14th International Symposium on Unsteady Aerodynamics, Aeroacoustics & Aeroelasticity of Turbomachines (ISUAAAT14)

Stockholm, Sweden
8-11 September 2015

ISBN: 978-1-5108-5934-0

Printed from e-media with permission by:

Curran Associates, Inc.
57 Morehouse Lane
Red Hook, NY 12571

Printed by Curran Associates, Inc. (2018)

ISUAAAT14 Organizing Committee
info@isuaaat.org

Additional copies of this publication are available from:

Curran Associates, Inc.
57 Morehouse Lane
Red Hook, NY 12571 USA
Phone: 845-758-0400
Fax: 845-758-2633
Email: curran@proceedings.com
Web: www.proceedings.com

Invited Speakers and Keynote

ISUAAAT14 Paper Sessions

Fan flutter : current and future challenges for engine manufacturers

Edouard De Jaeghere
SAFRAN-Snecma
Villaroche, FRANCE, edouard.dejaeghere@snecma.fr

Over the last decades turbofan engines have seen major changes concerning the fan stage design. The fan stage still represents a major issue for the next generation of engines. Highly loaded fan stage present transonic or low-supersonic velocities and are often prone to flutter at off-design conditions. Flutter is a highly undesirable and dangerous self-excited mode of blade oscillations that can result in high-cycle fatigue blade failure. Preventing flutter from happening in the operational range is therefore a matter of first importance for fan designers. As a consequence, continuous improvements have been made on fan flutter prediction tools. Among others, the massive access to CFD has brought new opportunities to evaluate the flutter risk. Steady and unsteady simulations have become standard practice to identify flutter and its root cause. An overview of these capacities as well as recent test results are proposed here.

However the next generation of turbofan engines brings new challenges as far as fan flutter is concerned: higher by-pass ratios, shorter and more integrated engines, new blade technologies and material... The effects of these evolutions need to be taken into account all along the fan design process. New steps forward need to be made on simulation tools to address these issues, both in aerodynamics and mechanics fields: non linear phenomena, mistuning effect, stronger coupling, faster simulations and multi-disciplinary optimization...

Transient Blade Row Methods to Predict Rotor-Stator Interaction and Assess Blade Flutter

For calculation of unsteady turbomachinery aerodynamics for aeroacoustics, forced response or flutter predictions it is necessary to account for the original frequencies. Hence, the original blade counts of the rotor and the stator, respectively, have to be considered for the unsteady numerical calculations. Sometimes the pitch-wise extent of multiple rotor passages match that of several stator passages. So the model can be reduced to a sector model, which features many passages or maybe is small enough to reduce computational costs to an acceptable level. But most turbomachinery stages do not have a common divisor of rotor and stator blade count, because of mechanical reasons. Hence the model cannot be reduced and full annulus simulations have to be performed. This is also valid, if the nodal diameter of blade flutter has no common divisor with the blade count. Additionally a different sector of the blade row has to be used for different nodal diameters. In order to overcome these requirements the "Transient Blade Row"-methods (TBR-methods) are implemented into ANSYS CFX. This numerical methods allow to account for the flow features of the full annulus whereas only one or two blade passages per blade row (depending on the rotor-stator-interface method) have to be considered for the numerical simulation. Therefore the time-inclination method of Giles and the Fourier method (phase-lag) of He were implemented. For flutter cases the Fourier method will be used in order to account for small and big interblade phase angles. The method allows to predict the aerodynamic damping for all interblade phase angles using a numerical model, which features just two rotor passages. To show the capabilities of the Fourier method to predict blade flutter, results of the calculations of the SCTF11 test case will be shown. Special attention will be put on the numerical effort of the simulations of the Fourier method in comparison to the full annulus calculations.

Proceedings of ISUAAAT14
Aeroacoustics & Aeroelasticity of Turbomachines
ISUAAAT14
September 7-11, 2015, Stockholm, Sweden

ISUAAAT14-IV3

FLUTTER GENERATION AND CONTROL IN CAPTIVITY

Roque Corral*
Industria de Turbo Propulsores S.A.
Senior Fellow in Computational Engineering
28108 Madrid, Spain
e-mail: roque.corral@itp.es

ABSTRACT

A free-flutter experiment of a stand-alone low-pressure turbine rotor blade in a high-speed rotating wind tunnel is described. The most relevant design decisions and solutions of a low-reduced frequency cantilever shrouded rotor blade are discussed also.. A strain-gauge-less test specimen underwent a experimental campaign with independent control of the shaft speed, pressure ratio and density. Measurements were obtained by means of Blade Tip Timing and unsteady pressure transducer probes located in the outer casing of the rig. The rotor blade was driven by design into flutter in a wide range of operating conditions.

The rotor blade was designed in such a way that masses could be mounted in the tip-shroud to create arbitrary mistuning patterns. Two mistuning patterns were tested, the first consisted of a classical alternate mistuning pattern designed to suppress flutter wheres a second pattern was designed to halve the baseline vibration amplitude. Both objectives were successfully achieved during the experiment.

INTRODUCTION

Most of flutter experiments fall under two categories. In the first an airfoil or a set of airfoils are forced to vibrate following a prescribed motion at a given frequency. These experiments, which cannot be properly called flutter experiments, are performed in a laboratory, in a sector or annular cascade, using a hardware specifically designed to be operated in the facility and that, from an aerodynamic and dynamic point of view, is a major simplification of a realistic bladed-disc. One or several blades may be relatively easy instrumented with unsteady pressure transducers, since these experiments are essentially non-rotating and feature scale-up airfoils. The outcome of these experiments is the unsteady pressure characterization of the vibrating airfoil, that is usually further post-processed to generate the aerodynamic damping of the configuration. These data are primarily used for the validation of aerodynamic codes. A classical collection of this type of date are the standard aeroelastic configurations [1].

The second category of experiments use engine-like hardware, operated either in a test bed or in a rig. Usually the hardware is instrumented with strain-gauges (both compressors and turbines) or tip-timing probes (mostly compressors) with the ultimate objective of clearing the hardware for certification and production. Obviously this hardware is not intentionally designed to flutter and therefore flutter signals, if measured at all, are obtained in unexpected operating points, or as a result of an error in the predictions. The use of dedicated hardware and experiments using engine parts is mainly restricted to compressors, where dedicated rigs are used to assess the stability margin and surge line of new designs. Simultaneously, these rigs are used as aeroelastic and dynamic rigs to map unexpected flutter and high response regions. (see for instance [2])

The main drawback of the latter type of experiments is its high level of complexity that induces a limited control of the test, and

*Also associate professor at the Department of Fluid Dynamics and Aerospace Propulsion of the School of Aeronautics and Space, UPM

the low level of instrumentation, due to the limitations in the number of dedicated measurement channels because both, the competition with other experiments running in parallel, and in many occasions, the smallness of the component size.

Free flutter dedicated experiments are rare. Recently Fransson et al. [3] conducted and academic free-flutter experiment in a linear low-speed academic cascade. They were able to modify the stiffness of the airfoils by changing the effective support length using a simple mechanism, and drive the airfoil to flutter with a vibration amplitude that exhibited a travelling-wave like shape. The lack of periodicity of the cascade limited the usefulness the experimental data, what were intended mostly for academic purposes. In the past some attempts were made to drive the EPFL annular cascade to flutter with little success. It is believed that the inherent cascade mistuning and friction precluded the acquisition of meaningful data.

Turbines operate at inlet conditions that are impossible to reproduce in practice in a cold flow rig, therefore engine and rig operating points may only be compared in non-dimensional terms. From a purely aerodynamic point of view only the exit Mach number and the Reynolds number need to be adjusted, whereas turbine geometry is usually scaled down to fit it in the rig. When aeroelastic phenomena need to be reproduced many more non-dimensional parameters need to be adjusted also and many times full similarity can not be completely achieved.

This lecture describes an experiment specifically designed to generate and measure flutter in a controlled environments, i.e.: a high subsonic wind-tunnel. An isolated turbine bladed-disc was designed to flutter, and measurements were obtained using tip-timing optical probes and casing flush-mounted unsteady pressure transducers. No instrumentation was installed in the rotor blade to avoid uncontrolled mistuning of them. Flutter was measured for different shaft speeds and Reynolds numbers with a reasonable matching of the numerical predictions. Once the baseline fluttering experiment was in place two mistuning patterns, that were designed to halve and suppress the vibration amplitude of the tuned case, were tested and the experimental results were found to be in good agreement with the analytical predictions as well.

It is believed that this is one of the few times that a clean free-flutter experiment has been ever conducted in a controlled environment and high quality measurements have been obtained for a cantilever turbine.

NOMENCLATURE

ADP	Aerodynamic Design Point
BTT	Blade Tip-Timing
c	Axial chord
H	blade span
IBPA	Inter-blade Phase Angle
k	$= \frac{\omega c}{U_{exit}}$ Reduced frequency
LE/TE	Leading/Trailing edge
LPT	Low-Pressure-Turbine
M	Mach number
ND	Nodal Diameter
PS/SS	Pressure/Suction side
$Re = \frac{\rho U_{exit} c}{\mu}$	Reynolds number
VIGV	Variable Inlet Guide Vanes
δ	Vibration amplitude
ρ	Density
ξ	Critical damping ratio
ω	Angular frequency
	Sub-scripts
e	Exit
0	Stagnation conditions

OVERALL DESCRIPTION OF WORK

Two high aspect ratio isolated rotor blades, one cantilever and another interlock, with the same aerodynamics, and the same disc, were tested in a rotating high-speed wind-tunnel at CTA. Here only the results of the baseline cantilever configuration are described in some detail, since due to the constrain of keeping the same aerodynamics in both rotors, it was not possible to make flutter the interlock configuration. The baseline cantilever configuration was further tested including mistuning masses. The data acquisition system was based on the tip-timing technique. BTT measurements and signal post-processing were performed by Rolls-Royce. The system is intrinsically non-intrusive and avoids the introduction of non-intentional mistuning associated to other types of instrumentation, such as strain gauges. The steady aerodynamic conditions at the inlet and outlet of the row were measured also. Aerodynamic damping characterization relied on measurements conducted in a sector annular cascade located at KTH. The target of the experiments was to obtain data for the validation of numerical tools. Two rotor blades were designed and manufactured keeping in mind the facility constraints and the maximization of the use of existing hardware.

Mechanical characterisation of the bladed-discs was performed it two steps. Firstly, a ping test of all blades was conducted (at Politecnico di Torino) and the frequency and mass of every individual blade were obtained. Secondly, modal testing of the bladed-disks was carried out, through excitation of the bladed-disc in

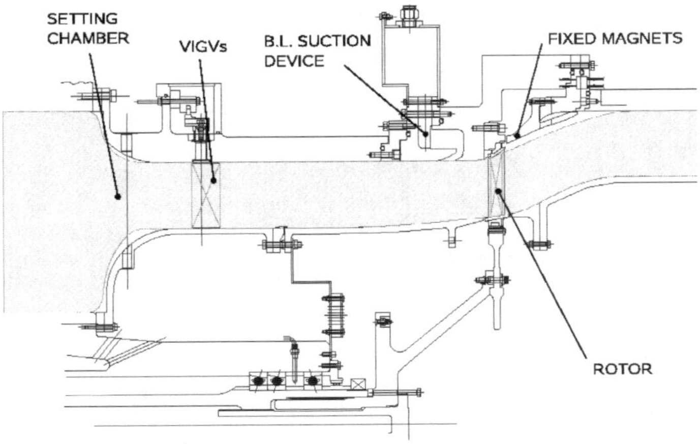

Figure 1. LAYOUT OF THE CTA WIND-TUNNEL USED FOR THE FREE-FLUTTER TEST

a in-vacuum rotating rig (at former Avio currently GE-Avio) to measure natural frequencies and derive mechanical damping from the BTT measurements.

The independent characterization of the aerodynamic and mechanical damping in simplified environments was intended to allow performing direct comparisons among different numerical tools and indirect comparisons with the single row aeroelastic experiment to increase the understanding of the whole problem.

Rotating Cold-Flow Rig

The main experiment consists in an isolated rotor blade subject to the incoming conditions of the facility IGVs, which are located more than 10 chords upstream of the testing article, therefore the rotor blade may be considered as an isolated rotor blade subject to weak perturbations associated to the blade-passing frequency of the upstream IGVs, whose chord is about 50% longer than that of the rotor blade. The boundary layer generated at the inlet duct is sucked one chord upstream of the rotor to avoid unrealistic effects in the secondary flow vortex system. A general layout of the wind-tunnel, which is located at the CTA in Zamudio (Spain), may be seen in Fig. 1. The stub shaft, the inlet and outer inner and outer casings where intended to be reused from a standard existing configuration of the facility in which a rotor blade, located upstream of the stator, is used as a wake generator while measurements are taken both in the airfoil surface and downstream of the stator. In our case an isolated rotor blade was tested and casings were redesigned to close the gap of the absent stator vane. No much margin was left in terms of the flow-path design.

The maximum mas-flow of the facility is about 20 kg/s, depending on the actual pressure ratio. Two compressors upstream and downstream of the test section allow an independent control of the pressure level, that determines the Reynolds number, and the pressure ratio across the rotor blade, that fix the exit Mach number for a given shaft speed. Rotor energy is dissipated using a water brake.

Inlet and outlet boundary conditions of the row were acquired using fast response five-hole miniaturized probes that are radially and azimuthally traversed across a distance equivalent to several rotor blade pitches. Stagnation temperature is also measured at the inlet. Blade Tip-Timing (BTT) probes were installed in the casing pointing radially to the rotor to measure the vibration amplitude of every individual blade. Finally flush-mounted unsteady pressure transducers were located one chord downstream of the rotor blade to correlate rotor vibration amplitudes with the unsteady pressure in the outer casing. More details of the facility may be found in [4].

Vibrating Annular Sector Cascade

Unsteady aerodynamics was characterized in a vibrating annular cascade, made up of five airfoils, located at KTH in Stockholm. Unsteady pressures associated to the vibration of the central blade were measured along the airfoil chord at three different span-wise positions. The influence coefficient technique was used to derive the work-per-cycle as a function of the inter-blade phase angle. The section chosen for the cascade corresponds to the 70% section of the 3D rotor blade tested at CTA modified to account for the stream-tube variation. This experiment allows an independent characterization of the aerodynamics, however, for the sake of brevity the results obtained in this campaign will not be described here.

In-Vacuum Rotating Dynamic Rig

A two-shaft mechanical rotating rig was used for the first time within the FUTURE project at Avio in Torino to dynamically characterise the bladed-disc in rotating conditions. Rotation not only slightly modifies the natural frequencies of the assembly at moderate shaft speeds, but significantly changes dry friction at the root attachment which, on the other hand, it is responsible of non-linearly saturate the vibration amplitude of aerodynamically unstable rotor blades and keep the vibration at a finite amplitude [5]. The test specimen is mounted in one of the shafts, whilst a set of magnets, forming a given ND, is mounted in an auxiliary structure that may be rotated independently of the blade-disc. The rig, therefore, allows the independent selection of the bladed-disc rotational speed, and the ND and frequency of the

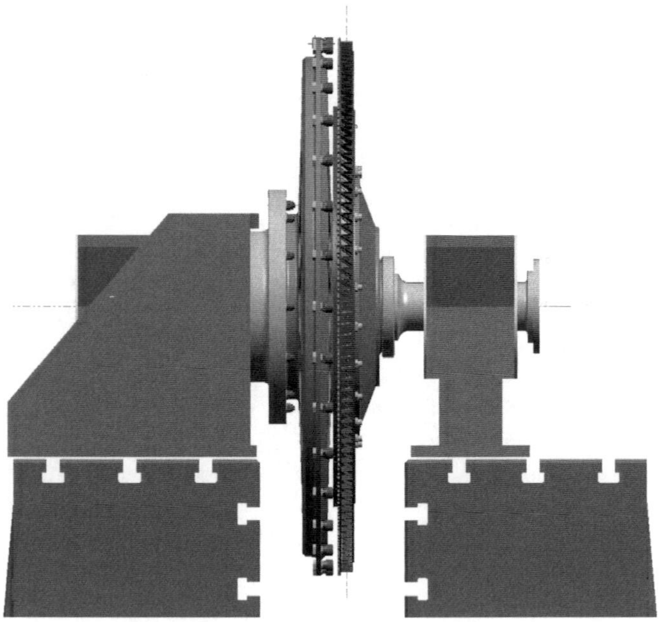

Figure 2. OVERALL DESCRIPTION OF THE STARGATE (AVIO) IN-VACUUM ROTATING RIG

excitation, which is generated using a set of magnets mounted both in the rotor blade and a auxiliary rotating supporting system. The excitation may be modulated varying the gap between the magnets located in the bladed-disc and the auxiliary structure.

ROTOR BLADE DESIGN

Design Intent

The reference stage aerodynamic design is representative of a state-of-the art LP turbine with high deflection (flow deflection around 100°), high lift blading and high sub-sonic exit Mach number conditions. To allow for an acceptable experimental cost, the turbine stage was downsized from engine scale, and a sub-atmospheric pressure environment was used. The main difficulty in designing the experiment is to reproduce all the non- dimensional parameters of the engine, including those related to the structural behaviour.

Aerodynamic non-dimensional parameters typical of LPTs are routinely simulated in cold flow rig testing; however, also identifying and reproducing non-dimensional aeroelastic parameters poses an additional challenge to the experiment. The main difficulties are linked to keeping the experimental cost down by downsizing the rotor blades to fit them in a relatively small wind-tunnel, and the pressure level (lower than in the engine as well).

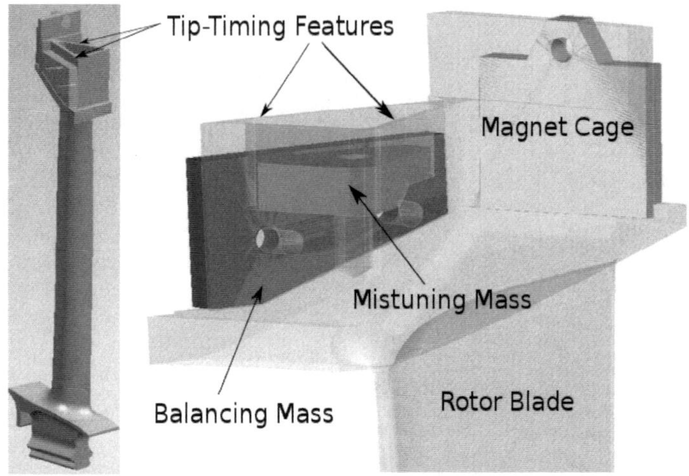

Figure 3. TIP-SHROUD CLOSED-UP

The new rotor blades were designed to match existing rig flow path, to minimize manufacturing costs by reusing existing casings, IGVs, and exit duct, as well as existing aerodynamic instrumentation (casing sensors and rakes).

The rotor aero-mechanical design was based on a reference stage aerodynamics and several iterations were conducted. The main problem was associated to induce an intentional aeromechanic instability, whilst ensuring that the vibration amplification would be large enough to be measured, but still small to avoid any risk during the operation of the rig. The baseline cantilever configuration was designed with a plain shroud, fined, blade tip to keep as much similarity as possibly with typical aeronautical LPT designs, avoiding clearance flow phenomena to become a key aero-feature, and, with respect to an interlocked shroud, (i) favor flutter because of the lower natural vibration frequencies and (2) avoid the uncertainties of interlocking-related frequency, damping and mistuning predictions. ITP was responsible for designing the baseline configuration.

The planned cold flow and mechanical test campaign was aimed at characterising both tuned and intentionally mistuned rotor configurations. Hence special provision for adding and removing masses had to be accounted for. Moreover a mechanism to excite the blades in the in-vacuum mechanical test campaign was needed.

Actual Design

Rotor Blade Design The rotor blade is representative of engine hardware, featuring a fir-tree attachment, shank, platform, airfoil and tip-shroud. However, the mechanical design of the tip-shroud was driven by several non-conventional constraints:

1. **Blade Tip-Timing features.** Optical BTT probes require the presence of sharp edges, to detect accurately the rotor blade passing. At some point it was decided to use straight probes pointing radially from the outer casing to the rotor blade tip-shroud. Two fence-like features covering the full length of the tip-shroud were added to ensure that the tip-timing probes may detect them. (see Fig. 3). Although in principle a single feature, such as that placed in the rig axis direction, would suffice to detect the vibration amplitude of the mode, a second one, inclined with respect the rig axis, was added to enable the possibility of detecting the axial displacement of the rotor blade as well. This can be useful to characterise the gap between the fixed and rotor blade magnets. As a result, a triangular-shaped cavity may be seen on the top of the tip-shroud.
2. **Magnet and magnet suppor**t. It is worth mentioning that several concepts were proposed at the early stage of the design to excite the rotor blades and characterise the damping of the bladed-disk in a in-vacuum spin rig at former AVIO. Finally, it was decided to use permanent magnets to excite the blade since this method was perceived as the simplest of all, for the purpose of the experiment (Magnet sizing and design of the excitation stem was leaded by Politecnico di Torino). In order to have the largest force per unit area a magnet-to-magnet concept was selected (i.e.: the force is exerted either by attraction or repulsion between two magnets) The magnet size was fixed by means of a characterization test of the force exerted by the magnets performed at POLITO. Several tests were conducted; specifically magnet versus magnet (M vs M) and magnet versus steel (M vs S) . The basic result is that the force exerted by the M vs M configuration doubles that of the M vs S, which was the expected result. Moreover, it was observed that the magnet specific force decreased very quickly with the magnet gap. It was decided to size the magnet to operate with a gap of a few mm in the M vs M configuration, since this is range in which the specific force varies smoothly, but still is large enough for the purpose of the experiment. The rotor blade was manufactured in aluminum and a cage to support the magnet, screwed to the rear part of the rotor blade, was designed (see Fig. 3). The magnet was mounted in the case by interference and glued for more security.
3. **Mistuning masses.** In order to modify the rotor blade frequency and implement mistuning patterns in the bladed-disk, provision was made to include the so-called mistuning masses in the rotor blade tip-shroud. The variability on the manufacturing of the rotor blade was expected to be below 1%, since there are machined in aluminum, however still a large variability was expected in the rotor blade frequency due to uncertainties in the attachment contact. To clearly distinguish the effect of the mistuning pattern on over the background scatter, the mistuning masses were designed to produce variations in the blade-alone frequency of about 4%. Mistuned masses were hosted between the two small fences used to trigger the optical signals of the BTT system (see Fig. 3) and screwed to the tip-shroud.
4. **Balancing mass** An additional problem found during the latest stages of the design was the fact that the rotor blade could not be simultaneously balanced for the wind-tunnel and in-vacuum tests because of the difference induced by the aerodynamic forces. The final decision was to balance the rotor for the vacuum conditions and to include an extra mass to minimise the rotor unbalance in the presence of aerodynamic forces at the ADP. The purpose was not to solve an integrity or life problem, but to ensure that the load distribution in the contact were as similar as possible in both tests. If the stress distributions are similar, it may be expected that the dissipation of both experiments will be comparable.The balancing mass was screwed to one of the fences used to trigger the BTT signal (see Fig. 3).

Finally it is important to remark that the axial retention of the rotor blade was implemented using lock plates to minimise other sources of friction different from that of the fir-tree attachment.

Aerodynamic Design The selected airfoil and boundary conditions are typical of an aeronautical low-pressure turbine. The flow-path was selected to minimize the modifications in the rig due to budgetary reasons. There are two types of rigs available at CTA from which to derive the FUTURE rig, namely, single stage rigs and three-stage rigs. Finally it was decided to derive the flow-path from the single stage rig concept. Using this concept it was possible to reduce the rig modifications to the outer and inner annulus. The exit Mach number, Reynolds number, velocity triangles and load distributions are representative of a low-pressure turbine, although some aerodynamic subtleties directed to reduce secondary losses have been omitted in the design for simplicity. We did not have a precise target for the Mach or Reynolds numbers since the design was entirely dominated by the blade dynamics and the flutter vibration amplitude. The rotor blade was designed with some tolerance to off-design conditions since it was expected that the rotor blade could be tested at different shaft speeds.

The design was entirely been performed inside the ITP design system which is a consistent suite of codes. This includes through-flow, 2D and 3D blading system together with 2D and 3D viscous analysis codes [6, 7, 8, 9]. However, since we ended-up with a single rotor blade, the aerodynamic design methodology plays a minor role in this design. It is worth mentioning

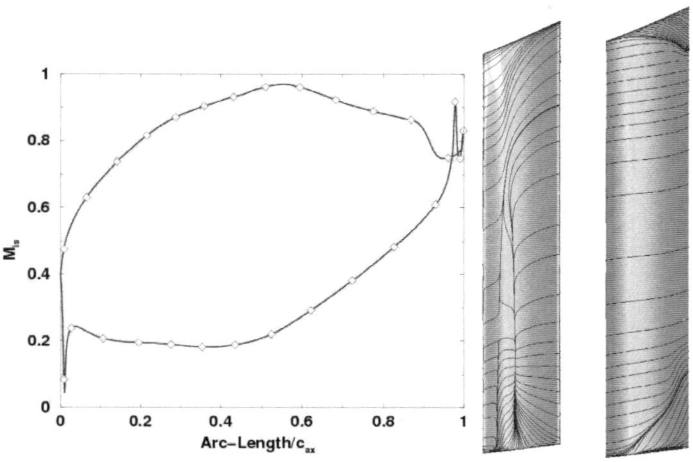

Figure 4. ISENTROPIC MACH NUMBER DISTRIBUTION OF THE 70% AIRFOIL SECTION (LEFT) AND SURFACE STREAMLINES OF THE PRESS SURE AND SUCTION SIDE OF THE AIRFOIL SURFACE (MIDDLE AND RIGHT RESPECTIVELY)

M_{in}	α_{in}	M_{exit}	α_{exit}	Re
0.42	36.5°	0.75	60°	1.2×10^5

Table 1. BOUNDARY CONDITIONS FOR THE DESIGN INTENT OF THE 70% SPAN OF THE CTA ROTOR BLADE

that three major iterations involving the detail design of the actual rotor blade and the corresponding aeroelastic analysis were conducted to define final airfoil.

The aerodynamic conditions at the 70% span section are displayed in Table 1. The exit Mach number is around 0.75, which is highly subsonic and higher than initially forecast at the earliest phases of the design. The hade angle of the annulus, both at the inlet and the outlet, does not present separation problems. The aerodynamic aspect ratio, $\Lambda = H/c_{ax}$, is $\Lambda = 5.3$. The number of rotor blades, N_b, was finally fixed to $N_b = 146$. The main aerodynamic parameters of the design intent of the 70% section may be seen in Table 1.

Figure 4 (left) displays the isentropic Mach number distribution of the 70% section of the airfoil for the ADP of the design intent. It may be appreciated that correspond with a typical high-lift, aft-loaded LPT section at the design point. The airfoil surface streamlines (middle and right of the same figure) show the penetration of the secondary flow system, that is confined to a small portion of the span, larger at the hub than at the tip of the airfoil, since the aspect ratio of the airfoil is high. This characteristic minimises the relevance of a detailed prediction of the secondary vortex system and therefore desensitizes the prediction of the flutter characteristic of the airfoil from the detailed aerodynamics.

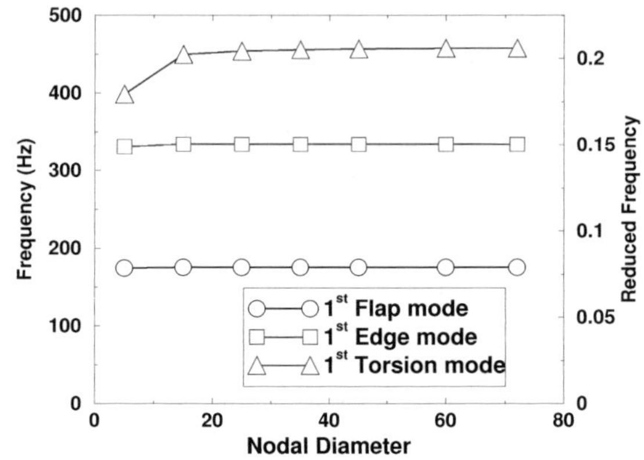

Figure 5. BLADED-DISC FREQUENCY CHARACTERISTICS

Aeromechanic Design The bladed-disc was designed with a very simple dynamics. Firstly, the disc is very stiff compared to the rotor blade and, secondly the first three modes (1^{st}flap, 1^{st}edge-wise and 1st torsion, ordered by frequency) of the rotor blade are well separated to avoid interactions among them. The rotor blade chord was minimised and fixed by manufacturing constraints to lower as much as possible the natural frequencies of the assembly. The frequency characteristics of the bladed-disc may be seen in Fig. 5 where it may be appreciated that the reduced frequency of the first three modes is very low.

The critical damping ratio, ξ, of the first three modes for the ADP is displayed in Fig, 6. It may be appreciated that the flap mode is highly unstable whereas the edge-wise and torsion mode are marginally stable. We intended by design to have only a single family of unstable modes although is worth mentioning that during many phases of the design we had slightly unstable edgewise or torsion modes. In the final design a single family of unstable modes remained, and during the testing only the instability of flap mode could be detected. The change of the aerodynamic damping with the incidence (shaft speed) is moderate while the change of ξ with the density is directly proportional.

Mistuning Pattern Design Another of the objectives of the test was to check the influence of mistuning on the flutter char-

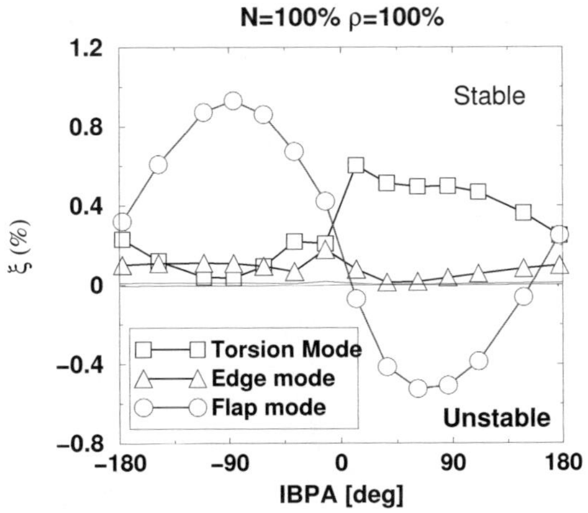

Figure 6. CRITICAL DAMPING RATIO OF THE THREE FUNDAMENTAL MODES AT $N = 100\%$ AND $\rho = 100\%$

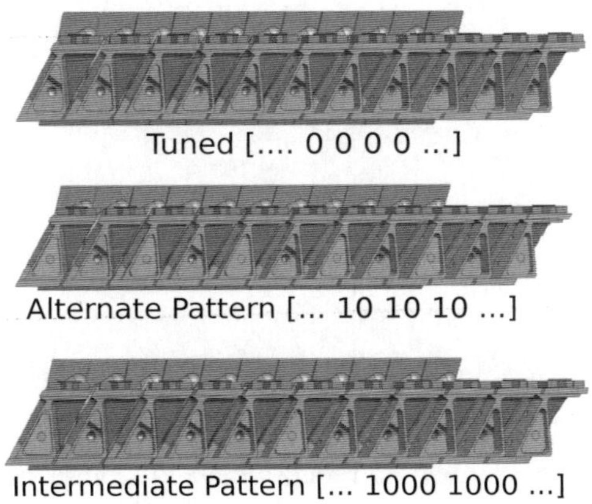

Figure 7. SUMMARY OF MISTUNING PATTERNS

acteristics of an unstable bladed-dics. Due to the random nature of natural mistuning, it is always difficult to obtain meaningful, conclusive data from mistuning experiments. The objective here was to design a conclusive realistic experiment such that the beneficial effect of mistunig on the stability of bladed-disc could not be discussed anymore, neither at a scientific technical level, nor at a management level, where indirect, not clean enough, evidences are much harder to sell.

The natural mistuning of the assembly was measured weighting and making a ping test of all the rotor blades. Mistuning masses were selected on the basis of highly detuning the bladed-disc in such a way that natural mistunig played a secondary role in the dynamics. Two mistuning patterns were selected [... 1 0 1 0 ...] and [...0 1111 0 1111 0 ...] where 1 means that a mass has been added in a particular rotor blade. The first pattern, [10],

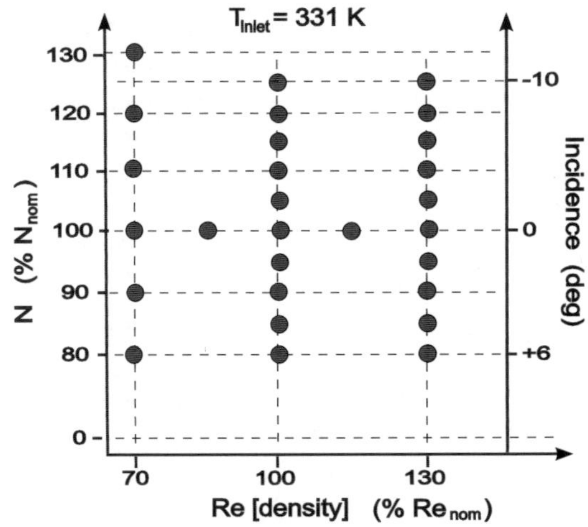

Figure 8. CTA TEST MATRIX SCHEME

corresponds with the classical alternate mistuning pattern. The mistuning mass was selected with the target of completely suppressing the flutter instability that had been previously created. The second pattern (derived by UPM), was designed with the objective of halving the vibration amplitude of the bladed-disc. The idea was that, if we succeed, firstly in making the rotor blade flutter, and secondly in suppressing the instability, it would be of a high added value to measure the response of a intentionally mistuned bladed-disc.

Mistuning patterns were implemented by screwing masses in the tip-shroud, as has been previously mentioned (see Fig. 7). Unfortunately, the so-called intermediate mistuning pattern was implemented differently in the CTA wind tunnel, where the pattern was [11110], than in the Avio STARGATE rig, where the actual tested pattern was [00001]. It may be shown that, surprisingly, both patterns are not fully equivalent.

EXPERIMENTAL RESULTS

Experimental Campaign and Operational Procedure

During the commissioning of the rig the incoming conditions to the rotor were measured and compared against the predictions. It was concluded that a correction of the operating point was needed since the IGVs of the facility were operating at maximum deflection and in highly off-design conditions, that were difficult to predict. Simulations were run to assess the impact of the new inlet boundary conditions and the shaft, that was slightly increased, in the steady loading of the blades.

The cantilever tuned configuration was tested first. The low shaft speed, low density points were tested first to gain confidence on the experimental set-up and ensure that the facility could be safely operated. Blade vibration was monitored on-line with a lite and quick BTT analysis system. Basically the mean vibration amplitude of the blades was monitored. From the very beginning there was a perceivable vibration level at the frequency of the first natural mode. The vibration amplitude increased with the density, as it could be expected. First results were very encouraging and the whole test matrix was measured.

During the test campaign the vibration amplitude of the blades was measured changing the shaft speed between 80% and 130% of its nominal value. Three sweeps of the density (Low=65%, medium=100% and high=130%) were carried out. Two additional points characterized by N=100% and a density between medium and high for the former, and low and medium for the latter, were measured also. In total, 28 test points were measured. A graphical illustration of the test matrix is presented in Fig. 8.

Immediately after the completion of the test cases for the tuned cantilever configuration, the testing of the mistuned test cases was carried out. In order to mount the mistuning masses, the rotor had to partially extracted from the facility. The whole assembly was unmounted rearwards and the mistuning masses screwed in the tip-shroud. This procedure was repeated for the two mistuning patterns. The same aerodynamic points that were tested for the tuned configuration were measured for the two mistuned configurations.

The facility was daily warmed up for one hour until a stabilised temperature was reached before before proceeding to acquire any data. Temperature stability was measured with T-type thermocouple inlet rakes. When the stationary temperature was reached, each measurement point was stabilized for a few minutes and then the data acquisition of the BTT system and the flush mounted pressure transducers was performed. The stability of the points was achieved by assuring that the control parameters were kept under a predefined threshold during the whole data acquisition process.

Inlet pressure rakes and fast response five-hole probes were used to measure total pressure, P_0, and Mach number, M. Shaft speed stability was reached by means of a water brake with a precision of ± 1 rpm. P_0 and M, could be independently varied as the facility is operated by means of a compressor and a vacuum compressor group, with which Reynolds and Mach number can be independently adjusted. A lite version of the BTT analysis system was used to guarantee the integrity of the rotor blade during the operation. All the acquired data were post-processed off-line. Aerodynamic data were post-processed in a daily basis, but BTT data were post-processed months after the completion of the test. This effectively means that it was not possible to drive the experiment using the full experience accumulated during the previous tests, since only the data coming from the lite BTT system were daily available.

.

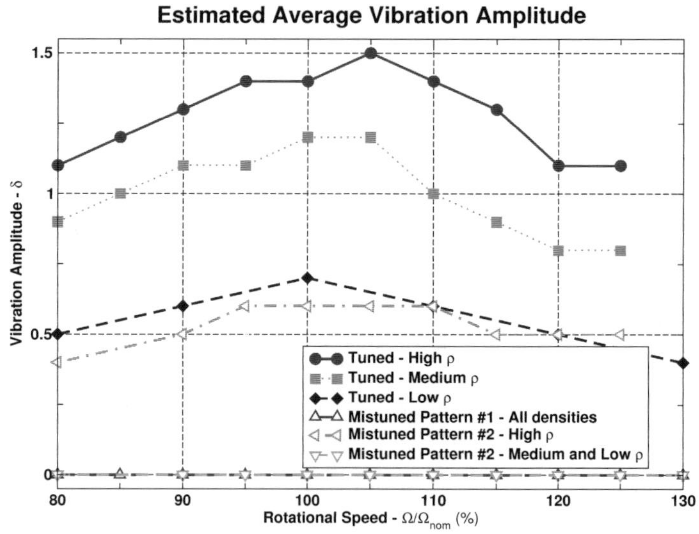

Figure 9. NORMALISED VIBRATION AMPLITUDE DERIVED FROM THE BTT DATA AS A FUNCTION OF THE SHAFT SPEED FOR DIFFERENT INLET DENSITY AND DIFFERENT MISTUNING PATTERNS.

Summary of Blade Tip-Timing Data

The summary average BTT data obtained from a on-line Lite BTT analysis system is shown in Fig. 9, where the normalised vibration amplitude is plotted as a function of the rotational speed. It may be seen that the vibration amplitude roughly scales linearly with the density, as it could be expected.

For the tune case the vibration amplitude grows with the shaft speed and then drops. The increase of the vibration amplitude with the rotational speed is something that could be expected, and that had been previously reported [10]. This trend is due to a reduction of the damping in the fir-tree attachment because of the pull load increase that prevents the relative motion between the rotor blade and disc. For very high pull loads, the attachment behaves as a perfect clamp, with zero dissipation, except that associated to the intrinsic material damping. The drop of the vibration amplitude with the shaft speed for the higher rotational speeds is not well understood yet.

For the alternate mistuning pattern (see Fig. 9 middle), the vibration is suppressed for all the density levels, as it was expected, since the mistuning masses were selected to completely suppress the instability. This is a clear, clean, experimental evidence of

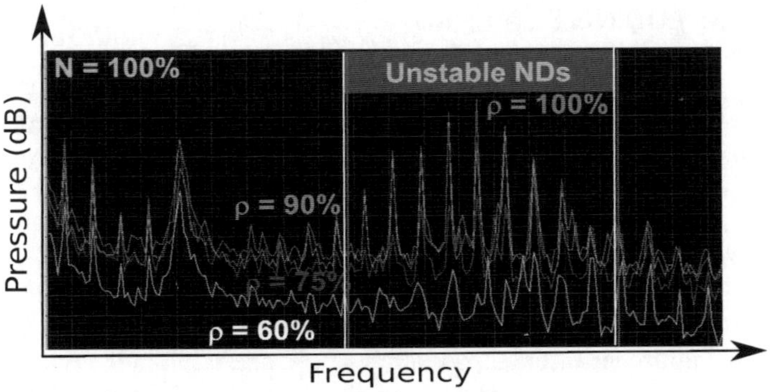

Figure 10. UNSTEADY PRESSURE TRANSDUCER DATA FOR THE TUNED CONFIGURATION AS SEEN IN THE CTA DISPLAYS

how alternate mistuning can be used to mitigate or completely remove flutter instabilities.

The second mistuning pattern (#2 in Fig. 9) was designed to approximately halve the vibration amplitude of the tuned case, and it may be seen that this pattern does a good job in reducing the vibration amplitude keeping the same trend with the shaft speed than the tuned case.

Summary of Unsteady Pressure Transducer Data

Twenty seven unevenly distributed unsteady pressure transducers were flushed mounted in the outer casing 1.15 chords downstream of the trailing edge of the rotor blade. Fourier transforms in time were performed for each individual transducer and then circumferentially averaged to obtain a mean representation of the harmonic content of the signal. Time signals were translated into nodal diameters using the natural frequency of the first mode obtained from a FEM analysis of the bladed-disk. Figure 10 displays the unsteady pressure transducer signals as they were seen in the CTA on-line BTT monitoring system at the nominal speed ($N = 100\%$). It can be appreciated a low-frequency noise caused by the facility, however at the frequency range of interest, a coherence signal, that coincides with the Doppler shifted frequency of the unstable modes, may be seen. The coherence is maintained for different levels of density. It can be appreciated that, since the shaft speed is constant, the frequency of the peaks is constant as well, while the intensity of the peaks depends on the density level. The higher the density is, the larger the excitation, as it could be expected. This was one of the most clear proofs that we had in our test a fluttering rotor blade.

Figure 10 shows a nice dependence of the unsteady pressure with the frequency. First of all the discrete mature of the signal is clearly identified and on the top of the background noise a clear

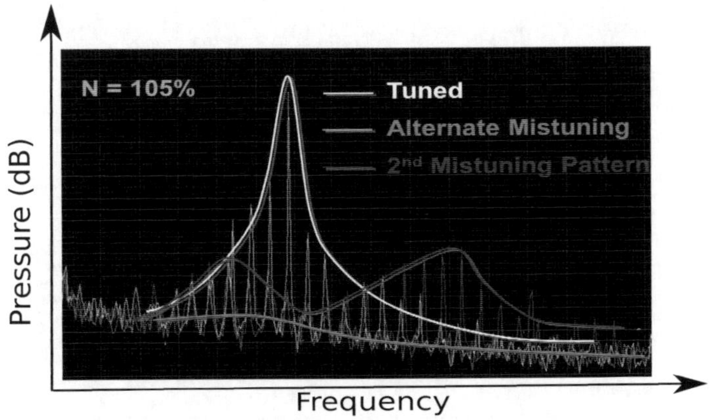

Figure 11. UNSTEADY PRESSURE TRANSDUCER DATA FOR THE MISTUNED CONFIGURATION AS SEEN IN THE CTA DISPLAYS

tone content of the signal may be appreciated. The highest peak corresponds to the Doppler shifted frequency of the most unstable nodal diameter, which is known by analysis to be around ND=22. The side bands represent the contributions from other ND which are unstable as well. The frequency window of non-zero amplitude corresponds very well with the predicted range of unstable nodal diameters translated in the frequency domain. This is a very clear indication that, as it had been conjectured before **(author?)** [5], there is a large number of unstable ND in a high aspect ratio cantilever low-pressure turbine.

Figure 11 displays a single pressure transducer signal for the tuned case and the two mistuning patterns (the enveloped of the signal has been plotted for the sake of clarity). Firstly, it may be appreciated that the alternate mistuned pattern does not give rise to any noticeable pressure signal, corroborating the results outlined previously coming from the BTT data. The intermediate mistuning pattern gives rise to a much lower unsteady pressure signal, but that is more scattered in frequency than the tuned case. This behaviour could have been anticipated, since mistuned configurations force, by design, the participation of more modes than in a tuned case. These NDs in the rotor frame of reference, give rise in a stationary frame of reference, and in the time domain, to a frequency scatter. The most unstable mode is split into two, with the corresponding side bands, indicating the existence of a multiplicity of unstable modes as well. The beauty of this plot is the high coherence of the unsteady pressure signals clearly distinguished from the facility background noise, that allows a direct identification of the ND participating on the dynamics of the bladed-disc.

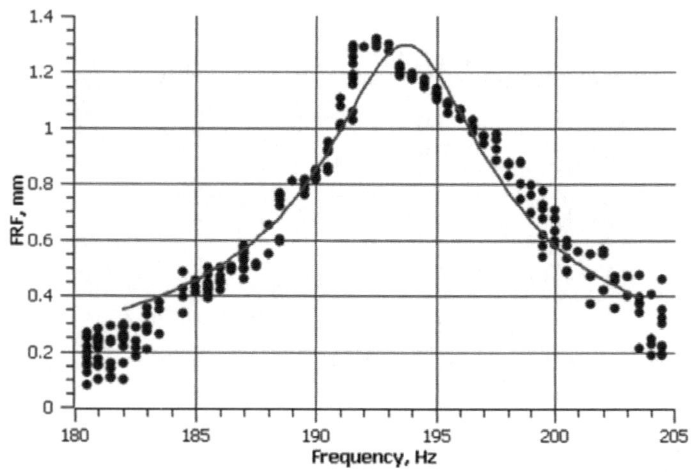

Figure 12. ROTOR BLADE VIBRATION AMPLITUDE AS A FUNCTION OF THE FREQUENCY

Summary of in-Vacuum BTT Data

An experiment devoted to dynamically characterise the bladed-disc was conducted at Avio in-vacuum rig known as STARGATE. The main objective of the experimental campaign was to derive the frequencies and damping of every individual blade and link them with the wind-tunnel results. With this aim in mind, the bladed-disk was rotated in vacuum and excited using magnets. The response of the system was measured and the data post-processed to derive friction damping. Different amplitudes, shaft speeds and nodal diameters were tested to characterize the non-linear nature of the phenomenon. Both tuned and mistuned configurations underwent this testing.

The non-linear nature of the saturation process due to the dissipation in the fir-tree attachment of the rotor blade with the disc was confirmed experimentally. This may be clearly seen in the non-symmetric resonant response of the rotor blade, characteristic of the response of non-linear friction dominated systems (see Fig, 12). An alternative way of highlighting the non-linear nature of the phenomenon is by looking at the dependence of the vibration amplitude of the bladed-disc in the wind-tunnel as a function of the shaft speed. It may be observed (see Fig. 9) that the amplitude first increases and then decreases with rotational speed. It has to be said that this behavior has not properly understood yet, but this variation of the amplitude with the speed is associated as well to a non-linear behaviour of the dissipation in the attachment..

CONCLUDING REMARKS

A whole set of experiments devoted to the aeromechanical characterization of a cantilever LPT bladed disc has been described. It has been demonstrated that by a careful selection of the non-dimensional parameters, it is possible to design a relatively simple, still highly realistic, experiment that contains all the basic ingredients of turbine classical flutter.

In spite of the fluttering conditions of the experiment, the vibration amplitudes of the blades allow a safe operation of the wind-tunnel, at stabilised fluttering conditions, and the acquisition of a valuable set of data. Although described in a much lesser extent than the aeroelastic experiment, it is though that a proper dynamic characterization of the bladed-discs is essential also to complete the picture of the problem.

We have experimentally demonstrated by means of a minimal fluttering unit that flutter can be suppressed, and even modulated, by using mistuning. Although the ultimate goal in real engines conditions is to minimise flutter vibration amplitude, if any, flutter modulation is a very valuable research tool.

Finally we have demonstrated that BTT measurements using optical probes pointing to the tip-shroud can be used to measure flutter, if appropriate features to trigger the laser signals are included in the shroud and that unsteady pressure transducers, located in the casing can be extremely helpful to understand the dynamics of fluttering bladed-discs.

ACKNOWLEDGMENTS

The author wish to thank the coordinator and deputy coordinator (Torsten Fransson and Damian Vogt), as well as the rest of the members of the Management Committee (Paolo Calza, Bob Elliot, Detlef Korte and Jan Ostlund) of the FUTURE Consortium for trusting on me to carry out this work and their support, patient and flexibility along the the development of this work. There are so many individuals across so many organizations (Avio, CTA, ITP, MTU, Polito, RR, UPM) that have contributed somehow to the success of this project that the list would so be long to be included here. My sincere gratitude to all of them, without of which it would have been impossible to carry out this work. In special, I would like to thank Fernando Partida and all the CTA team for all those nights of work operating the facility and their dedication. To all of you thanks. Finally, I would like to express my gratitude as well to Industria de Turbopropulsores S.A. for letting me to manage this project to write this paper.

This work has been funded by the FUTURE project of the European Commission under Contract ACP7-GA-2008-213414.

References

[1] Fransson, T. H. and Verdon, J. M., "Updated Report on Standard Configurations for Unsteady Flow," Tech. rep., KTH, 1992.

[2] Dodds, J. and Vahdati, M., "Rotating Stall Observations in a High Speed Compressor: Part 1 Experimental Study," *GT2014-25634*, ASME Turbo Expo 2014: Turbine Technical Conference and Exposition, Dusseldorf, Germany, June 2014.

[3] Lucio, M., Bergmans, J., Vogt, D., and Fransson, T., "A Remotely Operated Aeroelastically Unstable Low Pressure Turbine Cascade for Turbomachinery Aeromechanics Education and Training Remote Flutter Lab," *J. Eng. Gas Turbines Power*, Vol. 137, No. 3, October 2014, pp. 032507.

[4] Vazquez, R., Arsuaga, J. I. M., and Armananzas, L., "A Newr Transonic Test Turbine Facility," *XVI International Symposium on Air Breathing Engines (ISABE)*, ISABE, Cleveland, Ohio, USA, 2003.

[5] Corral, R. and Gallardo, J., "Nonlinear Dynamics of Bladed Disks with Multiple Unstable Modes," *AIAA Journal*, Vol. 56, No. 6, June 2014, pp. 1124–1132.

[6] Contreras, J., Corral, R., Fernandez-Castaneda, J., Pastor, G., and Vasco, C., "Semi-Unstructured Grid Methods for Turbomachinery Applications," *ASME Paper 2002-GT-30572*, Proceedings of the 47th ASME Gas Turbine and Aero engine Congress, Exposition and Users Symposium, Amsterdam, The Netherlands, June 2002.

[7] Corral, R., Escribano, A., Gisbert, F., Serrano, A., and Vasco, C., "Validation of a Linear Multigrid Accelerated Unstructured Navier-Stokes Solver for the Computation of Turbine Blades on Hybrid Grids," *AIAA Paper 2003-3326*, 9th AIAA/CEAS Aeroacoustics Conference, Hilton Mead, South Carolina, May 2003.

[8] Corral, R. and Pastor, G., "4. Parametric Design of Turbomachinery Airfoils Using Highly Differentiable Splines," *AIAA Journal of Propulsion and Power*, Vol. 20, No. 2, 2004, pp. 335–343.

[9] Burgos, M., Corral, R., and Contreras, J., "Efficient Edge Based Rotor/Stator Interaction Method," *AIAA Journal*, Vol. 41, No. 1, January 2011, pp. 19–31.

[10] Kielb, J. and Abhrari, R., "Experimental Study of Aerodynamic and Structural Damping in a Full-Scale Rotating Turbine," *J. Eng. Gas Turbines Power*, Vol. 125, No. 1, December 2002, pp. 102–112.

Unsteadiness and the Future of Aviation

by Alan Epstein
Vice President, Technology and Environment
Pratt & Whitney

Unsteadiness is at the core of modern aeropropulsion. It is the basis of energy exchange in turbomachinery. But it also generates many practically important phenomena that seriously challenge overall engine performance, life, and acceptability. Of course, I am referring to phenomena such as aeromechanics, noise and surge. These were problems in the very first jet engines and remain challenges in today's engines. Must this be the case in the future? Can new design paradigms reduce or eliminate such problems, or introduce new ones?

In order to improve commercial aircraft fuel burn, emissions and community noise, we are now moving into the era of Ultra-High Bypass ratio engines. Practical implementation of such concepts requires new engine architectures and new component design spaces. These new design approaches introduce new challenges and, perhaps surprisingly, obviates the need to solve some old ones. This means the solutions for tomorrow's engines will not necessarily evolve from the technical studies motivated by the challenges of legacy designs. There are new, rich opportunities for unsteady aero-sciences to deliver value to the aviation community.

This lecture explores the future of commercial propulsion and the role of unsteady aerodynamics, aeromechanics and acoustics in improving future air travel.

Proceedings of 14th International Symposium on Unsteady Aerodynamics, Aeroacoustics & Aeroelasticity of Turbomachines
ISUAAAT14
8-11 September, 2015, Stockholm, Sweden

I14-S1-1

MODE DEPENDENT SHOCK STRUCTURE EFFECTS IN LOW-PRESSURE TURBINE FLUTTER ANALYSES

Joshua J. Waite
Duke University
Durham, North Carolina, USA
joshua.waite@duke.edu

Robert E. Kielb
Duke University
Durham, North Carolina, USA
rkielb@duke.edu

ABSTRACT

As modern turbomachinery trends evolve toward higher Mach number and loading regimes, so grows the importance of physically understanding the onset of flutter as engineers race to keep pace with safe design of future generations of blades. This fact, along with recent experimental and numerical supporting evidence, has motivated the study of investigating the impact of load variations on the aerodynamic damping of turbomachinery blades. Looking beyond the most common parameters used in flutter-free design, the reduced frequency and mode shape, this study provides an initial look inside the complex, non-monotonic relationship between flutter and loading. A quasi-2D grid of Standard Configuration 4 is used to exhaustively examine the impact of both rigid body translations and rotations through various levels of load (e.g. subsonic/supersonic, passage-shock/shock-free). Critical reduced frequency is plotted as a function of mode shape, and several distinct modal regions are identified with vital qualitative distinctions in terms of their sensitivity to load and mode. It is shown that for a constant reduced frequency, certain mode shape regions may experience either increasing, decreasing, or non-monotonically changing stability as loading conditions advance. Therefore, it no longer adequate to consider only reduced frequency and mode shape in preliminary flutter design methods, as the loading condition may play a crucial role depending on the mode shape in question.

NOMENCLATURE

c	Blade chord
GF	Generalized force
h	Bending amplitude
k	$=\frac{\omega(c/2)}{V_2}$, Reduced frequency
M	Mach number
p	Pressure
PS/SS	Pressure/Suction side of airfoil
q	Dynamic pressure
P.A.	Pitching axis location
STCF 4	Standard Configuration 4
V	Relative flow velocity
W_{cyc}	Work-per-cycle
α	Pitching amplitude (rad)
β	Tangential Flow angle
η	Chord-normalized Tie-Dye ordinate coordinate
θ	Bend angle
ξ	Chord-normalized Tie-Dye abscissa coordinate
Π_c	Ratio of back pressure to critical choking pressure
Δz	Thickness of quasi-2D grid
σ	Inter-blade phase angle (IBPA)
Ξ	Aerodynamic damping coefficient
ω	$=2\pi f$, Angular frequency
Subscripts/Superscripts	
0	Stagnation
$*$	Critical choking condition
1, 2	Inlet, Exit

INTRODUCTION

Flutter may cause high-cycle fatigue and ultimate stress failure in high-aspect ratio low-pressure turbine (LPT) blades, especially when they are free-standing. As engineers, deep physical understanding of flutter helps prevent costly design delays, missed project deadlines, or even catastrophic failures of an active jet engine or power turbine. Thus far, the aeroelastic community has identified two parameters strongly correlated with LPT flutter [1–4]:

1. Reduced frequency, $k = \frac{\omega(c/2)}{V_2}$
2. Vibratory mode shape, e.g. torsion, bending.

Reduced frequency is defined here using the semi-chord definition, where $c/2$ is the semi-chord, $\omega = 2\pi f$ is the blade's angular oscillatory frequency in rad/s, and V_2 is the exit relative velocity, which is common to use for turbine blades (inlet relative velocity correlates better with fan and compressor flutter). Generally speaking, airfoils oscillating at lower reduced frequencies have a higher tendency to flutter. Therefore, the aeroelastic engineer strives to design blades with higher reduced frequencies to avoid unwanted vibrational instabilities.

The second well-known LPT flutter parameter, mode shape, can also dramatically change flutter margins [1, 3]. Due to their relatively high mass ratios, turbomachinery blades with adequate frequency margin between modes will flutter in a single mode as opposed to aircraft wings which are subject to mode coalescence [5]. Furthermore, since flutter normally occurs in one of the first few natural modes, these three-dimensional modes can be approximated as rigid body translations or rotations with varying spanwise amplitude. Combining this with the fact that LPT flow characteristics are remarkably linear [6], the stability of two-dimensional mode can be approximated by applying the principle of superposition to the 2D CFD of three fundamental rigid body mode shapes: two orthogonal translations and a rotation [1]. To illustrate this application, Panovsky and Kielb created a preliminary design tool called the "Tie-Dye" (TD) plot, which conveniently maps the critical reduced frequency, k_{crit}, which corresponds to neutral aerodynamic damping, as a function of mode. Figure 1 is an example of the TD plot for Standard Configuration 4 at a fixed exit Mach number of $M_2 = 0.90$. There are two orthogonal axes: 1) ξ - parallel to the chord-line (positive leading edge (LE) to trailing edge (TE)) and 2) η - perpendicular to the chord-line (positive pressure side (PS) to suction side (SS)). Each point on this map represents a pitching axis. For example, here are four points and their corresponding mode shape:

1. $(\xi,\eta) = (0,0) \rightarrow$ Pitching about the LE
2. $(\xi,\eta) = (1,0) \rightarrow$ Pitching about the TE
3. $(\xi,\eta) = (\infty,0) \rightarrow$ Translating/bending perpendicular to chord-line or "flex"
4. $(\xi,\eta) = (0,\infty) \rightarrow$ Translating/bending parallel to chord-line or "edgewise"

As the pitching axis varies throughout this four-chord window, k_{crit} changes by an order of magnitude. This simple fact underscores the strong correlation between mode shape and LPT flutter stability for a fixed reduced frequency, and sheds light on how harsh gradient regions, which we will soon explore in greater detail, should be intentionally avoided.

Not all flutter events are entirely explained by these well established parameters. Recently, evidence observed in engine

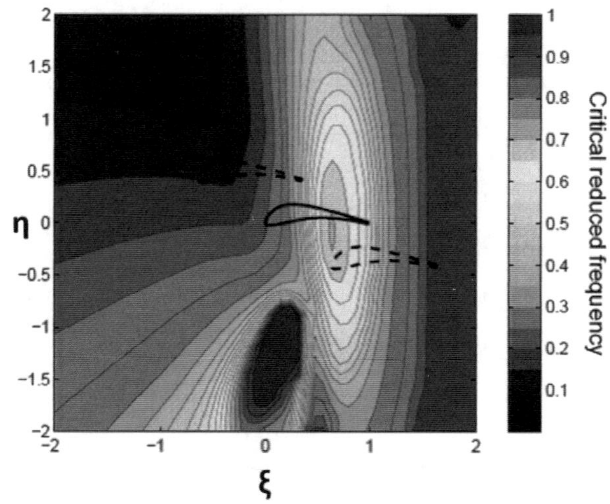

FIGURE 1: Tie-Dye plot of Standard Configuration 4: $M_2 = 0.90$ [7]. Neutrally stable reduced frequencies are shown for various rigid body pitching locations. Neighboring blades shown for cascade orientation. (ξ,η) normalized by the chord.

tests and via numerical methods have indicated steady loading combined with shock location are also key flutter drivers [8, 9]. This hot topic is more important than ever. The future of turbomachinery design demands lighter jet engines and higher energy-output from power turbines. Direct consequences of these trends are higher loading per stage and the present-day push into higher Mach number regimes rife with shock effects. With this, the authors predict that physically understanding the link between loading and flutter will soon be paramount. To address this concern, it is imperative to quickly learn how reduced frequency, mode shape, *and* loading conditions are all fundamentally interrelated in determining LPT flutter stability.

LOADING AND FLUTTER

The authors now examine previous efforts attempting to define the relationship between steady loading, shocks, and blade flutter. A major question is: how will higher blade loading and increased shock wave presence affect aerodynamic damping? The few studies addressing this topic focused on secondary flutter correlations to incidence angle, Mach number, or axial flow velocity [4, 7, 10, 11]. One early, notable study was completed by Cardinale in the 1980s [12]. Cardinale explored turbine flutter sensitivities in off-design conditions and found an unusual function of loading shown in the power turbine speed sweep seen in Fig. 2. By varying the bucket torque, approximate flutter boundaries were mapped as a function of torque and rotational speed. What is strange is that, for higher power turbine speeds, increasing the loading incited transitions into and back out of flutter as evidenced by the composite instability.

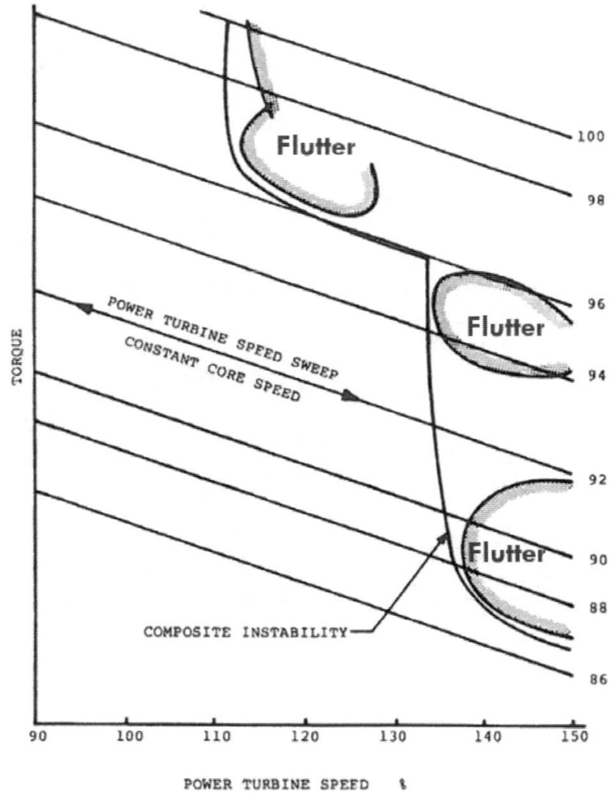

FIGURE 2: Investigation of torsion instability boundaries as a function of torque by Cardinale [12].

In 2002, Srivastava revisited the loading/flutter relationship after witnessing shocks abruptly influence work-per-cycle contributions on a fan blade [13]. Srivastava contended that raising fan's back pressure, therefore increasing the load, moves the shock forward where it also grows stronger and increases area of positive (destabilizing) work around the airfoil, especially in outboard stations of the fan blade. This research was followed up with a similar study by Srivastava in 2005, where it was concluded the inter-blade phase angle (IBPA) and shock location determined the influence of a shock wave on blade stability [14]. In the same study, Srivastava formulated mathematical expressions modeling flat plate work-per-cycle changes as a function of shock wave and IBPA. From these expressions, it can be logically inferred there exists a strong relation between shock structure, aerodynamic damping, and mode shape. Taking this into account, the authors numerically validated this hypothesis by running linearized unsteady CFD on a quasi-2D slice of Standard Configuration 4. By varying normalized back pressure (p_2/p^*) and fixing reduced frequency, the least stable damping was quantified over all IBPAs for four modes over this large range of loading conditions (e.g. exit Mach number, varied shock location).

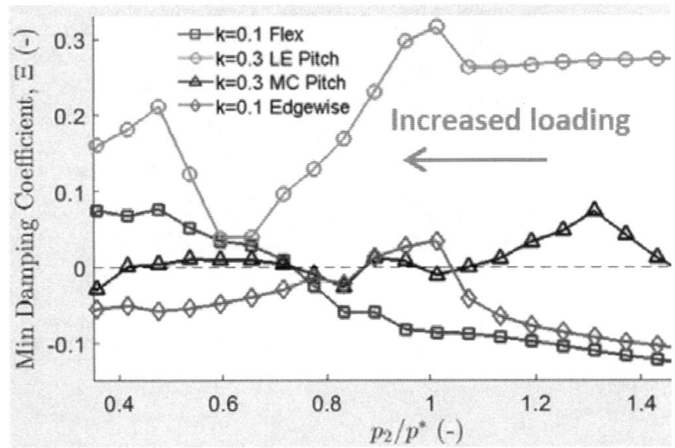

FIGURE 3: Minimum damping value (from 20 tested IBPAs) as a function of back pressure (normalized by critical choking pressure) for $k = 0.1$ flex bending, $k = 0.3$ leading edge pitching, $k = 0.3$ mid-chord pitching, and $k = 0.1$ edgewise bending.

These results are shown in Figure 3 [8]. One crucial finding was that increased loading does not monotonically decrease or increase stability, but rather the influence is indeed a strong function of mode. For example, Fig. 3 shows the flex mode (bending perpendicular to the chord) and leading edge pitch mode monotonically stabilizing and destabilizing, respectively, as the back pressure drops below the critical choking pressure, $p_2/p^* = 1$. As supporting evidence of mode/load sensitivity, the mid-chord rigid body pitching mode, which is a close approximation to a first-torsion mode, experiences six intersections with the neutral stability threshold (i.e. the dashed line) over this loading range.

The question must be answered why certain pressure ratios and loading conditions are detrimental for some modes, and beneficial for others' stability margins. An answer may be in the relationship between pitching axis and shock location. It is known that the proximity of the static back pressure to the critical choking pressure drastically changes the passage shock structure which in turn mutates the unsteady pressure amplitudes and phase [8]. While the pressure side of the airfoil is relatively unaffected, the suction side experiences significant changes in unsteady pressure amplitude and phase drastically altering flutter stability. In other words, it is no longer sufficient to consider only reduced frequency and mode shape. Loading and shock position must be considered simultaneously. An outline of the analyses to test these relations is now presented.

FIGURE 4: PHOTO OF STANDARD CONFIGURATION 4.

ANALYTICAL SETUP

The low-pressure turbine geometry to be used in the following flutter analyses is École Polytechnique Fédérale De Lausanne's (EPFL) Standard Configuration 4 (STCF 4). This annular, 20-blade, tuned, *non-rotating* LPT configuration, shown in Fig. 4, was experimentally tested for subsonic, transonic, and supersonic exit flows [15]. For the purpose of this paper, only the subsonic ($M_2 = 0.85$) baseline condition (Test Case 627) is used as a reference point. From this baseline, one of the following three control parameters is varied at a time: reduced frequency, mode, or back pressure.

All CFD results were collected using a validated frequency domain Reynolds-averaged Navier-Stokes (RANS) solver [16]. Since the shock motion's influence on work-per-cycle is linear for small blade oscillations [14], all unsteady CFD enforces small blade vibrational amplitudes below the linear limit. Experience working with LPT unsteady aerodynamics empirically supports this linearity assertion by Srivastava. In CFD tests, as the blade amplitude steadily increases low to high values, linearity holds damping nearly constant as a function of amplitude until unreasonably high vibration magnitudes (e.g., greater than HCF limits) are reached.

The computational domain is modeled in a viscous, quasi-2D sense. That is, the original 3D blade geometry is "sliced" near midspan creating a thinner grid with radial dimension $\Delta z = 8.057$ mm (0.3172 inches). Investigating numerous combinations of reduced frequency, mode shape, loading, and inter-blade phase angle, this simplified quasi-2D grid is necessary to decrease computational expense for the 24,245 CFD simulations required to generate the results that follow.

To control the loading, the exit static pressure, p_2, colloquially called the "back pressure", is fluctuated. Reducing back pressure increases the loading, flow velocity, and torque until fully choked conditions occur as shown in the steady CFD results of Fig. 5. The equation for the critical exit pressure when

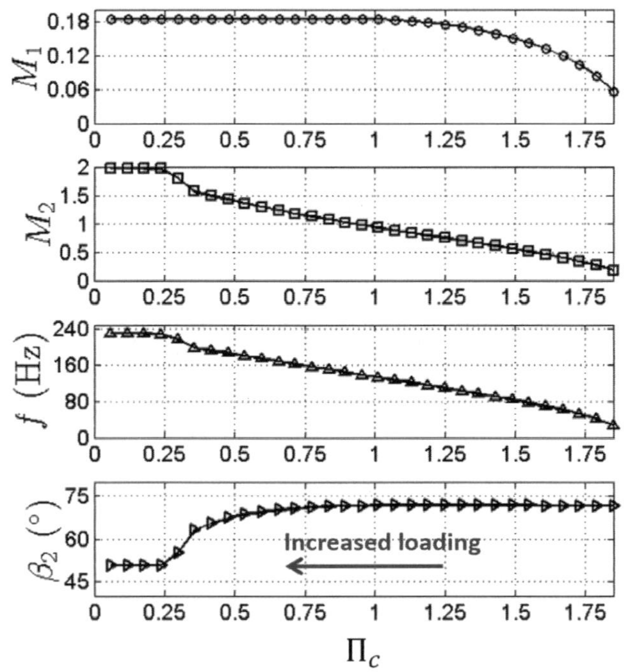

FIGURE 5: Variation of inlet/exit Mach, blade vibrational frequency, and exit flow angle as a function of normalized back pressure for constant reduced frequency, $k = 0.1$.

choking first occurs is given by [17]

$$p^* = p_0 \left[\frac{2}{\gamma+1} \right]^{\gamma/(\gamma-1)}, \tag{1}$$

where $\gamma = \frac{C_p}{C_v}$ is the specific heat ratio of the working fluid. When $p_2 \leq p^*$, it is impossible to increase the mass flow rate, and therefore M_1, by further reduction in p_2 alone. Therefore, the flow should be choked for $\Pi_c = p_2/p^* \leq 1$. For example, when $p_2/p^* = 1.01$, the suction side trailing edge is shock-free despite having a back pressure 1% higher than the critical choking pressure [8]. However, when the ratio slightly dips below this critical value of unity, the upstream mass flow can no longer increase, choking the flow and forming a visible shock on the aft suction side of the airfoil. On the airfoil pressure side, however, there are no visible unsteady pressure changes with reducing back pressure, indicating changes in stability due to shocks are dominated by suction-side effects.

For values of $p_2/p^* < 0.40$, the exit flow angle, β_2 drops significantly in Fig. 5. Flow turning abilities in this high loading region are greatly reduced. To isolate this potential effect from our current analyses, all results following are limited to values of $\Pi_c > 0.40$ where β_2 is relatively constant.

All other boundary conditions (i.e. angle of attack, inlet stagnation pressure, and total temperature) are held constant for this work. By adjusting p_2, the converged exit velocity, V_2, calibrates accordingly due to the change in downstream flow resistance. This change in exit velocity also alters the reduced frequency, k. However, to maintain consistency with the scientific method, the blade vibrational frequency is regulated proportionally to V_2 to guarantee constant reduced frequency is prescribed.

To vary the mode shape, three types of rigid body modes are tested: 1) Pure translational modes with various "bend" angles/directions 2) Pure pitching modes with pitching axes located on the chord-line (connecting the LE & TE) 3) Linear combinations of translations and rotations when constructing sample TD plots.

Finally, to quantify aerodynamic damping, a slightly different non-dimensional damping coefficient is used for bending and pitching modes, respectively, by [1]

$$\Xi_h = \frac{-W_{cyc}/\Delta z}{\pi q_2 h^2} \quad \& \quad \Xi_\alpha = \frac{-W_{cyc}/\Delta z}{\pi q_2 c^2 \alpha^2}. \tag{2}$$

Here, $W_{cyc} = \pi \text{Im}(\text{GF})$ is the work-per-cycle of the air on the blade (i.e. negative W_{cyc} is stable), Δz is the grid thickness, q_2 is the exit dynamic pressure, h is the translational/bending amplitude, α is the pitching angle in radians, and c is the chord length. Note that the complex generalized force (GF) is the surface integral around the airfoil of the dot product between the pressure and modal velocity.

BENDING MODE RESULTS

This section exhibits the relationship between bending modes and loading by investigating 31 equally-spaced translational rigid body modes. These "bend" angles vary in increments of $\theta = 6°$ and range from $\theta = 0° \rightarrow 180°$. Thanks to symmetry, the other half of the bend angles need not be considered. Figure 6 shows the sign and naming convention of the bending modes to come, and also provides a visual example of the $\theta = 90°$ "flex" mode. As a reminder, ξ is the chord-wise coordinate positive going from LE to TE, and is orthogonal to η, which is defined positive toward the airfoil suction side. For this study, bend angles, θ, are defined as positive starting from the $+\xi$ axis (edgewise translation) moving toward the $-\eta$ axis, or counter-clockwise as illustrated in Fig. 6.

Figure 7 shows the critical reduced frequency as a function of bend angle for eight selected critical pressure ratios. Each value of Π_c represents a fixed steady loading condition and shock structure. Note that three values of Π_c exceed the critical ratio of 1, and are therefore shock-free. This figure contains a wealth of information. First and foremost, each loading condition has a unique function of mode shape, and vice versa. For example,

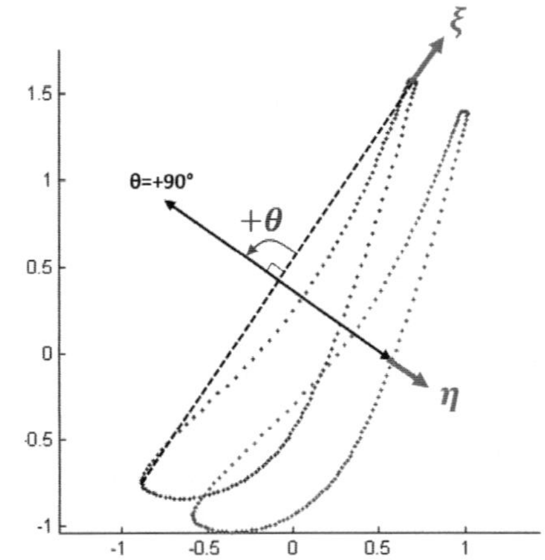

FIGURE 6: Depiction of the $\theta = 90°$ "flex" mode. The $+\theta$ sign-convention is shown to be counter-clockwise originating from the $+\xi$-axis.

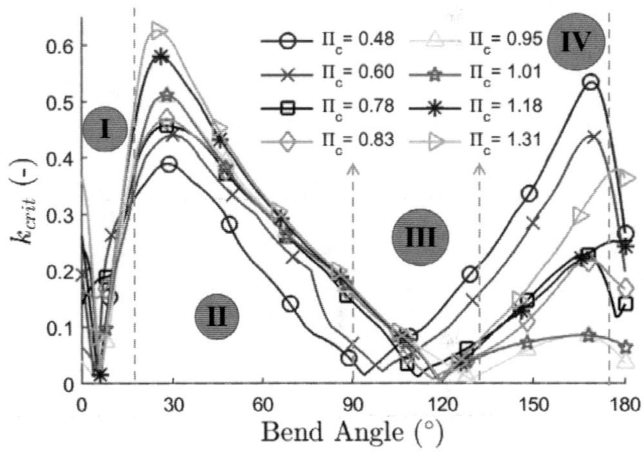

FIGURE 7: Critical reduced frequency as a function of every 6^{circ} bending direction for select loading conditions. Note: markers do not indicate refinement.

at $\theta = 30°$, loading is a stabilizing factor, whereas at $\theta = 165°$ loading destabilizes. Second, some bending modes are highly sensitive to the loading condition, whereas other regions maintain fixed k_{crit}. In an attempt to simplify the analysis of Fig. 7, the authors have divided the effect of loading on LPT flutter into four regions of bending modes as indicated by Roman numerals I-IV and the vertical dashed lines.

Region I covers approximately $\theta = -5° \rightarrow 20°$. These modes are close to edgewise translations. In this region, the effect of load variations is quite high; however, the trend with load is non-monotonic meaning it is difficult to predict if specific levels of loading are beneficial or detrimental to stability. The instantaneous rate or change of k_{crit} with respect to mode shape is the highest in Region I, indicating a high sensitivity to bend angle. Although, $k_{crit} < 0.4$ in the region for all loads signifying a favorable stability margin when compared to zones II & IV. The authors will note here that low reduced frequency is unfavorable, but low *critical* reduced frequency is quite desirable, as more stable operating conditions are in play.

Region II covers approximately $\theta = 20° \to 90°$. This wide range of modes includes translations moving in the direction of PS TE to SS LE all the way to the flex mode of $\theta = 90°$. In this region, higher loads, i.e. lower Π_c values, benefit from lower critical reduced frequencies, and are therefore considered more stable, especially at intervals $\theta = 20° - 40°$. Something interesting occurs between $\theta = 60° - 110°$. In this range, there is a noticeable loading insensitivity for values of $\Pi_c > 0.60$. Furthermore, k_{crit} roughly reduces to a linear function of bend angle.

Region III covers approximately $\theta = 90° \to 130°$. These are nearly "flex" bending modes, or plunge roughly perpendicular to the chord-line with a slight tendency going from SS TE to PS LE. This region is most stable and least likely to see flutter. While the bend angles of the local minima vary as a function of load, Region III offers the aeroelastic designer the most forgiving sensitivity to load and mode shape, as $k_{crit} < 0.2$ throughout.

Region IV covers approximately $\theta = 130° \to 175°$. Regions II and IV appear to behave in opposite manners, as the result of increasing the load in Region IV tends to destabilize (k_{crit} rises). This is especially true for values of $\Pi_c < 1$ where shocks are present.

Table 1 briefly summarizes the regional differences in terms of sensitivity to load variations, mode shape changes, and the range of k_{crit} values.

TABLE 1: Bending Mode Sensitivity Regions

Region	Sensitivity to Loading Changes	Reaction to Bend Angle	k_{crit}
I	high, non-monotonic	high: ↑ gradient	< 0.4
II	high, stabilizing	low: ≈ linear	< 0.6
III	low, non-monotonic	low: inflection	< 0.2
IV	high, destabilizing	low: ≈ linear	< 0.55

Fig. 8 plots the least stable non-dimensional damping coefficient and least stable IBPA as a function of bend angle. The reduced frequency is held constant at $k = 0.1$. The minimum damping is defined as the least positive or most negative damping value obtained testing 20 different IBPAs ranging from $\sigma = -162°$ to $\sigma = +180°$ in increments of 18°. Each intersection with the neutral stability dashed-line indicates $k_{crit} = 0.1$ for that specific mode/load combination, and can be verified in Fig. 7. Again, the observation holds that increasing the loading in the beginning of Region II near $\theta = 40°$ tends to stabilize, whereas the effect is reversed in Region IV near $\theta = 150°$ with the exception of the least loaded case of $\Pi_c = 1.31$. One interesting result, seen in Fig. 8, is the general trend defining the least stable IBPA varies proportionally with bend angle. There are of course small variations between loading conditions; however, relative to the macro trend, a conclusion is drawn that the least stable IBPA is a stronger function of bending mode direction and less of a function of load, with the exception of edgewise-like modes ($\theta \approx 0°$). Lastly, as these curves are quite similar in shape and grow in amplitude with larger pressure ratios, future work will include finding a proper scaling factor that collapse these curves for use in preliminary design.

PITCHING MODE RESULTS

This section displays the relationship between loading and rigid body torsion modes, or "pitching" modes. In all, 21 pitching axes (PA) spanning linearly from LE to TE in PA= 5% increments of chord are tested. Figure 9 illustrates an example of the mean and max rigid body deformation of the LE pitch mode, or pitching about 0% chord. Two other pitching axis locations, mid-chord (PA= 50%) and TE pitch (PA= 100%), are indicated for reference. Note the sample mode has exaggerated rotation. In the CFD, all modes rotate 2% of one radian, or roughly $\alpha = 1.15°$.

Figure 10 is much like its bending corollary of Fig. 7, but this time, k_{crit} is plotted as a function of pitching axis. For some combinations of mode shape and loading, the critical reduced frequency jumps off the plot above $k_{crit} = 1$. When this happens, it indicates only unstable conditions exist for that mode/load pair for $k \leq 1$. Once again attempting to simplify the interpretation of Fig. 10, the pitching mode range is divided into four qualitatively distinct regions.

Region I is defined by $0\% \leq PA < 46\%$. Generally speaking, this zone has the lowest average k_{crit}, and is therefore a good design region. The mode sensitivity is fairly low here, as $|\frac{dk_{crit}}{d(PA)}| << 1$ for most loads and pitching axes. The most stable pitching axis for each mode in this region is a function of load. For instance, $\Pi_c = 0.95$ is is least likely to flutter with a local minimum at PA= 45%, whereas the $\Pi_c = 1.31$ max stability point is near PA= 39%, and loading conditions $\Pi_c = 0.48, 0.60, 0.78, 0.83$ all have their lowest k_{crit} value with a leading edge pitch mode (PA= 0%). Therefore, while the mode sensitivity is low, the loading sensitivity is high as a small change in loading from $\Pi_c = 0.48$ to $\Pi_c = 0.60$ results in a 5-6x rise in

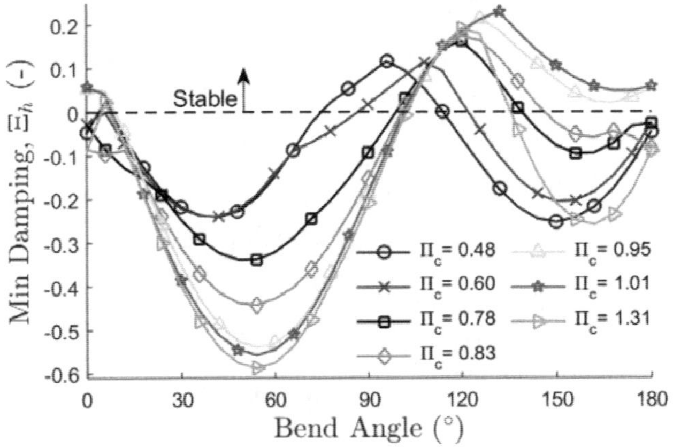

(a) Minimum Damping

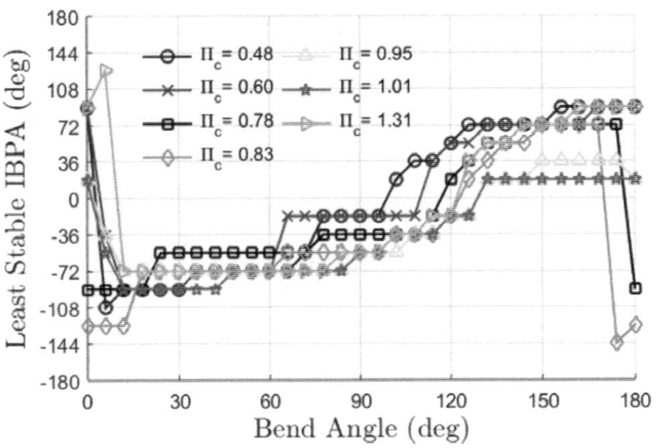

(b) Least Stable IBPA

FIGURE 8: Minimum non-dimensional damping coefficient and least stable IBPA for 20 IBPAs as a function of bend-direction and critical choking pressure ratio. Reduced frequency is held constant: $k = 0.1$.

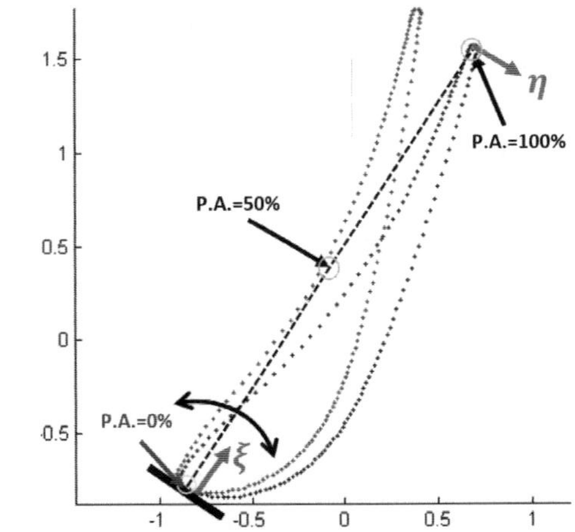

FIGURE 9: Sample sketch of the P.A.$=0\%$ "leading edge pitch" mode. Pitching-axis locations are also shown for mid- & trailing-edge pitch.

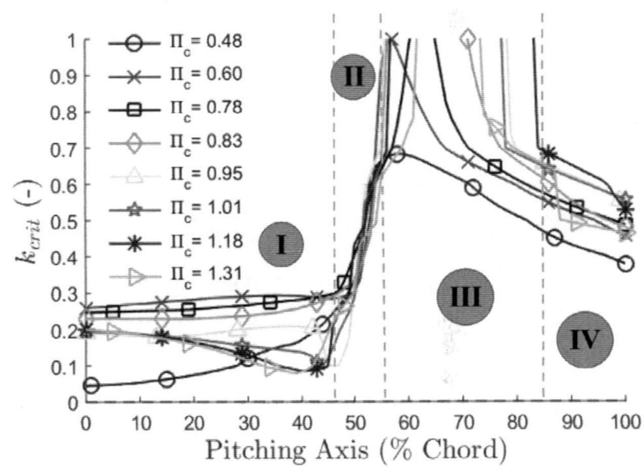

FIGURE 10: Critical reduced frequency as a function of chord-line pitching axis location (every 5%) for select loading conditions. Note: markers do not indicate refinement.

k_{crit} for PA=0%-25%.

Region II is defined by $46\% \leq PA < 55\%$. This zone is characterized by a very sharp destabilizing gradient in critical reduced frequency with respect to mode shape, i.e. $\frac{dk_{crit}}{d(PA)} >> 1$. Furthermore, k_{crit} is virtually independent of loading condition in Region II, as the mode shape solely dictates stability.

Region III is defined by $55\% \leq PA < 85\%$. This pitching region is the most unstable with very high k_{crit} values regardless of mode shape. Increased loading is favorable in Region III which is clearly visible near PA$= 80\%$.

Region IV is defined by $85\% \leq PA \leq 100\%$. Moving the pitching axis closer to the TE in this region improves stability. The critical reduced frequency ranges anywhere from $k_{crit} = 0.4 - 0.7$. For $\Pi_c < 1$, there is a somewhat monotonic relationship in that increased loading is improves stability.

High-level qualitative regional differences in terms of sensitivity to load variations, pitching mode changes, and the range of k_{crit} values are summarized in Table 2.

Figure 11 plots the least stable non-dimensional aerodynamic damping, Ξ_α, and the least stable IBPA as a function of chord-line pitching axis location for a constant reduced frequency of $k = 0.3$. The same 20 IBPAs tested in the bending mode results are retained for this pitching analysis. The fore-

TABLE 2: Pitching Mode Sensitivity Regions

Region	Sensitivity to Loading Changes	Reaction to Pitching Axis	k_{crit}
I	high, non-monotonic	very low	< 0.3
II	very low	high: $\approx$ linear	$0.1 - 0.6$
III	high, $\uparrow$ load stabilizes	high	> 0.5
IV	medium	low: $\approx$ linear	$0.4 - 0.7$

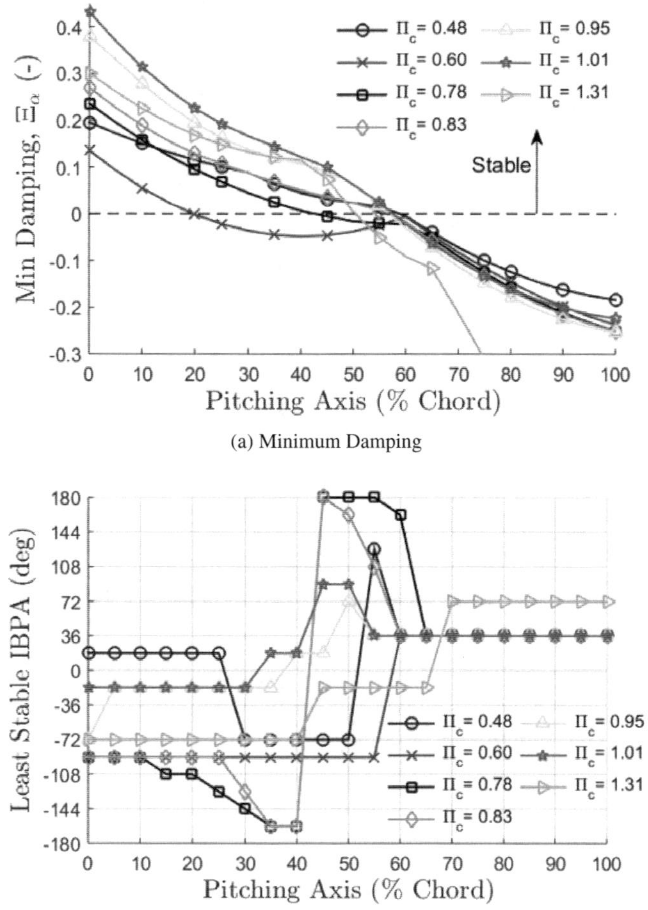

(a) Minimum Damping

(b) Least Stable IBPA

FIGURE 11: Minimum non-dimensional damping coefficient and least stable IBPA for 20 IBPAs as a function of chord-line pitching axis and critical choking pressure ratio. Reduced frequency is held constant: $k = 0.3$.

most pronounced feature of Fig. 11a is the generalization that moving the pitching axis aft is destabilizing. At first, this seemed to contradict the Region IV conclusions of Fig. 10. However, after generating data similar to Fig.11a this time using $k = 0.6$, the authors witnessed the tail end (90% $<$PA$<$ 100%) of all loading curves rise above the neutral stability line (i.e. $\Xi_\alpha = 0$) much like the $k = 0.3$, $\Pi_c = 0.60$ case does near PA= 60%. Therefore, one should be careful when making these types of generalizations when inspecting only one reduced frequency value. It is also worth noting, with the exception of the lowest loading condition $\Pi_c = 1.31$, the curves all seem to collapse between PA= 60% $-$ 70%.

Figure 12 now plots the minimum damping as a function of variable pressure ratio at a constant reduced frequency of $k = 0.3$. This alternative type of plot was created to provide an easier visualization tool of how specific modes change in aeroelastic stability as a function of load. For simple reference, the seven values of Π_c shown in Fig. 11a are indicated by gray, vertical dashed-lines. To analyze Fig. 12, the authors recycle the Region definitions of Fig. 10 and Table 2.

Region I modes appearing in Fig. 12 correspond to pitching axes of PA= 25%, 35%, & 45%. According to the conclusions of Table 2, there should be a low sensitivity to mode shape and a high sensitivity to load. As for the low sensitivity to mode, this is verified by noting how qualitatively similar the minimum damping curves are for the entire loading region. On the other hand, these three modes are all easily affected by Π_c variations. This assertion is especially accurate in the range $0.60 < \Pi_c < 0.90$ where there is a steep loss of stability with increasing loads. The authors hypothesize this is attributed to the passage shock moving farther aft as the back pressure drops, creating a larger distance, and therefore bigger moment-arm, between the shock and elastic axis thus amplifying the effect. The path of the suction side shock location (projected onto the chord-line) moving as a function of Π_c for STCF 4 is illustrated in Fig. 13. It can be seen that the shock moves almost linearly Π_c, as the position moves roughly 6%-chord aft for each pressure decrement of $0.1p^*$.

There is one Region II mode in Fig. 12 located at PA= 55% and is an interesting case. By now, there may be a misinterpretation of what is meant by "low" load effect in Table 2. While it may be true the absolute range of minimum damping values is relatively small for PA= 55% in Fig. 12, the impact of increasing the load initiates five distinct jumps across either side of the neutral stability line. This is the same type of intermittent flutter pattern seen by Cardinale in Figure 2. So, in Region II, while the impact of loading is "low" in terms of change in damping magnitude, the impact of loading is quite high and truly meaningful from the perspective of stability inflections.

The two Region III modes seen in Fig. 12 correspond to pitching axes of PA= 65% & 75%. As expected, these pitching axis locations are most unstable. At least for $k = 0.3$, increasing

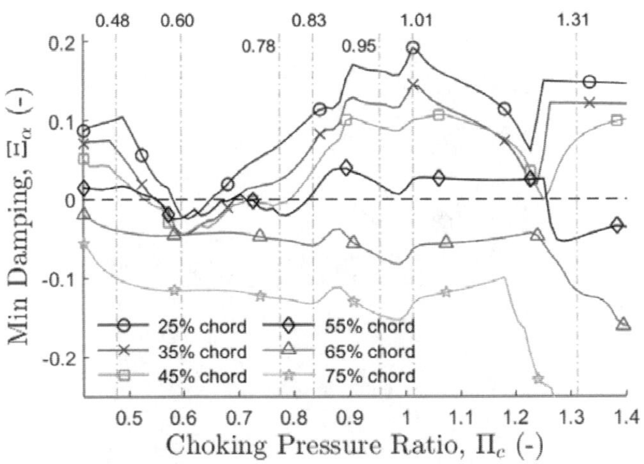

FIGURE 12: Minimum non-dimensional damping coefficient over all IBPAs as a function of choking pressure ratio for select pitching axes. Reduced frequency is held constant: $k = 0.3$.

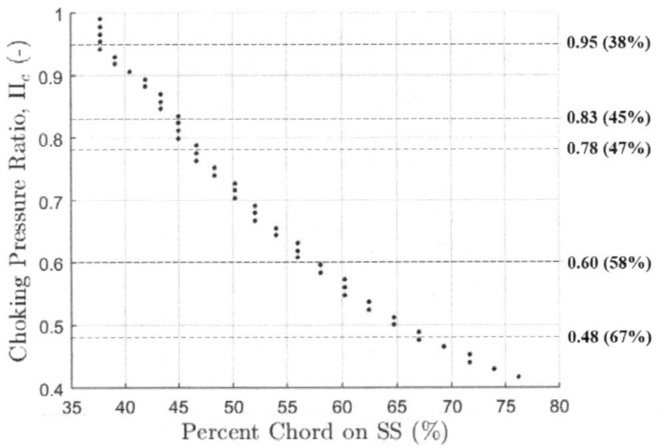

FIGURE 13: Evolution of shock location on the airfoil suction side as a function of pressure ratio.

loads is only helpful for either very low or very high values of Π_c. Between $0.60 < \Pi_c < 1.10$, the effect of pressure ratio and shocks carry minimal weight in changing stability.

Thus far, the analyses have been limited to rotational modes whose pitching axes are located on the chord-line. To offer an expanded look into how loading changes as a function of other pitching modes, seven Tie-Dye (TD) featuring distinct values of Π_c are shown in Figures 14-20. The layout and contour scale is identical between each TD plot. A black rectangle is placed in the center to indicate the LE/TE positions and provide the chord-length scale. Each plot has a pitching axis resolution of 200x200 spanning ± 2.5 chords in both the ξ- (parallel to chord from LE to TE) and η- (perpendicular to chord going from PS (down) to SS (up)) directions.

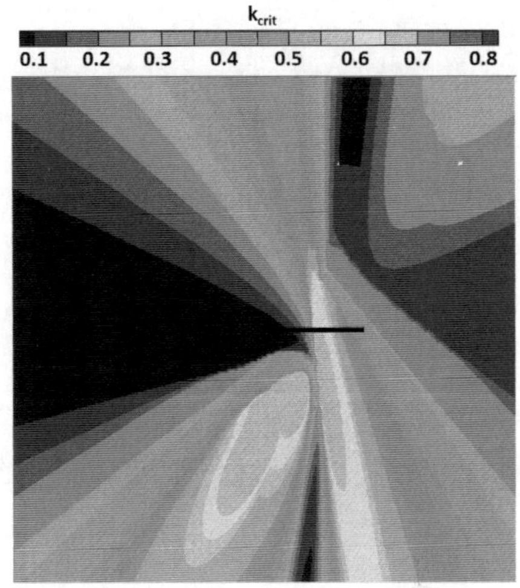

FIGURE 14: $\Pi_c = 0.48$, $M_2 = 1.42$ Tie-Dye plot: ± 2.5 chords.

Certainly, the sheer amount of data embedded within these seven TD figures is vast. Due to length constraints, individual analysis is left as an exercise to the reader. However, there are some high-level observations worth noting. Regardless of load, several areas in the TD plot visibly maintain their shape and relative stability level. A region that first comes to mind is the high k_{crit} region located approximately one chord below the LE (in the negative η direction). Pitching axes located in this region are quite unstable for normal levels of $k < 0.8$. Of course, an exception arises for the highest loading case in Fig. 14 where these pitching axes are slightly less likely to flutter with a $k_{crit} \approx 0.65$.

The second mainstay of the TD plots, regardless of load, is the high gradient region cutting through the airfoil. This undesirable region of low stability is essentially comprised of Regions II & III as defined above in Fig. 10. Region II can be interpreted as the quick transition from blue to green to red at approximately PA= 50% chord for each value of Π_c. Again, an exception exists for $\Pi_c = 0.48$, as this high loading condition demonstrates a greater stability tolerance.

The third and final region the authors will point out is the fairly stable (blue) region located in the upper left-hand corner of each plot. Although the exact area and value of k_{crit} slightly shifts as a function of load, this is without question the largest, consistently stable pitching axis zone. However, for once, the higher loads are not the most stable here as some intermediate (green) values of k_{crit} come into play.

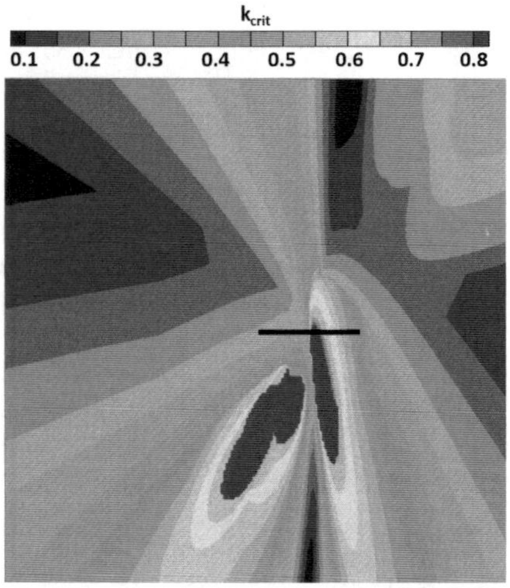

FIGURE 15: $\Pi_c = 0.60$, $M_2 = 1.30$ Tie-Dye plot: ± 2.5 chords.

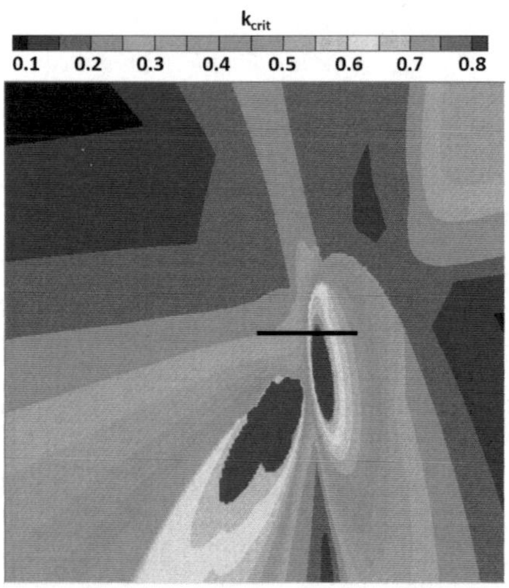

FIGURE 16: $\Pi_c = 0.78$, $M_2 = 1.18$ Tie-Dye plot: ± 2.5 chords.

In closing this section, the authors will note what is called the miracle of the Tie-Dye plot. When holding Mach conditions constant, the miracle of the Tie-Dye plot is that k_{crit} contours are relatively constant even for different blade geometries [7]. Therefore, despite investigating only one LPT geometry here, it is a reasonable hypothesis that many qualitative trends observed for the STCF 4 pitching and bending modes may also be readily applied to other LPT geometries.

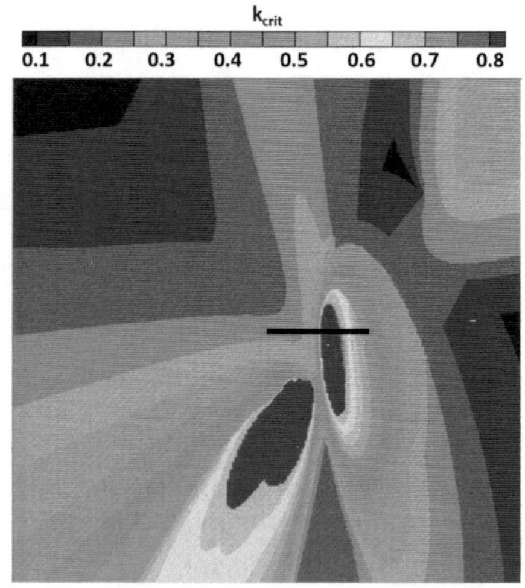

FIGURE 17: $\Pi_c = 0.83$, $M_2 = 1.07$ Tie-Dye plot: ± 2.5 chords.

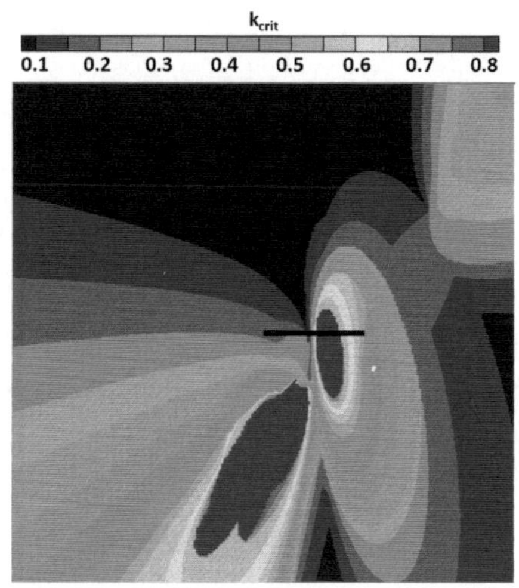

FIGURE 18: $\Pi_c = 0.95$, $M_2 = 0.97$ Tie-Dye plot: ± 2.5 chords.

FUTURE WORK

This study has presented evidence supporting the hypothesis that the manner in which loading affects flutter margin is a function of mode shape. There are three major follow-up steps needed in the immediate future: 1) Physical understanding of how loading/shocks influence stability 2) Generalization to other blade geometries and 3D mode shapes 3) Preliminary design tools.

Physical understanding is the most important next step. Previous studies have indicated that the shock strength, relative position of the shock to the elastic axis, and the variation in passage

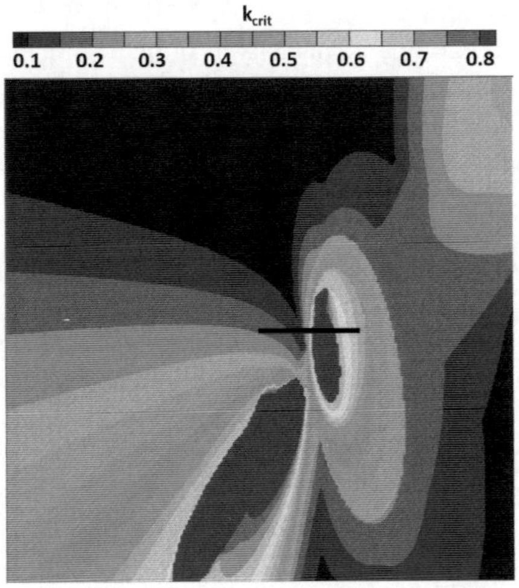

FIGURE 19: $\Pi_c = 1.01$, $M_2 = 0.93$ Tie-Dye plot: ± 2.5 chords.

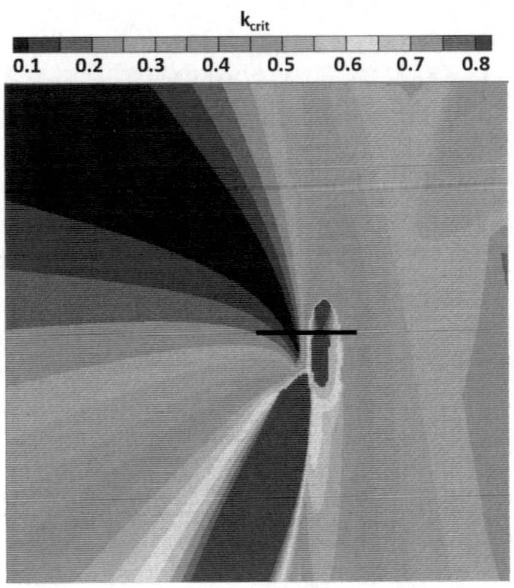

FIGURE 20: $\Pi_c = 1.31$, $M_2 = 0.70$ Tie-Dye plot: ± 2.5 chords.

throat as a function of mode and IBPA may all play a crucial role in solving the underlying physics. Figure 21 shows how the ratio of the max-to-min throat distance normalized by the mean throat distance varies as a function of mode shape and IBPA. For the bending modes of Fig. 21a, the largest variation in throat distance comes at $\theta = 100^\circ$, or nearly perpendicular to the chordline, when the adjacent airfoils are moving in anti-phase as one would imagine. For pitching modes in Fig. 21b, it is more complicated. The largest variations from the mean throat distance (located near 67% chord on the SS of STCF 4) are for higher values of IBPA and pitching between PA$= 0-25\%$ chord. Once the pitching axis moves aft of steady throat location, there is little change in mean throat for all IBPAs.

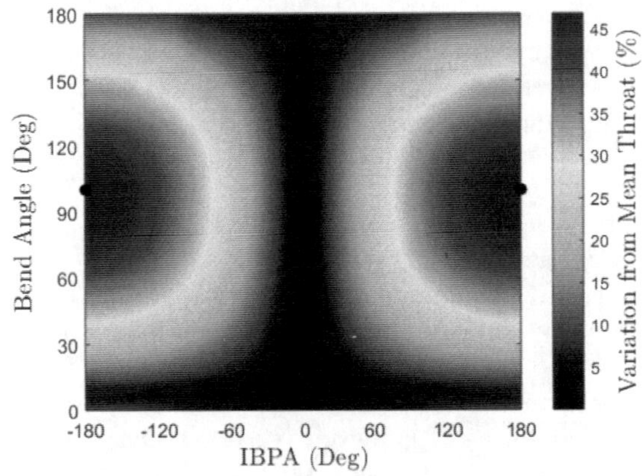

(a) Bending

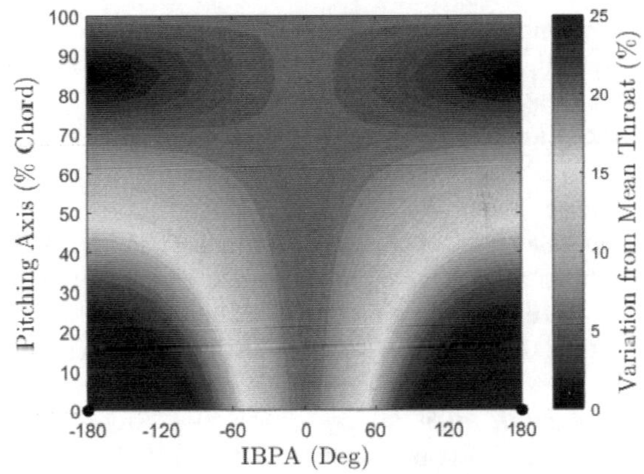

(b) Pitching

FIGURE 21: Visualization of peak-to-peak variation of minimum throat distance normalized by mean throat distance as a function of IBPA and mode shape.

Back in Fig. 11b, it was unmentioned that the most erratic region of least stable IBPA occurs between 30-70% chord. Perhaps unsurprisingly, this is the same region in which the shock exists for the five loading conditions below $\Pi_c = 1$ as seen in Fig. 13. The authors wish to formulate a mathematical expression that predicts work-per-cycle contributions as a function of loading condition, mode shape, and IBPA. This effort will first solve the problem using simplified airfoils before generalizing lessons learned to all possible airfoil design, in addition to creating a concrete set of preliminary design guidelines.

CONCLUSION

This work examined the effect of loading on LPT flutter stability as a function of bending and pitching rigid body mode shapes and reduced frequency. In addition, several important conclusions and observations are listed below:

1. The effect of loading on aerodynamic damping is a function of mode shape.
2. For Standard Configuration 4, k_{crit} variations as a function of bending mode and load can be divided into four regions:

   Bend Region I: $(-5^\circ < \theta < 20^\circ) \rightarrow$ High, non-monotonic load sensitivity and high sensitivity to mode shape. $k_{crit} < 0.4$
   Bend Region II: $(20^\circ < \theta < 90^\circ) \rightarrow$ High sensitivity that stabilizes with increased load and low, roughly linear sensitivity to mode. $k_{crit} < 0.6$
   Bend Region III: $(90^\circ < \theta < 130^\circ) \rightarrow$ Most stable region. Low sensitivity to load (non-monotonic) and mode shape. $k_{crit} < 0.2$
   Bend Region IV: $(130^\circ < \theta < 175^\circ) \rightarrow$ High sensitivity that destabilizing with increased load and low, roughly linear sensitivity to mode. $k_{crit} < 0.55$
3. For non-edgewise bending modes, the least stable IBPA is a strong function of bend direction, and less load dependent.
4. For Standard Configuration 4, k_{crit} variations as a function of pitching mode and load can be divided into four regions:

   Pitch Region I: $(0\% \leq P.A. < 46\%) \rightarrow$ Most stable region. High, non-monotonic load sensitivity and high sensitivity to mode shape. $k_{crit} < 0.3$
   Pitch Region II: $(46\% \leq P.A. < 55\%) \rightarrow$ Very low sensitivity to load and very high, linear sensitivity to mode. $0.1 < k_{crit} < 0.6$
   Pitch Region III: $(55\% \leq P.A. < 85\%) \rightarrow$ Least stable region. High sensitivities to both load and mode. Higher loads stabilize. $k_{crit} > 0.5$
   Pitch Region IV: $(85\% \leq P.A. \leq 100\%) \rightarrow$ Medium load sensitivity and low, linear sensitivity to mode. $0.4 < k_{crit} < 0.7$

ACKNOWLEDGMENTS

The authors wish to acknowledge the American Society for Engineering Education for supporting this research with the NDSEG Fellowship.

REFERENCES

[1] Panovsky, J., and Kielb, R. E., January 2000. "A design method to prevent low pressure turbine blade flutter". *Journal of Engineering for Gas Turbines and Power,* **122**, pp. 89–98.

[2] Tchernycheva, O. V., Fransson, T., Kielb, R. E., and Barter, J., 2001. "Comparative analysis of blade mode shape influence on flutter of two-dimensional turbine blades". *ISABE Paper ISABE-2001-1243.*

[3] Waite, J. J., and Kielb, R. E., 2014. "Physical understanding and sensitivities of low pressure turbine flutter". *Journal of Engineering for Gas Turbines and Power,* ***137 (1)***.

[4] Nowinski, M., and Panovsky, J., 2000. "Flutter mechanisms in low pressure turbine blades". *ASME Journal of Engineering for Gas Turbines and Power,* **122**, pp. 82–87.

[5] Clark, S. T., 2013. "Design for coupled-mode flutter and non-synchronous vibration in turbomachinery". *ProQuest Dissertations and Theses, 246.*

[6] Kielb, R. E., 2001. "CFD for turbomachinery unsteady flows - an aeroelastic design perspective". *AIAA Paper 2001-0429.*

[7] Meingast, M., Kielb, R. E., and Thomas, J. P., 2009. "Preliminary flutter design method for supersonic low pressure turbines". *ASME Paper GT2009-59177.*

[8] Waite, J. J., and Kielb, R. E., 2015. "The impact of blade loaload and unsteady pressure bifurcations on low-pressure turbine flutter". *ASME Paper GT2015-42857.*

[9] Corral, R., and Vega, A., 2015. "The low reduced frequency limit of vibrating airfoils. part i: Theoretical analysis". *ASME Paper GT2015-42440.*

[10] Carta, F. O., and Hilaire, A. O. S., 1980. "Effect of interblade phase angle and incidence angle on cascade pitching stability". *ASME Journal of Engineering for Gas Turbines and Power,* **102**, pp. 391–396.

[11] Széchényi, E., 1985. "Fan blade flutter - single blade instability or blade to blade coupling?". *ASME Paper 85-GT-216.*

[12] Cardinale, V. M., Bankhead, H. R., and McKay, R. A., 1981. "Experimental verification of turboblading aeromechanics". *56th Symposium of the AGARD Propulsion and Energetics Panel*(293).

[13] Srivastava, R., Bakhle, M. A., Keith, T. G., and Stefko, G. L., 2002. "Flutter analysis of a transonic fan". *NASA/TM-2002-211818.*

[14] Srivastava, R., and Keith, T. G., 2005. "Influence of shock wave on turbomachinery blade row flutter". *Journal of Propulsion and Power.*

[15] Bölcs, A., and Fransson, T. H., 1986. "Aeroelasticity in turbomachines comparison of theoretical and experimental cascade results". *Communication du Laboratoire de Thermique Appliquée et de Turbomachines.* EPFL, No. 13.

[16] Hall, K. C., Thomas, J. P., and Clark, W. S., May 2002. "Computation of unsteady nonlinear flows in cascades using a harmonic balance technique". *AIAA Journal,* ***40***(5).

[17] Shapiro, A. H., 1953. *The Dynamics and Thermodynamics of Compressible Fluid Flow*, Vol. 1. The Ronald Press Company.

Proceedings of the 14th International Symposium on Unsteady Aerodynamics, Aeroacoustics & Aeroelasticity of Turbomachines
ISUAAAT14
8-11 September 2015, Stockholm, Sweden

I14-S1-4

ROTOR BLADE FLUTTER IN LAST STAGE OF LP STEAM TURBINE

Romuald RZADKOWSKI, Artur MAURIN, Marcin DREWCZYNSKI, Leszek KUBITZ
Institute of Fluid-Flow Machinery
Polish Academy of Sciences
Fiszera 14, PL 80-231 Gdańsk, Poland
z3@imp.gda.pl

Vitally GNESIN, Luba KOLODYAZHNAYA
Department of Unsteady Gasdynamics and Aeroelasticity
Institute of Mechanical Engineering Problems
National Academy of Sciences of Ukraine
2/10 Pozharsky st., Kharkov,61046 Ukraine
gnesin@ipmach.kharkov.ua

Ryszard SZCZEPANIK
Air Force Institute of Technology
Ksiecia Bolesława 6,
01-494 Warszawa, Poland
ryszard.szczepanik@itwl.pl

ABSTRACT

This paper presents a numerical study of flutter in the last stage of a LP steam turbine blades using in-house codes and ANSYS CFX. The mode shapes and natural frequencies of the steam turbine bladed disk are calculated using an FEM model. The instability regions in 5 mode shapes and the distribution of the aerodamping coefficient along the blade length were shown in the case of harmonic blade oscillations with an assumed interblade phase angle.

INTRODUCTION

The design of blades in the last stage of steam turbines is the one of the most demanding engineering tasks. Increasing turbine efficiency has led to the designing of higher tip-to-hub ratios. Slender blading conforms to reliability requirements, such as high blade stiffness and a high first mode natural frequency. Unsteady prediction models for 3D non-viscous and viscous flutter have been discussed in literature over the last ten years (Huang 2006, Rzadkowski and Gnesin 2007, Gnesin at el. 2004, Sanvito at el. 2012)

Several high vibration amplitude problems have been reported regarding the slender last stage blading of commercial LP steam turbines. To solve this problem it was decided that the blades should be geometrically mistuned to prevent self-excitation or flutter. In this paper the 3D Euler and a 3D RANS solver (with the modified Baldwin and Lomax algebraic eddy viscous turbulence model and k-ε turbulence model) are used to calculate unsteady viscous and non-viscous flow through last stage LP steam turbine rotor blades, taking into account blade oscillations but excluding non-symmetrical pressure distribution behind the rotor blades caused by the exhaust hood.

NOMENCLATURE

p	pressure
C	damping matrix
F	unsteady aerodynamic forces
IBPA	interblade phase angle
K	stiffness matrix
M	mass matrix
T	temperature
ν	natural mode frequency
$\overline{L}$	dimensionless blade length

AERODYNAMICAL MODEL

Here the 3D transonic flow of an ideal and viscous gas through a multipassage blade row is considered. The flow equations are written for a three dimensional Cartesian co-ordinate system which is fixed to a rotating blade row. The numerical method uses the second order by time and coordinates an explicit finite-volume Godunov's type difference scheme with a moving H-O structured grid.

The numerical method uses the second order by time and coordinates an explicit finite-volume Godunov's type difference scheme with a moving H-O structured grid. The structure analysis used the modal approach and a 3D finite element blade model.

Hybrid H-H and O-O grids were used for each rotor passage. During the calculations, the outer H-grid of the rotor remained stationary, while the inner O-grid or H-grid was rebuilt in each iteration, in accordance with the blade motion.

To validate the numerical viscous code, the numerical results were compared with data measured for the 11th international standard configuration.

STRUCTURAL MODEL

The blade vibration formulation is based on a modal approach of the coupled problem (Gnesin et al. 2004). The dynamic model of the oscillating blade in linearized formulation is governed by matrix equation:

$$M\ddot{u}+C\dot{u}+Ku=F, \qquad (1)$$

where *M*, *C*, *K* are the mass, mechanical damping and stiffness matrices of the blade respectively; *u(t)* is the blade displacement; *F* is the unsteady aerodynamic forces vector, which is a function of blade displacement.

The first step of the modal approach consists of solving the problem of the natural mode shapes and eigenvalues without damping and in a vacuum. Then the displacement of each blade can be written as a linear combination of the first *N* modes shapes with the modal coefficients depending on time:

$$u=Uq=\sum_{i=1}^{N} U_i q_i \qquad (2)$$

where U_i is the displacement vector corresponding to i- th mode shape; $q_i(t)$ is the modal coefficient of i- th mode. Taking into account the equation (2) and the orthogonality property of the mode shapes the equation (1) can be written in form of:

$$I\ddot{q}+H\dot{q}+\Omega q=\lambda(t), \qquad (3)$$

where I=diag (1, 1, ... 1), H=diag (2 h_1, 2 h_2, ... 2 h_n), Ω =diag (ω_1^2, ω_2^2, ... , ω_n^2) are diagonal matrices; ω_i is i-th natural blade frequency; $\{\lambda(t)\}$ is the modal forces vector corresponding to the mode shapes, $h_i = \omega_i \xi_i$, where ξ_i is the i-th modal damping coefficient (Gnesin et al. 2004). Thus the dynamic problem Eq. (1) reduces to the set of independent differential equations relatively to modal coefficients of natural modes (Gnesin et al. 2004):

$$\ddot{q}_i+2h_i\dot{q}_i+\omega_i^2 q_i=\lambda_i, \qquad (4)$$

The equations of motion (4) can be solved using any standard integration method.

The modal forces λi are calculated at each iteration with the use of the instantaneous pressure field in the following way (Gnesin et al. 2004):

$$\lambda_i=\frac{\iint_{\sigma} p\overline{U}_i\cdot\overline{n}^{\circ}\,d\sigma}{\iiint_{v}\rho\overline{U}_i^2\,dv}. \qquad (5)$$

where p is the pressure along the blade surface.

NUMERICAL RESULTS

Numerical calculations have been carried out for last stage LP steam turbine rotor blades. The calculations were performed for harmonic oscillations of the blade row according to natural modes with the same amplitude and interblade phase angle (IBPA). The calculations were carried out for a nominal regime with the gasodynamic parameters of P_0 =12800 Pa, T_o=323 K at the inlet stage and P_2=2300 Pa at the outlet stage. The aeroelastic behaviour of the vibrating blade row was defined by the aerodamping coefficient D, which was equal to the negative work coefficient performed during one cycle of oscillations. The negative values of D corresponded to the transfer of mean flow energy to the blade (self-excitation), and positive values to the dissipation of oscillating blade energy to the flow (aerodamping).

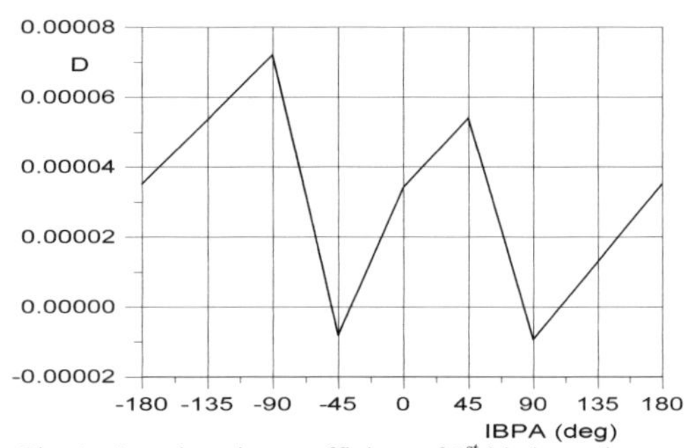

Fig. 1. Aerodamping coefficient of 1st blade mode, non-viscous model

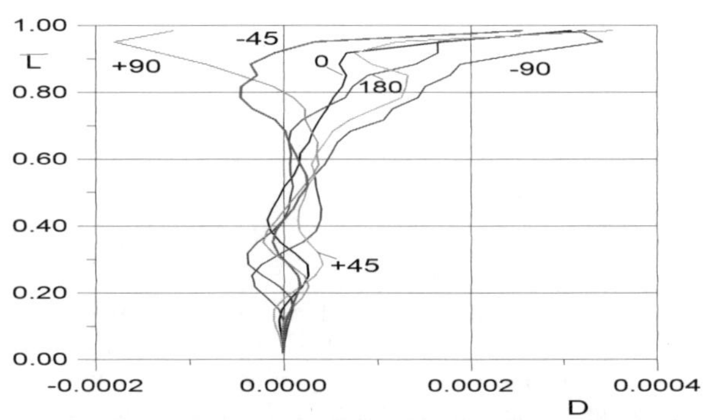

Fig. 2. Aerodamping coefficient along blade length of 1st blade mode, non-viscous model

The aerodamping coefficient calculated for a rotor blade at different interblade phase angles (IBPA) for the non-viscous code and the 1st mode showed that flutter appeared at a -45 deg and 90 deg IBPAs (Fig. 1). Fig 2 presents the aerodamping coefficient along the blade length for the 1st mode and various IBPAs. In the case of the 1st mode shape and IBPA=90 deg the flutter appeared in the region of 1 to 0.8 and 0.3 to 0.45 of the blade length and for IBPA=45 deg in the region of 0.7 to 0.95 and 0.3 to 0.4. For IBPA 180 deg flutter appeared in a small zone of 0.2 to 0.4 of the blade length, IBPA=-90 deg in the region of 0.1 to 0.3, IBPA=0 in the region of 0.3 to 0.5.

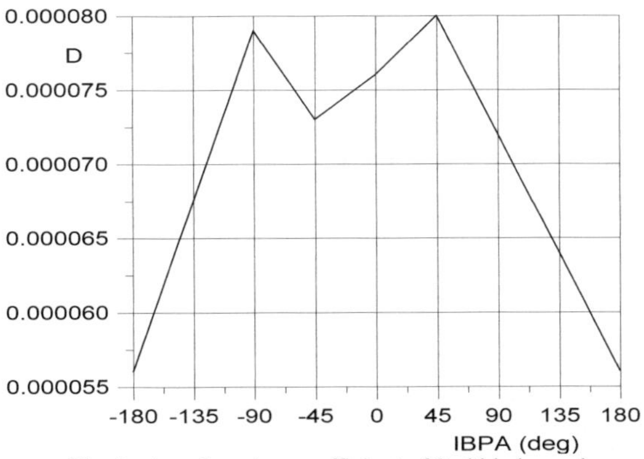

Fig. 3. Aerodamping coefficient of 2nd blade mode, non-viscous model

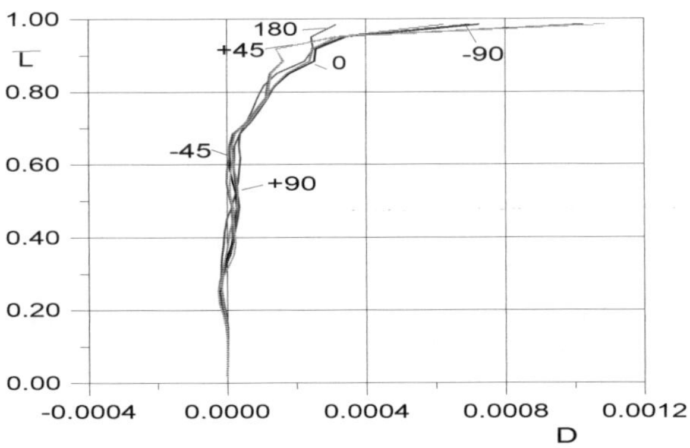

Fig. 4. Aerodamping coefficient along blade length of 2nd blade mode, non-viscous model

The oscillations in the 2nd mode, in which both bending and torsional motion occurred, were damped for all interblade phase angles (Fig. 3). For all the interblade phase angles, the flutter appeared in the small region of 0.1 to 0.4 of the blade length (Fig. 4).

It should be pointed out that oscillations in the 3rd mode are characterized by minimal values of an aerodamping coefficient near the IBPA of 45 deg. (Fig. 5). In the 3rd mode shape (see Fig. 6) and with IBPA = 90 deg the flutter appeared in the region of 0.74 to 1.0 of the blade length, whereas with IBPA=45 deg in the regions of 0.4 to 0.5 and 0.7 to 0.9 and 0.95 to 1 of the blade length.

The oscillations in the 4th mode were damped for all interblade phase angles (Fig. 7). In the 4th mode shape (see Fig. 8) and IBPA= 180 deg. the flutter appeared in the region of 0.9 to 1 of the blade length.

The oscillations in the 5th and 6th modes were damped for all interblade phase angles (Fig. 9, 10, 11, 12).

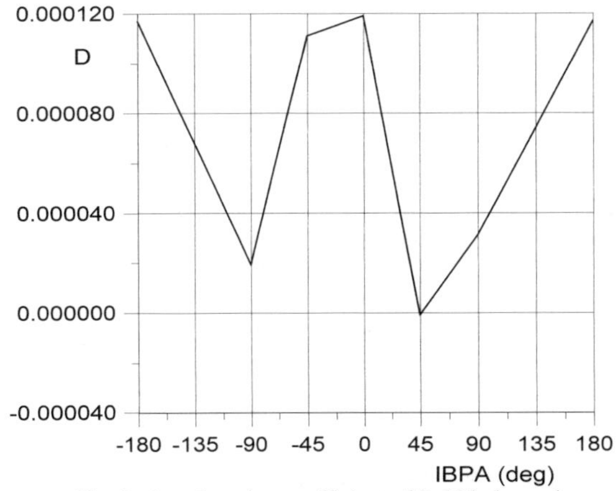

Fig.5. Aerodamping coefficient of 3rd blade mode, non-viscous model

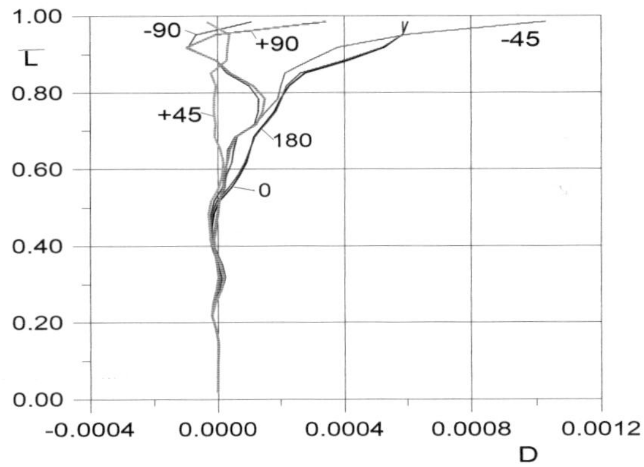

Fig. 6. Aerodamping coefficient along the blade length of 3rd blade mode, non-viscous model

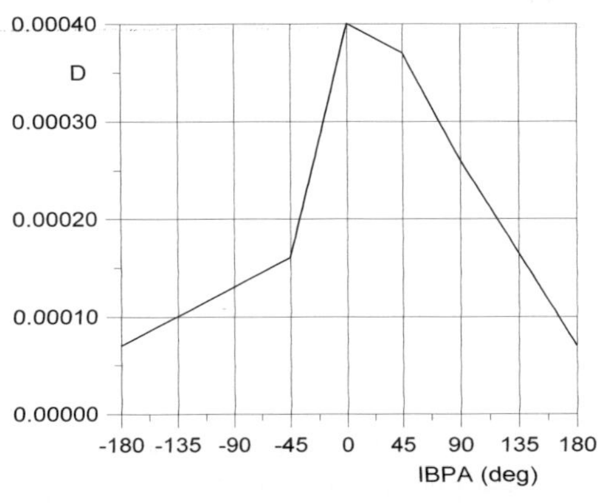

Fig. 7. Aerodamping coefficient of 4th blade mode, non-viscous model

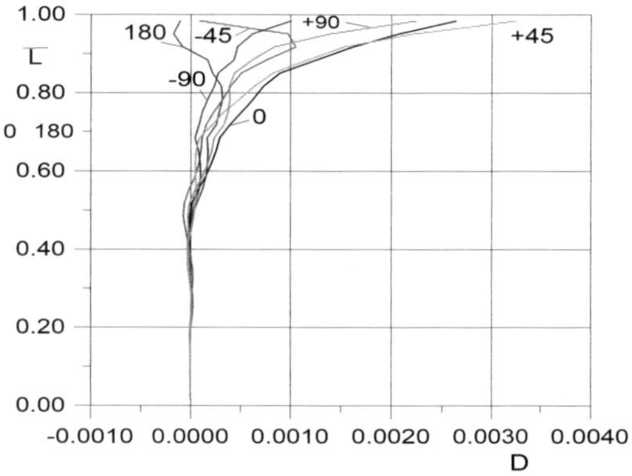

Fig. 8. Aerodamping coefficient along the blade length of 4th blade mode, non-viscous model

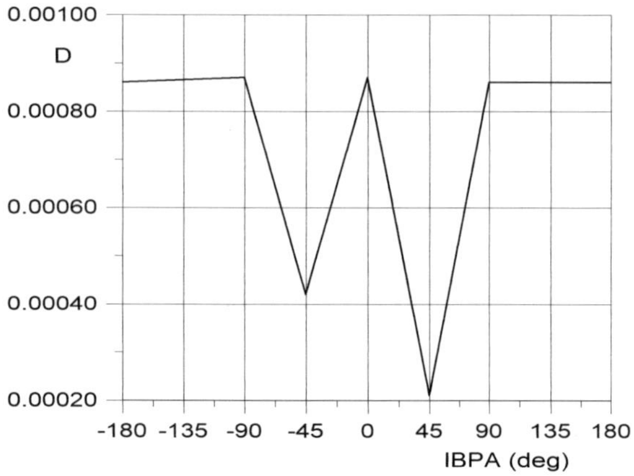

Fig. 9. Aerodamping coefficient of 5th blade mode, non-viscous model

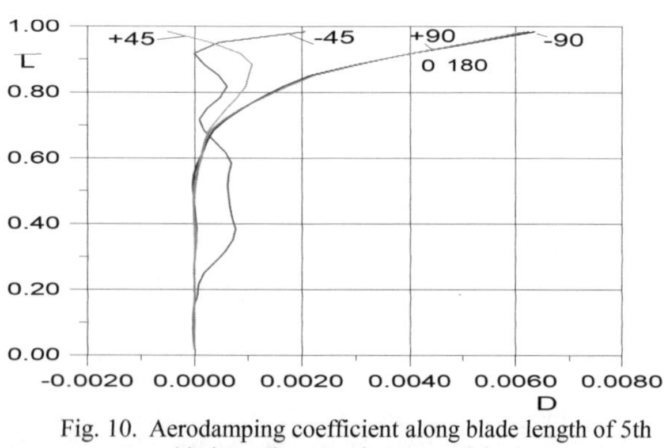

Fig. 10. Aerodamping coefficient along blade length of 5th blade mode, non-viscous model

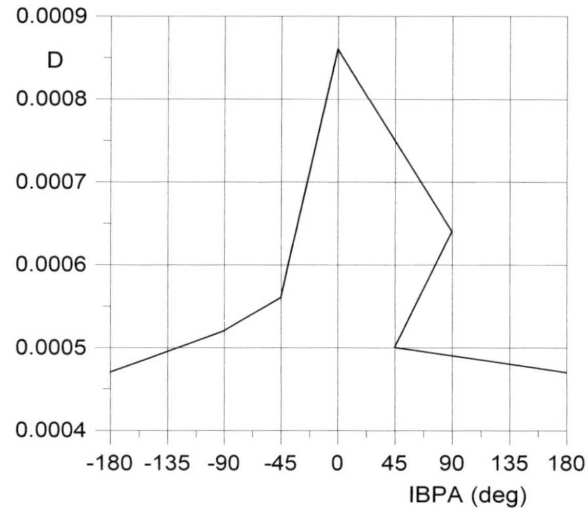

Fig. 11. Aerodamping coefficient of 6th blade mode, non-viscous model

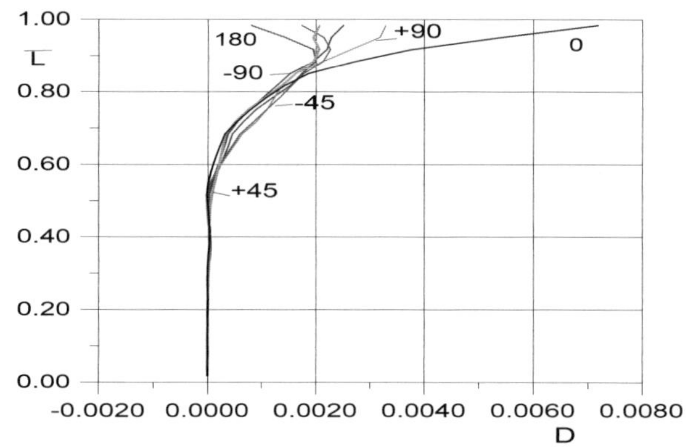

Fig. 12. Aerodamping coefficient along blade length of 6th blade mode, non-viscous model

In the 3D viscous calculation and Baldwin Lomax turbulence model flutter appeared with the 1st mode shape and IBPA = -90 deg (see Fig. 13), whereas at IBPA = +90 deg only a small area of instability appeared with no flutter. In this case there was little difference between the viscous and non-viscous calculations (Fig. 1).

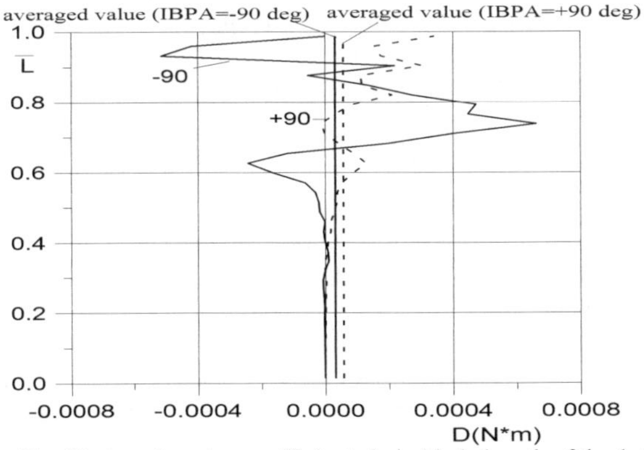

Fig. 13. Aerodamping coefficient along blade length of the 1st mode Baldwin-Lomax viscous model

In the 3D viscous calculation using the ANSYS CFX and TBR methods and the k-ε turbulence model, flutter appeared in the 1st mode shape with the IBPA from -100 deg. to -60 deg. (see Fig. 14). Fig 15 a, b present aerodynamic work along the blade surfaces with the IBPA =-90 deg. The oscillations in the 2nd, 3rd, 4th and 5th modes were damped in all the interblade phase angles

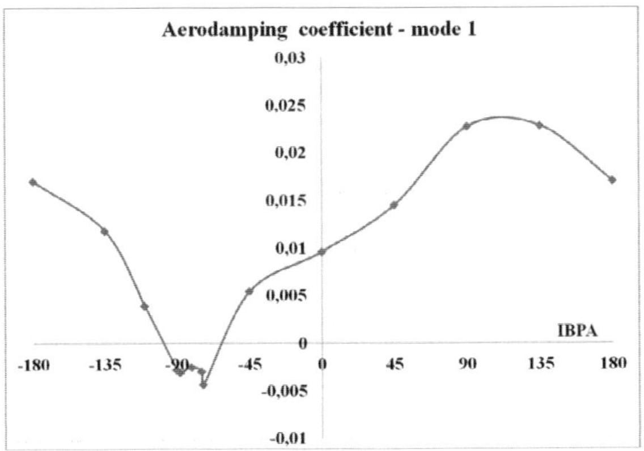

Fig. 14. Aerodamping coefficient of the 1st blade mode viscous model, TBR method

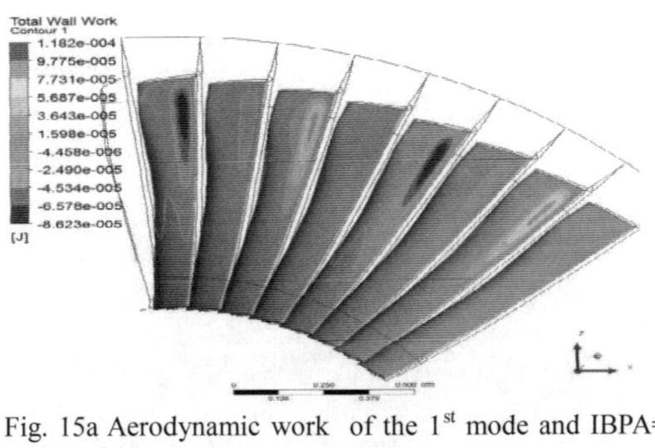

Fig. 15a Aerodynamic work of the 1st mode and IBPA=90 deg. suction side

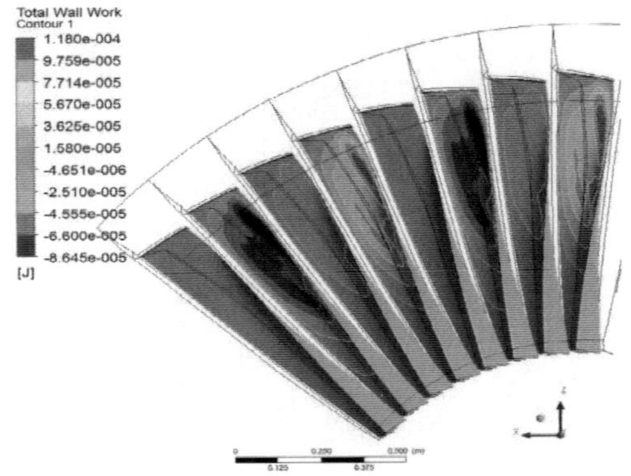

Fig. 15 Aerodynamic work of 1st mode and IBPA=90 deg. pressure side

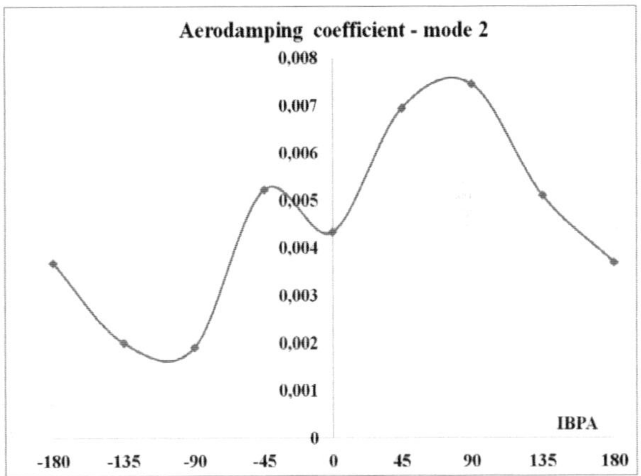

Fig. 16. Aerodamping coefficient of 2nd blade mode viscous model, TBR method

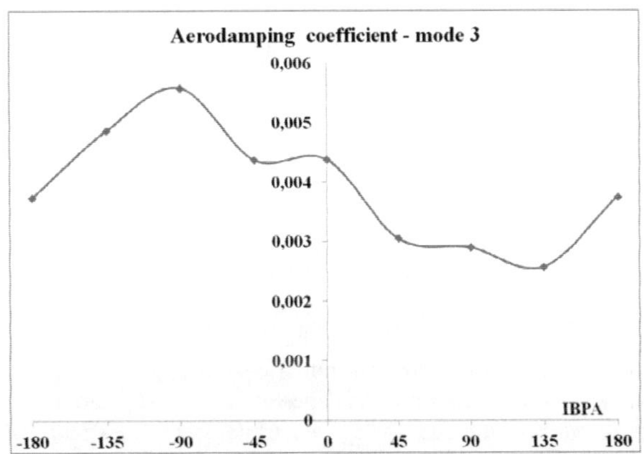

Fig. 15. Aerodamping coefficient of the 3rd blade mode viscous model, TBR method

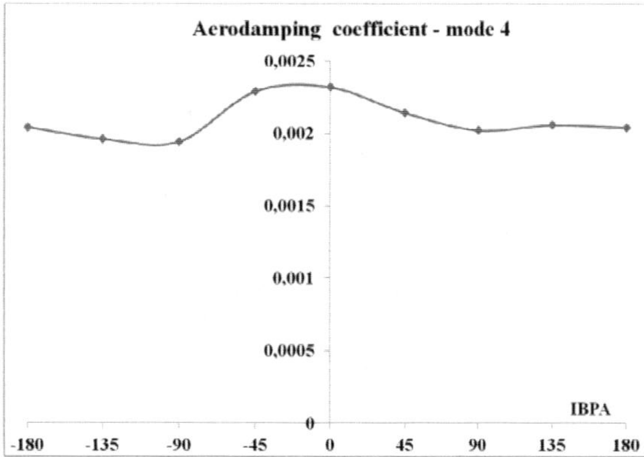

Fig. 16. Aerodamping coefficient of the 4th blade mode viscous model, TBR method

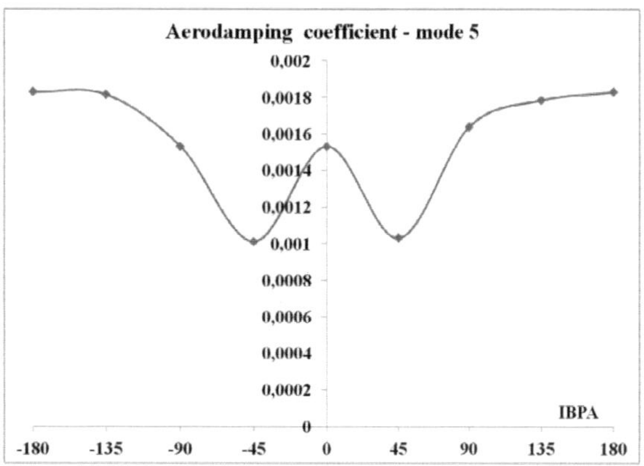

Fig. 17. Aerodamping coefficient of the 5th blade mode viscous model, TBR method

It is interesting to note that the sign of the aerodamping coefficient changes along the blade length. On a part of the blade the aerodamping coefficient is negative D<0 (the flutter condition), while on another part it is positive D>0 (the damping condition). These areas change with different mode shapes. The viscous calculations gave different results to those of the non-viscous calculations, but in both models the flutter appeared only in the first mode.

CONCLUSIONS

In the presented study, the simultaneous time domain method and the modal superposition method were used to determine the aeroelastic stability of last stage LP steam turbine blades. The influence of the natural modes on the aeroelastic blade behavior was numerically analysed.

Numerical calculations were performed for harmonic oscillations of the blade row in various natural modes with the same amplitude and various interblade phase angles (IBPA) using 3d viscous and non-viscous codes. It was found that flutter appeared only in the first mode. It is interesting to note that the aerodamping coefficient sign changes along the blade length and is dependent on the mode shape. In one part of the blade it is negative, while another part it is positive. These areas change, depending on the mode shape.

A numerical study of a steam turbine bladed disc with real, small mistuning shows that nodal diameters do not appear in the first four families associated with the first four mode shapes of a cantilever blade [Drewczynski et al. 2015]. In these modes only individual blades vibrate. This conclusion allows for new approaches to flutter analysis in the LP steam turbine rotor blades, where the IBPA should not be used. Instead the vibration of blades in the flow is modelled according to various mistuned bladed disc modes.

ACKNOWLEDGMENTS

The authors wish to thank NCBiR (project PBS1/B4/5/2012) for the financial support of this work. All numerical calculations were made at the Academic Computer Centre TASK (Gdansk, Poland).

REFERENCES

[1] Huang X.Q., Influence of Upstream Stator on Rotor Flutter Stability in a Low Pressure Steam Turbine Stage, *Proceedings of the Institution of Mechanical Engineers. Part A Journal of Power and Energy* 01/2006, 220(1), 25-35, 2006.

[2] Rządkowski R., Gnesin V., A 3D Inviscid Self-Excited Vibration of the Last Stage Turbine Blade Row, *Journal of Fluids and Structures,* 23, 858-873, 2007.

[3] Gnesin V., Kolodyazhnaya L., Rzadkowski R., A Numerical Model of Stator-Rotor Interaction in a Turbine Stage with Oscillating Blades, *Journal of Fluids and Structures*, 19(8), 1141-1153, 2004.

[4] Sanvito M., Pesatori E., Bachschmidt N., Chatterton S., Analysis of LP Steam Turbine Blade Vibration: Experimental Results and Numerical Simulations, 189-197, *10th International Conference on Vibrations in Rotating Machinery*, 11-13 Sept 2012, London, IMechE.

[5] Drewczyński M., Rzadkowski R., Maurin A., Marszałek P., Free Vibration of a Mistuned Steam Turbine Last Stage Bladed Disc, *Proceedings of ASME TURBO EXPO 2015,* June 15 – 19, 2015, Montreal, Canada, GT 2015-26011.

Proceedings of the 14th International Symposium on Unsteady Aerodynamics, Aeroacoustics & Aeroelasticity of Turbomachines
ISUAAAT14
8-11 September 2015, Stockholm, Sweden

I14-S1-5

TWO-DIMENSIONAL STEAM TURBINE FLUTTER TEST CASE

Paul Petrie-Repar
Royal Institute of Technology
Department of Energy Technology
S – 10044 Stockholm, Sweden
paul.petrie-repar@energy.kth.se

Christopher Fuhrer, Marius Grübel, Damian M. Vogt
University of Stuttgart, ITSM
Stuttgart, Germany

ABSTRACT

A new two-dimensional (2D) test case for steam turbine flutter is presented. Unlike previous 2D test cases, the geometry and flow conditions are similar to those of a typical steam turbine near the tip of the last stage. The flow near the tip is important for steam turbine flutter analysis because this is where most of the unsteady aerodynamic work is done that can lead to blade flutter.

The test case will be used to examine the periodic boundary condition, the grid motion, the grid convergence and the non-reflecting boundary condition of the linearized flow solver LUFT. It will be shown that the periodic boundary condition is working correctly and that the unsteady flow solutions are independent of the grid motion and location of the inlet and outlet.

Steady-state solutions assuming ideal gas and equilibrium wet steam for the test case calculated by LUFT and ANSYS CFX are presented. The agreement between the solutions is good. A steady-state solution which considers the finite growth rate in the water droplets (non-equilibrium wet steam) was calculated with ANSYS CFX. The non-equilibrium wet steam solution predicts that subcooling up to 30 K occurs in regions near the short blade passage (throat) where rapid expansion occurs and the shock on the suction side was 6% of chord further upstream. Unsteady flow simulations were performed using LUFT and ANSYS CFX to calculate the aerodynamic damping due to a bending mode. There were some differences in the LUFT and CFX solutions for ideal steam and equilibrium wet steam simulations. More solutions from other methods are required to determine the correct solution for this test case. The influence of non-equilibrium wet steam effects on the aerodynamic damping was also examined and some differences were observed, particularly on the aerodynamic work near the shock on the suction surface.

NOMENCLATURE

T_0:	total temperature
σ:	interblade phase angle (IBPA)
EQS:	equilibrium wet steam
NES:	non-equilibrium steam

INTRODUCTION

The turbine blades of the last stage of steam turbines for power generation are typically over one meter long. These long blades are susceptible to flutter because of their low structural frequency and supersonic tip speeds. Blade flutter is a concern for the manufacturers of steam turbines [1,2] and blade root cracking at the last stage of a steam turbine due to flutter has been reported [3].

Typically, steam condensation has occurred by the last stage. However, ideal gas is usually assumed when performing flutter analysis for steam turbines. A previous study [4] compared the results of a flutter analysis calculated assuming ideal gas and equilibrium wet steam. The blade geometry was taken from Standard Configuration 11[5] and the flow conditions were similar to flow conditions at the last stage in a typical steam turbine. The agreement between the ideal steam and equilibrium wet steam solutions was good if the polytropic index of the ideal gas was chosen such that it matched the polytropic index of the wet steam flow. The influence of non-equilibrium effects due to subcooling and the finite rate of water droplet formation was not considered.

The aim of the current study is to present a new 2D test case for steam turbine flutter. Unlike previous 2D test cases, the geometry and flow conditions are similar to those of a typical steam turbine near the tip of the last stage. The flow near the tip is important for steam turbine flutter analysis because this is where most of the unsteady aerodynamic work is done that can lead to blade flutter.

Firstly, the test case will be used to examine the periodic boundary conditions, the grid motion, the grid convergence and the non-reflecting boundary condition of the LUFT code. It will thereby be shown that the steady-state and unsteady flow solutions are independent of the inlet and outlet (far-field) boundary location. Secondly, the test case will be used to assess the effect of steam modeling on the steady-state and unsteady flow by applying ideal gas, equilibrium steam and non-equilibrium steam models using ANSYS CFX.

TEST CASE

The geometry of the test case consists of 2.5 mm thick flat plates with half circles at the leading and trailing edges. The stagger angle is 78°. The chord length is 222.5 mm and the pitch is 198 mm. The flow conditions are shown in Table 1. Two flow conditions with different inlet total temperatures were examined. The total flow conditions are superheated steam for both conditions. The isentropic Mach numbers at the turbine exit are 1.361 (T_0= 340 K) and 1.365 (T_0= 340 K) for equilibrium wet steam.

Inlet Total Pressure:	24 kPa
Inlet Total Temperature:	340 K and 355 K
Inflow Angle:	78 degrees
Exit Pressure:	9 kPa

Table 1: Test Case Flow Conditions

The polytropic index and steam quality (mass fraction of saturated water) as a function of Mach number for an isentropic expansion of equilibrium wet steam for the two flow conditions are shown in Figure 1. The flow remains superheated steam when expanding from rest until the Mach number reaches 0.265 when T_0 = 340 K and condensation starts when the Mach number increases above 0.265. When T_0= 355 K, condensation starts when the Mach number exceeds 0.675. There is a step change in the polytropic index of the flow when condensation starts. The calculated inlet Mach number for this test case is approximately 0.652, hence all of the expansion occurs after condensation when the inlet T_0 = 340 K and most of the flow expansion occurs after condensation has started for T_0= 355 K. As a result it was assumed that the ratio of specific heats was 1.11 for the ideal gas simulations. The absolute viscosity for the ideal gas simulations was assumed to be 1.1e-5 Pa.s.

The prescribed blade mode shape is a plunging mode (bending in 3D) in the direction normal to the chord at 100 Hz (reduced frequency 0.267).

METHOD

The coupled aerodynamic and structural dynamic (aeroelastic) system must be considered in order to analyze flutter. An aeroelastic system can only be considered flutter free if all possible eigenmodes of the aeroelastic system are stable. The aeroelastic eigenmodes for a turbomachinery row are the traveling wave modes if the following assumptions are made: each blade is identical (tuned blades), unsteady aerodynamic forces do not alter the structural deformation or frequency, each blade has only one degree of freedom (DOF), unsteady flow perturbations are linear with blade vibrations and the aerodynamic (and/or structural) coupling between blades is symmetric. The number of traveling wave modes is equal to the number of blades and each traveling wave mode is a pattern of motion with the blades moving with a constant arbitrary amplitude and with a constant phase shift between adjacent blades. This phase shift is referred to as the inter-blade phase angle and is equal to

$$\sigma = \frac{2\,\pi\,n}{N}$$

where n is the nodal diameter and N is the number of blades in the row. The nodal diameters are integers with the range

$$-{}^{N}\!/_{2} \leq n \leq {}^{N}\!/_{2}$$

A positive nodal diameter indicates that the direction of the phase (and the perceived direction of travel) of the blade vibrations in the blade row is in the same direction as the rotation of the rotor. For this test case the blades are not rotating and the positive phase direction is defined as the direction the blades would rotate if they were rotating.

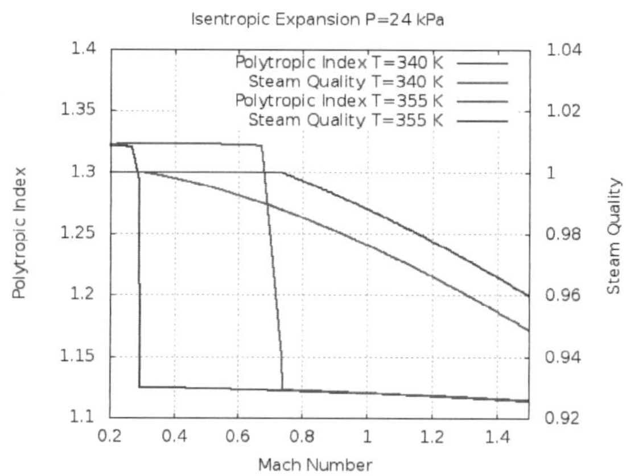

Figure 1: Isentropic expansion of wet steam for the two test conditions (total pressure 24 kPa)

The first task of a flutter analysis is to calculate the steady-state solution. The LUFT [6] code and the commercial flow solver ANSYS CFX were used to calculate steady-state solutions for the new test case. Both methods solve the Reynolds-averaged Navier-Stokes equations and used the Spalart and Allmaras turbulence model. Ideal steam and equilibrium wet steam (EQS) simulations were performed by both methods. Both methods used the formulae from the IAPWS-IF97 [7] to calculate the properties of the wet steam Non-equilibrium wet steam (NES) simulations were also performed with CFX. To model the condensation process the flow solver ANSYS CFX 16 is used. This model is based on an Eulerian-Eulerian approach and the interactions between the continuous and liquid phase are taken into account via source terms which are included in the conservation equations, see [8]. The steam properties are based on the IAPWS-IF97 formulation and the properties in the metastable region are evaluated by means of extrapolations from the superheated region. For all simulations, it is assumed that both phases share the same velocity field, which is valid for droplet sizes below 1 μm, see [9]. The applied model has been validated by means of nozzle and cascade flows as well as for turbine flows by [10] and [11].

For the calculation of the droplet formation the homogeneous nucleation rate

$$J = \frac{q_c}{1+\varepsilon}\sqrt{\frac{2\sigma}{\pi m^3}}\frac{\rho_v^2}{\rho_l}\cdot \exp\left(-\frac{4\pi\sigma r^{*2}}{3KT_v}\right)$$

is applied. In this equation, q_c is the condensation coefficient, ε is the non-isothermal correction from reference [12], σ is the surface tension of a flat water film, m is the mass of a water molecule, ρ_v and ρ_l are the densities of the vapor and the liquid, respectively, r* is the critical radius of a stable cluster of water molecules, K is Boltzmann's constant and T_v is the vapor temperature.

The subsequent droplet growth is modeled according to [9], where a correction factor based on the Knudsen number Kn is included to cover the free molecular regime of the flow

$$L\rho_l \frac{dr}{dt} = \frac{\lambda_v}{r}\frac{1}{1+3.18\,\mathrm{Kn}}\cdot \Delta T\left(1-\frac{r^*}{r}\right)$$

Here, L is the latent heat during condensation, dr/dt is the growth rate, λ_v is the thermal conductivity of the vapor and ΔT is the vapor subcooling.

The next step in the flutter analysis is to calculate the unsteady flow response to the blade motion for each traveling wave mode. Linearized unsteady flow simulations were performed by LUFT. The simulations were performed on a single passage with the interblade phase-angle applied at the periodic boundary and an exact 3D non-reflecting boundary condition was applied at the inlet and outlet. Unsteady flow simulations in the time domain were performed using CFX. Phase lag periodic simulations were not employed and the CFX simulations were run on multiple blade passages to match the interblade phase-angle, for example, in order to model σ=90°, four blade passages were used. The amplitude of the blade motion was 0.5 mm, to reach convergence for NES simulations up to 400 timesteps per period and seven run periods were necessary. All unsteady flow simulations were based on the unsteady RANS equations with the Spalart and Allmaras turbulence model.

The flutter stability of each traveling wave mode is assessed by calculating the work done by the unsteady pressure on the blade motion. This can be expressed as the aerodynamic damping which is equal to

$$\Xi = \frac{W_{aero}}{\pi\, b\, h_{max}^2 p_{ref}}$$

where W_{aero} is the aerodynamic work per cycle, b is the height of the blade, h_{max} is the maximum blade displacement and p_{ref} is the reference pressure which is set to the total pressure at the inlet minus the static pressure at the inlet.

While the aerodynamic damping is a measure of the overall stability, it does not show any detail about the underlying physics. This can be examined by looking at the local work coefficient on the blade. This can be calculated for simple mode shapes as,

$$w = \frac{-c\ p_{imag}\ \mathbf{h}\cdot\hat{\mathbf{n}}}{p_{ref}\ h^2{}_{max}}$$

where c is the chord length, p_{imag} is the imaginary (out of phase) component of the unsteady pressure, **h** is the vector of the local displacement on the blade and $\hat{\mathbf{n}}$ is the local normal.

MESHES

Flow simulations were performed on four different meshes. The standard mesh has 10285 cells and is shown in Figure 2. The inlet and outlet planes were 100 mm from the blade. This corresponds to 45% chord. A short mesh with the inlet and outlet planes 50 mm from the blade was also used. Two higher resolution meshes with the number cells increased successively by a factor of 4 were also used. Most of the presented results were run on the standard mesh unless specified otherwise.

	Standard	*Short*	*Fine*	*SuperFine*
Cells	10285	9829	39932	147412
Inlet Length (mm)	100	50	100	100
Outlet Length (mm)	100	50	100	100
Wall Cell Height (mm)	0.0175	0.0175	0.0155	0.01
First Cell y^+	1.2	1.2	1.1	0.62

Table 2: Mesh details

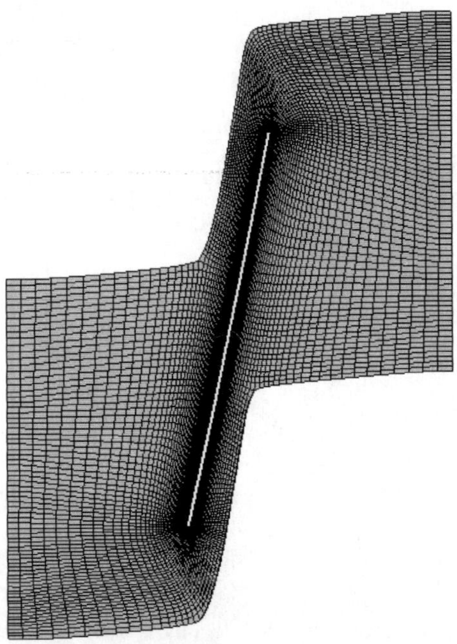

Figure 2: Standard (base) mesh for 2D steam turbine test case

GRID MOTION

The motion of the individual grid points of the CFD mesh has to be prescribed before performing the unsteady linearized flow simulations. The grid motion for the LUFT simulations was determined by solving a modified Laplace equation on the CFD mesh. The boundary conditions for the Laplace problem are: the mode shape is interpolated on to the CFD mesh of the profile, zero motion at the inlet, outlet and periodic boundaries and zero normal displacement at the hub and shroud (points can slide).

The stagger angle is very high for this test case and as a result the periodic boundary is very close to the leading edge on the suction side and to the trailing edge on the pressure side. This leads to high gradients in the prescribed motion which need to be at a maximum on the blade and zero on the periodic boundary. The high gradients in grid motion can be seen in Figure 3. Unsteady flow simulations were performed on a mesh based on the standard with three blade passages, in order to examine if these high gradients in the grid motion were affecting the accuracy of the unsteady flow solutions.

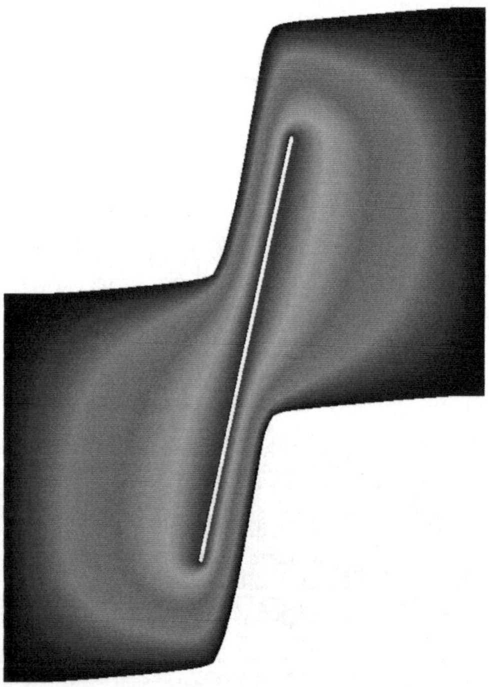

Figure 3: Grid motion in axial direction calculated on mesh with one blade passage

The grid motion calculated for a mesh with three blade passages for $\sigma = 0°$ and -160° are shown in Figures 4 and 5 respectively. It is necessary to calculate the grid motion for specific interblade phase-angles separately for the mesh with three blade passages because the blades in the mesh have different phases. When the grid motion is calculated for the mesh with three passages at $\sigma = 0°$ it can be seen that the gradient in the grid motion near the middle blade has been significantly reduced.

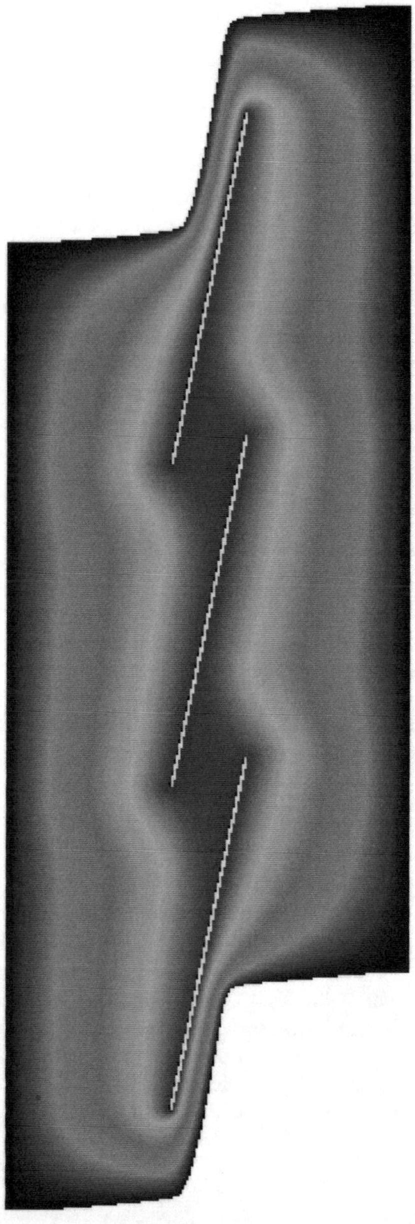

Figure 4: Grid motion in axial direction for $\sigma = 0°$ calculated on mesh with three blade passages

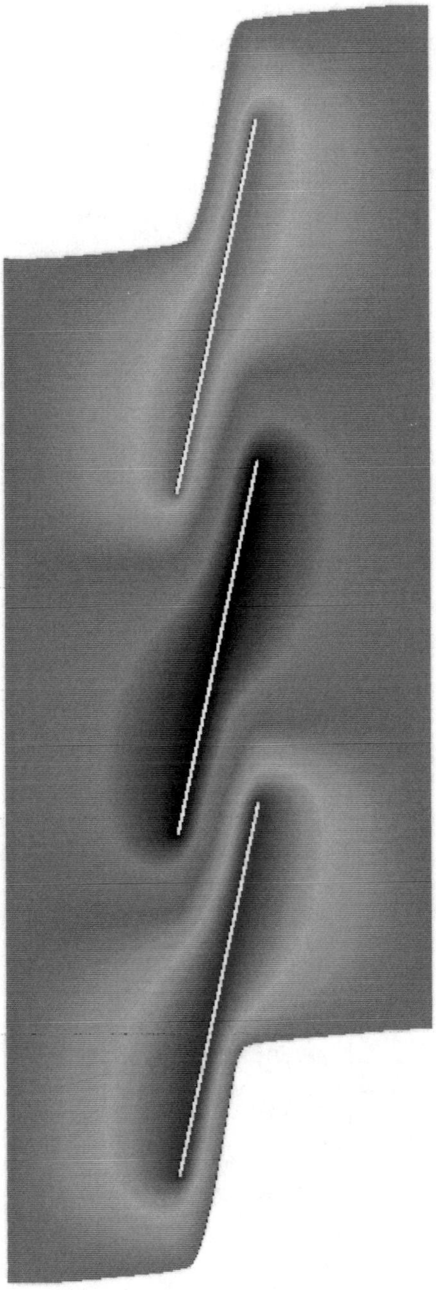

Figure 5: Grid motion in axial direction for σ = -160° calculated on mesh with three blade passages

The unsteady flow solutions calculated on the single passage mesh and three passage mesh for $\sigma = 0^{\circ}$ and -160° are shown in Figures 6 and 7 respectively. The work coefficient on the pressure and suction sides of the blades are shown. The influence of the shock can be seen on the suction side at approximately 35% chord. The work coefficient calculated on the three individual blades on the mesh with three blades are shown. The solutions agree very well except for a small discrepancy at the shock and downstream of the shock. This discrepancy is probably due to minor flow reflections from the outlet boundary and not due to the grid motion or the periodic boundaries. Note that the solutions calculated on the individual blades from mesh with three blades agree perfectly. It can be concluded from this good agreement that the high gradient in the grid motion is not affecting the accuracy of the unsteady flow solution and that the periodic boundary condition for the unsteady flow simulations has been applied correctly

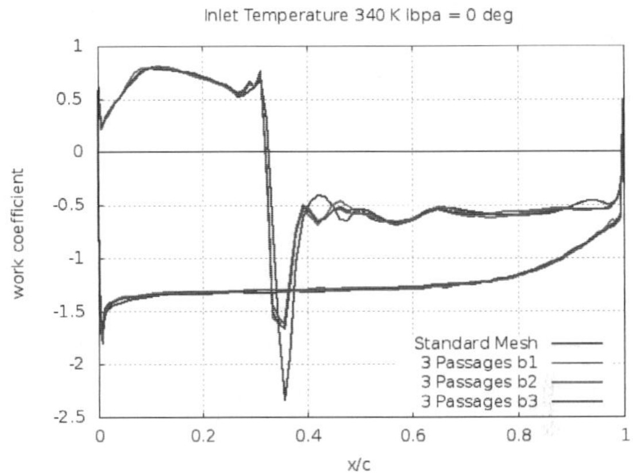

Figure 6: Work coefficient for $\sigma = 0^{\circ}$ calculated on standard mesh and mesh with three blade passages

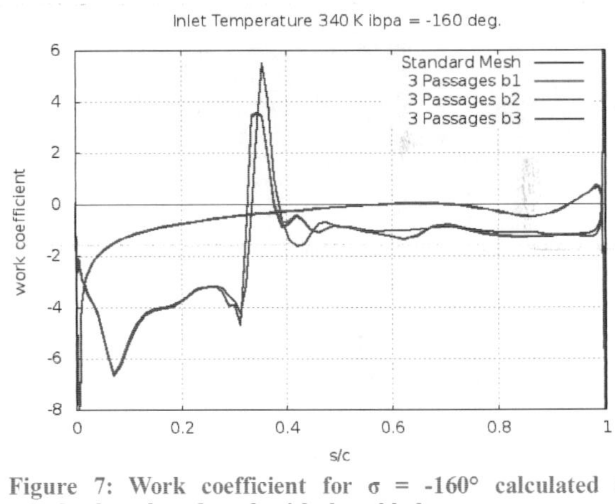

Figure 7: Work coefficient for σ = -160° calculated on standard mesh and mesh with three blade passages

STEADY-STATE RESULTS

The steady-state Mach number contour plot for the ideal steam solution from LUFT (short and standard mesh) for inlet $T_0 = 355$ K is shown in Figure 8. The flow is similar to the flow near the tip in a real steam turbine. The flow is sonic at the throat formed by the short blade passage and there are two oblique shocks starting at the trailing edge. One of these shocks forms a passage shock as the flow starts to expand after passing through the throat of the blade passage. The other shock points downstream and passes through the outlet plane. This shock de-accelerates the flow from the suction side. The solutions shown in Figure 8 were calculated on the short and standard mesh and it can be seen that the solutions are independent of the inlet and outlet location. A numerical Schlieren (magnitude of the gradient of density) is shown in Figure 9. This figure shows that the shock is exiting the outlet plane without reflection.

The steam quality (mass fraction of vapor) in the flow field from the LUFT EQS steady-state solution is shown in Figure 9. It can be seen that condensation is predicted to start as the flow enters the blade passage.

The blade loadings calculated by LUFT on the short and standard mesh from the ideal gas and equilibrium wet steam are shown in Figure 11. The calculated blade loadings are independent of the mesh and the steam model. The grid independence of the LUFT solutions is shown in Figure 12 (Note total temperature 340 K)

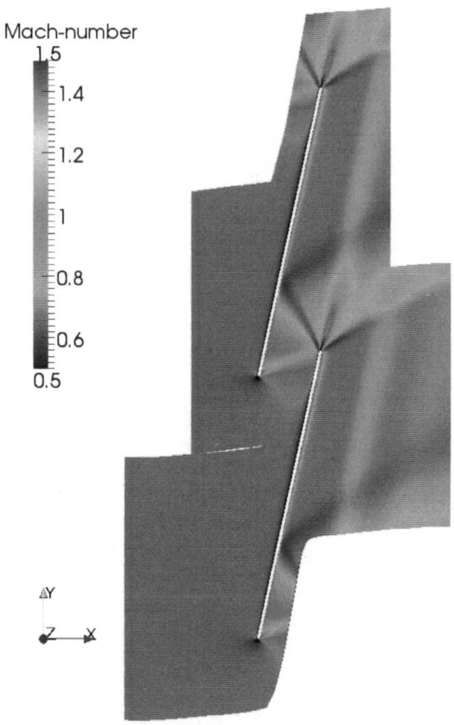

Figure 8: Mach number for ideal gas simulation with T_0 = 355 K

The LUFT and CFX blade loadings calculated on the standard mesh are compared in Figure 13. It can be seen that there is little difference between the CFX ideal and CFX equilibrium wet steam solutions. Also the CFX and LUFT solutions agree with some minor differences near the leading edge on the pressure side and near the shock on the suction side. However, there are some differences between the CFX EQS and CFX NES solutions. The pressure on the pressure side of the blade is 0.5 kPa lower on the pressure side for the NES solution and there are also some differences on the suction side near the solution. These differences are due to the finite rate of the droplet formation. The solution predicts that subcooling up to 30 K (Figure 14) occurs in regions near the short blade passage (throat) where rapid expansion occurs (gradient of Mach number is highest). The shock on the suction side is 6% of chord further upstream.

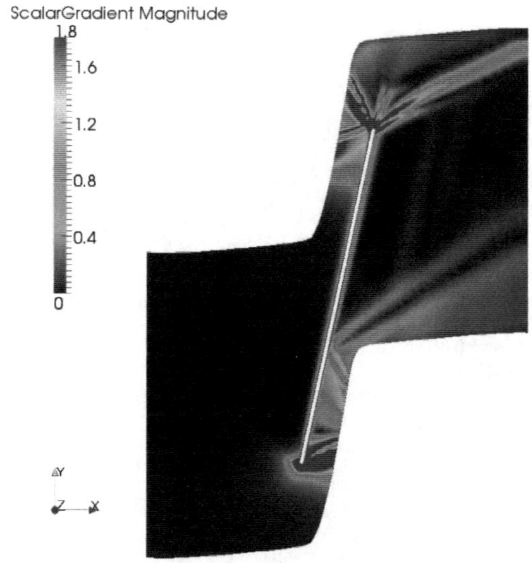

Figure 9: Numerical Schlieren of Steady-State Ideal Gas Solution from LUFT (T_0 = 355 K)

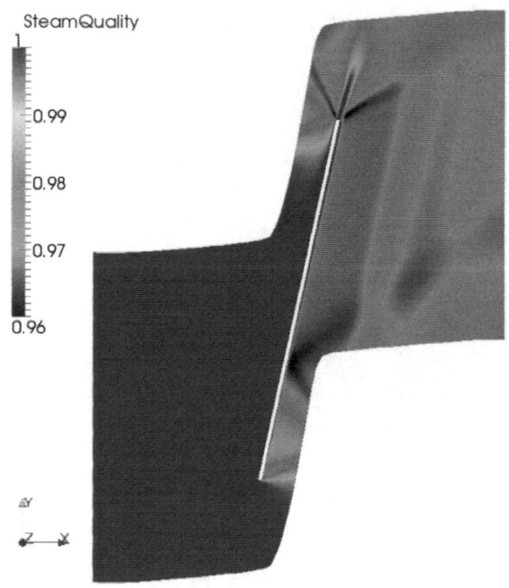

Figure 10: Steam quality from LUFT EQS solution calculated on the standard mesh (T0 = 355 K)

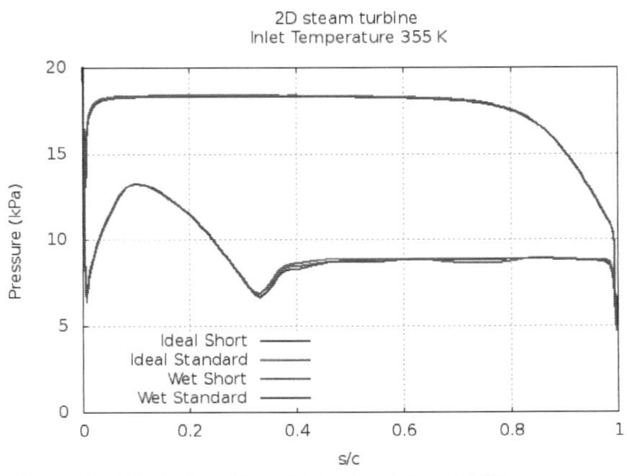

Figure 11: Blade Loadings Calculated by LUFT

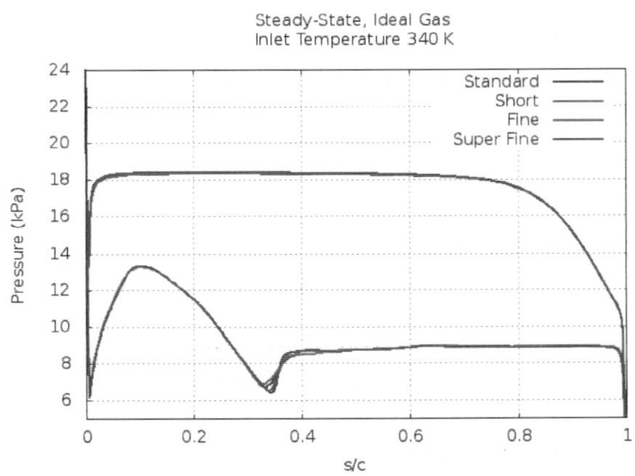

Figure 12: Blade Loadings calculated by LUFT for ideal steam on various meshes

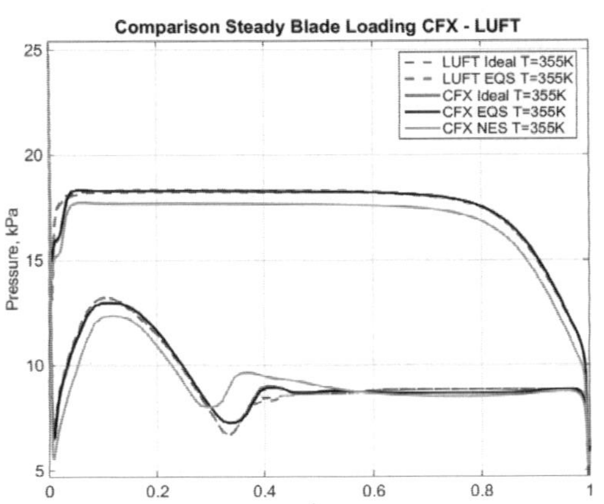

Figure 13: Comparison of blade loading from LUFT and CFX calculated on standard mesh

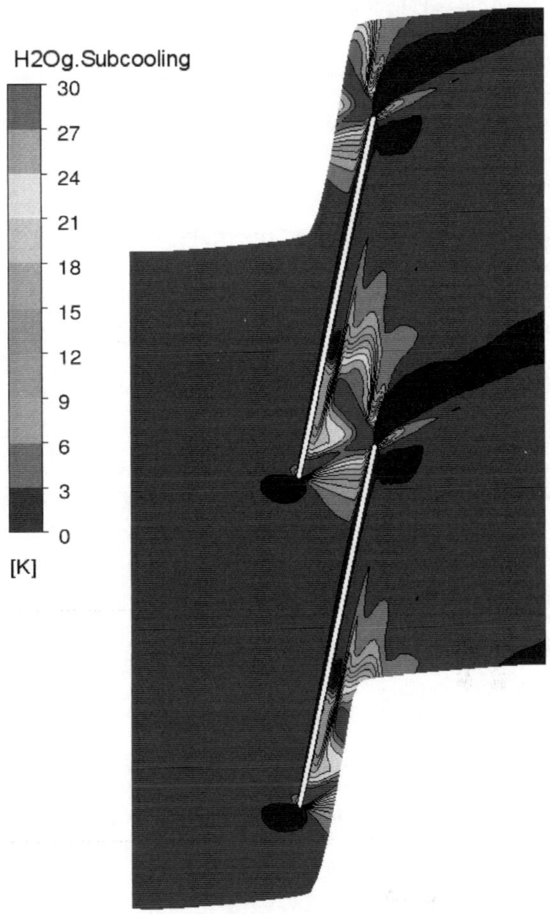

Figure 14: Level of subcooling in steady-state CFX NES solution

UNSTEADY FLOW RESULTS

The aerodynamic damping versus interblade phase-angle plot for the prescribed bending mode calculated by LUFT is shown in Figure 15. The total inlet temperature was 340 K and ideal steam gas was assumed. The damping values were calculated every five degrees on various meshes. It can be seen that the LUFT solutions are independent of the location of the inlet or outlet. There is also a good agreement with the damping values calculated in the higher resolution meshes, however, there are some small differences in the damping values calculated on the super fine mesh.

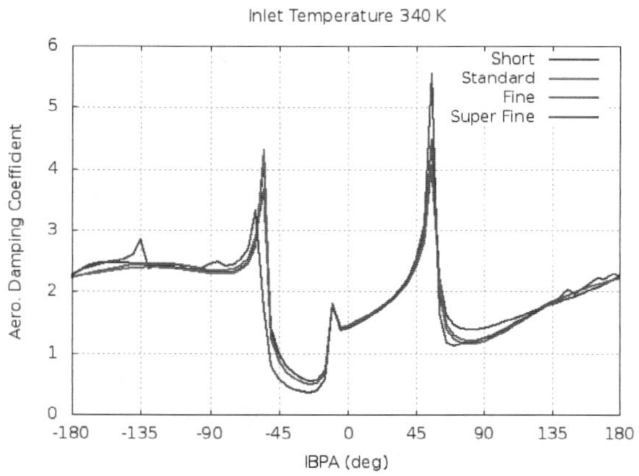

Figure 15: Aerodynamic damping calculated by LUFT on various meshes, ideal steam assumed

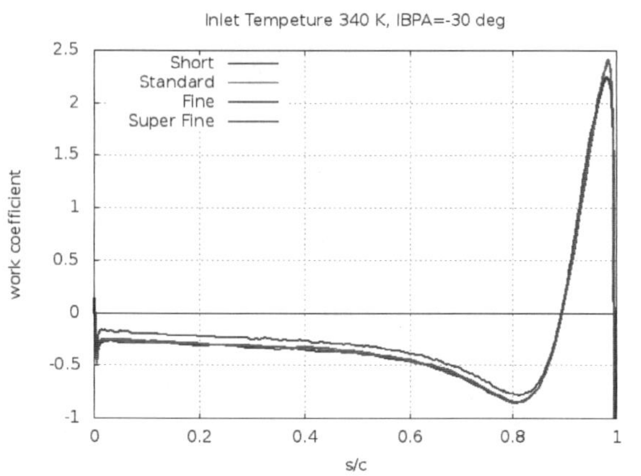

Figure 16: Work coefficient calculated by LUFT on pressure side for σ = -30° for inlet T_0 = 340 K

The work coefficients on the pressure side and suction side for σ = -30° for ideal steam with an inlet T_0 =340 K are shown in Figures 16 and 17 respectively. On the pressure side, there is a perfect agreement between all meshes except the super fine mesh where there is a small upward shift in the solution. The reason for this is unknown. On the suction side, there is a perfect agreement with all meshes upstream of the shock and there are some small differences downstream of the shock. It is speculated that these differences are due to minor unsteady flow reflections from the outlet. Overall the agreement is good and it is concluded that the LUFT solutions are independent of inlet and outlet location, and that grid convergence has been achieved.

The aerodynamic dampings calculated by LUFT and CFX for inlet T_0= 355 K are shown in Figure 18. The CFX solutions were calculated in the time domain for σ = -180°, -90°, 0°, 90°, and 180° on a mesh with 4 blade passages.. The peaks in the damping are due to acoustic resonance. They occur at +55.2° and -10.1° at the inlet and at -54.4° and -7.8° at the outlet. There is a difference in the location of the resonance points at the inlet predicted by the ideal gas model and EQS. This is because the ideal gas model assumes that the polytropic index is 1.11, however, the flow at the inlet is superheated steam (only just) and the real polytropic index is closer to 1.33. Also, it can be seen that there are some differences in the damping values predicted by the two codes. In order to investigate these differences further, the unsteady flow solution for σ = -90° was examined in more detail.

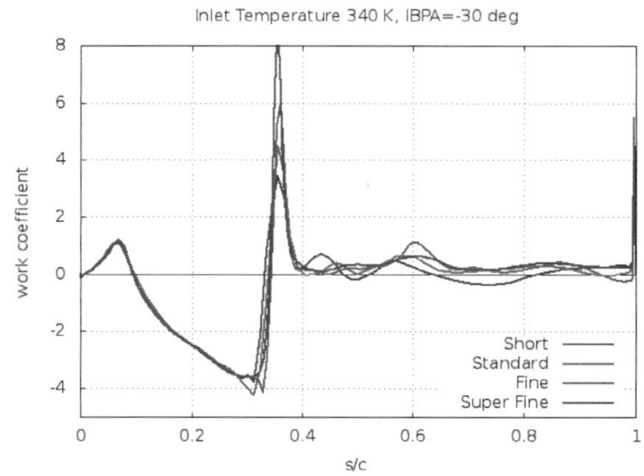

Figure 17: Work coefficient calculated by LUFT on suction side for σ = -30° for inlet T_0 = 340 K

The work coefficients on the pressure side for σ = -90° and inlet T_0 = 355K as calculated by LUFT and CFX are shown in Figure 19. The solutions were calculated on the short and standard mesh by LUFT and these solutions are independent of the inlet and outlet location. The LUFT solutions are calculated with ideal steam and equilibrium steam (EQS) and the CFX solutions are calculated with ideal steam, EQS and non-equilibrium steam (NES). The flow solutions on the pressure side appear to be independent of the fluid model. There are some minor differences near the leading edge but these differences are between the LUFT and CFX solutions.

The work coefficients calculated by LUFT on the suction side for the same case are shown in Figure 20. There are some minor differences at the shock and immediately downstream of the shock but in general the solution is independent of mesh and the steam model.

The work coefficients calculated by LUFT and CFX on the suction side with the ideal steam and EQS are shown in Figure 21. There are significant differences between the LUFT and CFX solutions downstream of the shock. The dominant acoustic waves at the outlet are cut-on (non-decaying) and cut-off (decaying) at the inlet for this case. The differences in the solutions may be due to unsteady flow reflections at the outlet within the CFX solutions. Note that it was not possible to calculate a reliable solution with CFX on the short domain.

The work coefficients calculated by CFX on the suction side with the ideal steam, EQS and NES are shown in Figure 22. Here we can see that the non-equilibrium wet steam effects is influencing the unsteady work. The differences are occurring near the shock and downstream of the shock.

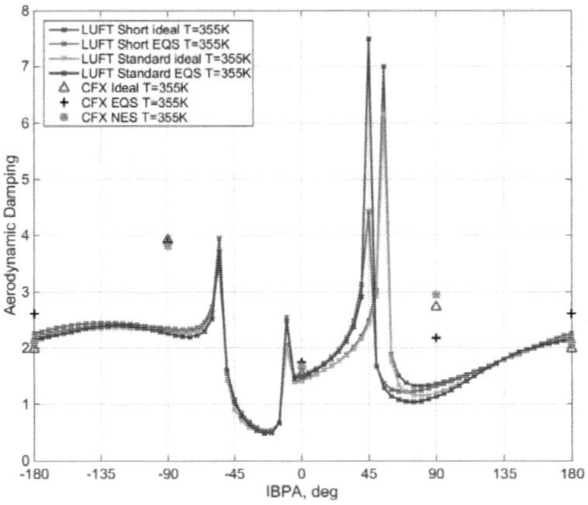

Figure 18: Aerodamping damping calculated by LUFT and CFX for inlet T_0= 355 K

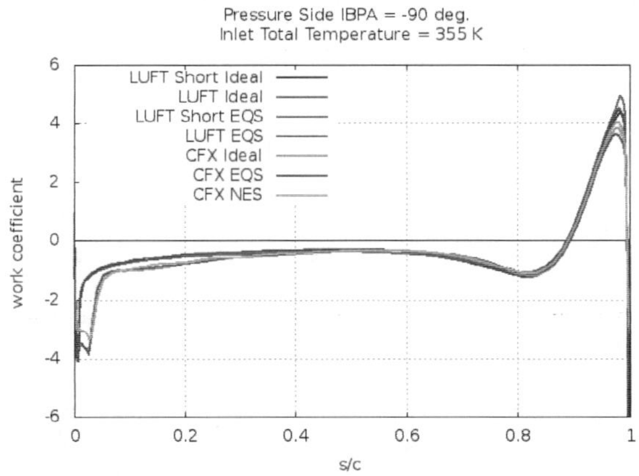

Figure 19: Work coefficient on pressure side for σ=-90° for inlet T_0= 355 K

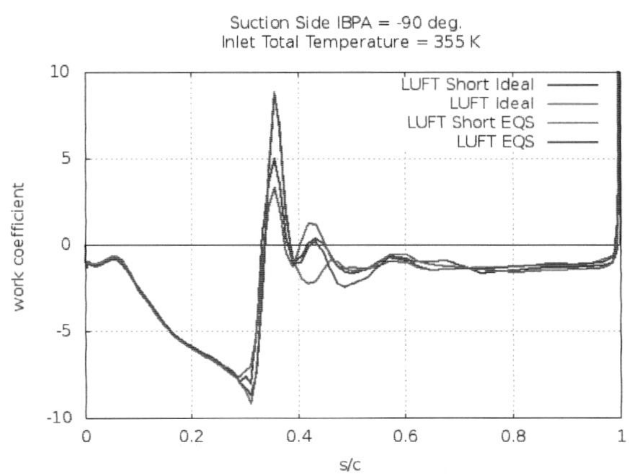

Figure 20: Work coefficient on suction side for σ=--90° for inlet T_0= 355 K calculated by LUFT

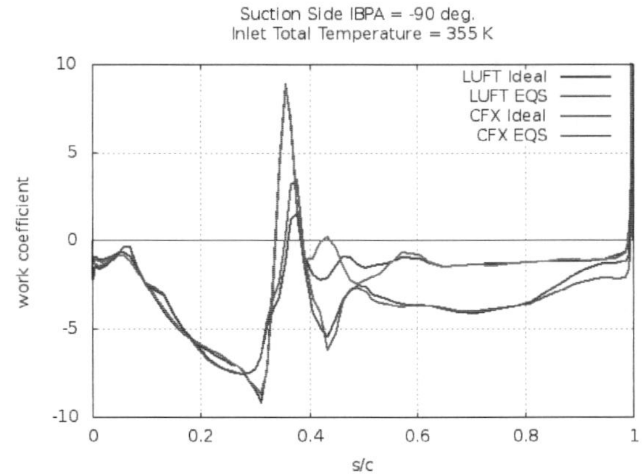

Figure 21: Work coefficient on suction side for σ = -90° for inlet T_0 = 355 K calculated by LUFT and CFX on standard mesh

The work coefficients on the pressure and suction side for σ = +90° calculated by LUFT and CFX on the standard mesh are shown in Figures 23 and 24. The differences are bigger on the pressure side and there is a reasonable agreement between the LUFT and CFX solutions on the suction side. For this case, the dominant acoustic waves are cut-on at the outlet and cut-off at the inlet. However, the rate of decay of the acoustic waves at the outlet is less (32% decrease in amplitude per 100 mm) compared with σ = -90° case (50% decrease).

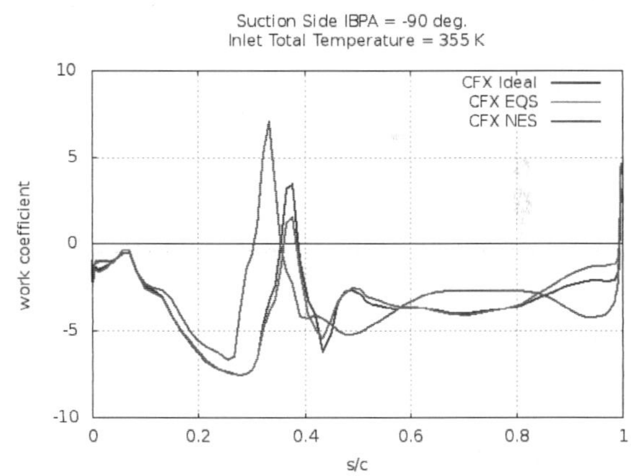

Figure 22: Work coefficient on suction side for σ = -90° for inlet T_0= 355 K calculated by CFX on standard mesh

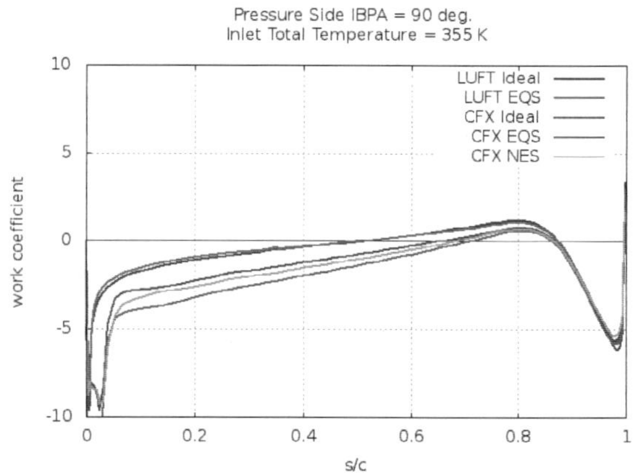

Figure 23: Work coefficient on pressure side for $\sigma = +90°$ for inlet $T_0 = 355$ K calculated by LUFT and CFX on standard mesh

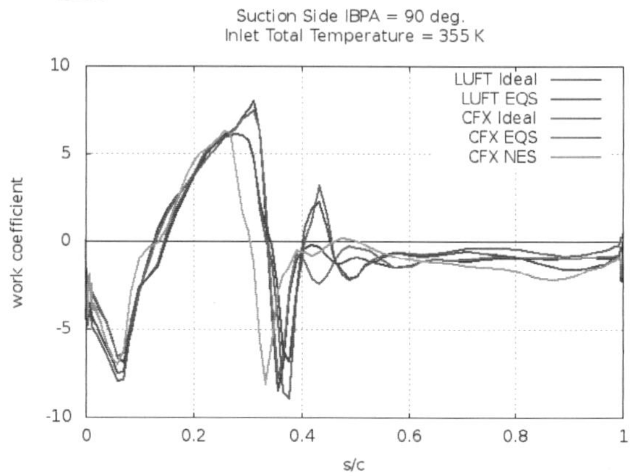

Figure 24: Work coefficient on suction side for $\sigma = +90°$ for inlet $T_0 = 355$ K calculated by LUFT and CFX on standard mesh

DISCUSSION

There were some differences in the solutions calculated on the super fine mesh by LUFT. For example, there is a jump in the aerodynamic damping at $\sigma = -135°$ (Figure 15) and there is a small positive shift in the work coefficient on the pressure side at $\sigma = -30°$ (Figure 16). The reason for this difference is unknown but it may be due to insufficient numerical dissipation in the simulation at the very high mesh resolution.

There are differences between the LUFT and CFX unsteady solutions for ideal steam and EQS. These differences can be seen in the aerodynamic damping plot (Figure 15) at $\sigma = -90°$ and $+90°$. These differences in damping are due to differences in the work coefficient on the suctions side for $\sigma = -90°$ and on the pressure side for $\sigma = +90°$. These differences may be due to unsteady flow reflections at the inlet and outlet within the CFX simulation. It was demonstrated that the LUFT solutions are independent of the inlet and outlet locations. However, we cannot guarantee that either solution is correct because we have no other reference solutions. Further solutions from other methods are required in order to gain confidence that we calculated the correct solutions.

The differences between the ideal and EQS solutions calculated by LUFT are negligible. This is also the case for the CFX solutions at $\sigma = -90°$, however, there is a small shift in the work coefficient on the pressure side at $\sigma = 90°$. This small shift does result in a significant difference in the calculated aerodynamic damping which can be seen in Figure 18. The reason for this difference is unknown.

There are differences between the EQS and NES solutions calculated by CFX. The main difference is the location of the shock on the suction side. It can be seen in Figure 13 that the shock is about 6 percent further upstream in the NES solution. The change in the shock position affects the unsteady work done near the shock. This can be seen in Figures 22 and 24.

CONCLUSIONS

A new 2D steam turbine test case has been established. The geometry is simple and fully described. Unlike previous open test cases, the stagger angle is high (78°) and the flow conditions are similar to those of the last stage of steam turbines.

Initial solutions for the test case calculated with the linearized flow solver LUFT and the commercial code CFX are presented. There were some differences in the solutions calculated by the two methods. The solutions calculated by LUFT were shown to be independent of the grid motion prescribed, the location of the inlet and the outlet boundaries and the mesh resolution. More solutions from other methods are required to gain confidence that the correct reference solution has been found.

Non-equilibrium wet steam (NES) steady and unsteady simulations were performed for $T_0 = 355$ K. At this condition, the steam is superheated at the inlet. The NES steady solution predicts that subcooling up to 30 K occurs in regions near the short blade passage (throat) where rapid expansion occurs and the shock on the suction side was 6% of chord further upstream. The change in the steady flow due to considering NES effects also affected the unsteady flutter work on the blade particularly near the shock on the suction side. It is planned to run NES simulations for cases where the steam has started to condense at the inlet ($T_0 = 340$ K) in the future. This work will examine the influence of the assumed water droplet size distribution at the inlet on the flutter analysis.

REFERENCES

[1] Rice, T., Bell, D., and Singh, G., 2009. "Identification of the Stability Margin Between Safe Operation and the Onset of Blade Flutter". Journal of Turbomachinery, 131(1), January, pp. 1–10

[2] Stüer, H., Schmitt, S., and Ashcroft, G., 2008. "Aerodynamic mistuning of structurally coupled blades". Proceedings of ASME Turbo Expo 2008: Power for Land, Sea and Air, ASME. GT2008-50204

[3] Masserey, P.-A., McBean, I., and Lorini, H., 2012. "Analysis and improvement of vibrational behaviour

on the ND37 A last stage blade". VGB Power Tech Journal(8)

[4] Petrie-Repar, P. J., 2012, "Establishment of a steam turbine flutter test case", In proceedings of 13th International Symposium on Unsteady Aerodynamics, Aeroacoustics and Aeroelasticity of Turbomachines, Tokyo, Japan

[5] Fransson, T., Jöcker, M., Bölcs, A., and Ott, P., 1999, "Viscous and inviscid linear/nonlinear calculations versus quasi-three-dimensional experimental cascade data for a new aeroelastic turbine standard configuration". ASME Journal of Turbomachinery, 121, October, pp. 717–725

[6] Petrie-Repar, P.J., McGhee, A.M., and Jacobs, P.A., 2007, "Three dimensional viscous flutter analysis of standard configuration 10", Proceedings of 2007 ASME Turbo Expo, GT2007-27800, Montreal, Canada

[7] Dooley, R. B., 1997. Release on the IAPWS industrial formulation 1997 for the thermodynamic properties of water and steam. Tech. rep., The International Association for the Properties of Water and Steam, Erlangen, Germany

[8] Gerber, A. G., 2008. Inhomogeneous Multifluid Model for Prediction of Nonequilibrium Phase Transition and Droplet Dynamics, Trans. ASME, J. Fluids Eng. 130(3):1-11

[9] Gyarmathy, G., 1962. Grundlagen einer Theorie der Nassdampfturbine (in German), Doctoral thesis, ETH Zürich, Juris Verlag Zürich

[10] Grübel, M., Starzmann, J., Schatz, M., Eberle, T., Vogt, D.M., 2014, "Two-Phase Flow Modeling and Measurements in Low-Pressure Turbines Part 1: Numerical Validation of Wet Steam Models and Turbine Modeling", J. Eng. Gas Turbines Power 137(4), pp. 042602

[11] Schatz, M., Eberle, T., Starzmann, J., Grübel, M., Vogt, D.M., 2014, "Two-Phase Flow Modeling and Measurements in Low-Pressure Turbines Part 2: Turbine Wetness Measurement and Comparison to CFD Predictions", J. Eng. Gas Turbines Power 137(4), pp. 042603

[12] Kantrowitz, A., 1951. Nucleation in Very Rapid Expansions, J. of Chemical Physics 19(9):1097-1100

Proceedings of the 14th International Symposium on Unsteady Aerodynamics, Aeroacoustics & Aeroelasticity of Turbomachines
ISUAAAT14
8-11 September 2015, Stockholm, Sweden

I14-S2-2

WAKE AND POTENTIAL INTERACTIONS IN A MULTISTAGE COMPRESSOR

Jing Li
Department of Mechanical Engineering and Materials Science, Duke University
Durham, North Carolina, USA
Jing.Li2@duke.edu

David Canon
Siemens Energy, Inc.
Jupiter, Florida, USA

Stefan Schmitt
Siemens Energy, Inc.
Mülheim, Germany

Fanny Besem[1]
Robert E. Kielb
Department of Mechanical Engineering and Materials Science, Duke University
Durham, North Carolina, USA

ABSTRACT
Resonant response prediction for aero-engines and power generation turbomachines is a critical aeromechanic design task. The most important part of such task is an accurate prediction of the unsteady flow field in the responding blade passage. When solving for the flow field, in order to reach the best compromise between prediction accuracy and computational cost, designers need to make careful decisions on what numerical method to use and what computational domain to model. However, there is a lack of industrial guidelines on this subject. This paper presents unsteady flow field and blade modal work predictions on the Purdue 3-Stage Axial Compressor using different CFD models, and gives recommendations on accurate resonant response predictions. The decoupled approach commonly employed in the industry is found to be inadequate and significantly underestimates response amplitude for cases investigated. The coupled approach should be adopted to resolve inter-row interaction. This is especially true when an acoustic cut-on mode due to inter-row interaction is present. When the excited row has neighbours of the same blade counts generating excitations at the same frequency, it is found that a coupled one-and-a-half stage domain is necessary.

NOMENCLATURE

1B	First bending
1T	First torsion
BC	Boundary condition
EO	Engine order
FEM	Finite Element Method
FSI	Fluid-structure interaction
IGV	Inlet guide vane
Q	Modal force
LE	Leading edge
TE	Trailing edge
ND	Nodal diameter
linTRACE	Linearized TRACE solver
hbTRACE	Harmonic Balance TRACE solver
P3S	Purdue 3-Stage (Compressor)
RANS	Reynolds-Averaged Navier-Stokes
PS	Pressure side
SS	Suction side
TPR	Total pressure ratio

1 INTRODUCTION
It is not always possible for a compressor design to avoid all resonance crossings in its whole operating range. When a potentially high-response crossing is identified on the Campbell Diagram, it is critical to accurately estimate the dynamic response to determine acceptability of the design. Numerical tools are often used to accomplish this.

Over the years, computational methods of varying degrees of sophistication are developed for simulating the unsteady flow events and predicting resonant response. They can be largely divided into fluid-structure-coupled and -decoupled methods [1]. The first method takes into account the fluid-structure interaction (FSI) where incoming disturbance to the blade passage (most commonly in the form of periodic flow from adjacent rows) creates an unsteady flow field around the blades and excites the blades to vibrate. This vibratory motion in turn modifies the unsteady flow field. To resolve FSI, flow and structural equations are solved simultaneously. Authors in [2] simulated wake excitation to a downstream rotor from IGV in a low pressure turbine with FSI and reported good agreement with experimental data. The study reported in [3] compared simulations on a transonic fan excited by inlet distortion with and without modelling FSI, and reported FSI effect to be strong on several mode shapes. However, this method is usually considered too computationally demanding to be used in an industrial environment, and thus is not discussed in this paper.

The fluid-structure-decoupled method is more commonly employed in industrial applications, and is adopted in this study. It assumes that the mode shapes and natural frequencies of the structure are independent of the flow. This simplifying assumption is usually found to be valid in turbomachines ([1]). Using this approach, the flow field is solved independently of blade motion to resolve the unsteady aerodynamic loading due to external excitations. This unsteady loading can then be used in combination with structural properties (typically obtained from FEM modal analysis) and aerodynamic damping values (usually obtained from separate CFD calculations with prescribed blade motion) to compute the response amplitude. The prediction accuracy of this method depends critically on the accuracy of the predicted unsteady flow, which can be solved in either a decoupled or coupled manner.

The decoupled approach is an efficient technique solving an isolated row with external excitations represented as unsteady boundary conditions. A steady-state (or time-averaged) flow solution is needed as its basis. This solution is Fourier decomposed at the boundary of the neighbouring row(s) to obtain spatial Fourier contents, which are applied to the excited row as unsteady boundary condition(s). Both linearized and nonlinear solvers could be used to solve the unsteady flow. However, any inter-row coupling effect is lost in the process, which in some cases could lead to inaccurate predictions.

Recently, there have been an increasing number of studies implementing coupled method where multiple rows are modelled simultaneously to capture inter-row interactions. Notable examples are [4] and [5]. In addition, in [6], resonant crossing of a high-speed compressor rotor blisk was predicted and was found to agree well with averaged experimental data. Multistage coupling effect was reported by [7] to be significant in an IGV-rotor-stator configuration in a transonic compressor, both experimentally and numerically. The study in [8] compared decoupled and coupled approaches, and found that the former may greatly overestimate or underestimate the response when it fails to capture a cut-on acoustic mode arising from the IGV-rotor interaction.

The coupled method is an improved approach compared to the decoupled method. However, it usually consumes more, if not much more, computational resources. The nonlinear Harmonic Balance method is found to be relatively computationally inexpensive compared to time-accurate and some other nonlinear frequency-domain methods ([9]). Using this technique, authors solved flutter ([10], [11] and [12]) and forced response ([12] and [13]) problems in multi-stage environments, and concluded that multi-stage effect could be significant in some cases.

2 PURDUE 3-STAGE AXIAL COMPRESSOR

This project is an effort within the GUIde 5 Consortium to address the Grand Challenges in Turbomachinery Forced Response and Non-Synchronous Vibration. This project is undertaken by a collaboration between Purdue and Duke Universities and is supported by partnering governmental agencies and industrial companies. In particular, resonant crossings in the Purdue 3-Stage (P3S) Axial Compressor are studied both experimentally and numerically. The P3S Compressor is a heavily instrumented facility housed at Purdue University. A schematic of the compressor flow path, blading and measurement stations is displayed in Figure 1. The flow path has a constant area throughout the rig with hub and tip diameters of 20 and 24 inches, respectively. Three rotor rows have declining numbers of blades: 36, 33, and 30. The IGV at the front of the compressor as well as the first and second stage stators each contains 44 vanes, while the last stage stator has 50 vanes. All stators are shrouded and have rotating seals. All rotors and the IGV are designed with NACA 65-series airfoil sections, while the three stators with double circular airfoil (DCA) profiles ([14]). The three rotors form one monolithic piece where the disk and blades are machined from one piece, eliminating energy dissipation mechanism at blade roots as usually found in conventional bladed disks. Inside the operating range, the second stage rotor, or simply R2, is excited at the 1T (1st torsion) mode at 3700RPM by the 44EO vane passing from both S1 and S2, and possibly also IGV further upstream. Recent studies on tuned and mistuned resonant response predictions at this crossing were reported in [15] and [16] and compared to experimental measurements documented in [17].

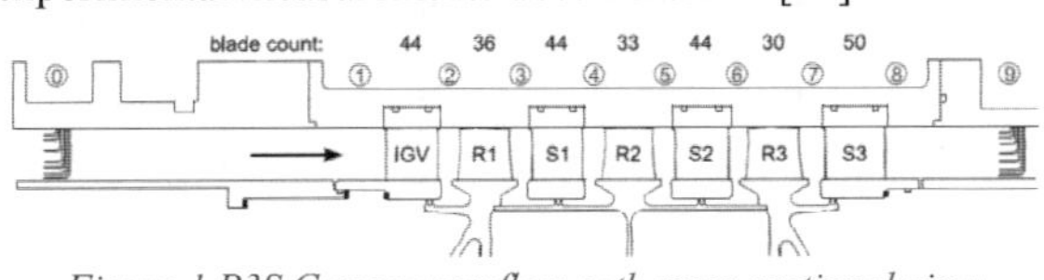

Figure 1 P3S Compressor flow path cross-sectional view, blading, and measurement staions (adapted from [15])

Recently, a new Stator 1 is designed with a reduced vane count of 38 ([15]), moving the resonance crossing of 1T on R2 from 3700RPM to 4284RPM as shown in Figure 2. The 1T mode shape is also displayed. This is to separate the S1 and S2 passing frequencies such that the influence of each stator row could be studied independently in R2. The dynamic pressure on R2 is also expected to be higher than when it is at a lower speed. This will produce larger disturbance and thus a stronger response. Therefore, the unsteady flow field and blade response are expected to be more measurable. The new S1 features increased axial chord length and reduced fillet size at the hub and the tip, and is designed so that the steady aerodynamics is largely maintained ([15]). Including the new S1 case in this study brings additional insight on the response prediction of blades experiencing excitations of multiple frequencies and how it compares with the nominal S1 case where two neighbouring rows provide the same excitation frequency.

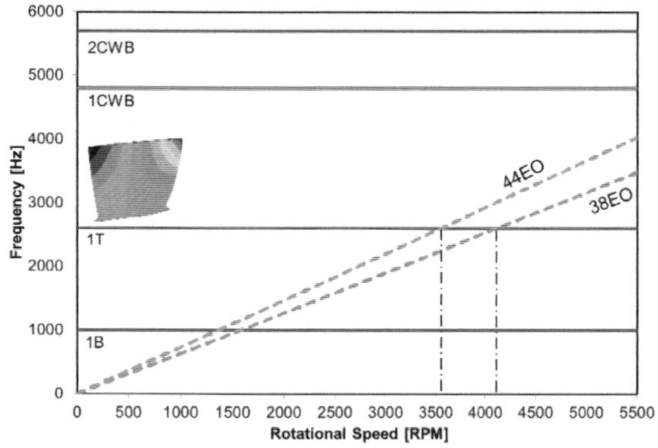

Figure 2 Campbell Diagram for R2

The mode shapes and natural frequencies of R2 estimated from an FEM modal analysis are shown in Figure 3. For the nominal S1 design, R2 is excited at 1T (mode 2) with -11ND. There is little coupling between blade and disk for this mode. The frequencies of adjacent 1B (first bending) and the 3rd mode are also sufficiently separate, therefore the coupling between modes can also be neglected. For the new S1 design, the 4284RPM resonance crossing occurs for 1T mode at -5ND. This falls in a veering region in the frequency versus nodal diameter map. For the purpose of this paper, the 11ND mode shape is adopted for modal force calculations for both crossings. Future work would carefully examine the 5ND 1T mode and its coupling with adjacent modes.

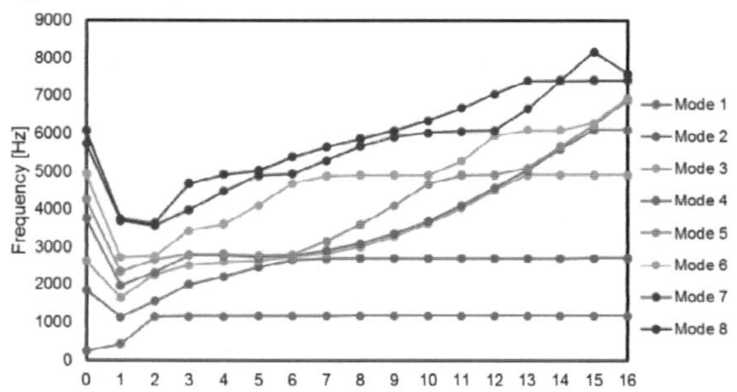

Figure 3 Frequency versus nodal diameter characteristics for the first 8 modes of R2

3 NUMERICAL SETUP

The TRACE solver developed at the German Aerospace Center (DLR) is used in this study. TRACE is a compressible, 3D RANS flow solver using a fully implicit algorithm ([8]). Recently, the Harmonic Balance method is implemented in TRACE ([18]), providing the capability of calculating highly nonlinear, unsteady flow at a relatively reduced computational cost. Only a single passage is needed to represent a row of arbitrary number of airfoils in a multistage configuration. One can specify explicitly what frequencies (or harmonics) to retain in each row. For this study, the primary adjacent-row passing frequency and its first two harmonics are included in the unsteady solution.

For a linearized decoupled calculation, the base steady-state solution is first obtained by ANSYS® CFX. On the exciting row, at its interface between the excited row, the flow quantities are Fourier transformed in the tangential direction at every radial location. They are perceived as disturbances of different harmonics to the excited row. The spatial harmonic corresponding to the excitation of interest is then applied as gust boundary condition to the linearized TRACE solver (linTRACE) where the unsteady flow is solved on the domain of the excited row. For a nonlinear coupled calculation, both exciting and excited rows are included and coupled directly. The steady-state nonlinear TRACE solver is first applied to provide an initialized flow field. The unsteady Harmonic Balance TRACE solver (hbTRACE) is then used to predict both the time-averaged and the unsteady flow fields.

The CFD cases using different modelling approaches and computational domains studied in this paper are listed in Table 1. Note that since the linearized solver we have access to currently lacks the capability of applying gust boundaries at both inlet and exit of the domain, we cannot simulate a case where R2 sees both up- and downstream excitations.

Table 1 Cases investigated in this study

Case #	Solver	CFD Domain	Note
1	linTRACE	R2	S1 disturbance applied as R2 inlet unsteady BC
2	hbTRACE	S1, R2	S1/R2 coupled
3	hbTRACE	S1, R2, S2	S1/R2/S2 coupled
4	linTRACE	R2	S2 disturbance applied as R2 exit unsteady BC
5	hbTRACE	R2, S2	R2/S2 coupled

The computational cost gradually grows from Case 1 to 3 with increasingly sophisticated numerical model. Case 1 would be considered adequate for many typical industrial applications. Designers would select the decoupled method with the assumption that the nonlinear interaction between exciting and excited rows is insignificant, and would also exclude S2 from the computational domain since the downstream potential disturbance is usually considered much smaller compared to the upstream viscous wake. A comparison between Cases 1 and 2 would reveal if there is any significant inter-row interaction that the decoupled approach cannot capture. A comparison between Cases 2 and 3 would yield whether including potential disturbance from the downstream stator will significantly alter the prediction. For the nominal S1 design, this additional excitation occurs at the same frequency of 44EO; for the new S1, this additional passing of 44EO is different from the 38EO resonant crossing. In addition, Cases 4 and 5 will be used to study the incident of S2 potential disturbance exciting R2, and how the results will differ between linear decoupled and nonlinear coupled methods.

Mesh used for this study is created by NUMECA AutoGrid™. The stator hub and shroud fillets, rotor hub fillet, as well as rotor tip gap are modelled. Each passage contains about 200,000 nodes; in total there are about 1,600,000 nodes in the 7-row model. The mesh size is considered adequate based on industrial experience, thus no mesh sensitivity study is performed. However, as highlighted in [2], it is anticipated that a finer mesh may better capture the viscous wake and frequencies of shorter wavelength. Menter SST k-ω turbulence model is applied for all unsteady simulations. Radial distributions of total pressure, total temperature, and flow angles in the axial-circumferential and axial-radial planes are specified at the inlet. Static pressure radial distribution is imposed at the exit.

4 RESULTS

4.1 P3S Compressor with Nominal S1

It is critical to ensure that all simulations are on the same or similar operating point for them to be comparable. This is determined by the boundary conditions. The importance of having the correct boundary conditions for an embedded compressor modelling is highlighted in [5]. It is also discussed in [19] that embedded rows in unsteady (in that case, transient) calculations may experience more flow blockage without the mitigation of mixing-plane and thus reduced mass flow rate. There are different criterions for matching operating points, and mostly commonly seen are the mass flow rate $\dot{m}$, or the parameter of total pressure ratio (TPR) over $\dot{m}$. In this study, both parameters are monitored by iteratively changing the exit pressure until its discrepancy from the experimental condition is reasonably small. For example, the exit pressure is reduced by 0.3% for Case 3 to match experimental $\dot{m}$ with only 0.01% and experimental $TPR/\dot{m}$ with only -0.15% discrepancy. Table 2 lists the predicted $\dot{m}$ and $TPR/\dot{m}$ of all unsteady cases compared against the experiment.

Table 2 Predicted operating points for Cases 1-3

Case	$\dot{m}$ [kg/s]	%Difference [-]	$TPR/\dot{m}$ [s/kg]	%Difference [-]
Exp. (S1/R2)	6.7762	-	0.1554	-
Case 1	6.7762	-	0.1561	0.47%
Case 2	6.7656	-0.16%	0.1554	0.03%
Exp. (S1/R2/S2)	6.7762	-	0.1549	-
Case 3	6.7774	0.01%	0.1547	-0.15%

The linearized, decoupled method relies on the steady-state solution as the basis, whereas the nonlinear solver actively calculates the time-averaged flow field. A comparison is plotted in Figure 4 on the steady-state (for Case 1) and the time-averaged (for Cases 2 and 3) blade surface pressure (normalized by P_{ref}= 1 [bar] or 101325 [Pa]) at 90% span. This spanwise location is chosen since the blade deflects dominantly near the tip at 1T mode and therefore the flow behaviour there is of most relevance. Although small, there is a noticeable difference between the steady-state and the time-averaged solutions. This can be partly explained by the use of different solvers (ANSYS® CFX for Case 1 and hbTRACE for Cases 2 and 3). This also reflects that the time-averaged flow out of a nonlinear unsteady simulation with inter-row coupling may differ from a steady-state solution with mixing-plane due to different modelling approaches.

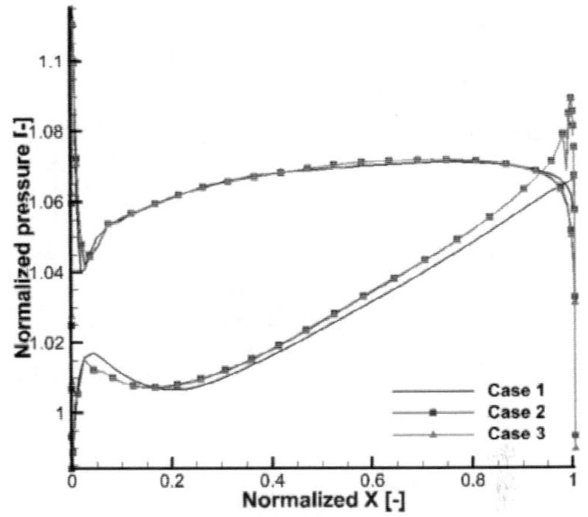

Figure 4 R2 blade surface normalized steady-state (Case 1) and time-averaged (Cases 2-3) pressure at 90% span

The normalized unsteady pressure amplitude on the R2 blade surface is shown in Figure 5. A comparison between Figure 5a and Figure 5b shows that the linear decoupled and nonlinear coupled predictions agree reasonably well when only the upstream perturbation is considered. Their contour patterns are similar, although Case 2 shows a relatively stronger level of unsteadiness on a large portion of PS, and closer to TE on the SS. When combined with the 1T mode shape, one can expect the aerodynamic modal force (and eventually the response amplitude) to be higher for Case 2 than Case 1. There is a much more pronounced difference of Case 3 compared to both Cases 1 and 2. A large, highly perturbed area occurs on the PS when S2 is included in the configuration. On the SS, the distribution of unsteady pressure amplitude is entirely changed with higher perturbation appearing near both LE and TE. The latter half of the SS adopts higher unsteady pressure, while the mid-chord region shows the opposite. Overall, Case 3 shows a higher level of unsteadiness than Case 2. This is consistent with experimental ([20]) and numerical ([21]) findings on vane clocking effects. In this case, S1 and S2 form a constructive interference such that collectively, they strengthen the R2 dynamic loading than when either S1 or S2 is acting individually.

The aerodynamic modal force amplitudes of 1T vibration predicted for three cases are shown in Table 3. Going from Case 1 to 2 results in a factor of 1.73 difference, while including S2 results in an additional factor of 1.41 which amounts to an overall factor of 2.44 difference between Case 1 and 3. This is a significant amount of difference. Although not shown here, aerodynamic damping is also available to be combined with the modal work information to calculate response amplitude for each case. The Case 3 result is indeed the closest match with experimental response measurement ([17]), albeit still underestimating the response amplitude. This indicates that if one relies on the linearized decoupled model, he/she will significantly underestimate the resonant response which could lead to bad design decisions.

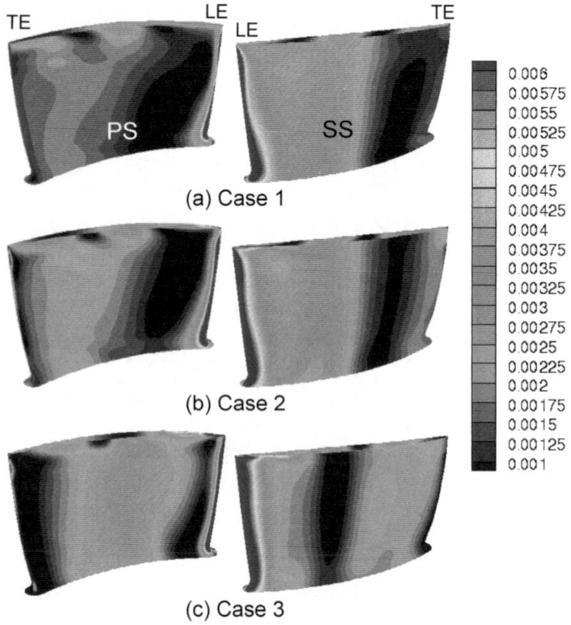

Figure 5 R2 blade surface normalized unsteady pressure amplitude contours of (a) Case 1, (b) Case 2, and (c) Case 3

Table 3 Aerodynamic modal force amplitudes of Cases 1-3

Case#	Q [N]	Q/Q_{Case1}[-]
1	0.0466	1.00
2	0.0808	1.73
3	0.1136	2.44

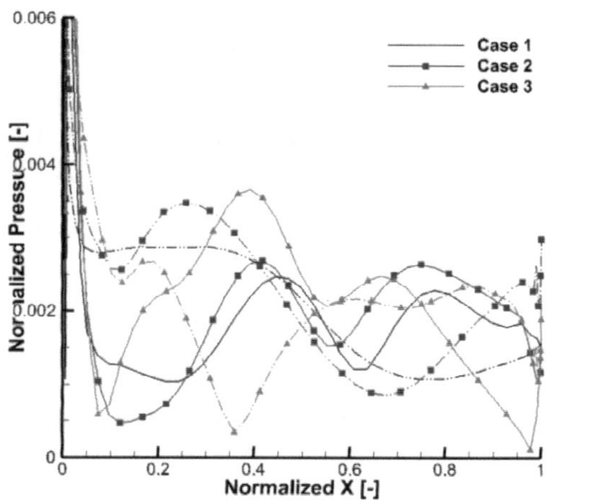

Figure 6 R2 surface normalized unsteady pressure amplitude at 90% span of Cases 1-3 (PS: solid line, SS: dashed-dotted line)

To take a closer look at the R2 surface unsteady loading of different cases, the 90% span is again chosen. Shown in Figure 6, the linear distribution is plotted in black and the nonlinear in red (Case2) and green (Case 3) curves. PS is drawn in solid and SS in interrupted lines. One could clearly see that Cases 1 and 2 have curves of similar shape in general, although the detailed amplitude distribution differs locally. For both, the SS has higher unsteady loading from the LE to about 40% chord. Around mid-chord, PS and SS adopt comparable unsteady loading levels. On and after about 60% chord, both cases exhibit higher unsteady loading on the PS. In contrast, Case 3 behaves qualitatively different, and displays a more complicated trend on PS and SS over the span of the chord. This can be explained by the interaction between S1-passing- and S2-passing which results in a different phase distribution compared to Case 2.

Figure 7 shows how the unsteadiness propagates through R2 passage at 90% span in terms of the real and imaginary parts of the normalized unsteady pressure. Cases 1 and 2 are shown to be more or less similar, although differences are also seen locally due to S1-R2 interaction in Case 2. One can clearly see a qualitative difference of Case 3 against Cases 1 and 2. Although relatively weak compared to the S1 excitation, the downstream perturbation from S2 is apparently perceived at the R2 domain exit. As it travels upstream, it "blends" with the perturbation due to S1 passing and together significantly modifies the unsteady pressure field.

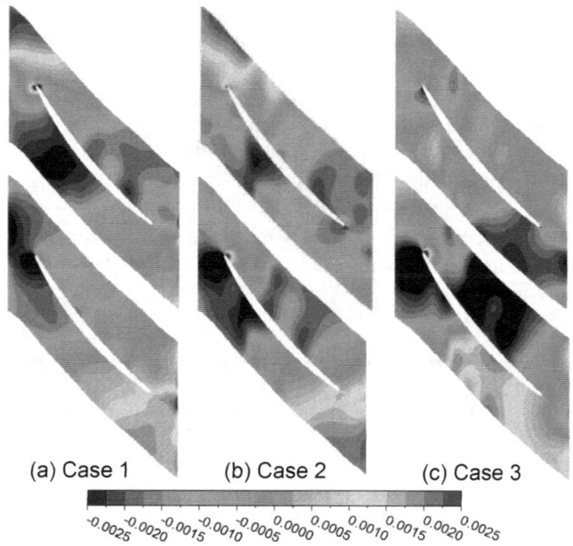

Figure 7 Real (top) and imaginary (bottom) parts of normalized unsteady pressure in R2 passage at 90% span of (a) Case 1, (b) Case 2, and (c) Case 3

Upstream stator wake is usually considered the primary source of excitation for a rotor in an embedded stator-rotor-stator environment, therefore we review how the S1 wakes are resolved in different cases. The normalized total pressure at the plane located axially mid-way between S1 TE and R2 LE is plotted in Figure 8. Note that Figure 8a is a steady-state solution with inter-row mixing plane present.

Quite differently, both Figure 8b and Figure 8c are time-averaged results of a nonlinear solver which couples each row in the configuration and calculates the time-averaged and unsteady flow fields simultaneously. Compared to Case 1, Case 2 predicts an apparently different S1 viscous wake. The wake, featured by locally low total pressure and thus high loss, is predicted to be wider along the entire span, and also deeper towards the endwalls. This means that the time-averaged flow solution with stator-rotor coupling modifies the steady-state wake characteristics. Comparison between Cases 2 and 3 reveals only a slight difference, indicating that the additional S2 has a small effect on the S1 exit wake.

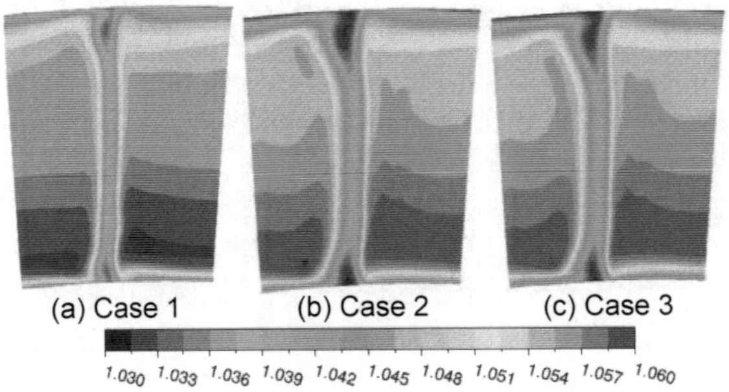

Figure 8 S1 exit normalized total pressure contours of (a) Case 1, (b) Case 2, and (c) Case 3

We discuss in detail three wake predictions at 80% span where experimental measurement is also available ([17]). Plotted in Figure 9 are the measured and predicted wake profiles in terms of normalized total pressure. Case 1 is in good agreement with the experiment in terms of both wake width and depth. Two nonlinear coupled solutions are both wider and deeper especially for Case 2, which is consistent with Figure 8. Similar observations are also reported in [19] where coupled time-marching calculations predict stronger S1 exit wake profiles than the decoupled model. It is not yet clear to the authors why Cases 2 and 3 do not compare to the experimental measurement as well as Case 1, given the fact that they represent the actual physical unsteady flow events more comprehensively than Case 1. One possible explanation could be that the mesh is not refined enough to yield reliable wake profiles, as also highlighted in [7]. Another possibility is the uncertainty in wake profile measurement. It has been reported in [22] that vane wake measurements could vary circumferentially by up to about ±25% on the 1st harmonic content.

The viscous wake forcing function to R2 can be obtained by Fourier decomposing the wake profile to obtain its Harmonic contents. Plotted in Figure 10 are the first five harmonics. They will translate to unsteady perturbations of 44EO and its first four harmonics in the rotating frame of reference. The experimental measurement shows a higher 2nd harmonic content compared to the 1st harmonic, while all CFD results show a monotonic decay. Again, same as observed with the wake profiles, Fourier contents of the decoupled case agrees better with the test data, while both nonlinear solutions over-predict by over a factor of three on the 1st harmonic (i.e. 44EO excitation). This increase in forcing function can largely explain the elevated unsteady blade loading predicted in Cases 2 and 3 as discussed in Figure 5.

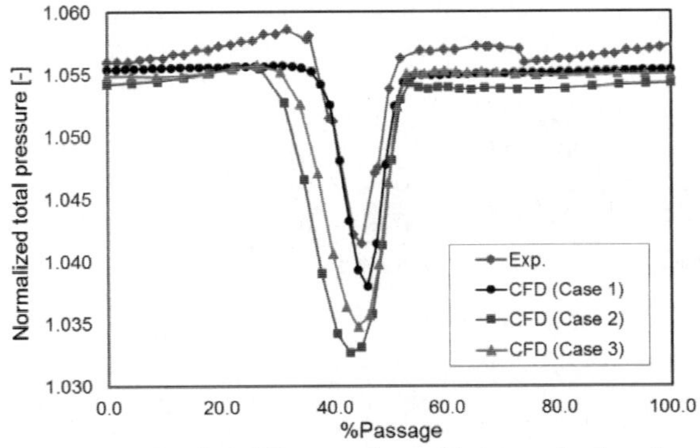

Figure 9 Predicted S1 exit wake profile (normalized total pressure) at 80% span compared with experimental measurement

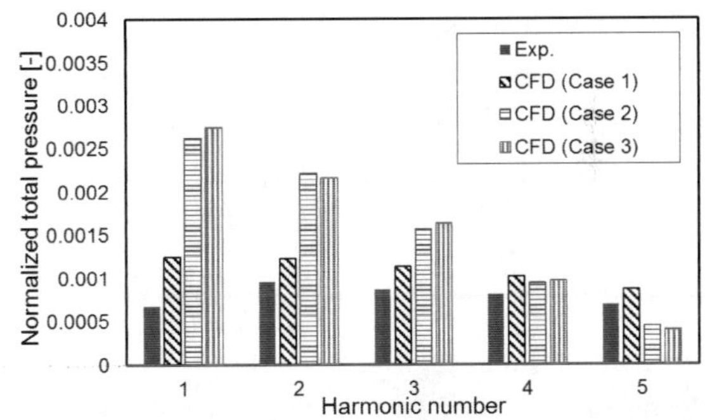

Figure 10 Fourier contents (first five harmonics) of predicted S1 exit wake profile (normalized total pressure amplitude) at 80% span compared with experimental measurement

To further assess S2 excitation, Case 3 is limited since the unsteady flow due to S2 excitation cannot be effectively separated from that due to S1 excitation. To this end, two additional simulations, Cases 4 (linear decoupled) and 5 (nonlinear coupled), are performed. When examining the unsteady pressure amplitude on R2 surface in Figure 11, we immediately notice a significant difference between two predictions. When R2-S2 interaction is present, the nonlinear solver predicts R2 surface unsteady pressure to be as high as 0.18% of 1 [bar] near the tip at about 30% chord length on the SS, marked by the red region on Figure 11b. In contrast, Case 1 is below approximately 0.08% of 1 [bar] anywhere on the blade surface. This huge difference is also reflected in the resultant modal force. According to Table 4, the modal forced amplitude of Case 5 is more than three times that of Case 4. This factor of three difference is also reflected in the predicted response amplitude, although not shown here.

We then investigate on both cases how the perturbation propagates in the R2 passage by observing the real and imaginary parts of the normalized unsteady pressure at 90% span as shown in Figure 12. The perturbation is perceived at the domain exit in both cases indicated by deep red and blue regions. However, its propagation differs dramatically between two cases. For Case 4, the excitation decays rapidly as it goes upstream. The unsteady field remains minimal on and upstream of R2 TE. In Case 5, the disturbance travels upstream without apparent decay with strong unsteady flow field throughout the R2 passage.

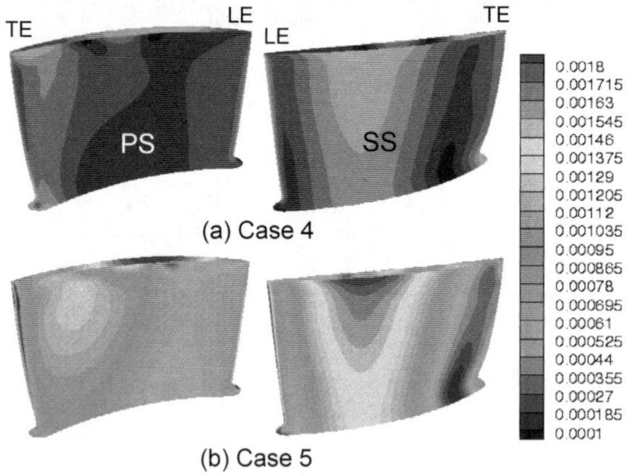

Figure 11 R2 blade surface normalized unsteady pressure amplitude contours of (a) Case 4 and (b) Case 5

Table 4 Aerodynamic modal force amplitudes of Cases 4-5

Case#	Q [N]	Q/Q_{Case4} [-]
4	0.0114	1.00
5	0.0356	3.13

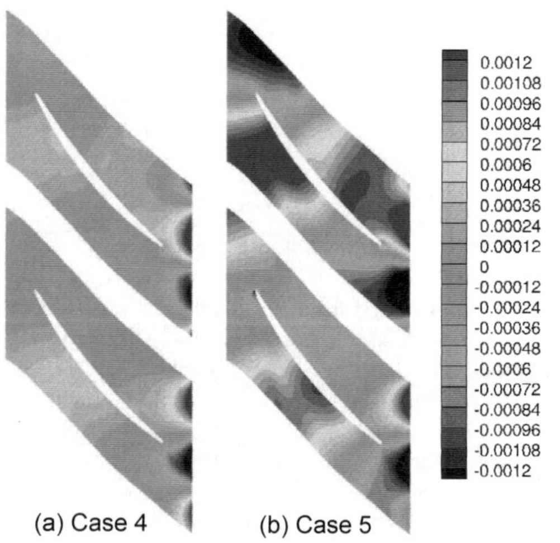

Figure 12 Real (top) and imaginary (bottom) parts of the normalized unsteady pressure at 90% span in R2 passage of (a) Case 4 and (b) Case 5

This decay vs. propagation explains the difference between Cases 4 and 5, and also in part between Cases 2 and 3. The nonlinear coupled model captures a cut-on acoustic mode due to R2-S2 interaction [23] which cannot be resolved by the decoupled approach. A similar case was reported in [8], where the presence of a cut-on acoustic mode due to IGV-rotor interaction results in huge difference between decoupled and coupled predictions. In light of this finding, designers are urged to make careful modelling decisions on cases involving both wake and potential excitations, especially when cut-on acoustic mode cannot be ruled out completely. It is unwise to exclude downstream potential source by assuming without justification that the potential disturbance is of secondary importance compared to the wake excitation. In addition, modelling inter-row coupling is imperative. The decoupled approach may considerably underestimate response, especially when a cut-on acoustic mode due to stator-rotor (or rotor-stator) interaction is present but fails to be captured.

4.2 P3S Compressor with New S1

In this section, the discussions extend to P3S Compressor with the new S1 design with reduced vane count of 38. Table 5 lists the designed and predicted $\dot{m}$ and $TPR/\dot{m}$ of each case. In order to reach the desired operating point within 1% discrepancy on $TPR/\dot{m}$, the original exit pressure distribution extracted from steady-state whole-rig simulation is multiplied by a factor of 99.2% in both Cases 2 and 3.

Table 5 Predicted operating points for Cases 1-3

Case	$\dot{m}$ [kg/s]	%Difference [-]	$\dot{m}/TPR$ [-]	%Difference [-]
Designed (S1/R2)	8.013	-	0.1283	-
Case 1	8.007	-	-	-
Case 2	8.194	2.26%	0.1291	0.67%
Designed (S1/R2/S2)	8.013	-	0.1332	-
Case 3	8.005	0.10%	0.1332	-0.74%

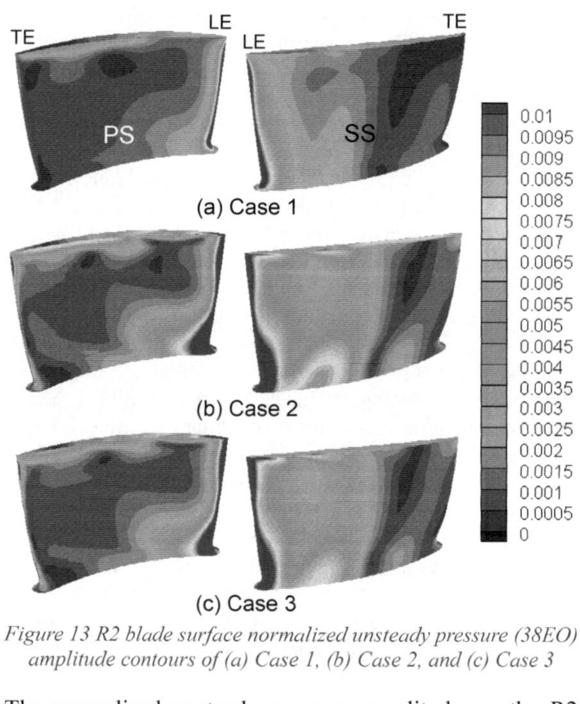

Figure 13 R2 blade surface normalized unsteady pressure (38EO) amplitude contours of (a) Case 1, (b) Case 2, and (c) Case 3

The normalized unsteady pressure amplitude on the R2 blade surface is plotted in Figure 13. The 38EO excitation provides an elevated level of unsteadiness compared with the nominal design (Figure 5). In general, all three cases have similar contour patterns, indicating that the primary unsteady flow event in R2 is due to S1 excitation, and that the decoupled approach is adequate to qualitatively capture the main characteristics. Compared to Case 1, Case 2 predicts an overall stronger dynamic loading, especially near the LE at lower span. This results in a larger modal force amplitude of 1.4 times of the linearized decoupled prediction, as shown in Table 6. Unlike the nominal design discussed in the previous section, here, the difference between Cases 2 and 3 is mild. This means that when up- and downstream perturbations have different frequencies, including the off-resonant excitation would bring limited improvement in prediction accuracy, which is also manifested in Table 6.

Figure 14 shows how the 38EO perturbation travels in the R2 passage in terms of the real and imaginary parts of the normalized unsteady pressure at 90% span. The flow disturbance is predicted to be considerably higher in Case 2 starting at the domain inlet and throughout the passage. This is consistent with the higher dynamic loading of Case 2 shown in Figure 13 and the resulting higher 1T modal force shown in Table 6. With the addition of 44EO S2-passing excitation, the 38EO unsteady flow field remains largely unchanged. Minor differences can be seen near the domain exit, but they do not significantly contribute to the dynamic loading.

Table 6 Aerodynamic modal force amplitudes of Cases 1-3

Case#	Q [N]	Q/Q_{Case1} [-]
1	0.0637	1.00
2	0.0898	1.40
3	0.0958	1.50

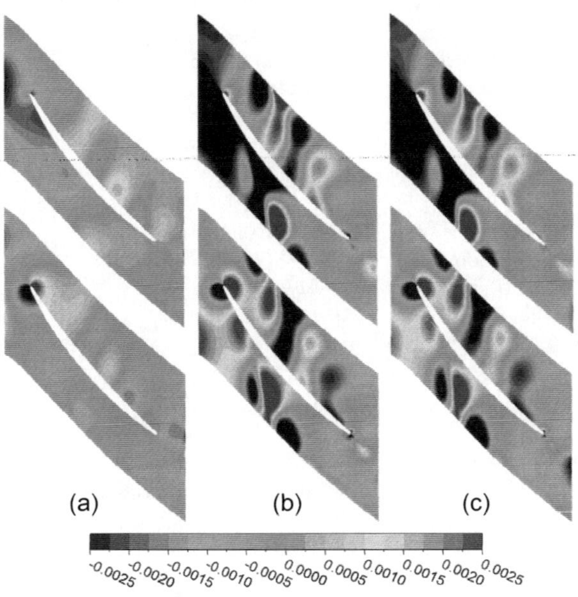

Figure 14 Real (top) and imaginary (bottom) parts of the normalized unsteady pressure (38EO) at 90% span in R2 passage of (a) Case 1 (b) Case 2, and (c) Case 3

The predicted S1 exit wake profiles are shown in Figure 15. Similar to what is previously observed, the coupled time-averaged solution is significantly different than the steady-state solution on which the linearized decoupled simulation is based. A detailed comparison is made in Figure 16 where the wake profiles at 90% span are plotted. Two coupled solutions are nearly identical to each other, indicating that including stator downstream of the stator-rotor pair in the computational domain barely alters the predicted S1 wake. They are both wider and deeper than the steady-state wake in Case 1. No general conclusion can be drawn at this point due to unavailability of test data, although experimental investigation will soon be carried out.

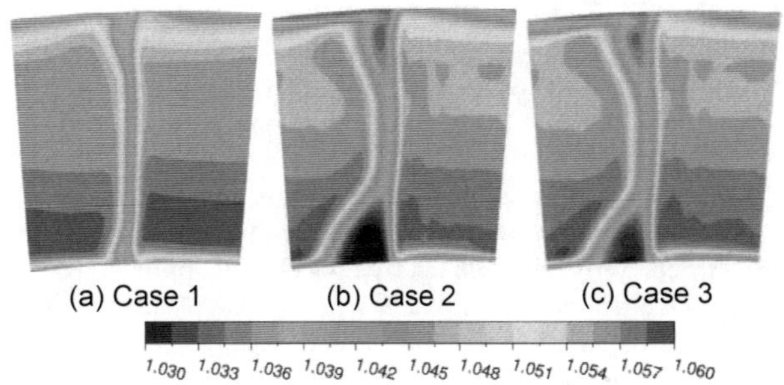

Figure 15 S1 exit plane normalized total pressure contours of (a) Case 1 (b) Case 2, and (c) Case 3

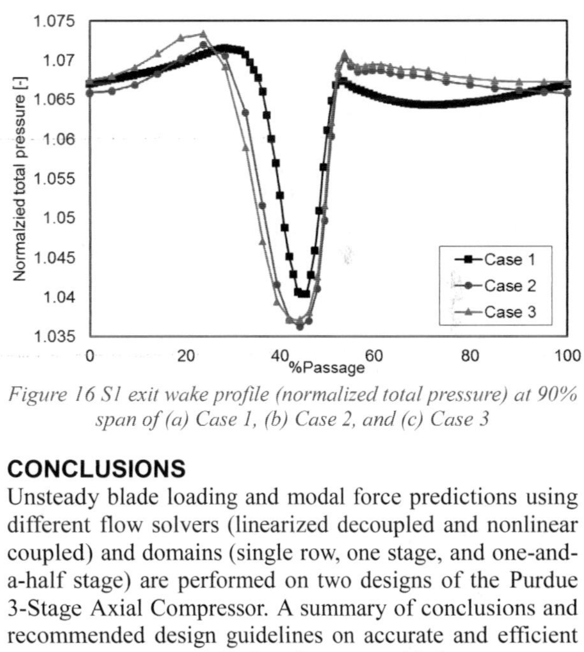

Figure 16 S1 exit wake profile (normalized total pressure) at 90% span of (a) Case 1, (b) Case 2, and (c) Case 3

CONCLUSIONS

Unsteady blade loading and modal force predictions using different flow solvers (linearized decoupled and nonlinear coupled) and domains (single row, one stage, and one-and-a-half stage) are performed on two designs of the Purdue 3-Stage Axial Compressor. A summary of conclusions and recommended design guidelines on accurate and efficient resonant response predictions is presented below:

1. Forced response problems in turbomachines can involve nonlinearities e.g. stator-rotor interaction, clocking, and other multistage effects, especially near resonance. Linearized decoupled approach is shown, for cases studied in this paper, to greatly underestimate the modal force by a factor of 1.7 (the nominal S1 design, from Case 1 to Case 2), thus should be avoided or at least used with great caution. It is recommended to model the coupled stator-rotor or rotor-stator domain with nonlinear

flow solvers, especially those applying efficient frequency-domain methods e.g. Harmonic Balance ([18]) to reduce computational cost.

2. An embedded row is disturbed by both its up- and downstream neighbours. When these disturbances occur at the same frequency, they can form either constructive or destructive interference that may significantly alter the unsteady flow field. In this case, it is recommended to include all three rows in CFD. When two disturbances are separate in frequencies, the resonant frequency would be of primary interest. The off-resonance disturbance is shown to only slightly modify the unsteady flow field at resonant frequency. In this case, a coupled stator-rotor or rotor-stator configuration resolving the resonant excitation event alone is found to be sufficient.
3. Cut-on acoustic mode due to inter-row interaction is an important unsteady flow phenomenon that can only be modelled by the coupled approach (or the decoupled approach with a carefully designed two-step method [8]). The coupled approach should always be used when there is a limited knowledge about the design and/or when acoustic cut-one mode due to inter-row interaction is likely to occur.
4. The accuracy of unsteady flow simulations can be highly dependent on the mesh quality. Viscous wake profile is hard to be fully resolved without a relatively fine mesh in the wake region. In addition, frequencies of shorter wavelength cannot be properly modelled without fine mesh. It is thus recommended to always perform a thorough mesh convergence study to remove any possible mesh sensitivity of the steady and/or unsteady flow solutions.

ACKNOWLEDGMENTS

This project is funded by the GUIde 5 Consortium. Dr. Nicole Key and her colleagues at Purdue University are gratefully acknowledged for providing experimental data. The authors would also like to thank Siemens Energy, Inc. for providing resources and support to complete this work and also for allowing the publication of this work.

REFERENCES

[1] S. Moffatt and L. He, "On Decoupled and Fully-coupled Methods for Blade Forced Response Prediction," *Journal of Fluids and Structures*, vol. 20, no. 2, pp. 217–234, 2005.

[2] A. I. Sayma, M. Vahdati, and M. Imregun, "Turbine Forced Response Prediction Using An Integrated Non-Linear Analysis," *Proceedings of the Institution of Mechanical Engineers, Part K: Journal of Multi-body Dynamics*, vol. 214, no. 1, pp. 45–60, 2005.

[3] C. Bréard, M. Vahdati, a. I. Sayma, and M. Imregun, "An Integrated Time-Domain Aeroelasticity Model for the Prediction of Fan Forced Response due to Inlet Distortion," *Journal of Engineering for Gas Turbines and Power*, vol. 124, no. 1, p. 196, 2002.

[4] S. Moffatt, W. Ning, Y. Li, R. G. Wells, and L. He, "Blade Forced Response Prediction for Industrial Gas Turbines," *Journal of Propulsion and Power*, vol. 21, no. 4, pp. 707–714, 2005.

[5] M. Vahdati, A. I. Sayma, M. Imregun, and G. Simpson, "Multibladerow Forced Response Modeling in Axial-Flow Core Compressors," *Journal of Turbomachinery*, vol. 129, no. 2, p. 412, 2007.

[6] M. Berthillier, M. Dhaninaut, F. Burgaud, and V. Garnier, "A Numerical Method for the Prediction of Bladed Disk Forced Response," *Journal of Engineering for Gas Turbines and Power*, vol. 119, no. 2, pp. 404–410, 1997.

[7] M. Richman and S. Fleeter, "Navier-Stokes Simulation of IGV-Rotor-Stator Interactions in a Transonic Compressor," 2000.

[8] H. Schoenenborn and G. Ashcroft, "Comparison of Non-linear and Linearized CFD Analysis of the Stator-Rotor Interaction of a Compressor Stage," in *Proceedings of ASME Turbo Expo 2014*, 2014.

[9] K. C. Hall, J. P. Thomas, and W. S. Clark, "Computation of Unsteady Nonlinear Flows in Cascades Using a Harmonic Balance Technique," *AIAA Journal*, vol. 40, no. 5, pp. 879–886, 2002.

[10] P. D. Silkowski and K. C. Hall, "A Coupled Mode Analysis of Unsteady Multistage Flows in Turbomachinery," *Journal of turbomachinery*, vol. 120, no. 3, pp. 410–421, 1998.

[11] K. C. Hall and K. Ekici, "Multistage Coupling for Unsteady Flows in Turbomachinery," *AIAA Journal*, vol. 43, no. 3, pp. 624–632, 2005.

[12] K. Ekici and K. C. Hall, "Nonlinear Analysis of Unsteady Flows in Multistage Turbomachines Using Harmonic Balance," *AIAA Journal*, vol. 45, no. 5, pp. 1047–1057, 2007.

[13] K. Ekici, K. C. Hall, and R. E. Kielb, "Harmonic Balance Analysis of Blade Row Interactions in a Transonic Compressor," *Journal of Propulsion and Power*, vol. 26, no. 2, pp. 335–343, 2010.

[14] R. D. Fulayter, "An Experimental Investigation of Resonant Response of Mistuned Fan and Compressor Rotors Utilizing NSMS," Purdue University, 2004.

[15] D. J. W. Monk, "A Computational Analysis of the Aerodynamic and Aeromechanical Behavior of the Purdue Multistage Compressor," Purdue University, 2014.

[16] F. M. Besem, R. E. Kielb, and N. L. Key, "Forced Response Sensitivity of a Mistuned Rotor From an Embedded Compressor Stage," in *Proceedings of ASME Turbo Expo 2015*, 2015, pp. 1–13.

[17] W. L. Murray III and N. L. Key, "Experimental Investigation of a Forced-Response Condition in a Multistage Compressor," *Journal of Propulsion and Power*, vol. 31, no. 5, pp. 1320–1329, 2015.

[18] C. Frey, G. Ashcroft, H.-P. Kersken, and C. Voigt, "A Harmonic Balance Technique for Multistage Turbomachinery Applications," in *Proceedings of ASME Turbo Expo 2014: Turbine Technical Conference and Exposition*, 2014.

[19] F. M. Besem, "Aeroelastic Instabilities due to Unsteady Aerodynamics," Duke University, 2015.

[20] Y. S. Choi, N. Key, and S. Fleeter, "Vane Clocking Effects on the Resonant Response of an Embedded Rotor," *Journal of Propulsion and Power*, vol. 27, no. 1, pp. 71–77, 2011.

[21] J. R. Salontay, N. L. Key, and R. D. Fulayter, "Investigation of Flow Physics of Vane Clocking Effects on Rotor Resonant Response," *Journal of Propulsion and Power*, vol. 27, no. 5, pp. 1001–1007, 2011.

[22] J. C. Methel, N. R. Smith, and N. L. Key, "Vane Wake Characterization for a Multistage Compressor," in *53rd AIAA Aerospace Sciences Meeting*, 2015, no. January.

[23] J. M. Tyler and T. G. Sofrin, "Axial Flow Compressor Noise Studies," *Transactions of the Society of Automotive Engineers*, pp. 309–332, 1962.

Proceedings of the 14th International Symposium on Unsteady Aerodynamics, Aeroacoustics & Aeroelasticity of Turbomachines
ISUAAAT15
8-11 September 2015, Stockholm, Sweden

I14-S2-3

TYLER-SOFRIN MODES IN AXIAL HIGH PRESSURE COMPRESSOR FORCED RESPONSE ANALYSES

Sven Schrape
Rolls-Royce Deutschland Ltd & Co KG
15827 Blankenfelde-Mahlow, Germany
sven.schrape@rolls-royce.com

Thomas Giersch, Jens Nipkau, Sina Stapelfeldt, Bernhard Mück
Rolls-Royce Deutschland Ltd & Co KG
15827 Blankenfelde-Mahlow, Germany

ABSTRACT

Integral aerodynamic excitations (i.e. forced response) in multi blade row turbomachinery are attributed to a multitude of excitation sources. In a common forced response analysis often only integral excitation orders directly connected to single blade row aerofoil counts and their sums and differences are considered.

For the blade row of interest various rotating pressure patterns can occur at a single frequency, which are generated by up- and downstream rotor-stator interactions. These spinning modes – termed Tyler-Sofrin modes - have been described by Tyler and Sofrin [1] from an aero acoustic point of view.

The present work considers Tyler-Sofrin modes in axial high pressure compressor forced response analyses. In the first part the effect of rotor stator interactions according to Tyler and Sofrin [1] is presented in detail and put into the context of a forced response analysis of an aerofoil dominated modeshape. This involves the calculation of affected frequencies and nodal diameters. The relevance of considering Tyler-Sofrin modes in the computation of the aerodynamic forcing is described based on the observation of varying blade individual forcing amplitudes over the circumference.

In the second part the results of numerical analyses of high pressure compressor multi blade row forced responses addressing the aforementioned set of problems are presented. The results are obtained using unsteady 3D CFD for the computation of the modal forces coming from classical and rotor-stator interaction sources. Vibration responses are calculated using a reduced order model employing a subset of nominal modes (SNM) [2], [3], [4] for a rotor blisk. Various implications of Tyler-Sofrin mode excitations are discussed for tuned and mistuned aerofoil vibrations.

INTRODUCTION

The ever growing demand for faster and less expensive development of aero engines has led to an increased amount of numerical simulations as part of engine certification programmes. In the course of that modern high pressure compressors of aero engines are being certified employing numerical analyses of aerofoil vibrations only reducing the amount of costly engine testing as much as possible. Here the challenge is not only the large number of simulations to be handled but also the growing complexity of the aeroelastic models. Vibration surveys conducted on a development engine include all possible excitation sources by nature. When it comes to numerical simulations of forced vibrations all possible sources of excitation have to be included in the simulation. Hence the numerical forced response models used nowadays not only include the direct up- and downstream blade rows but also those further upstream. This in turn poses new challenges with the formation of rotor-stator-interaction tones in such large setups. Such interaction tones, also called spinning modes, are the result of interfering unsteady pressure waves of rotating and non-rotating compressor parts.

The first to describe these interaction tones were Tyler and Sofrin in 1962 [1]. More details about the occurrence of Tyler-Sofrin modes are given in the next section. Ever since rotor-stator-interaction and its impact on forced responses have been the subject of numerous research activities. Besem et al. [5] worked on reproducing experimentally measured blade vibration amplitudes using a reduced order model and taking into account perturbations on measured blade frequencies, aerodynamic asymmetries and on the nodal diameter content of the forcing function. They found that additional nodal diameter content in the forcing function has a strong effect on the resulting blade vibration amplitudes. In 2005 Hall and Ekici [6] demonstrated that multi-stage effects can have a significant impact on aerodynamic loading. As a consequence they developed a three-dimensional time-linearised method which accounts for multi-stage effects by coupling the unsteady solutions corresponding to the different spinning modes at the blade row interface. Another approach to consider multistage effects on forced responses has been presented by Stapelfeldt and Di Mare [7]. They proposed a method for 3D unsteady viscous non-linear equations and were able to show considerable improvement in the computational effort needed to solve multistage forced response problems. Summarising the previous work it is indisputable that considering multistage interactions is vital for the accurate prediction of blade vibration amplitudes.

This paper addresses the impact of multistage effects on the aerodynamic modal forces acting on a compressor blade row. The unsteady simulations to compute the unsteady modal forces have been performed using a 3D unsteady Reynolds averaged Navier-Stokes solver with Spalart-Allmaras turbulence model developed by Imperial College London [8]. The code has been validated against numerous turbomachinery applications [9], [10], [11]. The result of the multistage forced response simulations with blade rows far upstream of the blade row of interest are shown in contrast to the results obtained including only the direct up- and downstream blade rows. The simulations have been performed for the stage 2 rotor of a 4.5 stage research compressor in preparation of future experimental blade vibration measurements.

The modal forces computed in these simulations have then been used to calculate blade vibration amplitudes for a tuned and mistuned rotor blisk with and without considering Tyler-Sofrin modes. The blade vibration amplitudes have been calculated using a reduced order model developed by [2] using a subset of nominal modes (SNM). As part of the structural mistuning investigations the identification of a worst case mistuning distribution has been performed by a genetic algorithm to generate a maximum amplification of the blisk response. Furthermore a probabilistic analysis of normal distributed blade individual frequencies and the resulting amplification factors have been carried out. Finally a comparison between forced response predictions based on measured mistuning distribution for models with and without Tyler-Sofrin mode excitations has been discussed.

NOMENCLATURE

1D	One dimensional
3D	Three dimensional
BLISK	Blade Integrated Disk
CFD	Computational Fluid Dynamic
CSM	Cyclic Symmetry Mode
DFT	Discrete Fourier Transformation
DLR	Deutsches Zentrum für Luft und Raumfahrt
EO	Engine Order
FE	Finite Element
HPC	High Pressure Compressor
LE	Leading Edge
ND	Nodal Diameter
NH	Mechanical Speed
OGV	Outlet Guide Vane
SNM	Subset of Nominal Modes
VIGV	Variable Inlet Guide Vane
B_R	Rotor blade count
B_S	Stator blade count
$\mathbf{D}$	Modal damping matrix
f	Frequency
$\mathbf{f}$	Modal force vector
$\hat{f}$	Amplitude of modal force
$\mathbf{G}$	Gyroscopic load matrix
k	Wave number
$\mathbf{K}$	Modal stiffness matrix
$\mathbf{M}$	Modal mass matrix
n	Disturbance index
p	Pressure
$\mathbf{q}$	Modal displacement vector
r	Radius
t	Time
y+	Dimensionless wall distance
$\mathbf{Z}$	Aerodynamic impedance matrix
$\boldsymbol{\varphi}$	Modal matrix
Φ	Phase angle
θ	Circumferential angle
Ω	Angular velocity
ω	Excitation frequency

OCCURRENCE OF TYLER-SOFRIN MODES

The fluctuating pressure field in an axial compressor (or turbine) due to a single row consists of a superposition of lobed patterns rotating with the row's rotational speed Ω. Assuming identical blades, the number of circumferential lobes in each pattern is an integer multiple of the row's blade count, B, such that the fluctuation can be mathematically described by a Fourier series:

$$p(r, \theta, t) = \sum_{n=1}^{\infty} a_n(r) \cos(nB(\theta - \Omega t) + \Phi_n) \qquad (1)$$

where the Fourier coefficients, a_n and Φ_n, are functions of radius since the compressor pressure field is typically not uniform from hub to tip. In a frame of reference which is fixed with respect to the engine, i.e. non-rotating, each of these fluctuations is associated with a blade passing frequency, $\omega = nB\Omega$.

In a multi-row environment, the pressure field contains multiple perturbations with different circumferential wave numbers and rotational speeds – so called spinning modes or Tyler-Sofrin modes. These perturbations interact. The rotor passing disturbance with circumferential wave number $n_R B_R$ rotating at angular velocity Ω_R, for example, becomes scattered, or modulated, through interaction with a stator pressure field consisting of circumferential wave numbers $n_S B_S$. This generates a new set of perturbations with circumferential wave numbers k:

$$k = n_R B_R + n_S B_S \qquad (2)$$

which appear with the rotor blade passing frequency $\omega_R = n_R B_R \Omega_R$ in the stator frame of reference, see [1]. If such a disturbance is reflected back into the rotor or travels on into another rotor row, its frequency in the relative frame of reference is shifted by $k(\Omega_S - \Omega_R)$, such that it appears with the stator blade passing frequency in the rotor frame of reference. By substituting appropriate rotational speeds, a similar pattern can be derived for rotor-rotor or stator-stator interaction. As a result a theoretically infinite number of perturbations may exist in a multi-stage compressor with circumferential wave numbers equal to linear combinations of the blade counts and angular frequencies equal to the blade passing frequencies and linear combinations thereof.

TYLER-SOFRIN MODES IN MULTI-BLADEROW COMPUTATIONS

The present investigations are focused on the Rig250 - a 4.5-stage research compressor from the DLR Cologne/Germany. The compressor consists of variable inlet guide vanes and variable stator vanes at stage one and two. In contrast to the previous rig builds the stage two of the current build has been equipped with a blisk while all other rotors remain conventionally bladed disks. The aerofoil geometry of the rotor two blisk is identical to the previous rotor blade geometry. No casing treatments are installed. At the design point the compressor pressure ratio is ~4.6, the reduced corrected massflow ~45kg/s. For details and a general arrangement it is refer to [12].

Based on numerical simulations this chapter addresses the impact of multistage effects onto aerodynamic modal forces acting on a compressor blade row. Therefore three dimensional flow computations have been carried out using the finite volume solver Au3D which was developed at the Imperial College London. The code solves the viscid, compressible, Favre-averaged, Navier-Stokes equations using the turbulence model of Spalart and Allmaras. A second-order implicit time marching algorithm has been applied for transient computations. Details can be read in [8].

Initial solutions for unsteady simulations are taken from steady state simulations based on a single passage, mixing plane, full HPC model from strut to OGV with periodic boundary conditions applied in circumferential direction. Associated blade counts are listed at Table 1.

Table 1 Blade counts Rig250

Blade Row	Count	Blade Row	Count
Strut	7	Rotor 3	38
VIGV	40	Stator 3	68
Rotor 1	23	Rotor 4	47
Stator 1	36	Stator 4	80
Rotor 2	28	OGV	80
Stator 2	48		

The CFD model consists of fillet radii, tip-, hub- and variable penny gaps based on the rig hardware, see Figure 1.

Figure 1: CFD grid example – semi-structured meshes are shown for the variable stator vane one and the rotor two aerofoil.

At the strut inlet radial profiles of total temperature, total pressure and velocity vectors have been prescribed. Steady state compressor characteristics were computed by closing a choked throttle at the HPC outlet for each point. Thereby ambient static pressure was fixed at the outlet boundary. Characteristics were computed at two speeds - 80%NH and 100%NH, see Figure 2.

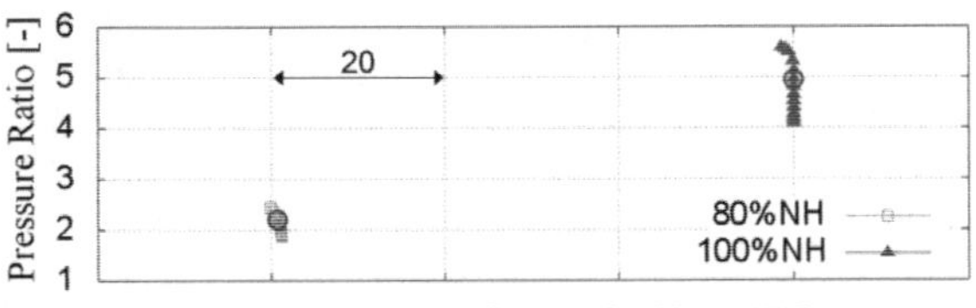

Figure 2: Computed compressor characteristics and chosen points for unsteady simulations.

For each speed a separate semi-structured grid for the full HPC was generated to capture a y+ value > ~30 with respect to the used wall functions. Figure 3 shows a full HPC steady state computation at 80%NH for example.

Figure 3: Compressor single passage mixing plane steady state CFD calculation at 80%NH - 11 bladerows from strut to OGV with ~3.5 mio grid points are shown with static pressure contours.

Transient, multistage simulations with sliding planes applied between adjacent blade rows have been compared to discuss the influence of Tyler-Sofrin modes onto modal forcings and blade vibration responses respectively. Thereby it has been focussed onto the rotor two blisk only. For each predicted speed two full assembly models were taken into account. Both models include the rotor 2 blisk of interest as the next to the last blade row. The first model is shown on the left hand side of Figure 4 with 3 rows including stator one, rotor two blisk and stator two. This model contains no Tyler-Sofrin modes generated by up- or downstream rotor-stator-interactions of the rotor two blisk.

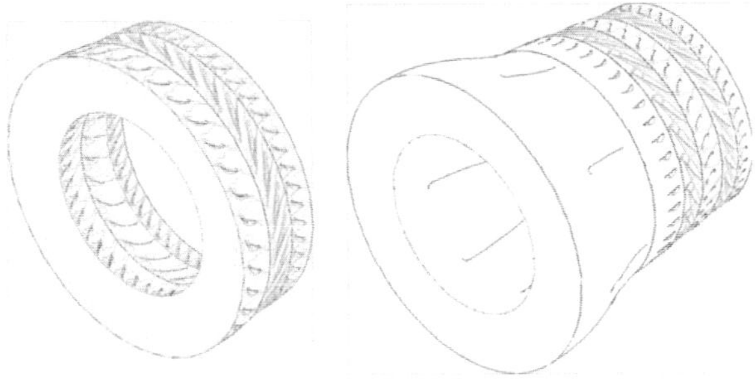

Figure 4: Full assembly CFD models at 100%NH - LHS: 3 row model from stator 1 to stator 2 with ~35 mio grid points, RHS: 6 row model from strut to stator 2 with ~55 mio grid points.

The second model consists of 6 rows starting from the strut up to stator two, see right hand side of Figure 4. Here the rotor one is expected to generate Tyler-Sofrin modes by interacting with the adjacent stator blade rows. The rotor two blisk modal forcings originated by these Tyler-Sofrin modes will be investigated. For reasons of time periodic modal forcings and to resolve the rotational frequency (EO1) all transient simulations ran at least about the duration of one rotor revolution. Thereby the modelled stator two engine order EO48 has been resolved with 150 time steps.

Within the present paper a higher aerofoil dominated mode of the rotor two blisk has been investigated. The modeshape of interest and corresponding resonance conditions are visible at Figure 5. The mode is excited by the directly upstream stator one with EO36 in the aliased nodal diameter ND 8. As the resonance of mode 11 occurs between 80%NH and 100%NH, a comparison of modal forcings between each CFD speed will be conducted.

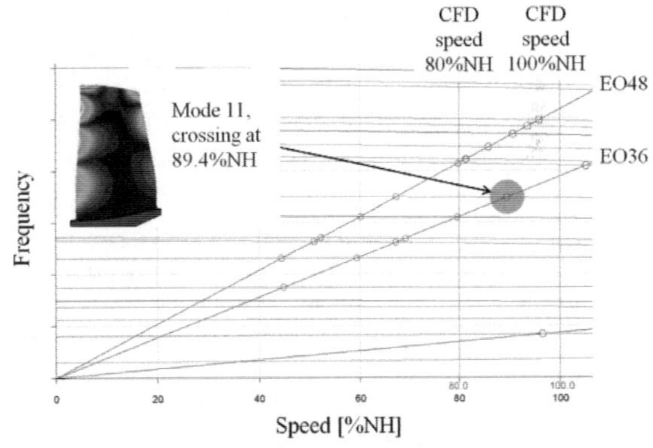

Figure 5: Campbell diagram of blisk rotor two for ND 8 with mode 11 (resultant displacement) in resonance at 89%NH.

During the transient rotor-stator-interaction simulations without considered aerofoil vibrations, unsteady pressures were recorded over time on all aerofoil surfaces over the circumference. Modal forces were subsequently predicted by considering the aerofoil modeshape of interest from a modal FE-analysis. Therefore the modeshape has been interpolated onto the corresponding CFD grids. Two different methods of modal force computations were used within the present paper. Both approaches can be transformed into each other.

A) Modal assembly force prediction:

For each nodal diameter the interpolated aerofoil modeshape has been expanded over the circumference to setup all assembly modeshapes. From these assembly modeshapes, time histories of modal assembly forces have been predicted using the unsteady pressure history from the full assembly CFD simulations. Subsequently a Discrete Fourier Transformation (DFT) delivers amplitudes and phase angles separately for each nodal diameter.

B) Blade individual modal force prediction:

As unsteady pressure histories has been recorded on every aerofoil surface over the circumference, blade individual modal forcings can be predicted using the interpolated modeshape of interest for every aerofoil without considering spatial nodal diameter information. A subsequent DFT delivers the blade individual modal forcing amplitude for every blade over the circumference which belongs to a specific excitation frequency.

At first the results of the 3 row CFD model will be discussed. Therefore Table 2 lists the expected rotor two blisk excitation sources, the corresponding engine orders and the aliased nodal diameters. These are known from the modelled stator blade counts, see Table 1.

Table 2: Expected rotor two excitation sources within 3 row models

Source	EO	ND
S1	36	8
S1 2nd harm	72	12
S1 3rd harm	108	4
S2	48	8
S2 2nd harm	96	12
S2-S1	12	12

The overall modal assembly forcing content (method A) of the 3 row CFD model at 80%NH and 100%NH related to mode 11 (ND8 of the blisk FE-model) are visualised by Figure 6 and Figure 7. The EO36 is marked by a black arrow. All forcings are normalised with the 80%NH ND8 forcing of EO36 of the 3 row model. The matrix plots show for every spatial nodal diameter of the rotor two blisk the modal forcing amplitude of each engine order EO and excitation frequency respectively. The dotted lines mark the zig-zag shaped relationship between the EO's and the aliased NDs of the rotor two blisk based on blade counts. Considering these plots at both speeds it has to be noted that resonant conditions for mode 11 are presumed only for EO36, see Campbell Diagram in Figure 5. All other forcings deemed to be off-resonant.

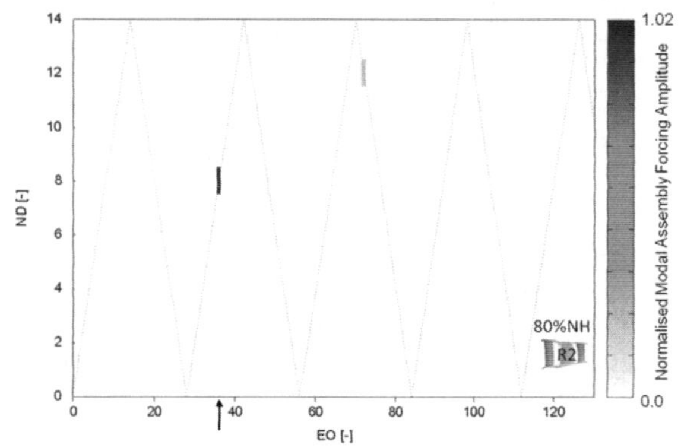

Figure 6: Normalised Modal assembly forcings of mode 11 at 3 row CFD model and 80%NH speed.

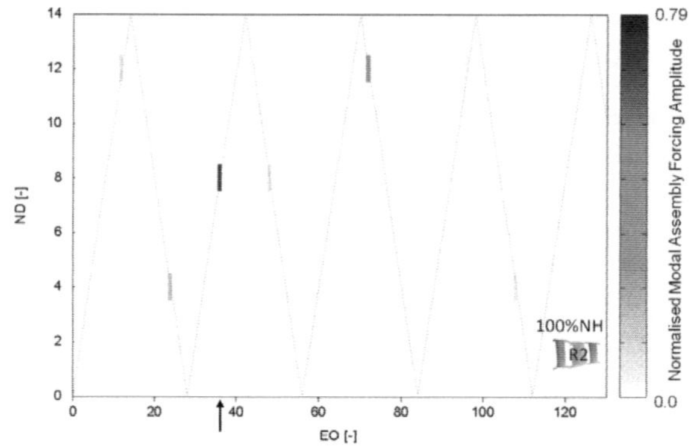

Figure 7: Normalised modal assembly forcings of mode 11 at 3 row CFD model and 100%NH speed.

All excitation sources from Table 2 become visible at 100%NH. In addition EO24 (S1 2nd harm – S2) and EO60 (S2 2nd harm – S1) show a small modal forcing. As expected, mode 11 is dominated by the direct upstream wake excitation of stator 1 and its second harmonic. For the predicted unsteady aerodynamics at 80%NH a distinct integral forcing of mode 11 was observed only for EO36, EO48 and EO72. EO36 shows the highest normalised model forcing at 80%NH speed with 1.0 while at 100%NH this forcing decreases to 0.73.

For discussing the 6 row model Table 3 lists the excitation sources from additionally modelled stator rows compared to the 3 row model excluding Tyler-Sofrin mode excitations. The overall modal forcing content of the 6 row models are shown in Figure 8 and Figure 9.

Table 3: Additional expected rotor two excitation sources within 6 row models compared to 3 row models without Tyler-Sofrin modes.

Source	EO	ND
Strut	7	7
Strut 2nd harm	14	14
VIGV	40	12
VIGV 2nd harm	80	4

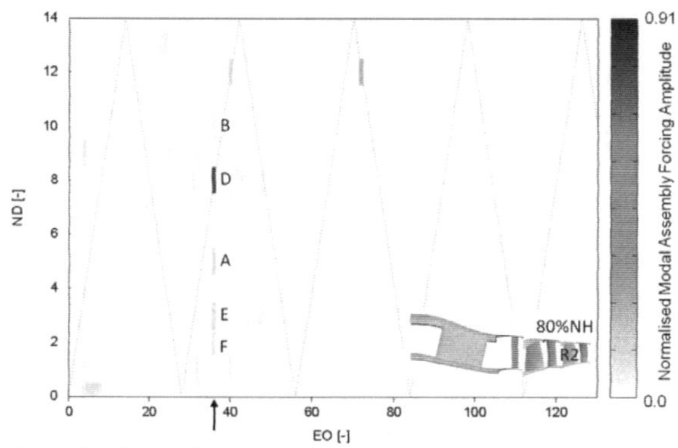

Figure 8: Normalised modal assembly forcings of mode 11 at 6 row CFD model and 80%NH speed.

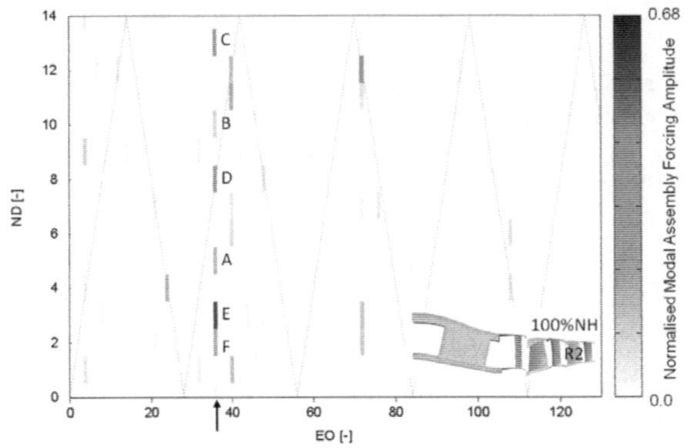

Figure 9: Normalised modal assembly forcings of mode 11 at 6 row CFD model and 100%NH speed.

Within the 6 row model generally the same excitation sources can be identified on the dotted zig-zag bands as shown for the 3 row model. This was expected as well as the occurrence of the additional 6 row excitation sources listed in Table 3. Only the VIGV EO40 contributes to a modal forcing of mode 11 at both CFD speeds.

As the EO36 is presumed to be in resonance with mode 11 (see Figure 5) the following discussion will be focussed on the EO36 only. It catches the eye that on both 6 row models rotor two blisk forcings occur in the EO36, which are located besides of the dotted zig-zag bands. These modal forcings are generated by Tylor-Sofrin modes which are part of the 6 row CFD solution as a result of the upstream interaction between rotor one and stator one. As the rotor one is not part of the 3 row model, these Tyler-Sofrin modes not exist therein. The occurrence of these spinning modes, the spatial wave numbers k and ratios of spinning speeds ω_k/Ω_{EO} can be predicted in accordance to [1] as discussed in the chapter before. Table 4 lists the results based on the EO36. Rotor two blisk forcings occur in different aliased NDs, which spin with individually speeds and hence excite the same EO as the originating stator wake. The rotating pressure field which is linked to the stator one in the rotating frame of reference undergoes a modulation of amplitude by the interacting rotor one. In Table 4 the IDs are given to easily identify Tyler-Sofrin mode excitations in the overall modal forcing plots.

Table 4: Tyler-Sofrin mode excitations of rotor two blisk based on the upstream interaction of rotor one and stator one, see equation (2). A negative ND indicates a backward travelling wave in the rotating frame of reference.

ID	n_S	n_R	k	ω_k/Ω_{EO36}	Aliased ND	Excited EO
A	1	-3	-33	-1.09	5	36
B	1	-2	-10	-3.60	10	36
C	1	-1	13	2.77	-13	36
D	1	0	36	1.00	-8	36
E	1	1	59	0.61	-3	36
F	1	2	82	0.44	2	36

For the unsteady aerodynamics at 100%NH the Tyler-Sofrin modes are more pronounced related to the directly upstream stator wake excitation in ND 8 as for 80%NH. The maximum normalised forcing at 100%NH has been observed in ND 3 with 0.6 while at 80%NH the ND 8 was maximum forced by the direct stator one wake excitation with 0.82.

Table 5: CFD predicted normalised modal assembly forcing amplitudes compared between all models.

		Normalised Modal Assembly Forcing Amplitude			
ID	Aliased ND	80%NH, EO36		100%NH, EO36	
		3 row	6 row	3 row	6 row
A	5	0	0.09 (Φ=-58°)	0	0.16 (Φ=-78°)
B	10	0	0.04 (Φ=12°)	0	0.09 (Φ=83°)
C	-13	0	0.01 (Φ=-78°)	0	0.31 (Φ=-75°)
D	-8	**1.0**	**0.82** (Φ=-155°)	**0.73**	0.34 (Φ=-51°)
E	-3	0	0.10 (Φ=142°)	0	**0.60** (Φ=-178°)
F	2	0	0.06 (Φ=-62°)	0	0.19 (Φ=-79°)
Sum		1.0	1.12	0.73	1.69

A comparison of computed assembly forcings between all models and speeds based on the stator one excitation EO36 is shown in Table 5 (maxima bolt marked). Four findings are stated:

- As seen for 100%NH, the maximum forcing of the 6 row models not necessarily occurs in the aliased ND 8 from the directly upstream wake excitation (stator one).
- Considering this ND 8 for both CFD speeds, the forcings observed by the 3 row models are higher compared to the 6 row models.
- The summation of 6 row model forcings (all NDs including Tyler-Sofrin excitations) amounts to higher forcings as computed for the 3 row models.
- While the 3 row model shows a smaller overall forcing for 100%NH compared to 80 %NH (0.73 < 1.0), the 6 row model behaves vice versa (1.69 > 1.12).

The modal forcings listed in Table 5 has been used in the chapter below to predict blade individual amplitudes of the rotor two blisk. It has to be noted, that the Tyler-Sofrin modes originating from the interaction between IGV and rotor one (EO40) is also visible and comprehensible for both CFD speeds as well as for the interaction between the 2nd harmonic of the stator one and the rotor one (EO72), see Figure 8 and Figure 9. Furthermore not necessarily all Tyler-Sofrin modes can be spotted within numerical simulations as they might be cut-off.

In the following the influence of the simultaneous occurrence of Tylor-Sofrin mode excited forcings in EO36 will be discussed for the rotor two blisk. The modal forcings act on different aliased NDs. These spin with individually speeds in a way that all of them excite the rotor two with the same EO and frequency respectively. In consequence a superposition of all nodal diameter contributions predicted by the 6 row model leads to an envelope of the blade individual modal forcings over the circumference which is fix in the rotating frame of reference. The envelope can be predicted by equation (3):

$$\hat{f}(\theta) = \sqrt{\begin{array}{l}\left(\sum_{i=1}^{ND_{max}}\left[\hat{f}_i \sin(ND_i\theta + \Phi_i)\right]\right)^2 \\ +\left(\sum_{i=1}^{ND_{max}}\left[\hat{f}_i \cos(ND_i\theta + \Phi_i)\right]\right)^2\end{array}} \quad (3)$$

Figure 10 shows the aforementioned behaviour for the EO36 at 80%NH. Therefore the blade individual forcings were predicted by method B), scaled to full assembly forcings for comparison reasons and plotted over the circumference. While for the 3 row model a constant blade individual forcing has been observed as expected for one travelling wave in ND 8, a rotor-fix envelope of a distinct modal forcing variation appears for the 6 row model. This behaviour could be reproduced semi-analytical by equation (3). Therefore forcing amplitudes and phase information of all contributed nodal diameters are taken from method A), see Table 5. The sum of the 6 row forcings marks a (theoretic) maximum which is not reached during the simulation due to the computed phasing Φ_i between the contributing NDs.

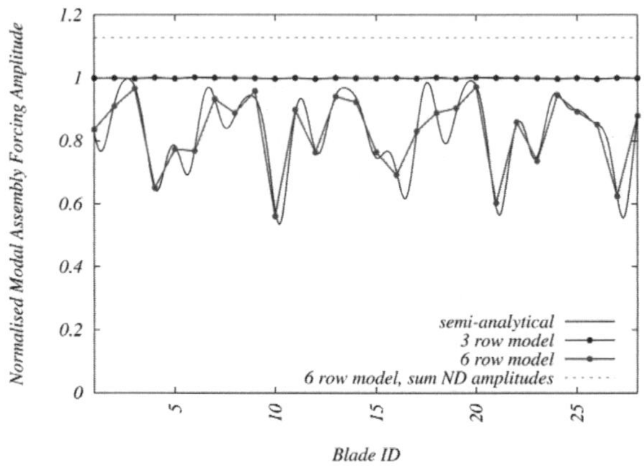

Figure 10: Blade individual modal forcings of mode 11 for EO36, 3 row model, 6 row model and semi-analytic function at 80%NH.

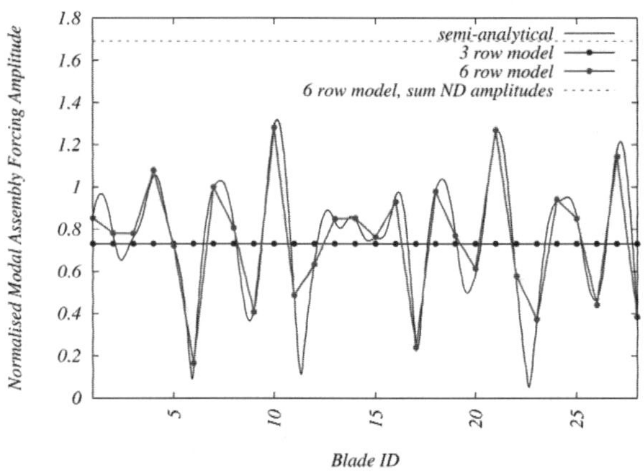

Figure 11: Blade individual modal forcings of mode 11 for EO36, 3 row model, 6 row model and semi-analytic function at 100%NH.

Figure 11 shows the results at 100%NH. Again the 3 row model computed a constant blade individual forcing. Compared to 80%NH this is considerably exceeded by the 6 row model due to the ND contributions. In sum these are higher than the 3 row forcing for 100%NH, but reaches not the theoretical value of 1.69 because of the computed phasing.

Figure 10 and Figure 12 show a distinct variation of the blade individual forcing - fix in the rotating frame of reference, which will be investigated with respect to the blade individual vibration responses in the following chapter.

TUNED FORCED RESPONSE COMPUTATIONS

For an integral structure as the discussed rotor two blisk, there is a significant blade to disk coupling yielding a variation of the eigenfrequencies for each nodal diameter. In addition, gyroscopic loads and aerodynamic coupling generate an additional variation of the eigenfrequencies for forward and backward travelling waves that differs for each distinct nodal diameter. Hence, if multiple nodal diameters are excited, the mechanical coupling between those has to be large enough to generate a rotor fixed amplitude variation for a certain resonance passing. To investigate the response of the rotor two blisk structure the subset of nominal system modes (SNM), introduced by Yang and Griffin [2] was utilized to study the impact of Tyler–Sofrin excitations on the blade response. The equation of motion reads

$$(\mathbf{M} + \Delta\mathbf{M})\ddot{\mathbf{q}} + (\mathbf{D} + \mathbf{G})\dot{\mathbf{q}} + (\mathbf{K} + \Delta\mathbf{K} + \mathbf{Z})\mathbf{q} = \mathbf{f}$$

where $\mathbf{M}$, $\mathbf{D}$ and $\mathbf{K}$ represent modal mass, damping and stiffness matrices. Gyroscopic loads $\mathbf{G}$ and aerodynamic coupling $\mathbf{Z}$ are neglected in this paper to decrease the complexity of the problem. For the tuned investigations modal mass $\Delta\mathbf{M}$ and stiffness mistuning $\Delta\mathbf{K}$ are neglected as well. The modal damping was set to a constant value for each of the participating modes and represents the expected aerodynamic and structural damping level. The blisk response is computed by solving for the modal deflection $\mathbf{q}$ on a certain excitation frequency ω with the right hand side modal force vector $\mathbf{f}$ in the frequency domain.

The advantage of the SNM is the ability to study the response of the structure in the modal domain by significantly reducing the degrees of freedom when starting from the cyclic symmetric finite element model, see [3], [4]. To study the aforementioned family of aerofoil modes 47 modes are used. These chosen modes have been highlighted in Figure 13. Two speeds have been marked for following discussions.

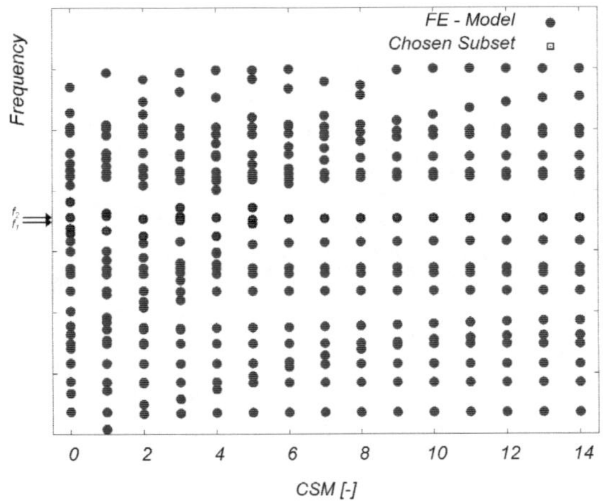

Figure 13: Frequency vs. nodal diameter diagram, showing choice of subset for discussed family of aerofoil modes.

At first the 80%NH forcing will be investigated, see Figure 14. As expected for a tuned structure excited by a travelling wave in one ND, the 3 row model computed a constant vibration level over the circumference. Applying a forcing of different NDs from the 6 row model, the predicted blade individual vibration levels are lower and subjected to a variation. This variation is linked to the variation of blade individual forcing (Figure 10) and depends on the structural coupling between all excited modes and their modal superposition. The coupling is mainly driven by the distance between the damped natural frequencies and the damping level itself.

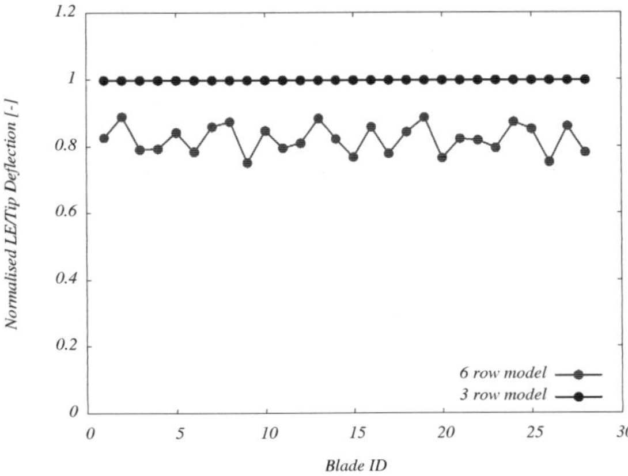

Figure 14: Predicted blade individual amplitudes with unsteady pressure force and aerodynamic damping from 80%NH.

Results for the 100%NH computations are shown in Figure 15. The 3 row model generates a constant blade individual vibration level at 40% normalised deflection of the 80%NH case because of a lower ND 8 forcing and higher aerodynamic damping. With respect to the 6 row model the higher aerodynamic damping increases the structural coupling compared to 80%NH.

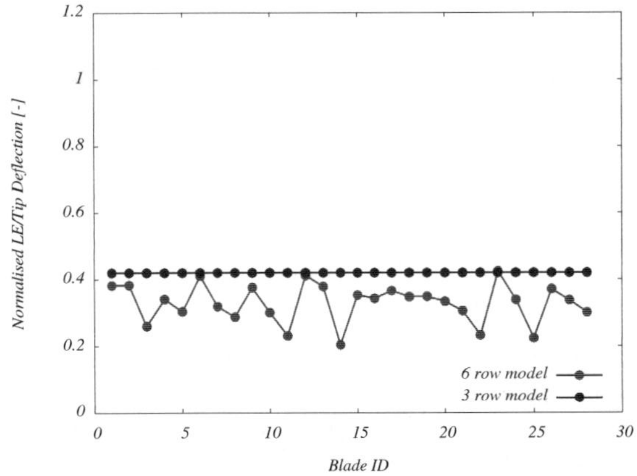

Figure 15: Predicted blade individual amplitudes with unsteady pressure force and aerodynamic damping from 100%NH.

In addition Tyler-Sofrin modes aliased also on high nodal diameter and are more pronounced related to the directly upstream stator wake excitation in ND 8, see Figure 9. This has been led to the distinct variation in blade individual forcing, see Figure 11. As a result an increased variation of blade individual vibration level occurred compared to 80%NH. Nevertheless, the most responding blade of the 6 row computation exceeds the largest response of the 3 row computation by only ~1%.

When taken a closer look to the amplitude vs. frequency diagram, see Figure 16, the response of a multi degree of freedom oscillator gets visible. Whereas the first peak is mainly related to the ND 2 and ND 3 response (marker f_1), the second peak is a superposition of the excited NDs 3, 5, 8, 10 and 13 (marker f_1), see Figure 13 for comparison. Again, the structural coupling between the modes depends on the distance of the damped eigenfrequencies and the damping level itself. Hence, a higher damping yields a stronger coupling of each excited mode.

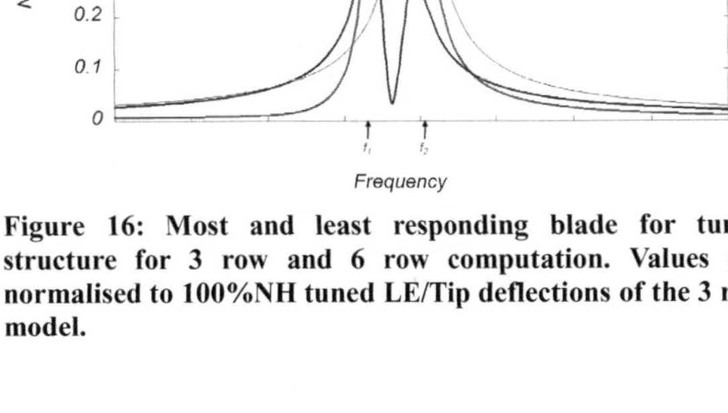

Figure 16: Most and least responding blade for tuned structure for 3 row and 6 row computation. Values are normalised to 100%NH tuned LE/Tip deflections of the 3 row model.

MISTUNED FORCED RESPONSE COMPUTATIONS

In real applications the perfect cyclic symmetry of the structure is destroyed by manufacturing tolerances and material inhomogeneity leading to different blade individual frequencies of the structure. This behaviour is described as mistuning and can lead to complex blade individual vibration responses on a rotating excitation. In addition, the mistuned response can supersede the maximum vibration level of the tuned response by a significant factor.

Within the structural model, stiffness mistuning is introduced by scaling the Young's Modulus of each blade i with a factor of (1+c_i) and transforming the blade stiffness matrix $\mathbf{K_b}$ to the modal domain with the nominal modal matrix $\boldsymbol{\varphi}$

$$\Delta\mathbf{K} = \sum_{i=1}^{B} c_i \boldsymbol{\varphi}_\mathbf{i}^T \mathbf{K}_{\mathbf{b,i}} \boldsymbol{\varphi}_\mathbf{i} \quad .$$

To analyse the impact of mistuning in interaction with complex force vectors containing several spinning modes, 3 different analyses have been carried out at 100%NH. First of all a measured mistuning pattern is analysed. The mistuning pattern was identified for the true hardware that is planned to run at an upcoming test campaign [13]. The identification process was carried out by the partner from the Brandenburg University of Technology Cottbus and follows the approach published in [14]. The second investigation aims on an identification of a worst case mistuning distribution that yields a maximum amplification of the structure. Therefore, a genetic algorithm was used to identify a worst mistuning pattern. The third object was to study normal distributed blade individual frequencies and the resulting amplification factors within a probabilistic investigation. Therefore, 1000 samples for a constant amount of standard deviation of blades individual frequencies are computed. This computation was done for 100 equally spaced standard deviations of blade individual frequencies.

The results of this analysis are shown in Figure 17. Taken a look at the probabilistic assessment the 6 row computation shows a larger variance and higher amplification factors as expected for a standard deviation of blade individual frequencies that is higher than 0.1%. In addition, the maximum computed amplification factors for each of the samples is higher for the 6 row computation than for the 3row computation. When assessing the optimised "worst case" solutions the behaviour is similar and the optimization has been found a higher amplification factor for the 6 row computation (~3.2) as for the 3row computation (~2.5). It is interesting to note that for a low level of frequency variation the model containing the forcing from the 6 row computation is less sensitive to mistuning than other one.

After all, the identified mistuning pattern yields amplification factors close to the expected values of both configurations and hence, will show different behaviour.

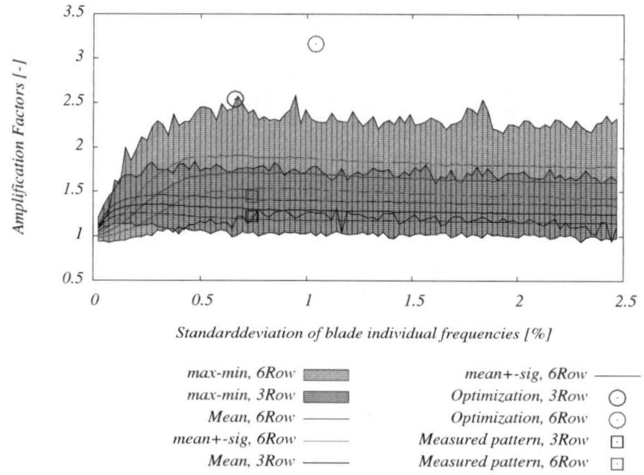

Figure 17: Mistuning analysis for modal force derived from 3row and 6row configuration at 100%NH.

If a direct comparison between computations and observed blade individual amplitudes is of interest the incorporation of spinning modes might be essential to derive the correct envelope. To highlight this issue the blade individual response for the mistuned setup is compared between both setups, see Figure 18. The shape of both envelopes considerably differs. In detail the most responding blade for the 6row computation differs from the one of the 3 row computation. Again the variance of the 6 row computed response is larger than, the of the 3 row model. For both models the mistuned response exceeded the maximum vibration of the tuned response.

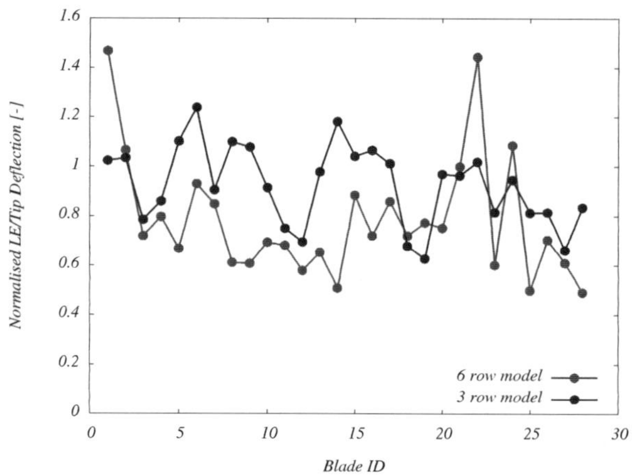

Figure 18: Predicted vibration response for measured mistuning pattern for both modal force vectors derived at 100%NH. Values are normalised to 100%NH tuned LE/Tip deflections of the 3 row model.

SUMMARY

Tyler-Sofrin modes have been computed within multi-bladerow CFD simulations for the stage 2 rotor of a 4.5 stage research compressor. These spinning modes can generate forcings of different nodal diameters at a specific excitation frequency as shown for an aerofoil dominated modeshape. In consequence a distinct variation of the blade individual forcing amplitudes over the circumference has been observed in the rotating frame of reference.

For multi-balderow setups including Tyler-Sofrin mode excitations (6 row models), the blade individual modal forcings can occur with higher amplitudes as computed by comparable CFD models excluding Tyler-Sofrin modes (3 row models). This depends on the contribution and superposition of Tyler-Sofrin modes at one excitation frequency. It has to be noted that the maximum modal forcing not necessarily occurs in the aliased nodal diameter from the direct up- or downstream excitation which is linked to the blade counts.

The modal forces computed by simulations with and without Tyler-Sofrin modes have been used to calculate blade vibration amplitudes for a tuned and mistuned rotor blisk. The blade vibration amplitudes have been calculated using a reduced order model developed by [2] using a subset of nominal modes (SNM).

For the tuned rotor blisk a variation of blade individual vibration amplitudes has been predicted for the 6 row model since Tyler-Sofrin mode contributions at various nodal diameters exist. Nevertheless, the computed blade individual vibration amplitudes based on the 3 row models tend to be higher or equal compared to the 6 row models. This behaviour is driven by the determined structural coupling of the blisk modes in combination with the observed distribution of modal forcings over different nodal diameters for the 6 row models.

As part of the structural mistuning investigations the identification of a worst case mistuning distribution has been performed by a genetic algorithm to generate a maximum amplification of the blisk response. Considering the 6 row model with contributing Tyler-Sofrin modes a ~30% higher amplification factor could be computed as for the 3 row model. Furthermore a larger variance and higher expected amplification factors have been predicted for the 6 row model (when standard deviation of blade individual frequencies > 0.1%) during probabilistic analyses. Finally a comparison between forced response predictions based on measured mistuning distribution for a 3 row and a 6 row model has been discussed. The envelopes of the blade individual vibration amplitudes differ considerably. Hence, the consideration of Tyler-Sofrin modes might be essential if blade individual vibration measurements have to be compared against forced response simulations.

ACKNOWLEDGMENTS

The investigations are part of the research project AeRoBlisk (FKZ: 20T0901C) which is funded by the German Federal Ministry of Economics and Technology. The authors gratefully acknowledge this financial support.

Furthermore the authors are grateful to Rolls-Royce Deutschland Ltd & Co KG for granting permission for its publications. Dr. Jeff Green and Dr. Caetano Peng from Rolls-Royce plc are greatly acknowledged for supporting this work.

Many thanks goes to the partner from the Brandenburg University of Technology Cottbus for the modelling support and detailed mistuning identification as well as to the DLR Cologne for providing the Rig data used for CFD modelling.

REFERENCES

[1] J. M. Tyler and T. G. Sofrin, "Axial Flow Compressor Noise Studies," *SAE Transaction* , vol. 70, pp. 309-332, 1962.

[2] M. T. Yang and J. H. Griffin, "A Reduced-Order Model of Mistuning Using a Subset of Nominal System Modes," *Journal of Engineering for Gas Turbines and Power,* vol. 124, no. 3, pp. 893-900, 2001.

[3] T. Giersch, B. Beirow, F. Popig and A. Kühhorn, "FSI-based Forced Response Analysis of a Mistuned High Pressure Compressor," in *Proceedings of the VIRM 10, Paper C1326/012*, London, UK, 11.-13. Sept. 2012.

[4] T. Giersch, P. Hönisch, B. Beirow and A. Kühhorn, "Forced Response Analysis of Mistuned Radial Inflow Turbines," *Journal of Turbomachinery,* vol. Vol. 135(3), 2013.

[5] F. M. Besam, R. E. Kielb and N. L. Key, "Forced Response Sensitivity of a Mistuned Rotor from an Embedded Compressor Stage," in *ASME Turbo Expo: Turbine Technical Conference and Exposition*, Montreal, 2015.

[6] K. C. Hall and K. Ekici, "Multistage Coupling for Unsteady Flows in Turbomachinery," *AIAA Journal* , vol. 43, no. 3, pp. 624-632, 2005.

[7] S. C. Stapelfeldt and L. di Mare, "Reduced Passage Method for Multirow Forced Response Computations," *AIAA Journal,* 2015.

[8] A. I. Sayma, M. Vahdati and M. Imregun, "An Integrated Non-linear Approach for Turbomachinery Forced Response Prediction. Part 1: Formulation," *Journal of Fluids and Structures,* vol. 14, pp. 87-101, 2000.

[9] M. Vahdati, A. I. Sayma and M. Imregun, "An integrated nonlinear approach for turbomachinery forced response prediction. Part II: Case studies," *Journal of Fluids and Structures,* vol. 14, pp. 103-125, 2000.

[10] M. Vahdati, A. I. Sayma and M. Imregun, "Multibladerow forced response modeling in axial-flow core compressors," *Journal of Turbomachinery - Transactions of the ASME,* vol. 129, pp. 412-420, 2007.

[11] C. Breard, M. Vahdati, A. I. Sayma and M. Imregun, "An Integrated Time-Domain Aeroelasticity Model for the Prediction of Fan Forced Response due to Inlet Distortion," *Journal of Engineering for Gas Turbines and Power,* vol. 124, no. 1, pp. 196-208, 2000.

[12] E. Johann and F. Heinichen, "Back to Back Comparison of a Casing Treatment in a High Speed Multi-Stage Compressor Rig Test," in *ISABE*, 2011.

[13] F. Popig, P. Hönisch and A. Kühhorn, "Experimental And Numerical Analysis Of Geometrical Induced Mistuning Technical Publication," in *Turbine Technical Conference and Exposition, GT2015-43272*, Montreal, Canada, June 15-19, 2015.

[14] A. Kühhorn and B. Beirow, "Measurement System in a Specific Environment Mechanical Measurement System Vibration Detection," Patent application number: 20100286934, http://www.faqs.org/patents/app/20100286934, 2010.

Proceedings of the 14th International Symposium on Unsteady Aerodynamics, Aeroacoustics & Aeroelasticity of Turbomachines
ISUAAAT14
8-11 September 2015, Stockholm, Sweden

I14-S2-4

VALIDATION OF A FORCED RESPONSE PREDICTION METHOD FOR DESIGN OF AXIAL COMPRESSORS

Steffen Hammer
GKN Aerospace
Trollhättan, Sweden

Marcus Lejon
GKN Aerospace
Trollhättan, Sweden

Pieter Groth
GKN Aerospace
Trollhättan, Sweden

Hans Mårtensson
GKN Aerospace
Trollhättan, Sweden

ABSTRACT

The linearized time-harmonic Navier-Stokes solver LINNEA is used to evaluate the forcing of a rotor blade in the transonic compressor Hulda, at part speed. Hulda is a 1½ stage demonstrator compressor consisting of a variable inlet guide vanes (VIGV), rotors, stators and outlet guide vanes (OGV). The forcing of the rotor is evaluated for different setting angles of the VIGV. The mesh study shows that, for this computational setup, a mesh independent solution was obtained in RANS simulation using CFX and in the linearized Navier-Stokes solver LINNEA for the same grid resolution.

The results obtained using LINNEA are compared to results from three URANS simulations using CFX. The first URANS case uses scaled geometries and rotational periodic boundary conditions. The second and third case use the true geometry in conjunction with rotational periodic boundary conditions: "profile transformation" and with time-lagged periodic boundary conditions: "time transformation" in CFX' nomenclature, which corresponds to time-inclined computational planes. LINNEA is run with two different boundary conditions at the inlet to the rotor domain, obtained from a RANS and URANS simulation, respectively.

The forcing is evaluated at engine orders 15, 30 and 45, corresponding to VIGV passing frequency and multiples thereof. The results show that the amplitude of the tangential force variation calculated with the different methods and boundary conditions, is within a 1.3% interval based on the total tangential force for the nominal (0 degree) VIGV setting and within a 3.3% interval for the VIGV setting angle of 30 degrees at engine order 15.

Using a linear time-harmonic Navier-Stokes solver to calculate forcing of a rotor blade interacting an upstream component is shown to be a possible and attractive alternative to URANS calculations. A linear harmonic simulation is less computationally demanding than a URANS simulation. The mean flow solution required by the linearized solver does not incur any additional computational effort since it is usually already available from general performance calculations. Further is shown that for URANS analyses a scaling of the model can be neglected since results obtained for a scaled model are in good agreement with results obtained for an unscaled model using profile transformation (PT) to handle the variation of pitch between components.

NOMENCLATURE

TBR	Transient Blade Row
RANS	Reynolds Averaged Navier Stokes
URANS	Unsteady Reynolds Averaged Navier Stokes
TT	Time Transformation
PT	Profile Transformation
EO	Engine Order
VIGV	Variable Inlet Guide Vane
OGV	Outlet Guide Vane
BC	Boundary Conditions
FF	Front Frame
RPM	Revolutions per Minute
FFT	Fast Fourier Transformation
TRN	transient

INTRODUCTION

The flow in axial turbomachines is inherently unsteady through the interaction between stationary and rotating blade rows. This will result in an unsteady excitation of blades and vanes. Depending on forcing amplitude and number of cycles, it may lead to high cycle fatigue. The two main sources for unsteady excitations are wake interaction and potential flow interaction. The wakes from upstream blade rows are approximately steady in its frame of reference. When the wakes interact with a downstream blade moving relative to them they will cause unsteady forces on these blades. Potential interaction is caused by the pressure field around the stators and rotors. The pressure in region between stator and rotor blade rows can be decomposed into one part which is uniform, a second part non-uniform and steady in the stator frame of reference and a third part non-uniform and steady in the rotor frame of reference. As the rotor blades move, the stator blades experience an unsteady pressure due to the non-uniform part locked to the rotors and the rotor blades experience an unsteady pressure due to the non-uniform part locked to the stators.

Two approaches are used in this paper, namely linear harmonic and non-linear.

Linear Harmonic is rapid and possible to use in the same time frame as the blade shape is designed, so sensitivities can be calculated. The present paper highlights the effects of how the incoming wake is calculated in order to show that the mixing plane boundary condition do not entirely distort the wake upstream of the interface. Exemplary the wake of a VIGV is shown in form of the static entropy for a RANS and a URANS simulation in Figure 1.

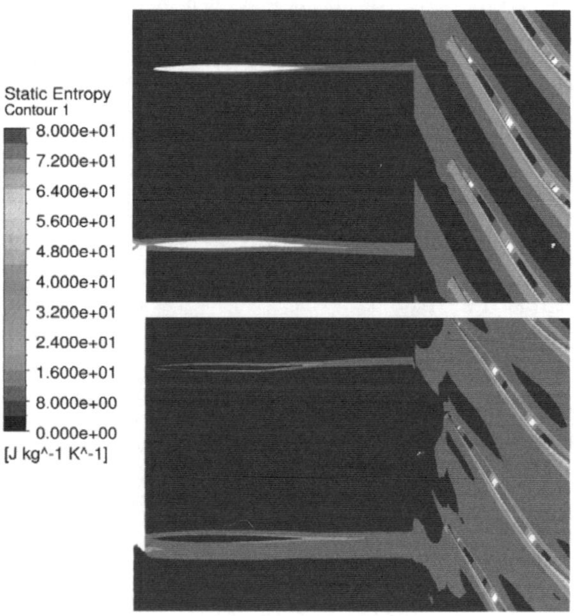

Figure 1: Static entropy at 75% span between VIGV and rotor domain. RANS calculation results (top) and URANS calculation results (bottom)

The non-linear approach is time consuming and more difficult in the context of convergence. By comparison to rig measurements, the test would run and let be settled for several hundred rotations and measurement follows during a substantial time, or by slowly increasing the speed to allow sampling. For the non-linear computation, only relatively few cycles can be afforded. Potentially the simulation covers more physics of the non-linear flow features so the development of the methods is important. The tests made serve to highlight the effects of choices available and allows an assessment of what we can expect in a real case on regular forcing.

In design it is important to have fast and reliable tools for predicting the unsteady forcing and the response to it. For that purpose a linearized solver may be an interesting and attractive alternative.

APPROACH

The general approach is to compare predictions of the forcing of a rotor blade due to an upstream guide vane. Two numerical solvers are used: the in-house solver LINNEA and the commercial solver ANSYS CFX 16.0. The results obtained by the linear time-harmonic Navier-Stokes solver LINNEA are compared to the Unsteady Navier-Stokes results obtained using ANSYS CFX 16.0.

As a test case the 1.5 stage transonic high pressure ratio compressor demonstrator Hulda presented in Figure 2 is used. The compressor consists of 5 Front Frame struts (FF), 15 VIGVs, 23 rotor blades, 51 stator vanes and 8 OGVs.

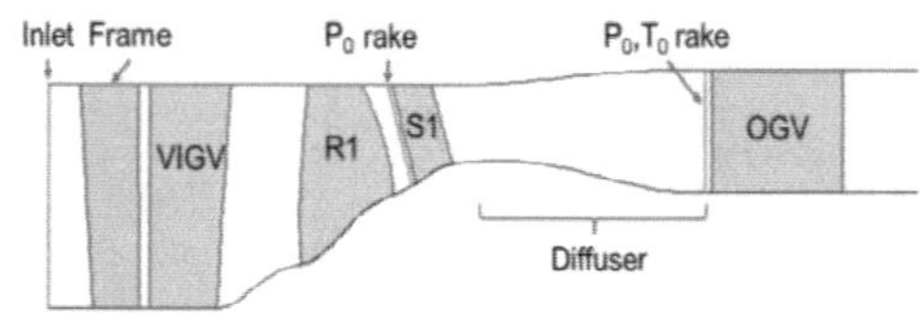

Figure 2: Hulda compressor layout [1]

The Hulda compressor is designed for a high total pressure ratio, above 2 at design speed. For this study two operating points are chosen, namely 0 and 30 VIGV angle at 63% speed. Measurement positions of total pressure and total temperature are shown in Figure 2 and experimental data collected at these positions are used here. For the comparison of the solvers the tangential force onto the downstream rotor is evaluated.

MESH STUDY

Since part of the study is based on an older study with a coarser mesh used in the scaled URANS, a mesh study on the complete compressor is performed to ensure mesh independency. For the further URANS and RANS simulations the OGV is omitted to reduce the model size and since this study is focused on wake interaction from front frame struts and VIGV's. For all computations the block structured mesh used, is generated by an in-house mesh generator (G3Dmesh). Four levels of refinement are investigated. The total number of cells in the computational grids used for the study is shown in Table 1. The arithmetic average y^+ values of the first grid node normal to the no-slip wall for both VIGV angles are presented Table 2 and Table 3.

The computational grid consists of a single passage for each component using a stage interface (mixing plane) in between, where the flow is mixed in the circumferential direction. An average static pressure is set at the outlet of the domain. The surfaces are set as adiabatic, no-slip walls.

The k-ε turbulence model is used with wall functions and the flow is assumed to be fully turbulent.

The inlet boundary conditions used in CFX are specified as a radial profile of total pressure, total temperature, k and epsilon. The radial profiles were obtained from rig measurements of the transonic compressor Hulda at 63% of design speed.

Table 1: Mesh study properties of the whole compressor

Config.	coarse	medium	fine	very fine
Cell count	215.1 K	534.7K	1726.4K	2593.6K

Table 2: Average y^+ values for 0 degree VIGV configuration

Config.	coarse	medium	fine	very fine
Average y^+ on blade surfaces	43	44	36	33
Average y^+ on shroud surfaces	92	75	81	35
Average y^+ on hub surfaces	97	70	88	43

Table 3: Average y^+ values for 30 degree VIGV configuration

Config.	coarse	medium	fine	very fine
Average y^+ on blade surfaces	41	37	34	32
Average y^+ on shroud surfaces	93	63	82	35
Average y^+ on hub surfaces	97	69	86	43

The normalized total pressure ratio and mass flow are shown for the nominal VIGV in Figure 3. It is shown that there is very little difference of the predicted total pressure ratio and mass flow for the medium grid and the two finer resolutions.

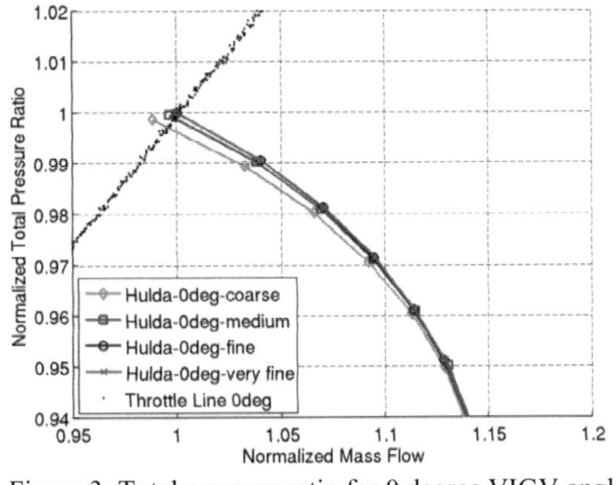

Figure 3: Total pressure ratio for 0 degree VIGV angle

The normalized polytropic efficiency versus mass flow for the nominal VIGV is shown in Figure 4. The maximum polytropic efficiency obtained on the fine grid has been used as to normalize the results on all other grids and all operating points.

Approximately 1% difference is shown for the medium and fine mesh in maximum total pressure ratio along the speed line, while very little difference is shown between the fine and very fine grid resolution. The mass flow at choke is shown to be relatively unaffected by the grid. From results presented, it is concluded that for the nominal VIGV, the mesh denoted as fine is sufficient for validation purposes.

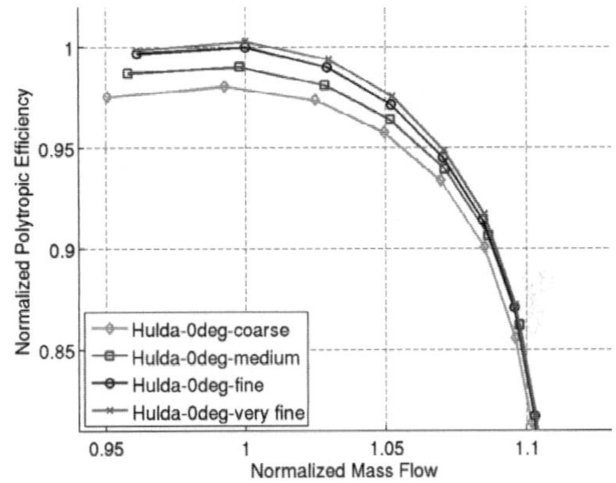

Figure 4: Normalized polytropic efficiency for 0 degree VIGV angle

The normalized total pressure ratio and mass flow are shown for the VIGV deflected 30 degrees in Figure 5. The performance in terms of total pressure ratio appears relatively insensitive to the computational grid variation done in this investigation.

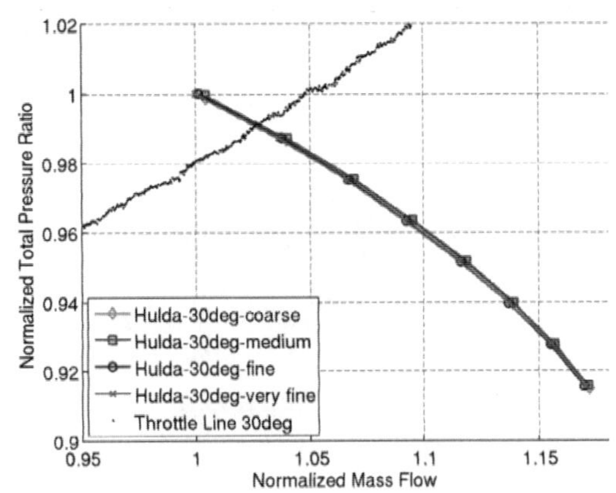

Figure 5: Total pressure ratio for 30 degree VIGV angle

The normalized polytropic efficiency for the same VIGV angle (30 degrees) is shown in Figure 6. The efficiency has been normalized in the same way as for Figure 4.

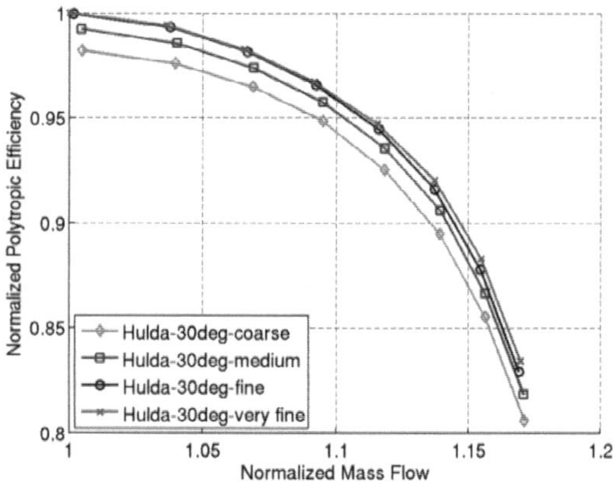

Figure 6: Normalized polytropic efficiency for 30 degree VIGV angle

The polytropic efficiency is shown to be more sensitive to grid variation than the total pressure ratio. The maximum polytropic efficiency is shown to increase by approximately 1% when the grid is refined from coarse to medium and increase again by approximately 1% when the grid is refined further to a fine mesh. A grid-independent solution is considered to have been achieved for the mesh denoted as fine. A mesh with this grid density will sufficiently resolve waves of 45EO. Thus this mesh resolution will be used in the following study.

CASE DESCRIPTIONS

The computational domain used for all RANS and URANS simulations consist of the FF-VIGV-R1-S1 domains while the OGV is omitted. A single blade passage is evaluated for the RANS computations, while for the URANS calculations a 72 degree sector is used. The use of a 72 degree sector for each of the components requires scaling of the components. 1 FF strut is used, followed by 3 VIGV blades, 5 rotor blades and 10 stator blades (Figure 7). The scaled model for the initial URANS simulation [2] uses a coarser mesh setup as seen in Table 4, while RANS and all other URANS simulations with a non-scaled geometry used the fine mesh setup as concluded in the mesh study.

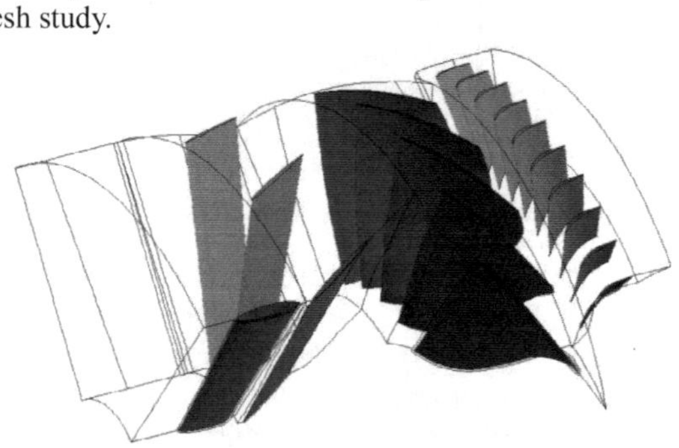
Figure 7: Computational domain with deflected VIGV

Table 4: Mesh properties for RANS and scaled URANS computations without OGV

	initial mesh before mesh study		fine mesh resolution from study	
Config.	scaled URANS (complete domain)	scaled URANS (single rotor passage)	RANS (complete domain)	RANS (single rotor domain)
Cell count	1300K	63K	1350K	260K

Operating point

Two speedlines at 63% design speed are calculated for each VIGV setting angle by a steady state simulation respectively. The speedline with a 354k grid refers to a previous study and the second speedline (1350K grid) is computed to ensure no major difference due to the increase mesh resolution. The speedlines can be seen in Figure 8 and Figure 9. Both RANS speedlines are computed using one passage per blade row. The 354k mesh uses the same mesh resolution used for the scaled URANS mentioned in Table 4, while the 1350k mesh represents the outcome from the mesh study. The backpressure for the URANS simulations is taken from the crossing point of the operating line with the RANS speedline. The time-averaged mass flow and total pressure ratio in the performed URANS simulations are additionally shown in the speedline plots to compare that the same operating point is considered in all analyses.

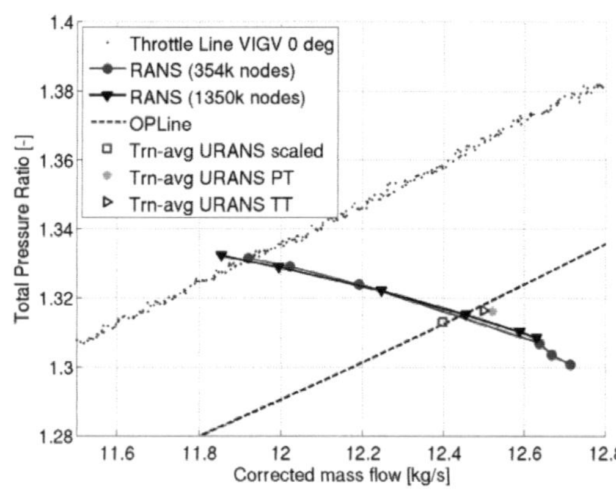

Figure 8: Speedline for 0 degree VIGV

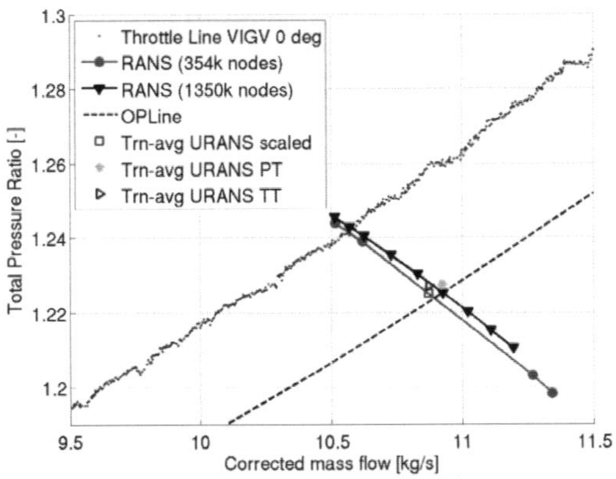

Figure 9: Speedline for VIGV deflected by 30 degrees

Six cases are used in the comparison study, namely 3 URANS performed in CFX and 3 cases performed in LINNEA.

Case 1

The first case is based on the scaled model used in [2] which uses a sector angle of 72 degrees to allow a matching pitch between the single stages. The blade numbers in the domain for the scaled model would result in a 25 instead of 23 rotors and 50 instead of 51 stators in the full annulus. The scaling of the blade rows will result in a scaling of the rotor blade forces.

The analysis type is set to transient in CFX.

Case 2

The second case uses an unscaled model but with the same number of blades in the domain as Case 1. This is not necessary for a profile transformation in CFX but chosen for comparison reasons. The further away from one the pitch ratio is, between neighbouring domains the more the simulated blade passing frequency will deviate from the true as will be seen below.

The analysis type is transient blade row (TBR) with profile transformation as described in [4] at VIGV – Rotor and Rotor – Stator interfaces. The Frame – VIGV interface is defined as a general Fluid – Fluid interface (Frame Change option none in CFX) as there is no pitch change between 1 frame and 3 VIGVs.

Case 3

The analysis type for Case 3 is set to transient blade row (TBR) with time transformation in CFX using an unscaled geometry. It is based on the concept time-inclined computational plane due to Giles [3]. The transformation model as described in [3] is applied at the VIGV and rotor interface. The Frame to VIGV interface is defined as a general Fluid – Fluid as in Case 2 and the Rotor – Stator interface is using a profile transformation. This case uses the same blade numbers in the domain as Case 2 since CFX requires a pitch change between 0.7 and 1.4 for a time transformation setup

The results are stored using a data compression of 7 Fourier coefficients in CFX and the "Moving Averages" setup option, where the Fourier coefficients are constantly accumulated and updated during the simulation.

A mid span view of the mesh used for Case 2 and Case 3 can be seen for the undeflected VIGV in Figure 10.

Case 4

For this case the unsteady VIGV exit flow of the scaled URANS of Case 1 is used as an input to the linearized solver (LINNEA). The mean flow however is taken from the RANS simulation. LINNEA calculates the mean forces on the rotor as well as the amplitude of the force variation for a specified engine order (EO).

Case 5

In this case the VIGV exit flow and the mean flow are taken from the unscaled RANS simulation performed in CFX and used as input to LINNEA.

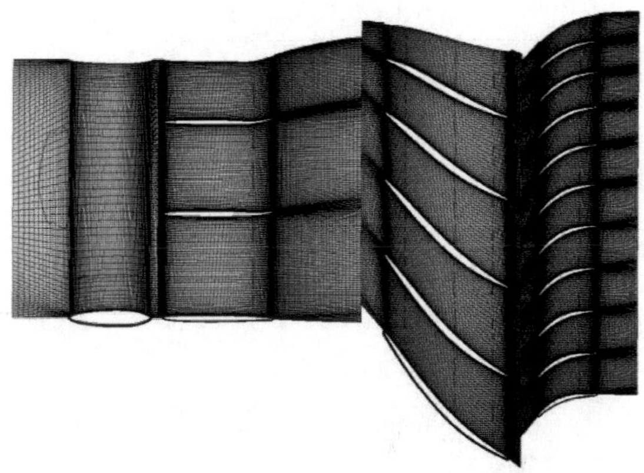
Figure 10: Mid span view of the used mesh model for undeflected VIGV

Case 6

Here the results from the unscaled URANS CFX simulation are used as input to the linearized solver. The difference here is the time averaged information of the time transformation simulation described in Case 3 are used for the mean flow and the VIGV exit flow.

RESULTS & DISCUSSION

The results of the unsteady nonlinear simulations are post-processed by extracting the pressures and calculating the tangential forces on the blade at every time step. Further a fast Fourier transform is applied on the time varying forcing from the URANS simulations to predict the amplitudes of the tangential forces. Of special interest are the forcing amplitudes of the VIGV, namely EO15, EO30 and EO45.

VIGV 0 degree

Figure 11 shows averaged tangential force amplitudes of the 5 rotor blades for Case 1 and a 0 degree VIGV angle. Since a scaled geometry is used in the simulation the forces are scaled back to original rotor blade count. It can be noticed that the magnitude of EO15 is about the same as EO5, which results from the front frame. The increase in forcing of the VIGV on the rotor (EO15) is nearly negligible. EO30 is about a quarter of EO15 and EO45 is close to zero.

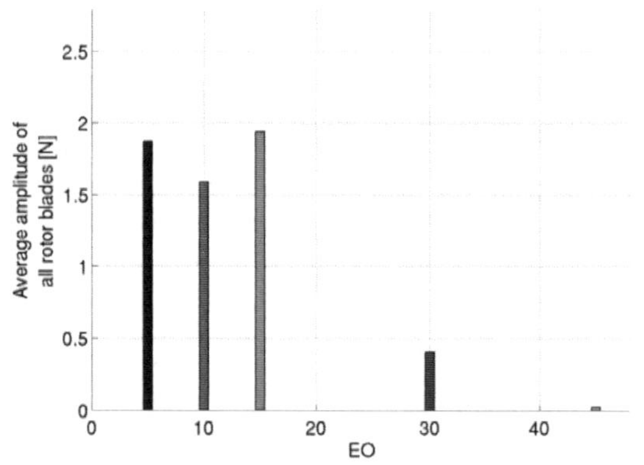

Figure 11: Case 1 – transient, Tangential force amplitude for 0 degree VIGV

For Case 2 using profile transformation, a frequency error proportional to the pitch ratio at each interface is recognized. This is in agreement with reference [4]. The pitch ratio between 3 VIGVs and 5 Rotors is 0.92. The ratio of the calculated frequencies to the expected ones can be seen in Table 5. The forcing amplitudes are taken at these shifted frequencies and plotted for their respective engine order in Figure 12.

Table 5: Frequency ratios for PT

EO with freq. [Hz]	frequency of PT [Hz]	freq. ratio
EO15: 3300	3039	0.92
EO30: 6600	6079	0.92
EO45: 9900	9118	0.92

The amplitude decreases at EO5, EO10 and EO15 compared to Case 1 presented in Figure 11. The influence of the VIGV however is comparable for EO30 and increased for EO45.

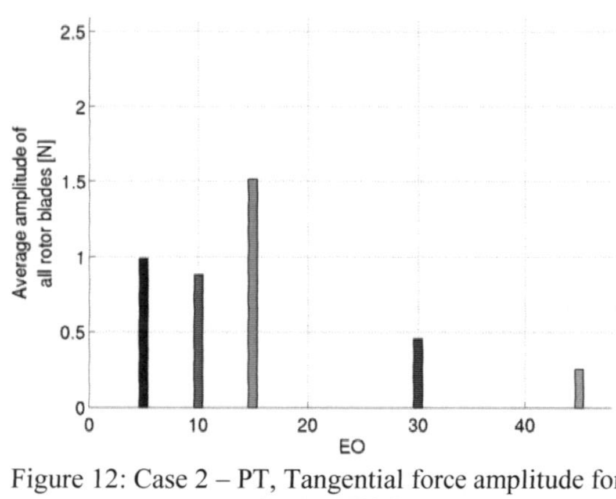

Figure 12: Case 2 – PT, Tangential force amplitude for 0 degree VIGV

The amplitudes for Case 3 are presented in Figure 13. The magnitudes of the amplitudes of EO15 and EO30 are reduced compared to Case 2, while EO45 shows the same amplitude compared to Case 2.

In the used setup it is not possible to show EO5 and EO10, this might be due to the number of Fourier coefficients used to store the data.

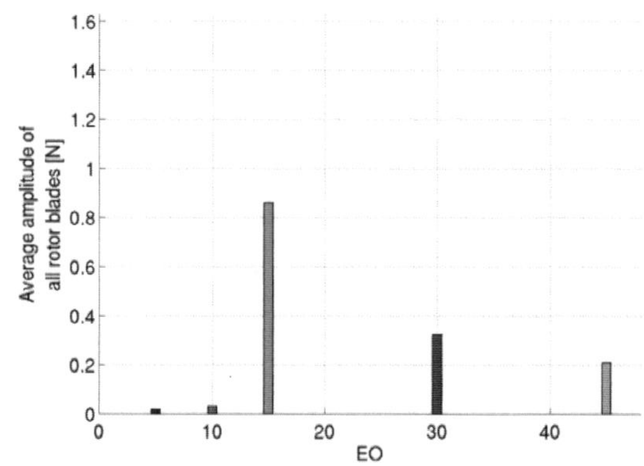

Figure 13: Case 3 – TT, Tangential force amplitude for 0 degree VIGV

Figure 14 presents the normalized tangential amplitudes for all 6 cases mentioned before, namely CFX and LINNEA, at EO15, EO30 and EO45. Each tangential amplitude is normalized by the total tangential force computed for Case 5 (LINNEA with RANS boundary conditions).

The range for EO15 of the different cases lies in an interval of 1.3% based on the total tangential force used to normalize the amplitudes. For higher engine orders this interval decreases to under 1% for EO30 and under 0.5% for EO45.

For Case 1 the results are rescaled by the ratio of the number of rotor blades from the scaled model to the true model. These results show good agreement with results from the profile transformation and the time averaged LINNEA results of Case 6 for EO15. According to these results a scaling of the geometry seems unnecessary. Being able to use the true geometry with profile transformation eliminates 1 extra preparation step before the computation and with it another error source.

For all engine orders LINNEA using the time averaged BC shows the highest amplitude of all cases.

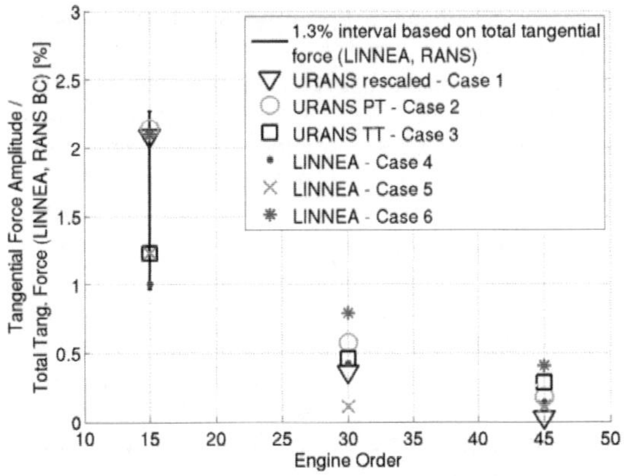

Figure 14: Tangential force amplitude normalized with total tangential force for VIGV not deflected

The computation time for each of the 6 cases at 0 degrees VIGV angle is listed in Table 6. Cases 2 and 3 are using about 70h using 37 and 50 cluster nodes respectively. Case 1 is computed within a third of the time by using 39 processors. LINNEA uses only 4 percent of the time of Case 3 while occupying 5 processors per calculated engine order. For this study the LINNEA simulations are done in parallel, which leaves at the same level but increases the number of processors by the number of calculated engine orders of interest, three in this case.

Table 6: Computation time per case for VIGV with 0 degree angle

analysis	# processors	approximate computation time
Case 1	39	23h
Case 2	37	69h
Case 3	50	70h
Case 4-6	5 per EO	3h

VIGV 30 degree

The results for a VIGV setting angle of 30 degree are presented below.

Figure 15 shows the tangential force amplitudes for Case 1. Notably the change in VIGV angle increased the forcing on the rotor for EO15 by about a factor of 3. The higher engine orders however stay at levels similar to that for the VIGV 0 degree configuration shown in Figure 11. Furthermore the magnitude at the engine orders 5 and 10 has reduced by about 20 and 60 percent, respectively.

The results for the profile transformation of Case 2 can be seen in Figure 16. The amplitude at EO15 is comparable to Case 1 seen in Figure 15. EO5 increased by about 60 percent, while EO10 more than doubled compared to the Case 1 with 30 degree VIGV angle. Additionally EO30 increased by a factor of 3 and even EO45 shows slightly higher amplitude.

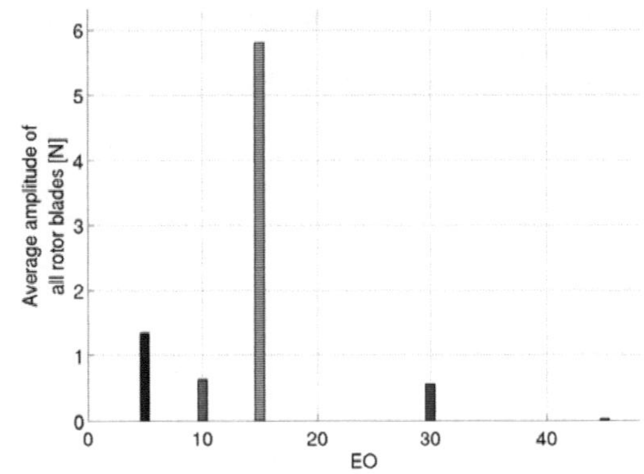

Figure 15: Case 1 – transient, Tangential force amplitude for 30 degree VIGV

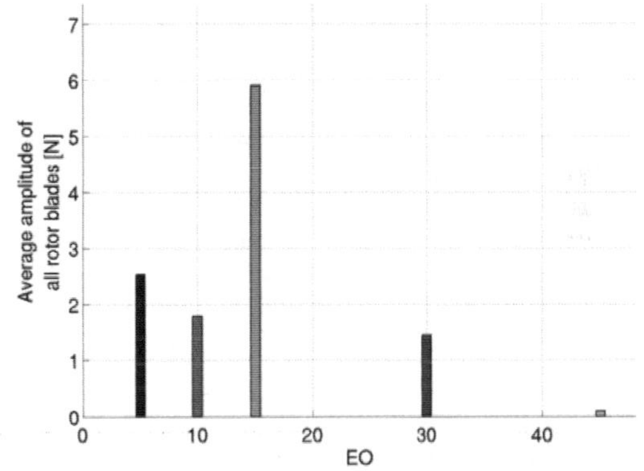

Figure 16: Case 2 – PT, Tangential force amplitude for 30 degree VIGV

In Figure 17 results of the time transformation are presented. Besides EO45 all other engine orders show lower amplitudes. EO15 and EO30 decrease by about 30 percent compared to the results of the profile transformation of Case 2 shown in Figure 16.

As for the 0 degree VIGV case EO5 and EO10 are not sufficiently detected in the FFT results.

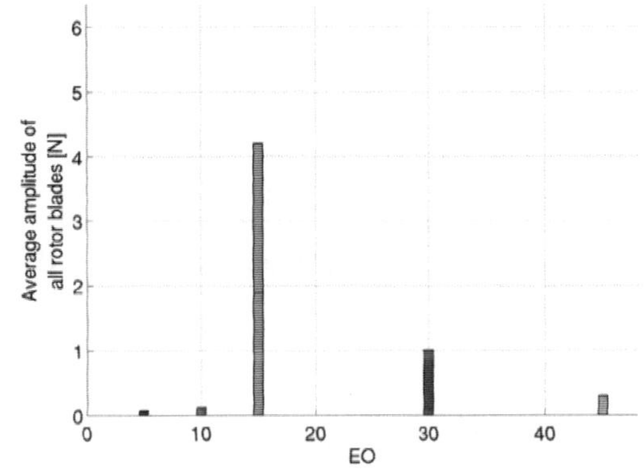

Figure 17: Case 2 – TT, Tangential force amplitude for 30 degree VIGV

Comparing the normalized amplitudes of the 6 cases for a VIGV angle of 30 degrees in Figure 18, the results at EO15 lie within an interval of 3.3% based on the total tangential force. At EO30 this error reduces to about 3% while at EO45 it is within 1%.

As for the 0 degree VIGV angle the rescaled Case 1 results are in good agreement with the results of the profile transformation. Here especially LINNEA results using the time averaged boundary conditions (Case 6) show the biggest difference at EO15. Further it can be seen that Case 6 has the highest normalized amplitudes at EO15 and EO30 for all the cases.

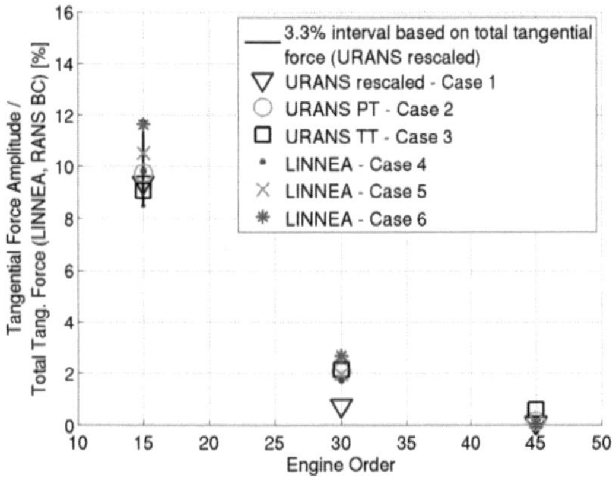

Figure 18: Tangential force amplitude normalized with total tangential force for VIGV deflected by 30 degrees

SUMMARY AND CONCLUSION

A comparison of different evaluations of the tangential forcing of the VIGV on a downstream rotor is presented. As a test case the 1.5 stage transonic demonstrator compressor Hulda is used.

URANS calculations are performed using CFX 16.0 as the commercial solver against a time-harmonic Navier-Stokes approach performed by the in-house code LINNEA.

The results show that the amplitude of the tangential force calculated with the different methods and boundary conditions, is within a 1.3% and 3.3% interval based on the total tangential force for a VIGV setting angle of 0 and 30 degrees respectively.

A major point in comparing the linear time-harmonic and non-linear solver is the computational cost during design. In this study all URANS calculations used more time and processors to reach a converged solution. Profile and time transformation computations used about similar time and processors while the transient simulation with scaled geometries needed only a third of the time.

The least amount of time and number of processors is used by the linear time-harmonic solver LINNEA. Although LINNEA needs to run for each engine order separately, the computations are done parallel and no extra time is needed. The initial RANS simulation is needed by both solvers so it has no impact on the cost difference.

With this, the linear time-harmonic Navier-Stokes solver presents itself as a possible and attractive alternative to URANS for predicting the forcing in axial compressor. A linear harmonic simulation is less computationally demanding than a URANS simulation. The mean flow solution required by the linearized solver is usually already available from general performance calculations and adds therefore no additional cost.

Furthermore, the results show that scaling the model can be neglected for forced response calculations since results from using the true geometry and a profile transformation between the interfaces are in good agreement. Eliminating the scaling of the geometry reduces the preparation time before the computation and with it an error source.

FUTURE WORK

During the evaluation of a time transformation some information is expected to get lost for EO5 and EO10 when using the Fourier reassembling of the data in CFX. The influence of higher number of Fourier coefficients for the reassembling is suggested, if a non-linear approach is chosen for other investigations.

The URANS calculations done for a 30 degree deflection angle of the VIGV displayed large fluctuations, and more time steps then used in the present study might be needed. Further studies are needed.

ACKNOWLEDGMENT

The support of ANSYS CFX is recognized.

REFERENCES

[1] Mårtensson H., Burman J., Johansson U., 2007, "Design Of the High Pressure Ratio Transonic 1½ Stage Fan Demonstrator Hulda", ASME Turbo Expo, GT2007- 27793.

[2] Mayorca, M. et al., 2012, "Uncertainty of Forced Responce Numerical Predictions of an Industrial Blisk – Comparison with Experiments", ASME Turbo Expo, GT2012-69534.

[3] Giles, M., 1988, "UNSFLO: A numerical Method For Unsteady Inviscid Flow In Turbomachinery", GTL Report No:195, MIT Cabmbridge, MA, 1988.

[4] Zori, L., Galpin, P., 2015, "Time Transformation Simulation of 1.5 Stage Transonic Compressor", ASME Turbo Expo, GT2015-43624.

[5] Einzinger J., 2015, personal communication, ANSYS Germany GmbH.

Proceedings of the 14th International Symposium on Unsteady Aerodynamics, Aeroacoustics & Aeroelasticity of Turbomachines
ISUAAAT14
8-11 September 2015, Stockholm, Sweden

I14-S3-3

DETAILED MEASUREMENTS OF THE UNSTEADY FLOW FIELD IN THE AREA OF ORIGIN OF ROTATING INSTABILITY

Mario Eck
Chair of Aero Engines
Department of Aeronautics and Astronautics
Technische Universität Berlin
Marchstr. 12-14, 10587 Berlin, Germany

Christine Tiedemann
Chair of Aero Engines
Department of Aeronautics and Astronautics
Technische Universität Berlin
Marchstr. 12-14, 10587 Berlin, Germany

Dieter Peitsch
Professor
Chair of Aero Engines
Department of Aeronautics and Astronautics
Technische Universität Berlin
Marchstr. 12-14, 10587 Berlin, Germany

ABSTRACT

This paper presents detailed flow structures developing in the area of origin of Rotating Instability (RI) in an isolated axial compressor stator with hub clearance. For these experiments, 59 unsteady pressure transducers are concentrated in an array upstream of the leading edge plane at a constant circumferential position relative to the compressor stator blades. A specially developed analysis method allows to identify characteristic trigger times, in which a so-called "modal event" excites for the dominant RI mode. With the knowledge about a huge number of stochastically distributed trigger times, it was possible to estimate quasi-instantaneous pressure contours over the whole circumference of the cascade. In addition, the analysis method allows to build up these global pressure contours at a high temporal resolution. As a result, modal flow structures can be individually assigned to the spectral components of RI.

Depending on the respective circumferential mode, pressure waves with different wave lengths move in circumferential direction. Alternating spots with decreased and increased pressure fluctuations cause abruptly changing inflow conditions to each compressor blade accompanied by an intense impact on the aerodynamic loading. Furthermore, the global patterns reveal a clear correlation between the orientation of the pressure waves and the inflow angle.

INTRODUCTION

A more accurate prediction of critical operating conditions in axial compressors requires detailed knowledge of an extensive range of unsteady flow phenomena being present at high blade loadings. Representing one of these flow phenomena RI predominantly occurs in axial compressors operating near their stability limit.

While the compressor system operates under stable conditions, RI can be detected by characteristic signatures in frequency spectra of single point measurements taken in the vicinity of the leading edge plane (1) (2). A spectral RI pattern arises with side-by-side peaks, each of which can be assigned to a circumferential mode. Regardless of whether the investigations are pursued in a rotor configuration or in stator configuration with hub clearance, these patterns comprise the same characteristics (3) (4) (5) (6). In general RI could be observed in axial compressors predominantly with large tip clearances, in shrouded configurations, in centrifugal compressors and even in low pressure turbines (7) (8) (9). Whereas the phenomenon RI appears persistently over an extended operating range, characteristic features like modal events are of intermittent nature (10). In the past, rotor coherent flow structures were mostly addressed to describe the formation mechanism of RI. Taking into account, that the occurrence of modal events is neither periodic nor rotor coherent, it is more suitable to concentrate the view at short periods of time in which modal events are clearly detectable.

A special triggering analysis enables to identify the stochastic modal events and in the following to represent the corresponding flow field in the area of origin of RI. Furthermore, the specially developed analysis method allows to estimate quasi-instantaneous pressure contours around the whole annulus and also at high temporal resolution. The purpose of this work is to accomplish a better physical understanding of the flow mechanism which can be assigned to the spectral components of RI and not to describe the formation of modal flow structures in their initial stage.

EXPERIMENTAL SETUP

The experimental investigation is pursued in an annular compressor stator cascade at the Department of Aeronautics and Astronautics at the Technische Universität Berlin (see Fig. 1).

TABLE 1: CSC FEATURES

CSC blades	20
Chord	30mm
Shroud diameter	240mm
Hub/Shroud ratio	0.72
Hub clearance	3% c
Reynolds number	3.0 x 10^5
Mach number	0.4
Incidence angle	+12.7 deg

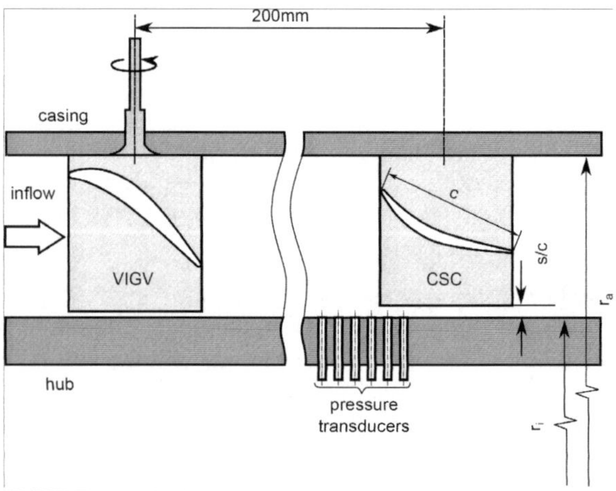

FIGURE 1: TEST SECTION

The test rig is designed as an open flow wind tunnel operated by a set of three radial compressors. The annular flow duct with a length of 1.27 m and a constant cross sectional area consists of two consecutive non-rotating blade rows, each with 20 blades mounted on the shroud. The first cascade (VIGV) turns the incoming axial flow in circumferential direction comparable to nozzle guide vanes. Thereby the inflow condition to the downstream compressor cascade (CSC; measurement domain) is provided by an intense swirl flow. By varying the stagger angle of the VIGV, the incidence angle to the compressor blades can be adjusted in a range of -5 deg up to +21 deg. The non-twisted, 2D-contoured compressor blades were designed aiming at flow separation sensitive characteristics. The gap between hub and compressor blade tip is 1 mm in height which corresponds to 3 % of the chord. Details of the test rig are summarised in Tab. 1.

SENSOR SETUP

An instantaneous hub wall pressure field measurement technique was developed, to give high spatial and temporal resolution of unsteady flow structures in a compressor stator with a minimal use of pressure sensors. For this purpose, 59 hub-mounted piezoresistive pressure transducers (Kulite) measure the static pressure distribution in an array upstream of the leading edge plane over one blade pith. The relative position between the array and the compressor stator blades remained constant during all measurements. Unsteady pressure measurements primarily taken inside one blade passage revealed that predominant pressure fluctuations dominate the flow field upstream of the cascade inlet (11). This finding led to the decision to place the array upstream of the leading edge plane.

In addition, 16 unsteady pressure sensors (A-P) are installed around the whole circumference (cf. Fig. 2) to capture the characteristic revolving pressure waves associated with RI. The axial position of the circumferential sensors is 1.5 mm upstream of the leading edge plane. All 63 signals were obtained simultaneously at a sampling rate of 24 kHz using a 64-channel signal amplifier with low-pass filtering at 10 kHz to avoid aliasing.

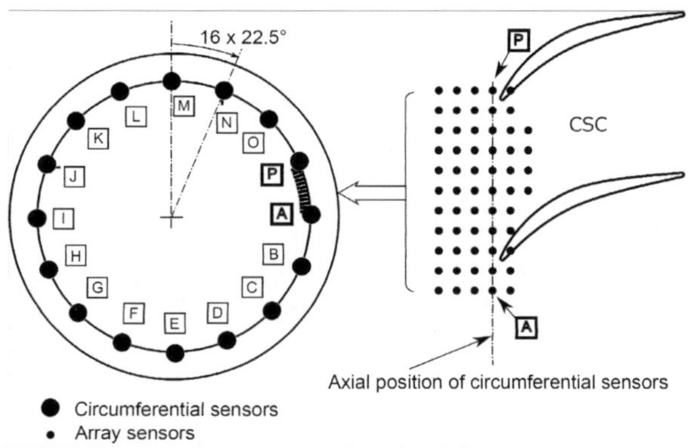

FIGURE 2: INSTRUMENTATION; SENSOR SETUP

RI CHARACTERISTICS

In this section, spectral and modal properties are analyzed to demonstrate the presence of RI together with all its well known specific characteristics at an operating point where the incidence angle is $i = +12.7$ deg.

Past experiences with the annular compressor stator cascade have revealed an extensive picture of the occurrence and intensity of RI over a wide operating range and varying Mach numbers. The detailed analysis in the present study focuses on an operating point with pronounced spectral RI characteristics at an inflow Mach number of $Ma = 0.4$. A predominant spectral RI signature can be observed by applying the power spectral density function to the signals from the 16 circumferential sensors. Figure 3 shows the mean power spectrum, coherence γ and phase response of various combinations of three selected circumferential staggered sensors A, C and F. The auto power spectra in the upper graph and also the coherence spectra in the centre exhibit a characteristic spectral signature with side-by-side peaks and an approximately constant frequency spacing between the adjacent peaks. The coherence indicating the propagation of distinct disturbances between two sensors with a certain offset decreases with increasing sensor distance. Phase information from associated cross spectra provide a stepwise phase distribution between 100 and 300 Hz. All findings indicate propagating pressure fluctuations with an azimuthal modal structure.

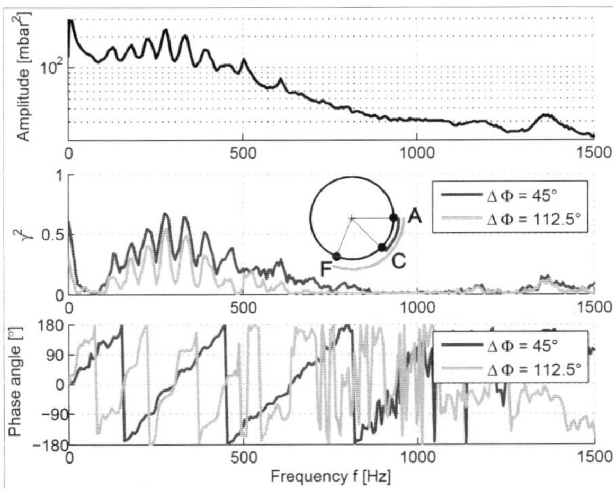

FIGURE 3: RI IDENTIFICATION WITH CIRCUMFERENTIALLY STAGGERED PRESSURE TRANSDUCERS

Furthermore, each peak within the comb structure can be assigned to a circumferential mode. Therefore, a mode decomposition technique was applied to the sensor ring in order to resolve the circumferential evolution of RI in frequency domain. The mode analysis has been already described precisely in Refs. (12) (13).

Using the 16 pressure transducers installed around the whole annulus an azimuthal mode analysis is feasible up to mode orders $m = \pm 7$ covering 2.9 passages of the stator row. Resulting mode amplitudes in a frequency range up to 1 kHz are shown in Fig. 4. Figure 4 clearly shows dominant azimuthal structures where adjacent peaks are of subsequent numbered modes with a dominant mode of order $m = 5$ at a frequency of 287 Hz. This finding representing typical RI characteristics could be observed in earlier studies at the presented compressor stator by Ref. (3) and also at rotor-stator configurations (4).

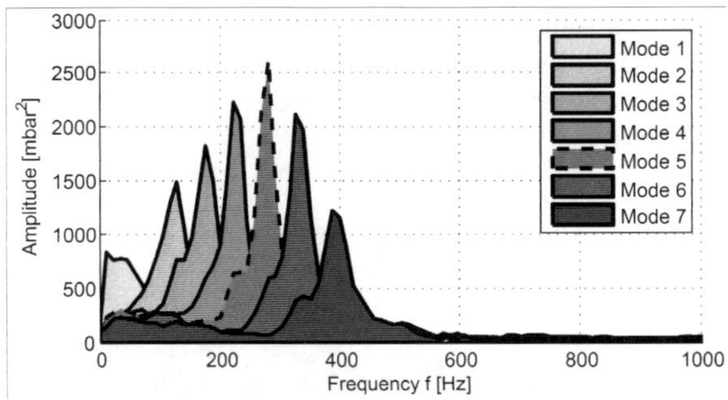

FIGURE 4: ANALYSIS OF CIRCUMFERENTIAL MODE AMPLITUDES

TRIGGERED MODE ANALYSIS

While RI appears continuously at high blade loadings, detailed characteristics of the phenomenon are of intermittent nature. Modal pressure distributions occur stochastically distributed in time and they are independent from each other (10). The highly temporal formation of modal pressure distributions with an increasing and subsequently decreasing mode amplitude is denoted as a "modal event". To understand which particular flow mechanisms can be assigned to the spectral components of RI it is instructive to analyse the short sequences in which a modal event is present.

For this purpose, a triggered data analysis enables the identification of trigger times where high amplitude bursts of different RI modes are existent. The corresponding instantaneous mode separation technique was developed by Kerschen and Johnston (14). In detail, a spatial Discrete Fourier Transformation analysis was used to calculate the time-resolved mode amplitudes $\tilde{A}_m(t)$ of order m at a time t.

$$\tilde{A}_m(t) = \frac{1}{N_\phi} \sum_{j=0}^{N_\phi - 1} \tilde{p}(\phi_j, t) \cdot e^{-im\phi_j}$$

For the detection of triggers within the time-resolved mode amplitude, thresholds for mode magnitude (Thr_m) and corresponding phase angles ($\phi_{Thr,m}$) are introduced. A trigger is found if the time-resolved magnitude exceeds the threshold level Thr_m. Furthermore, triggers are only taken into account if the related modal pressure distribution has the same circumferential position respectively phase ($\phi_{Thr,m}$).

$$|\tilde{A}_m(t_{trig})| > Thr_m \text{ and } arg\{\tilde{A}_m(t_{trig})\} = \phi_{Thr,m} \pm \phi_\Delta$$

As an example, the temporal distribution of the amplitude of mode order $m = 5$ is displayed in Fig. 5 with its real part (gray solid line) and its magnitude (black solid line). With a fixed threshold of $Thr_{m=5} = 1.3$ mbar the short time sequence contains three identified triggers. Around these triggers, the real part shows a harmonic oscillation with increasing and decreasing amplitudes denoting a modal event. Again, the occurrence and distribution of triggers is neither uniform nor periodic. While the phase angle $\phi_{Thr,m}$ was left constant at a value of 0 deg it was necessary to specify the threshold level Thr_m for each analysed circumferential mode gaining approximately 100 triggers. In general, triggers N of at least 100 are recommended for a reliable representation of stochastic quantities.

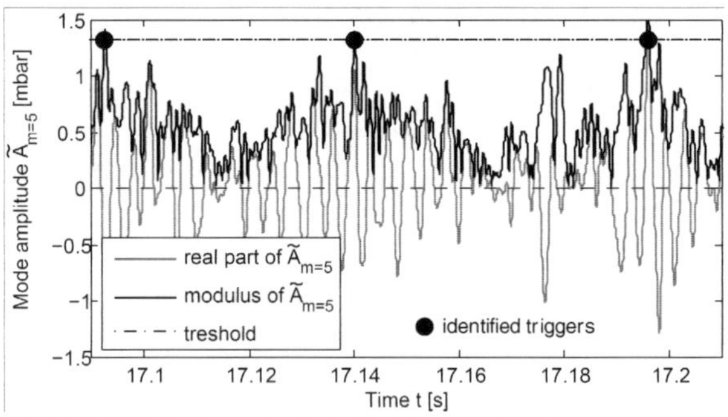

FIGURE 5: DETECTION OF MODAL EVENTS

STUDY OF MODAL EVENTS

The triggered mode analysis is performed for the dominating mode $m = 5$ to illustrate the corresponding modal pressure field measured by the array sensors in the vicinity of the leading edge plane. With a specified threshold level of $Thr_{m=5} = 0.9$ mbar, the resulting pressure fields are averaged over 105 triggers representing the unsteady flow field for one specific phase angle $\phi_{Thr,m=5} = 0$ deg. Thus, Fig. 6 shows an averaged snapshot of the pressure fluctuations being present at a modal event.

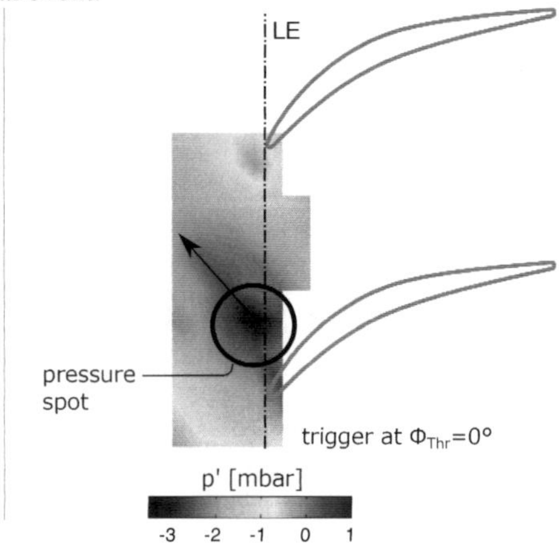

FIGURE 6: MODAL PRESSURE FIELD AT A SPECIFIC PHASE OF 0 DEG

The ensemble averaged pressure pattern is dominated by a pressure spot with decreased pressure fluctuations which has just passed the leading edge of the lower compressor blade. Comparable to observations made in tip leakage vortex investigations a thin trace of negative pressure fluctuations upstream of the pressure spot indicates the trajectory of a vortex system which is highlighted by a black arrow. Important to mention at this point, the trajectory is aligned perpendicular to the cascade inflow. In summary, the vortex is recognised by a low pressure core originating near the peak suction of the convex side of the compressor blade and extends counter the axial inflow direction out of the blade passage.

Having introduced the detailed flow field related to the dominating circumferential mode within this small region, we now move on to consider the same flow field both, over multiple blade passages and time-resolved.

ENSEMBLE AVERAGED PRESSURE CONTOURS

Summarising, the triggered mode analysis enables the identification of trigger times, at which a modal event is present. Now we define a sequence of length T around these trigger times (see Fig. 7). Equivalent sequences within the signal from the array sensors are further used for the subsequent analysis of the global flow field.

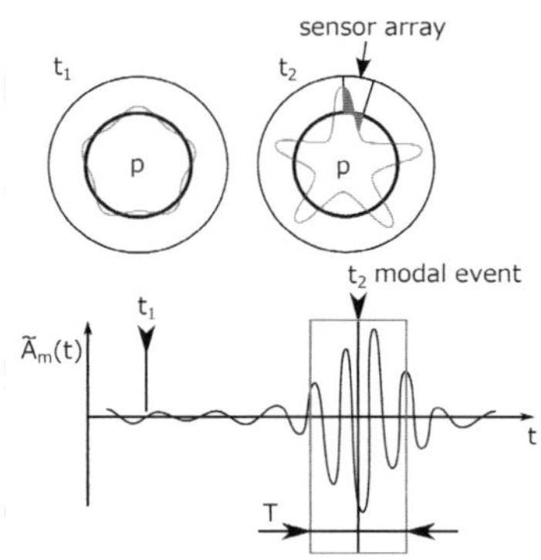

FIGURE 7: SEQUENCE AVAERAGING AT MODAL EVENTS

For this purpose we need information about the propagation velocity of the modal pressure waves. A cross correlation of the data from sensors **A** and **F** yields a propagation velocity of

$$u = \frac{\pi \cdot d}{\Delta t} = 30{,}59\ ^{m}/_{s}$$

A sufficient sensor distance was chosen in order to avoid any influence of the discrete blade distribution. The time offset of $\Delta t = 17.665$ ms leads to a rotation at $1/\Delta t = 56.609$ Hz which is equivalent to the frequency spacing of the adjacent peaks in Fig. 3.

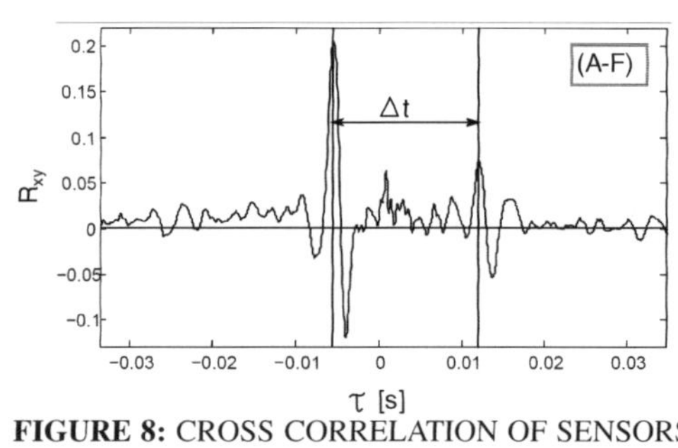

FIGURE 8: CROSS CORRELATION OF SENSORS **A** AND **F** FOR THE CALCULATION OF THE PROPAGATION VELOCITY

Using the blade pitch and the propagation velocity, it is possible to calculate the period of time Δt_s in which the pressure wave travelled the distance between two blade passages.

$$\Delta t_s = h/u$$

Assuming that the modal pressure wave remained unchanged within this period, approximately every 20th sample shows the characteristic flow field when it passed the respective neighbouring passage. The exact number of samples is calculated as follows.

$$\Delta s = \Delta t_s \cdot sf = 20.25$$

Starting with the detected trigger sample, the following samples with an interval of each $\Delta s = 20.25$ are appended in the negative direction of propagation in order to get a holistic picture of the modal pressure field over the entire cascade. By doing this, the samples are interpolated in time because of the odd sample distance Δs. Up to this step most of the pressure information was not included. If we start building the sequence exactly one sample later, the global pressure pattern can be analysed at the subsequent time. So the temporal resolution depends on the propagation velocity itself.

TIME-RESOLVED FLOW FIELD

In this section, we discuss the time-resolved static pressure contours for different mode orders. Beginning with mode order $m = 4$ the seven pressure maps depicted in Fig. 9 show the unsteady flow field extending over five blade passages each with a time interval of 0.125 ms.

By plotting the pressure fluctuations over five passages respectively a quarter of the entire circumference, exactly one wave length is represented for the analysed mode of order $m = 4$. Within the presented overall interval of 0.75 ms, moving pressure waves and related small-scale flow structures can be traced between two neighbouring blades.

At 0 ms the circumferential section is divided by alternately positive and negative pressure fluctuations. A global minimum of the pressure fluctuations is located around the leading edge of blade **4** indicating its increased aerodynamic load. In the following time step a small negative pressure spike arises near the peak suction of blade **4** (marked by the circle), and the aerodynamic load decreases. The dimension of the spike as well as its

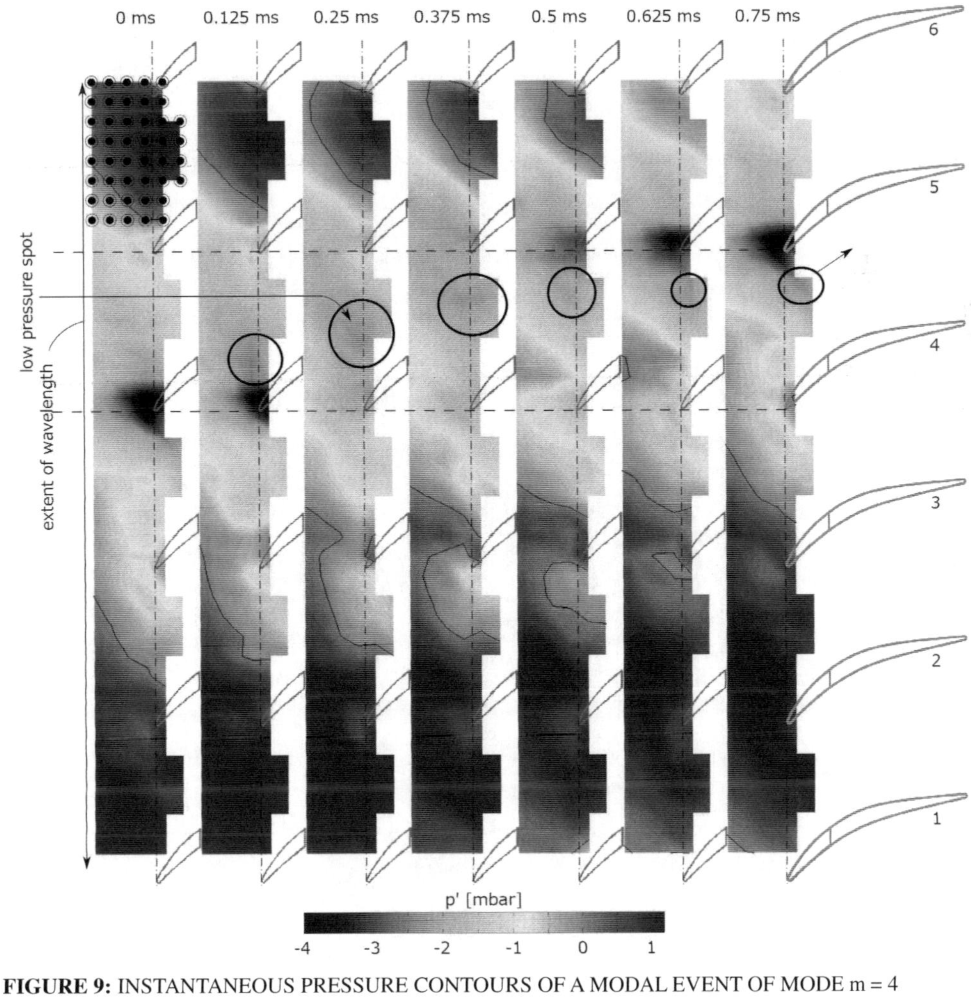

FIGURE 9: INSTANTANEOUS PRESSURE CONTOURS OF A MODAL EVENT OF MODE m = 4

intensity remains constant just over midway between blade **4** and **5**. Even before the spike impinges the neighbouring blade, a global minimum of the pressure fluctuations emerges at the leading edge of blade **5**. This tendency leads to the conclusion that the moving spike has an intense impact on the aerodynamic load of discrete compressor blades by provoking high incidence. In the course of the impact on the leading edge of blade **5** the spike seems to be slightly decelerated. After that, most part of the spike migrates downstream through the passage, as can be observed at $t = 0.75$ ms. However, based on this analysis method, the origin of spike formation cannot be determined even during the emergence of the modal event. For this purpose (not shown in this paper), an initial sample far before the trigger time was used for generating the global sequences. Furthermore, the circumferential extent of both, negative and positive pressure fluctuations varies slightly in time. Compared to the pressure pattern in Fig. 6, a characteristic vortex trajectory cannot clearly be identified for a circumferential mode of order $m = 4$.

As could be already seen in Fig. 4, pressure fluctuations exhibit various amplitudes depending on the mode order. Having studied the flow field for the mode of order $m = 4$, it is now useful to face the question of how the modal flow field develops for the dominant mode $m = 5$, with its highest pressure fluctuations.

The contour plots in Fig. 10 show the unsteady field which is represented by approximately 125 modal events. For triggering the modal events, the same phase of the time-resolved mode amplitudes was considered, to present the initial flow field at the same circumferential location as for the mode of order $m = 4$. In accordance to an increased mode order the extent of the wave length decreases while the magnitude of the pressure fluctuations has grows. Especially the flow structure belonging to the spike seems to have a more pronounced impact on the surrounding blades because the increased aerodynamic load is lasting for a longer period of time. Beginning at $t = 0$ ms, a thin trace of negative pressure fluctuations indicates a tangential vortex being shed from the suction side of blade **4**. While propagating in circumferential direction of the cascade the aerodynamic load decreases subsequently. The distinct trace disappears when the low pressure region impinges the neighbouring blade suggesting that most part of the vortex system dissipates in the course of hitting the leading edge. While one part of the low pressure region migrates downstream along the

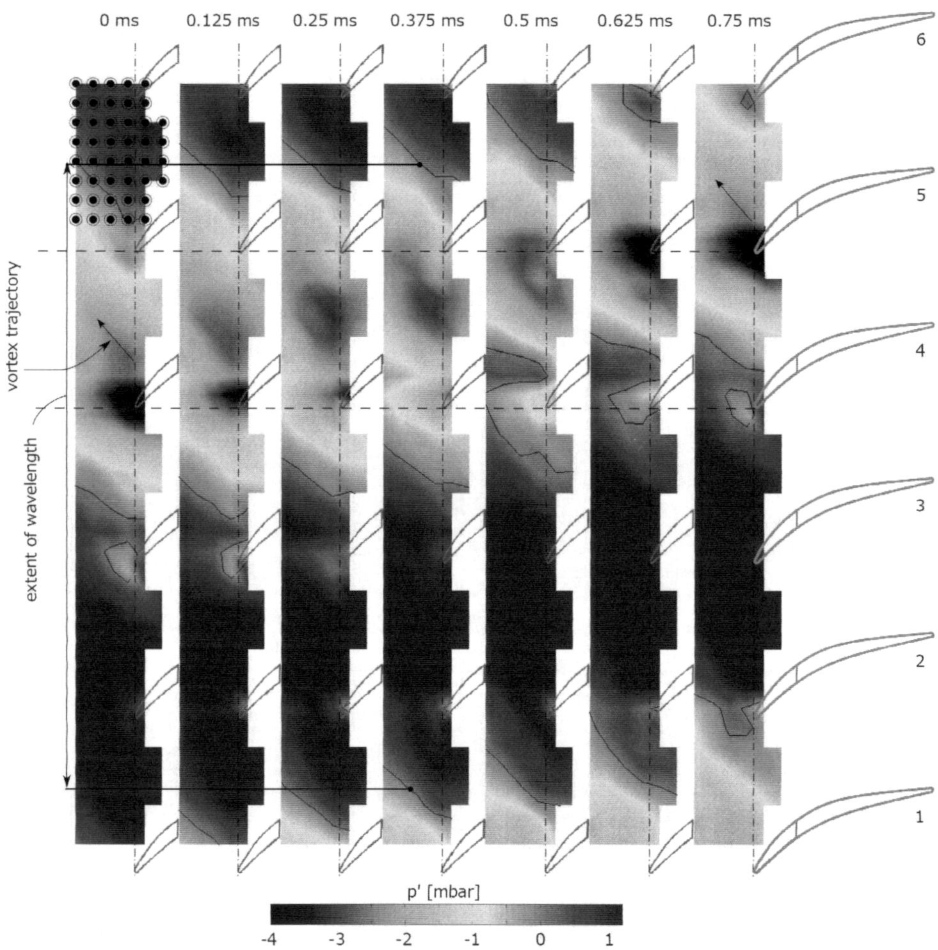

FIGURE 10: INSTANTANEOUS PRESSURE CONTOURS OF A MODAL EVENT OF MODE m = 5

pressure side, however, the other part seems to form a new vortex on the suction side. As can be seen at $t = 0.75$ ms, a thin trace of negative pressure fluctuations arises just *before* a low pressure spike detaches from the respective leading edge.

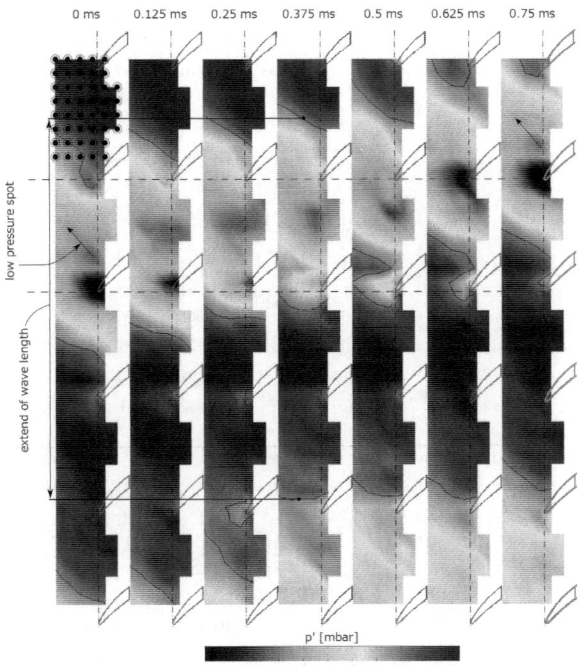

FIGURE 11: INSTANTANEOUS PRESSURE CONTOURS FOR MODE m = 6

Nevertheless, a comparison of Fig. 9 and 10 reveals that primary flow structures are independent of the respective mode order.
The alignment of the vortex trajectory and most of the pressure iso-lines suggest that the topology of the rotating flow structure is directly related to the inflow direction.

This trend becomes more apparent from the time-resolved pressure series of the mode of order $m = 6$ (see Fig. 11). In comparison to the pressure patterns of Fig. 10, the magnitude of the pressure fluctuations, especially of those to be found in the low pressure spot, decreased. This finding is in good agreement with the frequency-dependent mode amplitudes in Fig. 4. However, the characteristics of the moving spot, including its influence on the blade loading, are the same as to be considered in Fig. 9 and 10. Since each sequence comprises approximately 1.5 wavelengths for the presented mode order, the following low pressure spot is observed at the bottom of the last sequence ($t = 0.75$ ms) between blade **1** and **2**. The magnitude of fluctuations in the following pressure spot is lower compared to the spot in the upper passage indicating the short-half life of the modal flow structures. Again, the pressure contour plots are presented as a space-time diagram. If the rotating flow structures persist for multiple revolutions, its intensity is unchanged within this quasi-instantaneous view.

In order to verify this tendency it is instructive to analyse the quasi-instantaneous sequences covering the whole circumference of the cascade. Figure 12 shows the modal pressure patterns for the modes of order $m = 4,5$ and 6 at the same time-offset. The dashed lines represent the leading edge positions. The trigger array is positioned between leading edge **2** and **3**.

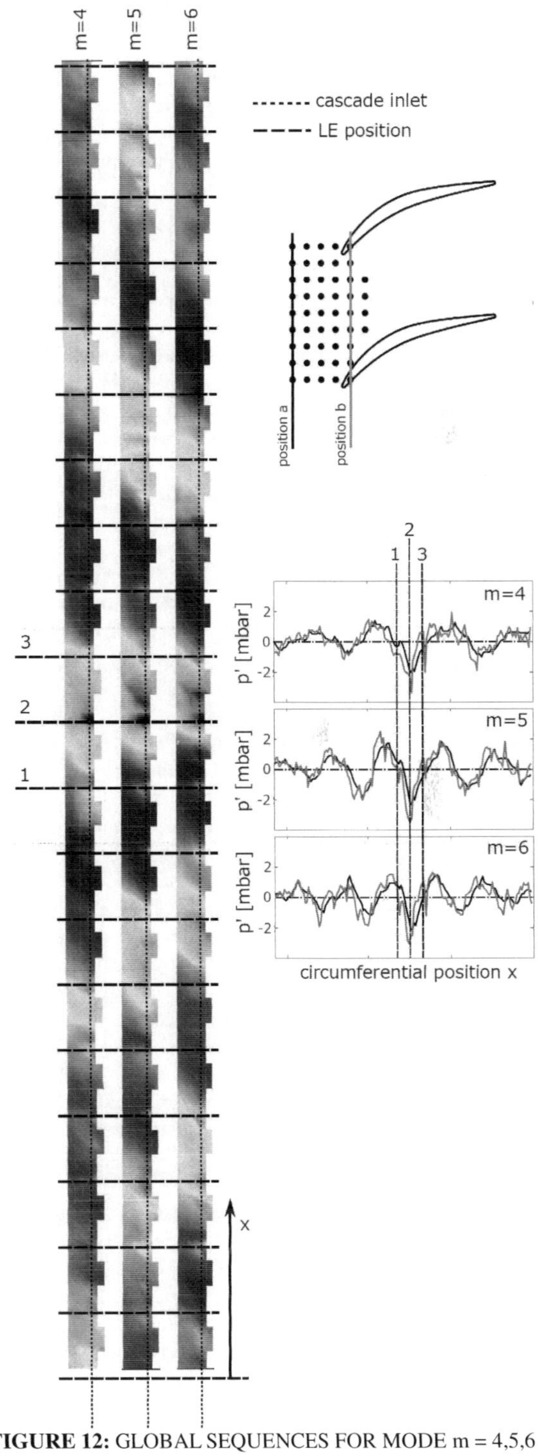

FIGURE 12: GLOBAL SEQUENCES FOR MODE m = 4,5,6

In each subplot, the number of global maxima matches the respective mode order denoting the special analysis procedure leads to appropriate pressure contour results. Further, maximum values of pressure fluctuations can be found around each trigger array with a decreasing trend on both sides. Thus confirming that a modal event does not even last for more than one revolution.

The subplots on the right side of Fig. 12 comprise the circumferential distribution of the pressure fluctuations being extracted from the pressure contours at different axial positions a and b which are schematically shown in the upper right of Fig. 12. For each mode order, the shift in x-direction between both distributions accentuate the orientation of the pressure waves as a characteristic feature of RI flow structures.

DISCUSSION OF THE RESULTS

Based on special analysis tools from acoustical investigations it is possible to obtain previously unseen information about the flow field which is directly related to the spectral components on Rotating Instability. In the past decades RI was widely investigated in terms of its influence on the compressor performance, noise production, and not at least to mention the description of the basic formation mechanism. Since then, the term "RI" was being contested in the sense that it would be merely a synonym for rotating stall.

For this reason the authors would like to emphasise some similarities and differences between RI and rotating stall. This shall be done considering two stand out stall inception patterns, namely the spike-type stall and the modal stall inception.

One route to rotating stall is represented by long length-scale perturbations which occur as modal oscillations and which can be described by a linear-to-nonlinear wave evolution (15). While the modal oscillation in its early form shows no stall cell, however, associated velocity fluctuations increase the local incidence on the blade causing it to separate. In general modal stall inception is present when the critical incidence is exceeded in the positively sloped compressor pressure rise characteristic (16). In contrast, RI has been observed in an axial fan stage even before reaching the pressure rise peak (4). The similarity to the modal stall inception is given by the fact that the modal pressure waves and its corresponding flow structures provoke high incidence on particular blades indicated by an extreme blade loading.

The second stall inception pattern, which is representing the one most common in axial compressors, is linked to short length-scale disturbances of large amplitudes. The so-called spike stall inception was observed in axial compressors with and without tip clearance as well as in centrifugal compressors. Typical spectral signatures of RI could be also identified at these compressor types as mentioned in the introduction. It has taken a longer time span and numerous research investigations to determine what kind of flow mechanism is related to the rotating disturbance which was later defined as "spike". The term "spike" bases on distinct low pressure spots which have been detected in the leading edge plane during the stall inception using arrays of unsteady pressure transducers over the rotor tip. The spike originates from a leading edge separation. During this process radial vorticity is shed from the suction side to form a discrete vortex filament that spans from the blade suction surface to the casing. While the radial vortex propagates in circumferential direction of the cascade it triggers a new separation on the leading edge of the neighbouring blade. The pattern subsequently grows in amplitude and length-scale before evolving into rotating stall (17) (18).

The quasi-instantaneous pressure patterns in the present paper also show discrete regions of pressure drop propagating around the annulus. Apart from the fact that these spots are arranged equidistantly around the circumference it is not clearly established what kind of flow structure causes these pressure drops. Similar patterns with patches of intense low pressure, also called as prestall flow irregularities, have been identified by Young et. al (19) in an isolated compressor stage operating with a negatively sloped pressure rise characteristic *before* stall. Their corresponding fourier transforms show characteristic spectral RI signatures with a broad band peak at 30% of the blade passing frequency (BPF) and a typical interaction of RI with the rotor (two peaks around BPF with smaller amplitudes; BPF-RI and BPF+RI). The signatures given in the paper suggests that they were observing RI. Hence, the so-called flow irregularities and the propagating spots presented in this paper might have the same origin.

For the common description of spike stall inception distinct spikes are merely defined as a characteristic feature of the stall inception process itself. By contrast, the low pressure spots which were obtained in the course on investigations on RI represent a persistent feature. They exist over a wide operating range where RI is present and not at a specific operating point.

But there is an interesting similarity between RI and rotating stall. Intermittent leading edge separations were identified as specific feature during the stall inception in the Cambridge low speed axial compressor by Pullan et al. (17) Modal events that are associated with RI also appear stochastically distributed in time. This fact could give an indication of the relationship between RI and rotating stall.

In case of RI predominant characteristics of the spike stall inception would persist *beyond* the inception process. In other words the transition from the state in which radial vorticity propagates around the annulus towards the final formation of rotating stall cells would be prevented. It is assumed that intermittent leading edge separations initialize a modal event wherein the separations are responsible for their stochastic occurrence. Before the resulting spikes evolve into fully developed stall cells their intensities are damped subsequently in order to finish the modal event.

Moreover, it should be clarified what flow mechanism contributes to the modal arrangement of finite disturbances around the whole circumference. For further discussion, the term "instability" plays an important role. It is hypothesised that flow instabilities cause the modal arrangement of disturbances by meeting circumferential resonant conditions for certain wavelength. Accordingly, intermittent leading edge separations evolve into propagating disturbances with temporarily increasing intensities if the corresponding resonant condition is met for one circumferential mode.

CONCLUSION

Using a combined setup of wall-flushed pressure transducers enabled the investigation of detailed unsteady flow structures which can be specifically related to the spectral components of RI. Therefore, a special analysis method was developed to identify trigger times where modal events occur. In the following the triggering method was used to analyse the characteristic time-resolved flow field in the region of origin of RI. The following conclusions can be drawn:

1. Global pressure patterns revealed harmonic pressure waves travelling circumferentially. The wavelength and the magnitude of the pressure fluctuations depend on the circumferential mode order. It has been demonstrated that each modal pressure wave exhibits temporarily increasing amplitudes. These amplitude bursts, not even persisting for more than one revolution, are denoted as modal events. Furthermore, modal events appear stochastically distributed in time.
2. Details of the modal flow fields manifested propagating spots of negative pressure fluctuations in the vicinity of the leading edge plane. Regardless of the involved flow structure the localised disturbances have an intense impact on the aerodynamic load of discrete blades. However, it is assumed that a narrow trace of negative pressure fluctuations upstream of the spot represents the trajectory of a vortex system which is aligned in perpendicular to the inflow direction and thus parallel to the hub wall. In general, primary flow structures which are involved in the development of RI, are independent of the respective circumferential mode order.
3. It is hypothesized that RI represents a special form of the spike stall inception without exceeding the transition into rotating stall. Accordingly modal events are triggered by leading edge separations which result in vortex formation and propagating around the annulus. Corroborated by the illustration of modal RI events the intensities of the propagating vortices increase temporarily without evolving into fully developed stall cells. During this process the convection, modal arrangement and not at least to mention the intensity of discrete vortices are influenced by an instability mechanism. So, the modal event is initiated if resonant conditions are met for certain wavelength. Summarising, RI contributes to a stall inception event.

NOMENCLATURE

$\tilde{A}_m$	time-resolved mode amplitude
c	chord
CSC	compressor stator cascade
d	hub diameter
h	blade pitch
i	incidence angle
m	circumferential mode order
Ma	Mach number
N	number of triggers
p'	pressure fluctuations
r	radius
sf	sampling frequency
Δs	sample interval
t	time
T	time sequence
u	propagation velocity
$VIGV$	variable inlet guide vane
ϕ	phase angle

ACKNOWLEDGMENTS

The presented work have been carried out within the framework of the research project "Flow Induced Acoustics in Turbomachines - The Rotating Instability" funded by the German Research Foundation (DFG). The discussions with our partners involved in this project, are gratefully acknowledged.

REFERENCES

1. **Kameier, F. and Neise, W.** Rotating Blade Flow Instability as a Source of Noise in Axial Turbomachines. *J. Sound Vib., 203.* 1997, pp. 833-853.

2. **Mailach, R., Lehmann, I. and Vogeler, K.** Rotating Instabilities in an Axial Compressor Originating From the Fluctuating Blade Tip Vortex. *Journal of Turbomachinery; No.123.* 2001, pp. 453-463.

3. **Pardowitz, B., Tapken, U., Sorge, R., Thamsen, P. U. and Enghardt, L.** Rotating Instability in an Annular Cascade: Detailed Analysis of the Instationary Flow Phenomena. *Journal of Turbomachinery; Vol. 136.* 2014.

4. **Pardowitz, B., Tapken, U., Neuhaus, L. and Enghardt, L.** Experiments on an Axial Fan Stage: Time-Resolved Analysis of Rotating Instability Modes. *Journal of Turbomachinery; Vol. 137.* 2015.

5. **Beselt, C., Pardowitz, B., van Rennings, R., Sorge, R., Peitsch, D., Enghardt, L., Thiele, F., Ehrenfried, K. and Thamsen, P. U.** Influence of clearance size on rotating instability in an axial compressor stator. *in Proceedings of the 10th European Conference on Turbomachinery Fluid Dynamics and Thermodynamics ETC 10.* 2013.

6. **Beselt, C. and Peitsch, D.** Influence of mach number and aerodynamic loading on rotating instability in an annular compressor cascade. *In Proceedings of ASME Turbo Expo 2012.* 2012, Copenhagen, Denmark.

7. **Raitor, T. and Neise, W.** Sound Genaration in Centrifugal Compressors. *Journal of Sound and Vibration; Vol. 314(35).* pp. 738-756.

8. **Truckenmueller, F.** *Untersuchungen zur Aerodynamisch Induzierten Schwingungsanregung von Niederdruck-Laufschaufeln bei Extremer Teillast; Ph.D. thesis.* [ed.] Universität Stuttgart. Stuttgart : s.n., 2003.

9. **Zhang, L. Y., He, L. and Stuer, H.** A Numerical Invesitigation of Rotating Instability in Steam Turbine Last Stage. *Journal of Turbomachinery; Vol. 135.* 2013.

10. **Pardowitz, B., Tapken, U. and Enghardt, L.** Time-Resolved Rotating Instability Waves in an Annular Cascade. *AIAA Paper No.2012-2132.*

11. **Beselt, C., van Rennings, R., Peitsch, D., Thiele, F. and Ehrenfried, K.** Experimental and Numerical Investigations of the Unsteady Endwall Flow in a Highly Loaded Axial Compressor Stator. *In Proceedings of ASME Turbo Expo 2014.* GT2014-25944, 2014.

12. **Liu, J. M., Holste, F. and Neise, W.** On the Azimuthal Mode Structure Of Rotating Blade Flow Instabilities in Axial Turbomachines. *AIAA Paper No. 96-1741.*

13. **Jürgens, W., Tapken, U., Pardowitz, B., Kausche, P., Bennett, G. J. and Enghardt, L.** Technique to Analyse Characteristics of Turbomachinery Broadband Noise Sources; AIAA Paper No. 2010-3979.

14. **Kerschen, E. and Johnston, J.** A Modal Separation Measurement Technique for Broadband Noise Propagating Inside Circular Ducts. *Journal of Sound and Vibration, Vol. 76(4).* pp. 499-515.

15. **Garnier, V., Epstein, A. and Greitzer, E.** Rotating Waves as a Stall Inception Indication in Axial Compressors. *Journal of Turbomachinery; Vol. 113(3).* 1991, pp. 290-301.

16. **Camp, T. and Day, I.** A Study of Spike and Modal Stall Phenomenon in a Low-Speed Axial Compressor. *Journal of Turbomachinery; Vol 120(3).* pp. 393-401.

17. **Pullan, G., Young, A. M., Day, I. J., Greitzer, E. M. and Spakovski, Z. S.** Origins and Structure of Spike-Type Rotating Stall. *Journal of Turbomachinery; Vol 137.* 2015.

18. **Weichert, S. and Day, I. J.** Detailed Measurements of Spike Formation in an Axial Compressor. *Journal of Turbomachinery; Vol. 136.* 2014.

19. **Young, A., Day, I. J. and Pullan, G.** Stall Warning by Blade Pressure Signature Analysis. *Journal of Turbomachinery; Vol. 135.* 2013.

Proceedings of the 14th International Symposium on Unsteady Aerodynamics, Aeroacoustics & Aeroelasticity of Turbomachines
ISUAAAT14
8-11 September 2015, Stockholm, Sweden

I14-S3-4

EFFECTS OF NOVEL CASING TREATMENT ON A LOW-SPEED AXIAL COMPRESSOR WITH CIRCUMFERENTIAL PRESSURE DISTORTION

Sun Dakun
BeiHang University
Beijing, China

Dong Xu
BeiHang University
Beijing, China

Li Fanyu
BeiHang University
Beijing, China

Sun Xiaofeng
BeiHang University
Beijing, China

ABSTRACT

Experimental research of circumferential total pressure distortion is conducted in a low-speed fan with emphasis on the compressor stability, and the stall margin improvement with a kind of stall precursor-suppressed (SPS) casing treatment. Some results on compressor stall margin enhancement and pre-stall behaviour under the restriction of circumferential total pressure distortion are presented. The results show that the SPS casing treatment can improve the compressor stall margin (6%-8%) in the condition of different distortion degrees. The experiment results also show that the mechanism of the stall margin improvement with such casing treatment is associated with delaying the nonlinear development of the stall precursor waves and weakening the unsteady flow disturbances in a compression system.

INTRODUCTION

Circumferential total pressure inlet distortion is one of the most important problem reducing compressor stability margin. As Cousins [1] mentioned, cross-winds cause a pressure loss on the windward side of a nacelle and high-angle take-off conditions can cause inlet distortion due to large boundary layer growth and possible flow separation at the nacelle lip. A great amount of work to characterize the variations and the performance effects of circumferential inlet distortion has been conducted [2-8] and various predictive methods have been developed [9-13]. However, further studies are required to extend compressor stall margin under the condition of circumferential inlet distortion.

The inchoate research about circumferential inlet distortion mainly aimed at the form and impact of the distorted inlet flow using various kinds of experimental methods or mathematical models. At that time, Hynes [9] carried out a "first-principle" approach to predicting the decrease in stall margin within the inlet distortion, and addressed a method for assessing the effects of circumferential pressure inlet distortion on compressor stability. But assessing these effects is not the real purpose and far from the demands of engine manufacturers about the compressor stability. So some concepts and methods of active stabilization or rotating stall control were received wide attention. Schalkwyk [2] described the first experimental validation of transfer function modeling and active stabilization for axial compressors with circumferential inlet distortion on a three-stage low-speed compressor, and his work successfully attempted to apply system identification and control theory to a compressor under the circumferential inlet distortion. After that, Spakovszky [3] used active feedback control attempting to stabilize rotating stall in a single-stage transonic axial flow compressor with inlet distortion, and his results were very promising for the future investigations and applications. While, with the development of advanced aircraft engine, the boundary layer ingesting engines have drawn more attentions with the potential of bringing significant reductions in fuel burn compared with traditional turbofan engines. For the sake of viability, the boundary layer ingesting engines must have the robust ability to work with the continuously inlet distortion which generally exists in the inlet flow due to large boundary layer growth and possible flow separation at the nacelle lip and directly impact the operating of fan. So Gunn [5] presented an experimental investigation of loss sources in a fan operating with continuous inlet distortion, by describing the fan flow field with inlet distortion and the transfer through the rotor and stator, his proposed the impacts of pressure inlet distortion on the performance of fan. Similarly, Mistry [6] investigated the effect of a 90 degrees circumferential distortion sector on the performance of a fan stage with two rotors, and they carried out the conclusions that a higher rotational speed second rotor can improve the performance of first stage under clean as well as distorted inlet. Based on these research, much extension of stall margin enhancement technology is required, and it will be the main object of this paper.

It is well known that casing treatments are an effective method to improve stall margin (e.g., [14-18]). The existing studies indicate that casing treatment can enhance the stall margin of both subsonic and transonic compressors in varying degrees along with some efficiency loss as cost. Up to now, very few works have been conducted on the effect of casing treatment in inlet distortion conditions. Madden et al (e.g., [19]) reported the combined effects of forward sweep and/or rotor casing treatment on a single stage military fan with tip-low total pressure radial inlet distortion. It is thought that the casing treatment could restrain the influence on surge trigger caused by inlet distortion and recognized the stability enhancement of casing treatment under the inlet distortion. It is noted that the casing treatment completely changed the original performance curves obviously and caused obvious efficiency loss in large mass flow rate case.

Recently, a kind of stall precursor-suppressed (SPS) casing treatment was proposed to affect the inception or evolution of the stall precursors to realize stall margin enhancement (e.g., [20-21]). It is obvious different from the traditional casing treatments aimed at the improvement of the blade tip flow structure. More importantly, the geometrical parameters of the SPS casing treatment can be optimized based on the flow stability model (e.g., [21-24]) instead of trial-and-error method. In fact, although there are many theoretical models available for predicting the flow stability related to isolated airfoil (e.g., [25-26]) and compressor cascade (e.g., [27-28), it is found that the three-dimensional flow stability model (e.g., [20-24]) is more suitable to study on such casing treatment. In addition, the casing treatment configuration has a low perforated ratio about 6%, which is far smaller than that in traditional casing treatments with over 50% open area ratio. The results show that the SPS casing treatment can obviously improve the stall margin, accompanied by less efficiency loss (e.g., [21]). Because of this, it is reasonable to judge that the SPS casing treatment may also influence the change of compressor pre-stall behavior caused by inlet distortion, and realize the purpose of stall margin enhancement. The previous experiments on the rotating inlet distortion showed the validity of the SPS casing treatment (e.g., [29]). Therefore, there is a strong motivation to extend such casing treatment to the case of circumferential inlet distortion.

In the present work, the experiments will be carried out to study the effect of the stall margin variation caused by the circumferential inlet distortion and the SPS casing treatment. All experimental work is conducted in a low speed compressor with varying degrees of circumferential distortion. The approach adopted is to adjust two distortion plates upstream of a low-speed compressor. Both time-mean and time-resolved measurements are recorded for this test rig with the results analyzed in terms of the given data processing approach. Comparisons are also made with the performance and stability limit corresponding to both solid casing and SPS casing treatment.

MOTIVATION AND SCOPE OF THE PAPER

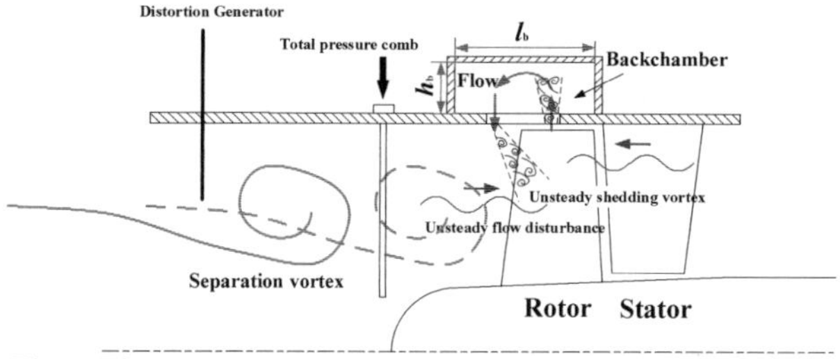

Fig. 1 The vortex-pressure wave interaction mechanism of the SPS casing treatment under the circumferential pressure inlet distortion

The major motivation of this article is to complete the experimental verification about the enhancement of SM by SPS Casing Treatment under different inlet conditions, and to further indicate the mechanism of SPS casing treatment. As mentioned in the introduction, previous experiments about the rotating inlet distortion has shown the validity of the SPS casing treatment (e.g., [29]), so extending the ability of such casing treatment to the circumferential inlet distortion situations is undoubtedly necessary. Besides, the circumferential pressure inlet distortion is a typical distortion form for advanced aircraft engine and truly different from rotating inlet distortion. First of all, the circumferential pressure inlet distortion can deliver an obvious total pressure loss which usually exists as a huge separation vortex moving and propagating. Secondly, this kind of inlet distortion is a broadband disturbance for the compressor system without particular frequency characteristic.

For this reason, if the mechanism of the stall margin improvement with SPS casing treatment is associated with delaying the nonlinear development of the stall precursor waves and the SPS casing treatment can weaken the energy of disturbances in broadband, such a novel casing treatment can also show an excellent effect under the circumferential pressure inlet distortion. As shown in Fig. 1, the distortion generator, usually a piece of metal plate, can generates a series of separation vortexes which will move downstream propagating. Then these vortexes will affect the rotor, change its operation point and other aerodynamic characteristics. At the same time, the separation vortexes can strengthen the energy of unsteady flow disturbance which may trigger the stall inception more easily and lead to rotating stall. So, if only the unsteady flow disturbance is not ridiculous strong, the vortex-pressure wave interaction can weaken the disturbance and enhance the stability of compressor.

Based on this understanding, the current work first carries out a method to generate and assess the circumferential pressure inlet distortion, and then delivery a series of experiments designed in different inlet conditions and different working speeds to prove the SPS casing treatment's ability of stall margin enhancement by using the steady characteristic of this test compressor. Finally, the dynamic stall and pre-stall signal are used to indicate the mechanism of the stall margin improvement with SPS casing treatment.

EXPERIMENTAL FACILITY, INSTRUMENTATION AND DATA ACQUISITION

The experimental rig employed for the present study is a low-speed, single-stage test compressor TA36. Fig. 2 shows the schematic of TA36 and measurement sections. It contains a one-stage fan (4 in Fig.2) with 20 blades, 300mm radium and 1.2067 pressure ratio; a fixed throttle cone (10 in Fig.2); an AC servo motor (7 in Fig.2) as the power of this rig; a distortion generator (Fig. 4) will be placed on position 11 in Fig. 3 and some pressure sensors (2, 3, 6 in Fig.2).

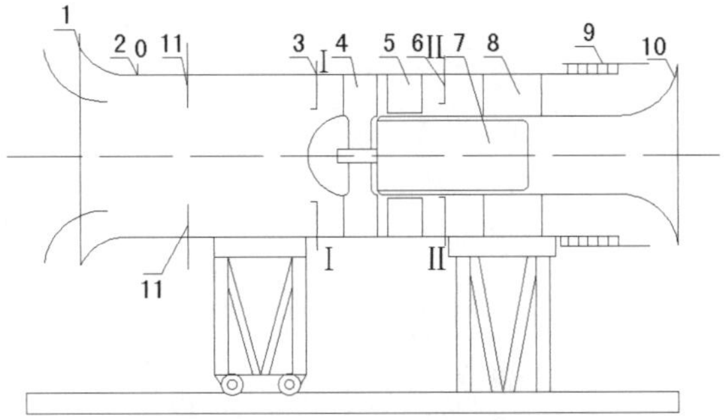

Fig. 2 Schematic of TA36 and measurement sections

Fig. 3 shows the SPS casing treatment which is installed upstream of rotor blade with about one-half overlapped area on the blade tip region. The open area ratio of perforated plate is 6.4%. The height of the annular back-chamber is 50 mm. The length of the back-chamber can be adjusted from 100 mm to 200mm. The length of the perforated plate is 75 mm.

Fig. 3 SPS casing treatment installed on TA36

A stagnation pressure distortion is created in the rig inlet by installing two plates at the location shown in Fig. 4. The positions of the plates can be adjusted separately by stepping motors. So this distortion generator can simulate circumferential inlet distortion in different intensities.

(a) Outside view (b) Inside view

Fig. 4 Circumferential Inlet Distortion Generators

The instrumentation of the TA36 rig mainly contains high-frequency response pressure sensors. The total pressure sensors are placed at the position of 3,6 in Fig. 2 to measure the inlet/outlet total pressure; and the static pressure sensors are placed on the casing wall at the position 2,6 in Fig. 2 to measure the inlet/outlet static pressure. Eight high-frequency response pressure sensors are mounted at the plane about half a chord length upstream of the rotor inlet to measure the static pressure perturbations, which are equally spaced around the circumference of the machine.

There are also 8 total pressure combs with 40 Kulite high-frequency response pressure sensors shown in the Fig. 5, those pressure sensors are used to measure the inlet pressure field structure and analyze the characteristic of inlet distortion. And because of the separation vortex is moving and propagating, the 8 combs must be placed really close to the rotor, in order to exactly reflect the rotor's inlet situation.

The data acquisition system is mainly a NI-PXI system and contains 6 acquisition devices up to 48 channels with an overall sampling rate of 100 kHz. The low speed fan TA36 is equipped with controllable bleed valves, which can accurately move the operating point near stall and quickly move away from stall. All the steady-state operating characteristics of the compressors are obtained using standard time-averaged instrumentation in these machines.

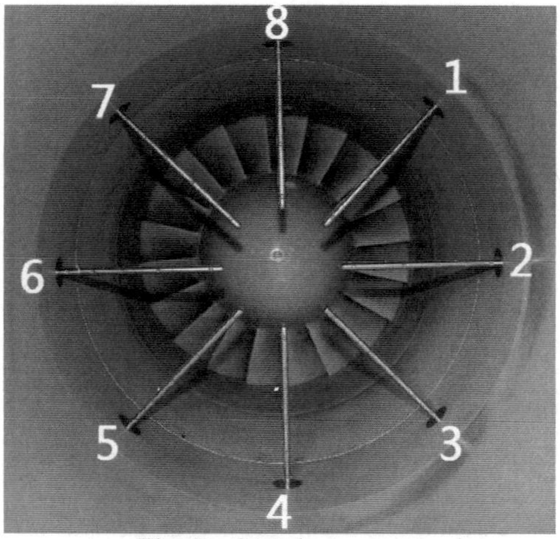

Fig. 5 8 total pressure combs

EXPERIMENTAL RESULTS

Degree of Inlet Distortion

In the present work, three kinds of total pressure inlet distortion (10%, 20% and 30%) are demonstrated and the degree of inlet distortion is expressed as a percentage of the distortion plate to the compressor inlet area. For example, 10% inlet distortion means that the distortion plate area covers 10 % of the compressor inlet area.

As mentioned above, there are 40 high-frequency response pressure sensors used to measure the inlet total pressure field, so the inlet distortion can be described by plotting a pressure contour. Fig. 6 shows the inlet total pressure field structure on 100% design working speed under the uniform inlet case, and the total pressure distributions are very uniform in the compressor inlet.

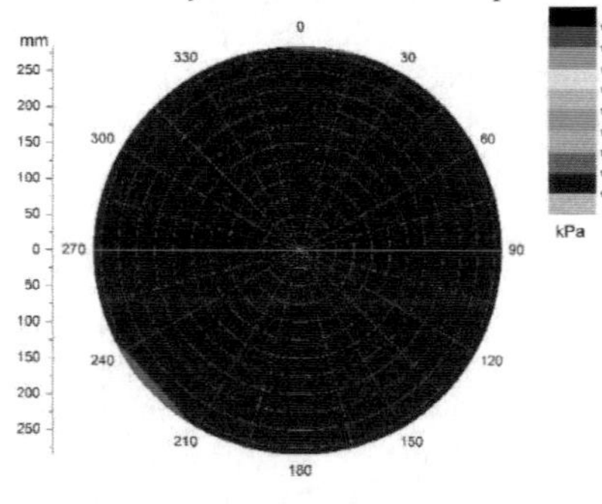

Fig. 6 Inlet pressure distribution Characteristics of low-speed fan under uniform inlet condition at design speed

Fig. 7-Fig. 9 show the inlet total pressure field structure on 80% and 100% working speed under the 10%, 20% and 30% inlet distortions, respectively. The distortion plate can make some total pressure loss compared with the atmosphere, and the higher the inlet distortion is, the bigger the total pressure value differences between high and low pressure districts become. Therefore, the distortion plate can generate different degree total pressure distortion to simulate the inlet flow conditions as mentioned in the introduction that the cross-winds can cause a pressure loss on the windward side of a nacelle and high-angle take-off conditions can cause inlet distortion due to large boundary layer growth and possible flow separation at the nacelle lip. While at the same time, in those figures, the low total pressure districts are to the right deviation because of the influence of fan rotor which is rotating in clockwise.

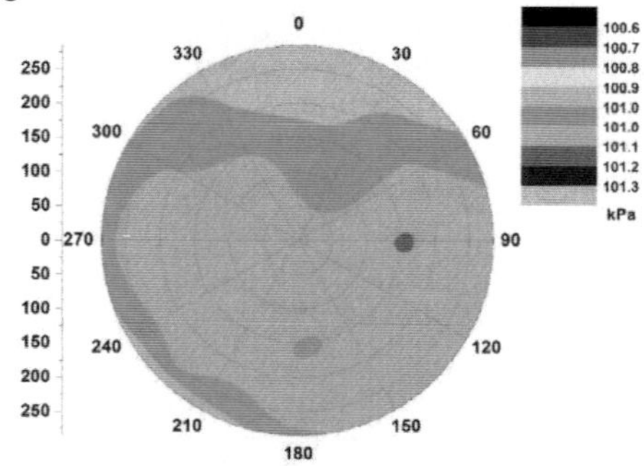

(a) 80% design working speed

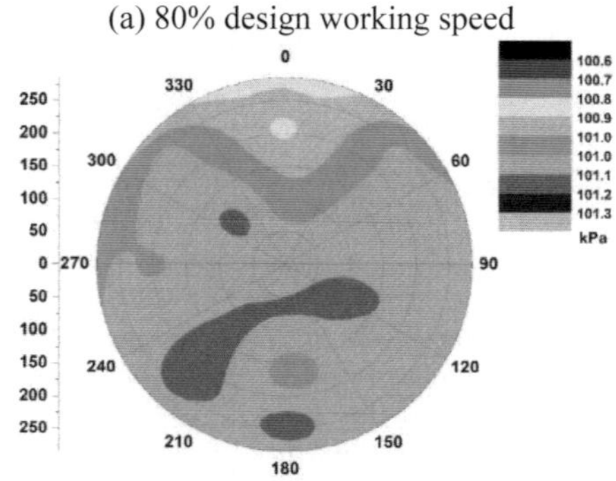

(b) 100% design working speed

Fig. 7 Inlet pressure distribution characteristics under 10% inlet distortion, 80% and 100% design working speed

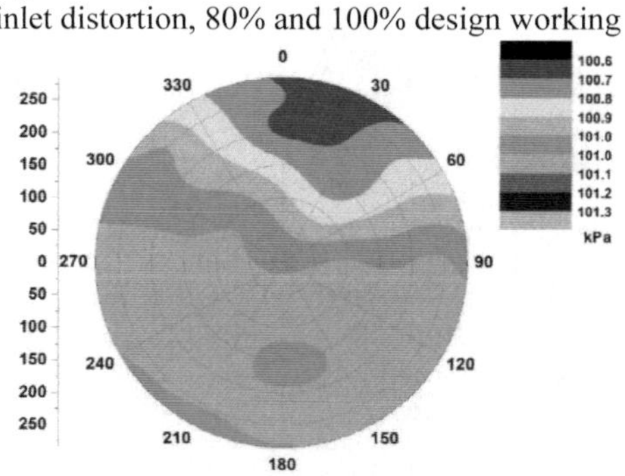

(a) 80% design working speed

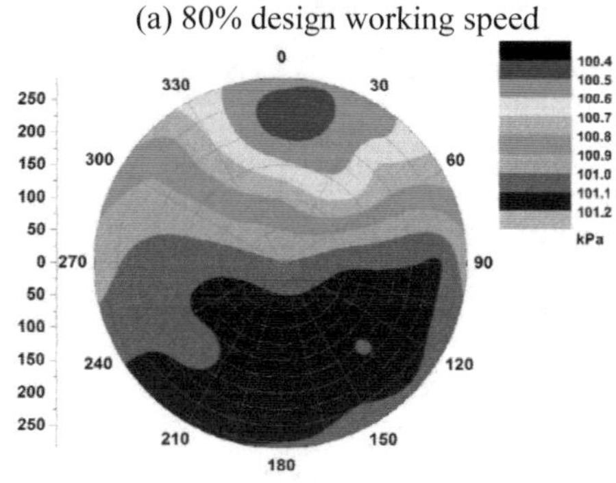

(b) 100% design working speed

Fig. 8 Inlet pressure distribution characteristics under 20% inlet distortion, 80% and 100% design working speed

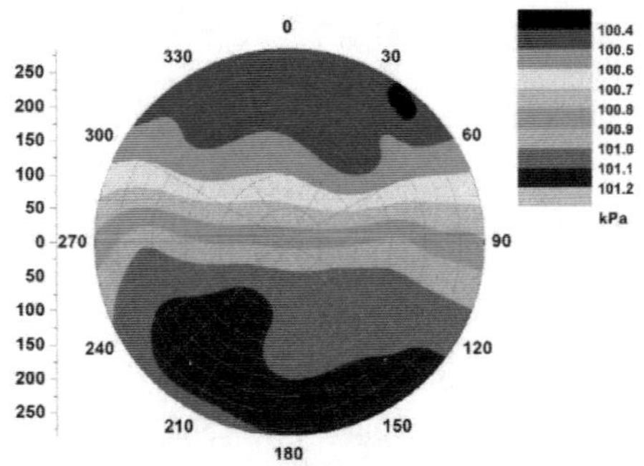

(a) 80% design working speed

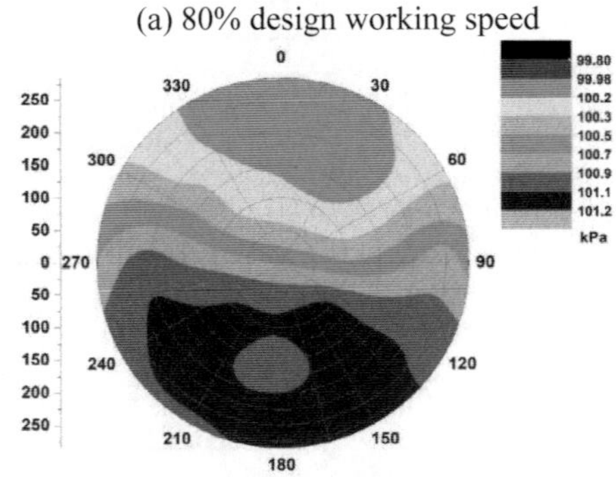

(b) 100% design working speed

Fig. 9 Inlet pressure distribution characteristics under 30% inlet distortion, 80% and 100% design working speed

Enhancement of SM by SPS Casing Treatment

The present work contains a series of experiments designed in different inlet conditions and different working speeds to prove the SPS casing treatment's ability of stall margin enhancement. In this paper, CT means using the SPS casing treatment with about 6% perforated ratio, SC means that the perforated plate is closed, which just corresponds to an ordinary solid wall casing. And the stall margin SM is defined:

$$SM = \left(\frac{\psi_s / \phi_s}{\psi_d / \phi_d} - 1\right) \times 100\% \qquad (1)$$

where subscripts s and d are parameters at stall point and design point respectively.

In order to assess the enhancement of SM by SPS casing treatment, the pressure rise curves and efficiency curves of this test compressor were carried out. As shown in Fig. 2, the test rig has a throttle cone which can control and adjust the flow subtly, so that the compressor characteristic can be described by curves within the steady static and total pressure sensors.

Fig. 10 shows the pressure rise curve under the 10% inlet distortion and 80% design working speed. In this figure, the SPS casing treatment can drop the stall point 1 under the inlet distortion of this degree back to the stall point 2 which is very close to the stall point without inlet distortion, make the stall margin loss caused by inlet distortion decrease from 7.22% to 1.30%. And Fig. 11 shows the pressure rise curve under the 10% inlet distortion and 100% design working speed. This time the SPS casing treatment can make the stall margin loss decrease from 7.09% to 0.51%, also shows an excellent ability in enhancing the stall margin.

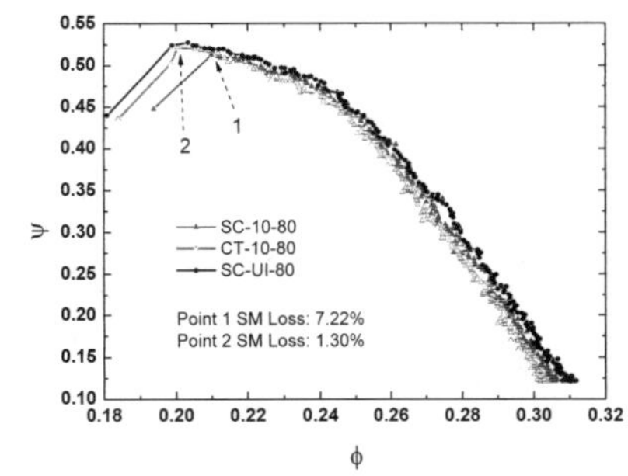

Fig. 10 The pressure rise curve under the 10% distortion, 80% design working speed

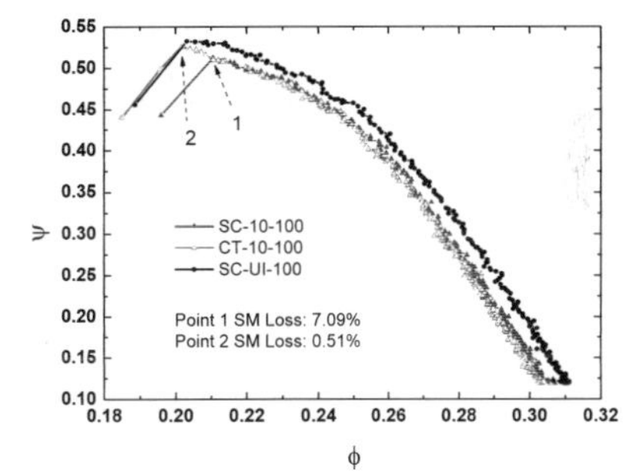

Fig. 11 The pressure rise curve under the 10% distortion, 100% design working speed

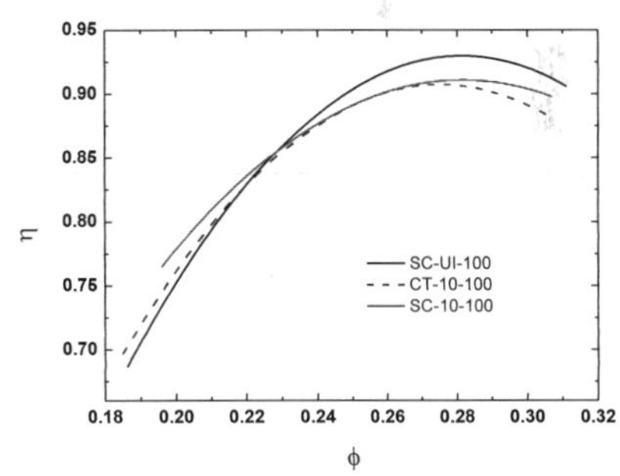

Fig. 12 Efficiency curves with 10% distortion under the rated speed

The efficiency curves with 10% inlet distortion under the 100% design working speed is shown in Fig. 12. The inlet distortion causes a big loss on the efficiency, and using thc SPS casing treatment also causes a little additional loss on efficiency. But in this figure, the efficiency loss caused by this SPS casing treatment is much smaller than that caused by the inlet distortion. In fact, using the static data, the additional efficiency loss caused by SPS casing treatment is less than 1%, which is almost within the error range.

If further increase the distortion plate inserting depth and the covering area ratio, the compressor's pressure rise characteristic is shown in Fig. 13. This figure shows different distortion degrees with two working speeds, in these cases, the inlet condition becomes worse and worse with the increasing of the distortion plate inserting depth and the covering area ratio, and the stall margin also becomes lower and lower. In spite of this condition, the SPS casing treatment still has the ability to enhance the stall margin. But the ability of SPS casing treatment may become limited when the distortion degree is big such as the results under the 30% inlet distortion. In that condition, the distorted inlet flow can cause about 25% stall margin loss, and the SPS casing treatment can only make up about 3%. While the optimistic fact is that there are nearly no aircraft engines having the opportunities to encounter such unfortunate and bad inlet conditions. So stall margin enhancement ability of SPS casing treatment still deserves a positive affirmation.

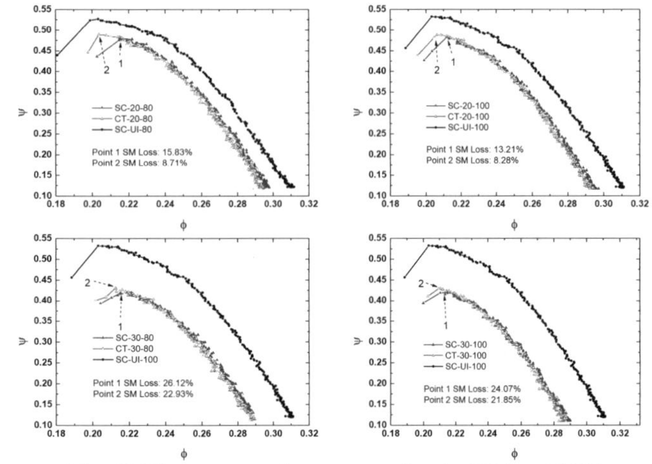

Fig. 13 The pressure rise curve under the 20% and 30% distortion, 80% and 100% design working speed

Fig. 14 shows the efficiency curves with three different distortion degrees under the rated speed. Compared with Fig. 12, it is found that no matter what kind of inlet distortion speed is, the additional efficiency change caused by SPS casing treatment is always so small as to be neglected.

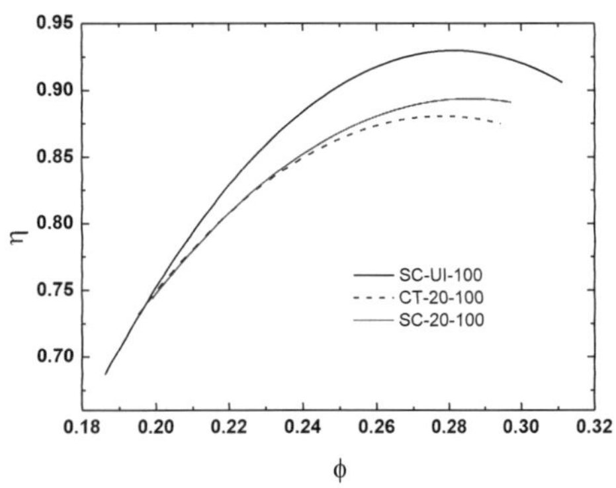

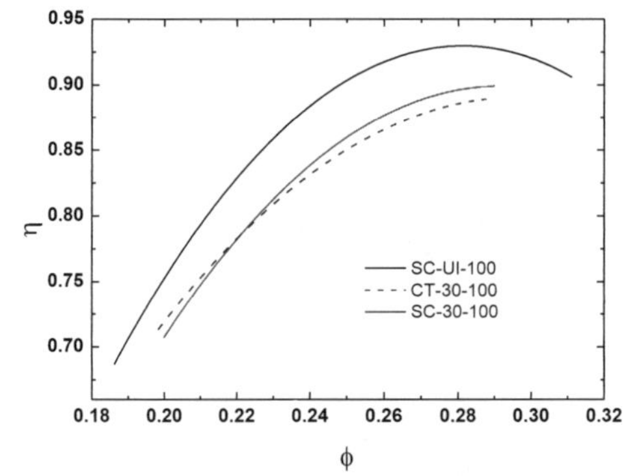

Fig. 14 Efficiency curves with 20% and 30% distortion under the rated speed

Experimental Results of Dynamic Pre-stall Signals

The experimental results shown previous and published recently (e.g., [27]) indicate that the mechanism of the stall margin improvement with such casing treatment is associated with delaying the nonlinear development of the stall precursor waves. And it is theoretically shown that the SPS casing treatment can not only delay the nonlinear development of the stall precursor waves but also can weaken the unsteady flow disturbance all the time (e.g., [22]). Its effectiveness under rotating inlet distortion has been checked in Ref.27, and in order to check the effectiveness of this mechanism under the circumferential pressure inlet distortion, a technical way to measure and analyze the pre-stall behaviour in the compressor system is carried out similarly. And the wall static pressure measured by eight high-frequency response pressure sensors (Kulite XT-190M) (3 in Fig.2)can be used to calculate and analyze.

Fig. 15 shows a typical compressor characteristic curve, and the method about how to plot a curve of our test rig has been clarified in the section of "Enhancement of SM by SPS Casing Treatment". And in order to observe the pre-stall behaviour, data are taken from the eight static pressure sensors while the compressor operating point moves from A to B very slowly. A is a near stall point, B is the stall point without using SPS casing treatment, and the entire time cost is about 2 seconds. So the flow range is equivalent for both CT and SC situations. The period of a rotor revolution is used as the unit of time, since this is a characteristic time scale for the phenomena.

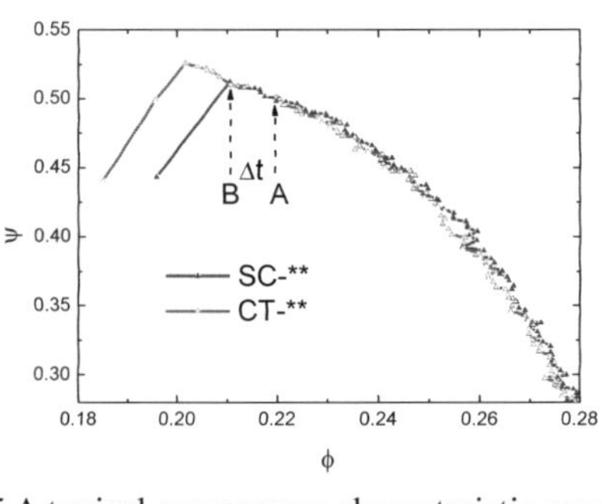

Fig. 15 A typical compressor characteristic curve shows the process of pre-stall observation

Fig. 16 and Fig. 17 show the PSD of the normalized frequency at 80% and 100% design working speed during a period of pre-stall, without inlet distortion, the normalized frequency is the ratio over shaft frequency. Because of the 3-D PSD figure can't show out the differences with or without SPS casing treatment obviously, so the time axis is cut down and the curves of different moments are plotted in one single moment.

Fig. 18 shows the PSD of the normalized frequency at 100% design working speed during a period of pre-stall within 20% inlet distortion. Compared with the cases without inlet distortion shown in Fig. 16-Fig. 17, the PSD amplitudes of disturbances in the range of 0.2 to 0.8 normalized frequency are much bigger which can more rapidly lead to the happening of compressor stall inception. According to these PSD results it is obvious that the SPS casing treatment can weaken the disturbances in broadband, and can also decrease the probability of the amplification of disturbance in flow.

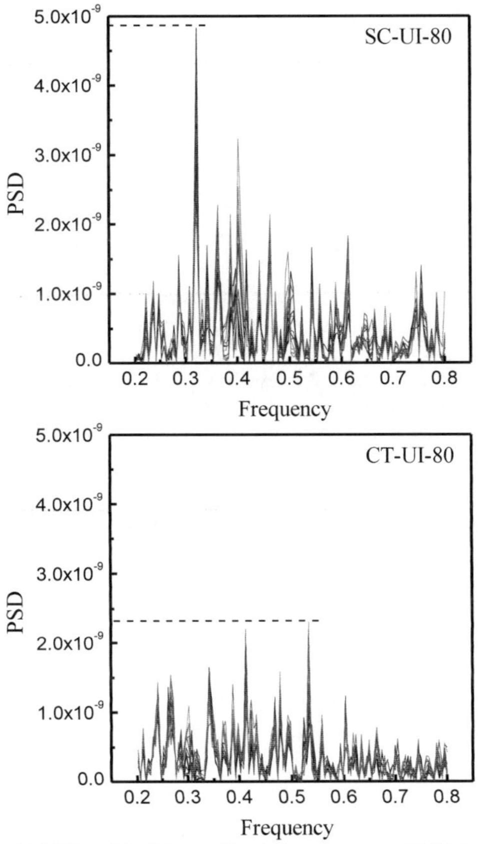

Fig. 16 PSD of the Normalized Frequency at 80% design working speed during a period of pre-stall, without inlet distortion

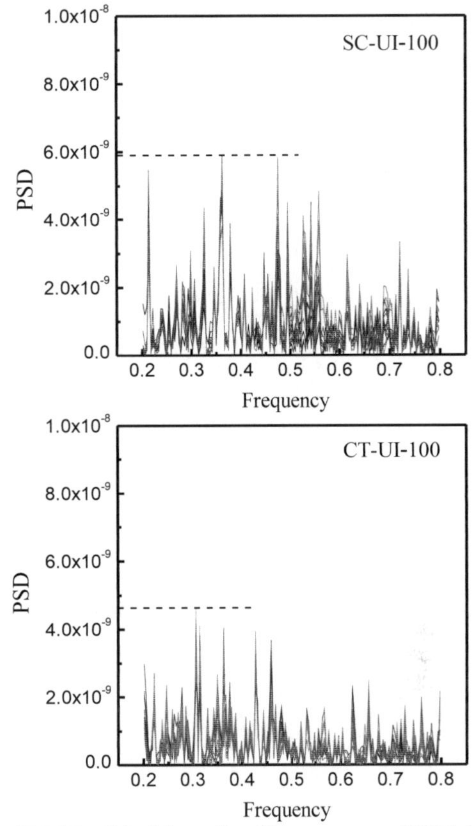

Fig. 17 PSD of the Normalized Frequency at 100% design working speed during a period of pre-stall, without inlet distortion

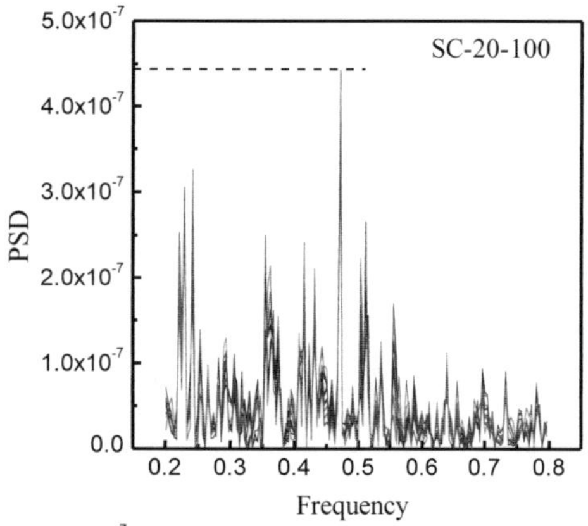

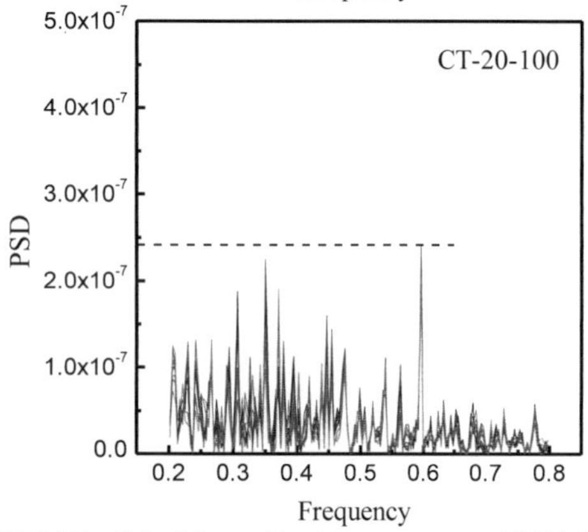

Fig. 18 PSD of the Normalized Frequency at 100% design working speed during a period of pre-stall, with 20% inlet distortion

CONCLUSIONS

This work presents various experimental data concerning the combined effect of the casing treatment under the restriction of circumferential pressure inlet distortion. The main results of circumferential pressure distortion on compressor stability with SPS casing treatment can be summarized as follows.

It is found that the circumferential pressure distortion completely changes the compressor performance. The stable operating range of TA36 is narrowed obviously and the stall margin is significantly reduced. And then the application of the SPS casing treatments can improve the stall margin without occurring obvious pressure-rise characteristics change. The inlet distortion causes a big loss on the efficiency, and using the SPS casing treatment also causes a little additional loss on efficiency under large mass flow rate conditions. But the efficiency loss caused by this SPS casing treatment is much smaller than that caused by the inlet distortion. In fact, using the static data, the additional efficiency loss caused by SPS casing treatment is less than 1%.

From the PSD of the normalized frequency results, the mechanism of the stall margin improvement with SPS casing treatment is probably to weaken the unsteady flow disturbances in broadband and delay the nonlinear development of the stall precursor waves.

ACKNOWLEDGMENTS

This work has been supported by National Natural Science Foundation of China (Nos. 51236001 and 51106154), and National Basic Research Program of China (No. 2012CB720201) and Aeronautical Science Foundation of China (No. 2014ZB51018) and Fundamental Research Funds for the Central Universities.

REFERENCES

[1] Cousins W T. History, philosophy, physics, and future directions of aircraft propulsion system/inlet integration. ASME Turbo Expo 2004: Power for Land, Sea, and Air. American Society of Mechanical Engineers, 2004: 305-320.

[2] C. M. van Schalkwyk, J. D. Paduano, E. M. Greitzer and A. H. Epstein, 1998, "Active Stabilization of Axial Compressors With Circumferential Inlet Distortion," ASME Journal of Turbomachinery, 120(3), pp. 431-439. DOI:10.1115/1.2841734

[3] Spakovszky, Z. S., Weigl, H. J., Paduano, J. D., Van Schalkwyk, C. M., Suder, K. L., and Bright, M. M., 1999, "Rotating Stall Control in a High-Speed Stage With Inlet Distortion, Parts II-Circumferential Distortion," ASME Journal of Turbomachinery, 121(3), pp. 508-522. DOI: 10.1115/1.2841345; DOI: 10.1115/1.2841346

[4] Hah, C., Rabe, D. C., Sullivan, T. J., Wadia, A.R., 1998, "Effects of Inlet Distortion on the Flow Field in a Transonic Compressor Rotor," ASME Journal of Turbomachinery, 120(2), pp. 233-246.

[5] Gunn E J, Tooze S E, Hall C A, et al. An experimental study of loss sources in a fan operating with continuous inlet stagnation pressure distortion. Journal of Turbomachinery, 2013, 135(5): 051002.

[6] Mistry C, Pradeep A M. Influence of circumferential inflow distortion on the performance of a low speed, high aspect ratio contra rotating axial fan. Journal of Turbomachinery, 2014, 136(7): 071009.

[7] Lesser, A., and Niehuis, R., 2014, "Transonic Axial Compressors with Total Pressure Inlet Flow Field Distortions," ASME Paper No. GT2014-26627.

[8] Lucas, J. R., O'Brien, W.F., Ferrar, A. M., 2014, "Effect of BLI–Type Inlet Distortion on Turbofan Engine Performance," ASME Paper No. GT2014-26666.

[9] Pearson H, McKenzie A.B. Wakes in axial compressors. Journal of the Royal Aeronautical Society, 1959, 63(583): 415-416.

[10] Mazzawy R S. Multiple segment parallel compressor model for circumferential flow distortion. Journal of Engineering for Gas Turbines and Power, 1977, 99(2): 288-296.

[11] Hynes T P, Greitzer E M. A method for assessing effects of circumferential flow distortion on compressor stability. Journal of Turbomachinery, 1987, 109(3): 371-379.

[12] Yao J, Gorrell S E, Wadia A R. High-Fidelity Numerical Analysis of Per-Rev-Type Inlet Distortion Transfer in Multistage Fans—Part I: Simulations With Selected Blade Rows. Journal of Turbomachinery, 2010, 132(4): 041014.

[13] Yao J, Gorrell S E, Wadia A R. High-Fidelity Numerical Analysis of Per-Rev-Type Inlet Distortion Transfer in Multistage Fans—Part II: Entire Component Simulation and Investigation. Journal of Turbomachinery, 2010, 132(4): 041015.

[14] Koch, C. C., 1970, "Experimental Evaluation of Outer Case Blowing or Bleeding of a Single-Stage Axial Flow Compressor," No. NASA CR-54592.

[15] Takata, H., and Tsukuda, Y., 1977, "Stall Margin Improvement by Casing Treatment—Its Mechanism and Effectiveness," ASME Journal of Engineering for Gas Turbines and Power, 99(1), pp. 121-133.

[16] Muller, M.W., Schier, H. P., Voges, M., and Hah, C., 2011, "Investigation of Passage Flow Features in a Transonic Compressor Rotor with Casing Treatments," ASME Paper No. GT2011-45364.

[17] Kroeckel, T., Hiller, S. J. and Jeschke, P., 2011, "Application of a Multistage Casing Treatment in a High Speed Axial Compressor Test Rig," ASME Paper No. GT2011-46315.

[18] Houghton T, Day I. Stability enhancement by casing grooves: The importance of stall inception mechanism and solidity. Journal of Turbomachinery, 2012, 134(2): 021003.

[19] Madden, D. S., West, M. A., 2005, "Effects of Inlet Distortion on the Stability of an Advanced Military Swept Fan Stage with Casing Treatment," ASME Paper No. GT-2005-68693.

[20] Sun, X., Sun, D., Liu, X., Yu, W. and Wang, X., 2014, "Theory of Compressor Stability Enhancement Using Novel Casing Treatment, Part I: Methodology," AIAA Journal of Propulsion and Power, 30(5), pp. 1224-1235.

[21] Sun, D., Liu, X., Jin, D., Gui, X. and Sun, X., 2014, "Theory of Compressor Stability Enhancement Using Novel Casing Treatment, Part II: Experiment," AIAA Journal of Propulsion and Power, 30(5), pp.1236-1247.

[22]Dakun Sun, Xiaohua Liu, and Xiaofeng Sun, "An Evaluation Approach for the Stall Margin Enhancement with SPS Casing Treatment," ASME Journal of Fluids Engineering, Vol. 137, 2015.

[23] Sun, X., 1996, "On the Relation between the Inception of Rotating Stall and Casing Treatment," 32th AIAA/ASME/SAE/ASEE Joint Propulsion Conference, No. AIAA-96-2579.

[24] Sun, X., Sun, D., and Yu, W., 2011, "Model to Predict Stall Inception of Transonic Axial Flow Fan/Compressors," Chinese Journal of Aeronautics, 24(6), pp. 687-700.

[25] Rusak, Z., and Morris, W. J., 2011, "Stall Onset on Airfoils at Moderately High Reynolds Number Flows," ASME J. Fluids Eng., 133(11), p. 111104.

[26] Morris, W. J., and Rusak, Z., 2013, "Stall Onset on Aerofoils at Low to Moderately High Reynolds Number Flows," Journal of Fluid Mechanics, 733, pp. 439–472.

[27] Sun, X., Liu, X., Hou, R. and Sun, D., 2013, "A General Theory of Flow Instability Inception in Turbomachinery," AIAA Journal, 51(7), pp. 1675-1687.

[28] Liu, X., Sun, D., Sun, X., 2014, "Basic Studies of Flow-Instability Inception in Axial Compressors Using Eigenvalue Method," ASME J. Fluids Eng., 136(3), p. 031102-1-9.

[29] Dong, X., Sun, D., Li F., Jin D., Gui X., and Sun, X., 2015, "Effects of Rotating Inlet Distortion on Compressor Stability with SPS Casing Treatment," ASME J. Fluids Eng.

Proceedings of the 14th International Symposium of Unsteady Aerodynamics, Aeroacoustics & Aeroelasticity of Turbomachines
ISUAAAT 14
8-11 September 2014, Stockholm, Sweden

I14-S3-5

STALL INCEPTION IN A TRANSONIC COMPRESSOR WITH INFLOW DISTORTION

Fabian Wartzek
Felix Holzinger
Christoph Brandstetter
Maximilian Jüngst
Heinz-Peter Schiffer
Technische Universität Darmstadt
Institute of Gas Turbines and Aerospace Propulsion
Darmstadt, Germany

Email: wartzek@glr.tu-darmstadt.de

ABSTRACT

Measurements of inflow distortions were performed in a transonic compressor stage. The inflow distortion was representative of separation in an engine inlet. Distortion generators were designed and measured to examine the impact of the size of the separation. The stage was heavily instrumented using steady and unsteady techniques. Steady measurements were performed to visualise the effect in the performance map. Unsteady measurements were taken to assess the stage's behaviour close to its stability limit and at stall onset. The flow was heavily disturbed before stall. Effects appeared that can usually only be seen during stall. The blades suffer from increased deflections due to the synchronous forcing resulting from the changing aerodynamic load around the circumference. The blade deflection values increased by a factor of two. Due to the highly disturbed flow before stall, the entry into stall is not as severe as for the reference case with no distortion. Increased vibration amplitudes are visible when the stage stalls. The behaviour of the stage at stall onset differs around the circumference. If the distortion is large enough the rotor partially stalls. A distinct stall cell over several passages does not appear before the whole stage stalls.

The radial extent of the distortion is more important than circumferential one. If the distortion is big enough an additional loss appears. This leads to highly inhomogeneous outflow from the stage, not just around the entire circumference but along the whole radius as well.

NOMENCLATURE

DTC	Darmstadt Transonic Compressor
N_{red}	ISA corrected rotor speed
ISA	International standard atmosphere
N_{mech}	Mechanical speed of the shaft
$\dot{m}_{red}$	ISA corrected mass flow
T_t	Total entry temperature in settling chamber
SC	Smooth casing - reference case without distortion
DG	Distortion Generator
MP20	Measurement position for the analysis at 20% axial blade chord
B10	120° distortion with 10% blade height
B10_60	60° distortion with 10% blade height
B10_90	90° distortion with 10% blade height
B5	120° distortion with 5% blade height
EO	Engine order - harmonics of rotational speed
$\pi_{t,30}$	Total pressure ratio of the whole stage
$\eta_{is,30}$	Isentropic efficiency of the whole stage
PE	Peak efficiency mass flow of reference
NS	Near stall operating point for each configuration
$p_{s,rel.circ}$	Static pressure relative to circumferential mean
LE	Leading edge
TE	Trailing edge
$\sigma_{pt,rel}$	Relative standard deviation of total pressure
BPF	Blade passing frequency

INTRODUCTION

Continuously rising ecological and economic requirements pressure the aircraft industry to further improve aircraft performance. During the design process, simplifications often need to be made to keep the process efficient. To enable safe operation of all parts, safety margins are defined, which restrict operation. One research goal is to enhance the accuracy of predictions to enable minimization of these margins.

For example, steeper angles of attack could reduce the perceived noise level in areas surrounding airports. The ability to withstand stronger crosswinds would reduce turn-around times and increase safety. Engine inlet lips need a certain thickness to guarantee a wide range without separation. Thinner lips would lead to smaller nacelles, saving weight. Additionally, this would enable more flexible placement of the engines, for example, more complete integration into the fuselage.

Inflow distortions indicate deviation from a homogeneous, steady and constant flow. For an engine, typically three types of distortion appear: pressure, swirl and temperature variations. Temperature variations usually appear only in military applications, i.e. ingested hot gases from weapon firing; pressure and swirl variations are the most common forms in civil aircraft engines.

Swirl distortions can appear in different forms (see, for example, [19] for their classifications). A bulk swirl is a constant angle deviation around the circumference. This would be produced, for example, by variable inlet guide vanes. The ingestion of a vortex can lead to local swirl variation. These vortices can appear if the engine operates close to the ground (see [10]) or due to installation effects (for example, a serpentine inlet duct, as shown in [3]).

Pressure variations can be classified further if distribution is taken into account. The most basic forms are radial (constant around the circumference) or circumferential (constant along the radius) variations. These forms can have varying strength or, for circumferential, a varying number or size around the circumference.

Unsteady effects have mostly been neglected in research. Inflow distortions in real applications originate from turbulent flow structures and are therefore intrinsic unsteady.

In the late 1960s, the investigation of inflow distortions and the reaction of the compression system became the focus of research. It was common practice to generate a steady and homogeneous total pressure drop as it was easy to create and measure. Using these simple distortion forms it was possible to derive analytical models, i.e. the parallel compressor model see [9], enabling prediction of compressor behaviour. From these simplified analyses the interaction between compressor and distortion could be explained [8]. Based on experimental results with generic distortion patterns generated by screens in the inflow, tolerable limits for distortions were defined. As real distortion patterns differ from the ones used to define these limits, additional safety margins are needed to ensure a safe operation of the engine under all circumstances.

An overview of the classification of inlet distortions, and their description and measurement is given by the SAE reports ARP1420 [17] and AIR1419 [16].

The intention of this work was to produce a more realistic distortion pattern and to investigate its interaction with a transonic compressor stage. Systematic variation of the distortion's topology assists understanding of the impact on rotor flow and the stage's behaviour. The difference between more realistic distortions and the available results from generic patterns can be analysed. Data from real inlet separations with an engine downstream are not publicly available. The testing of a real engine-inlet combination is expensive and is usually confidential. However, data on separations in inlets as through-flow models and data on engine testing with generic patterns are available. It is difficult to model realistic inlet separations, therefore the data provided can be used as generic test cases and the geometries used should cover a wide range of realistic patterns.

The accurate numerical prediction of distortion emergence and its interaction with the fan and compression system is challenging. Due to the lack of circumferential periodicity, full 360° setups are necessary. The effects are unsteady and turbulence models commonly used in industry suffer from difficulties in calculating the interaction between the transonic flow in the rotor and the highly turbulent flow created by the distortion.

Combining and improving the available approaches to enable the simulation of a full aircraft in realistic atmospheric conditions is the aim of the DFG[1] research group "FOR1066 - Simulation of Wing and Nacelle Stall". For an overview of their work see [13] and [12]. The work presented in this paper was part of this research project.

In a previous experiment using a different distortion generator, it was found that the rotor should be influenced at the tip (see [20]). This would be representative of an inlet separation. Therefore, distortion bodies were designed that mimic an inlet separation with different radial and circumferential extents. Investigations with a longitudinal vortex in the same stage were presented in [22]. The focus of the presented data is the flow of the compressor close to its stability limit and at stall onset when the rotor flow becomes unstable.

EXPERIMENTAL SETUP

The Darmstadt Transonic Compressor (DTC) was used for the investigations presented in this paper. The test rig is operated by the Institute of Gas Turbines and Aerospace Propulsion at the Technische Universität Darmstadt.

Test Rig

Figure 1 shows a sketch of the test rig. The rig is operated in an open-loop configuration. Air is sucked in from the atmosphere (in the upper left in the sketch). An inlet throttle enables the systematic variation of the inlet total pressure. Downstream of the throttle, the air is guided into the settling chamber. Several screens and honeycomb elements ensure a homogeneous inflow. A bell-mouth reduces the diameter to the outer diameter of the test section. The air passes the compressor stage, and down-

[1] German research foundation

stream of the stage, a radial diffuser. At the end of the diffuser an exit throttle enables the back-pressuring of the stage. A pneumatic valve is installed to quickly reduce the pressure to recover from unstable conditions. Further downstream, the air is guided back out to the atmosphere again.

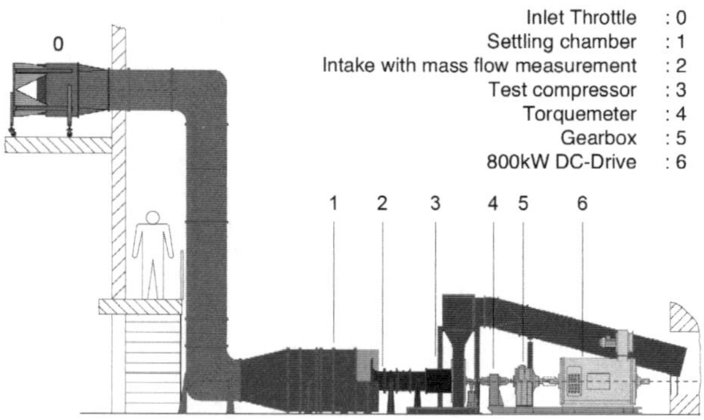

Figure 1: Sketch of the Darmstadt Transonic Compressor

The compressor is driven by an 800 kW DC drive. A gearbox alters the speed of the drive to a maximum rotational speed of 21,000 rpm. The shaft speed of the compressor and the torque are measured by a torquemeter. Further details of the test rig can be found in [18].

Table 1 summarizes some of the key technichal parameters of the test rig and the investigated stage. The stage used for the investigations was the "Rotor 1 - Stator 2" configuration. Previous results of this configuration without inflow distortions are shown in [11].

max. power	800 kW
max. torque	350 Nm
design mass flow	16 kg/s
design speed	20,000 rpm
outer diameter	0.38 m
hub-to-tip-ratio	0.51
rotor blades	16
stator blades	29

Table 1: Key parameters of the Darmstadt Transonic Compressor

Measurement Techniques

Measuring a circumferentially inhomogeneous flow requires a significant increase in effort. The missing periodicity of the flow makes measuring the whole circumference inevitable. The test rig was originally designed to take advantage of the periodicity, keeping the instrumentation fixed and traversing the inlet and outlet guide vanes. Therefore, it was necessary to traverse the inflow distortion 360°. This led to a drastic increase in measurement time. The stage was heavily instrumented to still enable detailed insight into the flow. The measurement matrix was reduced from 12 operating points to 2. The first point is the mass flow for maximum efficiency of the reference case. This operating point is representative of the engine in cruise conditions, where it should operate the most. The second operating point is the last stable operating point of each configuration individually. Therefore, the back-pressure was increased until stable flow broke down. The back-pressure was then slightly lowered so that stable operation of the compressor was possible, for the time needed to perform measurements.

The speed of the rotor was set to a fixed aerodynamic speed related to ISA conditions (288.15 K and 101325 Pa), according to Equation (1). The mass flow was related to ISA conditions using Equation (2). This enables the comparison of measurements taken under different ambient conditions.

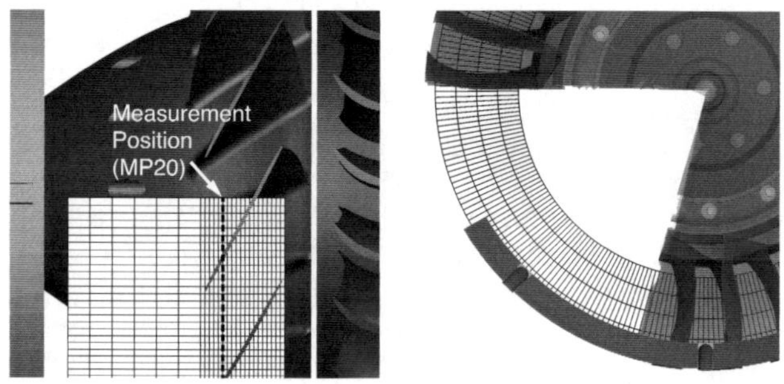

(a) Measurement grid for the unsteady wall pressure transducers

(b) Measurement grid for the unsteady exit pressure transducers

Figure 2: Sketch of the measurement grids (B10 is shown exemplary)

The focus of the present investigation was on the transient behaviour of the compressor stage and the unsteady phenomena so only a small part of the installed instrumentation was used. Figure 3a shows the measurement positions used in the test section. A detailed analysis of the steady flow phenomena at different operating points can be found in [21]. The measurement grids for the unsteady instrumentation are shown in Figure 2. To get an impression of the inflow distortion and the rotor tip flow with its characteristic structures (as the tip gap vortex or the shock), unsteady pressure transducers were flush mounted into the casing wall. KULITE XCS-062-5D transducers were used and sampled for 4 seconds with 500 kHz. This resulted in around 1300 measured revolutions. This amount is sufficient to get constant statistic values, such as mean value and standard deviation. The resulting resolution per passage was around 90 points. To reconstruct the whole 360° flow field, different relative positions between transducers and distortion were measured. 290 circumferential positions were chosen to get an accurate view of the structures. The resulting measurement grid and axial positions of the transducers, are shown in Figure 2a. Marked in this figure is the pressure transducer that will be used for the analysis (reffered to as MP20). Unsteady data on the stage's outflow were measured using a rake equipped with 5 KULITE XCS-062-25A transducers at distinct radial positions. The radial positions and the resulting measurement grid for the unsteady rake can be seen in Figure 2b.

By using a photoelectric barrier connected to the drive shaft, an accurate one-per-rev signal allows the data to be split into segments for each revolution. If the data for all these revolutions is averaged, the ensemble-average can be calculated using Equation (3). The relative standard deviation can be derived using Equation (4) by calculating the standard deviation over the revolutions and dividing it by the ensemble-averaged pressure.

This instrumentation was used for the transient measurements. The distortion was fixed at one position and the compressor continuously throttled. The data from all measurement techniques were recorded. Stall is potentially harmful to the rotor. Therefore, the process was repeated for only distinct positions of the distortion and not for all 290, as in the steady measurements. An accurate reconstruction of the 360° flow field is not possible.

$$N_{red} = N_{mech} \sqrt{\frac{T_{ISA}}{T_t}} \tag{1}$$

$$\dot{m}_{red} = \dot{m} \frac{p_{ISA}}{p_t} \sqrt{\frac{T_t}{T_{ISA}}} \tag{2}$$

$$\bar{p} = \frac{1}{N} \sum_{i=1}^{N} p_i \qquad N = \text{Number of revolutions} \tag{3}$$

$$\sigma_{rel} = \frac{1}{\bar{p}} \sigma = \frac{1}{\bar{p}} \sqrt{\frac{1}{N-1} \sum_{i=1}^{N} (p_i - \bar{p})^2} \tag{4}$$

A capacitive tip-timing system from FOGALE NANOTECH was used to monitor the blade's deflection and the tip gap. Six probes were positioned at three axial positions and various circumferential positions to enable measurement of all relevant modes. The sensors measure the blade tip gap and the time of arrival for all blades. By calculating time differences between the blades in conjunction with the speed, the deflection can be derived.

Stall onset is a random process, that depends on configuration and slightly differs every time repeated. This demands a possibility to synchronise not only the measurement techniques but also the events. The time between start of throttling and stall onset is always different. Hence, direct correlation of the data is difficult. To enable a comparison of different stall runs and measurement techniques, an indicator is needed that belongs to similar flow conditions. The speed of the DC-drive is kept constant by an automatic controller. When the rotor stalls, the change in load on the shaft happens so quickly that the speed increases for a short time. Therefore the signal of the photoelectric barrier was used. The shaft speed could be calculated from the time difference between two trigger events. The point where the first derivative of the speed is positive and the second derivative is at its maximum, is considered as Revolution 0. Comparison of different measurement techniques and stall events shows that this enables a repeatable and accurate determination of stall onset.

Distortion Generators

Four different distortion generators (DG) were investigated. The sizes of the DG are shown in Table 2. B10 is considered as baseline. For B10_60 the circumferential angle and for B5 the height was halved, respectively. A DG with an angle of 90° and 10% blade height suffered as intermediate size. The axial position of the downstream side, that faces towards the rotor, was kept at the same position. The upstream side was bevelled with an angle of 30° to suppress any additional separation on this side and an even more complex flow field.

	B10	B10_60	B10_90	B5
height [% blade span]	10	10	10	5
angle [°]	120	60	90	120
axial pos [axial chord]	2	2	2	2

Table 2: Dimensions of the used distortion generators

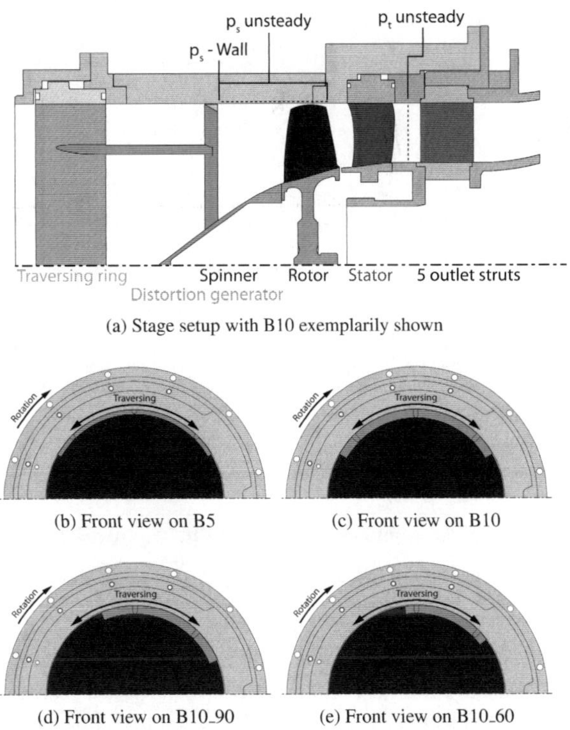

(a) Stage setup with B10 exemplarily shown

(b) Front view on B5

(c) Front view on B10

(d) Front view on B10_90

(e) Front view on B10_60

Figure 3: Sketch of the DG installed in the test section

Figure 3 shows the DG installed in the test section. Figure 3a shows the DG attached to a traversing ring upstream. This ring enables a 360° movement of the DG and therefore measuring of all relative positions between the DG and the measurement positions. To enable the integration of instrumentation into the casing close to the DG, they are not mounted directly on the traversing ring but by support struts. The rotation during all operating points, without touching the casing, is guaranteed by a 1 mm gap. The position of the DG is controlled by a magnetic displacement sensor directly attached to the traversing ring. This enables to eliminate the positioning error of the gear and the whole drive train.

RESULTS

The results of the measurements are analysed in this section. Starting with the last stable operating point, the data of the DG measurements are compared to the reference case. The data of the transient runs are then presented. To shorten the paper, the B10_90 data are not shown if no additional information is provided.

Global Behaviour

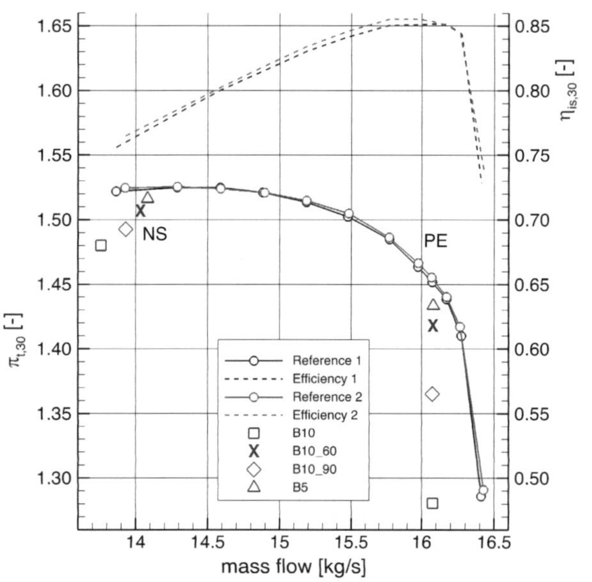

Figure 4: Performance map of the stage

The performance of the compressor stage gives a initial impression of the impact that the distortion generator had on the flow. For B10, a detailed analysis of the steady behaviour can be found in [21]. Figure 4 shows the performance map. For the reference case (SC), two measurements are shown to give an impression of the repeatability of the test rig. As explained previously, measurements with the DG are much more time consuming so only two points were measured: first, a constant mass flow, representing the peak efficiency point (PE) for the undistorted case; and second, each configuration was throttled to its last stable operating point (NS).

It is clear, that the loss in total pressure ratio at PE is significant. Comparing B5 and B10, which have the same circumference, B5 has only a slight impact on the overall performance. The same is visible at the stability limit. Even though B5 causes a slight loss in stability margin, the operating point for B10 lies completely off the reference case. Even though the NS mass flow for B10 is even lower than for the reference case, B10 causes the stability margin to decrease. The whole characteristic is shifted to lower total pressure ratios, but only slightly lower mass flows.

The other DG demonstrates more or less consistent behaviour. At PE, with the mass flow fixed, the total pressure ratio decreases with increasing angle of the DG. At NS, the shift of the operating point goes towards lower total pressure ratios and mass flows with increasing angle. However, the differences are clearly not linear. For example, the difference in total pressure ratio at PE for B10 is not twice the difference of B10_60. This is in contrast to available literature, which states that the influence of the DG stays more or less constant for a distinct angle (see, for example, [14]). These statements were related to radially constant distortions. They are no longer applicable if a big radial component exists.

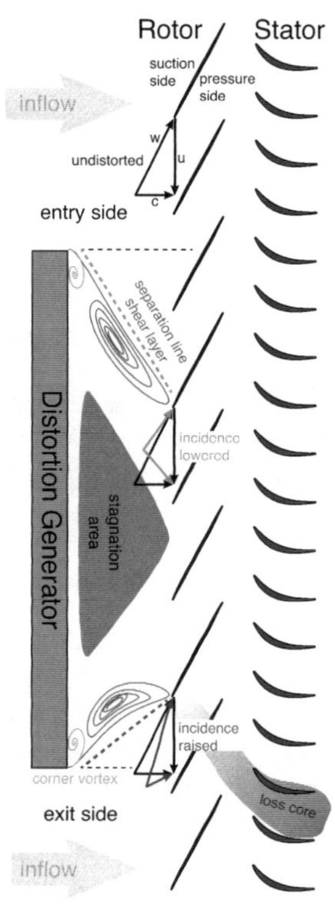

Figure 5: Flow model, taken from [21]

The values shown here are calculated by dividing the area-averaged total exit pressure by the inlet total pressure from upstream of the DG. This would be representative of an engine where an inflow distortion suddenly appears, for example caused by an inlet separation.

Figure 5 shows a flow model that was derived from the analysis of steady measurement data. Details on the development of the model are given in [21]. Different flow regimes can be identified. Close to the edges of the DG two corner vortices exist. The separation is limited by a separation line shear layer. Due to the high velocity gradients betwenn main flow and separation here high turbulence values exist. In the middle of the DG, a stagnation area appears. Here the turbulence is slightly reduced and a more or less steady separated flow topology forms. In direction of rotation of the rotor (from top to bottom in the figure) the first edge, where the blade enters the separation is referred to as "entry side". Here the incidence and the blade loading are reduced. The opposing end of the DG is referred to as "exit side". Here the incidence and the blade loading are increased. When the compressor is already operating at throttled conditions this increased loading can lead to the formation of an additional loss core. This depends on the DG geometry as will be shown in this paper.

Flow Field

An overview of the distortion can be derived from the static wall pressure taps. As the flow accelerates towards the rotor the pressure decreases due to the spinner (see Figure 3a). Across the rotor, the pressure is raised.

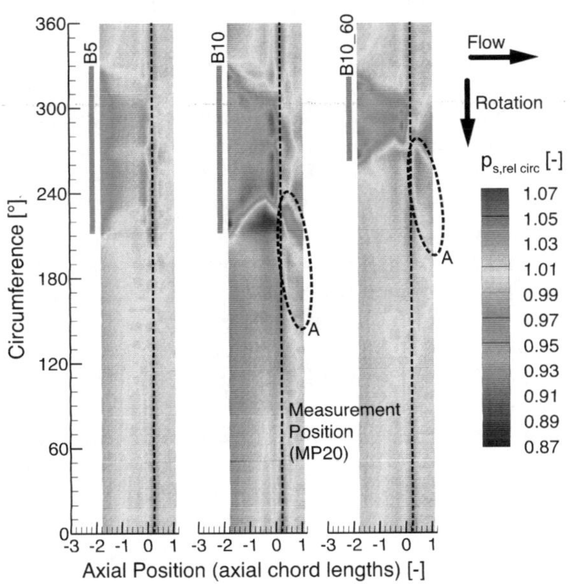

Figure 6: Steady static wall pressure for B5, B10, B10_60 (from left to right)

This axial gradient dominates the pressure field making the identification of circumferential variations difficult. By dividing the pressure values at each axial position by its circumferential mean, the axial pressure gradient is neglected. The result is shown in Figure 6. The rotor leading edge (LE) is located at axial position 0 and the trailing edge (TE) at position 1. The position of the KULITE transducer for the later analyses is also marked. This transducer is located at around 20% axial blade chord and is referred to as MP20 (see also Figure 2a).

The impact of the three distortion generators is clearly visible in the significant pressure drop. For B5, on the left, this drop is comparatively small. At the rotor LE, from around half of the DG up to the exit side the pressure is slightly increased. This variation is attenuated by the rotor, resulting in a more or less homogeneous flow at the TE.

When looking at B10, in the middle of Figure 6 this picture changes. The circumferential pressure variation caused by the separation is increased. At the exit side of the DG (A) a loss appears that is transported in the direction of rotor rotation. This loss appears only when throttling the compressor towards the stability limit. For further details, see [21]. This loss is present for B10_60 (A) as well, though with reduced intensity.

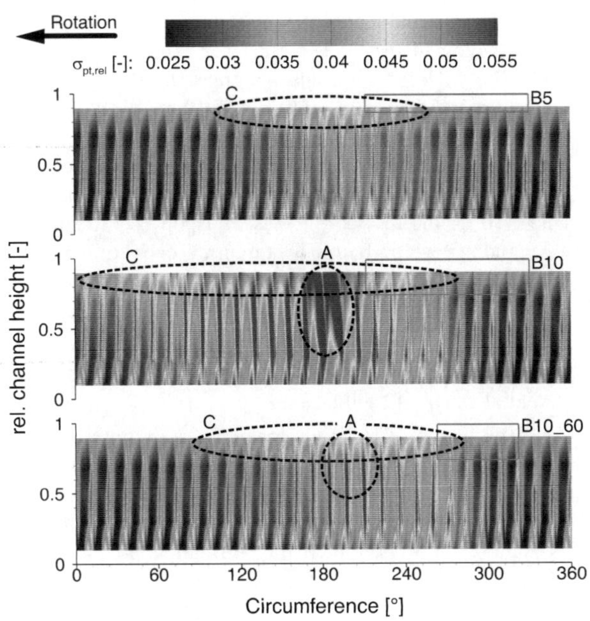

Figure 7: Relative standard deviation downstream stator for B5, B10, B10_60 (from top to bottom)

The reaction of the whole stage is investigated using the unsteady exit rake. The relative standard deviation (Equation (4)) gives an impression of the flow unsteadiness. Even though it is not directly connected to the turbulence, high values imply highly unsteady flow regimes that are usually caused by high turbulence. Determination of the total pressure loss is not possible

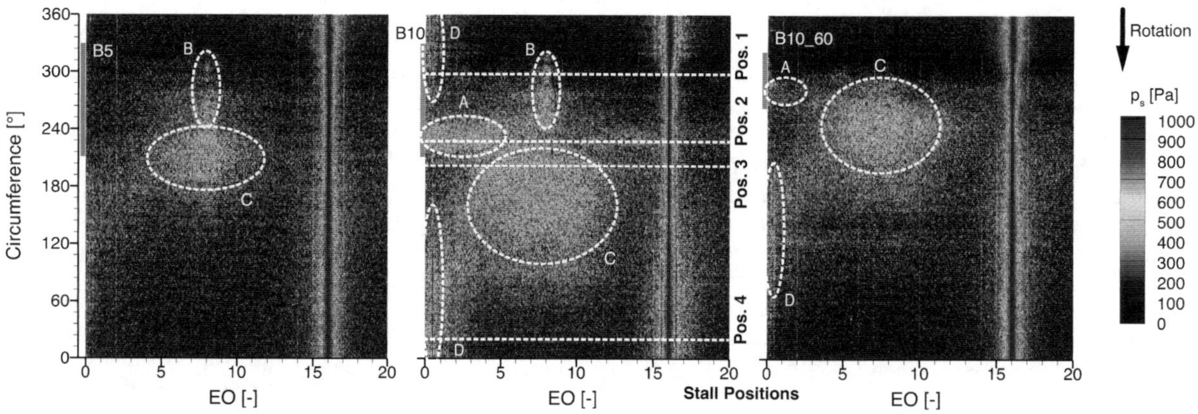

Figure 8: Frequency spectra around the circumference for B5, B10, B10_60 (from left to right)

for the same reasons. Figure 7 shows the results for the last stable operating point according to Figure 6.

As well as the differing circumferential impact of the distortions, the different radial distribution is also obvious. For all DG, increased values at the casing (C) are visible. This area is shifted in the direction of rotor rotation. The stator is designed with a bow to avoid hub corner separations. This bow results in high loading of the blade's midspan, especially on the suction side. This high loading leads to increased standard deviation values at around 50% channel height. B10 and B10_60 redistribute the flow, resulting in even greater unsteadiness. It seems that the flow on the suction side detaches. For B5 this effect is almost negligible. The stator wake is only slightly intensified. For B10 and B10_60 the loss core (A) is transported through the stator, even though it is shifted in the circumferential direction.

As already visible in the performance map, comparison of B5 and B10_60 shows that the impact of B10_60 affects not only a bigger part of the channel height, but also a bigger part of the circumference, especially when regarded relative to the circumferential extent of each DG.

The origin of the high unsteadiness is not apparent when looking at averaged pictures. One way to reveal more detail, is to look at the spectral content of the data. Therefore, at every circumferential position the data is transformed into the frequency domain using a fast fourier transformation (FFT). Dividing the frequency by the rotational speed gives the engine order (EO) as its harmonics. Consequently EO 16 represents the blade passing frequency (BPF) and EO 1 the rotational frequency.

As all unsteady pressure transducers cannot be shown at once, one was chosen as being representative, according to Figures 2a and 6. The resulting circumferential spectral map for this single axial position is shown in Figure 8. The DG are shown to give an impression of the position. The abscissa represents the frequency, not the axial position. Figure 6 shows that the separation becomes smaller from the DG to the rotor. This is due to the reduced pressure inside, which leads to bending of the streamlines. This bend leads to the separation becoming constricted. Therefore, effects that evolve at the edges of the DG are shifted in this picture.

The highest values appear at EO 16 and belong to the BPF. Increased fluctuations, with a frequency of around EO 8, are visible for all three DG. A lot of previous research shows that these fluctuations can be related to the tip leakage vortex (for example, [2], [4], [7] or [5]). The vortex increases in size and its trajectory moves towards the leading edge of the adjacent blade. When the vortex spills around the LE of the adjacent blade the pressure gradient in this passage is disturbed. A two-passage pattern forms, leading to frequency content around EO 8.

For B5, these values appear at the entry side (B) of the DG and subsequently increase, with the maximum close to the exit side (C). A spread in frequency can be seen, with slightly increased fluctuations between EO 4 and 12. The rest of the circumference appears undisturbed. This agrees with the result from the unsteady exit instrumentation in Figure 7.

Compared to the size of the DG, this effect appears before the rotor leaves the DG. This is due to the axial position of the shown transducer and the aforementioned constriction. This effect can be linked to the rotor leaving the separation. In Figure 7 a circumferential shift is visible further downstream, which is due to the deflection of rotor and stator. The maximum standard deviation visible belongs to this region as well.

With B10 installed, this effect is also visible. Additionally, two areas with increased values appear. At the exit side of the DG (A), fluctuations at lower frequencies (up to EO 5) can be seen over only a small part of the circumference. Looking at the trajectory of the low pressure trough, beginning at around 230° in Figure 6, this trough seems to be the source of these fluctuations. Following the trajectory further, the high values of unsteadiness in Figure 7 originate there as well. These values are not visible for B5 as the loss core does not form here. For B10 as well, when the rotor leaves the separation (C) an area of high fluctuations around EO 8 appears. This area is far more spread out for B10 than for B5.

The second area (D) with increased values can be seen at EO 0.5. This frequency usually belongs to the stall cell. Before stall, no fluctuations with this frequency can be observed for SC. For a small part of the circumference, from the middle of the DG at around 270° down to 170°, these fluctuations seem to disappear.

For B10_60 the patterns are more similar to B10, despite reduced amplitudes. On the exit side (A), higher fluctuations with lower frequencies are visible. This is in agreement with the weaker, but existing, loss core for this DG. When the rotor leaves the separation (C), a relatively wide spread frequency band around EO 8 with increased values appears. In contrast to B5 and B10, the small and relatively sharp frequency band with EO 8 at the middle of the DG (B, for B10 and B5 at around 280°) is not visible. It seems that the 60° angle of the DG is too small for these fluctuations to evolve. This agrees with earlier ones from [14], in which a critical angle of 60° was stated, but for a radial constant distortion. This value is not sufficient to classify this type of inflow distortion. This can also be seen when comparing B5 and B10_60. Whereas B5 with 120° shows nearly no influence, B10_60 has a small impact on stability.

Another difference are the fluctuations with EO 0.5 (D). They are visible only for a smaller part of the circumference, reaching from around 200° down to 60°. Compared to B10, the amplitude is reduced as well. Looking at these figures, it is hard to say where these fluctuations arise. A possible explanation could be that the rotor tip flow is heavily disturbed by the separation. When the rotor leaves the separation the loading reaches its maximum. This effect was explained in [21]. Another explanation for this overloading is given by [6], which takes into account the different time scales for convective and diffusive transports. If the rotor operates close to its stability limit, the blades are not able to withstand the high loading. Even though the compressor blades can endure higher loadings if they are unsteady, as in this case. This was shown by [15]. If the back pressure reaches a critical value, the tip gap vortex spills around the adjacent blade's LE. If this happens at a highly throttled state, it seems that the flow then breaks down. When the rotor reaches the entry side of the DG again, the load is reduced and the flow stabilises. As the flow around the circumference differs the formation of a discrete stall cell could not be observed. This stall cell first appears when the whole stage stalls.

Transient Data

Transient runs will now be analysed. The compressor was continuously throttled from stable operation until the flow collapsed and stall cells evolved. An overview of the behaviour of the compressor is given in Figure 9. The relative rotational speed of the rotor and the maximum measured deflection of all blades per revolution are shown. The relative speed is calculated by dividing the speed by the average speed of the first 100 revolutions. This enables the comparison of different stall runs, although the actual rotor speeds differ slightly. The rotor speed was kept at a constant reduced speed, according to Equation (1). The mechanical speed of the shaft differs, depending on atmospheric conditions. Blade deflection is the maximum value for all blades and all six sensors per revolution. This prevents deduction of the vibration's mode and gives an impression of how much the blades are deflected. Stall onset and revolution 0 are determined by the same algorithm as for the unsteady pressure transducers.

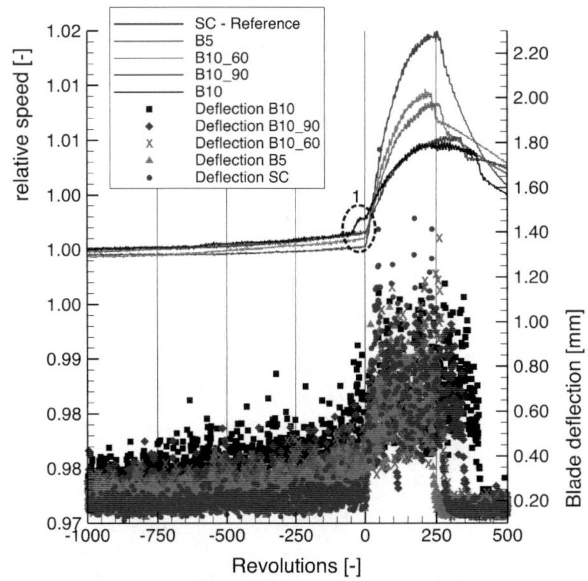

Figure 9: Speed and blade deflection for all configurations while compressor is throttled

The rotor speed reveals the increase caused by the loss of load when the compressor stalls. When comparing the different configurations, it appears that the rise in speed is attenuated with increasing distortion. Consequently, SC has the sharpest and highest speed slope. Speed stays constant for SC until the compressor stalls. With a DG installed, a slight speed rise is visible several hundred revolutions before stall.

When looking at B10 around 50 revolutions before stall (1), the speed rises by 0.3%. A small plateau is formed before the compressor actually stalls and the speed rises by around 1%. This pre-stall can also be seen in the wall pressure data. Even though the effect appears during many stall runs, the random character of stall onset means a reduced reproducibility. The duration of the plateau differs in every stall run and in some runs it does not appear at all.

The blade deflection has a similar pattern. With increasing inflow distortion, blade deflection increases. This seems reasonable as a bigger change in pressure leads to a bigger change in load and more severe synchronous vibration. When the compressor stalls, this changes: The blade deflection decreases with increasing inflow distortion. Although B10 shows the biggest values before stall, the rise in blade deflection is comparably small. The change in blade load could also be observed in unsteady numerical simulations of B10, as shown in [1]. The severest blade deflections during stall appear for SC. This can be explained by the change in the flow field. For B10 the distortion leads to high fluctuations, as seen in Figure 8. The difference to the values in stall is smaller. This further supports the theory that some stall-like effects are present before stall. For SC, the change in load on the blade is much more severe and harsh. The rotor builds up more and more pressure until suddenly the flow collapses, a stall cell appears and the pressure gradient across the blade breaks down. This change in pressure across the blade leads to aerodynamic forcing of the blade.

The transducer marked in Figures 2a and 6 is used to further analyse the flow physics. Transforming the pressure signal from time to frequency domain aids analysis of the unsteady behaviour. The raw pressure signal is highly fluctuating, making the underlying phenomena hard to identify.

The data were transformed into the frequency domain using a Fourier Transformation. As the data are unsteady they needed to be truncated into smaller parts. The resulting picture is known as a spectrogram. A Hanning window was, used with an overlap of 50%. The window length was set to 0.1 seconds, resulting in a frequency resolution of around 10 Hz. This resulted in a good compromise between time and frequency resolution.

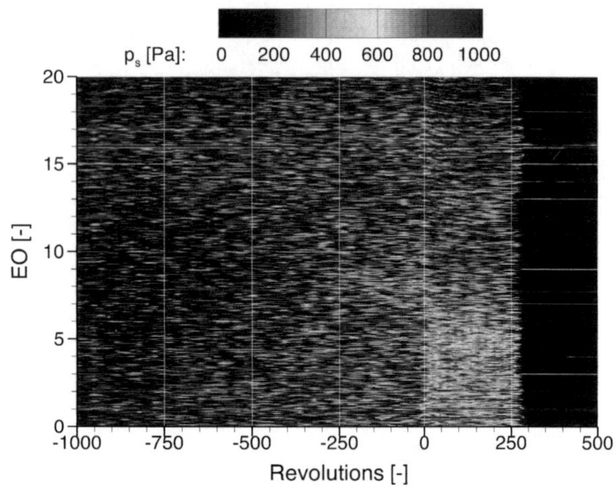

Figure 10: Spectrogram of the pressure while the compressor is throttled for SC (MP20 from Figure 2a)

The resulting spectrogram for SC is shown in Figure 10. The method to derive stall onset from speed variation works sufficiently well. The first significant change in the signal appears close to Revolution 0. Only the BPF is visible before stall. Previous research has identified the tip leakage vortex to form a two-passage pattern resulting in increased values at around half the BPF. The transducer location downstream of the LE explains that this pattern can hardly be seen in this figure. At around 300 revolutions prior to stall, a slight increase in the values at around EO 8 (half BPF) appears.

When the compressor stalls without inflow distortion, three phenomena appear:

1. At around EO 0.5 the stall cell rotating around the annulus shows relatively high values.
2. Increased values in the vicinity of EO 4 are visible. This could be linked to the size of the stall cell.
3. Beside the BPF, sub-harmonics appear with a difference of around 0.5, according to the stall cell rotational speed.

As could be expected when looking at Figures 8 and 9, this behaviour changes when a DG is installed. Figure 11 shows the spectrograms for four different stall runs with B10. The position of the DG is marked in the middle of Figure 8 (referred to as Stall Positions). B10 was chosen as it shows the most significant changes.

Looking at Position 1 in Figure 11a, the phenomena explained for the steady operating points in Figure 8 can be seen first. A relatively sharp frequency band at around EO 8 is visible 1000 revolutions before stall. The increased values at EO 0.5 appear around 250 revolutions before stall onset. For the edge of the separation in Figure 11b the frequency content changes. Mainly low frequency fluctuations below EO 8 can be seen. The amplitude of these fluctuations is already high. It seems that no stall precursors appear, in contrast the rotor is already stalled. At least the flow is highly turbulent and no well structured passage flow exists. When the rotor stalls, the amplitude values increase only slightly and the typical banded frequencies around BPF and EO 0.5 appear. The process towards stall is much smoother than for SC. This is in good agreement with the smoother speed rise for B10. When the rotor leaves the separation, at Position 3 in Figure 11c, the fluctuations that were visible at Position 2 are shifted to higher frequencies around EO 8. The loading on the blades is still high here, but the passage flow seems to be stabilised again. Therefore the blade tip vortex is strong and spills around the LE. A slight increase in EO 0.5 is visible at around 50 revolutions before stall. This behaviour is not as clear as for Position 1. When the rotor approaches the entry side of the DG (Position 4) in Figure 11d, a clear change can be seen. 500 revolutions before stall fluctuations with EO 0.5 appear. It is remarkable that for both Positions 1 and 4 high values at EO 0.5 appear even though they were two different measurements. Hence, there are two different stall runs. Sub-harmonics of the BPF are also present. The sub-harmonics, together with the EO 0.5 fluctuations, serve as stall indicators, as explained for SC.

Comparing Figures 8 and 11, it becomes apparent that the rotor is overloaded due to distortion, leading to a partial stall. This stall has similar effects to a classic stall, with different phenomena around the circumference. The fluctuations with EO 0.5 are an indicator that a partial stall cell is evolving. However ,it is difficult to determine whether the whole passage flow collapses, as seen for SC. As explained, a classic stall cell covering a large part of the circumference is not visible before revolution 0 , when the whole stage stalls. This variation of loading around the circumference matches the flow model in Figure 5.

CONCLUSION

Four distortion generators have been measured. As well as steady data to get the flow field and derive the performance of the stage, unsteady measurements were taken. These measurements were performed at both stable and unstable operating points. The impact of the distortion generator could be identified. When operating with inflow distortion the blades are exposed to increased deflection values caused by the change in aerodynamic load. This synchronous excitation leads to increased amplitudes

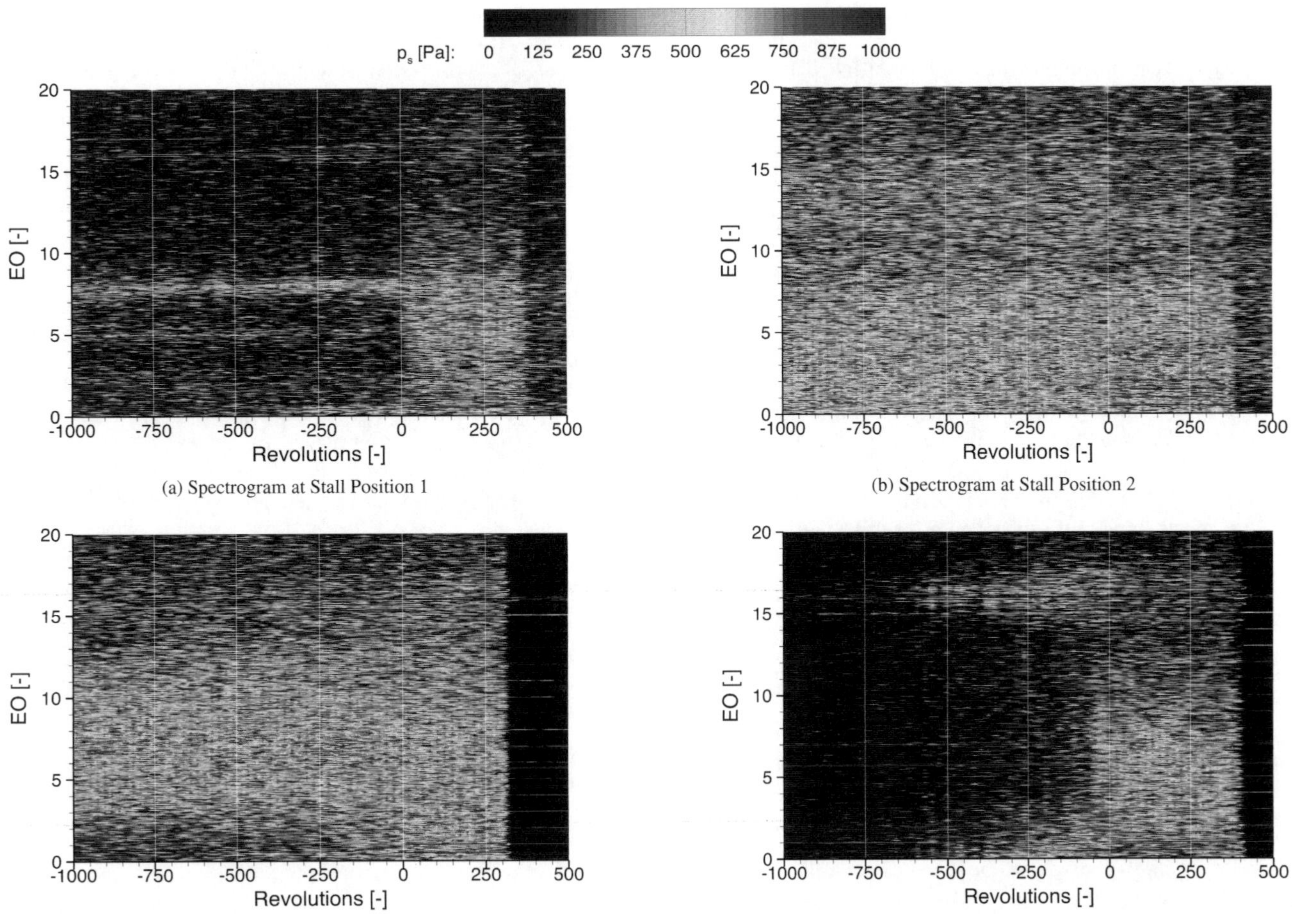

(a) Spectrogram at Stall Position 1

(b) Spectrogram at Stall Position 2

(c) Spectrogram at Stall Position 3

(d) Spectrogram at Stall Position 4

Figure 11: Spectrograms of transient stall runs for B10 at four different circumferential positions according to Figure 8

during resonance crossings. The high cycle fatigue limit could possibly be exceeded.

Classic descriptors based on screen measurements missed the impact change that occurred with increasing height of separation.The exceeding of a distinct height has a greater impact than a distinct circumferential angle. This should be taken into account in future definitions of safety margins. If the separation is smaller in height than a fixed limit the angle is not important. If a particularly height is exceeded the effects are only slightly dependent on the angle.

When classic stall descriptors are used parts of the circumference show stall-like behaviour due to local overloading of the blade. This happens before the flow collapses and the whole stage stalls. The resulting formation of additional loss cores leads to increased turbulence around the whole circumference and radius, which the following stages have to deal with.

Phenomena appear that are not captured by analytical models or classic screen measurements with steady and constant total pressure distortions.

ACKNOWLEDGMENTS

The members of the FOR 1066 research group gratefully acknowledge the support of the "Deutsche Forschungsgemeinschaft DFG" (German Research Foundation), which funded this research.

REFERENCES

[1] S. Barthmes, J. P. Haug, A. Lesser, and R. Niehuis. Unsteady cfd simulation of transonic axial compressor stages with distorted inflow. *Advances in Simulation of Wing and Nacelle Stall, Notes on Numerical Fluid Mechanics and Multidisciplinary Design*, Vol. 131:303–321, 2014.

[2] C. Biela, M. W. Müller, H.-P. Schiffer, and C. Zscherp. Unsteady pressure measurement in a single stage axial transonic compressor near the stability limit. *Proceedings of ASME Turbo Expo 2008*, GT2008-50245, 2008.

[3] R. V. Chima, D. J. Arend, R. S. Castner, J. W. Slater, and P. P. Truax. Cfd models of a serpentine inlet, fan, and noz-

zle. *48th AIAA Aerospace Sciences Meeting and Exhibit*, AIAA 2010-033, 2010.

[4] C. Hah, J. Bergner, and H.-P. Schiffer. Tip clearance vortex oscillation, vortex shedding and rotating instabilities in an axial transonic compressor roto. *Proceedings of ASME Turbo Expo 2008*, GT2008-50105, June 2008.

[5] C. Hah, M. Voges, M. Müller, and H.-P. Schiffer. Characteristics of tip clearance flow instability in a transonic compressor. *Proceedings of ASME Turbo Expo 2010*, GT2010-22101, June 2010.

[6] A. Lesser and R. Niehuis. Transonic axial compressors with total pressure inlet flow field distortions. *Proceedings of ASME Turbo Expo 2014*, GT2014-26627, 2014.

[7] F. Lin, J. Du, J. Chen, C. Nie, and C. Biela. Flow structures in the tip region for a transonic compressor rotor. *Proceedings of ASME Turbo Expo 2010*, GT2010-23025, June 2010.

[8] J. P. Longley and E. M. Greitzer. Inlet distortion effects in aircraft propulsion system integration. *Steady and Transient Performance Prediction of Gas Turbine Engines*, AGARD Lecture Series 183, 1992.

[9] R. S. Mazzawy. Multiple segment parallel compressor model for circumferential flow distortion. *ASME Journal of Engineering for Power*, Vol. 99:288–296, 1977.

[10] N. Mishra, D. MacManus, and J. Murphy. Intake ground vortex characteristics. *Proceedings of the Institution of Mechanical Engineers*, Vol. 000 Part G, December 2011.

[11] M. W. Müller. Untersuchungen zum Einfluss von Gehäusestrukturierungen auf die Stabilität und die Leistungsdaten eines transsonischen Axialverdichters. *Dissertation TU Darmstadt*, 2011.

[12] R. Niehuis, A. Lesser, A. Probst, R. Radespiel, S. Schulze, C.J. Kähler, F. Spiering, N. Kroll, F. Wartzek, and Schiffer, H.-P. Simulation of nacelle stall and engine response. *21st ISABE Conference 2013*, ISABE-2013-10135, September 2013.

[13] R. Radespiel, D.G. Francois, D. Hoppmann, S. Klein, P. Scholz, K. Wawrzinek, T. Lutz, T. Auerswald, J. Bange, C. Knigge, S. Raasch, P. Kelleners, R. Heinrich, S. Reuß, A. Probst, and T. Knopp. Simulation of wing stall. *43rd AIAA Fluid Dynamics Conference*, 2013-3175, June 2013.

[14] C. Reid. The response of axial flow compressors to intake flow distortion. *ASME 1969 Gas Turbine Conference and Products Show*, 69-GT-29, 1969.

[15] F. Roberts, G. A. Plourde, and F. Smakula. Insights into axial compressor response to distortion. *AIAA 4th Propulsion Joint Specialist Conference*, AIAA 68-565, 1968.

[16] S-16 Turbine Engine Inlet Flow Distortion Committee. Aerospace information report. *SAE Aerospace*, AIR1419 Rev. A, 2011.

[17] S-16 Turbine Engine Inlet Flow Distortion Committee. Aerospace recommended practice. *SAE Aerospace*, ARP1420 Rev. B, 2011.

[18] G. Schulze, C. Blaha, D. K. Hennecke, and J. M. Henne. The performance of a new axial single stage transonic compressor. *Proceedings of Int. Symposium of Air Breathing Engines 1995*, 1995.

[19] Y. Sheoran, B. Bouldin, and P. M. Krishnan. Compressor performance and operability in swirl distortion. *ASME Journal of Turbomachinery*, Vol. 134, July 2012.

[20] F. Wartzek, C. Biela, C. Pixberg, and Schiffer, H.-P. Modification of a compressor test rig for measuring the influence of inlet distortions on the compressor flow. *Third Symposium "Simulation of Wing and Nacelle Stall" Braunschweig*, 2012.

[21] F. Wartzek, C. Brandstetter, F. Holzinger, and Schiffer, H.-P. Response of a transonic compressor to a massive inlet distortion. *11th European Turbomachinery Conference, Madrid, Spain*, ETC2015-086, March 2015.

[22] F. Wartzek, F. Holzinger, C. Brandstetter, and H.-P. Schiffer. Realistic inlet distortion patterns interacting with a transonic compressor stage. *Advances in Simulation of Wing and Nacelle Stall, Notes on Numerical Fluid Mechanics and Multidisciplinary Design*, Vol. 131:285–302, 2014.

Proceedings of the 14th International Symposium on Unsteady Aerodynamics, Aeroacoustics & Aeroelasticity of Turbomachines
ISUAAAT14
8-11 September 2015, Stockholm, Sweden

I14-S4-2

FAN FORCED RESPONSE REDUCTION THROUGH BYPASS OUTLET GUIDE VANE OPTIMISATION

M. E. Rife
Imperial College London
London, UK

A. B. Parry
Rolls Royce
Derby, UK

L. di Mare
Imperial College London
London, UK

ABSTRACT

Optimisations of the bypass outlet guide vane cascade to minimise the upstream flowfield distortions caused by the pylon have been performed. Two cost functions for minimisation of the fan forced vibration are presented and compared. Both cost functions use a measure of upstream velocity distortion and downstream bulk swirl. The cost functions differ in that one targets a linear distribution of pressure coefficient on the suction side of the stator blades, while the other targets a prescribed loading distribution that is uniform from blade-to-blade. Geometry changes are applied through re-camber and re-stagger for the pressure coefficient linearity cost function and by orthogonal polynomials for the uniform blade-to-blade loading distribution cost function. Results are presented that indicate that losses through the cascade are lower for a cascade with uniform blade-to-blade loading at the cost of less reduction in flow non-uniformity at the fan face and residual bulk swirl aft of the stator cascade.

INTRODUCTION

It has long been known that bodies downstream of a stator cascade have a potential field that propagates upstream through the cascade of stators, often leading to significant circumferential variation of static pressure that negatively impacts the upstream rotor. The effect was first documented by Hemsworth [1] in a review of the lessons learned in the development of the TF39 high bypass ratio engine. In order to reduce the one per revolution vibration of the fan blades caused by the pressure field of the pylon, the pylon was integrated into the guide vane cascade and 5 different blade geometries were used in the cascade.

Rubbert *et al.* [2] conducted perhaps the first minimisation effort of the upstream pressure distortions by azimuthal modification of stator camber angles using a potential flow model. By attempting to minimise the variation of the mass flow through the cascade, the authors suggested that the upstream flow would be free from downstream pylon/strut induced distortions. The work showed that the primary means of obtaining this desired uniform mass flux around the annulus was through modifications of the stator blade trailing edge angles.

Potential flow solvers have been used by many researchers to evaluate and optimise the flowfield of a stator-strut interaction like that of a pylon behind a cascade of bypass outlet guide vanes (OGV).

Yokoi *et al.* [3] investigated the impact of stagger angle modifications on upstream pressure distortion and rotor vibration. It was shown that an optimised stagger angle distribution could almost eliminate the upstream pressure distortion at the rotor and provide a reduction in rotor vibration to around 40% of the vibration for the baseline, uniform stator design.

Cerri & O'Brien [4] conducted an optimization study of the cascade used by O'Brien *et al.* [5] using a potential model and numerical optimization techniques. By modifying stator stagger angle and strut azimuthal location only, a significant reduction in upstream pressure distortion was obtained. Two different objective functions were investigated, one being related to the amplitude of the upstream pressure distribution and the other being related to the integral of the upstream pressure distribution. The results of the study showed that the same stagger patterns for the stators was obtained from both cost functions, with only a small variation in strut azimuthal position.

Taylor & Ng [6] extended a Douglas-Neumann potential method to allow for cascades that move relative to each other. Using this method the authors were able to model an experimental rotor-stator-strut geometry and evaluate the unsteady pressure forces acting on the rotor caused by the downstream strut. Good agreement was found between the numerical and experimental results. The influence of strut-rotor spacing was evaluated using the unsteady potential model and showed a similar reduction in unsteady lift at the rotor with increased distance from the struts. Again, the results agreed well with the experimental data.

O'Brien *et al.* [5] performed an experimental investigation of the unsteady pressure disturbances at the rotor caused by downstream struts. The experimental results demonstrated that the effect of the struts was reduced by increasing the axial separation between the struts and the rotor. Additionally, a computational model of the cascade using a Douglas-Neumann singularity superposition method was shown to yield accurate results compared to the experimental data. While no attempt at modifying the cascade geometry was made by O'Brien *et al.*, the work has two key findings: with the current trends to reduce overall engine length to save system weight, means of reducing the influence of downstream struts and pylons on the fan are of increasing importance, and potential models of the flow provide an accurate yet computationally inexpensive means of modelling the geometry.

Potential flow models represent a middle ground level of fidelity for numerical simulation of the rotot-stator-strut flowfield. It is possible to further simplify the simulation by approximating the stator, and possibly the rotor, cascade using actuator disk theory.

Kodama [7] first demonstrated the use of an actuator disk method for the prediction of the performance of cascades subject to downstream static pressure distortions, such as those caused by the presence of a thick strut or pylon. The method provided good agreement with experimental data at transonic Mach numbers.

Kodama & Nagano [8] further validated the actuator disk model of Kodama [7]. Additionally, Kodama & Nagano used the actuator disk theory to arrive at an expression for flow angle at the trailing edge of a stator cascade that masks the pressure disturbance of downstream struts and pylons at the leading edge of the stator cascade. A non-uniform stator cascade geometry was obtained using the expression for the required stator exit flow angle. This cascade was then evaluated using a potential code and experimental model. A large reduction in the upstream pressure distortion was found with both the potential model and experimental test.

Parry [9] used a tandem fan-OGV actuator disk method to evaluate the upstream flow disturbance and optimise the OGV design. The tandem actuator disk method demonstrated a back-effect of the fan that reduced the pressure disturbance caused by downstream struts compared to the results without the fan. These results contrast with the findings of Cerri *et al.* [10] but were in agreement with experimental data. The optimisation procedure aimed to minimise the peak-to-peak fluctuation of the Mach number variation at the OGV leading edge by re-cambering the OGVs in groups. The limitation of up to 5 different blade profiles was imposed due to financial constraints. The perturbations at the OGV leading edge were shown to be reduced by up to 50% using 3 blades and up to 75% with 5 camber profiles.

The work of Parry [9] was later extended by Parry & Bailey [11] where a rotor-stator-strut-pylon configuration was considered using a tandem actuator disk model. The actuator disk model was validated against experimental data and showed very good agreement. While stator geometry optimisation was not considered by Parry & Bailey, it was shown that aligning the struts with the streamlines of the pylon-only flowfield reduced the flowfield disturbances to the levels of the pylon alone. It was then suggested that by increasing the stagger angles to twice the local streamline angle would completely eliminate the perturbation due to the presence of the pylon.

In contrast to the lower fidelity of the actuator disk theory approach, 2- and 3-dimensional computational fluid dynamics (CFD) approaches to the simulation and optimisation of the rotor-stator-strut problem have also been performed.

Chiang & Turner [12] used a 2-D CFD solver to evaluate the pressure field upstream of a row of stators ahead of thick struts. The authors noted an amplification of the upstream pressure distortion by the presence of a uniform cascade of stators ahead of the struts. By modifying the blade stagger angle circumferentially amplitude of the low frequency pressure disturbance was reduced by 92%. Additionally, it was shown that the blade-to-blade loading distribution remained more constant for the optimised geometry than in the baseline case.

Cerri *et al.* [10] used an improved 2-D potential model that included the rotors, stators, and struts simultaneously in conjunction with numerical optimisation techniques to minimise the influence of the downstream struts on the rotor. Differences between the stator alone, strut and stator, and rotor-stator-strut pressure distributions near the trailing edge position of the rotor were presented. It was shown that the rotor cascade tends to broadcast and increase the amplitude of the tone associated with the struts. Optimization of the stator cascade stagger angles and strut circumferential location provided a large reduction in the pressure distortion and thus the fan forced vibration.

Shahpar *et al.* [13] used 3 dimensional RANS simulations to optimise the bypass OGV cascade to minimise the fan forced vibration and cascade losses. The authors showed that by re-staggering all 52 blades in the cascade independently the upstream pressure variation could be reduced by 78%. However, it was noted that varying the stagger angle alone led to larger variation in incidence angle around the cascade, thus likely increasing the losses through the cascade.

Enoki *et al.* [14] investigated the effect of pylon spacing on the fan forced vibration. The work tested two different configurations of axial distance between the stator cascade and the pylon. Good agreement with experimental results was found. Additionally, a two-dimensional actuator disk model was compared to the experimental data, with good agreement.

The good performance of the three computational methods presented above makes them all suitable for OGV cascade optimisation. However, being considerably faster than both 2- and 3-dimensional CFD and providing flowfield details on the individual blades, such as pressure coefficient distributions, that are absent in the actuator disk method, the potential model offers a good compromise between accuracy and completeness of solution and available data and speed.

While the above surveyed works have generally attempted to minimise the pressure distortion at the fan face while maintaining acceptable levels of loss through the stator cascade, there have been a few studies that noted a relationship between the upstream pressure distribution and the blade-to-blade loading of the stator cascade. Chiang & Turner [12] found that when re-staggering the stator cascade to provide a uniform upstream pressure distribution the blade-to-blade loading was much more uniform. Jones *et al.* [15] also noted a more uniform blade-to-blade pressure distribution as a side effect of re-staggering the stator cascade to minimise the upstream pressure distribution. While both Chiang & Turner and Jones *et al.* identified a relationship with the stator cascade loading uniformity and the uniformity of the upstream pressure distribution, the loading uniformity was an unintentional result of optimisation.

Building on the work of Jones *et* al., Barton & Gentile [16] performed a comparison between the effectiveness of optimisations where the rotor exit plane pressure fluctuation and the stator cascade blade-to-blade loading uniformity were the cost functions. It was found that the cost function for upstream pressure variation yielded a larger reduction in the amplitude of the low frequency distortions to the pressure field near the rotor trailing edge than the cascade loading uniformity cost function.

The current work investigates the viability of blade-to-blade loading as a cost function for optimisation of a bypass outlet guide vane cascade. Two optimisation processes with different methods of geometry modification and cost functions, are presented. The paper will begin with a brief review of the potential model and optimisation routines. The results section begins with a description of a newly defined baseline blade that is subsequently used as the datum for the optimisations. This will be followed by a comparison of the geometry modification methods and cost function. Results will then be presented for the two optimised configurations based on the different cost functions.

NOMENCLATURE

CFD – Computational Fluid Dynamics
C_p – Pressure Coefficient
OGV – Outlet Guide Vane
RMS – Root Mean Square

METHODOLOGY

A potential model solver, based on the Douglas-Neumann singularity superposition method, is used to evaluate the flowfield around the pylon and OGV cascade in two dimensions. OGV blades are modelled as a series of vortices distributed along the blade camber-line. The thick pylon geometry is discretised and represented with lines of constant source and vortices at each panel. Solution of the system of equations is the obtained by applying a no-permeability condition on each of the panels and a Kutta condition at the trailing edges. Further details of the model are available in Annex A.

A compressibility correction similar to the Prandtl-Glauert transformation is employed. This allows for increased accuracy of results at higher Mach numbers. The correction is applied to the calculated pressure coefficients.

A numerical optimisation scheme is used to identify the geometrical configuration, within constraints, that minimises an objective cost function. The optimiser is based on IMSL libraries [17] and employs a constrained quasi-Newton optimisation procedure.

Two sets of optimisations have been performed, with differing methods of modifying the cascade geometry and different cost functions. The first method of modifying the geometry is by re-cambering and re-staggering the OGV blades. Re-cambering is achieved by modification of the camber-line internal control points defining the natural cubic spline used to create the camber-line. The control points are free to move only in the direction normal to the line connecting the leading edge and trailing edge points. Re-staggering is performed as a rotation about the leading edge of the blade. This method is depicted in Figure 1.

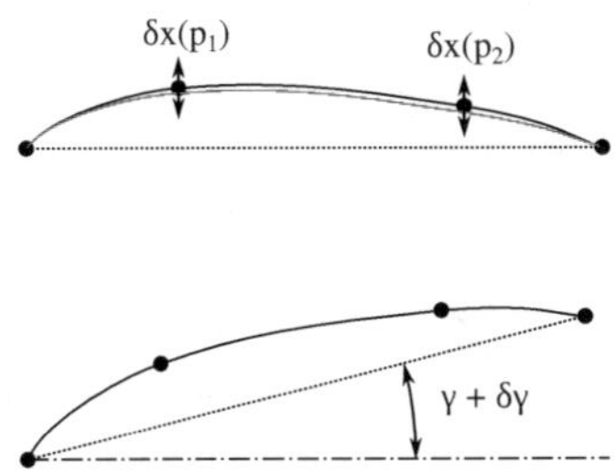

Figure 1 Re-camber and re-stagger methods

In order to minimise the number of variables in the optimisation process the modifications to the blade geometry are applied as a composition of Fourier harmonics. Up to 5 harmonics for each geometry parameter are used to build a new cascade.

The second method for geometry modification is through the use of orthogonal polynomials. The polynomials are used to describe the amplitude of the displacement in the local camber-line normal direction. Chebyshev polynomials provide a good basis for geometry modification because of their inherent orthogonality; a unique geometry can only be built using one combination of the polynomials. A visualisation of the geometry modifications to the baseline blade achieved from the first 5 Chebyshev polynomials is shown in Figure 2. It is seen from the figure that the first Chebyshev polynomial results in a rescaling and pitch modification of the blade, the second polynomial re-staggers the blade, and the higher order polynomials result in a re-cambering of the blade. Up to 5 Chebyshev polynomials were used for geometry modification.

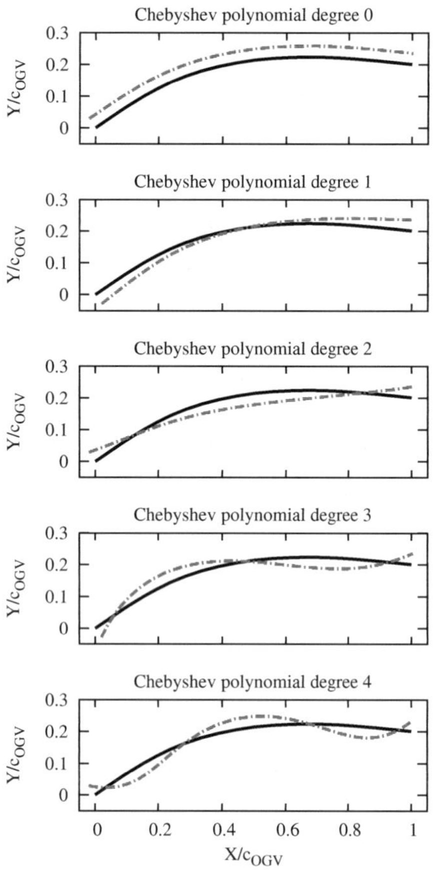

Figure 2 Chebyshev polynomial blade modification patterns.

Two formulations of objective functions were used for the optimisation, depending on the means of geometry modification. When re-cambering and re-staggering the cost function contains three main components. One is the root mean square (RMS) of the spatially varying part of the velocity profile 1 chord upstream of the cascade leading edge. The second is the sum of the squares of a measure of the suction side pressure coefficient linearity for each blade. Finally, they pylon circulation is included.

When geometry modification is performed using Chebyshev polynomials, the coefficients of Discrete Chebyshev Transforms of the loading distribution of each blade are used in the cost function. The difference between the coefficients for a uniform cascade with no pylon and the coefficients from the current tested geometry are used in the cost function. Additionally, the pylon circulation and upstream velocity RMS terms remain in the cost function.

The cost functions described above are a combination of three components. The pylon circulation component represents a measure of bulk swirl leaving the OGV cascade. Zero pylon circulation around a symmetrical pylon implies zero bulk swirl exiting the OGV cascade. The RMS of the spatially varying part of the upstream velocity profile acts as a measure of the fan forcing. The suction side pressure coefficient linearity term and the Chebyshev coefficients of the loading distribution on the blades are used as measures of the OGV load variation. In the first cost function only the desired linearity of the suction side pressure coefficient is specified while the amplitude is left unconstrained. In the second cost function, a specific loading distribution for each blade is targeted.

From the results of the potential model it is possible to obtain several inferred measures of loss. A measure of the diffusion ratio of the OGV cascade is obtained from the ratio of the maximum velocity on each blade and the upstream velocity. Additionally loss can be inferred from the incidence variation around the cascade [11]. These values are used after the optimisation to evaluate the loss characteristics of the cascade.

RESULTS

At the start of this work a baseline camber line was designed. The blade was designed for an isolated infinite cascade with a pitch to chord ratio of 1. As such no intake distortion or downstream struts or pylons were included in the geometry. The blade was designed to provide a minimum suction side pressure coefficient (C_p) near the leading edge of the blade and to reduce linearly along the chord. These criteria were applied over the central 90% of the chord as values at the leading edge and trailing edge require blade thickness to be refined. The criteria were specified as they provide similar performance benefits of prescribed velocity distribution and controlled diffusion blades [18]. The pressure side C_p was not considered in designing the blade as it was found that this was largely unaffected by changes to the camber line. The resulting blade geometry and pressure distribution are presented in Figure 3 and Figure 4.

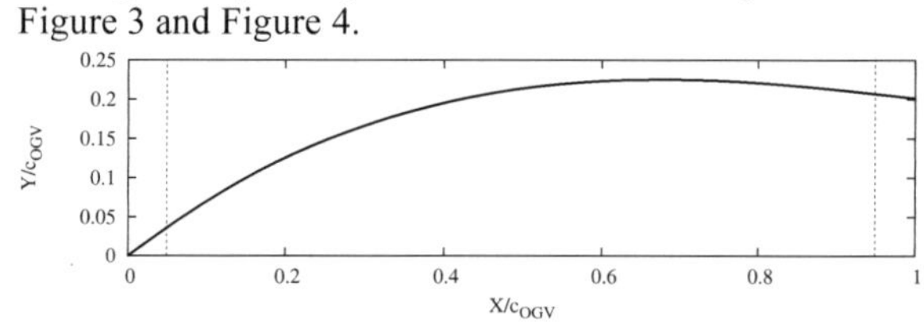

Figure 3 Baseline blade camber line geometry

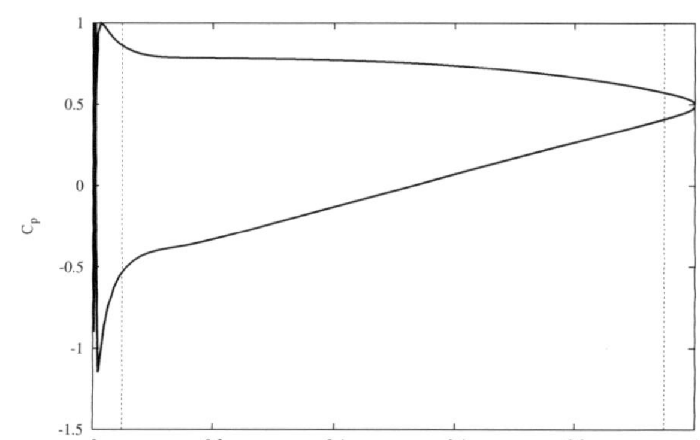

Figure 4 Baseline blade pressure distribution

A uniform cascade comprised of the baseline blades described above was used as the starting point for all further optimisations. While two sets of optimisations were completed, only the highest harmonic count results will be presented here, in the interest of brevity and clarity. The first series of optimisation was based on the re-staggering and re-cambering approach and the cost function was based on the upstream velocity distortion, pylon circulation and C_p linearity. The second series of optimisation used the orthogonal Chebyshev polynomials to modify the blade geometry and the cost function was based on the Chebyshev transform coefficients of the loading distribution of the blades, along with pylon circulation and upstream velocity uniformity. The resulting optimised geometries are plotted together in Figure 5. Overall, the changes to the geometry are quite subtle; little difference between the baseline and the re-staggered and re-cambered optimised geometry can be seen in the figure. When using Chebyshev polynomials to modify the geometry a small tangential displacement and scaling of some blades is visible; this is due to the degree zero Chebyshev polynomial, otherwise the cascades appear remarkable similar.

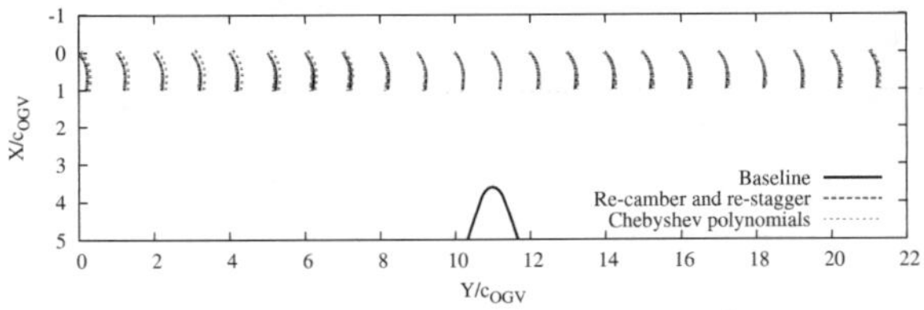

Figure 5 Optimised geometry comparison

To better understand the geometry variations caused by the optimisation, changes to the camber-line of each blade are shown in Figure 6 and changes to stagger angle are shown in Figure 7.

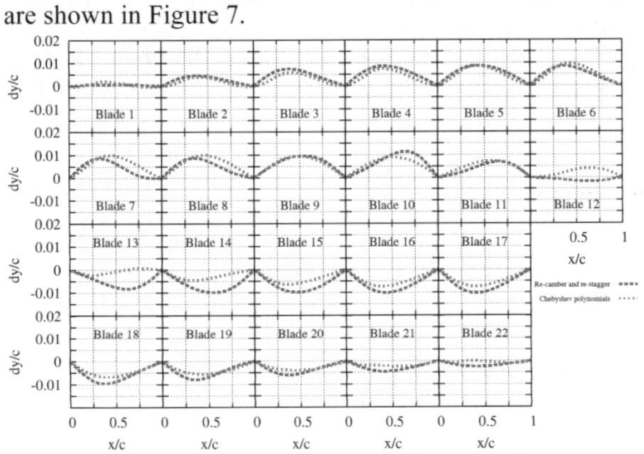

Figure 6 Optimised camber-line variation

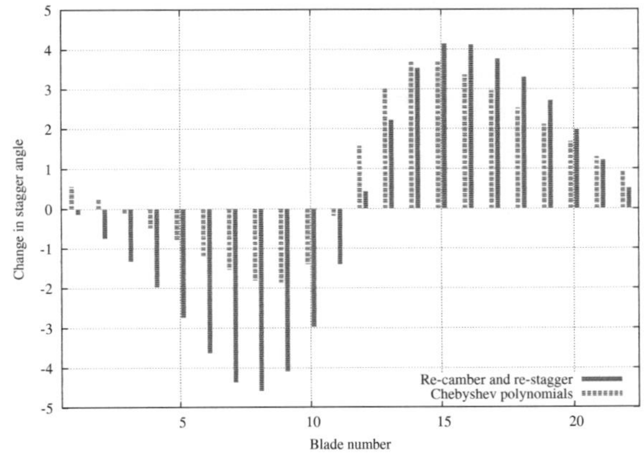

Figure 7 Optimised stagger angle variation

From Figure 6 it can be seen that both optimisations result in an increase in camber for blades 1 through 11. These blades are located to the left of the pylon centreline in Figure 5 and may be referred to as being 'below' the pylon later. Blades 13 through 22 (referred to as 'above' the pylon later) are shown to have the camber reduced by the optimisation. While the exact amounts of the re-cambering differ between the optimisations, it is clear that the optimum configuration has over-cambered blades located below the pylon and under-cambered blades above the pylon.

Looking at stagger angle variation in Figure 7 it is clear that both optimisations result in a clear cyclical variation of stagger angle. In this work the stagger angle is measured as the angle between the downstream engine axis and the blade chord-line, with positive values having a counter-clockwise direction. The stagger angle for blades below the pylon is generally reduced while it is increased for blades above the pylon. The exception being blades 1 and 2 in the Chebyshev polynomials optimisation, where the stagger angle is increased over the baseline angle. The re-camber and re-stagger optimisation results in a much more symmetric distribution of stagger angle around the annulus than the Chebyshev polynomial optimisation. However, the Chebyshev polynomial optimisation requires a lower amplitude of re-staggering. Overall, the changes to the camber and the stagger of the cascade are more subtle for the Chebyshev polynomial optimisation. However, it is important to remember the tangential shift and slight rescaling of a few blades below the pylon caused by the degree zero Chebyshev polynomial.

In order to evaluate the effect that the geometrical changes have on the flowfield the velocity distribution around the annulus at an axial location 1 OGV chord upstream of the baseline cascade leading edge is investigated in Figure 8. The figure shows the distribution of velocity amplitude and upstream swirl angle around the duct. It is clear from Figure 8 that both optimisations yield a significant reduction in the velocity, and hence pressure, distortion caused by the pylon. The use of Chebyshev polynomials to modify the geometry results in a large reduction in the low frequency distortion caused by the presence of the pylon. However, the re-camber and re-stagger optimisation is able to completely eliminate the distortion. These results show that the modified geometry of both cases can significantly reduce the fan forcing caused by the pylon.

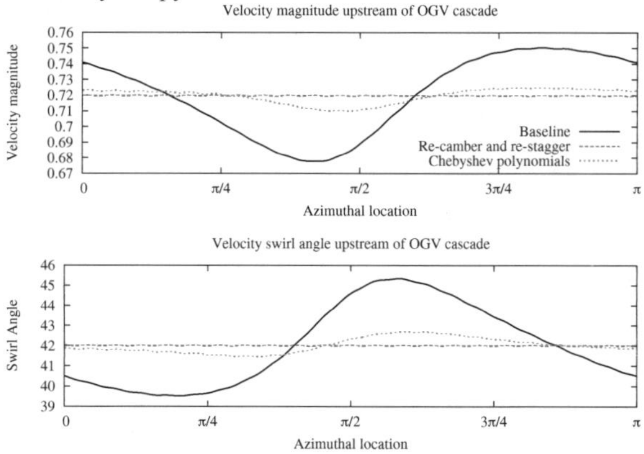

Figure 8 Upstream velocity profiles

In addition to reducing the flowfield distortion upstream of the OGV cascade, the cost functions for both optimisations include terms to reduce the pylon circulation and to attain a suitable loading distribution on the OGV blades. Attention is now turned to evaluation of these aspects of the optimisation. The minimisation of the pylon circulation is desired as it reflects a minimisation of the bulk swirl remaining in the bypass flow after the OGV cascade. Any residual bulk swirl downstream of the OGVs leads to increased losses and represents a lack of efficiency, as thrust that could be gained from purely axial flow is lost to the swirling motion. The swirl angle between the OGV cascade and the pylon is shown in Figure 9 for the baseline and optimised cases, as well as for the pylon alone in purely axial flow. The figure shows that the optimised cascade results in a near doubling of the swirl angle compared to the baseline case. This is a similar result to that predicted by Parry & Bailey [11]. In fact, the swirl angle distribution is much closer to the predicted ideal shape for the re-camber and re-stagger optimisation. The Chebyshev polynomial optimisation results in more resultant bulk swirl between the OGVs and the pylon.

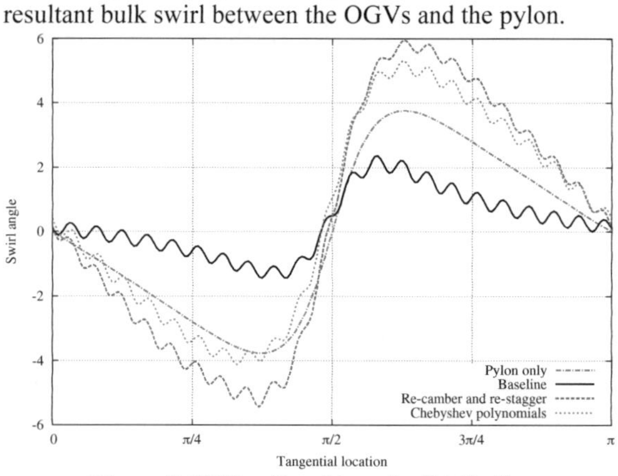

Figure 9 OGV exit swirl angle distribution

To evaluate the blade-to-blade loading variation and C_p distribution linearity, the loading for each of the OGV blades is plotted for the baseline case and the two optimisation runs in Figure 10, along with the baseline blade loading distribution. The figure shows the impact on blade loading around the cascade by the addition of a pylon behind the OGVs. Optimisation using re-camber and re-stagger with C_p linearity as a component of the cost function reduces the variation of the blade-to-blade loading. However, by aiming to get the loading on each blade back to the distribution for the baseline blade in isolation, almost all variation in loading around the cascade is eliminated. Reduction in the blade-to-blade loading variation is an important criterion in the design process as it helps to minimise the chances of blades stalling and choking of passages in the OGV cascade.

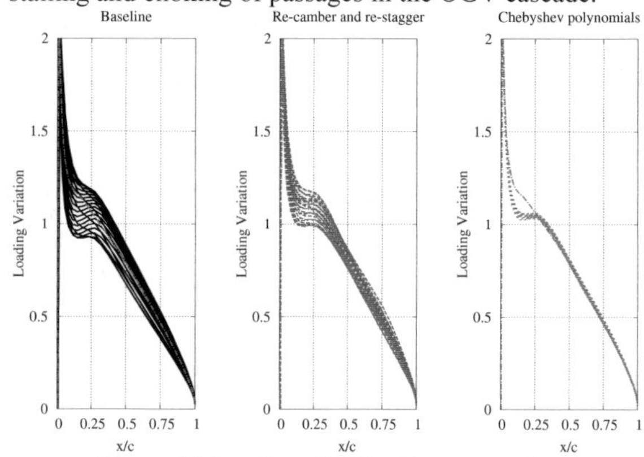

Figure 10 Loading distribution comparison

Further indications of OGV loss can be obtained from incidence angle variation and V_{max}/V. Distributions of these quantities around the annulus are shown in Figure 11 and Figure 12, respectively. Re-cambering and re-staggering the blades does not significantly alter the amplitude of the incidence variation of the OGV cascade, and provides only a modest reduction in the V_{max}/V distortion amplitude, and the mean value remains nearly unchanged, if not slightly increased. In contrast, the incidence and V_{max}/V distortion amplitudes are noticeably reduced in the Chebyshev polynomial optimisation. The reduction in both of these values represents a reduction in loss through the cascade [11] [18].

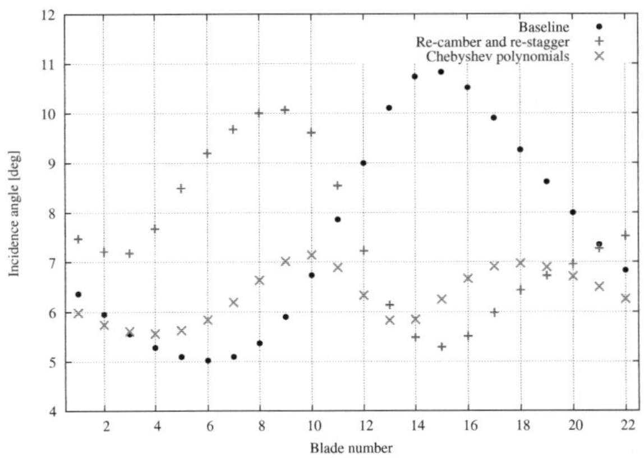

Figure 11 Incidence angle variation

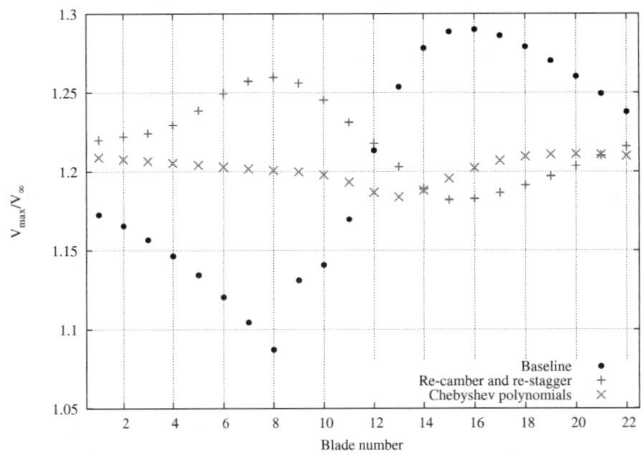

Figure 12 V_{max}/V variation

CONCLUSIONS

The results presented above demonstrate the ability of two optimisation approaches to modify the OGV cascade geometry in order to minimise the upstream flowfield distortion and cascade losses. One approach modified camber and stagger in an effort to minimise the upstream velocity distribution, pylon circulation, and generate a linear suction side pressure coefficient distribution across the OGV chord. The second approach used orthogonal polynomials to alter the blade geometry to minimise upstream velocity disturbance, pylon circulation and modify the suction side pressure distribution to adopt a prescribed shape.

It was shown that the re-cambering and re-staggering optimisation was able to provide the largest reduction in upstream velocity disturbance, and thereby provide minimum fan forcing. While the upstream flowfield distortion reduction was not as large in the Chebyshev polynomial geometry modification optimisation, a large and still acceptable reduction was found. It was found that the Chebyshev polynomial optimisation approach was able to create a cascade that had minimal loading variation around the cascade, and as a result, the variation of loss indicators such as incidence and V_{max}/V. Indeed, the Chebyshev polynomial optimised cascade demonstrated lower values for both incidence and V_{max}/V compared to the re-cambered and re-staggered geometry.

The geometry of both optimised cascades was found to have similar properties. OGVs located below the pylon, that is, on the windward side of the pylon centreline, were generally modified such that the camber was increased and the stagger angle reduced, compared to the baseline blades. Additionally, the blades located above the pylon, on the leeward side of the pylon centreline, were shown to have reduced camber and a greater stagger angle. These modifications to the blade geometry imply the need to direct the flow around the pylon. The re-staggering of the blades is likely the primary means of directing the flow; as there is no deviation in the current model, the trailing edge angle of each blade determines the flow exit angle. The changes to the stagger angle are then complimented with camber changes. It is expected that the re-cambering of the blades directs the leading edge into the oncoming flow, minimising the incidence and modifying the pressure distribution to the desired shape.

The target of one optimisation run was to modify the blade geometry in such a manner that the loading distribution became uniform from blade-to-blade and matched the designed distribution for an isolated uniform cascade comprised of the bespoke baseline blade. Use of this target in the cost function resulted in a sizeable reduction of the upstream velocity distortion. However, the real value of this approach was found in indicators of loss of the cascade. Compared to the baseline uniform cascade and the optimised cascade using C_p linearity as a component of the cost function, the minimised blade-to-blade loading approach resulted in a large reduction of the distortion amplitude of the variation of incidence and V_{max}/V around the cascade. Additionally, the blade-to-blade loading variation was essentially zero and matched closely to the target distribution. These results indicate that the optimised cascade will perform well. Furthermore, the low variation of V_{max}/V and incidence around the annulus will yield more predictable behaviour of the cascade, with minimised choking and stalling of cascade.

The approach is not without it's compromise, however. It was found that the upstream flowfield distortion was lower for the re-cambering and re-staggering optimisation approach, which used C_p linearity as a component of the cost function rather than blade-to-blade loading variation. Indeed, the optimum re-cambered and re-staggered cascade was able to completely eliminate the effects of the pylon upstream of the OGV cascade. Additionally, the bulk swirl of the C_p linearity approach was lower than the blade-to-blade loading method. However, the results above indicate that the use of blade-to-blade loading variation as an objective function can yield a large reduction in the upstream flowfield distortions caused by downstream bodies and minimised losses through the outlet guide vane cascade.

ACKNOWLEDGMENTS

This work was carried out as part of the NADIT programme, sponsored by Rolls-Royce plc and the UK Technology Strategy Board with grant number TSB 101368.

The authors gratefully acknowledge Rolls-Royce plc and the TSB for funding this work and granting permission for its publication.

REFERENCES

1. HEMSWORTH, M. C. **Development and Experiences of the First High Bypass Ratio Engine - TF39**. 11th Anglo-American Aeronautical Conference. London: [s.n.]. 1969.
2. RUBBERT, P. E. et al. **Concept and design of stators tailored to shield a fan from pressure disturbances arising in the downstream fan duct**. AIAA 10th Aerospace Sciences Meeting. San Diego: AIAA. 1972.
3. YOKOI, S.; NAGANO, S.; KAKEHI, Y. **Reduction of Strut Induced Rotor Blade Vibration with the Modified Stator Setting Angle**. Fifth International Symposium on Airbreathing Engines. Bangalore: [s.n.]. 1981. p. 61-1 - 61-7.
4. CERRI, G.; O'BRIEN, W. F. Sensitivity Analysis and Optimum Design Method for Reduced Rotor-Stator-Strut Flow Interaction. **Journal of Turbomachinery**, v. 111, p. 401-408, October 1989.
5. O'BRIEN, W. F.; REIMERS, S. L.; RICHARDSON, S. M. **Interaction of Fan Rotor Flow with Downstream Struts**. AIAA 8th Aeroacoustics Conerence. Atlanta: AIAA. 1983.
6. TARLOR, A. C.; NG, W. F. **Analytical Prediction of the Unsteady Lift on a Rotor Caused by Downstream Struts**. Gas Turbine Conference and Exhibition. Anaheim: ASME. 1987.
7. KODAMA, H. Performance of Axial Compressor with Nonuniform Exit Static Pressure. **Journal of Turbomachiner**, v. 108, p. 76-81, July 1986.
8. KODAMA, H.; NAGANO, S. Potential Pressure Field by Stator/Downstream Strut Interaction. **Journal of Turbomachinery**, v. 111, p. 197-203, April 1989.
9. PARRY, A. B. **Optimisation of Bypass Fan Outlet Guide Vanes**. International Gas Turbine and Aeroengine Congress and Exhibition. BIrmingham, UK: ASME. 1996.
10. CERRI, G. et al. **Optimization of Rotor-Stator-Strut Potential Flow Interaction Including rotor Feedback Effects**. International Gas Turbine and Aeroengine Congress and Exposition. The Hague: ASME. 1994.
11. PARRY, A. B.; BAILEY, R. H. **The use of Cyclic Variations in Strut Stagger to Reduce Coupled Blade-Vane-Strut-Pylon Interaction and System Losses**. International Gas Turbine & Aeroengine Congress & Exhibition. Orlando: ASME. 1997.
12. CHIANG, H.-W. D.; TURNER, M. G. Compressor Blade Forced Response Due to Downstream Vane-Strut Potential Interaction. **Journal of Turbomachinery**, v. 118, p. 134-142, January 1996.
13. SHAHPAR, S.; GIACCHE, D.; LAPWORTH, L. **Multi-Objective Design and Optimisation of Bypass Outlet-Guide Vanes**. ASME Turbo Expo. Atlanta: ASME. 2003.
14. ENOKI, T.; KODAMA, H.; KUSUDA, S. **Investigation of Fan Rotor Interaction with Pressure Disturbance Prodcued by Downstream Pylon**. ASME Turbo Expo. San Antonio: ASME. 2013.
15. JONES, M. G.; BARTON, M. T.; O'BRIEN, W. F. **The Use of Circumferentially Nonuniform Stators to Attenuate LP Compressor Rotor-Stator-Strut Aerodynamic and Mechanical Interactions**. International Gas Turbine and Aeroengine Congress & Exhibition. Birmingham, UK: ASME. 1996.
16. BARTON, M. T.; GENTILE, D. P. **The Use of a Circumferentially Nonuniform Stator to Attenuate Harmful Aerodynamic and Mechanical Interactions in an Advanced Mixed Flow Splittered Rotor/Tandem Variable Stator LP Compressor**. ASME Turbo Expo 2005. Reno, NV: ASME. 2005. p. 1-10.
17. IMSL Library Reference Manual. International Math and Sciences Libraries. Houston, p. ZXMWD-1-3. 1982.
18. CUMPSTY, N. A. **Compressor Aerodynamics**. Malabar: Krieger Publishing Company, 2004.
19. GOSTELOW, J. P. **Cascade Aerodynamics**. Oxford: Pergamon Press, 1984.
20. CHIANG, H.-W. D.; TURNER, M. G. Compressor Blade Forced Response Due to Downstream Vane-Strute Potential Interaction. **Journal of Turbomachinery**, v. 118, p. 134-142, January 1996.
21. KODAMA, H. Performance of Axial Compressor with Nonuniform Exit Static Pressure. **Journal of Turbomachinery**, v. 108, p. 76-81, July 1986.

ANNEX A

POTENTIAL FLOW SOLVER

The velocity field of an infinite row of vortices can be evaluated at any location, z, within the complex plane as [19]

$$\vec{w}(z) = \frac{\Gamma i}{2\pi} \frac{\pi}{\sigma} \coth\left(\frac{\pi}{\sigma}(\vec{z} - \vec{z_0})\right)$$

where $\vec{w}(z)$ is the complex velocity at z. The complex velocity is of the form $\vec{w}(z) = u - iv$ where u and v are the real and imaginary components of the velocity, respectively. Γ is the vorticity strength of the singularity, σ is the pitch of the cascade and $\vec{z_0}$ is the location of the vortex.

In order to obtain the circulation around each blade the blades can be treated as thin blades and represented by the camber line. The camber line is then discretised into linear panel with a vortex located at the one quarter chord location and a control point at the three quarters chord location of each panel. The aerodynamic influence coefficient matrix, $[A]$, is then generated, with elements a_{ij}. The influence coefficients represent the induced velocity at j caused by a unit strength vortex on panel I and is calculated as

$$a_{ij} = \vec{w_i} \cdot \vec{n_j}$$

The vortex strength is obtained by solving a linear system of equations in which the right hand side is the dot product of the the upstream velocity at infinity (inlet boundary condition) and the normal to the panel, i.e.

$$RHS_i = -q_\infty \cdot n_i$$

Then the equation for panel I is

$$\sum_i a_{ij} \cdot \Gamma_j = RHS_i$$

The complete system of equations for n panels of unknown vortex strengths, Γ_n, then has the form

$$\begin{bmatrix} a_{1,1} & a_{1,2} & \cdots & a_{1,n} \\ a_{2,1} & a_{2,2} & \cdots & a_{2,n} \\ \vdots & \vdots & \ddots & \vdots \\ a_{n,1} & a_{n,2} & \cdots & a_{n,n} \end{bmatrix} \cdot \begin{bmatrix} \Gamma_1 \\ \Gamma_2 \\ \vdots \\ \Gamma_n \end{bmatrix} = \begin{bmatrix} RHS_1 \\ RHS_2 \\ \vdots \\ RHS_n \end{bmatrix}$$

Solution of this system of equations yields the vortex strengths for each panel.

Thick bodies can be handled in a similar way. The velocity at any location, $\vec{w}(z)$, within the complex plane can be evaluated as

$$\vec{w}(z) = \ln\left[\frac{\sinh\left(\frac{\pi}{\sigma}(y - a)\right)}{\sinh\left(\frac{\pi}{\sigma}(y + a)\right)}\right]$$

where y is the tangential separation between the interrogation point and the panel and a is the length of the panel.

A matrix of aerodynamic influence coefficients, $[A]$, can be generated, taking the form

$$A = \begin{bmatrix} a_{1,1} & a_{1,2} & \cdots & a_{1,n} & \Sigma b_{1,j} & ny_1 \\ a_{2,1} & a_{2,2} & \cdots & a_{2,n} & \Sigma b_{2,j} & ny_2 \\ \vdots & \vdots & \ddots & \vdots & \vdots & \vdots \\ a_{n,1} & a_{n,2} & \cdots & a_{n,n} & \Sigma b_{n,j} & ny_n \\ a_{1,1}^t + a_{n,1}^t & a_{1,2}^t + a_{n,2}^t & \cdots & a_{1,n}^t + a_{n,n}^t & \Sigma\left(b_{1,j}^t + b_{n,j}^t\right) & ly_1 + ly_n \\ 0 & 0 & \cdots & 0 & s\frac{2\pi}{\sigma} & 1 \end{bmatrix}$$

This can be coupled with a vector of boundary conditions, G

$$G = \begin{bmatrix} -U_\infty nx_1 \\ -U_\infty nx_2 \\ \vdots \\ -U_\infty nx_n \\ -U_\infty lx_1 - U_\infty lx_n \\ V_\infty \end{bmatrix}$$

Where $a_{i,j}$ is the normal velocity at j caused by a unit strength source at panel i and $a_{i,j}^t$ is the tangential component of the velocity. Similarly, $b_{i,j}$ is the normal velocity at j caused by a unit strength vortex at panel i and $b_{i,j}^t$ is the tangential component of the velocity. s is used to denote the perimeter of the thick body. Values ny_i and nx_i represent the tangential and axial components of the unit vector of panel i, respectively, and ly_i and lx_i denote the tangential and axial components of the distance between the end points of panel i. U_∞ and V_∞ are the axial and tangential components of the velocity far upstream.

By inverting $[A]$ and multiplying by $[G]$ the source and vortex strengths at each panel on the thick body are obtained. More complex combinations of thin and thick bodies are easily solved together using one matrix of influence coefficients and a single vector of boundary conditions.

Proceedings of the 14th International Symposium on Unsteady Aerodynamics, Aeroacoustics & Aeroelasticity of Turbomachines
ISUAAAT14
8-11 September 2015, Stockholm, Sweden

I14-S5-1

ON THE VALIDITY OF DECOUPLED FLUTTER PREDICTION METHODS FOR COMPOSITE FAN BLADES

Christopher Chahine
Von Karman Institute for Fluid Dynamics
Rhode-Saint-Genèse, Belgium
chahine@vki.ac.be

Tom Verstraete
Von Karman Institute for Fluid Dynamics
Rhode-Saint-Genèse, Belgium
verstraete@vki.ac.be

Li He
Department of Engineering Science
University of Oxford
Oxford, United Kingdom
li.he@eng.ox.ac.uk

ABSTRACT

Composite fan blades are today increasingly used in high bypass ratio aero-engines due to their inherent weight advantage over metallic blades. However, decreased out-of-plane stiffness and reduced mass density increase their susceptibility to flutter. An accurate prediction of this aeroelastic instability therefore becomes a prime consideration in the design process. The classical energy method, which is based on the assumption of small aeroelastic coupling, is today the most widely applied flutter prediction technique. For composite fan blades the strength of aeroelastic coupling is uncertain and so is the applicability of the energy method. This paper presents an attempt to predict the aeroelastic coupling effects of a typical composite fan blade via the shift of the resonant frequency caused by the unsteady aerodynamic forces. To allow an assessment of the material impact on the frequency shift, the computations are performed with both, a full titanium and a composite blade. The results show that a significant amount of aeroelastic coupling is present for the titanium blade which is further increased by introducing a composite material. Values of the predicted frequency shift indicate that the energy method is likely to produce misleading results if applied for large aero-engine fan blades.

INTRODUCTION

Aero-engine manufacturers are increasingly introducing fan blades made of composite materials, notably carbon fiber reinforced plastics, to substitute their metallic counterparts for reduced engine weight and improved overall fuel efficiency. It is well known that fan blades of high bypass ratio aero-engines are prone to flutter due to geometrical features such as lean, sweep and low hub-to-tip radius ratios. With the introduction of composite materials the susceptibility to flutter compared to metallic blades is generally further increased due to a lower blade density (lower mass ratios) and decreased out-of-plane stiffness compared to metallic alloys. The onset of flutter can rapidly lead to harmful high amplitude vibrations with the potential result of complete component failure. It is therefore imperative that the flutter stability is accounted for during the design process.

In the past decades various computational methods have been developed for the prediction of flutter in turbomachinery applications. Reviews of these developments can be found in Bendiksen (1990), Marshall & Imregun (1996) and more recently Bartels & Sayma (2007) and Datta & Johnson (2008). A more specific overview on computational aeroelasticity methods for aero-engine fan blade assemblies is presented by Wu et al. (2004). The method that is by far the most frequently applied for turbomachinery flutter predictions today is the energy method proposed by Carta (1967). In this method the structure and the fluid are essentially treated in a decoupled manner, based on the assumption of negligible aeroelastic coupling which implies that the unsteady flow has no significant impact on the eigenmodes and eigenfrequencies of the blade. Thus, the unsteady flow field is computed with a prescribed vibratory blade motion, which is typically obtained from a modal analysis of the blade either in vacuum or with the steady state aerodynamic forces included, and the flutter stability is inferred from the work that the aerodynamic forces perform on the blade during one period of the forced oscillation.

It is generally accepted that the assumption of small aeroelastic coupling is valid for the majority of turbomachinery cases, as most applications have high stiffness and exhibit sufficiently high structural to fluid mass ratios with the result that the unsteady aerodynamic forces induced by the blade vibration are negligible with respect to the blade's inertia and stiffness forces. However, it is questionable whether these assumptions still hold true for large aero-engine fan blades made of carbon fiber reinforced plastics. An indication of the problem is given by Sadeghi & Liu (2005), who predicted the flutter stability of a two dimensional cascade with different mass ratios using both the decoupled energy method and a fully coupled aeroelastic code. Their results show that the energy method fails to correctly predict the onset of flutter if the mass ratios are small.

The purpose of this paper is to attempt a more specific assessment of the validity of the energy method for a composite fan blade representative for a typical modern high-bypass ratio turbofan engine.

CASE STUDY

The case study in this work is a transonic fan blade designed for a large high-bypass ratio turbofan engine. The blade is a wide-chord, low aspect ratio design which closely resembles a modern fan blade used in today's engines. However, no three-dimensional features (i.e. lean and sweep) were applied. An overview over some of the key design quantities of the blade is given in Table 1.

The bypass splitter was included in the design process in order to match the engine cycle design requirements for bypass and core flows. For computational efficiency, this part was removed from the fluid domain for the computations performed in this work.

FLUID ANALYSIS

The unsteady flow field resulting from the forced motion of the blade is computed using the commercial CFD code Numeca FINE/Turbo™. The non-linear harmonic method introduced by He & Ning (1998) and Ning & He (1998) is applied with an extension for moving meshes in Cartesian coordinates as reported in Debrabandere et al. (2013). Only the fundamental flow harmonic is considered in the computations, assuming that higher harmonics do not significantly contribute to the net energy transfer between the blade and the flow. A multi block structured mesh with an O4H topology consisting of about 1.2 million points is used to discretize the fluid domain. The eddy viscosity is computed using the one equation turbulence model of Spalart & Allmaras (1994). An overview of the discretized fluid domain along with the solid mesh is given in Fig. 1.

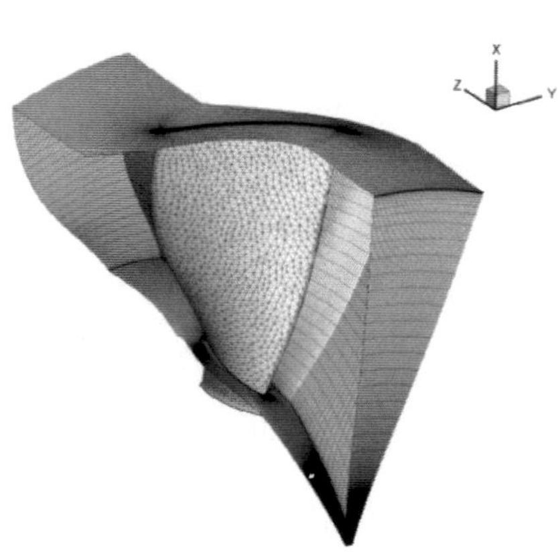

Figure 1: Overview of fluid and solid meshes used for the flutter analyses.

STRUCTURAL MODEL

The structural modelling of composite blades is distinctly more complicated than modelling blades made of metallic alloys. One reason for this lies in the anisotropy of the properties of composite materials, which is at the same time one of their biggest advantages over isotropic metals. In theory composite material properties can be tailored by adapting fiber angles, ply stacking sequences and ply thicknesses to give optimal support of a structure for a given loading condition. However, for design, analysis and manufacturing reasons, today's practical laminated fan blades exhibit quasi-isotropic material properties, meaning that the individual plies are stacked in such a fashion that the resulting laminate has globally isotropic in-plane properties. Although this approach clearly does not take full advantage of the capabilities of composite materials, the resulting laminate has usually still better specific strength and stiffness than a comparable metallic alloy, which is one of the driving factors behind the application of composites in aerospace applications. The usage of quasi-isotropic laminates has another advantage, in that it does enable the generation of a computational model with material characteristics that closely resemble those of a real composite blade.

In this work the fan blade is discretized with quadratic tetrahedral elements, cf. Fig. 1, and the material properties of a quasi-isotropic laminate made of a representative

Table 1: Fan blade design parameters.

Hub-to-tip radius ratio	0.35
Aspect ratio (height/mid-height chord)	1.7
Bypass ratio @ cruise	8.5
Number of blades	19
Tip relative inlet Mach number @ cruise	1.3
Tip diameter (m)	3.26

Table 2: Quasi-isotropic CFRP laminate properties.

E_x (GPa)	38.994
E_z (GPa)	11.793
G_{xz} (GPa)	3.2086
ν_{xy}	0.33375
ν_{xz}	0.30052
ρ (kg/m^3)	1594.3

carbon fiber reinforced plastic (CFRP) are applied, as reported in Akkerman (2002). The blade is modelled as a full composite blade without a honeycomb or foam core or any grid-stiffening features. The material properties used are summarized in Table 2. For each mesh element a local material coordinate system is defined, such that the out-of-plane component z is locally perpendicular to the closest point of the curved mid-surface of the blade.

A finite element solver (Dhondt (2004)) is used for the computation of the blade's modeshapes and eigenfrequencies. Centrifugal stiffening of the blade is taken into account by means of a non-linear geometric static analysis preceding the modal analysis.

Due to the discrepancy between the meshes used for the structural and CFD analyses an interpolation is required to transfer the modeshapes obtained from the structural analysis onto the CFD mesh in order to perform the computations with the blade vibration imposed. In this work a thin plate spline (TPS) radial basis function is applied for this purpose.

EVALUATION OF AEROELASTIC COUPLING

The aeroelastic system comprising the blade and the surrounding air flow is governed by the equation of motion

$$M\,\ddot{u} + C\,\dot{u} + K\,u = D, \qquad (1)$$

where M is the mass matrix, C is the damping matrix, K is the stiffness matrix and D represents the external aerodynamic forcing.

With the modeshapes $\Phi = [\phi_1, \phi_2, \dots, \phi_N]$ known from the finite element computations, the problem can be considerably simplified by transforming the set of N coupled equations of motion in physical coordinates given by Eq. (1) to a set of N decoupled equations in modal coordinates, hence allowing each equation to be solved independently. Such a transformation is possible by making use of the orthogonality of the modeshapes with respect to the mass and stiffness matrices. By defining a modal mass $m = \Phi^T M \Phi$ and a modal stiffness $k = \Phi^T K \Phi$, and decomposing the physical displacements as $u = \Phi \cdot q$, where q is the modal coordinate of the modeshape under consideration (only one in case of decoupled flutter analysis), the modal equation becomes

$$m\,\ddot{q} + c\,\dot{q} + k\,q = d, \qquad (2)$$

representing a single degree of freedom system for the unknown q. As the modeshapes are dimensionless, they are typically scaled for unity modal mass $m=1$ and modal stiffness $k = \omega_n^2$ which further simplifies the solution process. Because the modeshapes are not naturally orthogonal to the damping matrix, a modal damping is usually defined as

$$c = 2\xi\omega_n m \qquad (3)$$

where ξ is the damping ratio for the current mode.

Forces are transformed from physical to modal space by pre-multiplication with the transpose of the modeshape, hence the modal damping force is given by

$$d = \Phi^T D. \qquad (4)$$

One of the basic assumptions of the energy method is a fixed predetermined resonant frequency, i.e. that flutter occurs at one of the natural frequencies and eigenmodes of the blade. This assumption can be validated by investigating the amount of aeroelastic coupling in the system and more specifically the shift of the aeroelastic resonant frequency with respect to the natural frequency of the blade in the chosen eigenmode. If the frequency shift is small, say less than 1 %, the decoupled energy method might well be used to predict flutter with an acceptable accuracy.

In the energy method a harmonic motion of the blade in one of its eigenmodes ϕ, with an arbitrarily defined modal amplitude $\hat{q}_{CFD}$ (for instance 0.5% of chord, assuming the aerodynamic damping forces to depend linearly on the modal amplitude) and its associated eigenfrequency ω_n are imposed in the unsteady CFD computation for the determination of the modal aerodynamic damping forces $d(q)$. The work that the fluid does to the blade in one vibration period can then be determined by

$$W_A = \pi \sum_{j=1}^{N} [\hat{d}\,\hat{q}\sin(\psi)]_j, \qquad (5)$$

where N is the total number of blade surface cells of the fluid mesh and ψ is the phase shift between the modal damping force amplitude and the movement of the mesh cell. The aerodynamic damping is commonly defined as

$$D_A = -W_A, \qquad (6)$$

such that negative values of aerodynamic damping indicate an unstable situation where the fluid performs work to the blade. An equivalent viscous damping coefficient can be obtained by considering the work per period of a viscous damping force $\zeta_A \dot{q}$ which is given by $\pi\zeta_A 2\pi\omega\hat{q}_{CFD}$. Equating this expression with the aerodynamic damping yields

$$\zeta_A = \frac{D_A}{2\omega\pi^2\hat{q}_{CFD}}. \qquad (7)$$

In coupled flutter prediction methods the modal equation, Eq. (2), and the fluid equations are solved simultaneously to obtain the modal displacements q and the unsteady aerodynamic forces f. Since the bi-directional interaction between fluid and structure is accounted for, the aeroelastic resonant frequency and the associated vibration mode are a direct result of the computation.

However, in this paper a different approach is followed, which can be easily implemented also if no coupled solver is available. The method is based on an approximation of the frequency shift using a decoupled method. Similar approaches were reported by Marshall (1996), who used a harmonic balance method to derive an expression for the frequency shift based on the energy method, and later by Moffat & He (2005), who derived the frequency shift for forced response and flutter computations based on an emulation of a fully coupled method. Both methods were compared to results of fully coupled calculations and an acceptable agreement was found. The approach used in this paper is based on the latter method and will be further explained in the following.

Assuming a harmonic blade motion, $q = \hat{q}e^{i\omega t}$, and a harmonic damping force at the same frequency, $d = \hat{d}e^{i\omega t}$, with $\hat{q} \in \mathbb{R}$ and $\hat{d} \in \mathbb{C}$, the forces in the modal equation, Eq. (2), can be represented as vectors and plotted in the Argand diagram as shown in Fig. 2. Note that the modal blade displacement $\hat{q}$ is in phase with the inertial force amplitude $m\omega^2\hat{q}$. It is well known that only the component of the complex damping force that is out of phase with the blade displacement, $Im\{\hat{d}\}$, is able to do

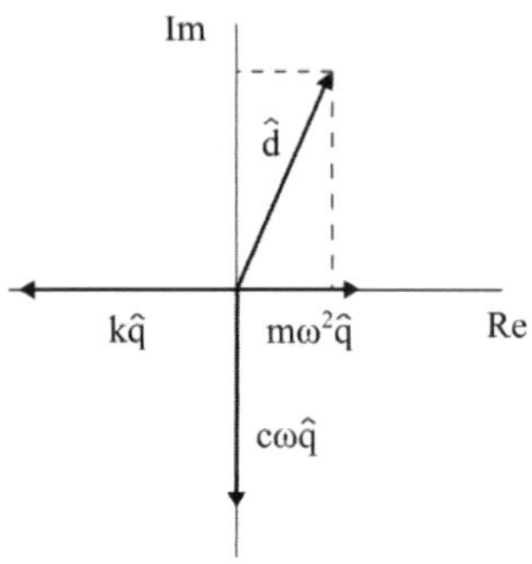

Figure 2: Argand diagram of the forces in the aeroelastic system.

work to the system and thus contribute to the aerodynamic damping. In contrast the damping force component that is in phase with the blade displacement, $Re\{\hat{d}\}$, is also in phase with the inertial and stiffness forces and therefore acts like an *added mass* in the aeroelastic system, which in turn affects the system's resonant frequency. As the energy method is based solely on the computation of the aerodynamic damping, the in phase component of the damping force is neglected, thus the method inherently assumes that resonance occurs at the blade natural frequency.

By applying Newton's second law to the real components of the forces we obtain

$$-k\hat{q} + Re\{\hat{d}\} + m\omega^2\hat{q} = 0. \tag{8}$$

An approximation of the resonant frequency of the system with the contribution of the added mass due to the damping forces included can thus be given as

$$\omega_{dyn} = \sqrt{\frac{k - Re\{\hat{d}\}/\hat{q}}{m}} \tag{9}$$

with $k = \omega_n^2$ and $m = 1$, as mentioned above. The amplitude of the complex modal damping force is obtained from the unsteady CFD computations as

$$\hat{d} = \frac{\hat{q}}{\hat{q}_{CFD}} \sum_{j=1}^{n} [\Phi_j \cdot (D_j^{Real} + iD_j^{Imag})] \tag{10}$$

where D_j^{Real} and D_j^{Imag} are the real and imaginary components of the physical complex damping forces.

RESULTS AND DISCUSSION

The operating point chosen for the evaluation of the frequency shift is 70 percent speed, near stall at sea level conditions with the fan blade vibrating in its first bending mode (1F). This is a condition where stall flutter typically occurs for fan blades, cf. Vahdati et al. (2001), Vahdati et al. (2011). Contour plots of this vibration mode for the composite fan blade and a geometrically identical reference blade modelled with material properties of Ti-6Al-4V alloy are shown in Fig. 3.

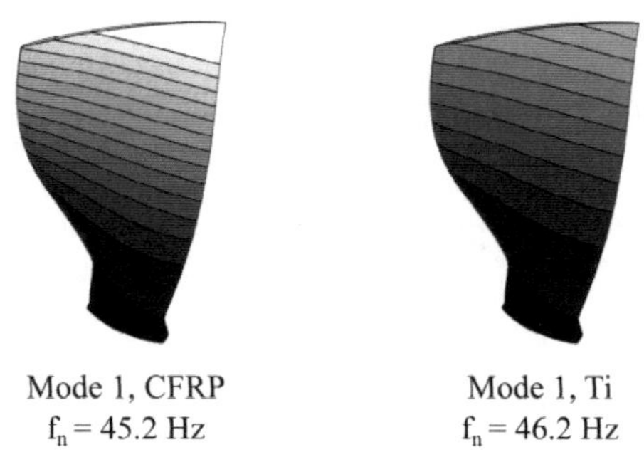

Figure 3: Contour plots of the mass normalized first bending modes of the composite blade (left) and the titanium blade (right) at 70 percent speed.

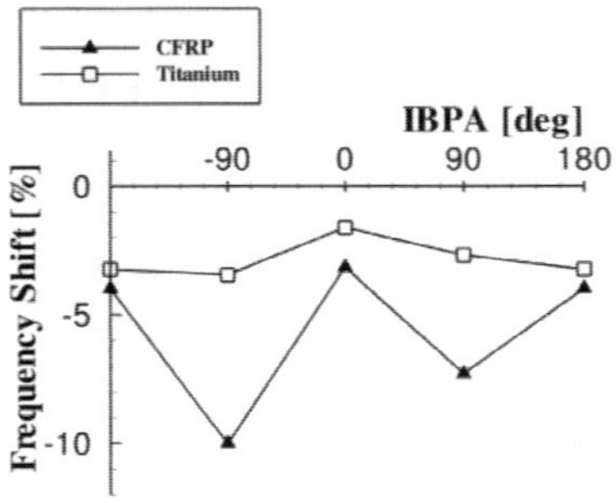

Figure 4: Frequency shift plotted versus the interblade phase angle.

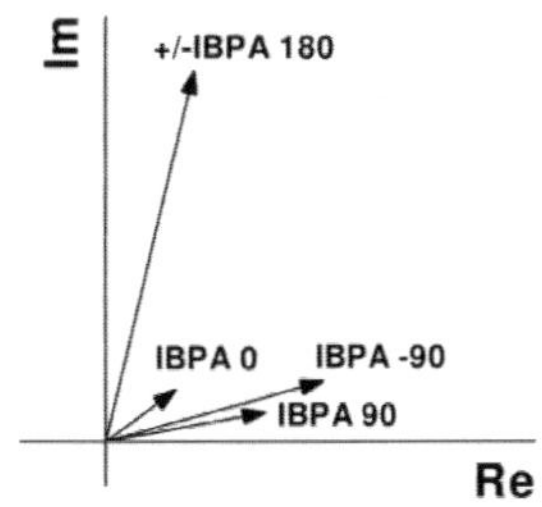

Figure 5: Argand diagram of complex modal damping forces at different IBPAs for the composite blade.

Frequency shift computations were performed for interblade phase angles (IBPAs) 0, -180, 180, -90 and 90 degrees according to the travelling wave assumption of Lane (1956). The results are summarized in Fig. 4, which depicts the frequency shift in percent plotted versus the IBPA. A negative frequency shift is obtained for all computations, indicating a reduction of the aeroelastic resonant frequency with respect to the blade's natural frequency due to the added mass effect. The values of frequency shift are overall significant for both materials under consideration. Values for the titanium blade range from -1.62 % at IBPA 0 to -3.45 % at IBPA -90. These are, already without the consideration of a composite material, values which point to a non-negligible impact of the unsteady aerodynamic forces on the aeroelastic response of the fan blade. Although a fixed rule for the applicability of the energy method depending on the frequency shift does not exist to the best knowledge of the authors, it is widely accepted that it needs to be applied with care for a frequency shift bigger than one percent, see e.g. Marshall (1996).

The substitution of the titanium alloy with a composite material leads to an overall increased magnitude of frequency shift, as initially expected at the outset of this work. The calculated ratios of the frequency shift of the composite blade over its corresponding value of the titanium blade listed in Table 3 show that the impact of the material substitution is highest at IBPAs of -90 and 90 degrees, where the frequency shift is almost tripled by the composite material. At this IBPAs also the highest magnitude of frequency shift is observed for the composite blade reaching a value of -7.3 % at IBPA 90 and -10.0 % at IBPA -90. The strong dependency of the frequency shift on the IBPA appears to be a reasonable result, as the effect of added mass originates from the phase shift between the unsteady aerodynamic damping forces and the blade motion, which can also be seen in Fig. 2. To explain this further, the complex modal damping forces for all computed IBPAs of the composite blade are plotted in the Argand diagram shown in Fig. 5. As can be seen, the modal damping force amplitude increases from 0 to ±180 degrees IBPA while its phase angle with the blade displacement (real component only) first decreases towards ±90 degrees IBPA and then increases again towards 180 degrees IBPA. This rather high phase angle is the reason why the real component of the modal damping force at 180 degrees IBPA is close to the value of 0 degrees IBPA, leading to comparable values of frequency shift despite the considerable difference in force magnitude between the two cases. For IBPAs of 90 and -90 degrees the modal damping force is dominated by its real component leading to the highest observed values of frequency shift as shown in Fig. 4.

Table 3: Impact of blade material on the frequency shift.

IBPA (deg)	$\Delta f_{comp}/\Delta f_{ti}$
-180	1.23
-90	2.90
0	1.95
90	2.71
180	1.23

CONCLUSIONS

In this paper a method for the computation of the aeroelastic resonant frequency of a fan blade and the associated shift with respect to the blade's eigenfrequency is presented and applied to a composite fan blade typical for a modern high bypass ratio turbofan engine. The obtained frequency shift is used as an indicator to assess the validity of the invariant vibration frequency assumption made in flutter predictions based on the energy method. A titanium blade of identical geometry is used as a reference case to enable an assessment of the material impact on the frequency shift.

Although the method applied is itself based on a decoupled approach, giving only an approximation of the frequency shift of the real aeroelastic system, the values obtained give rise to considerable doubt as to whether the energy method is an appropriate tool for the flutter prediction of large fan blades. Frequency shifts obtained for the titanium blade range from -1.62 % to -3.45 %, indicating an already significant amount of aeroelastic coupling at which the energy method should be applied with great care. Values of frequency shift are observed to globally increase over the entire range of interblade phase angles when the titanium material is substituted with a quasi-isotropic carbon-fiber reinforced-plastic. The predicted frequency shift for the composite blade ranges from -3.16 % to -10.0 %, giving a strong indication that the assumptions of the energy method are not valid anymore.

NOMENCLATURE

q	*Modal displacement*	*m*
Φ	*Modeshape matrix*	*[-]*
ϕ	*Single modeshape*	*[-]*
E	*Young's modulus*	N/m^2
G	*Shear modulus*	N/m^2
ν	*Poisson's ratio*	-
ρ	*Mass density*	kg/m^3
m	*Modal mass*	*kg*
c	*Modal damping*	*Ns/m*
k	*Modal stiffness*	*N/m*
d	*Modal damping force*	*N*
D	*Physical damping force*	*N*
M	*Mass matrix*	*kg*
C	*Damping matrix*	*Ns/m*
K	*Stiffness matrix*	*N/m*
ω_n	*Eigenfrequency*	*rad/s*
i	*Imaginary unit*	-
D_A	*Aerodynamic damping*	*J*
W_A	*Aerodynamic work*	*J*

ACKNOWLEDGMENTS

The research leading to these results has received funding from the European Union Seventh Framework Programme (FP7/2007-2013) under grant agreement no [316394] and the EPSRC.

REFERENCES

Bendiksen, O. O. (1990), Aeroelastic Problems in Turbomachines, AIAA paper 90-1157-CP

Marshall, J. G. & Imregun, M. (1996), A Review of Aeroelasticity Methods with Emphasis on Turbomachinery Applications, Journal of Fluids and Structures, Vol. 10, pp. 237-267

Bartels, R. E. & Sayma, A. I. (2007), Computational Aeroelastic Modelling of Airframes and Turbomachinery: Progress and Challenges, Phil. Trans. R. Soc. A, Vol 365, pp. 2469-2499

Datta, A. & Johnson, W. (2008), An Assessment of the State-of-the-art in Multidisciplinary Aeromechanical Analysis, AHS Specialist's Conference on Aeromechanics, San Francisco, CA, January 23-25, 2008

Wu, X. & Sayma, A. I. & Vahdati, W. & Imregun, M. (2004), Computational Techniques for Aeroelasticity and Aero-Acoustic Analyses of Aero-Engine Fan Assemblies, International Conference on Fans, London, UK, Nov 9th - 10th, 2004

Carta, F. (1967), Coupled Blade-Disk-Shroud Flutter Instabilities in Turbojet Engine Rotors, Journal of Engineering for Power, Vol. 89, No. 3, pp. 419-426

Sadeghi, M. & Liu, F. (2005), Computation of Cascade Flutter by Uncoupled and Coupled Methods, International Journal of Computational Fluid Dynamics, Vol. 19, No. 8, pp. 559-569

He, L. & Ning, W. (1998), Efficient Approach for Analysis of Unsteady Viscous Flows in Turbomachines, AIAA Journal, Vol. 36, No. 11, pp. 2005-2012

Ning, W. & He, L. (1998), Computation of Unsteady Flows Around Oscillating Blades Using Linear and Nonlinear Harmonic Euler Methods, ASME Journal of Turbomachinery, Vol. 120, No. 3, pp. 508-514

Debrabandere, F. & Tartinville, B. & Hirsch, C. & Coussement, G. (2013), Extension of the Non-Linear Harmonic Method to Flow Computations Around Moving and Deforming Structures, International Forum on Aeroelasticity & Structural Dynamics 2013, Bristol, UK, June 24th -26th, 2013

Spalart, P. R. & Allmaras, S. R. (1994), A One-Equation Turbulence Model for Aerodynamic Flows, Recherche Aerospatiale, No. 1, 1994, pp. 5-21

Akkerman, R. (2002), On the Properties of Quasi-Isotropic Laminates, Composites: Part B, Vol. 33, No. 2, pp. 133-140

Dhondt, G. (2004), The Finite Element Method for Three-Dimensional Thermomechanical Applications, John Wiley and Sons, Ltd.

Marshall, J. G. (1996), Prediction of Turbomachinery Aeroelasticity Effects Using a 3D Non-Linear Integrated Method, PhD Thesis, Imperial College of Science, Technology and Medicine, London

Moffat, S. & He, L. (2005), On Decoupled and Fully-Coupled Methods for Blade Forced Response Prediction, Journal of Fluids and Structures, Vol. 20, pp. 217-234

Vahdati, M. & Sayma, A. & Marshall, J. & Imregun, M. (2001), Mechanisms and Prediction Methods for Fan Blade Stall Flutter, Journal of Propulsion and Power, Vol. 17, No. 5, pp. 1100-1108

Vahdati, M. & Simpson, G. & Imregun, M. (2011), Mechanisms for Wide-Chord Fan Blade Flutter, Journal of Turbomachinery, Vol. 133, No. 4

Lane, F. (1956), System Mode Shapes in the Flutter of Compressor Blade Rows, Journal of Aeronautical Sciences, Vol. 23, pp. 54-66

Proceedings of the 14th International Symposium on Unsteady Aerodynamics, Aeroacoustics & Aeroelasticity of Turbomachines
ISUAAAT14
8-11 September 2015, Stockholm, Sweden

I14-S5-4

STALL FLUTTER ANALYSIS OF A TRANSONIC FAN BASED ON FULLY-COUPLED FSI AND AEROELASTIC MODE IDENTIFICATION

Atsushi Tateishi
School of Engineering,
The University of Tokyo
Tokyo, Japan

Toshinori Watanabe
The University of Tokyo
Tokyo, Japan

Takehiro Himeno
The University of Tokyo
Tokyo, Japan

Mizuho Aotsuka
IHI
Tokyo, Japan

Takeshi Murooka
IHI
Tokyo, Japan

ABSTRACT

An Aero-structure integrated approach for the cascade flutter simulations using the identification of aeroelastic modes were applied to the prediction of transonic stall flutter in an transonic fan. The flutter boundary obtained by FSI simulations fairly agreed in a qualitative sense on the higher speed lines. However, the simulations could not reproduce the end of flutter boundary on the lower speed lines. The existence of the detached shock leads flutter and sets the flutter boundary on the lower speed lines. The state of the shock was highly sensitive to the turbulence model, and transonic buffet can alter the steady flowfield around the shock. Aeroelastic modes can be identified accurately under stable flowfield. With the presence of strong flow instability near stall, the uncertainty of identification becomes large and modal identification cannot be applied.

INTRODUCTION

Flutter is one of well-known blade vibration phenomena which have to be prevented in aeroengine fans. Since it is experienced during the developing process of fans, compressors, and turbines, better prediction of the occurrence of flutter and the establishment of design criteria have been required so far. There have been reported many types of flutter in turbomachinery: classical flutter, subsonic/transonic stall flutter, choke flutter, supersonic unstalled flutter, and so on [1]. Among these types of flutter, stall flutter is one of the most important types in modern high by-pass ratio fans. It often occurs on part-speed, near-stall operating points, and it limits the operability of fans in lower mass flow rate conditions [2]. Examples of stall flutter can be found in literature [2-6]. In the reference [3] published in 1970's, the difficulty of predicting stall flutter was pointed out that there was no aerodynamic model for the analysis of stall flutter including the effect of separation on the blades.

Recent major approach for the evaluation of unsteady aerodynamic force in flutter analysis is based on computational fluid dynamics (CFD). Since 1990's, Reynolds-averaged Navier-Stokes equations have been applied to obtain motion-excited aerodynamic force and hence flutter analysis. Various computational approaches have been presented for flutter analysis: for example, forced blade oscillation approach with time-domain [7] or frequency domain solvers [8] and fluid-structure interaction (FSI) simulations [9]. Since CFD simulations provide detailed images of flow fields at near-stall conditions, it would be a powerful tool for analysing and understanding stall flutter phenomenon.

For the detailed modelling of aerodynamic-structural interaction and the improvement of accuracy of flutter analysis, the authors have been developed a new flutter analysis framework which is based on FSI simulations and system identification of vibrating cascade. It has been tested through two-dimensional multimode flutter problem [10] and the flutter analysis of NASA Rotor 67 fan [11].

In the present paper, the application of FSI simulation to stall flutter of a modern transonic fan is presented. After presenting the comparison of flutter boundary between experimental and numerical results, the mechanism of flutter inception and the applicability of modal identification will be discussed.

AEROELASTIC MODE AND ITS IDENTIFICATION

The equation of motion of cascade blades

Blade vibration in turbomachinery cascade is governed by the equation of motion below.

$$M\ddot{\boldsymbol{u}} + K\boldsymbol{u} = \boldsymbol{F}_{\mathrm{FR}} + \boldsymbol{F}_{\mathrm{SE}} \quad (1)$$

In Eq. (1), M and K are mass and stiffness matrices, $\boldsymbol{u}$ is displacement vector. Vector $\boldsymbol{F}$ represents external aerodynamic forces. Aerodynamic force can be divided into two terms: forced excitation force $\boldsymbol{F}_{\mathrm{FR}}$ which is not affected by the motion of the blade, and self-excited force $\boldsymbol{F}_{\mathrm{FR}}$ which is induced by the motion of the blade. Since flutter is unstable vibration caused by the coupling of the blade vibration and self-excited aerodynamic force, only the latter term $\boldsymbol{F}_{\mathrm{SE}}$ is of interest in this paper.

For the simplification of vibration analysis, the blade displacements are treated as a linear combination of lower structural modes. In this approximation, modal displacement vector $\boldsymbol{q}$ and modal force vector $\boldsymbol{f}$ are introduced to Eq. (1). Since the modal force $\boldsymbol{f}$ is considered as self-excited aerodynamic force, it can be expressed by the motion of the blade. When the blade displacement is small enough, the first order approximation can be applied to the modal force as follows.

$$\boldsymbol{f} \simeq A\boldsymbol{q} + B\dot{\boldsymbol{q}} \tag{2}$$

Matrices A and B are called aerodynamic force coefficient matrices. Modal form of Eq. (1) can be transformed into modal form as follows, introducing modal stiffness matrix Λ and the expression of linearized aerodynamic force expressed in Eq. (2).

$$\ddot{q} + \Lambda\dot{\boldsymbol{q}} = A\boldsymbol{q} + B\dot{\boldsymbol{q}} \tag{3}$$

Extending Eq. (3), a full set of equation of motion of the cascade can be written using vector form as follows.

$$\begin{bmatrix} \ddot{q}_1 \\ \vdots \\ \ddot{q}_{N_b} \end{bmatrix} + \mathrm{diag}(\Lambda) \begin{bmatrix} q_1 \\ \vdots \\ q_{N_b} \end{bmatrix} = [A_{ij}] \begin{bmatrix} q_1 \\ \vdots \\ q_{N_b} \end{bmatrix} + [B_{ij}] \begin{bmatrix} \dot{q}_1 \\ \vdots \\ \dot{q}_{N_b} \end{bmatrix} \tag{4}$$

System mode identification from a FSI result

If the matrices A and B are known, the stability of free response and mode shapes can be obtained. Generalized eigenvalue problem of Eq. (4) can be written as follows.

$$\lambda \begin{bmatrix} \psi \\ \lambda\psi \end{bmatrix} = \begin{bmatrix} O & I \\ -\mathrm{diag}(\Lambda) + [A_{ij}] & [B_{ij}] \end{bmatrix} \begin{bmatrix} \psi \\ \lambda\psi \end{bmatrix} \tag{5}$$

Here, λ is a complex aeroelastic eigenvalue, which includes the modal damping rate and frequency of aero-structure coupled system. Vector ψ is an aeroelastic eigenvector which describes the modeshape of the system, i.e. amplitude ratio and phase difference between degrees of freedoms. The aeroelastic damping rate μ_{AE} and frequency f_{AE} can be calculated from λ as follows.

$$\mu_{\mathrm{AE}} = -\mathrm{Re}(\lambda), \qquad f_{\mathrm{AE}} = |\mathrm{Im}(\lambda)|/2\pi \tag{6}$$

The occurrence of flutter can be judged by the sign of the aeroelastic damping rate. If the aeroelastic damping rate expressed by Eq. (5) is negative, blade vibration is unstable and amplified due to the energy input from the flow to the blade. Conversely, positive aeroelastic damping gives a stable aeroelastic mode.

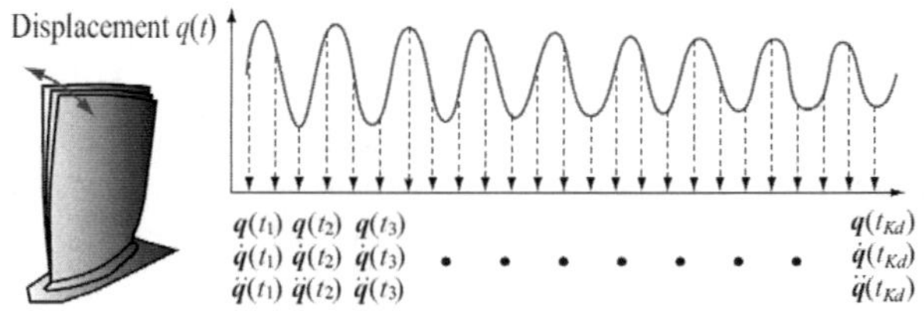

Fig. 1 Time-domain sampling of the blade vibration

To obtain the stability of aeroelastic system, matrices $[A_{ij}]$ and $[B_{ij}]$ need to be determined. In this paper, these matrices are directly identified from the history of free response of cascade which is obtained by FSI simulation. Figure 1 expresses sampling process of blade vibration from a FSI simulation. The identification process can be explained as follows.

(1) Run the FSI simulation of the cascade and sample modal displacement, velocity, and acceleration over different K_d timesteps.

(2) Matrices $[A_{ij}]$ and $[B_{ij}]$ are obtained by the least square method using sampled free decay history.

From these processes, the aeroelastic system of the cascade is identified based on the FSI result. The error of identification was monitored by the coefficient of determination R^2. The identification error in the aeroelastic damping rate and frequency was also estimated by the statistical way. Further detailed description in the modal identification procedure can be found in the literature [11].

NUMERICAL METHODS FOR FSI SIMULATION

All flutter simulations were conducted by an in-house FSI solver. Flutter simulations was conducted by solving unsteady flowfield and vibration of the blades simultaneously. Blade vibration was solved on the FEM solver with transferred instantaneous aerodynamic force from the CFD solver, and the CFD solver treats unsteady flowfield with given instantaneous blade deformation from FEM solver. Detailed information on these solvers is provided below.

CFD solver

Steady and unsteady flow simulations were performed with the structured finite volume solver of the compressible Reynolds-Averaged Navier-Stokes equations. To allow the blades and grid motions, arbitrary Lagrangian-Eularian formulation were employed. Time marching was conducted implicitly with Gauss-Seidel relaxation. The Eular implicit scheme was used for steady simulations, while a three-point backward difference scheme with inner iterations was used for unsteady simulations. The inviscid flux was evaluated by the SHUS scheme [12] with the third order MUSCL interpolation with Van-Albada's slope limiter. The viscous flux was calculated by the central difference. In the flutter simulation, Giles's one dimensional non-reflecting boundary condition [13] was applied on the inlet and outlet boundaries in order to reduce reflection of perturbation. The CFD code was parallelized both by MPI and OpenMP.

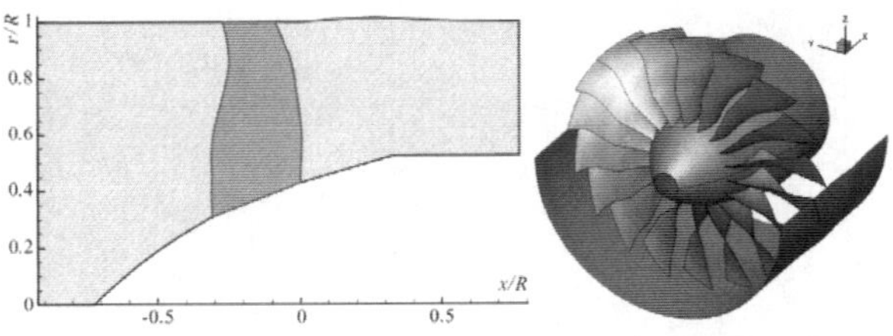
Fig. 2 Meridional and full-annulus view of CEFS1 fan

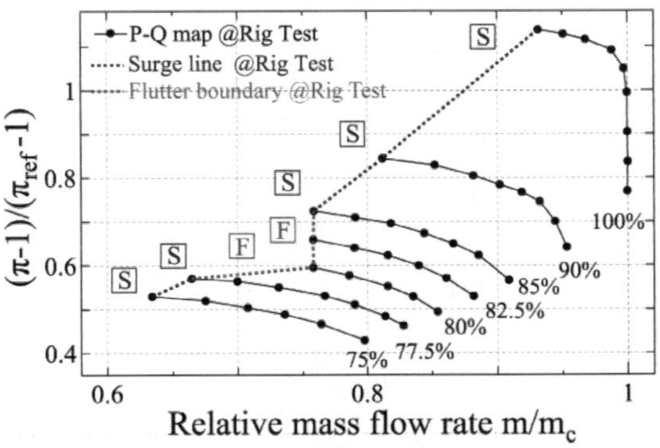

Fig. 3 Flutter bite on the fan characteristic map experienced in the rig test (S and F denote surge and flutter respectively)

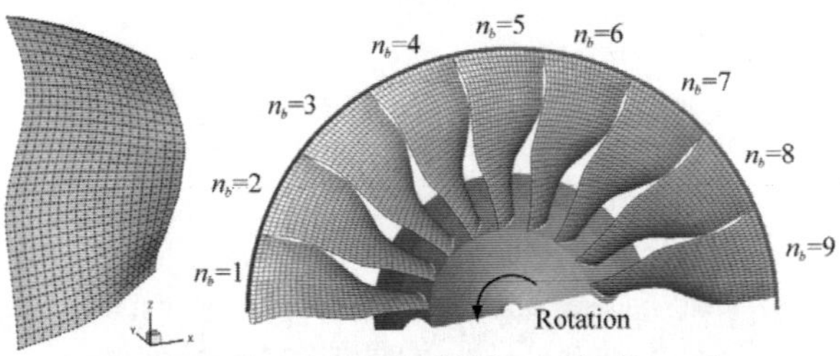

Fig. 4 FEM model of the blade and half annulus configuration for the flutter analysis

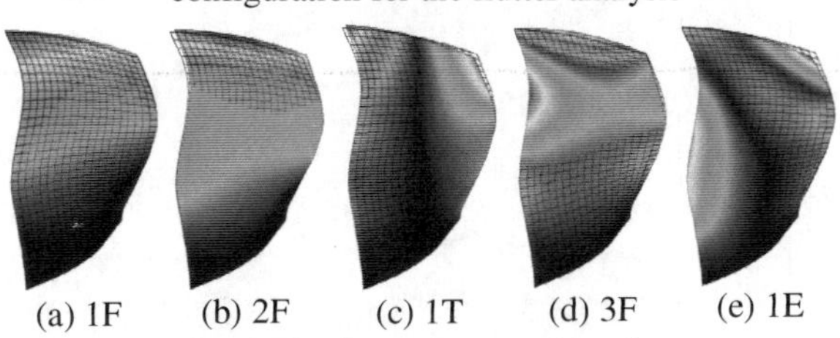
(a) 1F (b) 2F (c) 1T (d) 3F (e) 1E

Fig. 5 Five lower structural modes

Turbulence model

In this study, the Spalart-Allmaras (SA) one-equation model [14] was employed to estimate the eddy viscosity. With respect to the implementation of the SA model, the f_{t2} term, which controls the production term of eddy viscosity near the wall, were disabled for the flutter simulations. Latter of the paper the effect of this term to the CFD result is discussed in detail. Also, the effect of f_{t2} to the fundamental flat plate turbulent boundary layer is described in the Appendix part.

FEM, blade vibration, and FSI solvers

The blade structure was discretized by FEM using 20-nodes-brick elements. To analyse the deformation and structural modes of the blade under rotation, deformation was treated in a nonlinear manner. In the flutter analysis, structural modes were obtained by modal analysis and solved by Crank-Nicolson time marching scheme.

The blade motion and unsteady flow field were solved in a fully coupled manner with the implicit iterative algorithm. For spatial interpolation of solid displacement to the CFD cell node, moving least squares interpolation was used. This interpolation is very robust, and it gives spatially-smooth result even when large difference in mesh resolution exists between CFD mesh and FEM model. Aerodynamic force on the blade surface on the FEM model was transferred by the principle of virtual work.

TRANSONIC FAN RIG: ECO-CEFS1

The application target of FSI simulation is ECO-CEFS1 transonic experimental fan rig. This fan was developed and tested by IHI Corporation in the research and development of modern high by-pass fan which was aimed at achieving high efficiency and specific mass flow rate [15-17]. Figure 2 shows the meridional and full-annulus configuration of CEFS1. The fan has 18 blades in full annulus, and the blades were made of titanium alloy. The fan was blisk, and the structural damping was very low. Thus, the occurrence of flutter is governed by unsteady aerodynamic force characteristics.

Figure 3 shows the position of the flutter boundary in the fan operating map. The operation range with lower mass flow rate is limited by surge or flutter. Stall flutter was experienced at the operating points of lower mass flow rate on the speed lines of 80% and 82.5% design speed. On the other speed lines, flutter instability was not observed till the surge limit.

FSI setup and data reduction

Structure modelling. The FSI simulation was conducted to half-annulus model of the rotor. This configuration includes 9 blades, and as a result the nodal diameters (ND) of even number (ND=0, 2, 4, 6, 8) can be simulated. The FEM model of the blade for deformation and modal analyses is shown in Fig. 4. Manufactured blade shape was obtained from the design shape, and the running shapes and structural modes were obtained on each operating line. Figure 5 shows five lower structural modes of the blade. First flexure (1F), second flexure (2F), first torsion (1T), third flexure (3F), and edgewise bending (1E) are shown here. These five modes were taken into consideration in the flutter simulation.

CFD settings. Both steady and FSI simulations were conducted. Figure 6 shows CFD grids used in this study. Seven brocks were placed in one blade passage. The numerical grids were also placed in the tip clearance. The numbers of cells in chordwise, pitchwise, and spanwise directions are 120, 108, and 120, respectively. Total cell number is about 2.3 million points per one passage. Steady flowfield was obtained by local time stepping, while in the flutter analysis the time stepping was kept constant, and it took about 5000 steps for one blade oscillation of 1F mode. The CFD grid was morphed during the flutter

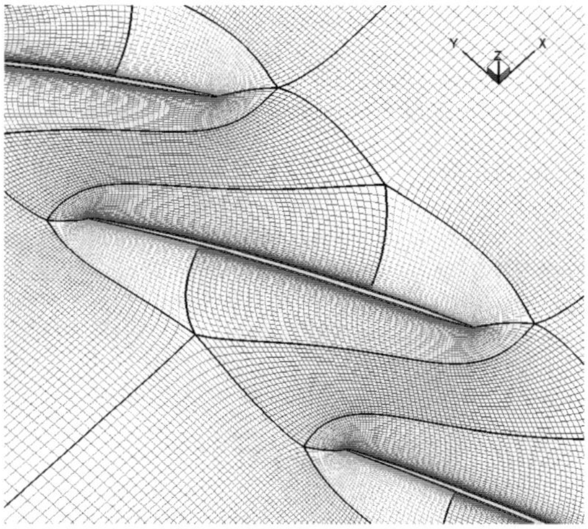

Fig. 6 CFD grid of CEFS1

simulations based on linear superposition of prescribed structural modeshapes.

Data reduction of FSI results. The aeroelastic modes and their eigenvalues were obtained based on the system identification procedure described in the previous sections. Figure 7 (a) shows an example of free decay history of blades (only five blades are shown). The free response of blade vibration was obtained by the FSI simulation. The time span for the modal identification was chosen to seven cycles of 1F oscillation. Fig. 7 (b) shows all identified aeroelastic modes from this transient signal. From modal identification, aeroelastic modes considering five structural modes can be obtained. Each mode has its own frequency and damping rate.

The flutter boundary was considered as zero-damping point on each speed lines. The mass flow rate and pressure ratio on the flutter boundary was estimated by linear interpolation based on identified damping rates from FSI simulations on two operating points across the zero-damping point.

FLUTTER ANALYSIS RESULTS

Steady flow results and characteristic map

Before the flutter simulation, steady flow simulations were performed in order to obtain steady flowfield around the flutter boundary. The range in the rotating speeds was varied from 70% to 90% of the design speed. Figure 8 shows the comparison of fan characteristic map between CFD and rig test results. In the operating points of higher mass flow rate, the total pressure ratio was calculated a little larger than experiment. However, overall, the CFD and test results were agreed fairly well. Flutter simulations were conducted on the 22 operating points shown by the bold circle. They are tagged as A', A, B, and C from choke side towards stall side on each speed line.

Flutter simulation results

The minimum aeroelastic damping rate within all identified aeroelastic modes was obtained on each FSI points. Before the evaluation of flutter boundary, the R^2 values of identification were evaluated in order to check the adequacy of the identified results. Table 1 summarizes

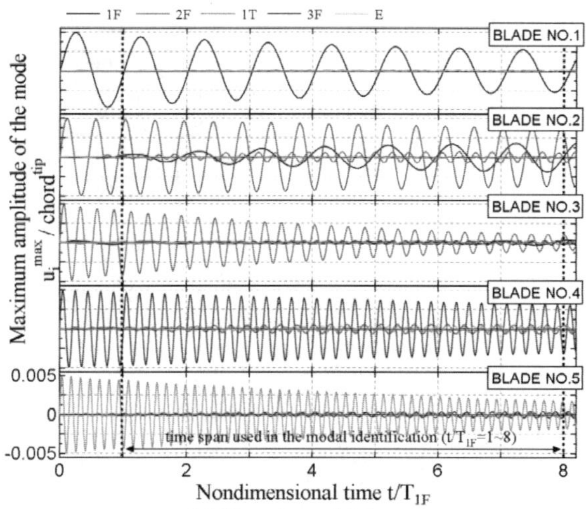

(a) Free decay history

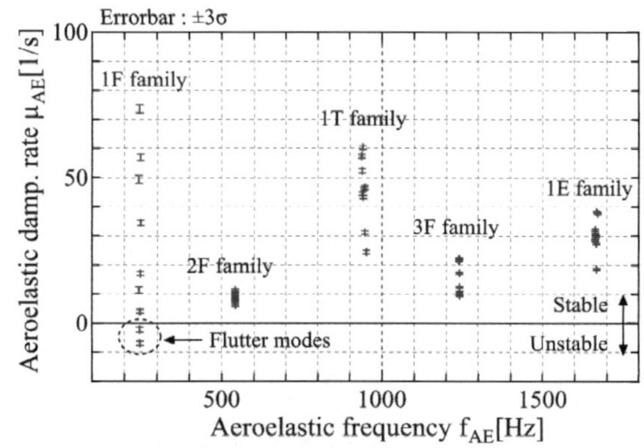

(b) All identified aeroelastic modes from the signal (a)

Fig. 7 An example of data reduction from FSI result

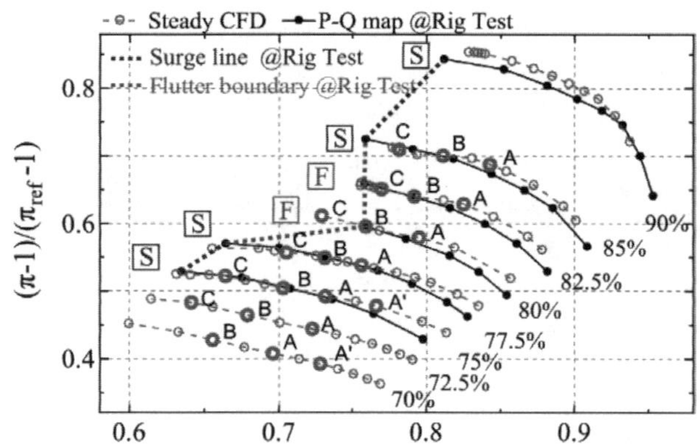

Fig. 8 Comparison of fan characteristic map between CFD and rig test results

the R^2 values that were obtained from the modal identification process on each operating point where the flutter analysis was conducted. The red line shows the position of flutter boundary, and (F) denotes the existence of negative damping mode. Also (†) denotes the operating point where the damping rate could not be determined accurately. The R^2 values were significantly high (over

Table 1 R^2 values of least-square identification on each operating points

OP/Speedline	70.0%N	72.5%N	75.0%N	77.5%N	80.0%N	82.5%N	85.0%N
A'	0.998	-	0.998	-	-	-	-
A	0.996	0.994	0.993 (F)	0.998	0.998	0.998	0.998
B	0.975	0.974	0.973 (F)	0.995 (F)	0.998	0.998	0.998
C	-	0.869 (†)	0.720 (†)	0.972 (F)	0.998 (F)	0.998 (F)	0.998

(F): operating points where negative damping mode was detected,
(†): operating points where the stability of blade vibration could not be distinguished due to large uncertainty,
red dashed line: flutter boundary, Green dotted line: inception of accuracy deterioration in the modal identification.

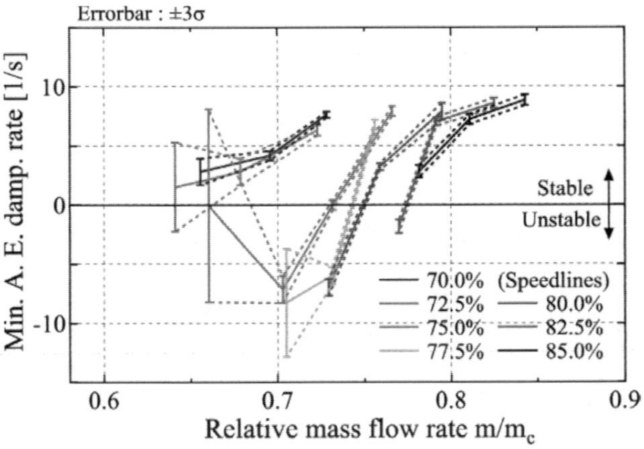

Fig. 9 Relationships between mass flow rate and the least stable damping rate

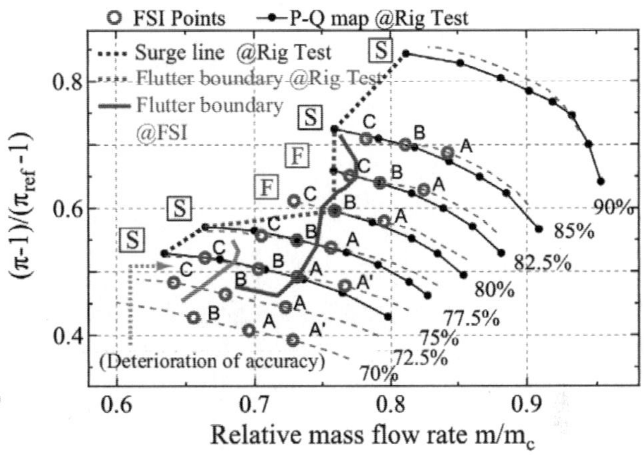

Fig. 10 Flutter boundary of 1F mode family calculated by FSI simulations

0.99) across the flutter boundary, so the aeroelastic modes could be successfully identified on these operating points. Besides, on the operating points C on the 72.5% and 75.0 speed lines, that are near-stall operating points on the lower speed lines, deterioration of R^2 were found. So the identified results should be treated carefully on these operating points.

Figure 9 shows the change in damping rate on each speed line as a function of mass flow rate. The error bar shows 99% confidence interval of identified damping rate. The common behaviour observed on all speed lines was that the damping rate decreases as the mass flow rate decreases. However, there can be seen some difference in their trend. In the speed lines from 75% to 82.5%, sharp decrement of damping rates was observed, and this behaviour results flutter inception. When the rotating speed was lower than 72.5%, this sharp decrement was not observed and damping rates were still positive even on the operating points of the lowest mass flow rate.

As for the uncertainty of the damping rates, the error bars were sufficiently small to determine flutter boundary on all speed lines. However, the error bars became extremely larger at the operating points nearest the stall side on the 72.5% and 75% speed lines. These behaviours of uncertainty that were observed in Fig. 9 were consistent with the trend of the magnitude of R^2 values in Table 1.

Based on the evaluation of damping rates and their uncertainties, the flutter boundary was obtained. Figure 10 shows the comparison of flutter boundary between FSI and the rig test results. In the speed lines of 80%, 82.5%, and 85.0%, the numerical result agreed qualitatively.

However, on the speed lines lower than 77.5%, FSI could not reproduce flutter boundary in the rig test. In these speed lines, flutter occurred in the simulations in spite of the fact that flutter was not observed experimentally. Thus, there might be some absences of important physics in the simulations to characterize stall flutter. In addition to this mismatch, the flutter boundary could not defined around points C on the 72.5% and 75.0% speed lines because of the deterioration of accuracy in identification. In order to explain the present numerical result, flow phenomena that lead to…

- the occurrence of flutter,
- the mismatch in the flutter boundary
- the large uncertainty of the identified damping rate

are discussed individually.

DISCUSSION: ON THE STEADY AND UNSTEADY FLOWFIELD AROUND THE FLUTTER BOUNDARY

Steady flowfield around the blade

The steady flowfield around the flutter boundary was observed in detail in order to understand key flow phenomena that are relevant to the occurrence of flutter. Figure 11 shows limiting streamlines on the suction side and relative Mach number fields on the operating points near the flutter boundary. In order to emphasize the difference in flow structures that relate to flutter, 72.5% and 80.0% speed lines are presented. Flutter did not occur on the 72.5% speed line, while flutter was observed on the 80.0% speed line both experimentally and numerically. On these speed lines, common flow features can be summarized as follows: (1) leading edge separation from the hub to the midspan, (2) radial migration of low speed

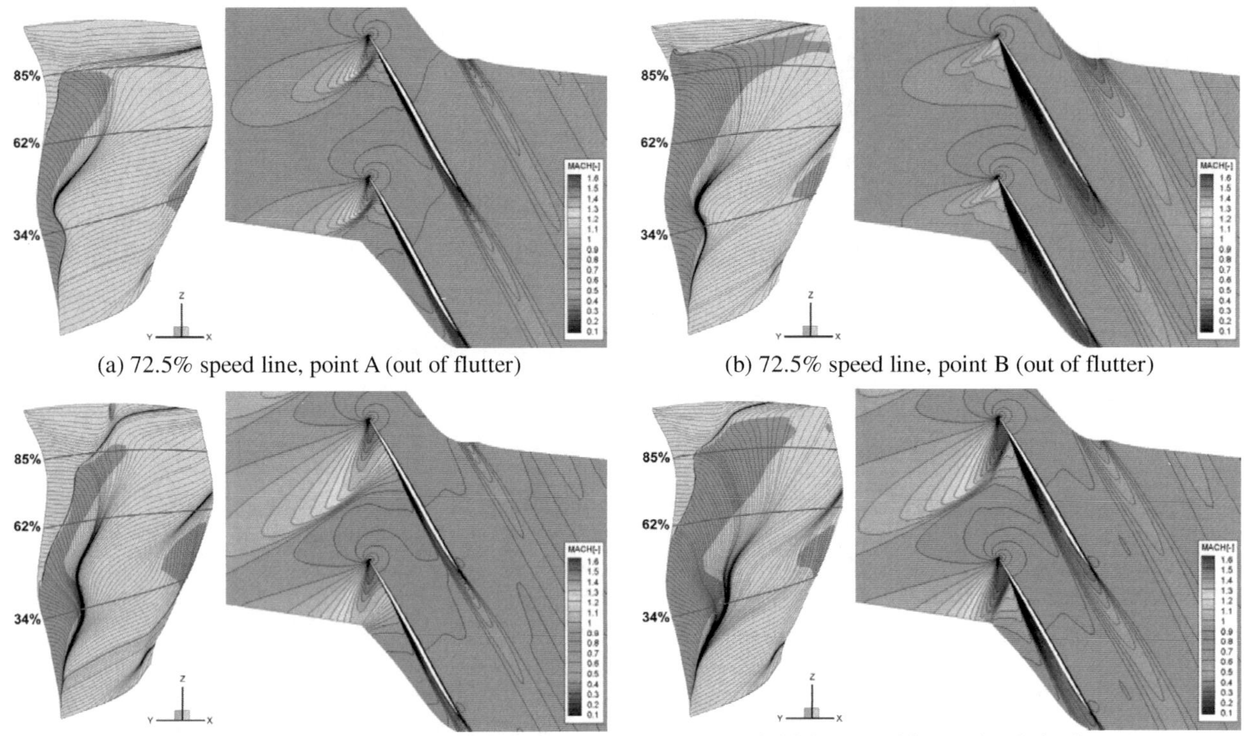

(a) 72.5% speed line, point A (out of flutter)

(b) 72.5% speed line, point B (out of flutter)

(c) 80.0% speed line, point B (out of flutter)

(d) 80.0% speed line, point C (in-flutter)

Fig. 11 Limiting streamline on the suction surface and relative Mach number field on the 85% span height. Light blue area shows axially-reversed area on the blade surface.

flow downstream of the leading edge separation, (3) detached shock near the leading edge from the midspan to the tip, and (4) interaction of migrated low speed flow and detached shock near the wall. And qualitative difference can be seen on the position of detached shock and the size of separated area across the flutter boundary.

On the 72.5% speed line, comparing Fig. 11 (a) and (b), it was observed that leading edge separation became larger as mass flow rate decreases and dominated the flowfield near the leading edge in almost all span height on the stall-side operating point. At the same time, the shock on the leading edge disappeared. This extinction of the shock might be related to the enlarged separation area migrated from lower spanwise positions.

On the 80.0% speed line shown in Fig. 11 (c) and (d), the enlargement of leading edge separation, migration, and the advancement of the shock towards upstream were observed in common with 72.5% speed line. However, the shock was still on the blade surface at the point C. Also, across the flutter boundary, the shock was strengthened due to interaction with enlarged separated area migrated from lower spanwise position.

Motion-excited component and its contribution to the aeroelastic instability

The existence and state of detached shock also affects to the blade excitation and hence flutter characteristics. In order to see the role of the existence of the shock to the flutter characteristics, local aerodynamic work which corresponds to the least stable aeroelastic mode was reconstructed from the history of blade surface pressure during FSI simulations. Local aerodynamic work LW is the indicator of blade excitation which is defined by time average of the product of the fluctuating pressure $\tilde{p}$, wall normal vector $\boldsymbol{n}$, and velocity of vibration $\boldsymbol{v}$ as follows.

$$\mathrm{LW} = \frac{1}{T}\int_{t}^{t+T}(-\tilde{p}\boldsymbol{n}\cdot\boldsymbol{v})dt \tag{7}$$

In this study, FSI simulation was used for the flutter simulation and the blade motion is not completely periodic. So, in order to obtain motion-excited component of pressure, the fluctuating pressure was linearly expanded by the motion of the blades with coefficients a_{ij} and b_{ij}, and these coefficients were determined by the least square method using time-domain signal from the FSI simulation on each cell surface on the blade. Here, the indexes i and j denote the number of blade and structural modes respectively.

$$\tilde{p} = a_{ij}q_{i,j} + b_{ij}\dot{q}_{i,j} \tag{8}$$

Once coefficients a_{ij} and b_{ij}, are obtained, local work which corresponds to the aeroelastic mode can be calculated by multiplying aeroelastic modeshape which was obtained from the previous modal identification.

Figure 12 shows local aerodynamic work distributions on the suction surface across the flutter boundary. Positive area of LW contributes to excitation of the blade, and

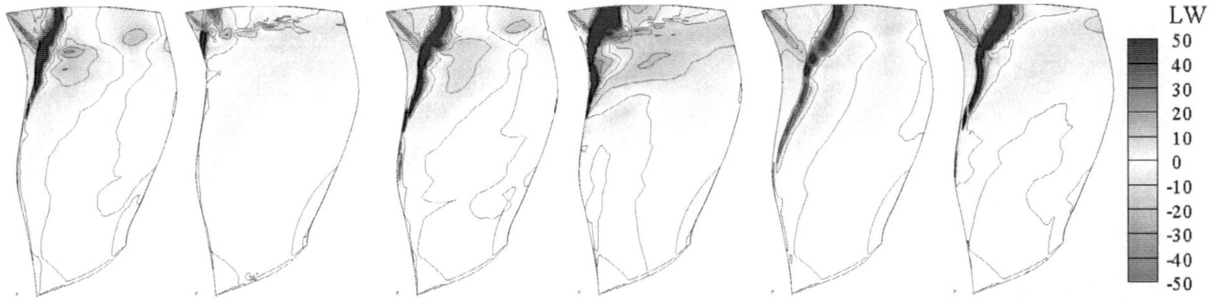

(a) 72.5%N, OP: A (out of flutter) (b) 72.5%N, OP: B (out of flutter) (c) 75.0%N, OP: A (on flutter boundary) (d) 75.0%N, OP: B (in-flutter) (e) 80.0%N, OP: B (out of flutter) (f) 80.0%N, OP: C (in-flutter)

Fig. 12 Local aerodynamic work distribution on the suction surface across the flutter boundary

negative area contributes to damping of the vibration. Three speed lines are shown in the figure: 72.5%, 75.0%, and 80.0%. In Fig. 10, the mismatch between experimental and numerical results was observed on the speed lines of 75.0% and 77.5%. The reason of this mismatch will be discussed in the following part with the consideration of flutter mechanism.

A common feature which can be seen on all speed lines presented in Fig. 12 is that the detached shock induces strong excitation force even the operating point is located out of the flutter boundary. On the 72.5% speed line where the extinction of the detached shock was observed numerically at the operating point of stall side, the extinction of excitation area which corresponds to the impingement area of the shock was also observed in Fig. 12 (a) and (b).

On the other hand, on the 80.0% speed line where the detached shock still existed on the operating point near stall, the excitation due to the shock was also present and strengthened in Fig. 12 (e) and (f). The result that the excitation under the shock was strengthened is consistent with the change in shock strength in steady flow fields shown in Fig. 11 (c) and (d). From the observation of local work and steady flowfield on 72.5% and 80.0% speed lines, it can be concluded that the existence of strengthened shock contributes to the occurrence of flutter.

The numerical result in Fig. 11 showed the mismatch in the flutter boundary on the lower speed lines. Figure 12 (c) and (d) show the local work distribution on the 75.0% speed line, where the mismatch exists. Point A in Fig. 12 (c) is just located on the flutter boundary, and the distribution of local work was similar to the distribution of Fig. 12 (a). In the local work inside the flutter boundary shown in Fig. 12 (d), however, it can be seen that the shock is still impinging on the leading edge of suction surface and contributes to strong excitation.

From these observations of local work, it can be concluded that the existence of detached shock on the leading edge on stall-side operating conditions leads the occurrence of flutter. As for the mismatch between experimental and numerical results, some differences of the state of the leading-edge separation and the shock might exist because the leading edge-separation dominates the flowfield near leading edge when the shock disappears.

Effect of turbulence model on the unsteadiness of transonic separated flow near the leading edge

On the operating points where the mismatch of flutter boundary exists, the shock was still on the blade surface similar to higher speed line(s) of 80.0% (and 82.5%). In order to obtain further information about this mismatch, the effect of minor change of turbulence model on the flowfield was studied.

The turbulence model which was used in the flutter simulations was the Spalart-Allmaras model without f_{t2} term. The position of f_{t2} in the model and its effect on the turbulent boundary layer is discussed in the Appendix part. An important thing on the inclusion of the f_{t2} term is that the wall boundary layer does not become "fully-turbulent". The fundamental test on the turbulent boundary layer showed that short laminar region appears on the leading edge when the f_{t2} term is activated. So it can be considered that the f_{t2} term acts as "uncontrollable" transition model or laminarization model on the leading edge.

The operating point A on the 75.0% speed line was chosen, and a steady flow simulation was attempted using the SA model with f_{t2} term similar to the previous simulations. However, the steady flow simulation could not converge due to the instability of flow around the leading edge. Thus, second-order in time, time-accurate simulation was conducted, and perfectly-periodic flow field could be obtained. It should be noted that the unsteadiness of the flow was caused purely by flow instability, since the blade motion was not included in this simulation.

Figure 13 presents the unsteadiness of the flow which appeared using the SA model with f_{t2} term. Figure 13 (a) shows the root mean squares (RMS) of blade surface pressure. Large pressure oscillation exists on the impinging area of the detached shock from 62% to 85% span height. The pressure variation was the highest at 75% span height.

Figure 13 (b) shows time history of the pressure on 3.3%, 4.6%, and 7.3% chord length downstream the leading edge at 75% span height. The horizontal axis is time reduced by oscillation period T_b. The pressure signal can be characterized by flat low pressure part and periodic spike-like signal. Similar signal has been reported by Lepicovsky, et al [18] in the result of wind tunnel test conducted in NASA. They explained that low pressure and high pressure parts correspond to supersonic and subsonic

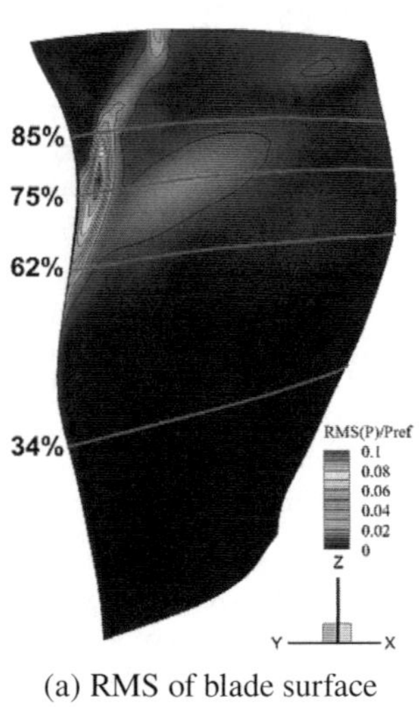

(a) RMS of blade surface pressure on the suction side

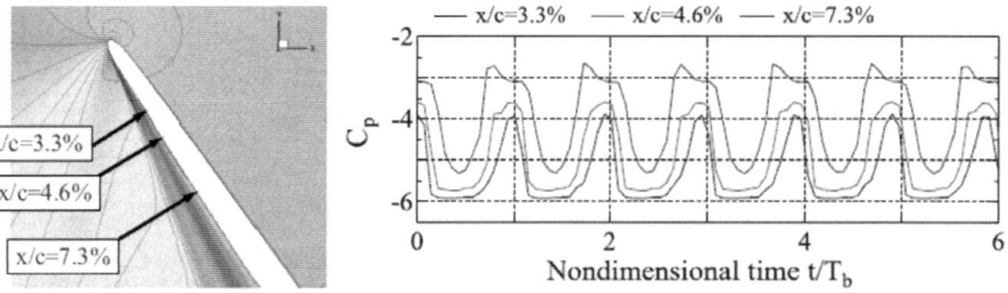

(b) Spike-like periodic pressure signal near the leading edge.

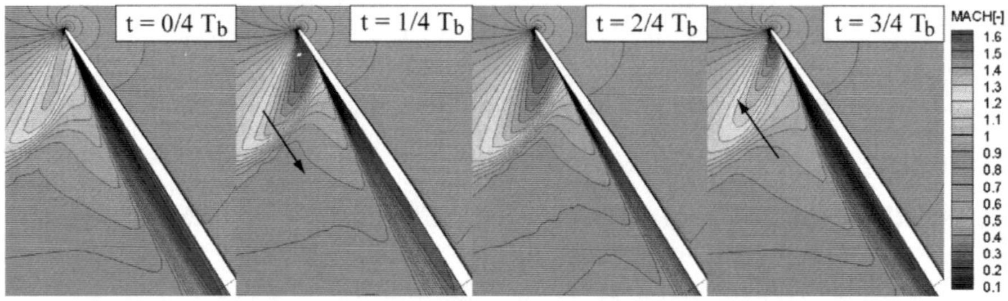

(c) Instantaneous Mach number distribution on the leading edge at 75% span height

Fig. 13 Unsteadiness of the detached shock obtained by a time-accurate CFD simulation at the operating point A on the 75.0% speed line with SA turbulence model including f_{t2} term

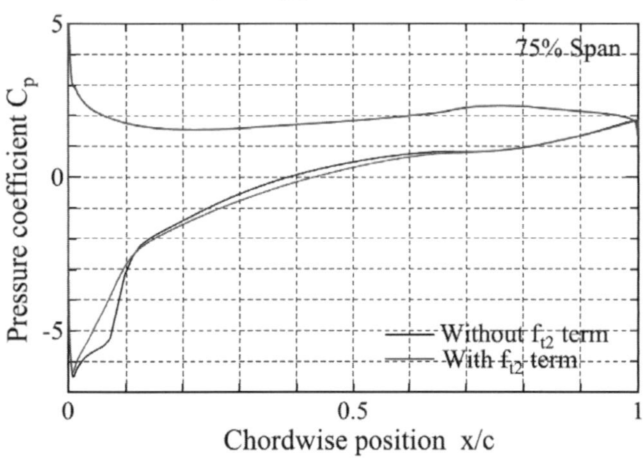

(a) Time-averaged blade surface pressure distribution

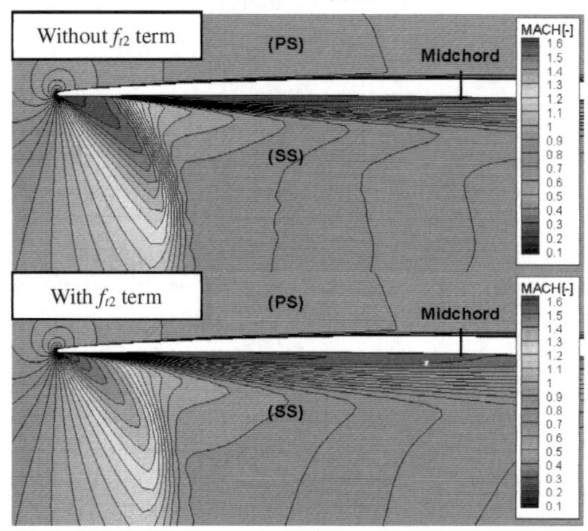

(b) Time-averaged Mach number distribution

Fig. 14 Effect on the unsteady shock motion on the time-averaged flowfield at 75% span height

flow respectively and there exists intermittent switching of separated subsonic / attached supersonic flow.

Figure 13 (c) shows the variation of relative Mach number distribution around the leading edge within one period of the pressure signal at 75% span height. The shock periodically oscillates near the leading edge. At the time of t = 0, the separated low-speed area exists on the leading edge, and the shock disappears. Then the shock appears and moves towards downstream with the decrement of separated area, and after the time of t = 2/4 T_b the shock travels towards the leading edge with increasing the separated area. This coupled motion of the shock and separated area is similar to transonic buffet which can be seen in the transonic supercritical airfoil, but in the present study, it occurred on the thin airfoil with sharp leading edge.

The inclusion of the f_{t2} term resulted in unsteady oscillation of the shock. This transonic buffet-like flow instability also affects to the time-averaged flowfield.

Figure 14 shows time-averaged blade surface pressure and relative Mach number distribution at 75% span height. The pressure distribution in Fig. 14 (a) indicates that the sharp pressure rise due to the impinging shock was disappeared and the pressure distribution became smooth in the case with f_{t2} term. It also could be confirmed from the Mach distribution in Fig. 14 (b). In the case without f_{t2} term, the shock directly impinged on the blade surface. On the other hand, in the case with f_{t2} term, the separation starts from the leading edge and the sharp gradient due to the shock significantly diffused in the chordwise direction.

The effect of buffet-like instability to the flutter characteristic is still not clear. However, if natural oscillation of the shock is completely independent to the blade vibration, only the steady flow should affect to the

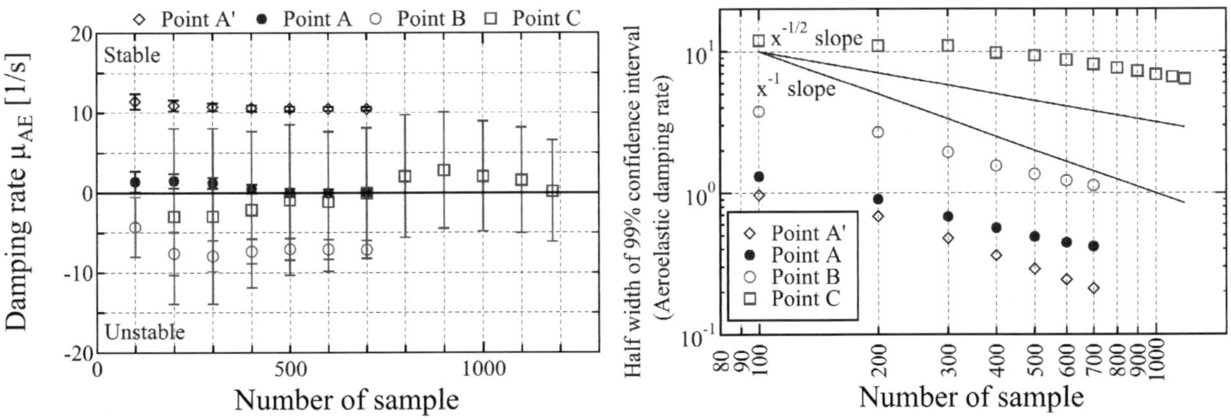

(a) Identified the least stable aeroelastic damping rate

(b) Half width of 99% confidence interval

Fig. 15 Convergence history of aeroelastic damping rate at the operating points on the 75.0% speed line

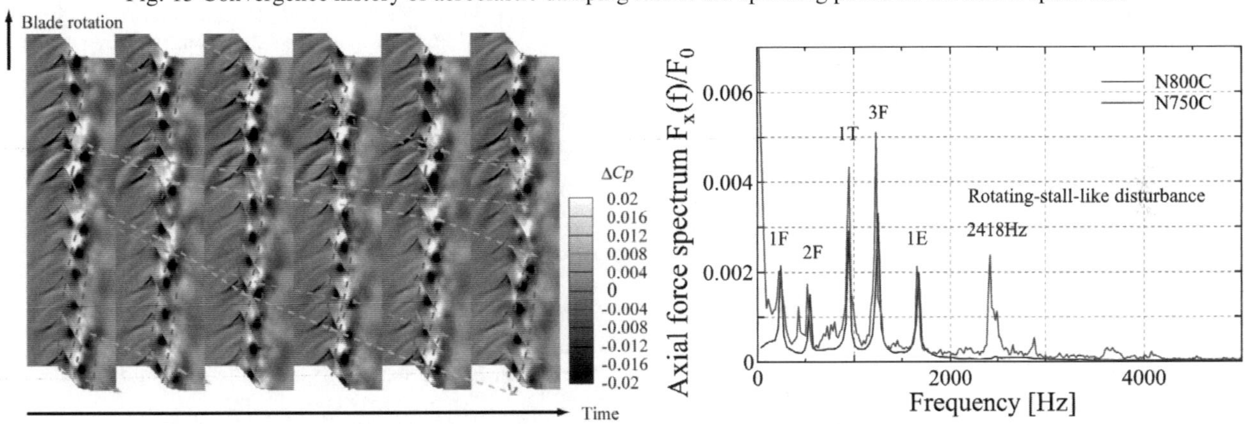

(a) Pressure fluctuation at 95% span height

(b) FFT of axial aerodynamic force acting the blades

Fig. 16 Unsteady flow observed at point C on the 75.0% speed line

motion-excited component of unsteady pressure. Considering the results that impinging area of the detached shock contributes to the excitation of blades and the buffet-like oscillation weakens the shock in the time-averaged flowfield, the flutter might be supressed when severe buffet occurs on the leading edge.

Flow instability and applicability of modal identification

At the end of the discussion part, the deterioration of the uncertainty in modal identification on the points C on the 72.5 and 75.0 speed lines is picked up as a topic to discuss the applicability of modal identification.

Figure 15 shows convergence history of identified aeroelastic damping rates on the 75.0% speed line. In the present simulations, 100 samples of instantaneous time per one cycle of 1F mode were used for the identification. Fig. 15 (a) shows identified the history of damping rate at point A', A, B, and C. The identified damping rate converged after 500 samples, and the error bars were sufficiently small on points A', A, B where the R^2 values were larger than 0.97. On the other hand, significantly large variation and error bar can be seen the point C. Fig. 15 (b) shows the history of the half width of the error bar. All histories were proportional to inverse of square root of sample number. So, in the statistic sense, when the number of samples are increased, the uncertainly of identification can be lessened. However, at the point C the initial uncertainty was very high compared to those on the other operating points.

This high uncertainty was directly related to the flow instability in the CFD simulations. Figure 16 shows the unsteadiness of the flow at point C on the 75.0% speed line. Figure 16 (a) shows pressure fluctuation during the FSI simulation at 95% span height, and different instantaneous snapshots are aligned in the horizontal direction. It can be seen that the low pressure area are clustered over 3 to 6 pitches and traveling toward the opposite direction of the blade rotation (in the rotating frame). This unsteady flow is similar to rotating stall and it might be independent to the oscillation of the blades.

Figure 16 (b) shows the FFT results of axial aerodynamic force acting on the blade at the point C on the 75.0% and 80.0% speed lines. On the 80.0% speed line where the R^2 value was sufficiently high, only motion-excited components corresponding to each blade modes can be observed. However, on the 75.0% speed line, the peak which corresponds to the rotating-stall-like flow superposed on the motion-excited components. Such signal which is independent to the blade excitation is considered as noise for the identification of the motion-excited component of unsteady aerodynamic force. Thus, on the applicability of modal identification technique to the FSI simulations, it can be said that it is difficult to identify aeroelastic modes accurately under strong flow instability. On the other hand, with clean and stable flowfield, accurate and low-uncertainty identification result can be obtained.

CONCLUSION

An Aero-structure integrated approach for the cascade flutter simulations using the identification of aeroelastic modes were developed and applied to the prediction of transonic stall flutter in an advanced transonic fan. The obtained flutter boundary from the FSI simulations were partially agreed with the experimental results on the higher speed lines. On the other hand, on the lower speed lines where flutter did not occur in the experiment, the FSI showed the occurrence of flutter. Steady flowfield and unsteady flowfield was observed in detail in order to clarify the reason of this mismatch and the applicability of modal identification technique to stall flutter analysis. Conclusions can be summarised as follows.

(1) The flow phenomena to be captured for the stall flutter simulations were recognized as: leading edge separation, radial migration of separated flow, detached shock, and these interactions near the wall.

(2) The existence of detached shock on the leading edge on stall-side operating condition leads flutter. On the lower speed lines, the flutter boundary ends where the shock disappears and leading edge separation appears.

(3) When the turbulence model was modified such that laminar region appears around the leading edge, transonic shock buffet occurred. Under the buffet, the detached shock in the time-averaged flowfield significantly diffused and sharp pressure rise disappeared.

(4) Aeroelastic modes can be identified accurately under stable flowfield. With the presence of strong flow instability, the uncertainty of identification becomes large.

ACKNOWLEDGMENTS

This work was supported by JSPS KAKENHI Grant Number 14J10312 and Research Fellowships for Young Scientists from JSPS. IHI developed CEFS1 fan under the contract with New Energy and Industrial Technology Development Organization (NEDO) as a part of "aircraft and space industry innovation program" and "energy innovation program" of Ministry of Economy, Trade and Industry (METI).

REFERENCES

[1] Platzer, M. F. and Carta, F. O. E. "AGARD Manual on Aeroelasticity in Axial-Flow Turbomachines Volume 1: Unsteady Turbomachinery Aerodynamics". AGARDograph No. 298, Vol. 1, 1988.

[2] Vahdati, M., Sayma, A. I., Imregun, M., and Simpson, G. "Influence of Intake on Fan Blade Flutter". ASME Paper GT2014-25859, 2014.

[3] Jeffers, J. D. and Meece, C. E. "F100 Fan Stall Flutter Problem Review and Solution". Journal of Aircraft Vol. 12, No. 4, pp.350-357, Apl. 1975.

[4] Weir, D. "Design and Test of Fan/Nacelle Models Quiet High-Speed Fan". NASA CR-2003-212370, Jul. 2003.

[5] Sanders, A. J., Hassan, K. K., and Rabe, D. C. "Experimental and Numerical Study of Stall Flutter in a Transonic Low-Aspect Ratio Fan Blisk". Transactions of the ASME, Vol. 126, pp. 166-174, Jan. 2004.

[6] Stargardter, H. "Subsonic/Transonic Stall Flutter Study Final Report". NASA-CR-165256, Pratt and Whitney Aircraft Group, Jun. 1979.

[7] Chuang, H. A. and Verdon, J. M. "A Nonlinear Numerical Simulator for Three-Dimensional Flows Through Vibrating Blade Rows". Journal of Turbomachinery, Vol. 121, No. 2, pp. 348-357, 1999.

[8] Hall, K. C., Thomas, J. P., and Clark, W. S. "Computation of Unsteady Nonlinear Flows in Cascades Using a Harmonic Balance Technique". AIAA Journal, Vol. 40, No. 5, pp. 879-886, May 2002

[9] Doi, H. and Alonso, J. J. "Fluid/Structure Coupled Aeroelastic Computations for Transonic Flows in Turbomachinery". ASME Paper GT-2002-30313, 2002.

[10] Tateishi, A., Watanabe, T., and Himeno, T. "Multimode Flutter Analysis Based on Time-Domain Fluid-Solid Interaction Simulation and System Identification". Proceedings of the 13th International Symposium on Unsteady Aerodynamics, Aeroacoustics and Aeroelasticity of Turbomachines, ISUAAAT13-S2-4, 2012

[11] Tateishi, A., Watanabe, T., Himeno, T., and Inoue, C. "Multimode Flutter Analysis of Transonic Fan Using FSI Simulation". ASME Paper GT2014-26702, 2014.

[12] Shima, E. and Jounouchi, T. "Role of Computational Fluid Dynamics in Aeronautical Engineering (No.12) - Formulation and Verification of Uni-Particle Upwind Schemes for the Euler Equations". Proceedings of the 12th NAL Symposium on Aircraft Computational Aerodynamics, NAL-SP 27, pp.255-260, 1994

[13] Giles, M. B. "Nonreflecting Boundary Conditions for Euler Equation Calculations". AIAA Journal Vol. 28, No. 12, 1990.

[14] Spalart, P. R. and Allmaras, S. R. "A One-Equation Turbulence Model for Aerodynamic Flows". Recherche Aerospatiale, No. 1, pp. 5-21, 1994.

[15] Murooka, T., Goto, S., Mizuta, I., and Kodama, H. "New Concept Design and Development of an Advanced Transonic Fan Rotor". IGTC2007 Tokyo, TS-053, 2007.

[16] Aotsuka, M., Tsuchiya, N., Horiguchi, Y., Yamamoto, K., and Nozaki, O. "Numerical Simulation of Transonic Fan Flutter with 3D N-S CFD Code". ASME Paper GT2008-50573, 2008.

[17] Aotsuka, M. and Murooka, T. "Numerical Analysis of Fan Transonic Stall Flutter". ASME Paper GT2014-26703, 2014.

[18] Lepicovsky, J., McFarland, E. R., Chima, R. V., Capece, V. R., and Hayden, J. "Intermittent Flow Regimes in a Transonic Fan Airfoil Cascade". NASA TM-2002-211375, Feb. 2002.

[19] Rumsey, C. L. "Apparent Transition Behavior of Widely-used Turbulence Models". Int. J. of Heat and Fluid flow, Vol. 28, pp. 1460-1471, May 2007.

APPENDIX: TRANSITION-LIKE BEHAVIOUR OF SPALART-ALLMARAS MODEL

The turbulence model employed in this study was the Spallart-Allmaras model and the sensitivity of f_{t2} term to the CFD result is shown in the discussion part. The governing equation of the model, major control functions and coefficients can be expressed as follows.

$$\frac{\partial\rho\hat{\nu}}{\partial t}+\frac{\partial\rho\hat{\nu}v_j}{\partial x_j}=c_{b1}(1-f_{t2})\hat{S}\rho\hat{\nu}-\rho\left[c_{w1}f_w-\frac{c_{b1}}{\kappa^2}f_{t2}\right]\left(\frac{\hat{\nu}_t}{d}\right)^2+\frac{1}{\sigma}\left[\frac{\partial}{\partial x_j}\left\{(\mu+\mu_t)\frac{\partial\hat{\nu}}{\partial x_j}\right\}+\rho c_{b2}\frac{\partial\hat{\nu}}{\partial x_k}\frac{\partial\hat{\nu}}{\partial x_k}\right]$$

$$\mu_t=\rho\hat{\nu}f_{v1},\qquad f_{v1}=\frac{\chi^3}{\chi^3+c_{v1}^3},\qquad \chi=\frac{\hat{\nu}}{\nu}$$

$$\hat{S}=\Omega+\frac{\hat{\nu}}{\kappa^2d^2}f_{v2},\qquad \Omega=\sqrt{2W_{ij}W_{ij}}$$

$$f_{t2}=c_{t3}\exp(-c_{t4}\chi^2),\qquad c_{t3}=1.2,\qquad c_{t4}=0.5$$

Here, ρ, v, $\hat{\nu}$, are density, velocity, and model variable related to the eddy viscosity, respectively. Coefficients cs and functions fs are model parameters and functions. The function f_{t2} located in the production and dissipation term is originally introduced to the model for the use of "trip term" to control the onset point of transition. Rumsey [19] explains about the role of f_{t2} that it controls the stability of numerical solution such that $\hat{\nu}=0$ to be a stable solution to keep the flow laminar on the upstream of transition point. It was also shown in [19] that short laminarized region appeared on the leading edge and the exclusion of f_{t2} (setting $c_{t3}=0$) removed this region and the boundary layer shows completely turbulent property.

As reported by Rumsey, the inclusion of the f_{t2} term brings the laminarized region on the leading edge with the present implementation to the CFD code used in this study. Figure A1 shows the effect of the f_{t2} term on the numerical result of basic subsonic turbulent boundary layer on the flat plate. In this case, the free stream was fully subsonic and the free stream Mach number was 0.059. The leading edge of the flat plate is located on x = 0. Here, the SA model with and without f_{t2} term are compared with respect to the onset of turbulent boundary layer. Figure A1 (a) shows spatial distributions of model variable $\hat{\nu}$ relative to free stream kinematic viscosity. The result obtained by the model without f_{t2} shows the production of $\hat{\nu}$ immediately after the leading edge. On the other hand, in the result obtained by the model with f_{t2}, low eddy-viscosity region can be seen near the leading edge. Figure A1 (b) shows the comparison of streamwise distributions of local skin-friction coefficient. Here, distributions of "Blasius" and "Turbulent BL" are given by following correlations as a function of streamwise Reynolds number Re(x).

$$C_f=0.664\mathrm{Re}(x)^{-0.5}\ \text{(Blasius)},\quad C_f=0.0576\mathrm{Re}(x)^{-0.2}\ \text{(Turbulent BL)}$$

Without f_{t2} term, the skin-friction follows the distribution of turbulent boundary layer from the leading edge, and this condition frequently referred as "fully-turbulent" computation. In the result with f_{t2} term, short laminar region which follows Blasius's distribution appeared from the leading edge to Re(x) = 1×10^5, and transition ends at Re(x) = 2×10^5. As can be seen from these results, the effect of the inclusion of f_{t2} term to the SA model in the present implementation brings laminarized region and transition-like behaviour around the leading edge.

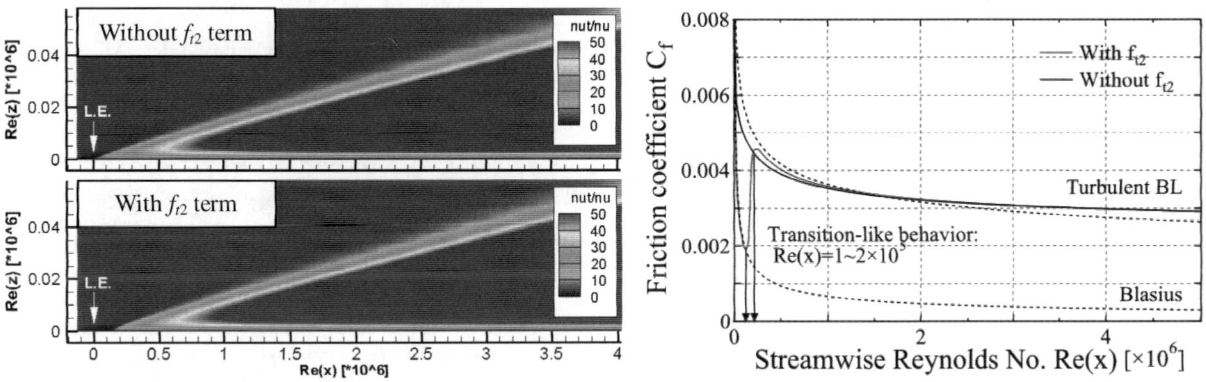

(a) Spatial distributions of $\hat{\nu}$ (b) Skin-friction coefficient distributions on the plate

Fig. A1 Effect of the f_{t2} term on the simulation of turbulent boundary layer on the flat plate

The effects of mistuning on Fan flutter

M. Vahdati
Imperial College London
Mechanical Engineering Department
Exhibition Road, London SW7 2AZ, UK
m.vahdati@imperial.ac.uk

L. Salles
Imperial College London
Mechanical Engineering Department
Exhibition Road, London SW7 2AZ, UK
l.salles@imperial.ac.uk

ABSTRACT

The aim of this paper is to study the effects of mistuning on fan flutter. The computations are based on a three-dimensional, whole assembly, time-accurate, viscous, finite-volume compressible flow solver. The unsteady flow cases are computed as Reynolds-averaged Navier–Stokes, with the basic assumption that the frequencies of interest are sufficiently far away from the frequencies of turbulent flow structures. The overall solution method is implicit, with second-order accuracy in space and time. A rig wide-chord fan blade, typical of modern civil designs, was used as the benchmark geometry for this study. There are 20 blades in the fan assembly. A random pattern frequency mistuning was used in most of analysis undertaken in this paper. The 1F mode shapes are the same for all mistuned blades. The objectives of this work are:

1. To check if the introduction of mistuning would bring the experimental and computed flutter boundaries closer
2. To explain how mistuning provide stability
3. Establish a relationship between mistuning and damping
4. Provide a data base which can be used in future for testing simple mistuning models

NOMENCLATURE

f	blade natural frequency, Hz
f_r	reduced frequency, fc/V
m	mass flow rate
m_{ref}	normalised corrected mass flow rate
p	static pressure
p_0	stagnation pressure
r	radius
r_t	rotor tip radius
V	velocity into rotor blades
Φ	inter-blade phase angle
θ	tangential coordinate
ω	rotor angular velocity
Ω	rotor non-dimensional angular velocity ($\omega 2\pi r_t/a_o$)

INTRODUCTION

It is well known that mistuning can have significant effect on fan flutter. The vast majority of turbomachinery blade mistuning studies have been conducted using only the effects of structural coupling. However, aerodynamic coupling between blades can have significant effects on the stability and blade forced response amplitudes. The significance of the aerodynamic coupling effects has been shown in recent literature. For example, Kielb et al. [1,2]and Miyakozawa[3] investigated the effects of aerodynamic coupling on both flutter and forced response using the single family of modes approach. The results suggested that the mistuning effects on flutter and forced response are dependent on the structural coupling levels. Campobasso et al. [4] conducted mathematical studies of flutter and forced response including the aerodynamic coupling effects assuming that the effect of the aerodynamic force is due to the blade motions of immediate neighbours. Sladojevic et al. [5] studied the correlation between the aerodynamic blade-to-blade coupling in mistuned bladed disk assemblies and the level of forced response. The analysis showed that the amplification factors were greatly increased due to aerodynamic coupling. He et al. [6] investigated the effects of aerodynamic coupling on both flutter and forced response problems using the tuned system modes to calculate the unsteady aerodynamic forces. For the case studied, aerodynamic coupling showed significant effects on the vibration amplitudes. All of these studies considered only blade frequency mistuning.

THE AERODYNAMIC COMPUTATION METHOD

The calculations in this paper were performed with the AU3D code written in Imperial College and developed over many years with support from Rolls-Royce [7]. The computations are performed using a three-dimensional, whole assembly, time-accurate, viscous approach. The unsteady flow cases are computed as Reynolds-averaged Navier–Stokes, with the basic assumption that the frequencies of interest are sufficiently far away from the frequencies of turbulent flow structures. The overall solution method is implicit, with second-order accuracy in space and time. The current computations use the one-equation Spalart–Allmaras turbulence model [8]. The parameters in Spalart–Allmaras have been adjusted on previous fans to get good agreement near the stability limit; the parameters are held constant in all the present work. The resulting CFD code has been used over the past 20 years for flows at off design conditions with a good degree of success [9, 10, 11].

TEST CASE AND STEADY FLOW

A rig wide-chord fan blade, typical of modern civil designs, was used as the benchmark geometry for this study. The rig was used extensively to study flutter. There are 20 blades in the fan assembly. More details of test case can be found in [9]. **Fig. 1** compares the calculated and measured speed line for the fan used in this report for a tip Mach number close to sonic. For this speed the non-dimensional fan speed is Ω =0.89. The pressure ratio and mass flow are seen to be in adequate agreement.

The speedline is terminated to the left by the instability line and to the right is shown the normal working line. It should be noted from **Fig. 1** that the instability both measured and calculated can either be stall/surge or flutter, with flutter occurring over a narrow range of fan rotational speed, referred to as 'flutter bite'. The calculations, based on the known frequency and mode of vibration show instability due to flutter at somewhat higher mass flow rate than those measured. The higher computed mass flow for instability is attributed to the presence of some mechanical damping and mistuning in the real fan which is wholly omitted in the present calculations. The trend is well predicted, notably the flutter bite (i.e.,instability at increased throttle area) around the sonic speedline. One of the aims of this work is to see if the introduction of mistuning would bring the experimental and computed flutter boundaries closer and the level of mistuning required for this purpose.

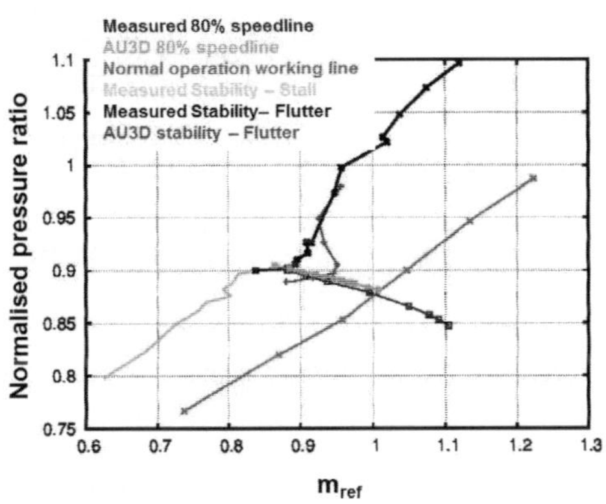

Fig 1. Performance plot for the model fan used in this investigation

It should be pointed out that the calculation in **Fig. 1** is for a complete fan, with outlet guide vanes and engine-section stators and that it includes an intake upstream of the fan. The intake was present for the measurements and the comparison must include the appropriate upstream boundary. The intake can decrease or increase stability, depending on the length, since the waves upstream of the rotor are reflected back from the intake plane and the phasing is critical [12]. Although the results in **Fig. 1** includes the intake to allow comparison with measurement, later calculations in this report assume no reflections, corresponding to infinitely long intake and exhaust ducts. The removal of reflection is achieved in the calculations by coarsening the computation mesh upstream and downstream of the rotor.

Fig. 2 shows the constant speed characteristic at Ω=0.89 for which flutter was observed during rig test.

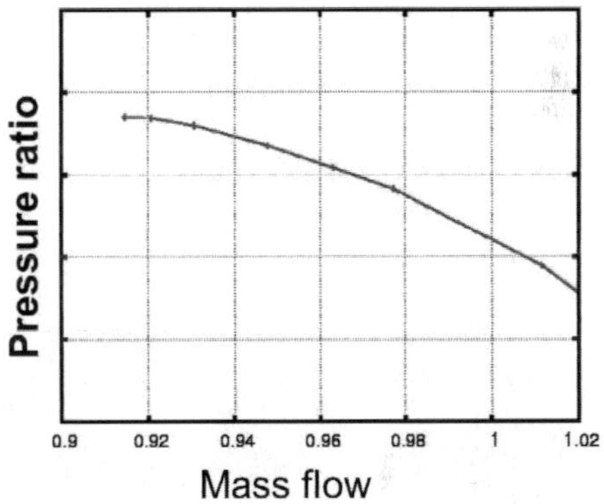

Fig 2: Constant speed characteristic at part speed

Fig 3a shows contours of isentropic Mach number in the blade-to-blade surface at 90% span for mass flow of m=0.93 when the relative velocity onto the rotor tip is close to sonic, Ω =0.89. There is a well-developed supersonic region downstream of the leading edge, terminated by a shock at about 20% chord. Fig. 3b shows the variation of static pressure along the blade at 90% height for different mass flows at Ω =0.89. It is seen from this plot that for all the operating points considered at this speed, the shock is expelled. However, the shock moves forward and becomes stronger as the mass flow decreases. Moreover, it can be seen from this plot that the pressure is nearly uniform along the pressure surface and nearly uniform downstream of the shock on the suction surface.

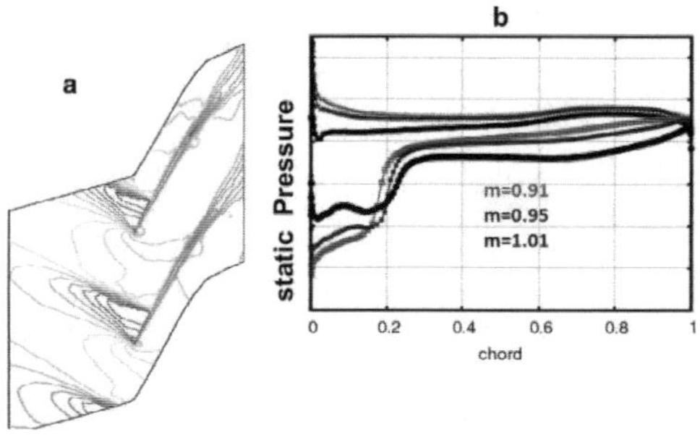

Fig 3: Typical view of steady flow at the speed. (a) Mach number contours at 90% height for m=0.93; (b) Static pressure variations along the blade at 90% height

MECHANICAL MODEL

The mode shape and frequency used here were obtained from an FE package and were interpolated onto the CFD grid. The mode shapes remain fixed during computation. It is an underlying assumption that the mechanical mode is unaffected by the aerodynamic forces. Only the 1st mode (1F), shown in **Fig. 4a**, is considered at this stage.

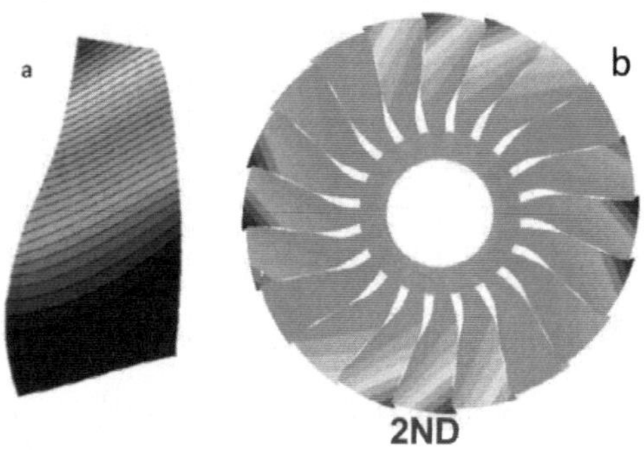

Fig. 4: (a) View of 1F mode shape for 1 blade, (b) Expaned 1F/2ND mode shape for whole assemblt

The usual approach in flutter computations is to expand the mode shape of **Fig. 4a** for all the blades and Nodal Diameters (ND) of interest. **Fig. 4b** shows the 1F/2ND expanded mode. This method of expansion will be referred to as ND expansion of mode shape in the rest of this paper. Unsteady flutter computations begin from a steady state solution by exciting all the nodal diameters of interest. The amplitude of motion is large enough to be well outside the noise range, but small enough for the unsteady flow to be represented as linear; the imposed tip displacements are typically of the order of 0.3% of blade chord at the tip. The time history of displacement of each ND is tracked to establish whether that ND is stable or unstable. The stability of each mode is determined from the time history of the modal displacement. **Fig. 5a** shows typical time history of displacement of a stable mode (black curve) and an unstable mode (red curve). Aerodynamic damping is evaluated by fitting a line to the modal displacement amplitudes. The slope of the line on a log scale yields the logarithmic decrement (logdec) with a positive value corresponding to positive aerodynamic damping and negative value corresponding to negative damping. **Fig. 5b** shows damping plotted against ND for mass flow of, m=09.3. It is seen from this plot that for this case 2ND mode is unstable.

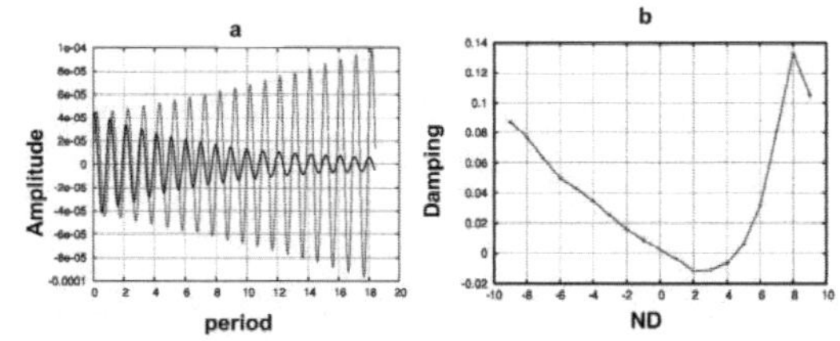

Fig 5: (a) Time history of displacement for a stable ND (black) and unstable ND (red); (b) Damping for m=0.93 as a function of ND

The approach shown above cannot be used for mistuned assemblies, as the expansion method used does not differentiate between blades (it can only distinguish between NDs), and hence one cannot assign different frequencies to blades. Therefore, in the absence of mode shapes for the whole assembly, a new approach for expansion of mode shapes is required. In the approach taken here, it is assumed that the fan assembly displacement is made of 20 different 1F modes, for which mode N corresponds to blade N having displacement while the displacement for all the other blades being zero. Modes 1, 5 and 13 are shown in **Fig. 6**. Therefore, with this approach, it is assumed that there is no structural coupling between blades and coupling is only through air. At time T=0.0, only blade 1 is excited and the time history of all the blades are tracked. This method of mode shape expansion will be referred to as blade expansion in the rest of the paper.

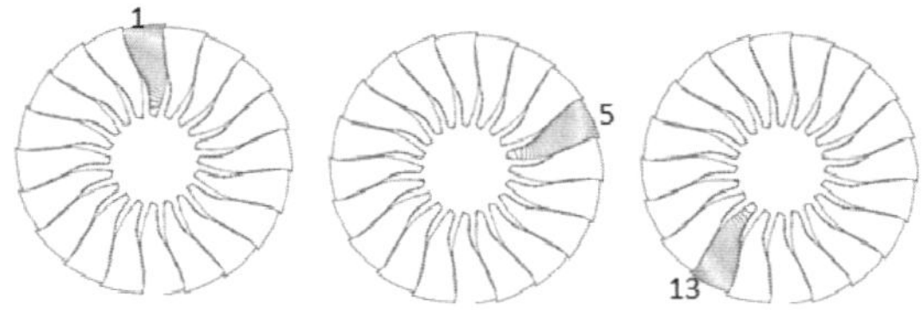

Fig 6: Alternative form of mode shape expnsion; blades with no contours have zero displacement in the mode

Before proceeding to mistuned computations, the blade expansion of the mode shape was checked. **Fig. 7** shows the time history of displacement of displacement for 4 blades around the assembly for a stable point (m=0.98) and an unstable point (m=0.93). It should be noted that at T=0.0 only blade 1 is excited.

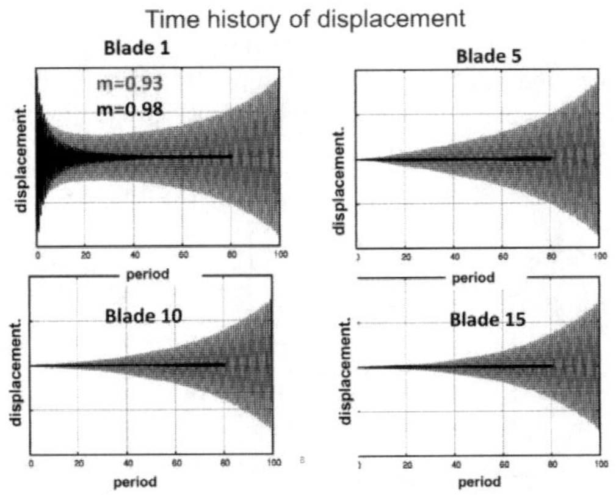

Fig 7: Time history of displacement for 4 blades around the annulus, m=0.93 unstable; m=0.98 stable

It can be seen from **Fig. 7** that initially (up to 10 periods of vibration) the time history of displacement for both mass flows is similar. However, for m=0.93 the initial disturbance of blade 1 is able to travel around the annulus and excite other blades and cause flutter, whereas for m=0.98 the initial excitation is unable to cause vibration on other blades. This can be clearly seen from **Fig. 8** which shows displacement plotted as a function of blade number for 5 instants of time during the motion.

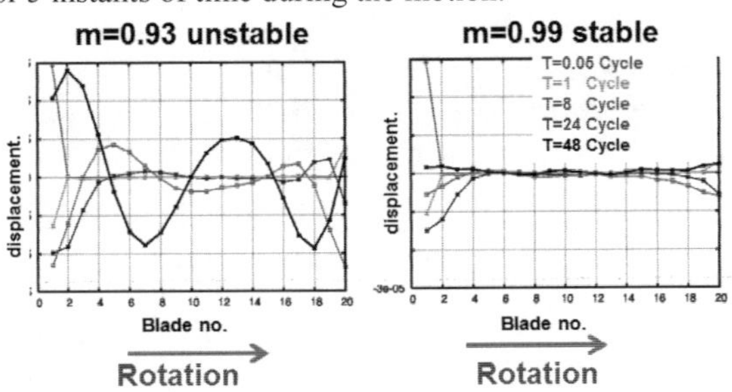

Fig 8: Time history of displacement as a function of blade number

It is seen from this plot that at T=0.05 cycle (red curve) and T=1 cycle (green curve) the displacement is much localised (around blade 1) and does not differ between the two mass flows. At T=8 cycles (blue curve) it is still hard to distinguish between the two mass flows cases, although for m=0.93more blades (19, 18 and 17) show vibration. At T=24 cycle (pink curve), the characteristic of vibration is totally different between the two mass flow cases and all the blades for m=0.93 show response. At T=48 cycles (black curve), a clear 2ND pattern is developed for m=0.93, whereas for m=0.98 all the vibration has died out. This can be clearly seen from **Fig. 9** which shows the time history of displacement after 90 cycles.. It can also be seen from this plot that, for m=0.93, the travelling wave moves by two blades in the direction of rotation in ¼ of cycle, which is consistent with a 2ND travelling wave pattern.

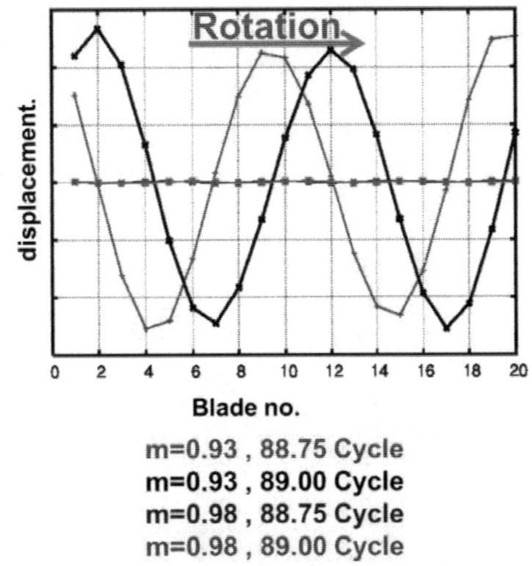

Fig 9: Time history of displacement as a function of blade number

Snapshots of unsteady pressure on the suction surface and the casing at two instants of time are shown in **Fig. 10**. It is seen from these plots that initially (T=0.05 Cycle), the disturbance is confined around blade 1 (Top dead centre) but after 80 cycles a clear 2ND pattern has developed. This clearly shows that aero coupling is sufficient to determine the least stable ND.

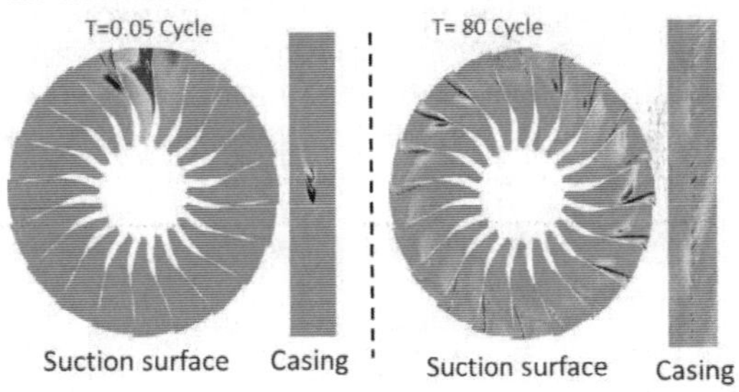

Fig 10: Instantaneous variations of unsteady pressure on the suction surface and the casing for m=0.93

Fig. 11 shows the comparison of damping between the ND and blade expanded mode shapes. For the ND expanded mode shapes, the damping curve is for a 2ND pattern. For the blade expanded mode shape computations, the unstable points at m=0.93 and m=0.91 converge to a 2ND pattern. As can be seen from time history plots of **Fig. 7**, obtaining damping from the time histories is not straight forward. In the strategy used in this work, for the stable points damping has been obtained from time history of blade 1 between 60-80 cycles, and for the unstable points damping has been obtained from time history of displacement over the last part of (80+ cycle) motion. For the unstable points, as the number of cycles increases the damping of all the blades converge to the same value. It is seen from this plot that, as expected, the two approaches give similar values for damping. The blade expanded mode shape strategy needs around 100 cycle of vibration for accurate computations of damping, whereas ND expanded mode shape strategy require around 10 cycle of vibrations. Therefore, blade expanded mode shape strategy is not advisable for tuned computations. Also shown in **Fig. 7**, is the overall damping of the blade 1 which can be interpreted as the sum of damping for all NDs.

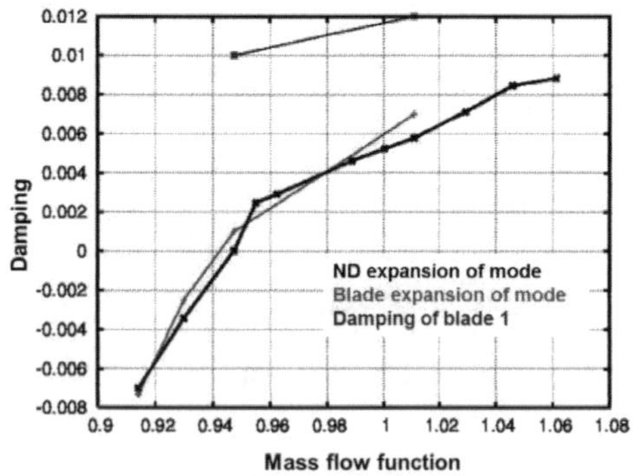

Fig 11: Comparison of damping between ND and blade expend mode shape.

MISTUNED COMPUTATIONS

The frequency mistuned pattern used for this study is shown in **Fig. 12**. The pattern has been obtained using a random number generator. The mode shape of blades is assumed to remain the same. In **Fig. 12** the y-axis is the amplitude of mistuning, which is defined as $(F_{max}-F_{min})/F_{ave}$, where F_{max}, F_{min} and F_{ave} are maximum, minimum and average frequencies respectively. For the case shown in **Fig. 12** the amplitude of mistuning is 6%.

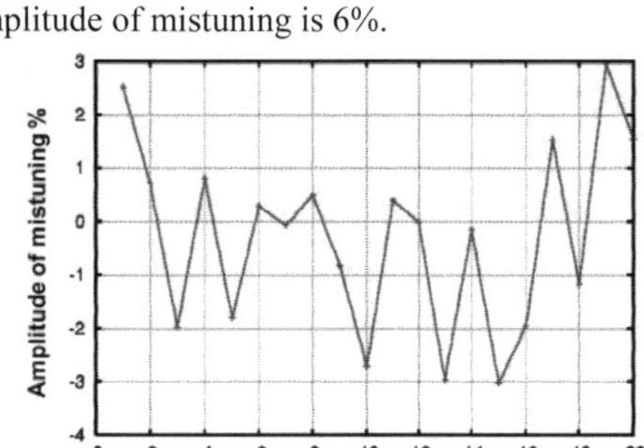

Fig 12: Frequency mistuning pattern used in this study

Initially, the computations were performed for unstable mass flow of m=0.93. **Fig. 13** shows the time history of displacement for 4 blades around the assembly for the tuned and the mistuned fan assemblies. It is seen that the addition of mistuning pattern of **Fig. 12** has made the blade stable. On comparison with **Fig. 7**, it is seen that the time history of mistuned assembly at m=0.93 is very similar to tuned system at m=0.98. This suggests that the introduction of mistuning stops the initial disturbance from propagating round the circumference.

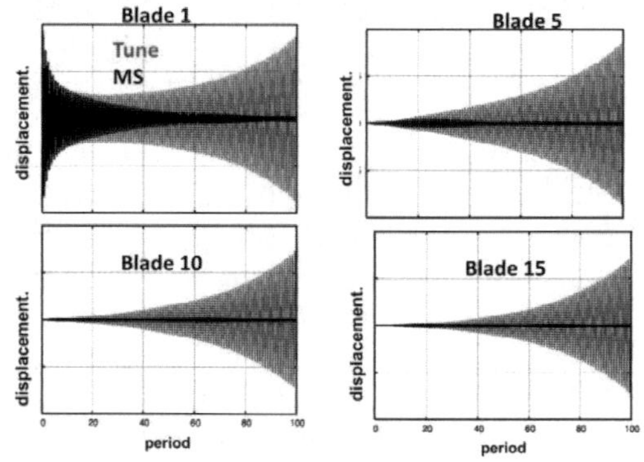

Fig 13: Time history of displacement for 4 blades around the annulus at m=0.93

This observation is further evident from the plots of **Fig 14** which shows the displacement plotted as a function of blade number for 5 instants of time during the motion for the tuned and the mistuned cases.

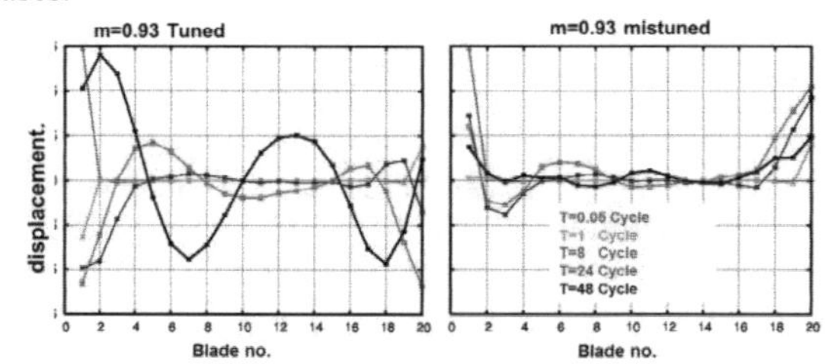

Fig 14: Time history of displacement as a function of blade number

The FFT of displacement around the circumference after 100 cycles for the tuned and mistuned systems is shown in **Fig. 15**. It is seen from this plot that for the tuned system the 2ND is the only mode present, whereas for the mistuned system there is response at in 1-4 NDs. It should be noted that the mistuned plot curve in **Fig. 15** has been multiplied by 5 so that it is visible in the plot.

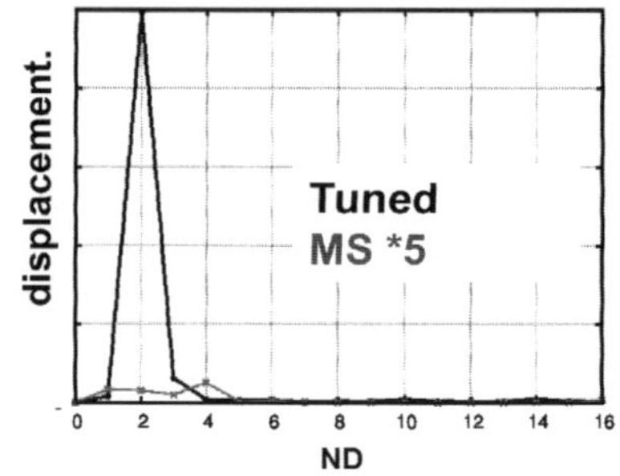

Fig 15: Fourier transform of displacement along the circumference

The time histories for mistuned system show a non-linear behaviour as can be seen from **Fig. 13** and **Fig. 16**. It can be seen from Fig. 16 that both blades respond in range of frequencies around the average frequency. Moreover, low frequency components are present in the response which is a due to interaction between blades.

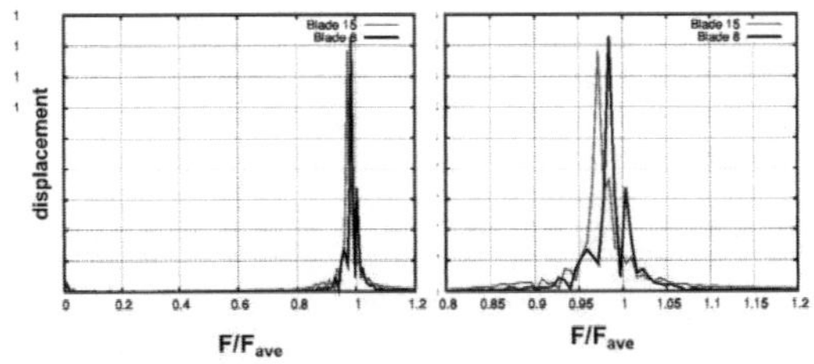

Fig 16: Fourier transform of time history displacement for two blade for mistuned case.

Fig. 17, shows the computed aero damping plotted against mass flow for a tuned system and the frequency mistuned assembly of **Fig 12**. In this plot a negative damping represents flutter. The results show that for the tuned system as the mass flow decreases, there is a sharp drop in aero damping and the blade flutters at mass flows below 0.95. However, for the level of mistuning used in these computations the blade remains stable at all the mass flows and the fan blade stalls before it flutters. As shown, above the time history for the mistuned cases is non-linear and hence obtaining a damping from a non-linear curve is quite approximate. The mistuned damping of **Fig. 17** has been obtained from time history of blade 1 between 90-100 cycles.

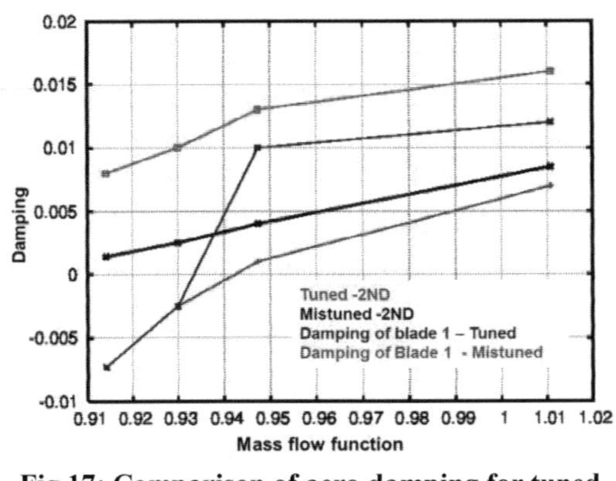

Fig 17: Comparison of aero damping for tuned and mistuned assembly

The effect of amplitude of mistuning on flutter at mass flows of m=0.91 and m=0.93 was considered next. **Fig 6** shows aero-damping plotted as a function of amplitude of mistuning. The amplitude of mistuning is defined as $(F_{max}-F_{min})/F_{ave}$, where F_{max}, F_{min} and F_{ave} are maximum, minimum and average frequencies respectively. For the pattern shown in **Fig. 12** the amplitude of mistuning is 6%. It should be noted that the same pattern of frequency mistuning is used in all the computations of this section and only the amplitude of mistuning (maximum frequency to minimum frequency) is changed. **Fig. 18** shows the aero damping of the fan assembly for m=0.91 and m=0.93 plotted against the amplitude of mistuning. The tuned system is shown as 0% mistuning level. It is seen from this plot that amplitude of mistuning has a significant effect on the aero damping of the assembly. For the pattern used, at m=0.91, a 3.2% amplitude is required for the blade to become stable, whereas for m=0.93 the amplitude is only 2%. Moreover, the aero damping of the blade asymptotes as the amount of mistuning is increased. Therefore, increasing mistuning will increase the flutter margin of a fan blade. A 2% mistuning level would bring the computed flutter boundary of **Fig. 1** much closer to the measured one.

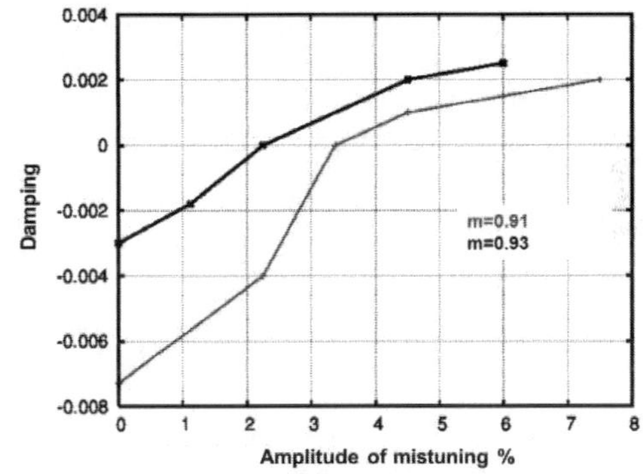

Fig 18: Aero damping as a function of amplitude of mistuning

Fig 19 which shows the displacement plotted as a function of blade number for 3 instants of time during the motion for an unstable mistuned case (m=0.91 and 1% amplitude of mistuning). It is seen from this figure that in the presence of mistuning instability will occur in more than one nodal diameter.

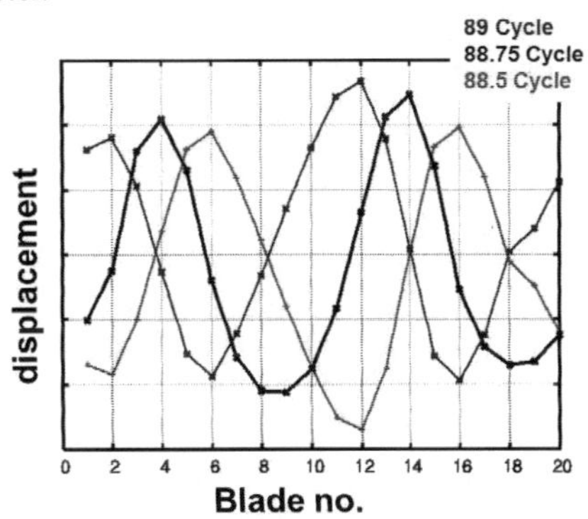

Fig 19: Time history of displacement as a function of blade number for an unstable mistuned case

Figure 20 shows the FFT of displacement around the circumference. It clearly shows the presence of 1, 2, 3 and 4 NDs is the response.

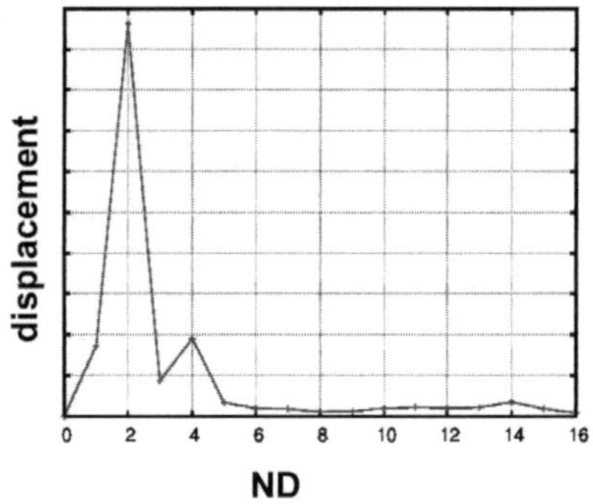

Fig 19: FFT of displacement around the circumference for an unstable mistuned case

CONCLUSIONS AND FUTURE WORK

The results clearly show that mistuning has a substantial effect on flutter stability of the fan blade and can increase the flutter margin of a fan blade significantly. Computed CFD results show that using a 2% random mistuning pattern will bring the computed and measured flutter boundary closer together. Initial results suggest that the amplitude of mistuning has much bigger influence on damping than the pattern of mistuning. Moreover, for points at lower flow, the amplitude of mistuning has to increase to guarantee blade stability. However, non-linear whole assembly, aeroelastic computations with mistuning require large amount of CPU time and faster models such as the one suggested in [1, 13] are required. In a future work, the results obtained in this work will be compared to those obtained from influence coefficient model of [1].

Acknowledgements

The authors thank Rolls-Royce plc for both sponsoring this work and allowing its publication. The first author also wishes to acknowledge support from UK TSB under the SILOET project. They gratefully acknowledge the contribution of their colleagues at Rolls-Royce plc: J. Green, R. Brooks, M. Wilson, and K. Johal.

REFERENCES

[1] Kielb, R. E., Feiner, D. M., Griffn, J. H., and Miyakozawa, T., 2004. , "Flutter of mistuned bladed disks and blisks with aerodynamic and FMM structural coupling". Proceedings of the ASME Turbo Expo 2004, Vienna, Austria, June 14–17, 2004, 6, pp. 573-579.

[2] Kielb, R. E., Feiner, D. M., Griffn, J. H., and Miyakozawa, T., March,2005. "Probabilistic analysis of mistuned bladed disks and blisks with aerodynamic and fmm structural coupling". 9th National Turbine Engine HCF Conference.

[3] Miyakozawa, T, 2008, "Flutter and forced response of turbomachinery with frequency mistuning and aerodynamic asymmetry", PhD thesis, Department of Mechanical Engineering and Materials Science Duke University

[4] Campobasso, M., and Giles, M., 2000. "Flutter and forced response of mistuned turbomachinery", Oxford University Computing Laboratory Report NA 00/20.

[5] Sladojevic, J., Petrov, E., Sayma, A., Imregun, M., and Green, J., 2005. "Investigation of the influence of aerodynamic coupling on response levels of mistuned bladed discs with weak structural coupling". Proceedings of the ASME Turbo Expo, 4, pp. 543 - 551.

[6] He, Z., Epureanu, B. I., and Pierre, C., 2007. "Fluid-structural coupling effects on the dynamics of mistuned bladed disks". AIAA Journal, 45(3), pp. 552 - 561

[7]Sayma, A. I., Vahdati, M., Sbardella, L., and Imregun, M., 2000, "Modeling of Three-Dimensional Viscous Compressible Turbomachinery Flows Using Unstructured Hybrid Grids," AIAA J., 38(6), pp. 945–954

[8] Spalart, P. R., and Allmaras, S. R., 1992, "A One-Equation Turbulence Model for Aerodynamic Flows," AIAA Paper No. 92-0439

[9] Vahdati, M., and Cumpsty, N. A., 2012, "Mechanisms for Wide-Chord Fan Blade Flutter," 13th International Symposium on Unsteady Aerodynamics, Aeroacoustics and Aeroelasticity of Turbomachines, University of Tokyo, Japan, Sept. 11–14, Paper No. ISUAAAT 13-I-5.

[10] Choi, M., Smith, N.H.S and Vahdati. M, 2013, "Validation of numerical simulation for rotating stall in a transonic fan", Journal of turbomachinery 135 (2), 021004

[11] J Dodds, J., Vahdati, M., 2015, "Rotating Stall Observations in a High Speed Compressor—Part II: Numerical Study" , Journal of Turbomachinery 137 (5), 051003

[12]Vahdati, M, Smith, N.H.S, Zhao, F.,2015, "Influence of Intake on Fan Blade Flutter", Journal of Turbomachinery 137 (8), 081002

[13] Kielb, R. E., Hall, K. C., Hong, E., and Pai, S. S., 2006, "Probabilistic Flutter Analysis of a Mistuned Bladed Disks," ASME Paper No. GT2006-90847, Barcelona, Spain, May 8-11, 2006

Proceedings of the 14th International Symposium on Unsteady Aerodynamics, Aeroacoustics & Aeroelasticity of Turbomachines
ISUAAAT14
8-11 September 2015, Stockholm, Sweden

I14-S6-1

INTENTIONAL ADJUSTMENT OF EXCITATION SOURCE PHASING - PHYSICAL INTERPRETATION OF AXIAL GAP VARIATION

Martin Enneking, Stephan Behre, Florian Fruth, Peter Jeschke

Institute of Jet Propulsion and Turbomachinery, RWTH Aachen University
D-52062 Aachen, Germany

ABSTRACT

The following paper presents the results of detailed unsteady numerical investigations on a 1.5 stage subsonic axial flow turbine. The main focus of this paper lies on the examination and consecutive evaluation of the unsteady forces acting on the rotor blade for three different blade row spacings. Thorough evaluation of the data showed potential to reduce the unsteady forces by means of the axial gap.

In a first step, time resolved blade forces were examined followed by an FFT analysis of the force vector for all gap configurations. As the gap size is enlarged from the smallest to the middle size the expected decrease of the circumferential force was found. However, contrary to intuitive consideration, a further increase of the gap from middle to largest gap size lead to an increase in the circumferential force, rather than a further decrease. Upon closer examination of the unsteady pressure sources generation, which ultimately results in the unsteady forces, an explanation for the phenomena was found and is presented in detail in this paper. First of all, the two sources of unsteadiness, the viscous wake effect and the inviscid influence of the first stator's potential field, are evident in the rotational frame of reference. In addition to these two classic pressure fluctuation sources a superposition effect of both was observed which explicitly explains the counter-intuitive behaviour of the middle gap configuration. Moreover, the time resolved simulations revealed that, depending on the axial gap size, both effects amplify or diminish the unsteady pressure fluctuations and result in higher and weaker unsteady forces, respectively. Additionally, examinations of related phase shift diagrams show that the identified superposition has a major impact on the unsteady blade forces and that it can be significantly modified by altering the axial inter row gap.

NOMENCLATURE

Symbols

F	Force	[N]
l	Axial gap length	[m]
L	First stator pitch	[m]
M_u	Circumferential Mach number	[-]
n	Number of time steps	[-]
$\bar{p}$	Time averaged static pressure	[Pa]
p	Static pressure	[Pa]
Δp	Unsteady static pressure	[Pa]
s	Chord length	[m]
t	Time	[s]
$\vec{w}$	Velocity vector	[m/s]
$\bar{\vec{w}}$	Time averaged velocity vector	[m/s]
$\Delta\vec{w}$	Unsteady secondary velocity vector	[m/s]
Δy	Distance in circumferential direction	[m]
x, y, z	Notation of coordinate axis	[-]
ω	Vorticity	[1/s]

Subscripts

A	Gap configuration A
ax	Axial
B	Blade, Gap configuration B
B1	Rotor blade
C	Gap configuration C
circ	Circumferential
rad	Radial
res	Resultant
RMS	Root mean square

Abbreviations

av	Average
BPP	Blade passing period
EO	Engine order
RANS	Reynolds-averaged Navier–Stokes

INTRODUCTION

More than a century of turbomachinery development has resulted in highly efficient conversion of energy. Having efficiency as a prime target, the structural integrity was mainly secured by over-engineering and basic beam models that could be solved analytically. In the case of forced response, the Campbell diagram [1] delivered already early in the past (since 1924) a general guideline for the design to assure structural integrity. For a more holistic development and optimization however, also the flow physics that has an influence on the excitation level of the blades had to be understood. Moreover, methods had to be developed to reduce the aerodynamic forcing levels without having an impact on the optimized blade profile.

The predominant means to lower unsteady forcing are, amongst others, the variation of the axial gap, clocking, blade count (ratio) and the stacking of the profiles. Due to the complexity of physical phenomena involved, most investigations focused on a single method. A summary of these methods can be found in [2]. For subsonic cases the increase of the axial gap has been widely investigated [3-8]. A general finding was the linear decay of the trailing edge wake and an exponential decay of the potential field [3, 4]. Moreover, Korakianitis [5] highlighted that a mere summation of these two effects for approximating the excitation amplitude is not sufficient due to non-linear effects. In fact, a minimum excitation level was found for an intermediate axial gap with a 2D compressible, inviscid rotor/stator-disturbance interaction program.

Due to the significant influence of the axial gap on aerodynamic forcing, it is essential to examine the findings of Korakianitis [5] in greater detail. Therefore, the current investigation uses highly improved numerical techniques and examines the flow physics and interaction of excitation sources in more detail. Understanding the physics of this nonlinear behaviour is a first step, which is essential for the base design and will, in a second step, allow for a more analytical design approach.

TURBINE GEOMETRY

The numerical investigations were performed on the 1.5 stage axial turbine test rig of the Institute of Jet Propulsion and Turbomachinery at RWTH Aachen University. The first stator is provided with a positive compound lean blade design and non-axisymmetric endwall contouring on the endwalls. The unshrouded rotor is likewise provided with an endwall contouring on the hub. For further details on the blade design, the authors would like to refer to [9]. The test rig additionally allows the investigation of three different axial gaps between the rows. The notation for each gap size is given in Table 2. A short description of the turbine's characteristics is given in Table 1. For further details on the 1.5 stage rig the authors refer to [10].

	Stator 1	Rotor	Stator 2
Number of blades [-]	33	44	66
Reynolds Number [-]	$8.1 \cdot 10^5$	$5.2 \cdot 10^5$	$3.3 \cdot 10^5$
Aspect Ratio [-]	0.95	1.4	1.51
Radial Tip Clearance [mm]		0.3	

Table 1: Turbine specifications

NUMERICAL SETUP AND METHODOLOGY

For all simulations the solver TRACE (Turbo Machinery Research Aerodynamic Computational Environment) in version 7.5 was used, which is a Reynolds-averaged Navier-Stokes solver based on a cell-centred finite volume approach and was developed at the German Aerospace Center (DLR) in cooperation with MTU Aero Engines. For turbulence closure a two-equation k-ω model was used. Transition was modelled on the first stator's pressure and suction side using the multimode model developed by Kozulovic [11]. The blade surfaces of the rotor as well as the second stator were calculated in a fully turbulent approach. However, for the present results, differences in transitional behaviour were found to be insignificant. The theory and methods of TRACE as well as code validation on the basis of experimental results can be found in [12], [13] and [14]. In the present work all computations were carried out on the RWTH Aachen University cluster.

To allow for computational grid independency and hence a better comparability of numerical results for each gap configuration, the meshing process for the blades has only been performed for the smallest Gap A. The two larger gaps B and C are simulated by moving apart the Gap A grid at the rows' interfaces. The generated clearances will then be filled by intermediate blocks consisting of non-sheared cells. This method has been successfully introduced by Restemeier [15]. The relative gap widths as well as the numbers of cells used for each configuration are summarized in Table 2. A relatively fine mesh resolution close to the endwalls provides $y^+ \approx 1$, such that the LowReynolds-Method can be applied on all surfaces.

	Gap A	Gap B	Gap C
$l_{Gap}/s_{B1,ax}$	0.26	0.61	0.97
Number of cells	$6.1 \cdot 10^6$	$6.4 \cdot 10^6$	$6.5 \cdot 10^6$

Table 2: Characteristics of the axial gap variation

The initial solution of each unsteady calculation was based on the preceding steady calculation. For all simulations, the inlet boundary condition values were chosen to resemble the experimental setup. At the outlet plane a constant back-pressure has been set. For the unsteady calculations, the moving and stationary blade rows are coupled by zonal interfaces. With reference to Figure 1, the computational domain was modelled using an eleventh fraction of the whole annulus of the turbine avoiding domain scaling or phase-lagged calculations. The time resolution was performed using 396 time steps per period and 40 inner iterations per physical time step.

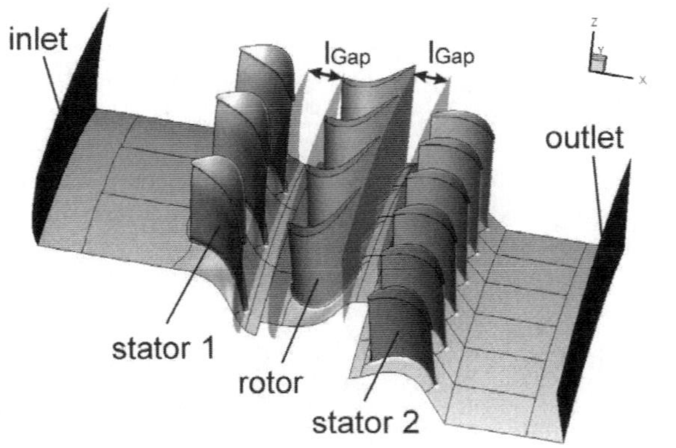

Figure 1: Computational domain of the turbine

RESULTS

Blade Forces for variable axial gaps

For each numerical time step, the rotor blade forces have been determined by integrating the pressure distribution over the suction and pressure side. For this consideration, the viscous shear stresses are negligible due to their small fraction of less than 0.1% of the blade forces. The time traces for all three gap variations for the axial (a), circumferential (b) and resultant force (c) are given in Figure 2. Due to the favourable blade count, three blade passing periods are shown in the rotational frame of reference. The resultant force is calculated from the force vectors acting in each spatial direction according to equation *(1)*.

$$|\overrightarrow{F_{res}}| = \sqrt{|\overrightarrow{F_{ax}}|^2 + |\overrightarrow{F_{circ}}|^2 + |\overrightarrow{F_{rad}}|^2} \qquad (1)$$

In addition to the unsteady force variations, the time averaged values are plotted as dashed lines. The values of the steady forces are normalized with Gap A's corresponding value and the results are summarized in Table 3.

	Gap A	Gap B	Gap C
$\overline{F_{ax}}/\overline{F_{ax,A}}$	1	1.003	0.997
$\overline{F_{circ}}/\overline{F_{circ,A}}$	1	1.002	0.995
$\overline{F_{rad}}/\overline{F_{rad,A}}$	1	1.001	1.001
$\overline{F_{res}}/\overline{F_{res,A}}$	1	1.002	0.998

Table 3: Normalized steady rotor blade forces

Examining both Figure 2 and Table 3, the overall magnitudes of all time averaged forces do not differ considerably. However, it should be noted here that the highest forces for all three spatial directions are observed for the Gap B configuration. Due to the insignificant differences between the variants, the averaged forces will be disregarded for the upcoming investigation.

Focusing next on the unsteady forces, the development of the normalized forces over time is shown together with their peak-to-peak amplitude values in Figure 2. Since the subsequent evaluation of the pressure fluctuations generating the unsteady blade forces is primarily two-dimensional, only the axial, the circumferential as well as the resultant blade forces are shown. Besides, an analysis of the radial forces' peak-to-peak amplitude turned out to be relatively moderate compared to the other components. For the sake of completeness, the corresponding values will be stated as follows: 1.4 N for Gap A, 0.5 N for Gap B and 0.4 N for Gap C.

A further consideration of Figure 2 reveals the periodicity of the blade forces due to the recurring first stator's potential field as well as its wake. With respect to the rotor frame of reference, the frequency can be determined to be 1925 Hz and will subsequently be indicated by EO33. It can further be noted, that the amplitudes of the unsteady axial forces show an expected behaviour and decrease with an increasing blade row spacing. The axial force amplitude is nearly halved from 14.7 N to 7.9 N when increasing the gap size from A to B and drops further to 3.0 N in case of the largest Gap C configuration. Moreover, a second superimposed oscillation of higher frequency is particularly visible for the smallest gap variation.

Considering the circumferential force amplitude change from Gap A to Gap B, a strong, but consistent, decrease is obvious. Comparing the two smallest gaps' ratios between the respective peak-to-peak values of the previously discussed axial force $(F_{ax,A}/F_{ax,B} = 1.8)$ and the circumferential one $(F_{circ,A}/F_{circ,B} = 7.6)$, the drop is even more apparent. However, focusing on the amplitude difference between Gap B and C, contrary to intuitive expectations, the amplitude of the circumferential force fluctuation rises by approximately 60%. Both anomalies explained above reveal the different behaviour of the middle gap. This can, as will be shown later, be attributed to a superposition effect originating from two pressure fluctuation sources leading to aerodynamic unsteadiness.

In contrast to the axial forces, the circumferential time traces of Gap A and C are mainly subjected to the first engine order EO33. However, the influence of another frequency of higher order is obvious for the middle gap. It is furthermore noticeable that the phase difference between the unsteady circumferential force for Gap A and C is approximately 0°. However the phase of the middle gap is shifted by nearly 180° meaning virtually half a blade passing period.

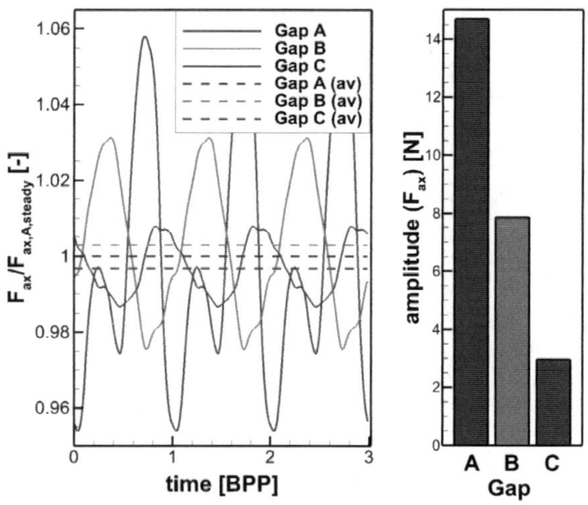

(a) Axial Force

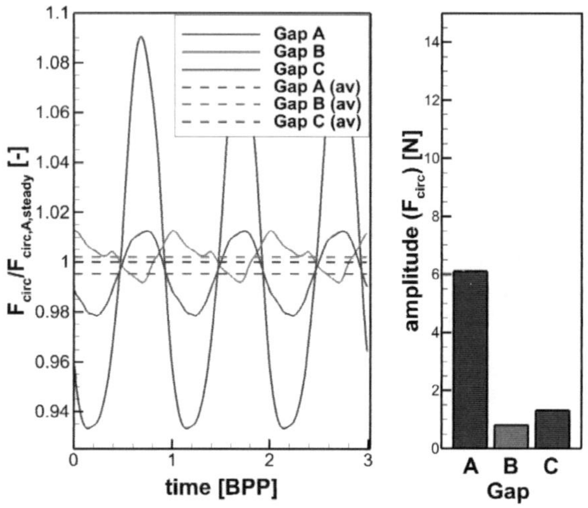

(b) Circumferential Force

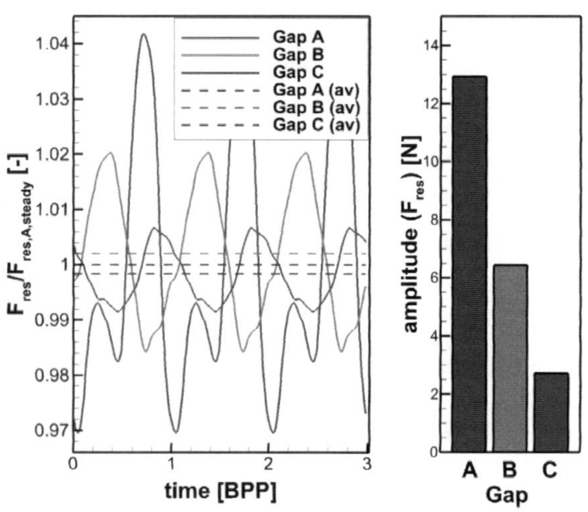

(c) Resultant Force

Figure 2: Unsteady rotor blade force developments (left) and peak-to-peak force amplitudes (right)

Due to the large amplitude of the axial force component, the smaller amplitude of the circumferential direction does not contribute significantly to the resultant force (see Figure 2 (c)). Due to the disproportionate amount of the axial force, the features seen for the circumferential force are not present in the resultant force. For that reason, for all following considerations, only the axial and circumferential fluctuations will be examined.

FFT Analysis of unsteady blade forces

An FFT analysis has been performed in order to examine the amplitudes of the characteristic frequency components of the unsteady forces. Figure 3 states the amplitudes of the first two harmonics, namely EO33 (1925 Hz) and EO66 (3850 Hz). Higher order harmonics will be disregarded for the further investigation due to their low impact with respect to the Campbell-diagram of the turbine rotor. Only the first two harmonics lead to intersections for relevant engine speeds.

Comparing the contribution of the two harmonics to the overall axial force, the unsteady variant of Gap A is the only configuration which is almost equally composed of both harmonic fractions. For the other two gaps, the unsteady axial force is dominated clearly by the first harmonic frequency.

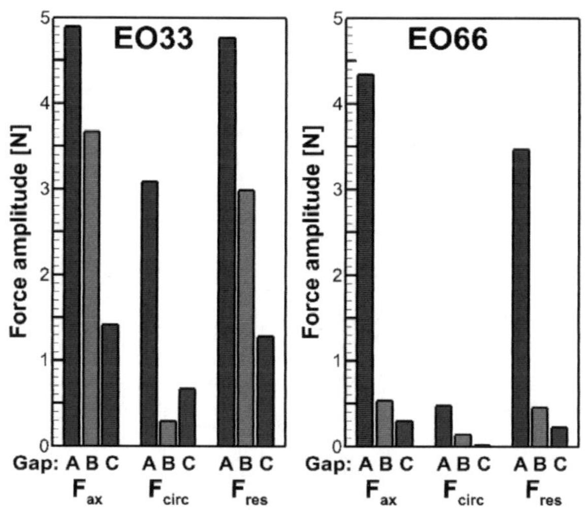

Figure 3: Amplitudes of unsteady rotor blade forces by means of FFT analysis for EO33 (left) and EO66 (right)

For the unsteady circumferential forces the results of the FFT analysis confirm the dependence of Gap A and C's amplitude on merely the first engine order EO33. In contrast to that, for the Gap B configuration the unsteady circumferential force amplitude is only halved for EO66 compared to EO33. This leads to the assumption that the second harmonic considerably influences the total circumferential force. With respect to the previous investigation, the atypical behaviour of Gap B (see Figure 2 (b)) is clearly obvious in the FFT analysis and predominately caused by the first EO33. Comparing the forces for Gap B and C related to EO33, the amplitude of

Gap C is about 220% of Gap B's amplitude. Since this effect cannot be confirmed by EO66, a flow mechanism - changing with the first harmonic frequency - must be responsible for the observed conspicuousness.

Since the unsteady forces are being generated by pressure fluctuations on the rotor blade surface, the reason for the behaviour can be found in the unsteady pressure field and its different sources.

Sources of pressure fluctuations

Initially, the sources of pressure fluctuations within the turbine will be discussed. To visualize the most pronounced effects, the pressure fluctuations within a blade-to-blade plane in the rotor are shown in Figure 4 at a random time step. Exemplarily, the unsteady pressure $\Delta p(t)$, defined as the difference between the instantaneous static pressure $p(t)$ and the time-averaged pressure $\bar{p}$, is illustrated for the smallest gap configuration A.

$$\Delta p(t) = p(t) - \bar{p} \qquad (2)$$

The first source of unsteadiness is shown in the left part of Figure 4. The pressure fluctuations' amplitudes recur with the periodicity of one stator pitch and move – considering the rotational frame of reference – upwards and can consequently be identified as the first stator's potential field. The rotor's potential field is not apparent in this image, due to the proposed definition of the unsteady pressure in equation *(2)*. In the rotating frame of reference, the rotor's potential field can be considered as stationary - disregarding the weak influence of the interaction with the first stator's potential field.

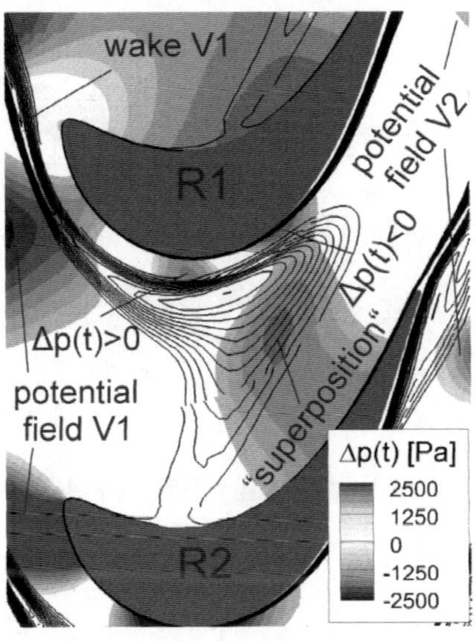

Figure 4: Blade-to-blade plane of the rotor with unsteady pressure contours and entropy lines (Gap A)

The pressure fluctuations caused by the potential field of the first stator propagate at the speed of sound in axial direction and their intensity decays exponentially, as was theoretically stated already in 1969 by Parker [16]. With respect to Figure 4, the impact of the first stator's potential field on the rotor blade surface pressure is obvious. Due to the exponential behaviour of the potential field, the front part of the suction side is particularly influenced by it.

Downstream of the rotor row, the second stator potential field is apparent. However, compared to the previously discussed potential field, the one from the second stator is significantly weaker by reason of its weaker blade loading. Consequently, the impact of the second stator potential field on the rotor blade pressure fluctuations is much weaker.

The second most important source of unsteady pressure fluctuations is also evident in Figure 4. The black contour lines show regions of high entropy, which reveal the instant position of the wake shedding from the upstream stator blades. At this time step, the visible wake forms its characteristic V-shape, after being chopped by the 2nd rotor blade R2 and distorted due to the circumferential pressure gradient within the rotor passage. Between the front part of the wake and the suction side of R1, a local minimum in unsteady pressure ($\Delta p(t) < 0$) can be observed. Upstream of the minimum, a local maximum ($\Delta p(t) > 0$) occurs between the wake's rear part and the suction side. In the upcoming paragraph, the described spots of local pressure unsteadiness will be identified as a direct consequence of the viscous wake migrating through the rotor passage. Another crucial source of blade pressure fluctuations, visualized in Figure 4 by the unsteady pressure ("superposition"), is located within the blade passage. It extends over the major part of the blade passage and influences especially the pressure side of the rotor blade R2. This source results from a superposition effect of the other two mentioned sources, the wake and the first stator's potential field. Section "Superposition Effect" of this paper is attributed to this effect.

In order to clarify the generation of the pressure spots close to the suction side, Figure 5 shows unsteady secondary velocity vectors along with the unsteady pressure contour of the rotor passage at the same time step as before. For the sake of understanding, red contour lines of high entropy indicate the first stator wake. The unsteady secondary velocity vectors $\Delta\vec{w}(t)$ are defined as the difference between the instantaneous velocity vectors $\vec{w}(t)$ at time t and their time averaged value $\bar{\vec{w}}$.

$$\Delta\vec{w}(t) = \vec{w}(t) - \bar{\vec{w}} \qquad (3)$$

The vectors reveal the unsteady flow behaviour under the absence of the mean velocity. With reference to Figure 5, the vectors within the wake point into the opposite direction of the mean flow representing the velocity deficit of the wake (see ① in Figure 5). Even considering the potential theory, the velocity deficit of the wake can be described as a parallel flow superimposed with two counter-rotating potential vortices. These circulations can be shown by means of the unsteady secondary velocity vectors, as visualized by ② in Figure 5. After the wake enters the blade passage, it is distorted by the pressure gradient and forms the characteristic V-shape. As a consequence, two counter-rotating vortices (③ and ④) are formed by the unsteady secondary velocity vectors. This vortex pattern has already been shown by Biester et al. [17] and experimentally by Stieger et al. [18] using Laser Doppler Anemometry.

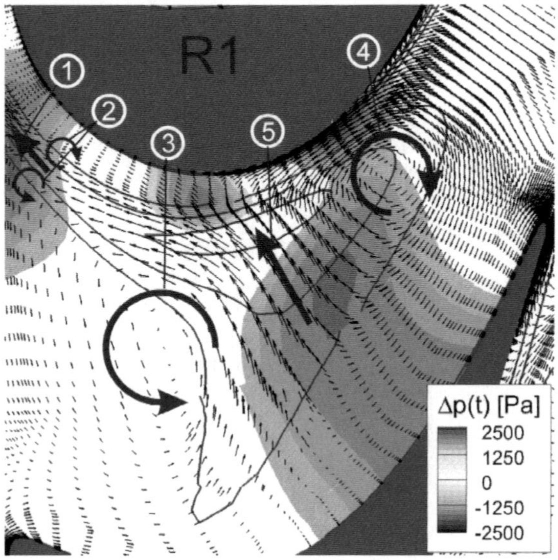

Figure 5: Blade-to-blade plane of the rotor with unsteady pressure contours, red entropy lines and unsteady secondary velocity vectors (Gap A)

When moving through the rotor passage, the wake's velocity deficit points towards the blade suction side (see ⑤). Considering the equilibrium of the circumferential blade-to-blade pressure gradient and the centrifugal force acting on the slower wake fluid, the wake experiences a stronger deflection towards the suction side compared to the free stream while moving through the passage. This fluid transport to the suction side (⑤) is commonly known as the "negative jet effect". Its overwhelming impact on low energy fluid is clarified by long unsteady secondary velocity vectors located between the aforementioned counter-rotating vortices. This effect is even amplified by the wake entering the rotor passage with negative incidence. The negative jet effect determines the intensity of both associated vortices.

As a consequence of the clockwise rotating vortex ④, the unsteady secondary velocity vectors in the vicinity of the suction side point in streamwise direction. Thus, the local velocity at this particular instant is much higher compared to the time averaged velocity. This in turn means that an acceleration of the flow took place right before the shown time step, in turn leading to a local pressure drop. However, the second vortex ③ and its corresponding vector field result in a lower velocity along the surface, due to a preceding deceleration of the flow, leading to a local pressure rise. The pressure drop as well as the pressure rise is obvious in both Figure 4 and Figure 5. Considering an entire period of the wake's migration through the passage, the local minimum and maximum in unsteady pressure are propagating with the wake's propagation speed along the blade surface from approx. 25% up to 70% relative axial coordinate.

The velocity deficit of the wake is responsible for the unsteady velocity field within the rotor passage (i. e. negative jet effect, two counter-rotating vortices). Its magnitude determines the strength of the associated pressure fluctuations on the suction side between 25% and 70% relative axial coordinate. In contrast to the wake-induced unsteady flow field, the other pressure fluctuations acting on the remaining regions of the suction surface are dominated by the first and additionally by the second stator's potential fields. Unlike the suction side, the pressure side is not directly faced towards the potential fields of the neighbouring rows and additionally it is less affected by the wake due to the pressure gradient within the passage. For that reason, the pressure fluctuations on the pressure side are not directly influenced by any of the individual effects. As will be shown in the upcoming part of the paper, the superposition of both effects plays an important role.

Evaluation of the three-dimensional distribution of blade pressure fluctuations

After investigating the sources of pressure fluctuations at an arbitrary blade-to-blade plane, subsequently, the pressure fluctuations along the blade's span are presented and the essential mechanisms, influencing their strength, are analysed. In Figure 6 the streamwise vorticity $\vec{\omega}_{SW}$ according to equation *(4)* is visualized at an axial plane within the rotor passage.

$$|\vec{\omega}_{SW}| = \frac{1}{|\vec{w}|} \cdot \left[\omega_x \cdot w_x + \omega_y \cdot w_y + \omega_z \cdot w_z\right] \qquad (4)$$

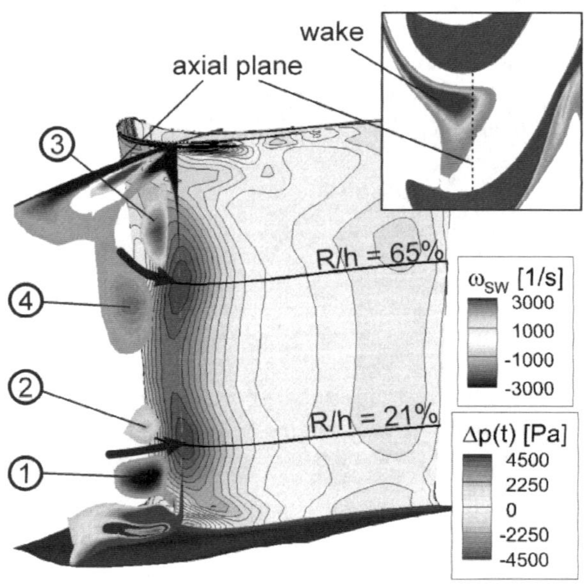

Figure 6: Streamwise vorticity at constant axial coordinate and unsteady pressure contours on the rotor blade suction side at the time indicated above right (Gap A)

In order to indicate the current position of the V-shaped wake, the illustration in the top right of Figure 6 depicts the location of the plane with respect to the wake. The red (maxima) and blue (minima) spots of vorticity within the axial plane indicate the existence of secondary vortices generated by the first stator. The strongest vortices can be identified to be the hub passage vortex ①, the hub trailing shed vortex ②, the tip passage vortex ③ and its mirror vortex ④ (see [19]). Considering the rotating direction of these vortices with respect to Figure 6, an enhanced fluid transport is evident at the two positions at 21% as well as 65% relative span, which are both marked by arrows. The increased circumferential velocity component amplifies the pronounced negative jet effect at these radial positions, which can be interpreted as longer unsteady secondary velocity vectors at position ⑤ of Figure 5. Combining both, Figure 5 and Figure 6, the direction of the secondary flow in the axial plane intensifies the negative jet effect leading in turn to higher vorticities of the two counter-rotating vortices and hence to higher unsteady velocities near the suction surface in Figure 5. Consequently, the velocity changes (deceleration and acceleration) and thus the changes in pressure must be stronger at these radial positions. This is confirmed in Figure 6 by the illustrated unsteady pressure field on the rotor suction side.

Pressure fluctuations for variable axial gaps

The preceding analysis of the rotor flow field and the sources of pressure fluctuations help to understand and to evaluate pressure fluctuations for different blade row spacing configurations. As mentioned above, three different axial gaps have been investigated by altering the gap between the rows equally. To visualize regions of significant pressure fluctuations on the blade surfaces, the RMS value of the unsteady pressure is calculated using the following equation *(5)*.

$$\Delta p_{RMS} = \sqrt{\frac{1}{n}\sum_{i=0}^{n} \Delta p(t_i)^2} \qquad (5)$$

Figure 7 presents the Δp_{RMS} distribution along the blade's suction and pressure side for all three gap configurations. Upon superficial consideration, it is evident that with increasing axial gap, the pressure fluctuations along the surfaces decrease. Keeping in mind the pressure fluctuations detailed above, the larger axial gap leaves the wake's velocity deficit more time to mix out leading to less intense deflection within the rotor passage. However, taking a closer look at the RMS distribution, further and much more interesting differences arise.

With regard to Figure 7 (a), (c) and (e), a characteristic section for every gap configuration can be identified. The region from approx. 25% to 70% relative axial chord length on the suction side is, as explained previously, influenced by the wake induced pressure fluctuations. Within this section, two spots of highest RMS values are present at 21% and 65% relative span. These spots correspond to the radial positions of the secondary flows arising from the first stator.

Focusing next on the leading edge region, in particular the region up to 20% axial chord, the RMS distribution of Gap A reveals a distinct influence of the first stator's potential field. Comparing all three gap configurations, the exponential decay at the leading edge can be seen.

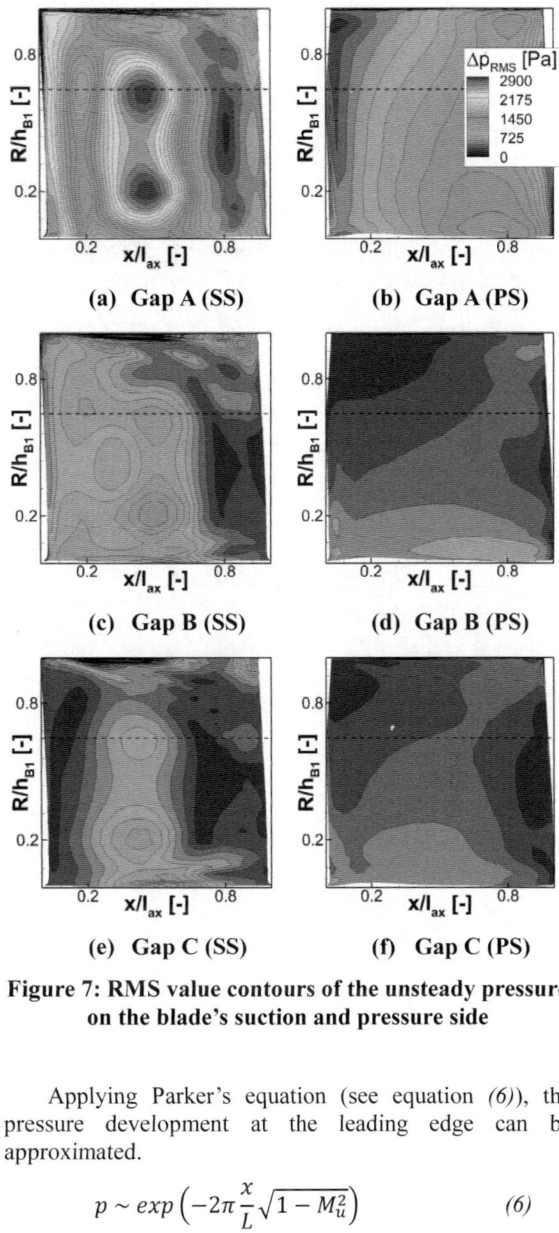

Figure 7: RMS value contours of the unsteady pressure on the blade's suction and pressure side

Applying Parker's equation (see equation *(6)*), the pressure development at the leading edge can be approximated.

$$p \sim exp\left(-2\pi\frac{x}{L}\sqrt{1-M_u^2}\right) \qquad (6)$$

The decay depends on the pitch L between neighboring stator blades as well as on the rotational speed of the rotor blades M_u. With increasing distance x between the stator and the rotor, the potential field's influence decreases, which is obvious for the largest Gap C, where the effects are not identifiable on the basis of the RMS values according to Figure 7 (e). Figure 9, for which the leading edge pressure amplitudes are extracted, clearly highlights this general trend. A comparison between equation *(6)* and the numerical results is shown in Figure 8.

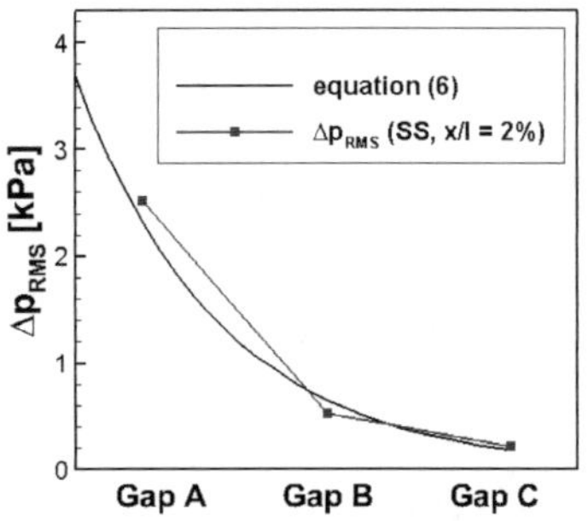

Figure 8: Comparison of theoretical decay of potential field and results of numerical simulation

Another distinctive feature applies to the suction side section between approx. 15% and 35% axial chord: the region is dominated by slight peaks of Δp_{RMS} for the smallest Gap A (see Figure 7 (a)) and the largest Gap C, yet by significantly stronger pressure fluctuations for Gap B. This difference is easier to see upon examination of Figure 9, illustrating the development of the RMS-values at one particular radial position (dashed lines in Figure 7). This feature can again – similarly to the pressure fluctuations seen on the pressures side – only be explained by a superposition effect.

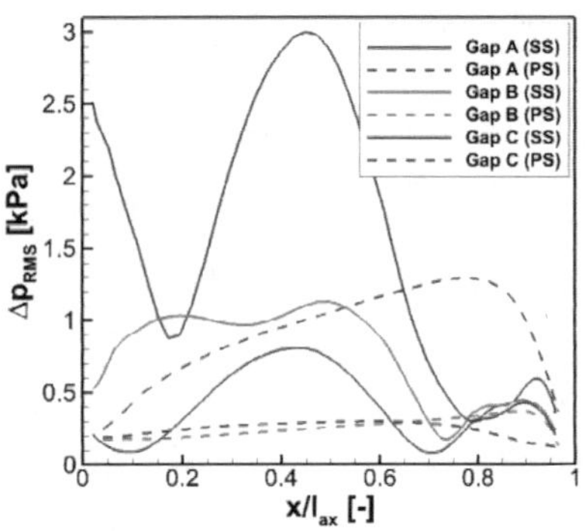

Figure 9: RMS values of unsteady blade pressure at 65% span for all gap configurations

Since both pressure variation sources cannot be separated, their simultaneous occurrence leads to a superposition. However, it is possible to modify the pressure fluctuations by changing for example the axial gap between the rows, as performed in this paper. In order to evaluate the effect of variable blade row spacing, the pressure fluctuations on the pressure side according to Figure 7 (b), (d) and (f) will be examined. In contrast to the suction side, it is apparent that the fluctuations are of larger spatial extension but of much smaller amplitude. As explained previously, this is due to the fact, that neither the wake nor the potential fields have any direct impact on the pressure surface.

Comparing all three gap configurations, it can be stated that the strength of the pressure fluctuations decreases with increasing gap width from Gap A to B. However, a further enlargement of the gap width from B to C leads to slightly more pronounced pressure fluctuations over a large proportion of the pressure surface, which is even more obvious in Figure 9. The reason for the different behaviour of the pressure fluctuations for Gap B and C can be explained using the idea of superposition.

Superposition Effect

Since both the wake and the potential field are highly unsteady phenomena, the superposition effect on the unsteady pressure field is best identified using instantaneous time steps. For all configurations two identic time steps with a phase shift of half the blade passing period are shown in Figure 10. Coloured contours indicate the unsteady pressure Δp and black contour lines are used again to highlight the location of the wake.

With respect to Figure 10, the two superposition effects (referred to as "I" and "II"), occurring for each configuration, are demonstrated for two distinct time steps. At time step 1 the wake impinges on the lower blade's pressure side and its deformation due to the pressure gradient has not yet begun. At time step 2, the wake is spread over the entire suction side. The green and red circles indicate an amplification and cancellation of both effects respectively.

Beginning the analysis with the smallest gap (see Figure 10 (a)), the unsteady pressure pattern located in front of the blade passage (time step 1: minimum, time step 2: maximum), caused by the first stator's potential field, has the same effective direction as the pressure fluctuations induced by the wake. In this turbine the same working direction of both fluctuation sources leads to an amplification zone, which extends over the entire blade passage up to the pressure side of the lower blade. Only due to the pressure amplification as a cause of the superposition effect I, significant unsteady pressure variations appear on the blade pressure surface. The same amplification is obvious not only for the second time step, but almost for the whole period, too.

The second phenomenon that occurs due to the interaction of different pressure fields is a reduction in pressure oscillations. As a consequence of an unsteady positive pressure maximum along the leading edge and its negative counterpart due to the viscous wake, both pressure variations extinguish each other. The location of cancellation (superposition effect II) is indicated by red circles in Figure 10 (a). The attenuation can also be seen in the averaged pressure fluctuations visualized in Figure 7 (a) between 15% and 35% axial chord. The same conclusions can be inferred for the second time step.

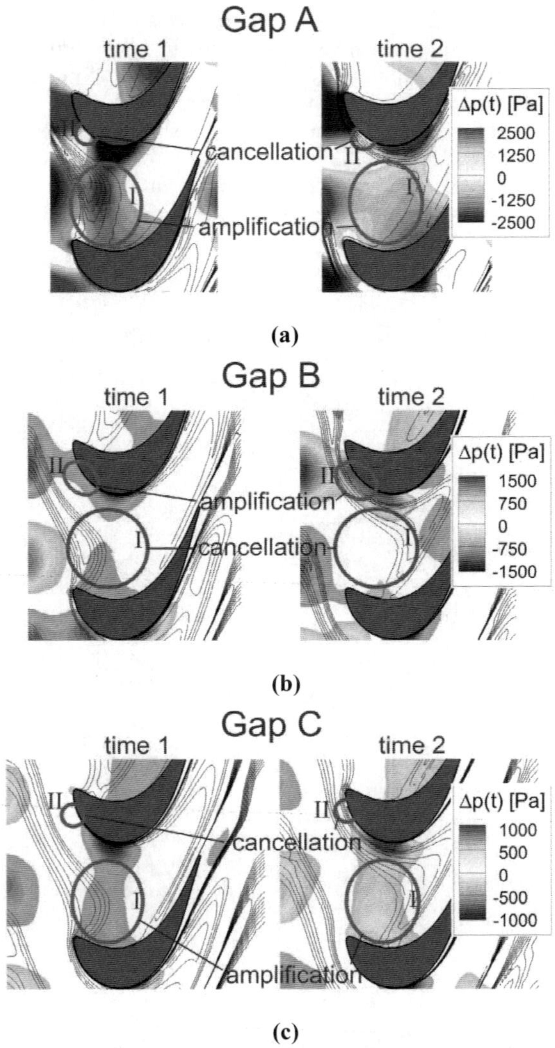

Figure 10: Unsteady pressure contours indicating the superposition effect I and II for all gap configurations

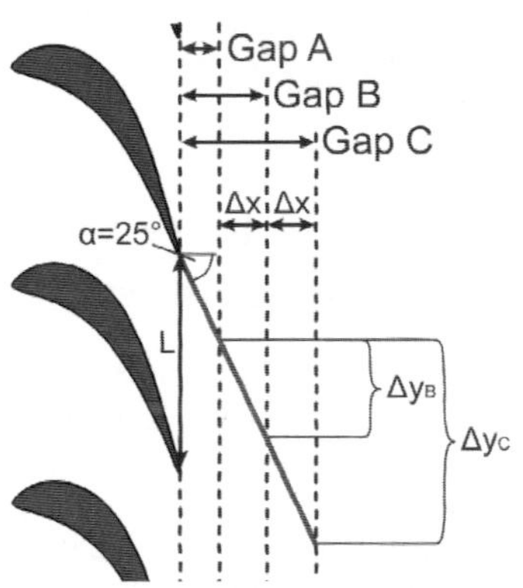

Figure 11: Schematic illustration to clarify the phase shifting of the unsteady pressure sources with variable axial gap

Considering next the Gap B configuration, the circumferential position of the first stator's potential field is shifted by half a stator pitch. The displacement is merely a geometric issue, which is explained by means of Figure 11. By enlarging the axial gap width from Gap A to Gap B, the stator wake reaches the rotor inlet plane at a different position in terms of the first stator's potential field. The difference in the circumferential direction can be determined as follows. Considering the stator outlet flow angle α and the increase in gap width Δx, for the given turbine case, the difference in circumferential distance Δy_B is calculated to be 0.48 relative to the first stator's pitch L. For the largest gap configuration, the distance Δy_C was computed to be 0.97 relative to the first stator's pitch L. Due to the accidental fact that the shift turns out to be almost exactly 50% of the stator pitch in the Gap B case, the amplitudes of the stator potential field have been reversed. Hence, the amplification that was observed for the smallest gap transforms into a cancellation and vice versa. Therefore, the unsteady pressure fluctuations increase between 15% and 35% axial chord along the front part of the suction side (marked with a green circle) and decrease along the pressure side (marked with a red circle). Referring again to Figure 7 (c), the higher pressure variations on the suction side seen here, can be identified to be caused by the superposition of the potential field and the wake.

Finally, the Gap C configuration will be investigated. As previously described, the phase shift is nearly one, leading to almost identic inflow conditions for the rotor compared to the smallest configuration. Hence, similar superposition effects I and II are to be expected. With reference to Figure 7 and Figure 9 the amplification of the unsteady pressure variations within the blade passage leads to increased fluctuations on the pressure side despite the larger gap size. The suction side is only subjected to the wake induced pressure fluctuations. The very weak pressure fluctuations up to 20% axial chord of the rotor's suction side (see Figure 7 (e)) can be traced back to two effects: firstly, the pressure variation of the first stator's potential field has the opposite working direction compared to the wake induced pressure variation. Secondly, its amplitude is very low due to the exponential decay with the axial distance.

Link between unsteady blade forces and pressure fluctuations

In the following chapter, the link between the observed unsteady pressure fluctuations – the noticed cancellation and amplification – and the unsteady blade forces presented initially in Figure 2 and Figure 3 is established. In order to interpret the development of the unsteady axial and circumferential 1D blade forces, it is necessary to determine a correlation between the sources of pressure variations and their impact on the axial or circumferential force component respectively.

Referring to the preceding results, four pressure fluctuation sources influencing both the blade suction and pressure side have been identified. Depending on the impact location along the rotor profile, the pressure variation sources lead to unsteady forces with preferred directions. The local extensions of the presented effects are visualized in Figure 12 (bottom) and labelled from ① to ⑤. Additionally, the preferred direction is stated as well.

1. The first stator's potential field, influencing the leading edge region ①, primarily leads to unsteady forces in axial direction.
2. The wake predominantly influences the suction side between 25% and 70% axial chord (section ②) leading in turn to mainly unsteady circumferential blade forces.
3. Referring to Figure 10, the superposition effect I results in an area of amplification or cancellation, depending on the gap size. In the case of an amplification, the occurring pressure fluctuations act on the pressure side and modify the circumferential blade force in the front and mid region of the blade (up to 60% axial chord, section ③). Additionally, particularly at the rear part of the pressure side, (downstream of 60% axial chord, section ④), the amplification effect leads to axial blade forces. In the case of a cancellation, the superposition effect I reduces the unsteady pressure in this regions and leads in turn to a diminished influence on the unsteady forces.
4. The superposition effect II likewise leads to an amplification and cancellation. The former results in both, axial and circumferential blade forces due to pressure fluctuations on the suction side between 15% and 35% axial chord (section ⑤). The latter exhibits no influence on the unsteady forces.

The influence of the second stator potential field on the unsteady blade forces is relatively weak and will be disregarded for the succeeding investigations.

First of all, the larger circumferential force amplitude of Gap C in comparison to Gap B results from the amplified unsteady pressure within the blade passage due to the superposition effect I. This leads to stronger pressure fluctuations in section ③ for Gap C, which generate increased blade forces in circumferential direction. The same holds true for the Gap A configuration, explaining the large difference in peak-to-peak amplitude of the circumferential forces $\left(F_{circ,A}/F_{circ,B} = 7.6\right)$ compared to Gap B. The cancellation effect observed for Gap B enhances the difference between both amplitudes even further. Hence, by considering the superposition effect within the blade passage, the variation of the blade row spacing allows to modify the magnitude of the unsteady circumferential blade forces.

In order to identify whether or not the unsteady pressure modification, due to axial gap variation, leads to an enlargement or reduction of the forces, the upcoming consideration will refer to the time profiles of the forces seen in Figure 2. For that purpose, the phase shift of the unsteady pressure variations acting along the blade profile are investigated exemplarily at 65% span. Therefore, an FFT analysis of the time resolved unsteady pressure is performed for each cell along the profile shown in Figure 12 (bottom). The resulting phase angles over the axial coordinate, associated with EO33, are given in Figure 12 (top) for all three gap configurations and both pressure and suction side.

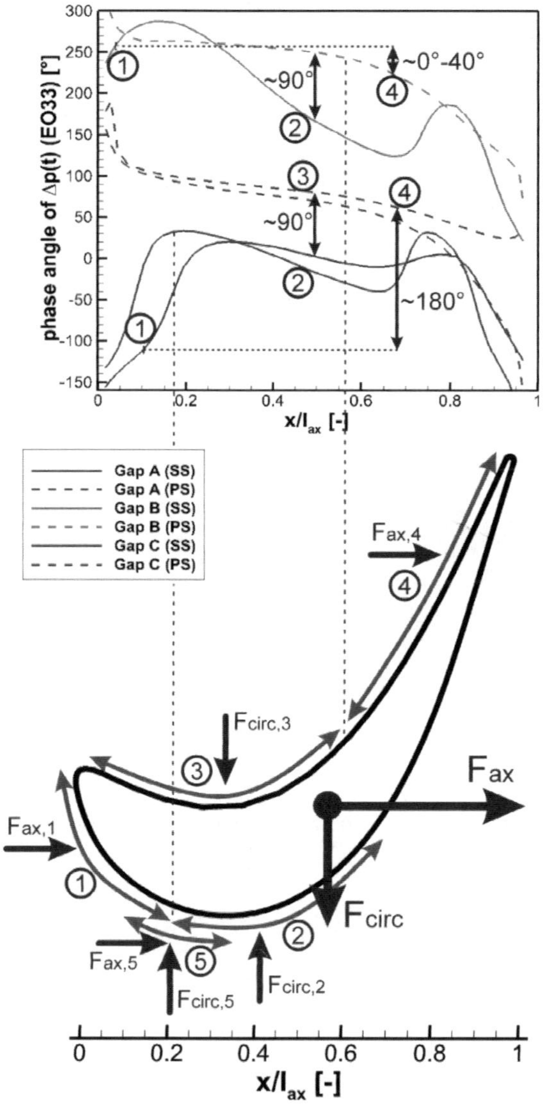

Figure 12: Phase angle (EO33) of unsteady blade pressure at 65% span (top) and blade areas of influence (bottom)

On the one hand, the phase angle diagram contains information about the propagation direction of the unsteady pressure. Here, a positive gradient of the phase angle in positive axial direction indicates an upstream movement of the unsteady pressure signal and vice versa. Hence, the pressure fluctuations due to the first stator's potential field move along section ① upstream from approximately 20% axial chord to the leading edge. Additionally, the downstream propagation of the wake induced pressure fluctuations between 25% and 70% axial chord is confirmed.

On the other hand, the phase shifts can be used to provide further insight into the correlation between the local pressure fluctuations along the pressure and suction side. First of all, the resemblance of the Gap A and C's phase angles is obvious due to the previously mentioned phase shift of nearly one blade passing period (see Figure 11). The corresponding sources of unsteadiness determining the circumferential forces, as explained above, are the wake induced pressure fluctuations (section ②) and the superposition effect I (section ③). According to Figure 12, the difference in phase angle of the pressure und suction side is about 90°, meaning that both pressure fluctuations act successively in the same direction. This results in a relatively high amplitude for Gap A and C. That difference in phase angle between these two sources does not correlate with the varying axial gap. Therefore it is rather a function of the blade count ratio than of the axial gap.

Taking next a closer look to the time resolved axial force component of the smallest gap according to Figure 2 (a), the corresponding sources of unsteadiness are the potential field (section ①) and the unsteadiness due to superposition effect I influencing the pressure side in section ④. With reference to Figure 12 (top), the difference in phase angle of the prescribed phenomena is approximately 120° to 180°. Consequently, both sources contribute to the axial force with a phase shift of about half a blade passing period. Additionally, the pressure fluctuations detected on the blade surfaces comprise frequencies of higher order as well. These two reasons lead to two distinct maxima in the development of the unsteady axial force. One is located at 0.25 BPP because of the positive acting of the potential field and a further maximum at 0.7 BPP due to the positive unsteady pressure on the basis of the superposition effect I. The inversely working pressure fluctuations lead in turn to the overall relatively moderate amplitude of the unsteady axial force for the smallest gap configuration compared to the middle gap.

However, the phase diagram for Gap B distinguishes significantly. The superposition effect I influencing the unsteady pressure at the pressure side (section ④) is rather weak. Nevertheless, it acts in axial direction almost in phase (approx. 0° to 40°) with the superposition effect II influencing the leading edge region of the suction side (section ⑤). The simultaneously acting pressure variations result in a rather high amplitude of the axial force for Gap B. This is the reason why the difference in peak-to-peak amplitude between Gap A and B is much larger for the circumferential $\left(F_{circ,A}/F_{circ,B} = 7.6\right)$ than for the axial forces $\left(F_{ax,A}/F_{ax,B} = 1.8\right)$.

CONCLUSION

Detailed unsteady RANS simulations for a 1.5 stage axial turbine geometry have been conducted to investigate the effect of inter-blade row gap variation on the rotor blade force fluctuations. Since the unsteady forces are being generated by pressure fluctuations on the rotor blade, the unsteady pressure field and its different sources were described and examined. Subsequently, the findings have been used to interpret the characteristic force developments over time. The main results are summarized as follows:

1. The following sources for the observed unsteadiness on the rotor have been identified: the potential field of adjacent blade rows and the first stator's wake. In addition to these well-known effects, a superposition of both sources has been observed. Depending on the axial gap and the associated phase shift of the sources, two distinct areas of the unsteady pressure field either experienced amplification or cancellation. On the basis of the presented results, Gap A behaves similar to Gap C, due to the mutual phase shift of nearly one blade passing period.
2. The numerical results showed a 60% increase in the amplitude of the unsteady circumferential force for the largest Gap C relative to Gap B. The analysis revealed that the superposition effect leads to intensified pressure fluctuations on the pressure side for Gap C causing the observed higher values.
3. For the intermediate gap, the same superposition effect reduced the pressure fluctuations on the pressure side leading to small amplitudes in the circumferential force fluctuation. For the other two gaps, the effect amplifies the force fluctuations in circumferential direction. This is the reason why the difference in circumferential force amplitude between Gap A and B is rather high $\left(F_{circ,A}/F_{circ,B} = 7.6\right)$.
4. Different phase shifts of the sources influencing the unsteady axial forces were figured out for Gap A and C compared to Gap B. For the latter, the two associated sources act nearly in phase, while they are acting with a phase shift of 180° for the Gap A configuration. Therefore, the difference in axial force amplitude is relatively low $\left(F_{ax,A}/F_{ax,B} = 1.8\right)$.
5. The strength of rotor pressure fluctuations in spanwise direction is intensified by the secondary vortex system generated by the first stator.

In the future, the numerical analyses will be validated by experimental investigations in the 1.5 stage turbine rig to validate the results found. It is intended to detect the rotor blade pressure fluctuations by means of piezoresistive pressure transducers distributed at specific locations along the suction and pressure surface.

ACKNOWLEDGMENTS

The investigations were conducted as part of the joint research program AG Turbo 2020. The work was supported by the Bundesministerium für Wirtschaft und Technologie (BMWi) as per resolution of the German Federal Parliament under grant number 03ET2011F. The authors gratefully acknowledge AG Turbo and MTU Aero Engines AG for their support and permission to publish this paper, as well as DLR Cologne for all support in using the TRACE solver. The responsibility for the content lies solely with its authors.

REFERENCES

[1] Campbell, W., 1924. "Protection of Steam Turbine Disk Wheels from Axial Vibration". General Electric Company.

[2] Fruth, F., 2013. "Reduction of Aerodynamic Forcing in Transonic Turbomachinery - Numerical Studies on Forcing Reduction Techniques". PhD thesis, KTH Stockholm.

[3] Lambertz, J., Wallmann, T., and Gallus, H. E., 1980. "Vergleich der Messergebnisse Dynamischer Schaufelkräfte in Axialen Turbomaschinen mit Berechnungsverfahren Verschiedener Autoren". VDI-Report, No. 381.

[4] Gallus, H. E., Grollius, H., and Lambertz, J., 1981. "The Influence of Blade Number Ratio and Blade Row Spacing on Axial-Flow Compressor Stator Blade Dynamic Load and Stage Sound Pressure Level". 81-GT-165, ASME Turbo Expo.

[5] Korakianitis, T., 1992. "On the Prediction of Unsteady Forces on Gas Turbine Blades: Part 2 - Analysis of the Results". Journal of Turbomachinery, Vol. 114, Issue 1.

[6] Jöcker, M., Hillion, F. X., and Fransson, T. H, 2001. "Numerical Unsteady Flow Analysis of a Turbine Stage with Extremely Large Blade Loads". GT2001-0260, ASME Turbo Expo.

[7] Chang, D., Tavoularis, S., 2013. "Effect of the Axial Spacing between Vanes and Blades on a Transonic Gas Turbine Performance and Blade Loading ".International Journal of Turbo Jet-Engines, Vol. 30, Issue 1.

[8] Chen, T., Patel, K., and Millington, P., 2012. "Combined Effects of both Axial Gap and Blade Count Ratio on the Unsteady Forces of a Steam Turbine Stage ". GT2012-68874, ASME Turbo Expo.

[9] Pöhler, T., Niewöhner, J., Jeschke, P., and Gündogdu, Y., 2014. " Investigation of Non-Axisymmetric Endwall Contouring and 3D Airfoil Design in a 1.5 Stage Axial Turbine Part I: Design and Novell Numerical Analysis Method". Proceedings of ASME Turbo Expo 2014, GT2014-26784.

[10] Restemeier, M., Jeschke, P., Gündogdu, Y., and Gier, J., 2011. "Numerical and Experimental Analysis of the Effect of Variable Blade Row Spacing in a Subsonic Axial Turbine". Proceedings of ASME Turbo Expo 2011, GT2011-45637.

[11] Kozulovic, D., 2007. "Modellierung des Grenzschichtumschlags bei Turbomaschinenströmungen unter Berücksichtigung mehrerer Umschlagsarten". PhD thesis, Ruhr-Universität Bochum.

[12] Franke, M., Kügeler, E., and Nürnberger, D., 2005. "Das DLR-Verfahren TRACE: Moderne Simulationstechniken für Turbomaschinenströmungen". Deutscher Luft-und Raumfahrtkongress (Friedrichshafen).

[13] Nürnberger, D., and Greza, H., 2002. "Numerical Investigation of Unsteady Transitional Flows in Turbomachinery Components Based on a Rans Approach". Flow, Turbulence and Combustion, 69, pp. 331–353.

[14] Kügeler, E., Weber, A., and Lisiewicz, S., 2001. "Combination of a Transition Model with a Two-Equation Turbulence Model and Comparison with Experimental Results". Proceedings of the 4th European Conference on Turbomachinery, Fluid Dynamics and Thermodynamics.

[15] Restemeier, M., 2012. „Einfluss des Schaufelreihenabstandes auf Strömung und Wirkungsgrade in einer subsonischen Axialturbine". PhD thesis, RWTH Aachen.

[16] Parker, R., 1969. "Relation between Blade Row Spacing and Potential Flow Interaction Eeffects in Turbomachines". Proceedings of the Institution of Mechanical Engineers, SAGE Publications, 184(7):1–8.

[17] Biester, M., Henke, M., and Seume, J., 2012. „Unsteady Wake-Blade Interaction: A Correlation between Surface Pressure Fluctuations and Loss Generation". Proceedings of ASME Turbo Expo 2012, GT2012-69616.

[18] Stieger, R. D., and Hodson, H. P., 2005. „The Unsteady Development of a Turbulent Wake Through a Downstream Low-Pressure Turbine Blade Passage". Journal of Turbomachinery, Vol. 127, pp. 388-394.

[19] Walraevens, R. E., 2000. „Experimentelle Analyse dreidimensionaler instationärer Strömungseffekte in einer 1 1/2-stufigen Axialturbine". VDI-Verlag.

Proceedings of the 14th International Symposium on Unsteady Aerodynamics, Aeroacoustics & Aeroelasticity of Turbomachines
ISUAAAT14
8-11 September 2015, Stockholm, Sweden

I14-S6-2

INFLUENCE OF COMBUSTION CHAMBER DEFECTS ON THE FORCED RESPONSE BEHAVIOR OF TURBINE BLADES

Thomas Hauptmann
Institute of Turbomachinery and Fluid Dynamics (TFD),
Leibniz Universität Hannover
Hannover, Germany,
hauptmann@tfd.uni-hannover.de

Paul Christ, Jens Aschenbruck
Formerly at Institute of Turbomachinery and Fluid Dynamics (TFD), Leibniz Universität Hannover
Hannover, Germany

Christoph Hennecke, Friedrich Dinkelacker
ITV, Leibniz Universität Hannover
Hannover, Germany

Joerg R. Seume
TFD, Leibniz Universität Hannover
Hannover, Germany

ABSTRACT

Combustion chambers of aircraft engines are exposed to oxidation, corrosion, and high thermo-mechanical loads. Accordingly, they are particularly vulnerable to damage and defects. Such defects of combustion chambers can introduce a non-uniform flow and a distortion at the downstream turbine inlet. This distortion at turbine inlet can lead to forced response excitation of the downstream turbine blades.

In this study, forced response excitation due to three different inlet temperature profiles is numerically investigated using a fluid-structure interaction (FSI) approach previously validated in a five-stage axial air turbine. The focus of this study is the aerodynamic excitation due to the combustion chamber defects and the influence on the excitation of the downstream turbine blade row.

The results show that the distortion at turbine inlet creating by the combustion chamber can lead to a significant increase of the excitation of the downstream blades. The present distortion due to the combustion chamber with defects leads to a threefold higher vibration amplitude compared to normal operating conditions. These high amplitudes cause high dynamic stress levels in the blades and reduce the life cycle of the blades.

INTRODUCTION

The overhaul of jet engines, also called regeneration, is of high interest because it generates 8% of the operating cost of an airplane [1]. The main cost factor at overhaul is caused by the airfoils which incur approximately 50% of the cost. This cost is mainly caused by the replacement of worn blades from the high-pressure turbine, because they are one of the most highly loaded parts and thus subject to substantial wear. For this reason, the collaborative research center (CRC) 871 "Regeneration of Complex Capital Goods" aims to develop the scientific basis for the overhaul of jet engines. One main goal is to save as many of the worn components as possible [2].

One approach in the CRC 871 is to investigate and reduce the aerodynamic excitation which occurs due to defects in the flow path. These excitations can be introduced by defects in the combustion chamber. The objective of this paper is to determine the excitation of turbine blades and the vibration amplitude subjected to inlet temperature profiles. The influence of the distortion at turbine inlet on the vibration amplitude of a high pressure turbine (HPT) is numerically investigated for three different inlet conditions.

In [3] various degradation effects on a stationary gas turbine were modelled and studied the variation in overall efficiency and power. The authors showed that a combustor with damaged fuel nozzles does not necessarily change the combustion efficiency, but rather distorts the temperature distribution at the turbine inlet. Previous simulations of mixing of temperature non-uniformities within turbines resulting from the combustion concentrated on the effects of combustor hot streaks on the first turbine stages as presented in [4], [5], and [6]. Typical defects of a combustion chamber can appear due to deviations of the fuel supply and the distribution of the secondary air supply which is shown in [7]. Such defects were numerically investigated in [8]. It was proven that combustion chamber defects can introduce a non-uniform flow field and a distortion of the temperature distribution in the downstream turbine stages and throughout the hot gas path well into the exhaust jet. This distortion is one source of synchronous vibrations in gas turbines.

This excitation source can propagate through the turbine stages, causing an excitation of the downstream turbine blades. This is in contrast to the wake interaction or the potential flow interaction between the blade rows where the neighbouring blade row is mainly influenced. A temperature distortion can lead to a forced response excitation of a low pressure turbine large enough to cause high cycle fatigue (HCF) failures presented in [9]. In [10] and [11] the influence of hot streaks from burner-can induced excitation of low pressure gas turbine blades is numerically and experimentally investigated. In this research the vibration levels excited by the burner cans was estimated and it was shown that the effect of a distortion at turbine inlet is non negligible.

As the HPT blade is arranged directly after the combustion chamber the mechanical and thermal loads of the blades are directly influenced by the combustion chamber. Hence, studies also focused on the influence of temperature distortion at turbine inlet on the HPT stage ([12], [13] and [14]). In [13] the influence of different temperature distortion on the low engine order excitation of HPT blades was investigated. In contrast to [13], the objective of this paper is to determine the excitation of turbine blades and the vibration amplitude subjected to different turbine inlet temperature profiles, resulting from numerical simulations with a combustion chamber in the example of a non-uniform fuel supply of the fuel spray nozzles. Temperature profiles of combustion chamber without defect and with a faulty fuel supply to one of the fuel nozzles in the combustor are compared with a uniform temperature profile.

OUTLINE

In the first part of this paper, the numerical approach to simulate the combustion chamber, with and without defects due to wear, is presented. By simulating different defects subjected to the deviation of the fuel supply, different outlet conditions were determined.

In the second step, the different temperature profiles were interpolated on the inlet of numerical model of the turbine stage. Three different temperature profiles are numerically analyzed using CFD. The focus is set on the impact of the inlet temperature distortion on the excitation of the turbine blades.

Finally, forced response analyses are performed subjected to the three different temperature profiles. The unidirectional fluid-structure approach previously validated experimentally in [15] is used to predict numerically the vibration amplitude.

SIMULATION OF THE COMBUSTION CHAMBER

In this study the influence of the combustion chamber on the aeroelastic behavior of the first stage HPT rotor blades is investigated. Hence, the numerical model was restricted to the combustion chamber, the first stage of the HPT and an outlet pipe. The simulations are separated into two parts. The flow and reaction inside the combustor are modeled with ANSYS Fluent. Subsequently, the simulation of the HPT was performed with ANSYS CFX with the combustion chamber outlet conditions as inlet boundary condition. A schematic sketch of the numerical setup is shown in Fig. 1.

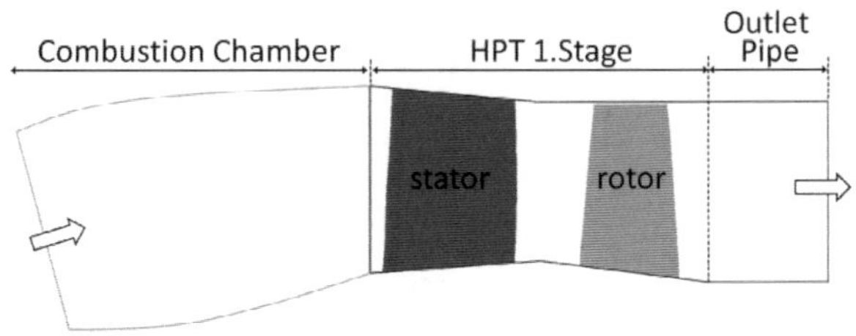

Figure 1: Sketch of the numerical setup in the meridional view

In a first step, the combustion chamber was simulated to determine the turbine inlet temperature profile. The reacting flow in the combustor is simulated using ANSYS Fluent 13.0. Here, turbulence is modeled with the transition SST-model (see [16] and [17]). For combustion, the non-premixed combustion model with the Mixture Fraction Approach is used under the simplifying assumption that the instantaneous thermo-chemical state of the fluid is related to a conserved scalar quantity. Herein the interaction of turbulence and chemistry is calculated with an assumed-shape Probability Density Function (PDF) that considers 20 different species (see [18]). The PDF-table calculates compressibility effects and a non-adiabatic energy treatment have been used. As fuel, gaseous jet fuel A has been chosen. The process of spray breakup and evaporation are neglected, as the purpose of the work so far is to show the general procedure of the approach. The combustor considered is an annular combustion chamber which is supplied with air/fuel mixture by 20 fuel spray nozzles. For this geometry, an unstructured numerical mesh of about 43 million cells was used in order to capture the rather complex nozzle geometry. The geometry of nozzle and combustor is depicted in Fig. 2.

Figure 2: Geometry of combustor and nozzle [8]

The results of the combustion chamber without defect at the design point of the aircraft are shown in Fig. 3. It is shown that a normal operating combustion chamber causes temperature non-uniformity at turbine inlet itself and is dependent on the structure of the burner flame. This distortion is intensified by a defect of a fuel spray nozzle. This is the reason why a failure of a spray nozzle between two normal operating combustors was investigated. A defect of a spray nozzle leads to a failure of the flame at this combustor. This influences the temperature distribution of the adjacent combustor. The pattern of the temperature distribution persists at a lower average temperature level. Hence, a failure of a combustion chamber has an intensified influence on the non-uniformity of the temperature profile at turbine inlet. The profiles of the combustion chamber outlet outlet are both shown in Fig. 5. These profiles are used as inlet condition for the numerical investigations of the HPT turbine. The influence of both cases on the vibration amplitude of the downstream turbine blades were investigated using a forced response analysis, which is described in the following sections.

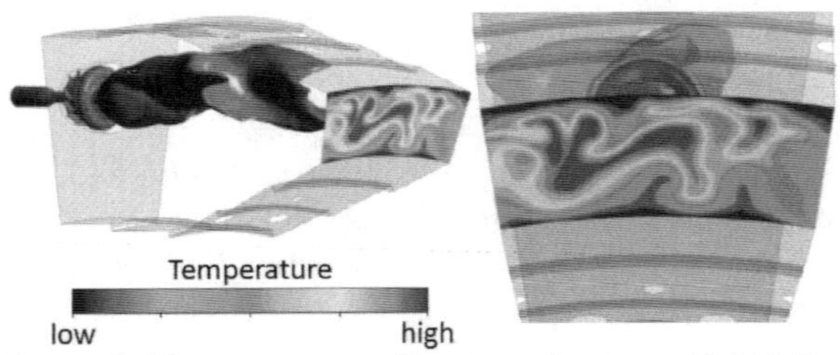

Figure 3: Temperature profile at combustor outlet at the design point [7]

SIMULATION OF THE TURBINE STAGE

Combustion chamber defects generate a non-uniform flow field at the outlet of the chamber. These distortions enter directly into the inlet of the downstream HPT. Therefore, the excitation of the turbine blades can be increased due to these distortions at the inlet of the HPT. The influence of these distortions on the vibration amplitude of a high pressure turbine blade is numerically determined for three different inlet conditions using a unidirectional fluid-structure interaction approach. The following cases are numerically investigated and compared to each other in terms of the vibration amplitude:

- a uniform temperature profile,
- a non-uniform profile without combustion chamber defects
- and a turbine inlet profile in case of failure of every second combustion chamber.

As described in the last section the breakdown of a burner has the highest impact on the distortion of the flow field. Thus, the breakdown of a burner is investigated because it can be assumed that this condition leads to the highest excitation. Not only defect burners are generating non-uniformities at the HPT inlet, but also a combustion chamber without defects which was also shown in the last section. Hence, the influence on the excitation of operation based distortions is investigated as well. The distortions are then compared with a uniform inlet condition.

For the analysis of the excitation mechanisms due to combustion defects, the first stage of a high pressure turbine is simulated by CFD. This stage of a typical aircraft engine consists of 40 vanes and 64 blades. To identify critical operating conditions the eigenfrequencies and engine orders (EO) are shown in the Campbell-Diagram in Fig. 4. In this Campbell-Diagram the eigenfrequencies and the rotational speed are normalized on the frequency and on the rotational speed respectively of the operating point. The operating point is the design point of the turbine and is close to the maximum rotational speed. At this operating point the first eigenfrequency is in resonance with the 10th EO as depicted in Fig. 4. As the combustion chamber of the analyzed aircraft engine consists of 20 burners, the 10th EO can occur in case of a failure of every second combustor. Although this is not a realistic event, it can be expected that the general excitation mechanism is applicable for approaches with a smaller critical EO.

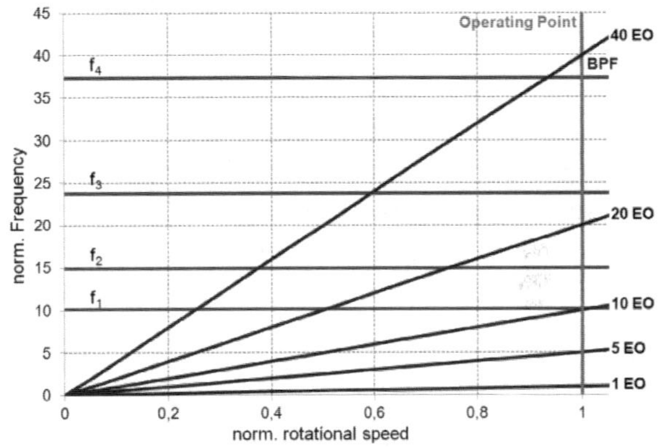

Figure 4: Campbell-Diagram of the HPT rotor blade

Numerical Setup

The numerical investigations of the influence of the combustion chamber on the aerodynamics and the vibration amplitude of the turbine blade were conducted subject to three different inlet temperature profiles. The inlet profiles were obtained from the simulations with the combustion chamber presented in the previous section and are shown in Fig. 5. The cases of a uniform temperature profile, a non-uniform profile without combustion chamber defects and a turbine inlet profile in case of failure of every second combustor were compared regarding the vibration amplitude.

The model depicted in Fig. 6 was used for the numerical investigations. The temperature profiles of the combustion chamber without defects and with a defect of every second combustor were interpolated on the inlet of the CFD model. Besides the temperature profile, the velocity direction was interpolated as well. For the uniform profile, a constant temperature was assumed which is by a factor of 0.94 lower than the real temperature in the engine at design point. As the vanes and blades of the HPT usually have to withstand high temperatures above the melting point of the materials, a cooling flow is applied in the real engine. To consider this cooling flow, the constant temperature at the inlet of the uniform profile was set to be a factor of 0.94 lower. Through the same reasoning, the massflow rate was increased for this operating point and defined with a constant value at the inlet of the CFD model for all three cases. The defined mass-flow rate is by a factor of 1.15 higher than in the real engine. The outlet boundary condition was set to a constant area average static pressure in all three cases.

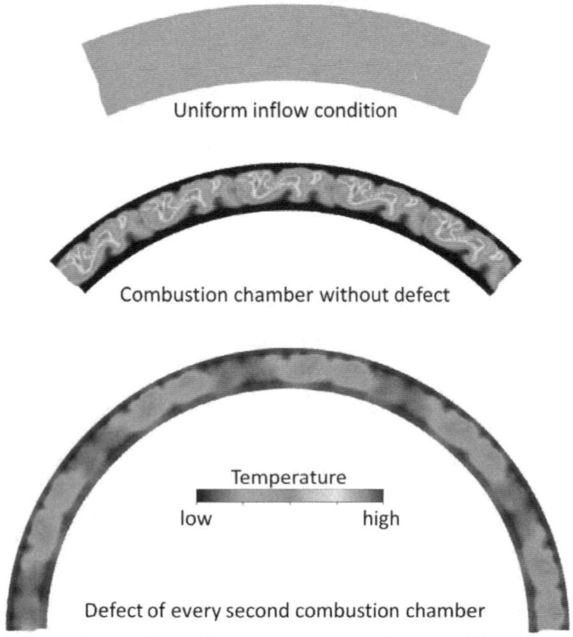

Figure 5: Investigated temperature profiles at turbine inlet

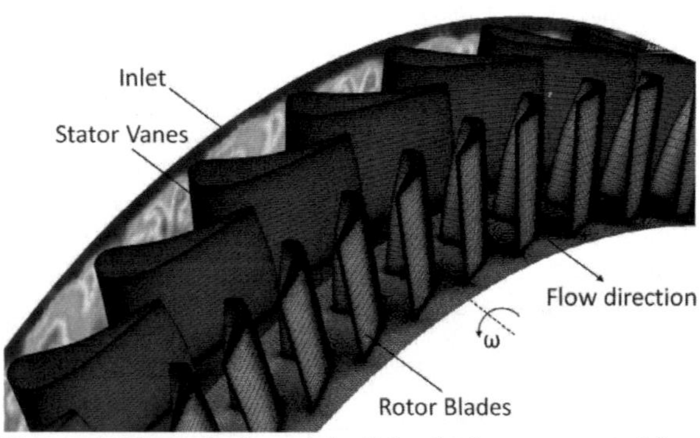

Figure 6: Numerical model of the high pressure turbine stage

The three inlet profiles lead to different pitches which have to be simulated in order to ensure an equal pitch of the domains for the unsteady simulations. The simulated passages for all cases are summarized in Table 1. In all cases rotational periodicity was defined at the circumferential boundaries of the segment. In the steady simulations, the rotor-stator interface was defined as a frozen rotor interface. The frozen rotor interface was replaced by a transient rotor-stator interface in the unsteady simulations. The unsteady computations were simulated with 16 time steps per pitch and with the steady simulation as an initial solution. The numerical model was discretized with 4.3Mio elements for the uniform case and 17.2Mio elements for the case of a defect of every second combustor. All simulations of the HPT stage were performed with ANSYS CFX 14.5 using the SST turbulence model. The turbulence intensity was set to 20% because in high pressure turbines the turbulence intensity is higher compared to the standard turbulence intensity in turbomachines.

Table 1: Simulated passages of the first HPT stage

	Uniform profile	Profile without defect	Profile with defect burners
Passage in °	45°	90°	180°
Vanes	5	10	20
Blades	8	16	32
Burners	2.5	5	10
Mesh elements	4.3Mio	8.6Mio	17.2Mio

For the numerical investigations it is important that the numerical error caused by the mesh resolution is negligible. Hence, a mesh study was performed. In the mesh study the parameters of main interest like total pressure loss coefficient or the performance were determined. These parameters were evaluated by the grid convergence index (GCI) defined in [19]. The diagram is exemplary shown in Fig. 7 for the performance of the rotor and the total pressure loss coefficient at 50% span height of the rotor stage. The mesh study proves that the results of the medium and fine mesh are almost independent regarding the total pressure loss coefficient and performance. Consequently, the medium mesh size was used for the investigations of the HPT stage. The mesh quality of the HPT stage is in a good agreement with the recommended quality for numerical simulations. The meshes have a grid quality of a minimum angle above 42.7° and a maximum aspect ratio below 500.

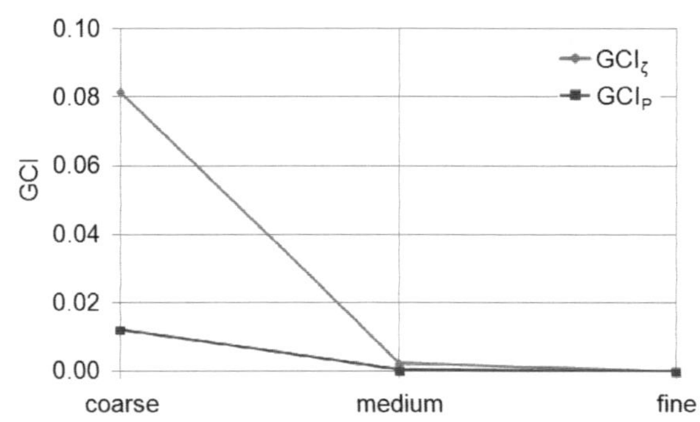

Figure 7: Grid convergence index of total pressure loss coefficient and performance

Aerodynamic Results

The focus for the aerodynamic results was set on the analysis of the total pressure loss ζ coefficient behind the stator of the first HPT stage. The total pressure loss coefficient is defined in Eq. (1) with the total pressure $p_{tot,in}$ at inlet and static pressure $p_{stat,in}$ at inlet.

$$\varsigma = \frac{p_{tot,in} - p_{tot}}{p_{tot,in} - p_{stat,in}} \quad (1)$$

Besides the potential effect, the wakes of the vanes are one of the main excitation mechanisms of the downstream rotor blades. The wake behavior can be analyzed with the total pressure loss coefficient which is depicted in Fig. 8 as a function of the pitch. The first 10 pitches are shown in this chart to indicate the trend of the wake behavior. The coefficient is evaluated at 80% span behind the stator vanes, indicated by the red line in the sketch above the chart. In Fig. 8 it is shown that the uniform inflow condition has an almost constant wake behavior. The spacing between each wake is about 1.03 pitches. As expected, the wake behavior of the uniform inflow condition is influenced only by the stator vanes.

In contrast to that, the outlet condition of the combustion chamber has a significant influence on the wake behavior by the combustion chamber (see Fig. 8). The normal operating condition with a temperature profile caused by a combustion chamber without defect shows that every second maximum of the total pressure loss coefficient is lower by 0.89%. The 20th combustor leads to a decrease of the total pressure loss coefficient in an alternating distribution and influence the wake behavior of the 40 vanes.

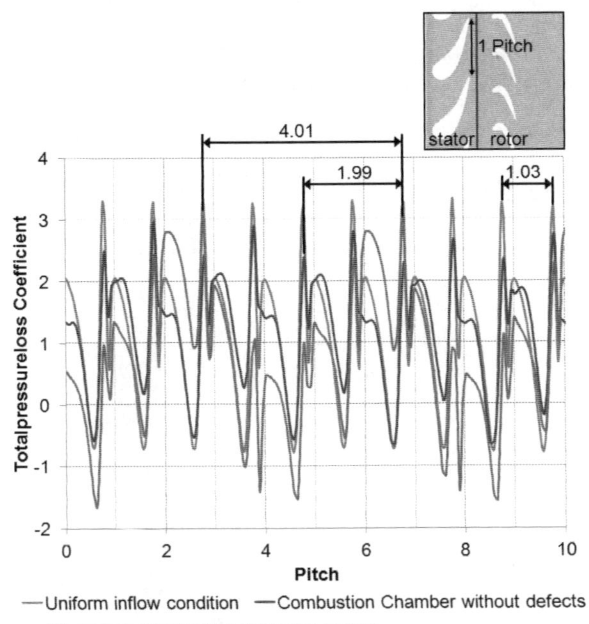

Figure 8: Total pressure loss coefficient behind the stator vane at 80% span

For the case with the temperature profile of the case with the alternating combustor failure, the distribution of the total pressure loss coefficient repeats every 4 pitches. This causes a variation in the wake behavior behind every 4th stator vane. Hence, the distortion of the temperature profile due to a defect or failure of a combustion chamber leads to a significant influence of the aerodynamic behavior behind the stator vanes. Consequently, the combustion chamber with or without defect has a different influence on the excitation of the downstream rotor blades which is investigated in the following section.

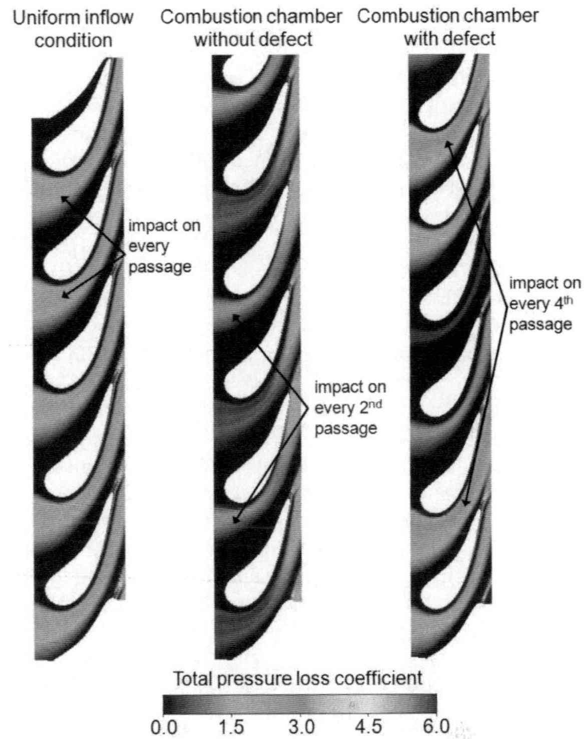

Figure 9: Contourplot of total pressure loss coefficient in the stator domain at 80% span

The results of Fig. 8 are confirmed by the results of Fig. 9. In Figure 9 a contour plot of the total pressure loss coefficient at 80% span is visualized of each case. In these plots the influence of the inlet distortion on the vane passages is clearly shown. The uniform inflow condition has the same impact on every passage. In contrast, the case of the combustion chamber without defect has a repeating impact on every second passage due to the 20 combustor. Hence, the excitation behind the stator vanes should also differs from the uniform inflow which was also examined in Fig. 8. In these contour plots it can also be seen that the case of the alternating combustor has the highest influence on the disturbance of the flow in the passages of the vanes. The defect of the combustion chamber has an impact on every fourth passage which confirms the results of Fig. 8 where the distribution of the total pressure loss coefficient repeats every 4 pitches.

The results of the total pressure loss coefficient show that the influence of the inlet distortion has a significant effect on the flow field. Furthermore, the wake behaviour differs between the three cases. Hence, it can be assumed the excitation mechanism of the downstream rotor blades is also different. Based on these results, the next step is to investigate the influence of the aerodynamic excitation on the vibration amplitudes subjected to the temperature distortion by performing a forced response analysis.

FORCED RESPONSE ANALYSIS

In the following section the influence of the combustion chamber on the vibration amplitude of the high pressure turbine blade is investigated based on the aerodynamic investigation of the preceding section. For this purpose a forced response analysis was conducted. A unidirectional fluid-structure interaction (FSI) approach is used which is experimentally validated and showed a good agreement between the experimental and numerical results (see [15]).

Forced Response Method

The vibration amplitude of the turbine blades was predicted numerically by performing a forced response analysis. The approach of the unidirectional FSI is shown in Fig. 10. After performing steady and unsteady CFD calculations, the unsteady surface pressures are extracted. Subsequently, the unsteady surface pressures are mapped from the CFD mesh on the FEM mesh. The unsteady surface pressures are then transformed into the frequency domain by performing a Fast Fourier Transformation (FFT). A harmonic analysis is conducted with the pressure loads for the main excitation frequencies resulting from the FFT. In the last step, the responses of the harmonic analysis are superimposed.

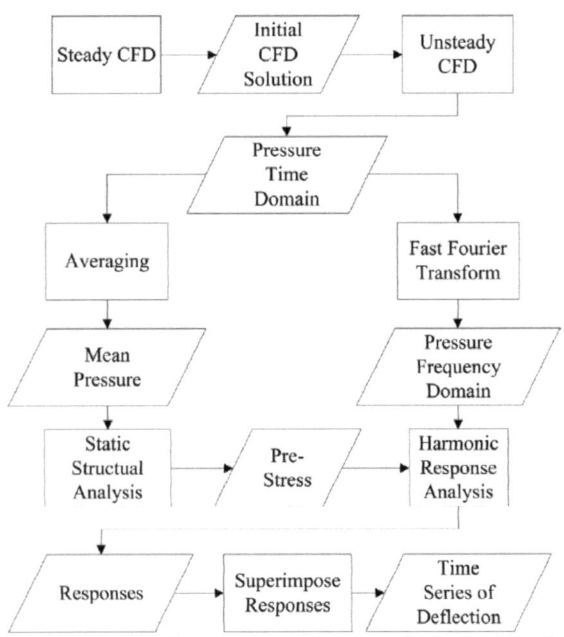

Figure 10: Unidirectional FSI approach by using the harmonic analysis [15]

By using this approach it is possible to estimate the vibration amplitudes of the turbine blades depending on various inlet conditions. This approach was performed with a damping ratio of 0.001. This damping ratio did not consider the aerodynamic damping and was assumed because it is a typical value of turbine blades in turbomachines. The vibration amplitudes which are determined in the forced response analysis will be compared relative to each other.

Forced Response Results

The impact of the temperature distortion at turbine inlet on the amplitude of the pressure integrated over the blade surface is investigated in the first step. For this purpose, the integrated pressure amplitudes are evaluated for the engine orders with highest occurring amplitudes and are shown in Fig. 11. The amplitudes are normalized on the amplitude of the case with the combustion chamber without defects at the 40th EO. The case with a defect of the combustion chamber results in an excitation of the 10th EO. As the profile of this case has the highest distortion at turbine inlet, this case has also the highest impact on the excitation of the blade and results in high integrated pressure amplitudes. The pressure amplitude is 11% higher than the pressure amplitude for the normal operating condition at the blade passing frequency.

High pressure amplitudes occur at the 20th EO for the case with combustion chamber without defect. This excitation is caused by the distortion of the 20 combustor Compared to the blade passing frequency excitation, this pressure amplitude is 5.6% lower. For the case with the alternating combustor failure, the amplitude at the 20th EO is significant lower and about 30.1%. The uniform case has a remarkable excitation of the blade at the blade passing frequency (40th EO). This excitation of the blade passing frequency in all three cases is non-negligible and results in high pressure amplitudes. For EO below the blade passing frequency, the pressure amplitudes for the uniform temperature profile are negligible and just the profiles with combustion chamber have a significant influence.

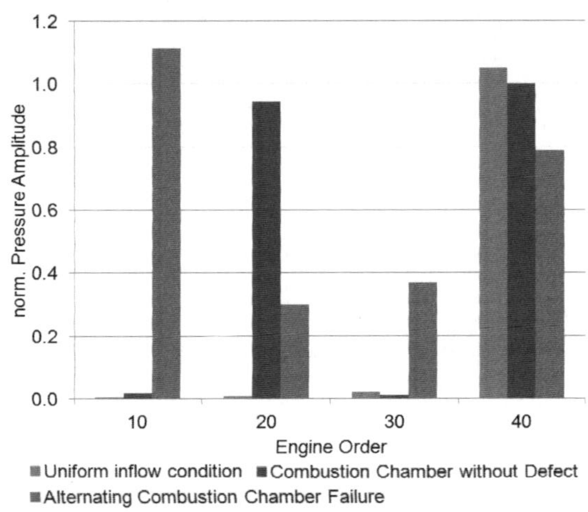

Figure 11: Blade-surface Pressure amplitude integrated over the blade surface for different engine orders

Identical to the results of the integrated pressure amplitudes, the vibration amplitudes are significantly influenced by the inhomogeneous flow field which results from the combustion chamber. Compared to the uniform inflow condition, additional excitation mechanisms are detected due to the combustion chamber. In Fig. 12, the normalized vibration amplitude subjected to the excited engine order is presented at the tip trailing edge of the rotor blade. The amplitudes are normalized on the amplitudes of the case of combustion chamber without defects at the 40th EO which correspond to the blade passing frequency (BPF). In the case of the uniform inflow condition, detectable vibration amplitudes only occur at the blade passing frequency. These amplitudes are by a factor of 4 lower than the amplitudes in the case of the combustion chamber without defects. An additional excitation at the 20th EO for the non-uniform profile without defects results from the twenty combustor. A further amplification of the vibration amplitude can be induced by the combustion chamber with defects which are depicted in Fig. 12. The 10th EO, which is close to the first eigenfrequency of the turbine blade, occurs due to a failure of every second combustor (see Fig. 12). The first eigenfrequency shows the highest displacement at the rotor tip trailing edge. Hence, this resonance induces higher vibration amplitudes compared to excitations at other frequencies. The amplitude for this case is by a factor of 3.8 higher than the case of normal operating conditions at the blade passing frequency. The examination of the vibration amplitudes also shows that distortion of the combustion chamber without defects leads to a higher excitation of the rotor blade compared to the uniform profile. The results of the normalized vibration amplitude at the rotor tip trailing edge show a similar trend as the integrated pressure amplitudes in Fig. 11.

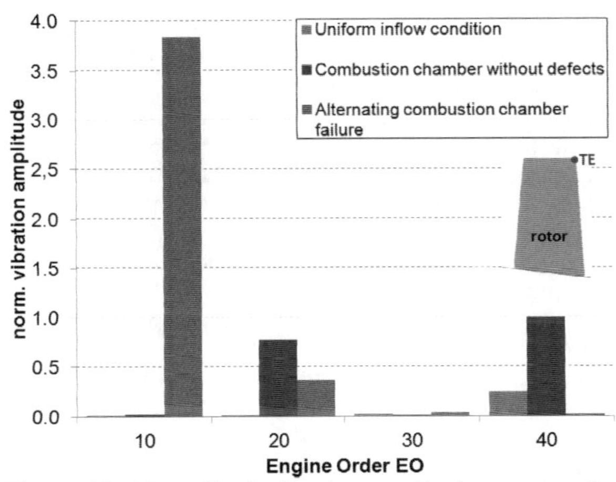

Figure 12: Normalized vibration amplitude at rotor tip trailing edge

The influence of the three inlet temperature profiles on the superimposed vibration amplitude is depicted in Fig. 13. The superimposed vibration amplitude is normalized for the case of normal operating conditions in the combustion chamber. The superimposed amplitude of the case with combustion failure is by a factor of 3 higher compared to the combustion chamber without defect (see Fig. 13). This amplification is due to the excitation of the first eigenfrequency by the 10th EO which causes high vibration amplitude. Nevertheless, the superimposed vibration amplitude of the combustion chamber without defect is by a factor of 4 higher compared to the uniform inflow conditions. Consequently, the distortion of the inlet temperature profile leads to a further amplification of the vibration amplitude at the rotor tip trailing edge. The excitation frequencies due to the uniform temperature profile have negligible influence on the vibration amplitude which in turn also causes lower superimposed vibration amplitude.

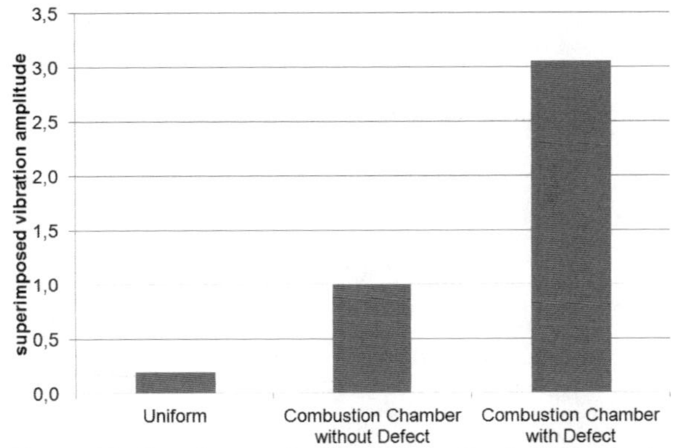

Figure 13: Superimposed vibration amplitudes at rotor tip trailing edge

These results show that the excitation due to the combustion chamber because of the distortion at turbine inlet has to be considered in the design of the rotor blades. Distortions at turbine inlet directly influence the vibration amplitude and, therefore, the dynamic stress level of the downstream turbine blades.

CONCLUSIONS

The influence of combustion chamber outlet conditions on the excitation of a high pressure turbine blade is numerically investigated using a unidirectional FSI method. The following outlet conditions were investigated and compared regarding the vibration amplitude:

- uniform temperature profile,
- a non-uniform profile without combustion chamber defects,
- and a turbine inlet profile in case of failure of every second combustor.

The prediction of the vibration amplitude using the FSI approach shows an increase of the superimposed vibration amplitudes in case of a normally operating combustion chamber by a factor of four compared to the uniform temperature profile. A further amplification of the amplitude by a factor of three is caused by a failure of every second combustor. This amplification results from the excitation of the first eigenfrequency of the rotor blade by the 10th EO. Based on these results, such temperature distortions due to the combustion chamber or combustion chamber defects should also be considered in the design process of turbine blades and during the regeneration process in order to avoid unacceptably high dynamic stress levels in the blades.

A future step will be the investigation of multistage effects subjected to combustion chamber defects. The investigation of the vibration amplitude by including further turbine stages is of particular interest.

NOMENCLATURE

f	frequency
D	damping ratio
$\dot{m}$	mass flow rate
p	pressure
rpm	rotor speed
T	Temperature
ζ	total pressure loss coefficient

Indices

dyn	dynamic
in	turbine inlet
out	turbine outlet
stat	static
tot	total

Abbreviations

BPF	blade passing frequency
CFD	computational fluid dynamics
CRC	collaborative research center
EO	engine order
FSI	fluid-structure interaction
GCI	grid convergence index
HPT	high pressure turbine
LE	leading edge
PDF	probability density function
TE	trailing edge

ACKNOWLEDGMENTS

The authors kindly thank the German Research Foundation (DFG) for the financial support to accomplish the research projects C4 "Regeneration-induced Variances of Aeroelastic Properties of Turbine Blades" and A4 "Influence of Combustion Chamber Defects on the Exhaust Jet" within the Collaborative Research Center (CRC) 871. Furthermore, the authors thank ANSYS for providing CFX in an academic license and the Leibniz Universität Hannover IT Services (LUIS) for the computational resources provided.

REFERENCES

[1] Rupp, O., 2001. "Instandhaltungskosten bei zivilen Strahltriebwerken". Deutscher Luft- und Raumfahrtkongress 2001, Hamburg DGLR-2001-008

[2] Aschenbruck, J., Adamczuk, R. and Seume, J., 2014. "Recent Progress in Turbine Blade and Compressor Blisk Regeneration". *Proceedings of 3rd International Conference on Through-life Engineering Services*, November 4-5 2014, Cranfield, England

[3] Kurz, R., and Brun, K., 2001. "Degradation in gas turbine systems". *Journal of Engineering for Gas Turbines and Power,* ***123***(1), p. 70.

[4] Basol, A. M., Jenny, P., Ibrahim, M., Kalfas, A. I., and Abhari, R. S., 2009. "Hot streak migration in a turbine stage: Integrated design to improve aerothermal performance". *Journal of Engineering for Gas Turbines and Power,* ***133***, pp. 1–10.

[5] Qingjun, Z., Fei, T., Huishe, W., Jianyi, D., Xiaolu, Z., and Jianzhong, X., 2008. "Influence of hot streak temperature ratio on low pressure stage of a vaneless counter-rotating turbine". *Journal of Engineering for Gas Turbines and Power,* ***130***(3), p. 031901.

[6] Qureshi, I., Beretta, A., and Povey, T., 2011. "Effect of simulated combustor temperature nonuniformity on HP vane and end wall heat transfer: An experimental and computational investigation". *Journal of Engineering for Gas Turbines and Power,* ***133***(3), p. 031901.

[7] Hennecke, C., Frieling, D., Dinkelacker, F., 2013. „Influence of Combustion Chamber Disturbances to Patterns in the Hot Gas Path". Proceedings of the European Combustion Meeting 2013.

[8] Adamczuk, R., Buske C., Roehle, I. Hennecke, C., Dinkelacker, F. Seume, J. R., 2013. „Impact of defects and damage in aircraft engines on the exhaust jet". Proceedings of ASME Turbo Expo 2013, June 3-7 2013, San Antonio, Texas, USA, GT2013-95079

[9] Manwaring, S. R., Kirkeng, K. L., 1997. „Forced Response Vibrations of a Low Pressure Turbine Due to Circumferential Temperature Distortions". Unsteady Aerodynamics and Aeroelasticity of Turbomachines, Proceedings of the 8th International Symposium, 14-18 September 1997, Stockholm, Schweden.

[10] Mayorca, M. A., Bladh, R., Ozturk, U., 2013. "Estimation of burner can-induced excitation levels in industrial gas turbine". Proceedings of ASME Turbo Expo 2013, June 3-7 2013, San Antonio, Texas, USA.

[11] Mayorca, M. A., Torres, A., Vsevolod, K., Bladh, R., 2015. "Estimation of Burner-Can Induced Excitation Response of an Industrial Low Pressure Turbine Blade – Validation against Prototype Test Data". Proceedings of ASME Turbo Expo 2015, June 15-17 2015, Montréal, Canada.

[12] Bréard, C., Sayma, A., Vahdati, M. And Imregun, M., 2002. "Aeroelasticity Analysis of an Industrial Gas Turbine Combustor using a simplified model". Journal of Fluid and Structures, Vol. 16 (No.8), pp. 1111-1126; *doi:10.1006/jfls.2002.0466*

[13] Bréard, C., Green, J., and Imregun, M., 2003. "Low-Engine-Order Excitation Mechanisms in Axial-Flow Turbomachinery". Journal of Propulsion and Power, Vol. 2003 (No.19), pp. 704-712.

[14] He, L., Menshikova, V., Haller, B.R., 2004. "Influence of Hot Streak Circumferential Length-Scale in Transonic Turbine Stage". Proceedings of ASME Turbo Expo 2004, June 14-17 2004, Vienna, Austria.
[15] Aschenbruck, J. and Seume, J. R., 2015. „Experimentally Verified Study of Regeneration-Induced Forced Response in Axial Turbines". *ASME Journal of Turbomachinery, Paper No: TURBO-14-1177; doi 10.1115/1.4028350, Vol. 137(3)/031006*
[16] Menter, F. R., 1994. "Two-equation eddy-viscosity turbulence models for engineering applications". AIAA Journal, 32(8), Aug., pp. 1598–1605.
[17] Menter, F. R., Langtry, R. B., Likki, S. R., Suzen, Y. B., Huang, P. G., and Volker, S., 2006. "A correlation-based transition model using local Variables - Part I: model formulation". Journal of Turbomachinery, 128(3), July, p. 413.
[18] Sivathanu, Y., and Faeth, G., 1990. "Generalized state relationships for scalar properties in nonpremixed hydrocarbon/air flames". Combustion and Flame, Jan.
[19] Salas, M. D., 2006. "Some observations on Grid Convergence". NASA Langley Research Center, Hampton, VA 23681-2199

Proceedings of the 14th International Symposium on Unsteady Aerodynamics,
Aeroacoustics & Aeroelasticity of Turbomachines
ISUAAAT14
8-11 September , 2015, Stockholm, Sweden

I14-S6-3

FORCED RESPONSE VIBRATION AMPLITUDE OF MISTUNED BLADED DISKS WITH AERODYNAMIC DAMPING

Carlos Martel
ETSI Aeronáutica y del Espacio
Universidad Politécnica de Madrid
28040 Madrid, Spain

Luo Cuiping
School of Aerospace and Engineering
Beijing Institute of Technology
Beijing 100081, China

ABSTRACT

Mistuning can dramatically increase the forced response of a bladed disk. In the case of damping coming from aerodynamic effects the situation is more complicated, and the effect of mistuning can be the reduction of the maximum response below that of the tuned case. In this paper the Asymptotic Mistuning Model (AMM) methodology is used to analyze this situation. The AMM is a reduced model that is systematically derived from the complete mistuned bladed disk dynamical model using a perturbative procedure based on the smallness of the mistuning and the damping. Several different forcing cases are analyzed, and the results from the AMM are quantitatively compared with those from the numerical simulation of a complete mistuned bladed disk. The AMM gives information about the key mechanisms involved in the forced response of mistuned cyclic structures, and about the maximum amplification of the vibration amplitude that the system can experience.

INTRODUCTION

The term mistuning refers to the small differences among the (theoretically perfectly identical) blades in real turbomachinery bladed disk. These small undesirable random imperfections result from manufacturing, assembling and material tolerances, as well as in-service wear, and are essentially unavoidable.

Mistuning strongly influences the dynamical response of the bladed disk. In the case of a forced response excitation, it typically leads to a considerable amplification of the vibration amplitudes of individual blades, which can significantly exceed the corresponding vibration amplitude of the tuned structure (i.e., without imperfections). These increased response levels result in a strong decrease of the fatigue life of the blades, and thus have a considerable negative impact on the safety, operability and readiness of aircraft turboengines.

There has been a lot of research since the 1970s in the effect of mistuning (see, e.g, the review [1]), and it is well known that mistuning typically increases the amplitude of the forced response by a factor that can be between 2 an 5.

But recently, the possibility of having a reduction (by a factor of 2 or more) of the forced response vibration amplitude by the effect of mistuning has been reported [2, 3]. The reason for this change in the effect of mistuning relies in the fact that the damping considered is now aerodynamic damping, which has different values for the different travelling wave (TW) modes of the tuned structure. Mistuning couples the TWs and therefore can produce an increase of the effective damping that translates directly into a reduction of the vibration amplitude. This situation is particularly relevant for the case of BLISKS (integrally bladed disks with no blade-disk interfaces), where practically all the damping comes from the gas flow. In the cases considered previously, the structural damping was taken to be uniform, and the effect of mistuning was always to produce an increase of the forced response vibration levels.

The analyses available in the literature perform statistical explorations and optimizations on full FEM models and on reduced order models to gain information about the reduction of the vibration amplitude due to mistuning [2, 3, 4, 5].

The purpose of this work is to asymptotically derive consistent, simplified mistuning models by exploiting the fact that both, the mistuning and the aerodynamic effects (damping and frequency correction), are small quantities. Similar asymptotic mistuning models (AMM) have been already successfully applied to the analysis of the effect of mistuning on the flutter characteristics of unstable rotors [6], and to the computation of the amplification forced response vibration amplitude [7, 8]. The resulting AMM will be used to to identify the physical mechanisms that play a role in the effect of reduction of vibration by mistuning. We will also use the AMM to look for the mistuning patterns that produce maximum vibration reduction, and to find bounds for the maximum response amplification and reduction that can be obtained when aerodynamic effects are included. The results from the AMM will be also checked against those from a FEM model of a simplified rotor.

NOMENCLATURE

$\mathbf{M}$	mass matrix
$\mathbf{K}$	stiffness matrix
$\Delta\mathbf{M}$	mistuning mass matrix
$\Delta\mathbf{K}$	mistuning stiffness matrix
$\mathbf{F}$	external forcing
$\mathbf{X}$	displacement vector
ω	forcing frequency
r	forcing engine order
N	number of sectors
m	number of DOF per sector
X_j	sector j displacements
M	in-sector mass matrix
M_c	sector-to-sector coupling mass matrix
K	in-sector stiffness matrix
K_c	sector-to-sector coupling stiffness matrix
F	in-sector external forcing amplitude
ΔM_j	sector j mass mistuning matrix
ΔK_j	sector j stiffness mistuning matrix
$\mathbf{P}$	TW modes matrix
$\mathbf{A}$	TW mode amplitudes vector
Z_{kj}	tuned TW mode shape
ω_{kj}	tuned TW mode frequency
ΔM_k^F	mass mistuning Fourier k harmonic
ΔK_k^F	stiffness mistuning Fourier k harmonic
Δ	mistuning matrix in TW basis
ω_0	resonant TW mode frequency
ω_a	active TW mode frequency
$D, D_{j'j}$	mistuning coefficients in the TW basis
$m(j)$	mistuning amplitude distribution
K_{mist}	stiffness mistuning matrix
$\mathbf{M}_{\text{aero}}$	matrix of aerodynamic forces
η_a	aerodynamic frequency correction
ξ_a	aerodamping
ND	number of nodal diameters
EO	engine order of the forcing
TW	traveling wave

MISTUNED BLADED DISK FORCED VIBRATION

In this section we give a brief explanation of the derivation of the AMM; for a more detailed description see [6, 7, 8].

The starting point are the general FEM model equations for the aeroelastic vibration of a mistuned bladed disk subject to a harmonic forcing with frequency ω,

$$(\mathbf{K}+\Delta\mathbf{K}-\omega^2(\mathbf{M}+\Delta\mathbf{M}))\mathbf{X}=\mathbf{M}_{\text{aero}}(\omega)\mathbf{X}+\mathbf{F}, \qquad (1)$$

where the response of the structure is also assumed to be harmonic with frequency ω. The vector $\mathbf{X}$ contains the displacement amplitudes of all independent DOFs of the structure, and is arranged by sectors

$$\mathbf{X}=\begin{bmatrix} X_1 \\ \vdots \\ X_j \\ \vdots \\ X_N \end{bmatrix}, \qquad (2)$$

with the vector X_j containing the displacements of the m DOF associated with sector j, and N is the total number of sectors of the bladed disk.

The tuned mass $\mathbf{M}$ and stiffness $\mathbf{K}$ matrices are cyclic matrices given by

$$\mathbf{M}=\begin{bmatrix} M & M_c & 0 & \cdots & M_c^{\mathrm{T}} \\ M_c^{\mathrm{T}} & M & M_c & \cdots & 0 \\ & & \ddots & \ddots & \\ M_c & 0 & \cdots & M_c^{\mathrm{T}} & M \end{bmatrix}, \quad \mathbf{K}=\begin{bmatrix} K & K_c & 0 & \cdots & K_c^{\mathrm{T}} \\ K_c^{\mathrm{T}} & K & K_c & \cdots & 0 \\ & & \ddots & \ddots & \\ K_c & 0 & \cdots & K_c^{\mathrm{T}} & K \end{bmatrix}, \qquad (3)$$

where the $m\times m$ sector stiffness and mass matrices K and M are symmetric and the coupling between adjacent sectors is represented by the coupling stiffness and mass matrices K_c and M_c.

The forcing from the wakes of the previous row is assumed to take the form of a pure traveling wave, with frequency ω and engine order r, that is,

$$\mathbf{F}=\begin{bmatrix} F e^{\mathrm{i}(\frac{2\pi r}{N})1} \\ \vdots \\ F e^{\mathrm{i}(\frac{2\pi r}{N})j} \\ \vdots \\ F e^{\mathrm{i}(\frac{2\pi r}{N})N} \end{bmatrix} + \text{c.c.} \qquad (4)$$

The matrix associated with the linearized aerodynamic forces due to the vibration of the blades, $\mathbf{M}_{\text{aero}}$, is a block-sector cyclic matrix for the tuned configuration, and depends only on the vibration frequency ω. The full computation of $\mathbf{M}_{\text{aero}}$ would involve: (i) the computation of the stationary flow around the blades, and then (ii) for each given ω value the computation of the linearized Navier-Stokes around the base flow equations for all linearly independent blade displacements. This task is extremely costly and simply not realizable for any practical realistic bladed disk configuration, but the matrix $\mathbf{M}_{\text{aero}}(\omega)$ is formally kept in the formulation in order to progress with the AMM derivation.

Finally, the mistuning is included in the model (1) through the mistuning stiffness $\Delta\mathbf{K}$ and mass $\Delta\mathbf{M}$ matrices, that take the form

$$\Delta\mathbf{K}=\begin{bmatrix}\Delta K_1 & 0 & \cdots & 0\\ 0 & \Delta K_2 & \cdots & 0\\ & & \ddots & \\ 0 & 0 & 0 & \Delta K_N\end{bmatrix}, \Delta\mathbf{M}=\begin{bmatrix}\Delta M_1 & 0 & \cdots & 0\\ 0 & \Delta M_2 & \cdots & 0\\ & & \ddots & \\ 0 & 0 & 0 & \Delta M_N\end{bmatrix}, \quad (5)$$

$$\text{wit}\Delta\text{h} \quad \textstyle\sum_{j=1}^{N} \Delta K_j = 0,\ \sum_{j=1}^{N} \Delta M_j = 0, \quad (6)$$

$$\text{and} \quad \Delta K_j = \Delta K_j^{\mathrm{T}},\ \Delta M_j = \Delta M_j^{\mathrm{T}} \quad \text{for} \quad j = 1 \ldots N. \quad (7)$$

Note that we are not considering mistuning in the coupling between sectors, and that, in order to avoid any ambiguities, we have set the sector to sector mistuning average to zero (see eq. (6)), which is equivalent to define the tuned sector as the average of the mistuned sectors.

The system (1) is transformed from the displacements description, $\mathbf{X}$, to the tuned traveling wave amplitudes, $\mathbf{A}$, using the linear change of variables :

$$\mathbf{X} = \mathbf{P}\mathbf{A}, \qquad \mathbf{A} = \begin{bmatrix}A_1\\ \vdots\\ A_k\\ \vdots\\ A_N\end{bmatrix} \quad (8)$$

where

$$\mathbf{P} = \frac{1}{\sqrt{N}}\begin{bmatrix}P_1 e^{i(\frac{2\pi 1}{N})1} & \ldots & P_N e^{i(\frac{2\pi N}{N})1}\\ \vdots & & \vdots\\ P_1 e^{i(\frac{2\pi 1}{N})j} & \ldots & P_N e^{i(\frac{2\pi N}{N})j)}\\ \vdots & & \vdots\\ P_1 e^{i(\frac{2\pi 1}{N})N} & \ldots & P_N e^{i(\frac{2\pi N}{N})N}\end{bmatrix} \quad (9)$$

and the block matrices P_k of the matrix $\mathbf{P}$ contain all tuned elastic eigenmodes Z_{kp} corresponding to nodal diameter equal to k:

$$P_k = \left[Z_{k1}|Z_{k2}|\cdots|Z_{kp}|\cdots|Z_{km}\right]. \quad (10)$$

The transformed system takes the following form:

$$\left(\begin{bmatrix}\Omega_1^2 - \omega^2 I & \ldots & 0\\ \vdots & \ddots & \vdots\\ 0 & \ldots & \Omega_N^2 - \omega^2 I\end{bmatrix} + \Delta\right)\begin{bmatrix}A_1\\ \vdots\\ A_N\end{bmatrix} =$$
$$= \mathbf{P}^{\mathrm{H}}\mathbf{M}_{\text{aero}}(\omega)\mathbf{P}\begin{bmatrix}A_1\\ \vdots\\ A_N\end{bmatrix} + \mathbf{P}^{\mathrm{H}}\mathbf{F} \quad (11)$$

where the identity matrix I and the Ω_k^2 matrices come from the transformed tuned mass and stiffness matrices, respectively. The Ω_k^2 matrices just have the squares of the eigenfrequecies ω_{kp} for nodal diameter k on the diagonal:

$$\Omega_k^2 = \begin{bmatrix}\omega_{k1}^2 & 0 & \ldots & 0\\ 0 & \omega_{k2}^2 & \ddots & \vdots\\ \vdots & \ddots & \ddots & 0\\ 0 & \cdots & 0 & \omega_{km}^2\end{bmatrix} \quad (12)$$

As expected, the tuned elastic part becomes strictly diagonal in the TW basis.

The forcing $\mathbf{F}$ has a traveling wave form with a nodal diameter $k = r$, and consequently its left multiplication by the matrix $\mathbf{P}^{\mathrm{H}}$, will result in a block vector with null entries, except for the block corresponding to the engine order r (recall that the TWs with different number of nodal diameters are orthogonal):

$$\mathbf{P}^{\mathrm{H}} \cdot \mathbf{F} = \begin{bmatrix}\mathbf{0}\\ \vdots\\ P_r^H \cdot F\\ \vdots\\ \mathbf{0}\end{bmatrix} \quad (13)$$

The linearized aerodynamic flow around the blades has also cyclic symmetry because the aerodynamic flow is the same on each sector. This leads, again formally, to a block diagonalization of the aerodynamic matrix $\mathbf{M}_{\text{aero}}$ in the traveling wave basis:

$$\mathbf{P}^{\text{H}}\mathbf{M}_{\text{aero}}(\omega)\mathbf{P} = \begin{bmatrix} M_1(\omega) & 0 & \dots & 0 \\ 0 & M_2(\omega) & \ddots & \vdots \\ \vdots & \ddots & \ddots & 0 \\ 0 & \cdots & 0 & M_N(\omega) \end{bmatrix} \quad (14)$$

And finally, the mistuning correction matrix Δ, that contains all the mistuning terms is defined as:

$$\Delta = \mathbf{P}^{\text{H}}(\Delta\mathbf{K} - \omega^2\Delta\mathbf{M})\mathbf{P} \quad (15)$$

To further understand the structure of this matrix it is convenient to expand in discrete Fourier series the mistuning distribution along the N sectors of the bladed disk (see (5))

$$\Delta K_j = \sum_{k=1}^{N} \Delta K_k^{\text{F}} e^{\text{i}(\frac{2\pi k}{N})j} \quad \text{and} \quad (16)$$

$$\Delta M_j = \sum_{k=1}^{N} \Delta M_k^{\text{F}} e^{\text{i}(\frac{2\pi k}{N})j} \quad \text{for} \quad j = 1 \dots N \quad (17)$$

The k Fourier coefficients ΔK_k^{F} and ΔM_k^{F} indicate the amount of harmonic wave with k nodal diameters that is contained in the mistuning pattern, and verify the following conditions that are just direct consequences of (6) and (7).

$$\Delta K_N^{\text{F}} = 0,\ \Delta M_N^{\text{F}} = 0 \quad \text{and} \quad \Delta K_k^{\text{F}} = \overline{\Delta K_{-k}^{\text{F}}},\ \Delta M_k^{\text{F}} = \overline{\Delta M_{-k}^{\text{F}}},$$
$$\Delta K_k^{\text{F}} = (\Delta K_k^{\text{F}})^{\text{T}},\ \Delta M_k^{\text{F}} = (\Delta M_k^{\text{F}})^{\text{T}}, \quad \text{for} \quad k = 1 \dots N \quad (18)$$

After inserting the above expressions (16) and (17) into the mistuning correction matrix Δ, it takes the form

$$\Delta = \begin{bmatrix} 0 & \Delta_{12} & \dots & \Delta_{1N} \\ \Delta_{21} & 0 & \dots & \Delta_{2N} \\ \vdots & \vdots & \ddots & \vdots \\ \Delta_{N1} & \Delta_{N2} & \dots & 0 \end{bmatrix} \quad (19)$$

where each block is related to the Fourier components of the block mistuning matrices ΔK_j and ΔM_j by:

$$\Delta_{kk'} = P_k^{\text{H}}(\Delta K_{k-k'}^{\text{F}} - \omega^2 \Delta M_{k-k'}^{\text{F}})P_{k'} \quad (20)$$

The idea of the AMM methodology is to take into account the fact that the mistuning, the forcing and the aerodynamic effects are just small corrections of the elastic tuned dynamics, and apply an asymptotic perturbative procedure to obtain a reduced model of the system (see [6, 7, 8]). Near a resonance, the only TW modes that have to be retained are those with frequency close to the tuned resonant frequency (active modes). The rest of the modes (passive modes), with frequencies that are well apart from the resonant frequency, have very small amplitudes and can be neglected in a first order approximation.

The AMM is basically the reduced model that results from keeping only the active modes in system (11), which obey equations of the form

$$(\omega_a^2 - \omega^2)A_a + \sum_{a' \neq a} \delta_{aa'}A_{a'} - [M_k(\omega)]_{aa}A_a = F_a, \quad (21)$$

where A_a is the amplitude of the active mode considered, and $A_{a'}$ stands for the amplitudes of the rest of the active modes. The different terms in the above equation can be further simplified: (i) close to the tuned resonant frequency ω_0, both the modal frequency and the forcing frequency can be expanded as $\omega_a = \omega_0(1 + \Delta\omega_a + \dots)$ and $\omega = \omega_0(1 + \Delta\omega + \dots)$, and the first coefficient can be written as

$$(\omega_a^2 - \omega^2) \cong 2\omega_0^2(\Delta\omega_a - \Delta\omega + \dots), \quad (22)$$

(ii) the mistuning coupling coefficients $\delta_{aa'}$ can be also approximated as

$$\delta_{aa'} = Z_a^H \left(\Delta K_{a-a'}^F - \omega^2 \Delta M_{a-a'}^F\right) Z_{a'}$$
$$\cong Z_a^H \left(\Delta K_{a-a'}^F - \omega_0^2 \Delta M_{a-a'}^F\right) Z_{a'}, \quad (23)$$

(iii) the aerodynamic coefficient $[M_k(\omega)]_{aa}$ can be also expressed in terms of the frequency correction, η_a, and the aerodynamic damping, ξ_a,

$$[M_k(\omega)]_{aa} = -2\omega_a^2(\eta_a + \text{i}\xi_a), \quad (24)$$

and (iv) the forcing term F_a is zero everywhere except at the active mode that is being directly forced, where it takes the value

$$F_a = 2\text{i}\omega_a^2\xi_a \quad (25)$$

in order to make the maximum amplitude of the tuned forced response equal to 1.

The AMM is finally constructed just by putting together the equations for the active modes to build a reduced linear system of the form

$$\begin{bmatrix} d_{a'} & & D & \\ & d_a & & \\ & & \ddots & \\ D^{\mathrm{H}} & & & d_{a''} \end{bmatrix} \begin{bmatrix} A_{a'} \\ A_a \\ \vdots \\ A_{a''} \end{bmatrix} = \begin{bmatrix} 0 \\ 2i\omega_a^2\xi_a \\ \vdots \\ 0 \end{bmatrix} \tag{26}$$

where $A_{a'}, A_a, \ldots, A_{a''}$ are the amplitudes of the active traveling wave modes, the diagonal terms on the left hand side are given by

$$d_a = 2\omega_a^2(\Delta\omega_a - \Delta\omega) + 2\omega_a^2(\eta_a + i\xi_a), \tag{27}$$

and the off-diagonal mistuning elements in D are defined by the following expression:

$$D_{aa'} = Z_a^{\mathrm{H}}(\Delta K_{a-a'}^{\mathrm{F}} - \omega_0^2 \Delta M_{a-a'}^{\mathrm{F}}) Z_{a'} \tag{28}$$

The AMM indicates that only the active modes are involved in the mistuned response of the system, and that the effect of mistuning is to couple TW with different nodal diameters: the TW with k nodal diameters is coupled to the TW with k' nodal diameters precisely through the $(k-k')$ Fourier mode of the mistuning distribution. A step by step derivation of the AMM and a more elaborated comments on its implications can be found in [6, 7, 8].

AMM DESCRIPTION OF THE FORCED RESPONSE OF A MISTUNED ROTOR

The rotor model is a simplified configuration with $N = 21$ sectors, which is represented in Fig. 1 together with the mesh used for the sectors. The FEM model has approximately 3600 DOFs/sector, and the forced response of the complete rotor ($\sim 75\,10^3$ DOFs) can be computed in a reasonable time. Fig. 2 shows the elastic vibration characteristics of the tuned rotor.

The matrix of linear aerodynamic effects considered produces a tuned aerodynamic correction that is the same for all family modes. The resulting small frequency correction and aerodynamic damping are shown in Fig. 3. All damping values are positive, therefore the rotor is stable (no flutter present), and the TW mode with lower damping is that with $ND = 5$.

The mistuning effects are modeled including a small perturbation proportional to the blade stiffness of each sector. Using the nomenclature of the previous section, the mistuning matrices

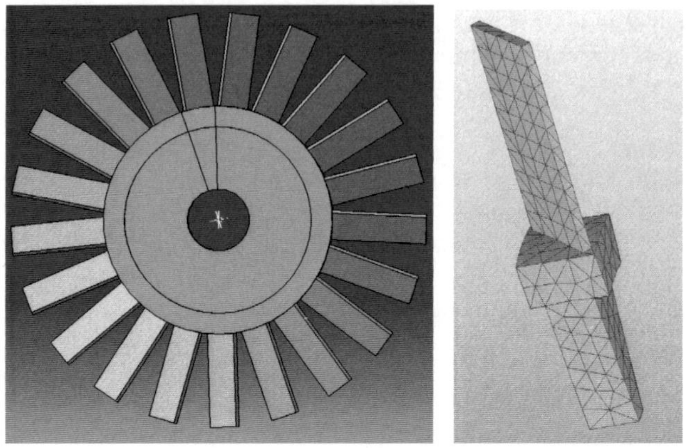

Figure 1. Left: simplified rotor. Right: sector mesh.

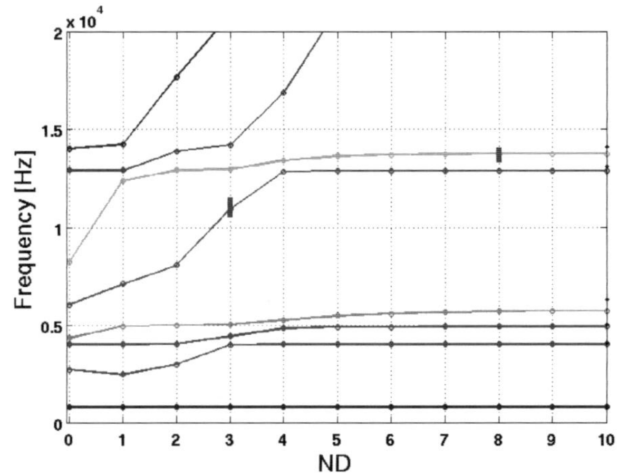

Figure 2. Natural vibration frequencies of the tuned rotor. Thick red lines indicate the forcing cases EO = ± 3 and EO = ± 8.

are of the form

$$\Delta\mathbf{M} = 0, \quad \Delta\mathbf{K} = \begin{bmatrix} m(1)K_{mist} & 0 & \ldots & 0 \\ 0 & m(2)K_{mist} & & \vdots \\ \vdots & & \ddots & 0 \\ 0 & \ldots & 0 & m(N)K_{mist} \end{bmatrix},$$

where K_{mist} is the sector stiffness mistuning matrix that only includes the blade stiffness

$$K_{mist} = \begin{bmatrix} K_{bb} & 0 \\ 0 & 0 \end{bmatrix},$$

and the small numbers $m(1), \ldots, m(N)$ represent the sector-to-sector mistuning distribution around the rotor.

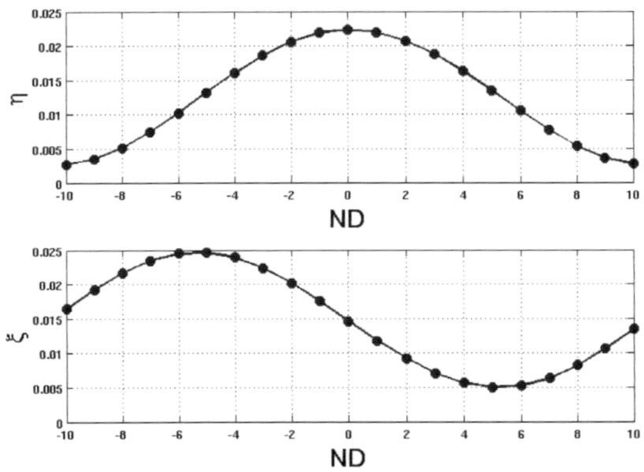

Figure 3. Aerodynamic frequency correction η (top) and damping ξ (bottom).

And finally, the forcing applied to the rotor is modeled as a TW with engine order EO and frequency ω

$$\mathbf{F} = \begin{bmatrix} F e^{i(\frac{2\pi}{N}\text{EO})1} \\ F e^{i(\frac{2\pi}{N}\text{EO})2} \\ \vdots \\ F e^{i(\frac{2\pi}{N}\text{EO})N} \end{bmatrix} e^{i\omega t} + \text{c.c.},$$

with the following in-sector structure

$$F = \begin{bmatrix} 1 \\ 0 \end{bmatrix},$$

which corresponds to constant forcing equal to 1 at all blade DOFs and no forcing on the disk.

Forcing with EO=3

The first case corresponds to forcing with EO$=3$ near the resonance frequency of the fifth family, see Fig. 2.

If we look close to the tuned resonant frequency, it is clearly seen that this mode is well separated from the rest. The only possible active modes are the TW with 3 nodal diameters (the one directly excited by the forcing) and its symmetric TW corresponding to -3 nodal diameters.

For this isolated mode forcing configuration, the AMM can be written in its simplest possible form (see [6, 7, 8]), with only two symmetric active TW,

$$\begin{bmatrix} -\Delta\omega + (\eta_+ + i\xi_+) & D \\ \bar{D} & -\Delta\omega + (\eta_- + i\xi_-) \end{bmatrix} \begin{bmatrix} A_+ \\ A_- \end{bmatrix} = \begin{bmatrix} i\xi_+ \\ 0 \end{bmatrix}, \tag{29}$$

where $\Delta\omega = (\omega - \omega_0)/\omega_0$, ω_0 is the resonant tuned frequency corresponding to the TW of the fifth modal family with 3 nodal diameters (see Fig. 2), the sub-indexes + and - stand for the TW with $ND = 3$ and $ND = -3$, and the effect of mistuning is contained in the off-diagonal coefficient (see eq. (28))

$$D = Z_+^{\mathrm{H}}(\Delta K_6^{\mathrm{F}} - \omega_0^2 \Delta M_6^{\mathrm{F}}) Z_- / (2\omega_0^2). \tag{30}$$

Note that only the harmonic with wavenumber 6 of the sector-to-sector mistuning distribution appears in the AMM formulation; the rest of the harmonics have no effect on the response of the system.

The AMM model above has been used to perform a numerical optimization to find the maximum response amplification (relative to the tuned case) for each value of the mistuning. The results are shown in Fig. 4. As the mistuning is increased, the amplification first shows a slight growth and then decays to a value of approximately one half of the tuned value. This strong reduction of the maximum amplification of the forced response due to mistuning is a very interesting result that was recently reported in [2].

The mechanism behind this reduction of the vibration amplitude can be easily explained by just looking at the AMM equations (29). The coupling due to the mistuning of the $ND = 3$ TW (which has smaller damping, see Fig. 3) with the $ND = -3$ TW (with larger damping) results in an increase of the damping and, therefore, the response is reduced. This effect competes with the capability of the mistuning for increasing the response (as it happens in the uniform damping case, see [7]), producing the initial small maximum of the curve.

The AMM formulation is so simple in this case that it can be solved explicitly. This allows to derive closed form expressions for the global maximum amplification

$$A_{max} = \frac{1 + \sqrt{\xi_+(\frac{1}{\xi_+} + \frac{1}{\xi_-})}}{2}, \tag{31}$$

(which generalizes that obtained in [7]), and for the maximum amplification for large mistuning

$$A_{max} = \frac{2\xi_+}{\xi_+ + \xi_-}. \tag{32}$$

The values from these two expressions are also plotted as thick red lines in Fig. 4, and they show a perfect agreement with the optimization results.

The full FEM problem has also been solved for the mistuning patterns

$$m(j) = \varepsilon \cos(6(2\pi j/N)), \tag{33}$$

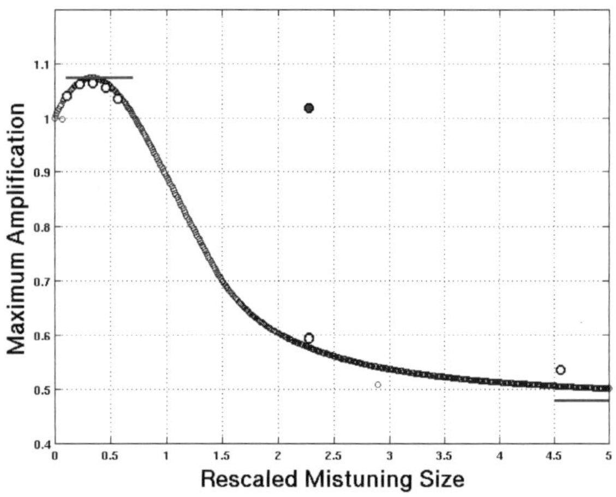

Figure 4. EO$=3$ forcing. Maximum amplification vs rescaled mistuning size $|D| * 0.01$ from the AMM. Red lines correspond to the global maximum amplification value (31) and to the large mistuning value (32). The dots correspond to solutions of the full FEM model of the mistuned rotor.

with $\varepsilon = 0.0005, 0.001, 0.0015, 0.002, 0.0025, 0.01, 0.02$. The results are represented in Fig. 4 as white dots, and are very similar to the AMM results. Another full FEM computation has been performed for a mistuning pattern of the form

$$m(j) = \varepsilon \text{ rand}(j), \tag{34}$$

where $\varepsilon = 0.01$, and rand(j) is a uniform random distribution with values in the range $[-1, 1]$ whose harmonic 6 has been removed (Fig. 5). In this case, the resulting amplification is very close to 1 (see red dot in Fig. 4), indicating that, as predicted by AMM methodology, this mistuning pattern has almost no effect in the forced response of the system.

Forcing with EO=-3

The situation is completely similar to the previous case, but the forcing is now acting on the $ND = -3$ TW, which has higher aerodynamic damping (see Fig. 3). The resulting AMM formulation is the exactly the same as in (29) but with the forcing appearing in the equation for A_- with the value $i\xi_-$.

The results of the numerical optimization of the response amplification computed using the AMM are shown in Fig. 6. The behavior is now completely different from the previous case: as the mistuning level is increased there is first a strong growth of the maximum amplification, and then it becomes stabilized at a value slightly above 1.5. The amplification is now much higher than in the case of constant damping, which is limited to approximately 1.2 [7]. The strong amplification is again caused by the

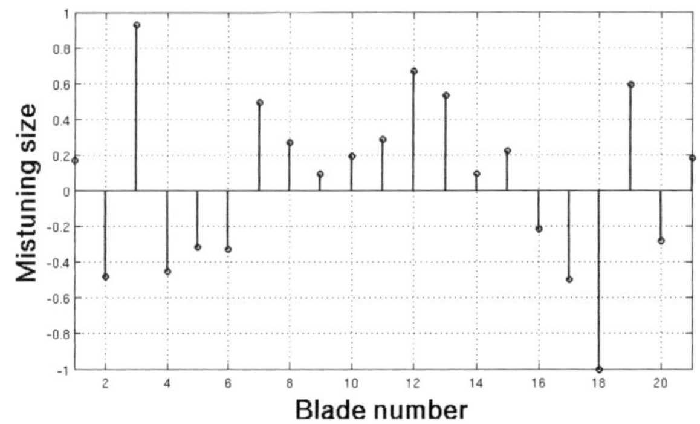

Figure 5. Random mistuning pattern $rand(j)$ with zero harmonic 6 content.

mistuning induced coupling that, in this case, results in a reduction of the effective damping that boosts the amplitude of the forced response.

The expressions for the global maximum amplification and for the maximum amplification for large mistuning can be obtained from (31) and (32) just by interchanging the subindexes $+$ and $-$. The resulting values happen to be very similar in this particular configuration and are plotted with thick red lines in Fig. 6.

The numerical results from the full FEM model with harmonic mistuning pattern given by (33) are again in good agreement with AMM results (white dots in Fig. 6). And again, the random mistuning pattern with zero content in the harmonic 6 (34) leaves the tuned response almost unchanged (red dot in Fig. 6).

Forcing with EO=8 and EO=-8

Two more forcing cases are considered with EO$=8$ and EO$=-8$, and with a frequency close to that of the sixth modal family (see Fig. 2).

The AMM includes a total of 10 active modes with frequencies very close to that directly excited by the forcing, and with nodal diameters $ND = 6, 7, 8, 9, 10, -10, -9, -8, -7, -6$. The AMM formulation is a system of 10 equations for the amplitudes of the active modes given by (26).

The resulting maximum amplification of the forced response computed from the AMM as a function of the mistuning size is shown in Fig. 7 and Fig. 8. The mistuning size is now measured as the norm of the vector D that contains the AMM mistuning coefficients (eq.(28)) divided by $(2\omega_a^2)$. Both curves grow from the tuned value, even for the case EO$=8$ that has a damping three times smaller than that of EO$=-8$ (see Fig. 3). This is due to the fact that the mistuning can now couple with more active TWs, generating a stronger increase of the amplitude that overcomes the effect of the higher damping of the other TWs. The maxi-

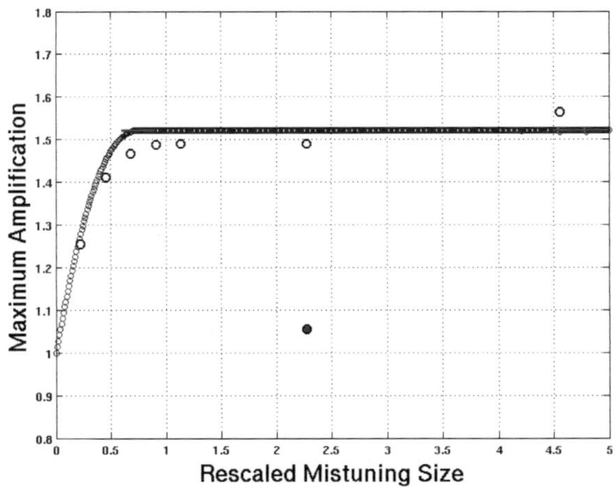

Figure 6. EO$= -3$ forcing. Maximum amplification vs rescaled mistuning size $|D| * 0.01$ from the AMM. Red lines correspond to the global maximum amplification value (31) and to the large mistuning value (32). The dots correspond to solutions of the full FEM model of the mistuned rotor.

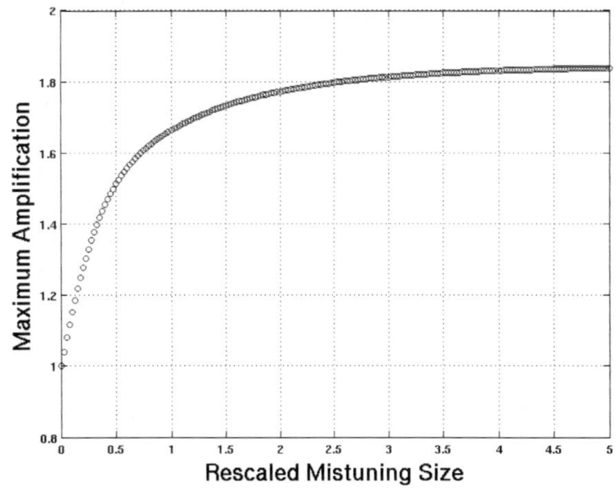

Figure 7. EO$= 8$ forcing. Maximum amplification vs rescaled mistuning size $||D|| * 0.01$ from the AMM. Red line corresponds to the global maximum amplification value (35).

mum amplification obtained is approximately 1.85, smaller than that of the constant damping case $(1+\sqrt{10})/2 = 2.08\ldots$ [7]. On the other hand, in the case EO$= -8$ the damping is higher, and the maximum amplification reaches a much larger value of approximately 2.7.

The global maximum amplification for the general case of N_a active modes can be also obtained in closed form:

$$A_{max} = \frac{1 + \sqrt{\xi_a (\sum_{a'} \frac{1}{\xi})}}{2}, \tag{35}$$

where ξ_a is the damping of the mode that is directly forced, and the sum goes over all active modes. The process for computing the expression above is completely analogous to that used in [9] for the case of all modes active, and also in [7] for a reduced number of active modes. The value for the global maximum amplification from the expression above is marked as a thick red line Fig. 7 and Fig. 8.

It is worth to mention that if there are many active modes present then the system requires higher levels of mistuning to be able to couple all modes and approach the global maximum amplification value (compare Figs. 4 and 6 with Figs. 7 and 8). The AMM properly captures the high sensitivity of the system to the presence of small mistuning imperfections. If the mistuning is further increased then more modes (not considered initially as active modes) can be coupled, and higher amplification levels can be probably achieved, but, of course, this regime is not covered by the AMM description whose validity is restricted to small mistuning values.

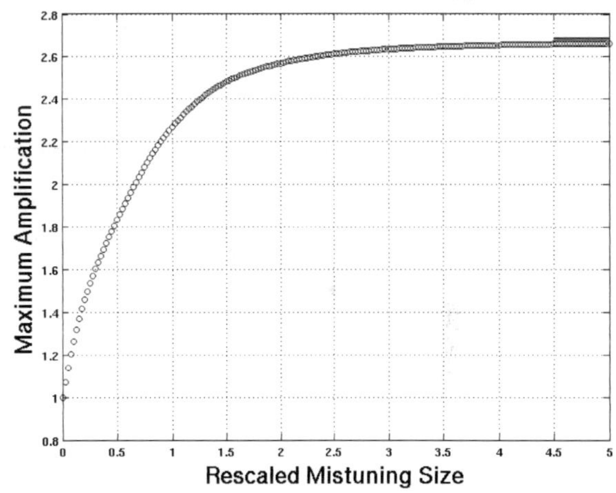

Figure 8. EO$= -8$ forcing. Maximum amplification vs rescaled mistuning size $||D|| * 0.01$ from the AMM. Red line corresponds to the global maximum amplification value (35).

CONCLUSIONS

The AMM [6, 7, 8] has been applied to the analysis of the mistuning effects on the forced response of a stable rotor with damping coming from the aerodynamic effects. The AMM is systematically derived from the complete mistuned bladed disk dynamical model using a perturbative procedure based in the smallness of the mistuning and the damping. The resulting AMM is a very reduced model that allows to draw the following remarks about how mistuning operates in this configuration.

1. There are two main effects involved: (i) the amplification of the response due to the mistuning coupling of the TW modes (as in the case of uniform material damping), and (ii) the resulting effective damping that is a combination of the different dampings of the TWs coupled by the mistuning.
2. In the case of forcing an isolated TW mode with aerodamping (ξ_+) smaller than that of its counter-rotating TW (ξ_-), the maximum amplification can be reduced below 1 (that is, below the tuned response), to the value of

$$A_{max} = \frac{2\xi_+}{\xi_+ + \xi_-} < 1.$$

   If, on the other hand, the aerodamping of the forced TW is higher ($\xi_+ > \xi_-$) then the maximum amplification becomes larger than in the tuned case.
3. In the case of forcing a TW mode that is not isolated and has nearby modes with similar frequencies, then there is typically no reduction of the maximum amplitude. The mistuning couples the TW forced with other active TWs, generating a stronger increase of the amplitude that overcomes the effect of the higher damping of the other TWs.
4. The global maximum amplitude amplification can be computed from the AMM:

$$A_{max} = \frac{1 + \sqrt{\xi_a (\sum_{a'} \frac{1}{\xi})}}{2},$$

   this expression is a generalization of that obtained in [7]:

$$A_{max} = \frac{1 + \sqrt{N_a}}{2},$$

   for the case of uniform damping.

ACKNOWLEDGMENTS

The authors wish to thank O. Khemiri for providing us with the FEM model of the rotor. This work has been supported by the Spanish Ministerio de Educaci6n y Ciencia under grant DPI-2013-46403-R, and by the Universidad Politécnica de Madrid under grant GI1514120024.

REFERENCES

[1] M.P. Castanier and C. Pierre. Modeling and analysis of mistuned bladed disk status and emerging directions. *Journal of Propulsion and Power*, 22(2):384–396, 2006.

[2] E.P. Petrov. A method for forced response analysis of mistuned bladed disks with aerodynamic effects included. *Journal of Turbomachinery*, 132:062502–1–10, 2010.

[3] E.P. Petrov. Reduction of forced response levels for bladed disks by mistuning: Overview of the phenomenon. *Journal of Turbomachinery*, 133:072501–1–10, 2011.

[4] B. Berirow, T. Giersh, A. Kühhorn, and J. Nipkau. Forced response analysis of a mistuned compressor blisk. *Journal of Turbomachinery*, 136:062507–1–13, 2014.

[5] F.M. Bessem, R.E. Kielb, and N.L. Key. Forced response sensitivity of a mistuned rotor from an embedded compressor stage. *Proceedings of the ASME Turbo Expo 2015*, GT2015-43032, 2015.

[6] C. Martel, R. Corral, and J.M. Llorens. Stability increase of aerodynamically unstable rotors using intentional mistuning. *Journal of Turbomachinery*, 130:011005, January 2008.

[7] C. Martel and R. Corral. Asymptotic description of maximum mistuning amplification of bladed disk forced response. *Journal of Engineering for Gas Turbines and Power*, 131:022506, March 2009.

[8] O. Khemiri, C. Martel, and R. Corral. Forced response of mistuned bladed disks: Asymptotic description and fem validation. *AIAA Journal of Propulsion and Power*, 30:397–406, 2013.

[9] D.S. Whitehead. Effect of mistuning on the vibration of turbomachine blades induced by wakes. *Journal of Mechanical Engineering Science*, 8(1):15–21, 1966.

Proceedings of the 14th International Symposium on Unsteady Aerodynamics, Aeroacoustics & Aeroelasticity of Turbomachines
ISUAAAT14
8-11 September, 2015, Stockholm, Sweden

I14-S7-1

EVALUATION OF FAST NUMERICAL METHODS FOR TURBOMACHINERY BLADE FLUTTER ANALYSIS

Lorenzo Pinelli, Francesco Poli,
Juri Bellucci, Matteo Giovannini,
Andrea Arnone
Department of Industrial Engineering
University of Florence
Via di S. Marta 3, 50139 Florence, Italy
Lorenzo.Pinelli@tgroup.unifi.it

ABSTRACT

Different numerical methods for blade flutter analysis were employed to evaluate the flutter stability of a typical low pressure turbine bladerow tested in the context of the European research project FUTURE. This aeroelastic test-case was extensively studied during the project by a large number of partners, becoming a suitable benchmark to compare numerical techniques for flutter prediction.

In this study, two different uncoupled aeroelastic methods (time-linearized and non-linear) were used to assess the flutter stability. The results highlight a very good agreement between the two approaches even when the non-linear code was applied with some approximation strategies (non viscous endwalls and wall functions).

The two analyzed bladerow configurations (cantilever and interlock) show unstable and stable flutter behavior respectively, and the numerical flutter predictions are in agreement with the experimental evidences obtained during the project.

Finally the two methods were also evaluated in terms of computational time. This is due to the fact that flutter assessment methods are more and more used during the turbomachinery design, even during the bladerow optimization.

INTRODUCTION

Greening aviation implies the design of highly efficient and environmentally friendly aero engines. Reduced component counts and decreased weight of mechanical parts are trends which make the propulsion "greener". This request pushes the engine manufacturers towards aggressively lightweight blade designs, which are more and more prone to forced response and flutter instability problems. For this reason, the need for fast and accurate numerical methods to be employed during the engine design process is now fundamental to avoid unwanted high cycle fatigue failures.

In this context, the European project FUTURE (Flutter-Free Turbomachinery Blades) has given a concrete opportunity to enhance the predicting capability of numerical aeroelastic simulations by exploiting the acquired database of high quality flutter measurements. This research project is focused on the two primary components of turbomachines where the flutter phenomenon is observed: Low Pressure Compressors (LPC) and Low Pressure Turbines (LPT).

The unsteady response of a vibrating cascade (responsible for the flutter occurrence) can usually be analyzed with different methodologies, (e.g. linear, time-linearized, harmonic balance or non-linear methods). The first aeroelastic numerical methods, developed at the beginning of the 1970s, were based on linear approaches where all the equations are linearized [1, 2]. Since the 1990s, time-linearized methods have been developed: according to these approaches, the flow is decomposed into a non-linear steady flow plus a small-perturbation harmonic unsteady flow [3–8]. In recent years thanks to the growth in computer technology, non-linear methods have become a viable alternative and several non-linear (uncoupled and coupled with structural solver) approaches have been implemented [5, 9–14]. Although computationally expensive, these methods are able to include non-linear effects into aeroelastic analysis. All these methods are well known in the literature and some comparisons among them have been already carried out by several authors [15, 16].

Within this paper, the flutter results obtained by different numerical techniques are compared in order to highlight the pros and cons of each methodology. The selected FUTURE testcase for this numerical activity is the LP turbine blade tested at the Centro de Tecnologías Aeronáuticas CTA (Bilbao, Spain). This bladerow was designed to experience flutter instability (free-flutter) for the cantilever configuration and this behavior was confirmed during the experimental campaign. On the other hand, the interlock configuration showed a stable behavior. Moreover, the excellent agreement among the flutter results provided by the partners involved in the project makes this test-case particularly attractive for validation and comparison purposes.

The flutter stability of this blade for the cantilever and interlocked configurations was computed with two different codes: the time-linearized aeroelastic and aeroacoustic solver Lars [7] and the unsteady CFD Traf code [14] (with moving mesh and phase lagged periodicity conditions), thus comparing a linearized with a non-linear approach.

Since non-linear simulations, although more accurate in particular flow conditions (with presence of shock waves or detached flow regions), are more computationally expensive in terms of time and memory usage, a few simplified models (Thin Layer approximation and Wall Functions model) are used to make the computational effort more comparable with time-linearized analyses. One of the aims of the paper is to assess the effect of such approximations on flutter analysis.

NOMENCLATURE

a modal amplitude
b blade span
c velocity
CFD Computational Fluid Dynamics
CSD Computational Solid Dynamics
E kinetic energy
h harmonic index
$\imath$ imaginary unit
LE Leading Edge
m modal mass
Ma Mach number
N number of blades
nd number of nodal diameters
NS Navier-Stokes
p pressure
PS Pressure Side
SS Suction Side
T vibration period
TE Trailing Edge
TL Thin Layer
WF Wall Functions
ν frequency
ξ energetic damping coefficient
Σ blade surface
ϕ IBPA Inter Blade Phase Angle
ω angular frequency

NUMERICAL METHODS

The two uncoupled aeroelastic methods (time-linearized and non-linear) used in the following to perform the flutter assessment of the two cascades require similar pre-processing and post-processing.

First of all, the blade modeshape (usually coming from a modal analysis solution) has to be transferred to the blade surface within the CFD domain. Then, the two solvers are able to evaluate the unsteady pressure response over the blade surface while the bladerow vibrates in a traveling wave manner. The unsteady pressure is finally integrated during an oscillation period over the blade to obtain the aerodamping parameters.

Modeshape transfer

A modeshape transfer technique is used to interpolate a real or a complex modeshape from the modal analysis to the blade surface within the CFD mesh. Usually the CSD and CFD employ different discretization grids and a nearest point interpolation strategy is used.

Once the modeshape was transferred to the CFD blade surface it is important to verify if the mode is correctly reproduced on the CFD surface, especially when simplified models as inviscid endwalls (hereafter named Thin Layer) and Wall Functions are used to lighten the computational domain. Figure 1 confirms the correct transfer of the cantilever modeshape to the Navier-Stocks mesh and also to the coarser grid (Thin-Layer + Wall Function).

The blade motion is the source of unsteadiness for the problem and, obviously, the whole fluid domain (where Naiver-Stokes equations are numerically integrated by the solvers) has to be deformed following the bladerow vibration.

For the linearized method the grid perturbation is generated through a three-dimensional elliptic procedure. This procedure generates a very smooth distribution of mesh node displacements, yet it does not ensure that the volume of the mesh cells remains positive during the deformation, especially where the displacements are larger. This is not a problem for the linearized method where the node displacements are used within the governing equation of the conservative variable perturbations.

On the other hand, the non-linear method employs the whole grid deformation to actually rebuild the computational domain at different equally-space instants over the oscillation period. For this reason the volume of each cell must be positive during the whole deformation in order to avoid cell interlacement. For this reason, the grid deformation (in case of non-linear analyses) is created by using a different procedure. An algebraic method distributes the largest deformations where the biggest mesh elements are located; thus maintaining the shape of the smallest elements (located, for instance, within the boundary layer or near the blade clearance). Anyway, as the blade oscillation is imposed, the grid deformation can be verified in advance before running the non-linear flutter simulations.

Another aspect which deserves to be discussed for the non-linear simulations concerns the choice of the blade os-

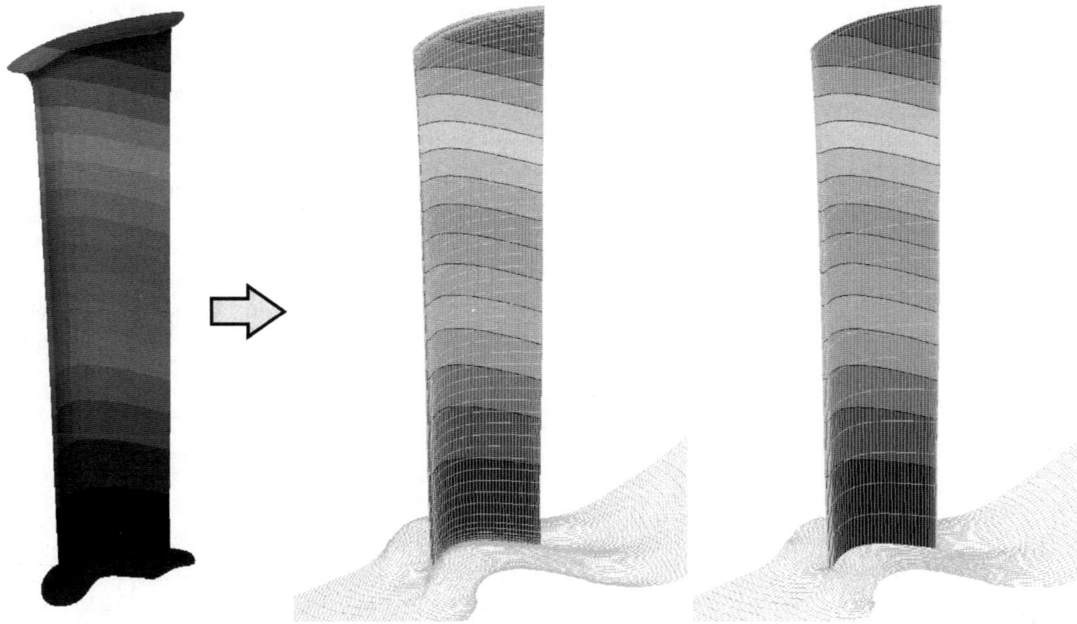

(a) Modal analysis (b) NS mesh (c) TL+WF mesh

Figure 1. Cantilever configuration: real part of the grid perturbation magnitude (nodal diameter = ± 36)

cillation amplitude. It is well known that linearized solutions can be scaled with the blade amplitude, whereas non-linear effects may change depending on the amplitude of the surface motion. As these analyses are usually focused on the flutter occurrence, the blade oscillation is chosen quite small in order to be representative of typical limit cycle amplitudes.

Linearized method

The aeroelastic and aeroacoustic three-dimensional solver named Lars has been developed at the Department of Industrial Engineering (University of Florence) to perform flutter analyses and tone noise evaluations on axial turbomachine blade rows. This solver implements a time-linearized computational method and is designed to work together with the Traf steady/unsteady aerodynamic solver. All the details of this method can be found in previous publications by the authors [6, 7].

As already mentioned, small time-sinusoidal blade vibrations are assigned to the aeroelastic calculation as input data. In order to represent vibrations in traveling wave form, while limiting the computation domain to one single angular pitch of the blade row, phase-lagged conditions are applied to periodicity boundaries. In order to reduce spurious reflections at inlet and outlet boundaries, three types of non-reflecting conditions are implemented (one-dimensional, two-dimensional at each radius, and three-dimensional radial-mode-based conditions). Two-dimensional conditions are used in this study.

Non-linear method

The Traf three-dimensional steady/unsteady aerodynamic solver has also been extended to simulate unsteady flows around vibrating blades, thus implementing an aeroelastic non-linear method. The non-linear uncoupled method for flutter analysis is here briefly described. The comprehensive method description and validation are included in a previous article [14].

Time-sinusoidal blade vibrations are again assigned to the aeroelastic calculation as input data, but in the non-linear method, the mesh perturbation is used to actually deform the grid at each time step, rather than to derive the linearized governing equation. The non-linear unsteady flow equations are integrated on the deforming mesh by applying the same dual time-stepping algorithm used for aerodynamic unsteady simulations.

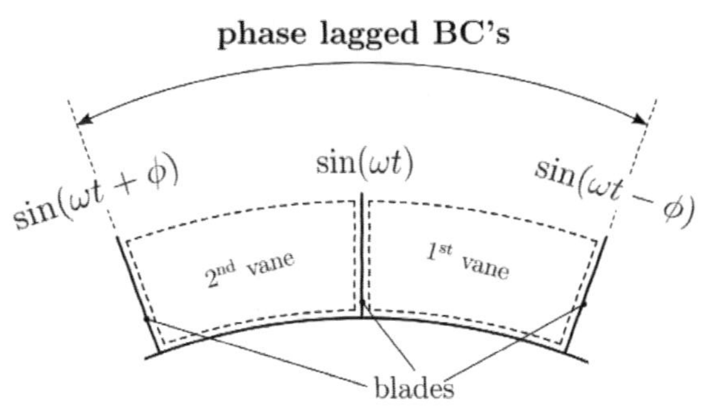

Figure 2. Sketch of the Phase Lagged method application

Traveling waves are represented on a single-angular-pitch domain by applying phase-lagged conditions to periodicity boundaries. Their implementation in the non-linear method is definitely less trivial than in the linearized one: in the time domain, Fourier transforms are used to assign quantities on periodicity boundaries with appropriate time shifts, while limiting the memory storage requirements. As shown in Figure 2, two tangential blocks can be simulated in order to speed up the convergence [17].

Flutter stability assessment

Since in uncoupled methods blade vibrations are input data, rather than a result of the computation, flutter stability is estimated by checking the sign of the energetic damping coefficient ξ (also known as critical damping ratio) applied by the fluid onto the blade during one vibration cycle. This quantity is defined as follows:

$$\xi = \frac{-L}{8\pi E} = \frac{-L}{2\pi m\omega^2 a^2} = \frac{-L}{8\pi^3 m\nu^2 a^2} \tag{1}$$

where E is the blade average kinetic energy, m is the blade modal mass, a is the modal amplitude, ω is the angular frequency and ν is the frequency of the vibration.

A positive critical damping ratio means that energy is transferred from the blade to the fluid flow and the cascade oscillation is damped, hence a stable configuration. On the contrary, if the damping is negative, the unsteady aerodynamic pressure adds energy to the blade motion and the cascade is unstable. The different aeroelastic behavior mainly depends on the local phase shifting between unsteady pressure and blade vibration.

To obtain the aerodynamic work, the blade surface unsteady pressure is numerically integrated over the entire blade for a cycle of oscillation:

$$L = \int_t^{t+T}\int_\Sigma (-p)\vec{N}\cdot\vec{c}_{\text{blade}}\,\mathrm{d}\Sigma\,\mathrm{d}t \tag{2}$$

where Σ is the blade surface, T is the vibration period, p the unsteady pressure, $\vec{N}$ is Σ outgoing normal unit vector and $\vec{c}_{\text{blade}}$ is the blade velocity.

The linearized method solves the first harmonic of the blade pressure response and the aerodynamic work is directly computed in the frequency domain, whereas the non-linear solution also includes the higher harmonics and the aerodynamic work is numerically integrated in the time domain. In this LPT environment, where the flow is subsonic and strong non-linear phenomena (shock wave, detached flows, etc...) are absent, the unsteady pressure response content basically consists of the first pressure harmonic contribution and, for this reason, the critical damping ratio obtained from the two presented methods is very similar as shown in the following of the paper.

To further compare these two approaches, a Fourier analysis of the results coming from the non-linear code was performed and the first harmonic of the unsteady pressure was compared with the linearized solution.

$$\overline{\overline{p}}^{(h)} = \frac{1}{N_{div}}\sum_{k=1}^{N_{div}} p_k e^{-\imath 2\pi nk/N_{div}} \tag{3}$$

The analyzed pressure time history p_k is equally spaced, and it was noticed that around 20 instants during the oscillation period are enough to correctly compute the aerodynamic work and to extract the first pressure harmonic.

The last quantity used in the following comparisons is the energetic damping coefficient surface density defined as follows:

$$\rho_\xi = \frac{dL_\delta}{8\pi E\frac{d\Sigma}{\Sigma}} = \frac{dL_\delta}{2\pi m\omega^2 a^2\frac{d\Sigma}{\Sigma}} = \frac{dL_\delta}{8\pi^3 m\nu^2 a^2\frac{d\Sigma}{\Sigma}} \tag{4}$$

This local quantity is useful to pinpoint the blade surface areas which contribute to the stability or instability of the whole bladerow for a given nodal diameter. Blade distributions of this value usually give valuable information for the free flutter blade design.

RESULTS

Two high aspect ratio aeronautical low-pressure turbine rotor blades sharing the same aerodynamic surface were designed to experience flutter instability in a rotating high-speed wind tunnel located at the CTA. The first rotor blade has plane not-contacting tip-shrouds and it is referred to as the cantilever configuration whereas the second features a z-shape shroud where there is a contact with the neighboring blades and it is named the interlock configuration.

This is the first time that a bladed-disk has been specifically designed to flutter under controlled conditions. The major technical difficulty was to design a rotor blade within the design space of an existing rig whose vibration amplitude were high enough to be measurable but not to compromise the safe operation of the rig and the facility.

During the experimental campaigns, these two rotor blades were tested in several configurations (tuned or with different mistuning patterns) at different operating conditions. The tuned cantilever configuration showed free-flutter vibrations, whereas the tuned interlock one resulted stable although designed to flutter. The steady and flutter results collected in this paper are referred to the design point (OP1) for the cantilever and interlock tuned configurations.

	NS	TL	TL + WF
dimensions	$244\times60\times72$	$244\times60\times16$	$244\times40\times16$
total cells	$\approx 1.13\times10^6$	$\approx 0.27\times10^6$	$\approx 0.18\times10^6$

Table 1. Grid dimensions

Steady results

Before starting the flutter simulations it is necessary to perform the steady state analysis for the selected operating point (OP1). The steady solution is employed in a different manner by the two aeroelastic codes: the time-linearized method uses this field as mean frozen background to compute the small time-sinusoidal perturbations of the conservative quantities; whereas the non-linear approach employs the steady solution as field initialization.

For this study, three different steady computations with $k-\omega$ turbulence model in high-Reynolds formulation were carried out:

1. viscous computation (full Navier-Stokes)
2. inviscid endwalls (Thin-Layer)
3. inviscid endwalls (Thin-Layer) + Wall Functions

This is due to the fact that the two simplified approaches (with Thin-Layer and Wall Function approximations) allow the use of coarser grids: the dimensions of the different grids are listed in Table 1.

As it can be seen in Figure 3, which depicts the steady Mach number at the blade midspan from the full Navier-Stokes solution, the rotor shows a typical LPT subsonic flow field.

Finally the three mean aerodynamic solutions were also compared with each other, highlighting a very good agreement in terms of static pressure over the blade surface. Figure 4 shows the pressure distribution at the blade midspan. For the further flutter analyses, it is essential that the steady solutions are as close as possible to avoid possible discrepancies due to different mean flow conditions.

Figure 3. Steady solution (OP1): Mach number contours (NS solution)

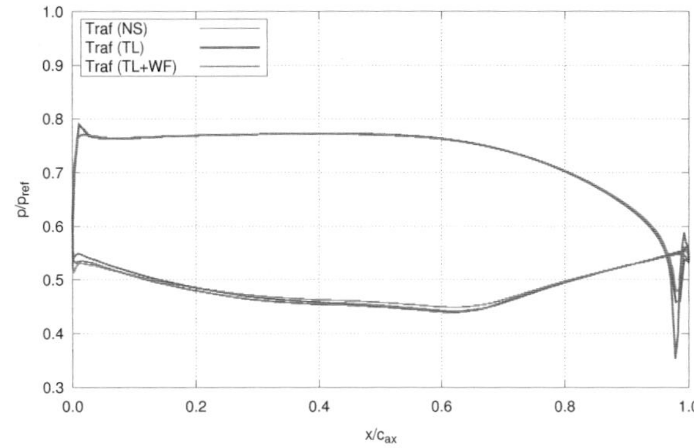

Figure 4. Steady solution (OP1): static pressure distribution at midspan

	NS	TL	TL + WF
linearized (Lars)	✓	✗	✗
non-linear (Traf)	✓	✓	✓

Table 2. Matrix of flutter computations

Flutter results

In order to compare the two aeroelastic methods with different levels of approximation, it was decided to analyze the two tuned configurations as reported in Table 2.

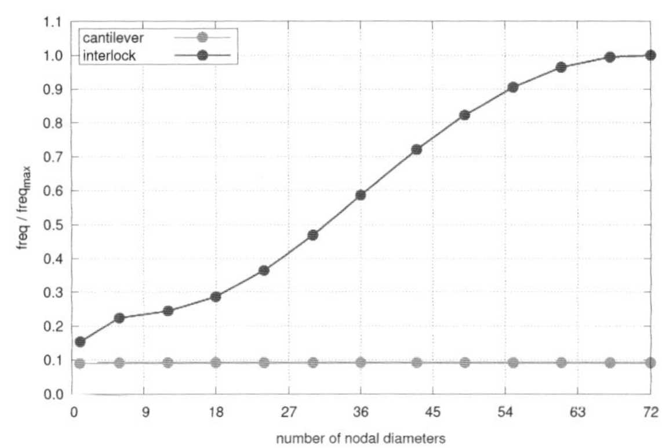

Figure 5. Normalized modeshape frequency

The time-linearized method was applied in the full Navier-Stokes configuration, while three configurations of the non-linear analysis were run: full Navier-Stokes, Thin Layer and Thin Layer + Wall Functions.

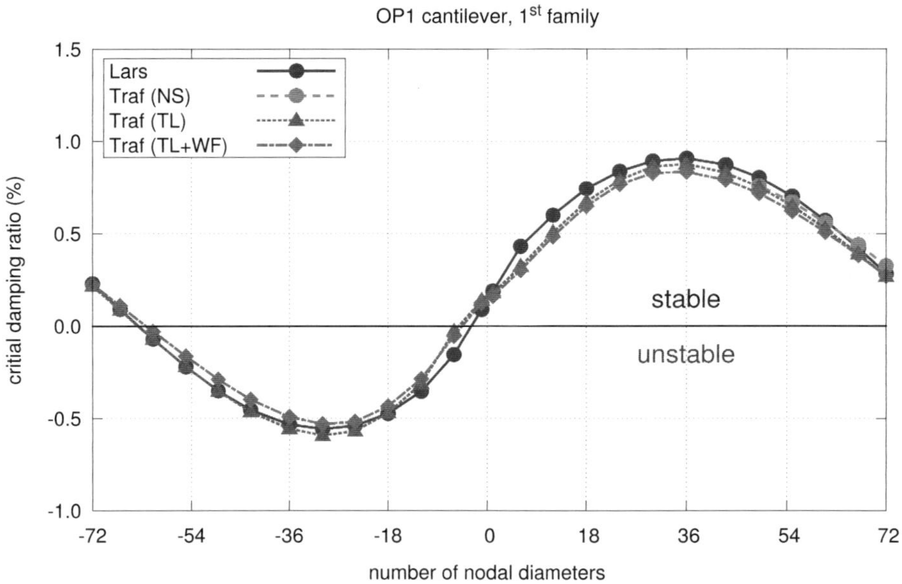

Figure 6. OP1 Cantilever configuration: critical damping ratio vs. number of nodal diameters

The flutter analyses for cantilever and interlock configurations were focused on the first rotor mode (1st family) and a selection of nodal diameters was computed in order to obtain the critical damping ratio curves. The trend of the modeshape frequencies are reported in Figure 5. As can be seen, the interlock configuration shows a relevant coupling due to the presence of the z-shape shroud, whereas the cantilever rotor has very low coupling and the frequency remains almost constant over the different nodal diameters.

In addition to the main comparisons in terms of aerodamping curves (which show an excellent agreement), the unsteady pressure response on the blade surface is compared over the analyzed nodal diameters. Finally, some distributions of energetic damping coefficient surface density are shown as well.

Cantilever configuration The cantilever configuration was designed to experience flutter at all the operating points and the lack of mechanical coupling results in real modeshapes with similar deformations.

The aerodamping curve obtained for this rotor configuration (see Figure 6) shows a sinusoidal trend. This means that the unsteady blade response essentially contains the contribution of the fist harmonic and the aerodynamic work is generated by the auto-influence of the blade and by the effect of the two adjacent blades. As far as the flutter assessment is concerned, the four approaches are in a very good agreement and highlight an unstable region for a wide range of negative nodal diameters. The nodal diameters included in this unstable range are self-excited and extract energy from the flow. It is likely that the self-excited nodal diameters grow in amplitude until reaching a limit cycle where the mechanical damping equal the negative aerodamping. The limit cycle condition cannot be predicted by uncoupled methods.

Experimental campaigns proved the flutter instability of this cantilever configuration. Tip timing and acoustic measurements confirmed that there is a range of negative unstable nodal diameters. The most unstable one correspond to $nd = -21$. As it can be seen in Figure 6, the minimum of the numerical aerodamping curve is located near to this value.

Moreover, to further confirm the accuracy of the numerical flutter predictions, it is important to add that also the numerical predictions of other European partners within the project (not shown here) are in excellent agreement with the presented results.

Figure 7 and 8 compare the amplitude and phase of the first pressure harmonic at midspan for all the computed nodal diameters. The agreement among the results of the four approach under investigation is very satisfactory. As can be seen in Figure 7, the pressure amplitude is symmetrical with respect to the $nd = 0$ line and the maximum of the pressure response is located on the blade suction side where the mean velocity gradient are higher. This small area of the blade surface is main responsible of flutter behavior of the bladerow (see also Figure 9).

On the other hand, the pressure phase (referred to the blade motion) has a crucial role to define the energy transfer from fluid to blade or vice versa. As shown in Figure 8, the phases on the blade suction side (where the highest pressure response is located) have opposite sign between the upper (positive nd) and the lower (negative nd) part of the plot. This explains the overall flutter behavior already shown in Figure 6

To gain a deeper insight in this phenomenon it is useful to visualize the amplitude and phase of the pressure first

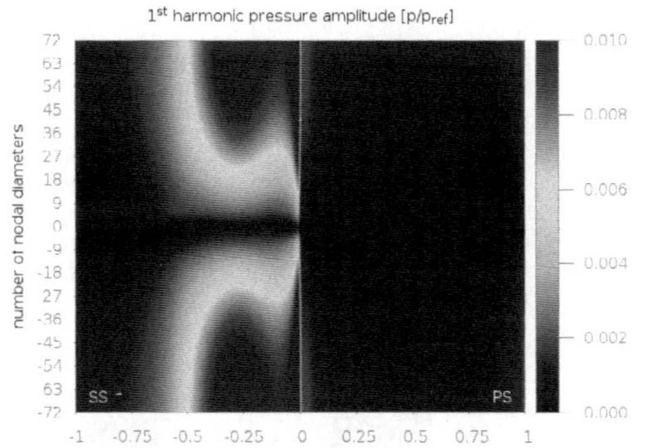

(a) Lars

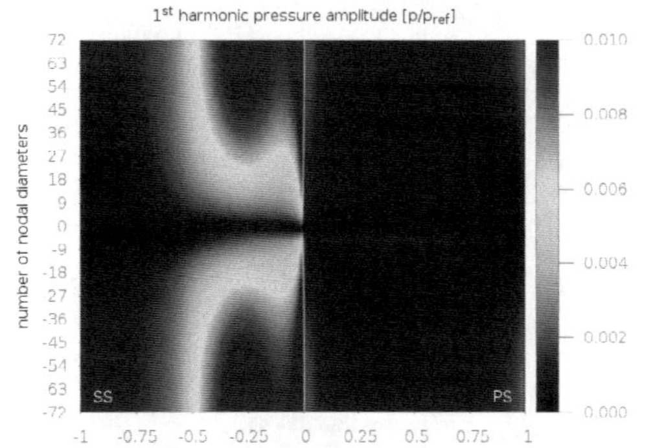

(b) Traf (NS)

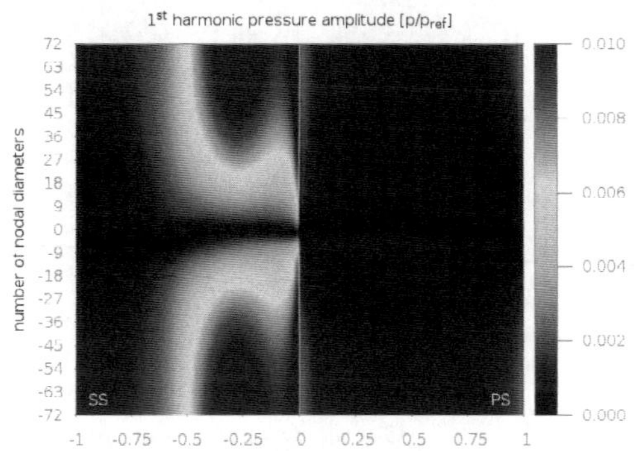

(c) Traf (TL)

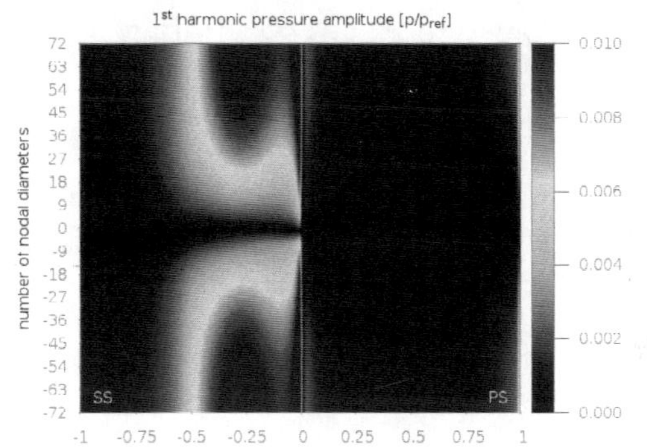

(d) Traf (TL+WF)

Figure 7. OP1 cantilever configuration: blade unsteady pressure: amplitude of 1st harmonic at the 50% span for the computed IBPAs

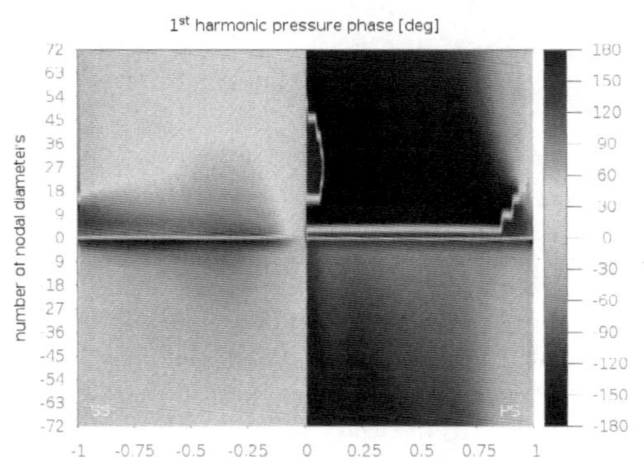

(a) Lars

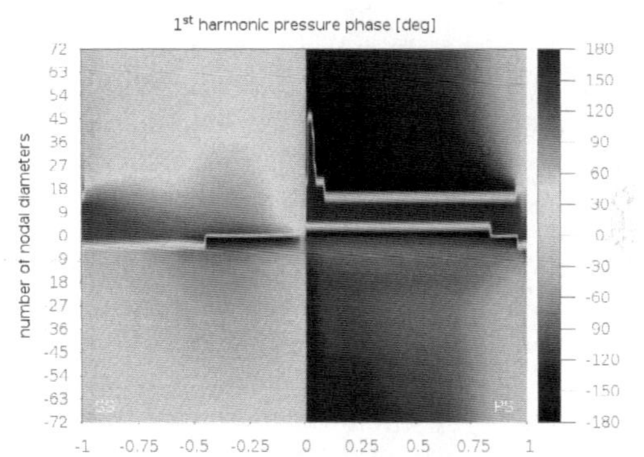

(b) Traf (NS)

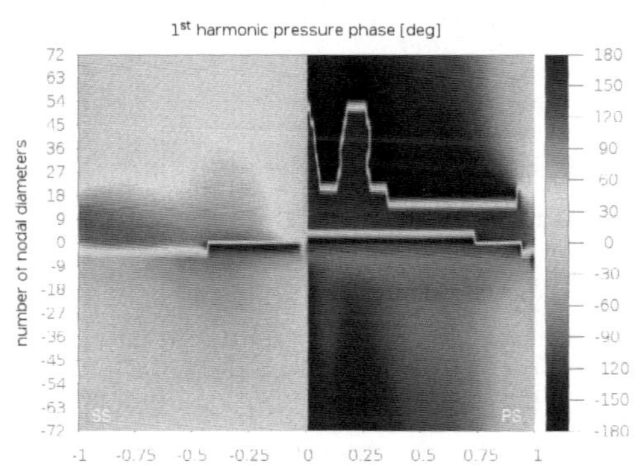

(c) Traf (TL)

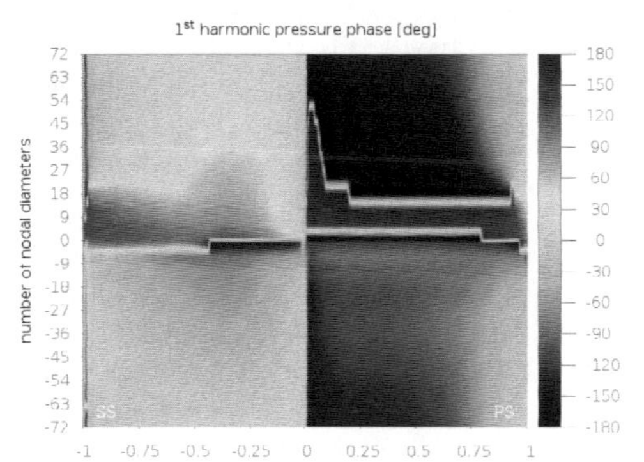

(d) Traf (TL+WF)

Figure 8. OP1 cantilever configuration: blade unsteady pressure: phase of 1st harmonic at the 50% span for the computed IBPAs

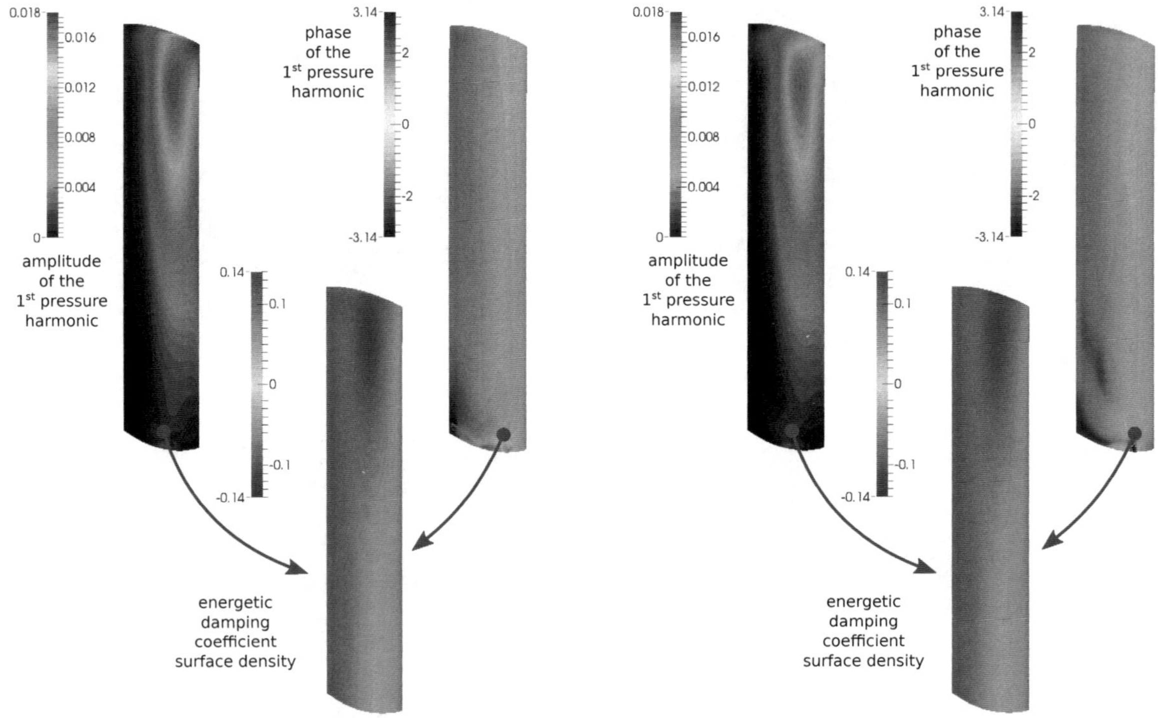

(a) nodal diameter = -36 (unstable)

(b) nodal diameter = +36 (stable)

Figure 9. OP1 cantilever configuration: blade distributions of aeroelastic quantities (from NS non-linear method)

harmonic over the blade surface for a given nodal diameter. Figure 9 compares the blade response to the two physical traveling waves related to the nodal diameters $nd \pm 36$. The $nd = -36$ shows an unstable behavior, while the positive one (spinning in the opposite direction) is stable. The amplitude distribution of the first pressure harmonic is very similar in both cases: there is a maximum on the blade suction side near the tip and in this zone the energy exchange is maximum (see the distribution of the energetic damping coefficient surface density at the bottom of Figure 9). As already pointed out, the phase plays here the fundamental role and negative phases results in a positive aerodynamic work. On the other hand, the phase is less important when the unsteady response of the blade is low, as happens at nodal diameters near to zero. In this cases the aerodynamic work is near zero as the pressure response is very weak.

Interlock configuration The tuned interlock configuration exhibited a stable behaviors during the experimental campaigns. This is in agreement with the computed aerodamping curve which is always positive. As in the previous configuration the tested methods gave a similar trends over all the nodal diameters: the only differences are located around $nd = +12$ (see Figure 10).

The stable behavior of this configuration is basically due to the presence of the shroud at the tip. This z-shape connection changes the modeshapes and, most important increases their frequencies (see again Figure 5). Higher frequencies with same aerodynamic usually mean stable configurations. Again, the other partners of the project numerically confirmed the stable behavior of this configuration, yet some discrepancies (not shown here) were found in the shape of the aerodamping curve. This was caused by some differences in the blade modeshapes, as different contact models at the blade tip may significantly alter the modal analysis results.

Concerning the unsteady pressure response, the four methodologies produced very similar trends (see Figure 11). As in the previous case, the highest pressure response is on the suction side near the blade leading edge where the phase is positive (producing positive local damping). Than, moving along the suction side to the trailing edge the phase jumps to negative values resulting in a negative local damping. This is clearly visible in the energetic damping coefficient surface density reported in Figure 12: after a blue zone at the blade LE it can be seen a red unstable area near the blade TE.

This figure also shows the excellent agreement among the employed approaches. In this cases the maximum energy exchange is located around blade midspan as the present of the shroud reduces the blade oscillation at the tip.

In conclusion both the studied testcases showed a very good agreement between the numerical results and the experimental evidences. This, on one hand further validates the two aeroelastic solvers and on the other hand confirmed the accuracy of fast simplified method (TL and WF) for LPT environments. In light of the computational time which will be discussed in the following section, this conclusion is very

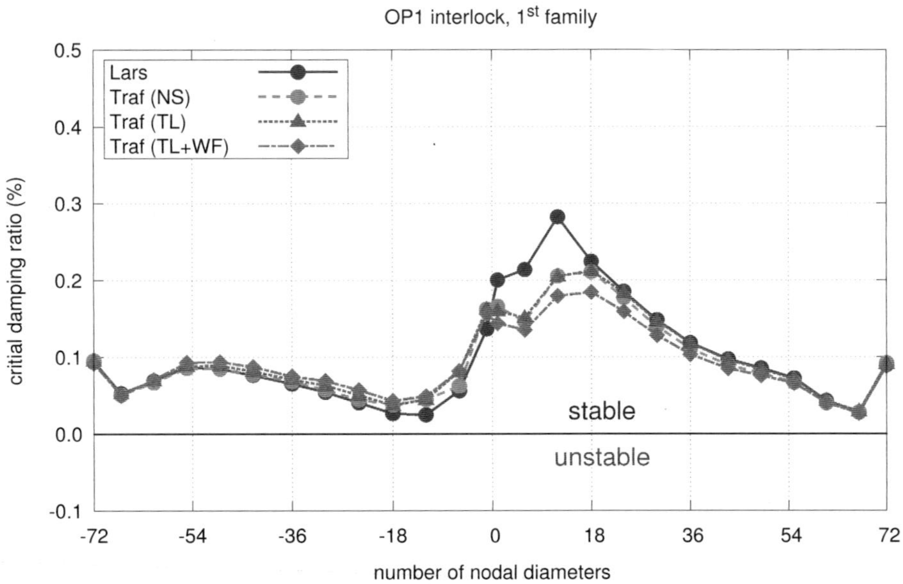

Figure 10. OP1 interlock: critical damping ratio vs. number of nodal diameters

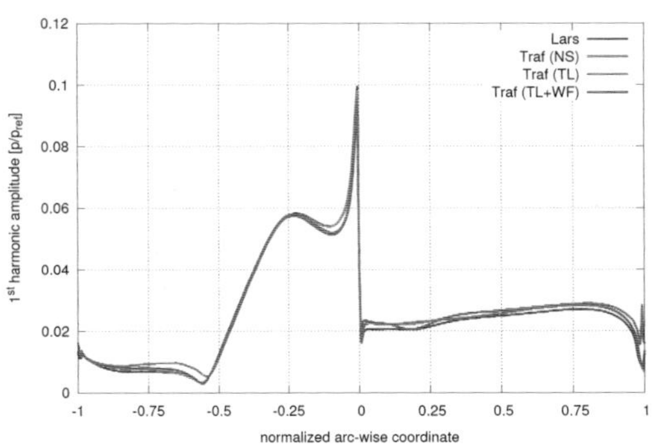

(a) amplitude

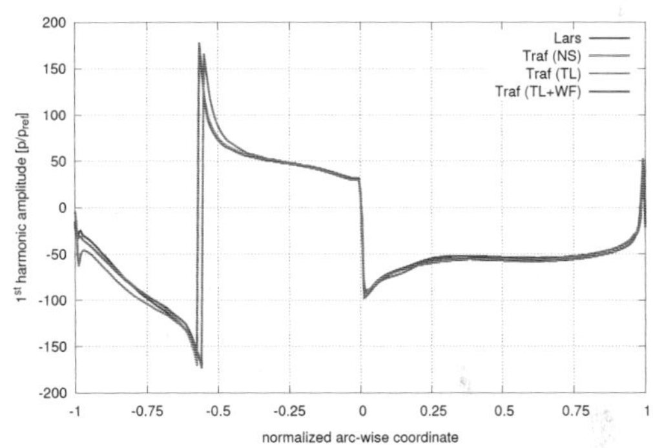

(b) phase

Figure 11. OP1 interlock configuration: blade unsteady pressure: amplitude (left) and phase (right) of 1st harmonic at the 50% span ($nd = -72$)

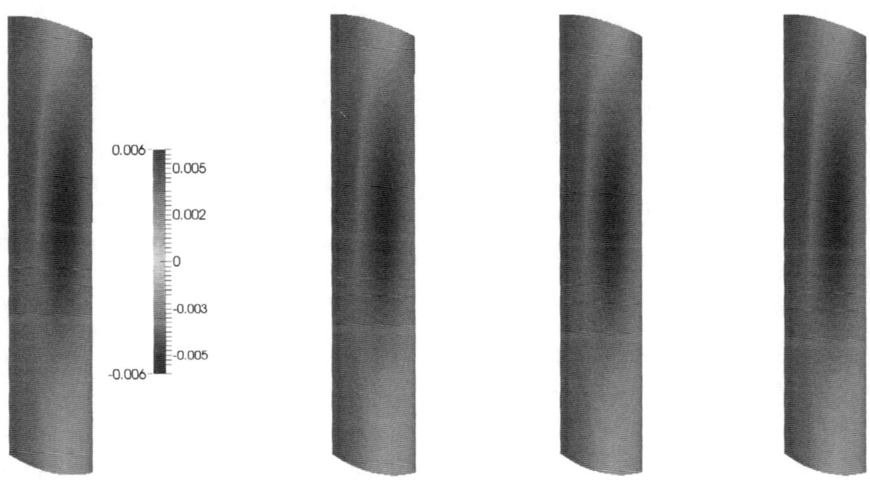

(a) Lars (b) Traf (NS) (c) Traf (TL) (d) Traf (TL+WF)

Figure 12. OP1 interlock: energetic damping coefficient surface density (nodal diameter = -72)

	CPUs	grid dimension	time instants	convergence	total time [s]
Lars (NS)	16 cores (16 OMP)	$\approx 1.13\times10^6$	–	1500 iterations	$\approx 2.9 \times 10^3$
Traf (NS)	8 cores (2 MPI $\times$ 4 OMP)	$\approx 1.13\times10^6$	80	15 periods	$\approx 2.5 \times 10^4$
Traf (TL)	8 cores (2 MPI $\times$ 4 OMP)	$\approx 0.27\times10^6$	80	15 periods	$\approx 3.8 \times 10^3$
Traf (TL+WF)	8 cores (2 MPI $\times$ 4 OMP)	$\approx 0.18\times10^6$	80	15 periods	$\approx 1.7 \times 10^3$

Table 3. Computational time

important for the application of these methods within an optimization strategy which can take into account also flutter issues.

Computational time

Optimization strategies require fast tools in order to investigate large design spaces, and for this reason, flutter verifications are usually not taken into account during aerodynamic optimization procedures. Nevertheless, simplified aeroelastic methods, as the ones presented in this article, may become suitable to be included within an optimization process.

As the accuracy of the presented methods was proven in a LPT environment, it is important to quantify the computational effort, especially in terms of time requirement. Table 3 shows the comparison of the computational times. These time requirement are referred to a single IBPA computation on Intel® Xeon® E5-2680 v2 2.8 GHz CPU.

It is worth mentioning here that both codes are parallelized: the linearized solver takes the advantage of an OpenMP parallelization, while the non-linear method implements both MPI (Message Passing Interface) and OpenMP parallelism. As can be seen in the Table 3 the computational time of the simplified approaches (and especially of the non-linear TL+WF model) may be considered attractive for optimization purposes.

CONCLUSIONS

Four different uncoupled aeroelastic strategies were applied to investigate the flutter response of two LPT rotor row configurations. These two configurations (named cantilever and interlock) were designed and tested in the context the European research project FUTURE. The experimental activity was carried out at the Centro de Tecnologías Aeronáuticas (Bilbao, Spain).

During the experimental campaigns, the cantilever configuration exhibited an unstable flutter response, whereas the interlock one was stable for all the possible nodal diameters. These experimental evidences confirmed all the numerical predictions reported in this article, further validating the two aeroelastic solvers (a time-linearized method in the frequency domain and non-linear approach in time domain) employed in this study.

The two codes were used with different levels of approximation with the aim of demonstrating the applicability of fast and simplified approaches to predict flutter occurrence in a LPT environment. The obtained flutter results are in excellent agreement among the different approaches in terms of the unsteady pressure response on the blade during the imposed oscillation, thus confirming the accuracy of the results coming from the simplified approaches. The curves of the critical damping ratio show a very good correspondence as well.

Finally, the computational time required by the different approaches was also compared. This analysis highlights the capability of the simplified models (especially the non-linear TL+WF approach) to provide a very fast response in terms of flutter occurrence prediction. Such conclusions are very important in the perspective of introducing flutter verifications within a typical aerodynamic optimization process.

ACKNOWLEDGMENTS

This research effort was funded by the European Commission under the Seventh Framework Programme (grant agreement number 213414, project FUTURE - Flutter-Free Turbomachinery Blades). The permission to publish is gratefully acknowledged.

REFERENCES

[1] Verdon, J., 1987. "Linearized unsteady aerodynamic theory". *In AGARD Manual on Aeroelasticity in Axial Flow in Turbomachines,* ***AGARD-AG-298***.

[2] Whitehead, D., 1987. "Classical two-dimensionnal methods". *In AGARD Manual on Aeroelasticity in Axial Flow in Turbomachines,* ***AGARD-AG-298***.

[3] Hall, K., and Clark, W., 1993. "Linearized Euler predictions of unsteady aerodynamic loads in cascades". *AIAA Journal,* ***31***, pp. 540–550.

[4] Hall, K., and Lorence, C., 1993. "Calculation of three-dimensional unsteady flows in turbomachinery using the linearized harmonic Euler equations". *Transactions of the ASME, Journal of turbomachinery,* ***115***, pp. 800–809.

[5] Chassaing, J., and Gerolymos, G., 2000. "Compressor flutter analysis using time-nonlinear and time-linearized 3-d navier-stokes methods". In 9th International Symposium on Unsteady Aerodynamics, Aeroacoustics and Aeroelasticity of Turbomachines.

[6] Arnone, A., Poli, F., and Schipani, C., 2003. "A method to assess flutter stability of complex modes". In 10th International Symposium on Unsteady Aero-

dynamics, Aeroacoustics and Aeroelasticity of Turbomachines.

[7] Poli, F., Gambini, E., Arnone, A., Poli, F., and Schipani, C., 2006. "A 3d time-linearized method for turbomachinery blade flutter analysis". In 11th International Symposium on Unsteady Aerodynamics, Aeroacoustics and Aeroelasticity of Turbomachines.

[8] Chenaux, V., and Grueber, B., 2015. "Aeroelastic investigation of an annular transonic compressor cascade: Numerical sensitivity study for validation purposes". ASME Turbo Expo, June 15–19, Montreal, Canada.

[9] McBean, I., Liu, F., Hourigan, K., and Thompson, M., 2002. "Simulations of aeroelasticity in an annular cascade using a parallel 3-dimensional Navier-Stokes solver". *ASME paper GT–2002–30366*. ASME Turbo Expo, June 3–6, Amsterdam, The Netherlands.

[10] McBean, I., Hourigan, K., Thompson, M., and Liu, F., 2005. "Prediction of flutter of turbine blades in a transonic annular cascade". *ASME Journal of Fluid Engineering,* ***127***, November, pp. 1053–1058.

[11] Vahdati, M., Sayma, A., and Imregun, M., 1999. "Case studies in turbomachinery aeroelasticity using an integrated 3D non-linear method". In *Lecture series programme on Aeroelasticity in axial flow turbomachines*. Von Karman Institute for fluid dynamics, May.

[12] Doi, H., and Alonso, J., 2002. "Fluid/structure coupled aeroelastic computations for transonic flows in turbomachinery". *ASME paper GT–2002–30313*. ASME Turbo Expo, June 3–6, Amsterdam, The Netherlands.

[13] Cinnella, P., and Palma, P. D., 2004. "A numerical method for turbomachinery aeroelasticity". *Transactions of the ASME, Journal of turbomachinery,* ***126***, pp. 310–316.

[14] Pinelli, L., Poli, F., Arnone, A., and Schipani, C., 2009. "A time-accurate 3D method for turbomachinery blade flutter analysis". In 12th International Symposium on Unsteady Aerodynamics, Aeroacoustics and Aeroelasticity of Turbomachines (ISUAAAT). September 1–4, London, UK, paper I12-S8-3.

[15] May., M., Mauffrey, Y., and Sicot, F., 2011. "Numerical flutter analysis of turbomachinery blading based on time-linearized, time-spectral and time-accurate simulations". In 15th International Forum on Aeroelasticity and Structural Dynamics (IFASD). June 26-30. Paris, France.

[16] Poli., F., Pinelli, L., and Arnone, A., 2015. "Aeroelastic stability analysis of a non-rotating annular turbine test rig: a comparison between a linearized and a nonlinear computational method". In 22nd International Congress on Sound and Vibration, (ICSV22). July 12-16, Florence, Italy.

[17] Giovannini, M., Marconcini, M., Arnone, A., and Bertini, F., 2014. "Evaluation of unsteady computational fluid dynamics models applied to the analysis of a transonic high-pressure turbine stage". *Proceedings of the Institution of Mechanical Engineers, Part A: Journal of Power and Energy,* ***228***(7), pp. 813–824.

Proceedings of the 14th International Symposium on Unsteady Aerodynamics, Aeroacoustics & Aeroelasticity of Turbomachines
ISUAAAT 14
8-11 September, 2015, Stockholm, Sweden

I14-S7-2

TURBINE PROFILE OPTIMIZATION TO MAXIMIZE AERODYNAMIC DAMPING

Christian Peeren
Siemens AG
Mellinghofer Strasse 55
Mülheim a.d. Ruhr, Germany
christian.peeren.ext@siemens.com

Stefan Schmitt
Siemens AG
Mellinghofer Strasse 55
Mülheim a.d. Ruhr, Germany
schmittstefan@siemens.com

Konrad Vogeler
TU Dresden
George-Bähr-Strasse 3c
Dresden, Germany
konrad.vogeler@tu-dresden.de

ABSTRACT

Quasi-two dimensional (Q2D) aerodynamic damping optimizations are carried out for different gas turbine blades with the goal to examine the influence of the turbine airfoil shape on the flutter stability. The selected cases are the 11th Standard configuration (a public literature test case) under subsonic operating conditions and a transonic gas turbine blade with a simple shock pattern.

The first part of the paper deals with the importance of the unsteady pressure phase since it directly determines local excitation or damping in case of real mode shapes. The amplitude of the unsteady pressure has in contrast just a scaling influence on the aerodynamic work. Geometrically designing both, the phase and the amplitude, would help to improve stability of the blade. The key question, which is raised at the end, is: Can the unsteady pressure be actively designed to obtain a beneficial flutter performance?

This question is answered in the second part where the two turbine geometries are aeroelastically optimized for different aeroelastical settings, here mode shape and Eigen frequency. The optimization is used to minimize or maximize the aerodynamic excitation of the original airfoil, while parameters like the blade loading and reduced frequency are kept constant. The latter case is used to better identify phenomenas which worsen flutter. For the subsonic dase the change of the flutter performance is very limited with respect to the baseline. However, great optimization potential is found for the transonic blade and the modified shock position is seen as one of the key features for the improved flutter perfomance. It is concluded that blades with much unguided turning and shocks can unveil a great impact on flutter if a local redesign is applied.

NOMENCLATURE

IBPA Interblade Phase Angle
LE, TE Leading, trailling edge
FTW, BTW Forward, backward travelling wave
SS, PS Suction, pressure side
x_{real}, x_{imag} Real, Imaginary part of x
x Chordwise coordinate
c Chord length
t Pitch
r_x, r_y x,y-component rotational center coordinate system
α_0 Amplification factor
f Blade Eigen frequency
$\vec{\phi}$ Mode shape vector, $\vec{\phi} = a_c \pm ia_s$
$\vec{n}$ Steady blade normal vector
U Flow velocity
M Mach number
k Reduced frequency, $k = 2\pi\frac{cf}{U_1}$
σ Interblade phase angle
W_{cyc} Work per cycle done by the blade
Λ Aerodynamic damping
w_{aero} Real part of the local aerodynamic work
$\widehat{p}$ Pressure amplitude
Φ_P Unsteady pressure phase

INTRODUCTION

The trend to more aggressive aerodynamic blade design for higher efficiencies and mass flows is one demand the designer is confronted with during the design of next generation turbomachinery. In the case of large power generation gas turbines the resulting blade structure shows thin profiles for the transonic flow regime and high aspect ratios resulting in a slender structure which is especially prone to aeroelastical phenomena. This is in particular flutter. Flutter is an aerodynamically self-excited vibration at blade Eigen frequency, where the blade absorbs energy from the fluid leading to strong vibrations which cause high cycle fatigue and in the worst case severe damage of the engine.

There are several numerical ways to predict flutter and the methods can be roughly separated in time domain methods and frequency domain methods. While time domain methods have the advantage, that they can better cover nonlinear effects like shocks for large geometrical deformation, they lack in computational speed and are expensive compared to frequency domain methods. See [1], [2] and [3] for flutter calculations, which have been performed in the time-domain for different blades. This paper uses the linear frequency domain method to predict flutter. This method linearizes the governing equations around their steady state (linearized perturbed Navier-Stokes Equation) and assumes that the flow perturbation due to the imposed airfoil displacement is harmonical, see e.g. [4] or [5]. The response of the flow field is then computed by solving the complex algebraic system for every possible interblade phase angle (IBPA), σ, which models the effect of forward and backward travelling waves [6]. Physically interpreted the complex unsteady pressure, which is obtained from the solution of the system, can be decomposed into its amplitude and phase. The amplitude is a measure for the local distortion while the phase indicates the propagation through the flow domain. To assess if the blade is damped or excited the accumulated energy per cycle is calculated and usually normalized by the maximum vibration energy yielding the aerodynamic damping Λ. Plotted for every σ the typical shape of those damping curves resembles a S-shape with a defined minimum and maximum. By definition a negative damping value indicates excitation and positive damping value acts stabilizing.

For a safe design, the minimum damping value must not be negative. This is sometimes inevitable and requires countermeasures. These are traditionally either reduction of blade loading or increasing the mechanical damping. The first option is in contrast to an aggressive blade design. The second option leads to expensive mechanical measures like under platform dampers, shrouded blades and snubbers and impacts the mode shape and Eigen frequency of the blade. In e.g. [7] and [8] the influence of the mode shape is discussed and it was found that there is an high influence on the stability behavior of the blade. [7] shows this influence in form of the so called tie-dye plot, a preliminary design method, where the rotational center coordinates r_x/c and r_y/c are plotted with the corresponding critical reduced frequency, k, where flutter occurs. The rotational center is another way to define the modeshape $\vec{\phi}$ for a two-dimensional airfoil and represents a fixed position around which the airfoil is perfoming a pivoting motion with a defined amplitude α_0. In [9] this prelimary design concept is extended for airfoils under transonic operation condition.

In this paper the focus is put on the influence of the airfoil geometry to examine whether a local geometrical redesign of the airfoil shape can help to remedy flutter excitation. The mode shape, reduced frequency and steady loading are held constant. With a constant hold reduced frequency and steady loading the question is answered, how a purely aeroelastically optimized minimum excitation (MinEx) airfoil featuring the same aerodynamical performance does look like. Efficieny considerations are not made since the aeroelastical optimization potential is in the focus of this paper. It further helps to understand in what way the unsteady pressure distribution is designable by modifying the geometry. The same approach is used for the opposite case, the maximization of the excitation. The idea behind the maximum excitation (MaxEx) airfoil is to identify effects which worsen flutter. Furthermore, the difference between the aeroelastical best and worst design can be interpreted as potential for a fixed steady aerodynamic task. In order to reduce the complexity, especially for the geometry optimization, all calculations are quasi two-dimensional.

Two representative airfoil geometries, one for subsonice and the other one for transonic operating conditions, are optimized and compared with each other. Additionally, the reduced frequency and rotational center are modified for both geometries and the optimization repeated to see the impact on the geometry and the aeroelastical optimization potential. The first regarded airfoil geometry is the Eleventh Standard Configuration from the set of standard configurations for unsteady aerodynamics and aeroelastics of turbomachinery under subsonic flow conditions [10] at mid span, denoted as case A. The transonic case, denoted as B, is a representative section of a stationary gas turbine.

MODIFIABILITY OF THE UNSTEADY PRESSURE

Key to alter the flutter perfomance of an airfoil is the unsteady pressure, which is caused by the motion of the airfoils. Before proceeding to the idea to geometrically design the unsteady pressure, it is neccessary to discuss the sources of the unsteady pressure and their geometrical modifiability. These sources can be roughly separated into three effects. First, there is the motion of the stagnation point that locally causes a pressure disturbances at the leading edge region. Second, the mutual interaction of the reference blade with its neighbours. Last, there are gasdynamic effects like shocks. The geometrical modification of the stagnation point does not have a significant influence on the associated unsteady pressure distribution. A fundamental influenceability is assumed for the second and third effect. An example for the modifiability of the second effect is a redesign of the channel region and throat area. It is obvious that this will come along with a changed unsteady pressure distribution. An example for the gasdynamic effect is a fishtail shock pattern, whose lower branch impinges on the suction surface of its neighbor blade. An e.g. specific geometrical modification of the impinging location might help here to reduce excitation. There are also simpler shock patterns, like shown further down (case B), where the effect of the shock is only confined on the blade itself. In general, shocks can transfer large amounts of unsteady energy into the structure like shown in [11] and are critical for the stability behavior.

Note that a good indication for mutual interaction with the neighbour blade is, when a local effect does have a certain variation with the IBPA. If the variation is weak one can generally assume that the effect is mainly induced by the blade itself. A global indication for the interaction of the reference blade with its first neighbours is the spreading, the difference between the maximum and minimum damping value. A first guess for the stability of the blade itself is the mean value of all aerodynamic damping values. Finally, there is the convection of all mentioned disturbances, whose crucial influence parameter is the reduced frequency, and its mutual superposition, which can make it difficult to identify the different single mechanisms.

UNSTEADY PRESSURE DESIGN FOR REAL MODE-SHAPES

In this section the link between the unsteady pressure and aerodynamic work is established. It is shown that the unsteady pressure phase is critical for local stability. To obtain the unsteady pressure the linearized perturbed Navier-Stokes equations for the given Eigen mode and interblade phase angle are solved. The energy, which is accumulated for one vibration cycle by the blade surface A, is

$$W_{cyc} = \int_A \underbrace{\pi p_{imag}(\vec{n} \cdot \vec{\phi})}_{=w_{aero}} dA. \tag{1}$$

Note that in equation 1 a real mode shape without imaginary components is assumed [12], $\vec{\phi} = \vec{a}_c$. Rewritting the imaginary part of the complex pressure in terms of its amplitude, $\widehat{p}$, and phase, Φ_p, gives

$$w_{aero} = \pi \widehat{p} \sin(\Phi_p)(\vec{n} \cdot \vec{\phi}) \tag{2}$$

for the local aerodynamic work distribution w_{aero}. The amplitude of the complex pressure is usually made dimensionless with $c_{up} = \frac{\widehat{p}}{\alpha_0(p_{t1} - p_1)}$, the steady dynamic pressure and the amplitude factor of the deformation. By definition a positive value of w_{aero} stands for local excitation and damping otherwise. In case of a value equal zero the regarded location acts neutrally stable. As seen in equation 1, w_{aero} is not a direct function of the steady pressure and is only dependent on the harmonic unsteady part. The integral value W_{cyc} is usually normalized with the vibrational energy, $K_E = \frac{1}{2}M\omega^2$, with M for the modal mass and ω for the modal frequency,

$$\Lambda = -\frac{W_{cyc}}{2K_E} \tag{3}$$

and so the damping value Λ for the given σ is obtained. Plotting Λ over the different σ gives the typical damping or flutter curve. In equation 2 some observations can now be made, assuming a given $(\vec{n} \cdot \vec{\phi})$ distribution. First is that the sign of w_{areo} is only dependent on the phase of the pressure. Second is that the pressure amplitude has a scaling influence on the local aerodynamic work. Hence the influence of Φ_p is critical for local stability. Both are functions of the geometry, mode shape, steady-state solution, the reduced frequency and the interblade phase angle. The inner product $(\vec{n} \cdot \vec{\phi})$ is a pure function of the blade geometry and mode shape and is seen as static part since a constant mode shape is assumed. It will be later shown that the change of $(\vec{n} \cdot \vec{\phi})$ has turned out to be small and thus the changes of the unsteady pressure are merely responsible for an improved or worsened flutter performance. The primary question, which is raised in this paper, is:

- Is it possible to design the unsteady pressure distribution by modifing the blade geometry so that the damping is maximized?

In above sections first thoughts were already made about the geometrical modifiability of the unsteady pressure. It was recognized that especially effects, which arise from the mututal interaction of the blades with each other, must have a dependency on the airfoil geometry. To answer this question the minimum damping value of the baseline configuration is optimized towards the stable direction by modifying the blade geometry and the changes of the unsteady pressure observed. Focus of this analysis is whether the aeroelastically improved geometry is achieved by an altered behavior of $\widehat{p}$ or Φ_p or both.

METHODOLOGY

For the optimization the CFD tool chain, consisting of a NACA like blade parameterization, a meshing process, steady-state evaluation and unsteady time-linearized flutter solver, is linked with an optimizer. Due to the assumption of a constant mode shape and Eigen frequency no structural solver is needed. The assumption of a constant mode shape is justified since the exciting energy entries for freestanding blade flutter occur mainly at the upper part of the blade. This is the part where the optimization is carried out. The mode shape for a fundamental mode is, however, dominated by the thick hub part. Modifying the upper thin sections will have an influence on the overall blade mode shape but it will not completely change the original configuration. It will rather express itself in a minor change of r_x and r_y. Assuming a center of rotation, which does, referring to the tie-dye plot of [7], lie in a region with small stability gradients, will have an neglectable stability influence if disturbed.

The blade parameterization uses an inhouse software and the meshing process Autogrid (v.9.1.2). Steady-state and time-linearized flutter calculations are done with TRACE, LinearTRACE, (v.8.2) respectively. The Dakota framework (v.5.2) is used as optimization software [13] and as optimization strategy a surrogate based local optimization (SBLO) approach is opted. The surrogate model is a fast to evaluate low fidelity representation of the unknown real function [14]. For its construction it requires a certain amount of expensive real function evaluations which are dependent on the type of model and the desired accuracy. In this work Kriging is used to create a surrogate model out of the real function evaluations. Those are obtained by a prior executed design of experiments (DOE) where Latin hypercube sampling (LHS) is used [15] to ensure a space-filling sampling plan that covers as much details as possible from the real function behavior. Subsequently, the optimization is executed on the surrogate model with an evolotionary strategy. To conclude, the optimization is performing the following three steps:

I.) Sampling with LHS around its initial (or from a iteration step before) location within the given trust region and construction of the meta model (surrogate)
II.) Find minimum with the selected optimization strategy, compare the minimum of the surrogate with the minimum of the true model
III.) Depending on accuracy: Repeat step one and two at new point with a expanded or contracted trust region or abort optimization.

As optimization constraints the outlet velocity and outlet flow angle is allowed to vary within a small tolerance of $\pm 1\%$ with respect to the baseline's value and $\pm 0.1^{\circ}$, respectively. This ensures that the flow turning of the profile is fixed as well as the reduced frequency. For all subsonic cases (A1 to A5) 180 designs samples are used per surrogate while having 12 geometrical design parameters. The transonic cases (B1 to B5) have due to the simpler droplet shape only 7 design parameters and use therefore only 150 samples per surrogate.

GEOMETRICAL FLUTTER OPITMIZATION

As mentioned in the introdution above, the flutter optimization is conducted for two geometrical different baseline airfoils. The first airfoil has subsonic operating conditions (case A) and is a typical example for a section in a subsonic turbine blade configuration. The second airfoil geometry is a representative section of a stationary gas turbine, experiencing transonic flow conditions with a simple shock pattern at the TE (case B). The original center of rotation coordinates and reduced frequencies are treated in case A1 and B1. Case A2 and B2 have the same r_x and r_y coordinates as A1 and B1, but are run at half nominal reduced frequency. The remaining cases A3-A5 and B3-B5 exhibit modified r_x and r_y coordinates with the nominal reduced frequency of case A1 and B1, respectively. Figure 1 gives an overview of the examined cases. Due to the definition of the mode shape of case A1, see [10] a real plunging motion normal to the chord, the associated representation as center of rotation is a r_x/c part of infinity and a r_y/c of zero. That is why case A1 and A2 can not be found in figure 1. The baseline mode shape of the transonic case B1 is a real plunging rigid body motion with an overlaid pitching component which can be described by a rotational center upstream of the blade. For each case two

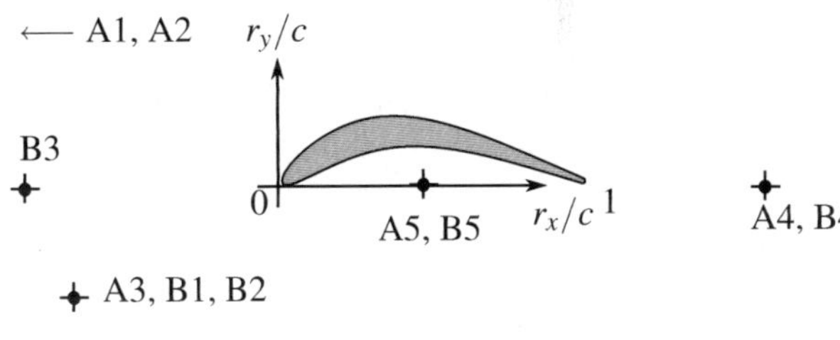

FIGURE 1: Blade coordinate system and sketched rotational center for the different cases. r_x and r_y define the rotational center. Case A1 and A2 are not seen here because the rotational center lies at $r_x/c = \pm\infty$ and $r_y = 0$.

optimizations are carried out: an aeroelastically improving optimization where the flutter exciation is minimized (MinEx) and a contrary one where flutter exciation is maximized (MaxEx). The damping curve, airfoil geometry and isentropic Mach number on the blade surface are then plotted for the three different airfoil versions, Baseline, MinEx and MaxEx. For some selected cases the unsteady pressure distribution is additionally discussed.

Also mentioned in the introdution chapter is that the difference between MinEx and MaxEx can be interpreted as aeroelastical optimization potential Π_{pot} for a given steady aerodynamic task. Mathematically, the following equation 4 is obtained

$$\Pi_{pot} = \frac{\Lambda_{min}^{MinEx} - \Lambda_{min}^{MaxEx}}{\left(\Lambda_{max}^{MaxEx} - \Lambda_{min}^{MaxEx}\right)}. \quad (4)$$

Here the spreading, the difference of the maximum and minimum damping value, of the MaxEx airfoil version is used to make the different cases compareable with each other. Figure 2 shows the aeroelastical optimization potential for the subsonic and transonic cases introduced above. Clearly seen is that the

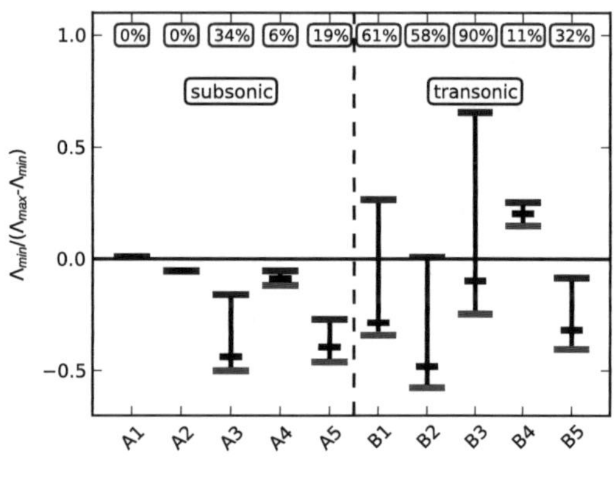

FIGURE 2: Optimization Potential for different cases. The line between MinEx and MaxEx is Π_{pot}, the black line between is the baseline.

subsonic case A1 and A2 have no potential. Cases A3-A5, where the center of rotation has been modified, reveal a potential of 34% and 19% but are still unstable since they lie underneath $\frac{\Lambda_{min}}{\left(\Lambda_{max}^{MaxEx} - \Lambda_{min}^{MaxEx}\right)} = 0$. More potential is found for the transonic cases with up to 90% at case B3. Except for case B5 all transonic cases have been optimized in such a way that the MinEx airfoil version does not have excition anymore ($\Lambda_{min}^{MinEx} > 0.0$).

RESULTS SUBSONIC CASES A1-A5

In the next chapters the subsonic cases are discussed first, starting with the baseline optimization A1, followed by the influence of a changed reduced frequency A2 and finished by the influence of the rotational center A3-A5. This analysis is analougsly repeated for the transonic cases.

BASELINE OPTIMIZATION (A1)

Case A1 is the aeroelastical optimization for the 11th standard configuration at subsonic flow conditions as described in [10]. It is neither possible to significantly improve nor worsen flutter, figure 3a. Analyses of the unsteady pressure barely showed any difference compared to the baseline distributions. The very small improvement in case A1 originates from the increased positive incidence, figure 3b and 3c which caused a flow situation close to separation around the LE. The same effect is found in case A3 and more details are discussed there as well as the changes of the unsteady pressure distribution.

INFLUENCE OF THE REDUCED FREQUENCY (A2)

Case A2 is the same configuration like case A1 just with the half value of the nominal reduced frequency. Similar to case A1, does case A2 not show any geometrical impact, figure 3e, on the flutter curve figure 3d. One surprising result is that the resulting airfoil geometries for an optimization targeting to minimize the excitation (MinEx) figure 3e compared to the resulting geometries for an optimization targeting to maximize the excitation (MaxEx) are "switched" between the cases A2 and A3, figure 3h. A plausible explaination for this effect could be that the flutter performance is weakly correlated or rather indepedent of the airfoil geometry for the regarded Eigen mode. That is why case A1 and A2 show practical no flutter potential even if the airfoil geometry is strongly altered like in case A2. For the subsonic case it is conluded that the reduced frequency has no impact on the aeroelastical potential.

INFLUENCE OF THE CENTER OF ROTATION (A3-A5)

The center of rotation of case A3 is identical with case B1, a translational plunging motion with an overlayer pitching component. In case A5 the airfoil is perfoming a pure pitching motion at midchord. The changed center of rotation of case A3 and A5 mainly impacted the spreading, figure 3g and 4d. At first sight, this is an indication that especially the interaction with the neighbour blades is geometrically influenceable for this cases. For both cases similar mechanisms in the unsteady pressure were found. The same is observed for case A4, where the center of rotation lies downstream of the blade. Hence, only case A3 is discussed in detail here.

Figure 3g shows the resulting damping curve for the baseline, MinEx and MaxEx optimization. Comparing the baseline and MaxEx less difference is noted for the flutter performance, though the geometrical change, seen in figure 3h, shows strong

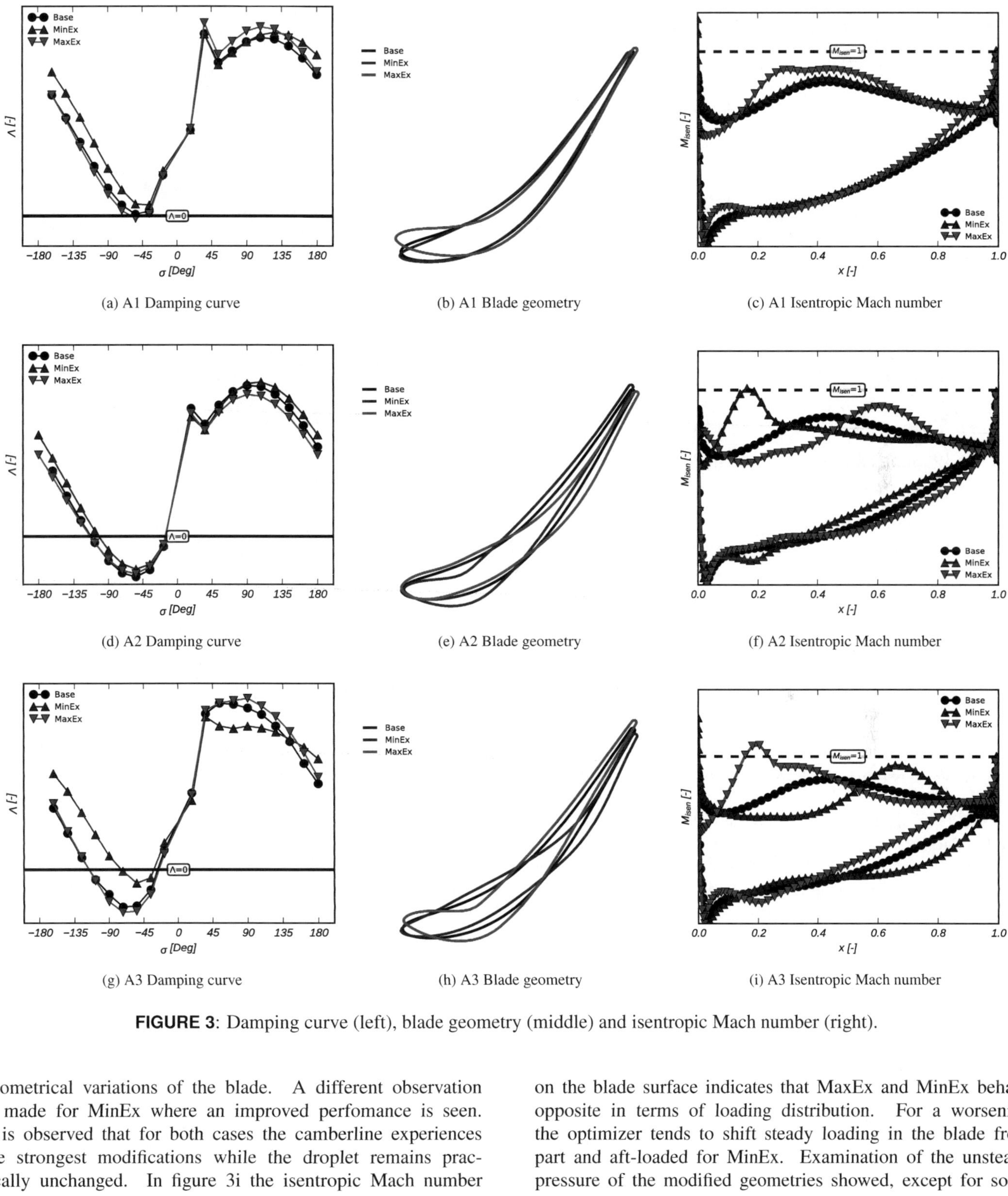

(a) A1 Damping curve (b) A1 Blade geometry (c) A1 Isentropic Mach number

(d) A2 Damping curve (e) A2 Blade geometry (f) A2 Isentropic Mach number

(g) A3 Damping curve (h) A3 Blade geometry (i) A3 Isentropic Mach number

FIGURE 3: Damping curve (left), blade geometry (middle) and isentropic Mach number (right).

geometrical variations of the blade. A different observation is made for MinEx where an improved perfomance is seen. It is observed that for both cases the camberline experiences the strongest modifications while the droplet remains practically unchanged. In figure 3i the isentropic Mach number on the blade surface indicates that MaxEx and MinEx behave opposite in terms of loading distribution. For a worsening the optimizer tends to shift steady loading in the blade front part and aft-loaded for MinEx. Examination of the unsteady pressure of the modified geometries showed, except for some

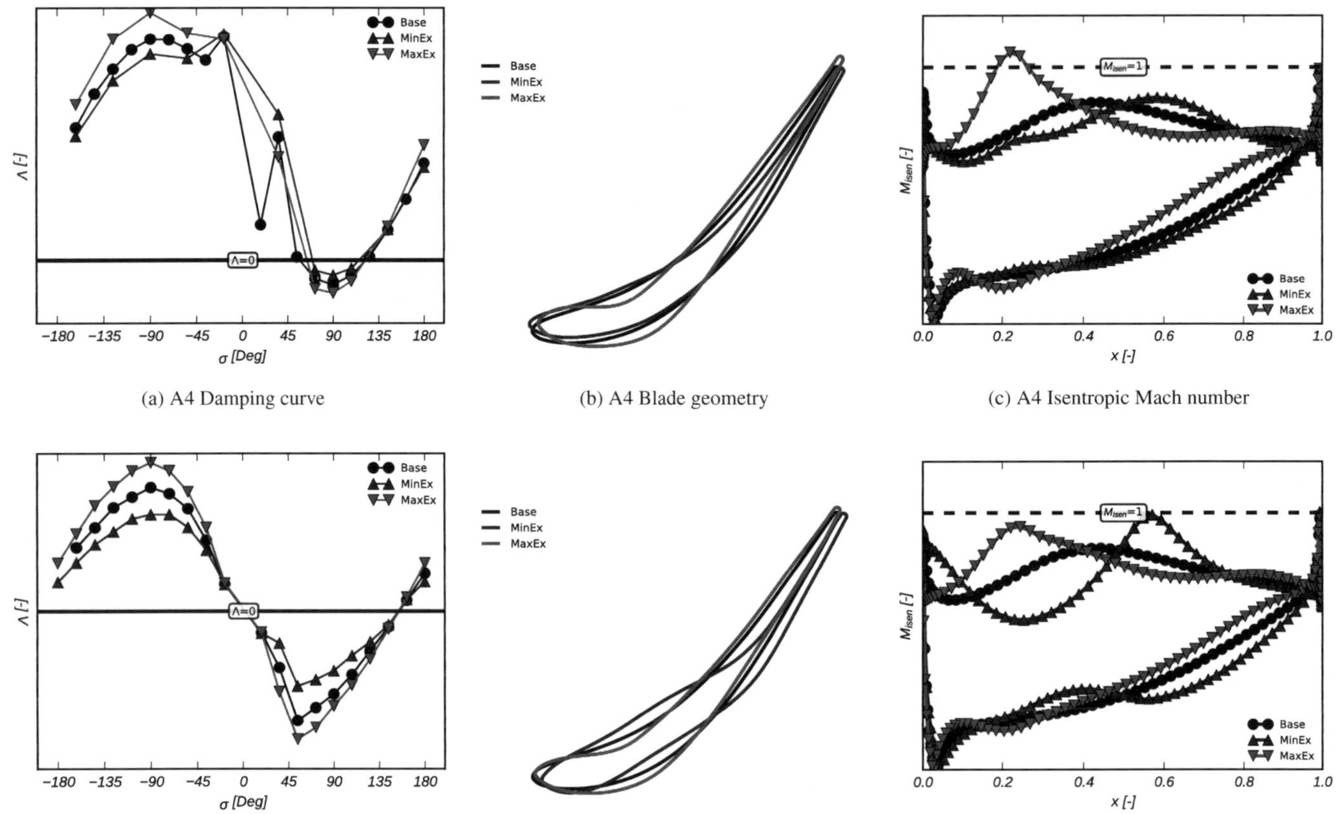

(a) A4 Damping curve (b) A4 Blade geometry (c) A4 Isentropic Mach number

(d) A5 Damping curve (e) A5 Blade geometry (f) A5 Isentropic Mach number

FIGURE 4: Damping curve (left), blade geometry (middle) and isentropic Mach number (right).

local effects, no fundamentally changed behavior compared to the baseline. This is recognized in figure 5 where w_{aero}, Φ_p and $\widehat{p}$ are plotted over the airfoil surface for the critical IBPA, $\sigma = -72.00°$. The mentioned local effects are seen for both airfoil versions on the SS. MaxEx exhibits a local exciting work peak at x=0.2, while MinEx reduces excitation by having a less distinct crest compare to the baseline around the same location. For MinEx the reason is found in a decrease of $\widehat{p}$, figure 5c and a pressure phase, which has moved closer to neutral 180°, figure 5b. Reasons for this change of MinEx are identified in figure 3i where loading is taken away from the front part. The flow does here experience less forced turn through the channel geometry, which is nothing else than a reduction of the neighbor blade influence what seems to be beneficial for the unsteady pressure. Around x=0.6 where the steady loading is rising a dip in Φ_p is seen and analogously an increased interaction with the neighbor is a plausible explaination for this phenomena. The MinEx $\widehat{p}$ peak close to the LE is accounted for by the increased positive incidence, which even cause supersonic Mach numbers for a short time. In combination with the corresponding Φ_p, which is also here very close to neutral 180°, there is no visual impact on w_{aero}, though. Referring to the MaxEx w_{aero} at x=0.2, the reasons is quickly found when considering $\widehat{p}$ at the same position. Also here the source of the disturbance originates from a two-dimensional effect, because when regarding the steady isentropic Mach number again, it is observed that the flow turns supersonic for a short time again. This time the reason is the airfoil geometry which features its maximum camber at this position. Apart from this local effect, MaxEx practically does not differ from the baseline distribution. For both airfoil variants it was in general not possible to completely alter the unsteady pressure behavior. The observed local effects have a minor impact on the flutter perfomance. Referring to the question above if the unsteady pressure is designable the answer is no in this case regarding the drastic geometrical change both airfoil variants have been experiencing. This conclusion is, of course only valid for the regarded operational point. Comparing MinEx with the MinEx solution of case A1 and A2 the pitching

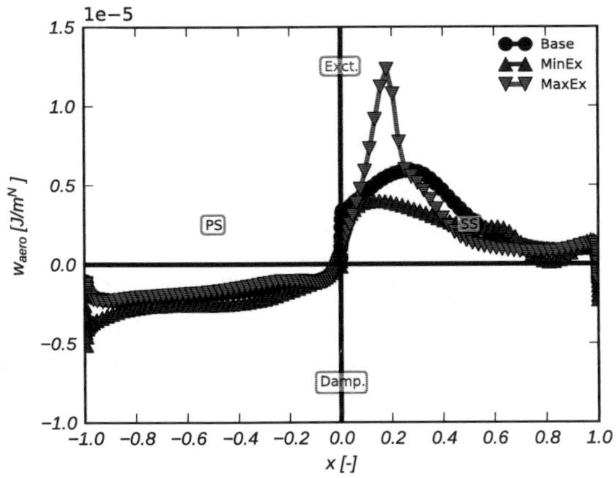

(a) w_{aero} on the blade surface, $\sigma = -72.00°$

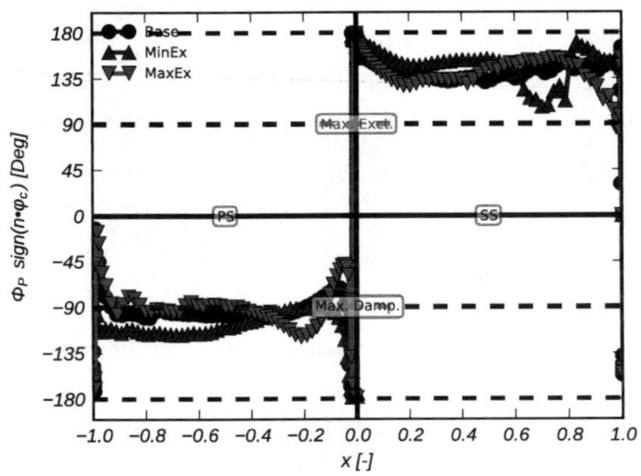

(b) Φ_p on the blade surface, $\sigma = -72.00°$

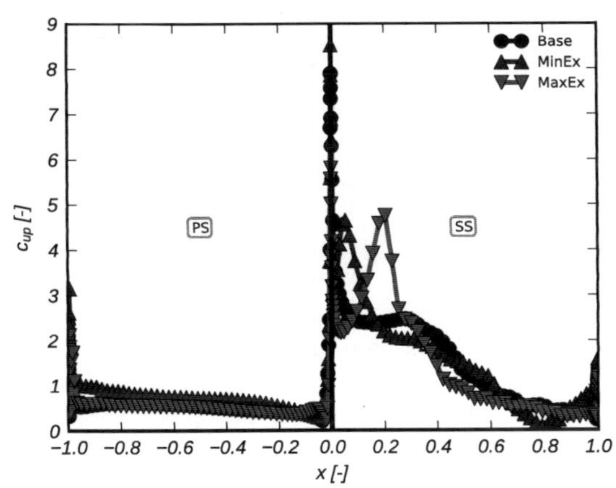

(c) $\hat{c}_p$ on the blade surface, $\sigma = -72.00°$

FIGURE 5: Real part of the specific aerodynamic work and unsteady pressure distribution for $\sigma = -72.00°$ case A3

component seems to influence the geometrical optimization potential. This is coincides with the findings of [7] and emphasis that the mode shape determines the aeroelastical flutter potential.

Not mentioned so far and last point being discussed for case A3 is the change of the inner product $(\vec{n} \cdot \vec{\phi})$ shown in figure 6. The impact of the modified geometry is, also here, very small and partically does not differ from the baseline. The same was observed for all other cases, that is why the change of $(\vec{n} \cdot \vec{\phi})$ is not further discussed in this paper. Above's results showed that a

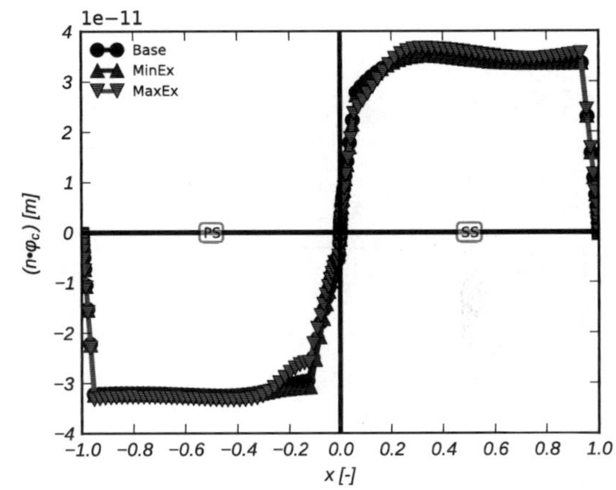

FIGURE 6: $(\vec{n} \cdot \vec{\phi})$ distribution case A3

changed rotational center impacts the aeroelastical optimization potential. This is in particular the pitching component in the airfoil mode shape. The effect of the modified mode shape is seen in the geometrically flutter optimization where the spreading and hence the mutual interaction between the blades is influenced.

RESULTS TRANSONIC CASES B1-B5

The following sections discuss the results of the transonic cases B1-B5 in the same manner like for the subsonic cases.

BASELINE OPTIMIZATION (B1)

The optimization of case B1 indicates aeroelastical potential, figure 9a. For the minimum excitation airfoil version it is possible to shift the damping curve in the completely stable region. The modification of the geometry remains moderately and affected only the camber line of the profile, figure 9b. The maximum excitation airfoil variant reveals a geometrical modification in the opposite direction. Here the airfoil has become

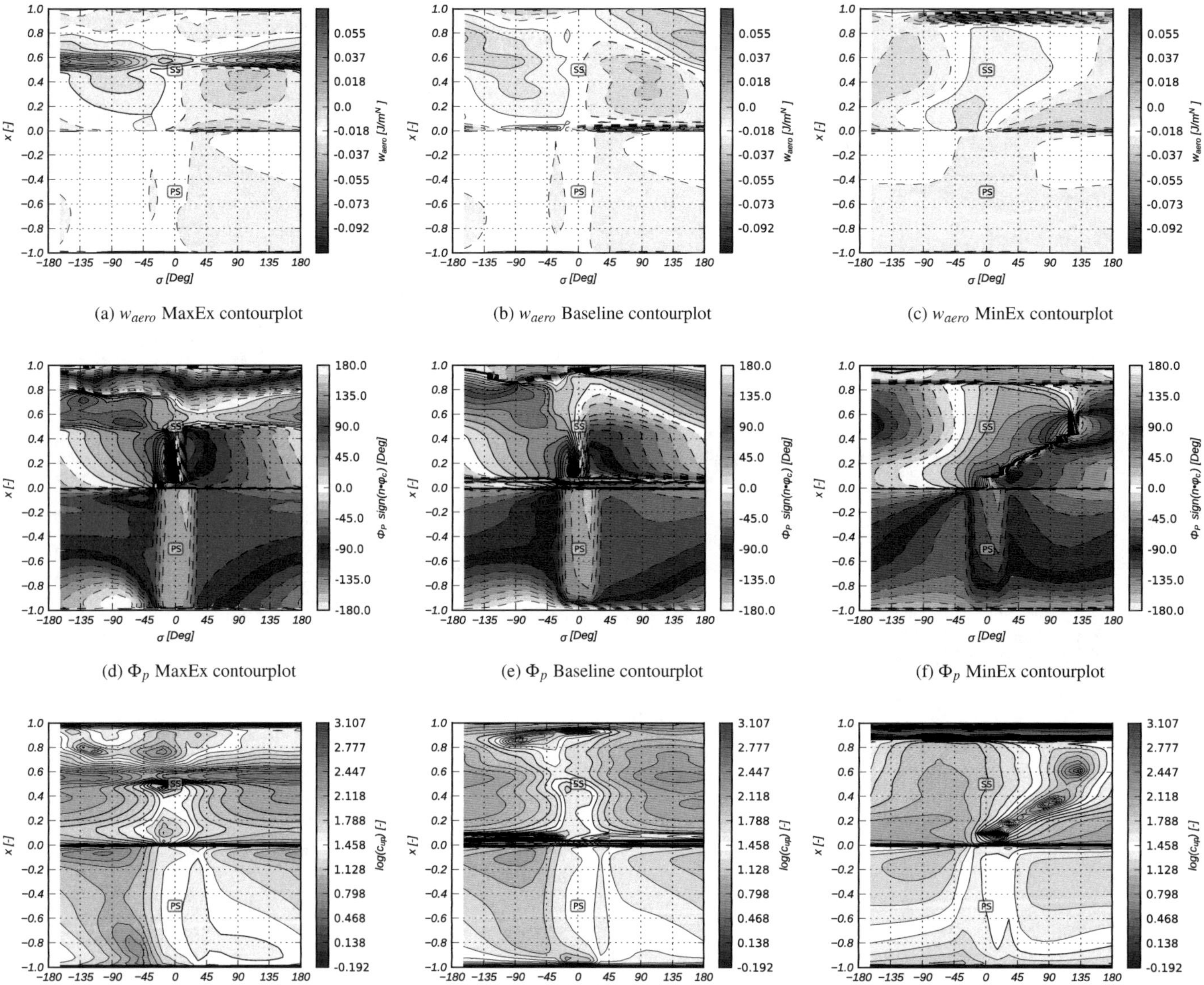

(a) w_{aero} MaxEx contourplot

(b) w_{aero} Baseline contourplot

(c) w_{aero} MinEx contourplot

(d) Φ_p MaxEx contourplot

(e) Φ_p Baseline contourplot

(f) Φ_p MinEx contourplot

(g) $\log_{10}(c_{up})$ MaxEx contourplot

(h) $\log_{10}(c_{up})$ Baseline contourplot

(i) $\log_{10}(c_{up})$ MinEx contourplot

FIGURE 7: Case B1: w_{aero}, $\Phi_p\,\mathrm{sign}(\vec{n}\cdot\vec{\phi})$ and $\widehat{p}$ over the blade surface for all interblade phase angles and different reduced frequencies. Red: exciting work/phase, Blue: damping work/phase. The log10 function is applied to the pressure amplitude plots to make changes better visible.

less cambered with a strong turning close to the LE and a flat-plate like geometry downstream. The variation of the damping curve shows in contrast to the MinEx case less difference to the baseline and the worsening effect remains very small. Regarding the change of the steady loading, figure 9c, MaxEx shifts loading to the front in contrast to MinEx where the blade receives an aft-loaded trend. In order to identify from where the stabilizing potential arises the local aerodynamic work and unsteady pressure is plotted in the next figures. Before it is mentioned that the unsteady pressure features, in contrast to the A cases introduced prior, substantial changes for this airfoil. That is why another graphical presentation way is used to display the results of all unsteady quantities. In figure 7 w_{aero}, Φ_p and $\widehat{p}$ is plotted as function of the spanwise coordinate and the IBPA as con-

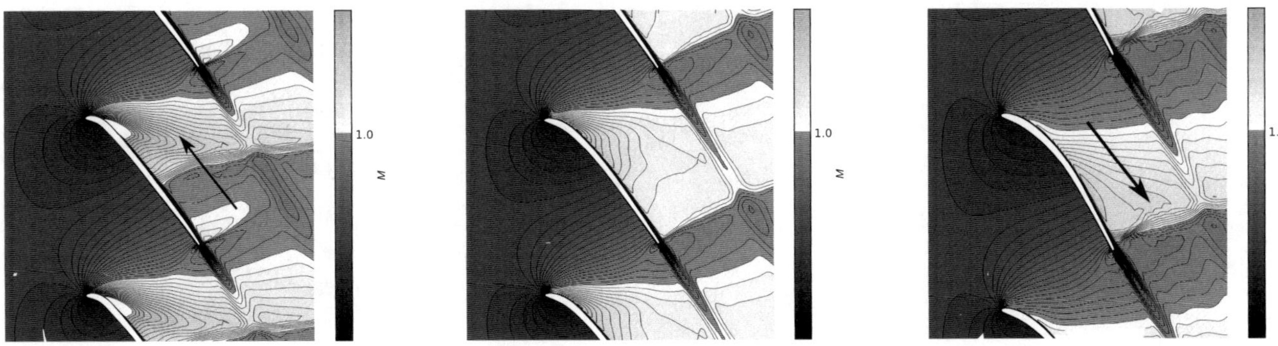

(a) Mach MaxEx contourplot

(b) Mach Baseline contourplot

(c) Mach MinEx contourplot

FIGURE 8: Case B1 Mach contourplots. Yellow: Supersonic regions, Blue: Subsonic regions. Clearly seen: The optimizer moves the shock either in regions with high interaction with the blade neighbours (MaxEx) or away from them (MinEx).

tourplot. This allows to additionally identify the behavior for different IBPA on a more global level. For the subsonic cases the strong resemblance with the baseline made this kind of plot obsolete, thus the classical presentation way was used. Regarding w_{aero} it is seen that for MaxEx as well as the MinEx optimized version the shock on the SS exhibits a certain key function for stability. Whereas the shock contributes exciting work in the MaxEx case, it causes a strong entry of stabilizing work for MinEx. Analyzing the unsteady pressure at the shock location of MinEx, fig. 7f and 7i, reveals that Φ_p has a phasing close to the neutral $-180°$ but is still in the stable phase sector. Physically interpreted the MinEx version shifts the shock in regions with less interaction by making the airfoil more cambered figure 8. The shock is more isolated from the neighbour blades and only disturbed by the reference blade itself. That is seen in the weak and almost constant variation of Φ_p with the IBPA at shock location. As reason for the strong increase of the magnitude of $\widehat{p}$ the increased Mach numbers upstream of the shock are identified, figure 7i. Both effects explain the damping work entry in figure 7c. The opposite case takes place for the MaxEx version of case B1 where the optimizer shifts the shock upstream in a region with increased interaction with the blade neighbour resulting in destabilizing work at the shock location. The Φ_p behavior, figure 7d, which is practically unstable for all IBPA, reaches even Φ_p values for maximum excitation (90°). There is a clear variation with σ and thus interference with the neighbour blades. Combined with a slightly increase pressure amplitude, fig. 7g, the exciting work in figure 7a is explained. Interestingly MaxEx shows an almost constant broad band of stabilizing Φ_p downstream of the shock, what is however, not critical for the aero. work due to the small amplitude. Another very interesting effect is observed in the front part of all three versions from the LE till midchord and for the MinEx even up to $x = 0.8$. Whereas the baseline and MaxEx look quite similar here, MinEx shows a shift in BTW direction which is linear increasing with the chordwise coordinate. The same is seen for the pressure amplitude.

INFLUENCE OF THE REDUCED FREQUENCY (B2)

Case B2 shows similar to case B1 great aeroelastical optimization potential, figure 9d. The damping curve for the MinEx optimized version of blade exhibits a reduced spreading of about 79% referred to the baseline what indicates that the optimization drove, globally considered, the influence of the neighbour blades to a minimum. The opposite case is found for the MaxEx case where the spreading increased by 13%. Further studies of the aerodynamic work and unsteady pressure revealed that the optimization potential arises from the same mechanisms like case B1. In principal the geometry of case B2 shows for the MinEx as well as MaxEx optimization identical results like for case B1. The reduced frequency does thus have not a critical influence on the optimization potential.

INFLUENCE OF THE CENTER OF ROTATION (B3-B5)

Case B3 has a modified center of torsion and reveals the largest amount of aeroelastical optimization potential. The effects from where this potential arise are similar to case B1 and B2 and so not discussed here. Analougs results are observed for case B5, figure 10d to 10e, where the center of rotation lies at mid chord. Case B4 has a rotational center downstream of the blade. Two details are very different compared to the other transonic cases. This is the minimum damping value for both optimized versions, figure 10a, which does in contrast to the B cases discussed prior only show a slight difference to the baseline and that the geometrical solution behave the opposite way,

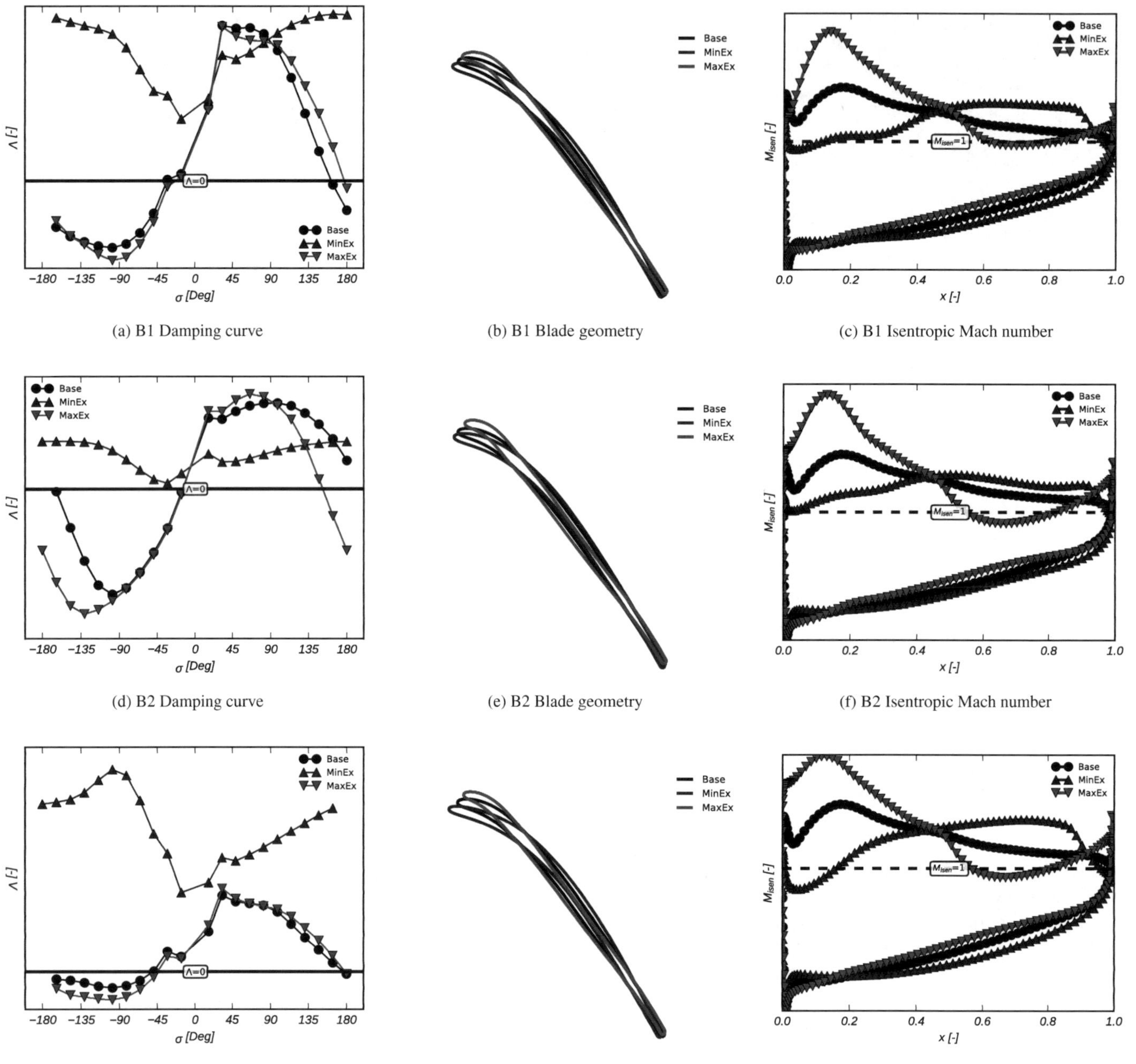

(a) B1 Damping curve

(b) B1 Blade geometry

(c) B1 Isentropic Mach number

(d) B2 Damping curve

(e) B2 Blade geometry

(f) B2 Isentropic Mach number

(g) B3 Damping curve

(h) B3 Blade geometry

(i) B3 Isentropic Mach number

FIGURE 9: Damping curve (left), blade geometry (middle) and isentropic Mach number (right).

figure 10b. MinEx is front-loaded while MaxEx is aft-loaded, what was the typical MinEx solution for all other B cases. Analyzing the unsteady pressure of MaxEx at the shock location revealed a similar mechanism like in case B3 for MinEx. For MinEx, the MaxEx solution of case B3, the shock, has, however, a vanishin g influence on the aerodynamic work. All improving effects arise from the airfoil front part. Similar to the subsonic case the transonic case shows that the aeroelastical optimization potential is dependent on the regarded mode shape.

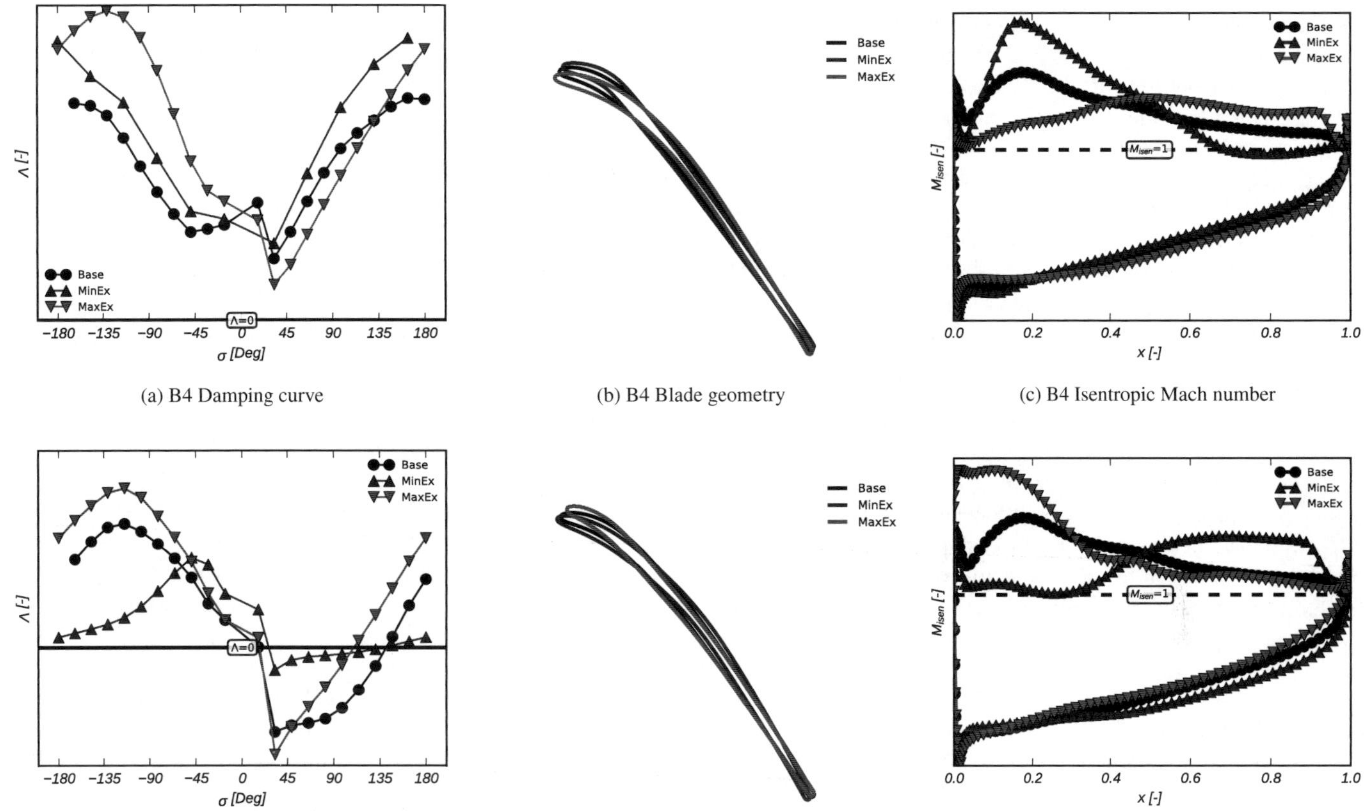

(a) B4 Damping curve (b) B4 Blade geometry (c) B4 Isentropic Mach number

(d) B5 Damping curve (e) B5 Blade geometry (f) B5 Isentropic Mach number

FIGURE 10: Damping curve (left), blade geometry (middle) and isentropic Mach number (right).

CONCLUSIONS

In this paper it has been showed that the unsteady pressure, which decides about the stability for real mode shapes, can be actively designed by modifying the geometry. Key for the local stability is the unsteady pressure phase, while the unsteady pressure amplitude is of scaling nature only. Both are functions of the interblade phase angle, steady-state flow field, Eigen mode and airfoil geometry. The key question which is raised is whether it is possible to actively design the unsteady pressure in such a way that a beneficial flutter perfomance is attained.

The considered baseline geometries are the Eleventh Standard Configuration under subsonic flow conditions and the section of a stationary transonic gas turbine blade. For the subsonic cases (A1-A2), the baseline and baseline with half nominal reduced frequency, no optimization potential and effect of the airfoil geometry is observed. The cases with the altered center of rotation (A3-A5) revealed a slight potential, which is based on a reduction of the interaction with the neighbour blades. Analyzing the changes of the unsteady pressure showed, that there is no substantial difference from the baseline distribution to speak from an aeroelastical designability for this case. Reasons for this limited potential could be the, compare to the transonic case, small pitch to chord ratio, which results in a relatively long channeled region between the blades. With the imposed optimization constraints, here mainly the fixed outlet flow angle the optimizer was not able to fundamentally change the geometry.

This looks differently for transonic cases (B1-B5), which strived independent of the aeroelastical settings, to the same solution: An aft-loaded airfoil design where the shock is moved away as far as the optimization constraints allow from the zone of interaction with the airfoil neighbour. The resulting airfoil is more cambered an has higher Mach numbers upstream of the shock, what also yielded in a higher pressure amplitude at the shock location. Due to the reduced interaction, however, the pressure phase does show less variance with the IBPA and lies in the stable phase sector close to $-180°$. The work entry is, hence, damping for all IBPAs. Another main driver behind this improvement is the new Φ_p and $\widehat{p}$ distribution at the blade front part, which originates from the changed steady blade loading and cause compensation of exciting and damping fields. Those two effects are found for all transonic cases. A change of the reduced frequency showed a minor effect on the optimization potential, while different rotational centers had a signifcant impact. It is concluded that the aeroelastical optimization of transonic blades can reveal high optimization potential and that the unsteady pressure is geometrical designable. Based on above observations, the following design recommendations are made:

• The unsteady pressure can be actively designed. Effects occur in both, the unsteady pressure amplitude and phase.
• In case of a simple transonic shock pattern: The Shock should be moved out of region of interaction. Less interaction with neighbour blades cause a decreased dependency on the IBPA.
• The aeroelastical airfoil optimization showed that it is possible to improve or worse flutter. Aeroelastical optimization potential does not neccessarily exist, though, and is dependent on the regarded flow case as well as mode shape.

REFERENCES

[1] Sadeghi, M., and Liu, F., 2001. "Investigation of Non-Linear Flutter by a coupled Aerodynamics and Structural Dynamics Method". *Paper No. AIAA 20010573 ,39th AIAA Aerospace Sciences Meeting and Exhibit 2001, Reno NV.*

[2] Siden, L., 1991. "Numerical Simulation of Unsteady Compressible Flows Applied to a Blade Flutter Analysis". *ASME Paper No. GT91-203.*

[3] He, L., and Denton, J., 1994. "Three Dimensional Time Marching Inviscid and Viscous Solutions for Unsteady Flows around Vibrating Blades". *Journal of Turbomachinery,* ***116***, pp. 469–476.

[4] Kersken, H., Frey, C., Voigt, C., and Ashcroft, G., 2010. "Time-Linearized and Time-Accurate 3d RANS Method for Aeroelastic Analysis in Turbomachinery". *ASME Paper No. GT2010-22940.*

[5] Clark, W., and Hall, K., 2000. "A Time-Linearized Navier-Stokes Analysis of Stall Flutter". *Journal of Turbomachinery,* ***122***, July, pp. 467–476.

[6] Hall, K. C., 1997. "Linearized Unsteady Aerodyamics". In Presented at the Von Karman Institute for Fluid Dynamics Lecture Series Programme on Aeroelasticy in Axial Flow Machines.

[7] Panovsky, J., and Kielb, R. E., 2000. "A Design Method to Prevent Low Pressure Turbine Blade Flutter". *ASME Journal of Engineering for Gas Turbines and Power,* ***122***, pp. 89–98.

[8] Waite, J., and Kielb, R. E., 2014. "Physical Understanding and Sensitivities of LPT Flutter". *ASME Paper No. GT2014-25133.*

[9] Meingast, M., Kielb, R. E., and Jeffrey, P. T., 2009. "Preliminary Flutter Design Method For Supersonic Low Pressure Turbines". *ASME Paper No. GT2009-59177.*

[10] Fransson, T. H., Jöcker, M., Bölcs, A., and Ott, P., 1999. "Viscous and Inviscid Linear/Nonlinear Calculations Versus Quasi-Three-Dimensional Experimental Cascade Data for a New Aeroelastic Turbine Standart Configuration". *Journal of Turbomachinery,* ***121***, pp. 717–725.

[11] Srivastava, R., and Keith, T., 2004. "Shock Induced Flutter of Turbomachinery Blade Row". *ASME Paper No. GT2004-53479.*

[12] Hall, K. C., Kielb, R. E., and Thomas, J. P., 2006. *Unsteady Aerodynamics, Aeroacoustics and Aeroelasticity of Turbomachines*. Springer.

[13] Adams, B. M., Ebeida, M. S., Eldred, M. S., Jakeman, J. D., Swiler, L. P., Bohnhoff, W. J., Dalbey, K. R., Eddy, J. P., Hu, K. T., Vigil, D. M., and Lara E. Bauman, P. D. H., 2009. *Dakota. A Multilevel Parallel Object-Oriented Framework for Design Optimization, Parameter Estimation, Uncertainty Quantification, and Sensitivity Analysis. Version 5.4 Users Manual*. Livermore, CA, USA.

[14] Queipo, N. V., Haftka, R. T., Shyy, W., Goel, T., Vaidyanathan, R., and Tucker, P. K., 2005. "Surrogate-Based Analysis and Optimization". *Progress in Aerospace Sciences,* ***41***, pp. 1–28.

[15] Forrester, A. I. J., Sobester, A., and Keane, A., 2008. *Engineering Design via Surrogate Modelling*. John Wiley & Sons Ltd., University of Southampton, UK.

Proceedings of the 14th International Symposium on Unsteady Aerodynamics,
Aeroacoustics & Aeroelasticity of Turbomachines
ISUAAAT14
8-11 September 2015, Stockholm, Sweden

I14-s7-3

FLUTTER AND FORCED RESPONSE INVESTIGATION OF A TRANSONIC AXIAL TURBINE STAGE USING FLUID-STRUCTURE INTERACTION SIMULATION

Iroizan Ubulom
University of New South Wales (UNSW)
Canberra, ACT, Australia

Andrew Neely
University of New South Wales (UNSW)
Canberra, ACT, Australia

ABSTRACT

The prediction of the mutual interaction between the unsteady flow phenomena and the oscillating blades is essential for improved turbine performance and component efficiency. Previous studies have shown this interaction in the form of energy transfer (self-excitation and flutter) due to the coupled relation between the aerodynamic, inertia and elastic forces acting on the blades. This invariably will lead to unstable material behaviour and structural fatigue failures. Thus the exclusion of the mutual interactions of the structural blade motion and aerodynamic field may not accurately represent turbomachinery fatigue and flutter failures. These problems are characterized by strong instability, energy exchange and strong aeroelastic coupling. This study adopts a method of fluid-structure interaction modelling to combine the unsteady aerodynamics and structural domain discretization to investigate the aeroelasticity phenomenon of a transonic turbine stage. The stage flutter-prone regions were predicted for various inter-blade phase angles (IBPA). The first bending mode was shown to be most unstable at 0 IBPA as compared to the second and third vibration modes. For the effect of unsteady forced response due to upstream wakes, the 36th engine order (EO) excitation loads were used to estimate stress amplitudes including the added effect of thermal and rotational loads. The result showed the dominant effect of thermal effects on the estimation of the stress distribution across the blade profile. The results of the flow-field and heat transfer through the stage were also compared against experimental data showing significant shock-boundary layer effects.

INTRODUCTION

Fatigue failures in the form of flutter instabilities and forced response excitations are major structural concerns in the operation of gas turbine engines. The overall structural integrity of the engine system depends to a large extent on the ability of the machine to withstand both steady and vibratory stresses, [1]. In addition, a certification requirement is that the engine withstands the specified design steady stress and measured vibratory stress level. It has been noted that approximately

55% of the jet engine mishaps with total damage or loss of aircraft worth over 1$ Million are due to fatigue failures [2]. A similar case was reported by [1] who stated that nearly 70% of engine failure was attributable to high-cycle fatigue failure.

These problems arise due to the extreme operating conditions the blades operate in. These are mostly driven by safety and economic demands for improved performance and low operating cost. Thus, from an economic standpoint, the onus is on the manufacturers to design components for high efficiency, drastic weight reduction achievable through reduced engine size, improved fuel consumptions etc. These demands invariably translate into the need to design turbine components with higher blade loading, closer axial spacing, high temperature operation etc. As such, the material strengths are pushed to their aeromechanical limits with increased risk of high-cycle fatigue failures. Considering a stator-rotor stage, the reduced axial gaps between the blade-rows often lead to strong inter-blade row effects, where the incoming flow disturbances in the form of wakes from upstream components often distort the pressure fields around the blades. In addition to this are the intra-blade row effects where shockwaves and unsteady wakes impinge on the boundary layers of the blade in the forward staggered position. These phenomena often lead to significant aeroelastic instabilities (flutter) as the unsteady flow disturbance frequency closely matches the natural frequency of the blade structures and thus fatigue failures [3].

While fatigue is a phenomena common to both the compressor and turbine stages, flutter has mostly be studied as a compressor-related problem, [4]. Recent studies reported in [5] have shown this to be an issue of concern to turbine stages equivalently. Aeroelasticity as a domain of fatigue problems is herein defined as the study of the mutual interaction between the inertial, elastic and aerodynamic forces acting on blades exposed to an airstream. These related fields were represented by [6] as a triangle of forces including inertial, elastic and aerodynamic forces. However, as noted by [7], in the turbine stages with the added thermal effect, a tetrahedral dimension would often be termed 'aerothermoelasticity' due to the high temperature effects. The problem becomes much more complicated considering the added effect from transonic operations with high exit Mach numbers. As noted by Whitehead [8], for turbine blades unlike compressors, they work in flow velocity well in excess of the critical value which causes additional concern for flutter and vibration failures. This current study therefore investigates the problem of turbine stage flutter using a coupled numerical approach of fluid-structure simulation, for a test case with significant temperature and transonic viscous flow effects.

Over the years, numerical approaches have shown to be viable tools in the understanding of fatigue and flutter instabilities. Marshal and Imregun [7] provided a review of the related numerical methods in turbomachinery aeroelasticity including fluid only aerodynamic computation procedures. They noted previous numerical predictions had considered these problems from two separate domains of fluid and solid computations, with little or no interaction effects being considered. For such case where the feedback effect of the structural dynamics motion on the fluid domain is ignored, it is often regarded as a decoupled approach. On the other hand, studies have have turbomachinery aeroelastic and fatigue problems as highly nonlinear and therefore, are characterised by strong couplings between the fluid and structural domains, which thus require a coupled interaction. As noted by Moffat and He [3], a decoupled simulation would assume a linear aerodynamic damping for which case the blade mode shapes are not affected by the aerodynamic loading. They had compared two predictions using a decoupled and a fully coupled approach. Their result showed the forced response prediction from the coupled method to be significantly lower than the decoupled method. The difference between the two approaches was attributed to the effect of aerodynamic damping present in the coupled method. Thus the induced pressure acting on the blade resulted in a modal damping force which lags the blade motions.

In light of fatigue life estimation, [9] had presented a fatigue life estimation of a 2D airfoil under the influence of high cycle fatigue. This was mostly a parametric study to investigate the influence of major design features on the blade aerodynamic response. The rotor wakes were modelled as rows of Karman Vortices superimposed on an inviscid uniform flow. Their result showed a stable flow induced vibration. This was explained in light of the fact that although the blades in the cascades were under continuous Karman vortices excitation, the level of induced vibration instability was limited by the effect of fluid damping. It is precisely the damping term that limits the amplitude of oscillation of the central blade which resulted in stable flow-induced vibration. It should however be noted that the flow condition here was assumed to be incompressible, two-dimensional and inviscid. Similarly, for the structural blade solutions, they modelled a two-degree of freedom model with assumed plunging and pitching modes of oscillation. Gnesin [10] adopted a nonlinear time marching method to investigate aeroelastic instability on an oscillating blade row. They adopted an unsteady Euler solver to describe the aerodynamic equations. The blade motion was described as a linear combination of mode shapes while the modal coefficients were obtained as solution variables. For their results, they showed the interaction of blade modes and aeroelastic blade response to be essentially nonlinear often resulting in blade limit oscillation.

On turbine related flutter investigations, Sayma and Imregun [11] reported a whole assembly flutter prediction of the low pressure turbine stage. The flow was assumed inviscid with the steady-viscous term accounted for using a distributed loss model. The loss model invariably represents the effect of shear stress by a distributed friction force with the resulting flow being inviscid, but not isentropic. [12], have shown that for a turbine stage, strong viscous effects are able to cause a significant change in the aerodynamic damping of turbomachinery. They had compared the predictions using an Euler code to that from Navier-Stokes solution. While the Euler code prediction showed a damped oscillation for the case considered, the Navier-Stokes solution predicted an excited vibration mode. Roque and Vasco [13] compared the aerodynamic damping of an LPT configuration to that obtained when pairs of blades are forced to vibrate as rigid body modes similar to that obtained from a welded-pair assembly. Thus, the reduction in thickness in blades and disk coupled with increased aspect ratio were noted as likely cause of flutter mechanism in turbine blade stages.

While the above cited works have shown the susceptibility of turbine stages to flutter, very little information exist on the effect of coupled field interactions on the flutter boundaries. While the motion of the structure is controlled by the inertial forces, the structural damping terms, and the excitation from the thermal and aerodynamic stresses are able to significantly modify the global model stiffness matrices. This therefore requires an understanding of the mechanisms of the coupled interactions between the fluid and the vibrating blades. Such are essential for accurate fatigue life estimation. This study uses a coupled fluid-structure simulation to investigate preliminary flutter and forced response behaviour of a turbine blade with added thermal effects.

In a coupled interaction simulation as formulated here, at each solution time step, the aerodynamic field receives an update of displacement and conservative vector variables for the position of the solid boundary. The requirement of such is that the interpolation algorithm must be property conserving in order not to introduce nonphysical dynamics into system. A commercial CFD code with an advanced mesh morphing capability was used in this study. The regions of motion on the fluid-solid interface were specified as displacements boundaries relative to an undeformed initial mesh. This enables the preservation of the mesh quality especially at the inflated boundary layers close to the airfoil wall boundary.

AERODYNAMIC FLOW MODEL

A finite-volume code was used to predict the viscous interactions in the fluid domain. The integral governing equations were discretised as fluxes through a 2-blade row passage. This finite volume technique was initially developed by Jameson et al [14] and later applied on turbomachinery related problems by Denton [15]. The full unsteady Reynolds-averaged Navier-Stokes (URANS) was used with full retention of the viscous terms. The time-averaging of the turbulence quantities are modelled using the SST model. The spatial discretisation was done using the cell-centered scheme together with fourth-order dissipation. The solution was marched in time using the four-step Runge-Kutta approach with convergence accelerated using a time-consistent multi-gird approach. For the related flow boundary conditions; the inlet total pressure and temperature were specified while at the outlet a non-reflective outlet speed was modelled. By using a logarithmic wall function, the blade and endwalls were specified as nonslip boundaries with reasonable wall function resolutions. The phase-lagged periodic boundary conditions at the blade upper and lower sidewall boundaries were modelled using the Fourier transformation approach.

Turbomachinery flutter prediction methods are either done using a time-domain approach or frequency domain approach. For a typical nonlinear time-marching method, they are noted to provide a more robust and comprehensive modelling capability to simulate the unsteady flow features in turbomachines. The method is limited by the required running time to achieve solution convergence to a steady-state. The use of a Fourier transformation approach enhances solution speeds by reducing the number of variables stored for each timestep. For this work, two-blade row passages were modelled while maintaining solution accuracy close to that of a full-stage simulation by the application of phase-shifted periodic boundary conditions. This functionality was proposed by He [16] as a shape correction approach with the time-wise variation in periodic boundary variables approximated using the Fourier series transformation. As such, few orders of Fourier coefficients were stored at the sampling interface instead of a whole transient history of the flow variables. The coefficient storage functionality therefore reduces the computational storage memory thus speeding the solution time. Although this was implemented in the time domain solution, variants of this approach are also used in the frequency domain analyses methods such as harmonic balance methods of [17]. They adopted a frequency domain non-linear harmonic method for forced response prediction of turbine blades of an industrial turbine engine. With only two blade passages modelled in this work, it is assumed the flow variable at the upper periodic boundary and the lower periodic boundary satisfy the phase-shifted (time-lagged) condition. This relationship was represented following that given by He [18] as;

$$U_{Lower}(x,t) = U_{upper}(x,t-\Delta t)$$

Where the time-step value $\Delta t(= {}^{\sigma}/_{\omega})$ with σ as the IBPA, and ω as the blade vibration frequency. By assuming a temporal periodicity, a flow variable $\bar{X}(x,t)$ can be decomposed into a time-averaged flow variable part, and an unsteady part.

To ensure the conservation of the variables at the upper and lower periodic boundaries, the Fourier equations were represented in the form [19];

$$U_{lower}(x,t) = \bar{X}(x,y,z) + \sum_{n=1}^{N_i} [A_n(x)sin(\theta) + B_n(x)\cos(\theta)]$$

$$U_{upper} = \bar{X}(x) + \sum_{n=1}^{N_i} \left[A_n(x)sin\big(n(\beta)\big) + B_n(x)cos\big(n(\beta)\big)\right]$$

Where $\theta = (n\omega t)$ and $\beta = n(\omega t + \sigma)$. The above equations represent the approximation of the Fourier terms assuming both spatial and temporal periodicity. Where N_i represents the order of Fourier series. The terms x,y,z are the three-dimensional coordinate and t is the physical time. The IBPA is represented by the term σ while $\omega(\frac{rad}{s})$ is the angular frequency of the unsteady disturbance. A more detailed representation can also be found in [19] and [20].

This case was used for the validation of the numerical codes used in this study. Based on the design condition of the stage, the initial predictions for the code validation was done for the stator and rotor blades, including the heat transfer Stanton number prediction. In order to properly resolve the wakes from the upstream stator impacting on the rotor, including the shock-waves at the exit flow conditions, a multiblock meshing approach was used. The inner viscous layer in the boundary layer was modelled with the O-type mesh while the H-type mesh was used in the far-field domain areas. An interface mixing plane was specified at the stator outlet and the rotor inlet using a constant velocity approximation. An enhanced wall treatment enabled the resolution of the boundary layer with a reasonable y+ specification. The spatial and temporal error reduction was ensured following the grid convergence index of Roache et al (1986) [21]. A direct order of convergence was obtained by running three solutions with a grid refine ratio of 1.5 while observing the asymptotic behaviour of the flow variable

For the structural model, the basic equation of motion was represented in the form

$$[M]\ddot{X} + [C]\dot{X} + [K]X = F(t)$$

Where M is the mass matrix, C is the system damping (mechanical and aerodynamic), K is the stiffness matrix and X represents the displacement of the body due to the source of excitation F(t). The excitation force was modelled as a function of time using the simple harmonic motion function $\{F(t)\} = \{F\}\varrho^{i\omega t}$. Similarly, the resulting sinusoidal response can also be represented in the form $\{X(t)\} = \{X\}\varrho^{i\omega t}$. By introducing a mass normalised mode shape and a scaling factor, the structural equation of motion can be represented in the form;

$$-\omega^2\psi^*[M]\psi + i\omega\psi^*[C]\psi + \psi^*[K]\psi = \psi^*F(t)$$

For which case the term ψ is the mass normalised mode-shape. By multiplying with a scaling term, x, the vector displacement can be obtained as $X = \psi.x.\varrho^{i\omega t}$. The definition of mass normalisation used here is similar to that found in [22] where;

$$\psi^*[M]\psi = 1, \quad \psi^*[C]\psi = 2\xi\omega_n, \quad \psi^*[K]\psi = \omega_n^2$$

The term ξ is the damping term and ω_n is the natural frequency of the model. The damping due to the excitation source has both imaginary and real component. The imaginary part is decomposed to the aerodynamic damping term. Thus a positive damping factor means the blade is imparting energy to the flow, while a negative value signals a destabilizing mode depending on the magnitude of the mechanical damping term. The work done by the aerodynamic force for a given cycle of blade vibration is estimated in the term $\psi^*F(t)$.

For the structural equations, the blades were discretised in a finite element solver where the local displacements and loads are conveniently represented at nodes. A solid mesh element with 20 degree of freedom was used on the solid blade model with prescribed displacement and rotational velocity as boundary conditions.

Three isolated modes with natural frequencies (ω_n); 4437Hz, 6145Hz and 10321Hz were investigated for this study. This due to the fact that for turbomachinery blades, the large mass of the blade compared to the mass of fluid in the blade volume causes an insignificant variation of the aerodynamic coupling between the modes, thus the modes can be considered as vibrating in a single degree of freedom which is either torsion or bending, Whitehead, [8]. A free-vibration analysis was solved initially with the predicted initial mode-shapes interpolated on the CFD mesh to investigate aeroelastic flutter boundary for each mode.

TEST CASE

The test case considered for this analysis was the vane blade interaction case. The experimental measurements were originally reported in Dunn et al [23]. In their work, they presented a time-averaged and phase-resolved measurement of the pressure and heat-flux for two cases with varying vane positions. The related stage parameters for the case considered in this work are shown in Table 1. This was a transonic case with stator exit Mach number of 1.121. The unsteady aerodynamics and heat transfer measurement data for this case were also reported in Rao [24] [25].

Table 1: Vane Blade Interaction Turbine Stage Parameters

Parameters	Stator	Rotor
Rotor Speed (RPM)	----	11400
Inlet Mach number	0.164	0.483
Exit Mach number	1.121	1.054
Number of Blades	30	45
Chord, (mm)	66	46
Inlet Total Temperature (K)	522	---
Inlet Total Pressure (kPa)	303.4	---
Mean Radius, (mm)	241	239
Tip Radius, (mm)	265	265

In the work by Dunn [23], the measurements were obtained from a shock-tunnel facility with the flow generated through a short-duration source with heated and pressurised air. The turbine stage configuration was reportedly constructed by the Allison Gas Turbine Division. The two configurations considered were known as open and closed vane positions. For this work, the flow conditions similar to the latter were considered. This had a stage pressure ratio of 3.4 with a turbine weight flow of 10.20 kg/s. The corrected shaft speed was 11400 rpm for a test duration of approximately 35-45 ms. A more detailed explanation of the measurement facility and the test flow conditions can be found in the cited works.

For the heat transfer prediction, a wall temperature boundary condition was specified on the blade walls for both rotor and stator blades, similar to that reported in [25]. The surface finish was resolved by specifying a roughness factor of 2μm. This was to model the effect of flush mounting of the heat flux gauges which were installed to surface finish of 2.5 x 10^{-6}m. It was also reported that surface inserts were mounted at the leading edge position to capture the high gradient heat transfer region. Non-reflective boundary conditions were specified at the inlet and outlet while a phase-lagged periodic boundary was used to reduce the computational domain to two passages. The full-stage isolated blade rows were modelled for the code validation while only the rotor passages were modelled for the flutter prediction.

RESULTS

The predicted steady-state pressure coefficients are shown in figures 1 and 2 for the rotor and stator blades respectively. Also shown are the measured data from the experimental work of [24]. This was performed for the closed vane position with a transonic exit flow condition. From the predicted fluid-flow results, the effect of shock wave in the passage resulted in significant unsteady flow behaviour on both the rotor and stator blades. This unsteadiness can be seen in the figure 1 and 2 plot at the wetted chord distance across the suction surfaces of both blades. The regions between 0 to 1 (wetted chord distance) represent the suction surface, while the region from 0 to 0.5 is the pressure surface for both airfoils. For the rotor blade, while a relatively steady pressure profile was maintained on the pressure surface, a reasonable close match with the experimental data was predicted at this region. The code shows an under prediction at the suction surface region closed to the trailing edge. The effect of the shockwave reflections was more predominant on the suction surface. This unsteady shock effect could be seen in the Schlieren plot of Figure 3, with both in-passage shock and an oblique shock emanating from the trailing edge region of the neighbouring blades. A similar effect was noticed for the stator blades which also showed significant flow unsteadiness on the suction side. The shock was stronger in the stator domain given the higher Mach number of its exit flow as compared to that of the rotor. This effect is explained further in the Stanton number prediction plot of Figure 4 and 5.

Thus, noting that shock waves are irreversible and adiabatic, their effect was much noticed as they introduce some form of entropy (dissipation) in the flow. Such high entropy values can used to evaluate the shock strength based on the Prandtl-Meyer shock function. For the stage considered, although the local Mach numbers are high, the generally predominant oblique shock waves often resulted in reduced ratio of predicted static pressure more specifically for the rotor domain with significant pressure loss.

The Schlieren plot, Figure 3 obtained through density gradient contours clearly shows the movement of the shocks across the passages. These are for the flow passage stator and rotor blades, with the axis of rotation being the Z-axis. Region-A on the stator blade shows the in-passage shock impinging on the boundary layer. This invariably reduces the thickness of the boundary layer in this region with another reflected shock at the blade trailing edge. Regions B and C are the shock reflections in the rotor passage. While B is the point where the passage shock is convected downstream in the rotor domain, the reflection of the shock at point C is also shown to weakly impinge on the suction surface of the neighbouring blade. Such effects will be strongly influenced by the inter-blade row spacing and the stagger angle of the airfoils.

While a detailed flow field analysis is outside the scope of this work, there was also noticed an endwall secondary flow features in the rotor domain. This resulted in the separation of the flow at about 10-15% span position from the hub region. The flow was purely three dimensional as the flow seems attached at the upper airfoil spanwise regions. A recirculating region of dead air cell was noticed at the lower hub separation region. Thus further analysis of this case will be needed to qualitatively state the interactions of the shockwave reflections in the passage and the secondary horseshoe vortex predominant at the hub region. Added to this complexity is the three-dimensional nature of the tip leakages which was not considered in the present work.

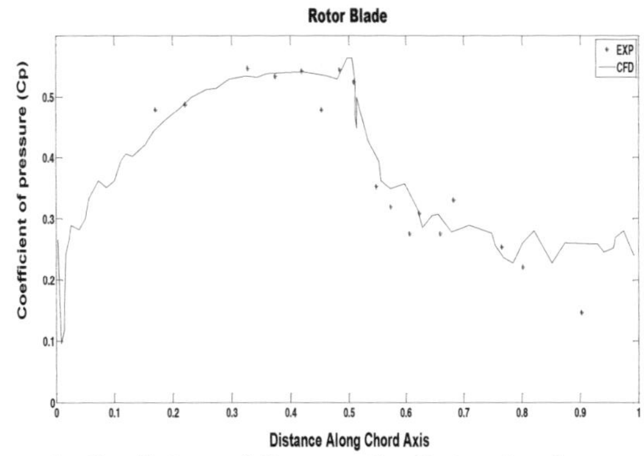

Figure 1: Coefficient of Pressure Prediction for the rotor blade

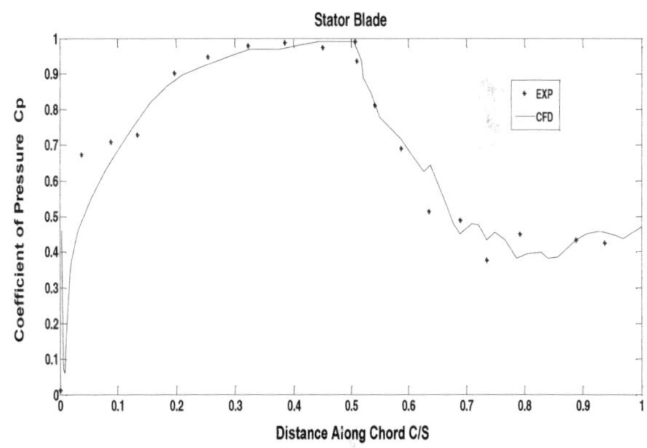

Figure 2: Coefficient of Pressure prediction for the stator blade

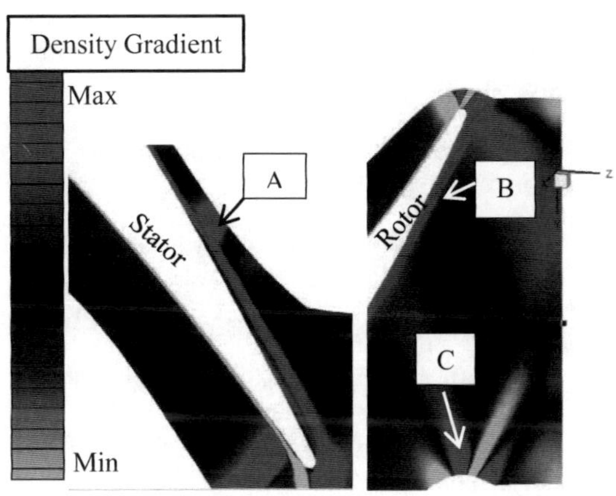

Figure 3: Numerical schlieren plots at trailing edge regions of stator and rotor showing regions of Shock-wave reflections

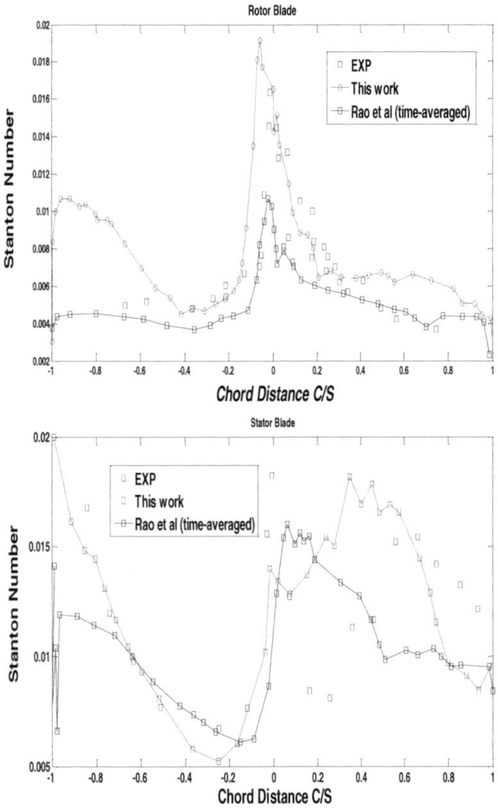

Figure 5: predicted Stanton number distribution for the rotor and stator blades respectively

Figure 5 shows the computed Stanton number distribution for the rotor and stator blades respectively. Also plotted were the 2-D predictions reported in Rao et al [25]. They had compared the Navier-Stokes prediction using an explicit Runge-Kutta scheme for both time-averaged and unsteady predictions. As shown in the figure, the effect of the three-dimensional flow features on the stage is quite noticeable for the rotor blades as well as for the stator blade. The wetted distance -1 and 1 in this case being the trailing edges while the leading edge is at position 0. The predictions are shown for the blade midspan position. The pressure side is to the left of the curves while the suction side is the region between points 0 to 1. As reported in Rao et al [25], the dip noticed at about 0.1 wetted distance position on the suction side of the stator was the laminar-turbulent transition region which was not clearly captured in this predictions, although the flow specified as fully turbulent. The elevated predictions of heat transfer at region 0.2-0.4 were associated with the freestream turbulence. On the rotor blade however, the impinging shock from the upstream stator causes the localised increase in the Stanton number predicted at the leading edge. The coming wakes from the upstream however, was shown to not result in much unsteadiness in the passage but rather assumed a stabilising role. This is evident by the much better agreement with experimental data noticed at this point. In general, the capability of the numerical code is shown by the close match with experiment while the strong shock effect in the stator vane passage resulted in the mismatch noticed over suction side regions. As noted earlier, the increasing Stanton number at the stator 40-50% suction surface could be attributed to the shock effect impinging on the boundary layer. On the pressure surface, close to the trailing edge, the impinging shock emanating from the neighbouring blade trailing edge causes the increasing heat transfer in those regions coupled with the flow approaching a Kutta condition at the trailing edge. In conclusion, the plots show a strong correlation of the predicted blade loadings with experimental data thus giving an informed capability of the code used in this study. The related influence of such unsteady flow field including the shock reflections on the blade aeroelastic instabilities are reported in the following sections.

FLUTTER PREDICTION

For flutter related issues, the blade instability is usually defined in terms of the aerodamping coefficient. This is defined as the negative of the blade work coefficient during one cycle of oscillation. While the aerodynamic damping coefficient could conveniently describe the energy exchange between the flow and the blade, its sign interpretation in this case is used as the condition for self-excited vibrations. It should also be noted that the blade response will also depend on the blade kinetic energy and the flow acting on the blade. For a typical mode of vibration, the aerodynamic damping coefficient was estimated using the relation;

$$\int_{t=i}^{\Delta t}\int \frac{p\vec{v}.\vec{n}dadt}{m\omega^2 A_{max}^2}$$

Where the numerator is the work coefficient term defined as the energy imparted on the blade by the flow for each vibration cycle. The denominator denotes the blade energy estimated as a function of the blade mass, angular frequency and the surface area. The blade pressure and velocity were integrated over the area to estimate the work coefficient. A similar expression for the torsion mode was given by Gnesin [26]. For this work, two isolated rotor passages were modelled for the flutter instability prediction using the Fourier domain transformation. The instabilities were predicted for the first three vibration modes of the blade. Figure 6 shows the estimated aerodynamic damping ratio for various inter-blade phase angles. The ratio of the bending to torsion frequency considered in this case was 0.72, thus signifying a strong coupling between the bending and torsion modes of vibration. The blade elastic axis and C.o.G position was located at about 50% of the chord position. For an interblade phase of 180^0, the two blade passages are assumed to deform in opposite directions with amplitudes computed as a result of the oscillations. The IBPA are also related to the nodal diameters by the correlation; $IBPA = {}^{2\pi n_d}/_{N_R}$ where n_d is the nodal diameter number and N_R is the number of rotor blades.

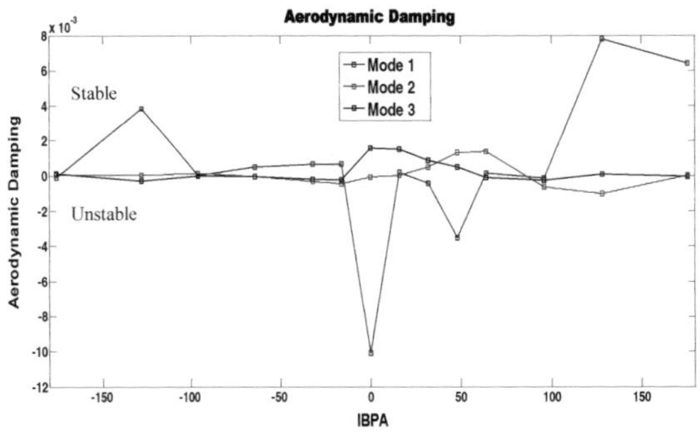

Figure 6: Blade aerodynamic damping vs IBPA

The plot shows the blade to be mostly stable in the backward traveling wave modes where the estimated aerodynamic blade damping is positive. The plots are shown for the first three vibration modes of reduced frequencies (K) defined by $K = {}^{\omega(c/2)}/_{v}$ where v is the outlet velocity, c is the blade length and ω is the angular frequency of vibration of the blades. It signifies the ratio of the time taken for a given fluid particle to convect through the blade passage to that which the rotor moves pass one upstream stator passage. Thus, the higher blade modes will have increased reduced frequencies as compared to lower vibration modes. A decreasing value of the reduced frequency will tend to destabilise the flow and vice versa. As shown in the plot, the lower vibration frequency (mode 1) seems to be the most unstable of the three modes. The negative IBPA values representing the backward traveling wave modes while the positive IBPA are the positive traveling wave. The region of highest instability however is predicted at the neutral traveling wave mode of 0 IBPA. Another region of aerodynamic instability was noticed at the 50^0 IBPA for the first mode, while low damping coefficient for mode 2 was noted at approximately 130^0 IBPA. The higher reduced frequency value of 2nd and 3rd vibration modes tends to stabilise the blades aerodynamically. Flutter instabilities were more likely for mode 1 depending on the magnitude of the structural damping. The variation of the predicted aerodynamic damping for the various IBPA also signifies that flutter instability for turbomachinery blade rows has a strong aerodynamic coupling as opposed to wing flutter. Thus the blades vibrate circumferentially in a traveling wave manner clearly determined by the nodal diameter lines. While the reduced frequency of the model will be considered higher than typical critical values ($K < 0.5$), the case is still prone to flutter instability. These instabilities can be attributed to the nonlinear effects of shocks predicted within the passage. Thus, as the speed of flow increased across the passage, the vibration characteristics alter gradually at amplitudes that is controlled by the nonlinear flow effects.

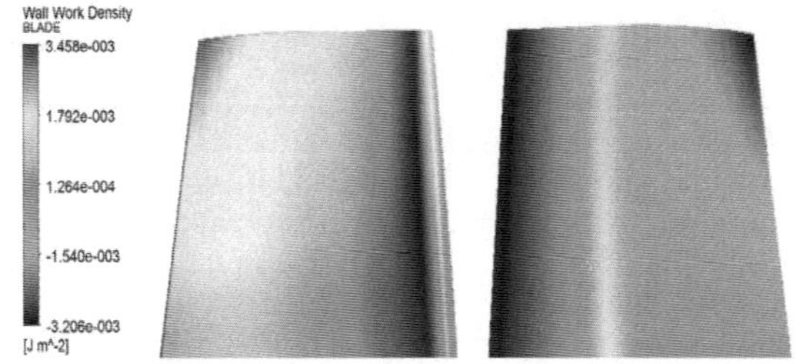

Figure 7: Wall work density for the pressure and suction sides

In most cases of turbomachinery flutter analysis, the effect of shock wave on the pressure amplitude variation has been noted as the leading cause of flutter instabilities. As noted by [27], a transonic flow across an airfoil blade will create thin boundary layers around the profile. While such effects on a typical compressor airfoil will cause a free transition shortly after the leading edge, for a turbine blade, this is known to offset transition at the point of maximum velocity. Similarly, Srinivasan [1] had noted that with the resulting effect of shockwaves on thermodynamic properties of the blades such as density and pressure, the decreasing effect of the shockwave reflections on pressure terms will reduce the stage unsteady work and thus possibly lead to an unstable mode. Figure 7 shows the work per unit area of the blade for the pressure and suction side of the blade at IBPA of 0^0 in the forward traveling wave mode. It shows the region of the blade with the highest expansion work per unit area. Regions of local maximum and minimum were predicted for the trailing edge tip location of the both surfaces. This is also the region of maximum displacement of the airfoil. The figure also represents the spanwise distribution of the loading on the airfoil. The airfoil is highly loaded in the upper half span location on the pressure side. From figure 6 as shown earlier, this is relatively a stable depending on the magnitude of the structural damping. Thus flutter instability will then be a function of the energy balance between the unsteady forcing by the fluid, inertial and damping forces of the structure as well as the elastic forces of the structure which acts to counteract the unsteady pressure force.

While the study shown here only considers the rotor blade flutter without accounting for any effects of the upstream stator, recent studies have shown the significant alteration of the flutter stability regions of low pressure turbines due to effects of an upstream blade row. He and Quang [20] have investigated the influence of upstream stator on a LP steam turbine flutter. This was done to investigate the effect of inter-blade row on the acoustic wave reflections [20]. Similarly, Waite and Kielb et al [28] investigated the effect of back pressure on flutter instability of the STCF-04 configuration. Further details on the influence of operating conditions on LP flutter can also be found in [29].

FORCED RESPONSE AND FATIGUE ESTIMATION

To investigate the effects of the unsteady aerodynamic fields on blade fatigue life, the resolved Fourier transforms of the pressure amplitude are shown in figure 8. For typical blade fatigue problems, HCF normally arises when the vibration levels induced by the flow disturbances exceed the material's endurance limit. While current industrial approaches would identify possible resonances using the Campbell diagram, it is virtually impossible to avoid all resonances for all the blade's natural frequencies. The plot shows the unsteady work amplitude variation for the 1st mode at 0 IBPA. This is necessary to consider given that for low excitation frequencies it is likely that the fundamental acoustic mode (IBPA=0) will propagate. As frequencies increased, depending on the acoustic resonance wave, higher mode frequencies are also prone to propagate, [28]. A potential damage region was noted for the higher engine order excitations. For instance, at 36th stator passing period the EO excitation frequency ($f = {}^{EO * \omega}/_{2\pi}$) estimated as 6840 Hz was closed to the 2nd vibration mode frequency of the blade although at much lower amplitude. The significant peak shown in the figure was away from the first 3-vibration modes frequency

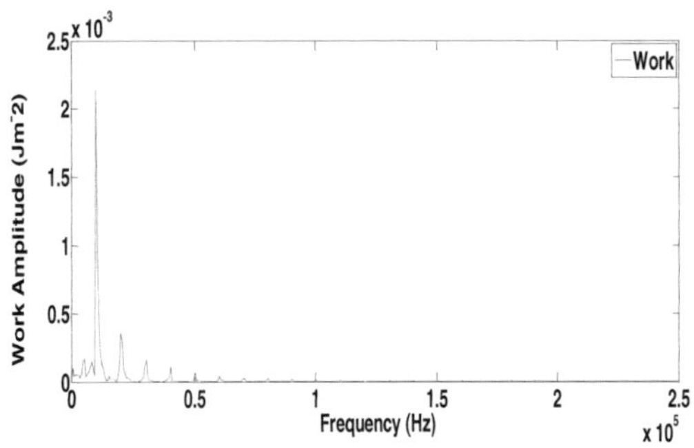

Figure 8: Fourier transform of blade unsteady work

The related stress magnitude was estimated using a coupled fluid-interaction code with added thermal effect in a fully coupled simulation. A finite element code was coupled to a finite volume fluid code with information (displacement, forces, temperature and heat flux) exchanged at the fluid-solid interface. The code was used to model the static loads due to blade rotation at constant speed value of 11400 rpm while the dynamic loads were coupled through time-step iterations to the finite element solver. It was expected that with the added static loads due to temperature, rotation, and unsteady aerodynamic forces, they would significantly reduce the ability of the material to sustain more dynamic stresses. The added rotation and thermal effects however resolved to a static nonlinear problem of the form

$$F_r + F_t = [K].X$$

Where X is the static displacement of the blade, [K] is the stiffness matric while F_t, F_r, are the thermal, rotational forcing functions which will also in addition to the unsteady aerodynamic load alter the stiffness matrix in the blade energy equation. These forces generate a deformation of the structure dependent on the coordinate and associated mass of the system. The resulting deformation was feedback into the fluid dynamics to re-calculate the aerodynamic and convective heat transfer coefficients. Thus for each solution time-step, the deformation was computed and the displacement matrix of the structural equation updated. Also, the aerodynamic damping term from the aerodynamic fluid code was added to the damping matrix while the structural damping was estimated based on the elastic properties of the blade

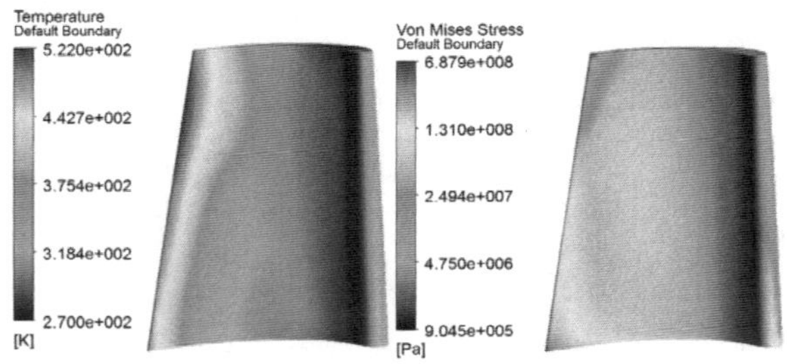

Figure 9: Temperature and Von Mises stress on rotor blade

Figure 9 shows the predicted stress and temperature contour plots on the rotor blade. While the aerodynamic pressure loads would often induce bending moments at the leading and trailing edge of the blade, the predicted directional displacements were predominant in the tangential span-wise direction indicating the dominant influence of the rotational acceleration on the blade. However, with closer axial gap machines, the aerodynamic load is expected to transfer a significant amount of energy to the blade. A local hotspot was predicted for the trailing edge due to the reduced thermal mass of the blade at this region. While the effect of rotational load was evident in the high stress noticeable at the hub region, the predominant stress was mostly from the thermal heat conduction. This is clearly visible by the spanwise distribution of the stress across the pressure mid surface area. Although not shown here for conciseness, the rotational load was the dominant stress term at initial solution time-step while the effect of the thermal load was noticeable as the blade metal temperature evolved approaching an adiabatic wall condition.

CONCLUSION

A coupled fluid interaction study was used to investigate the flutter and forced response of the blade typical of a turbine stage. From the flutter investigation it was shown that the most susceptible mode was the first vibration bending mode at neutral traveling wave mode. The flutter characteristics of the blade, while strongly dependent on the inter blade phase, were also affected by the reduced frequency term. Although the estimated reduced frequency value of the case considered were quite high (due to low mass and aspect ratio), the flutter instabilities were mainly due to shock reflections which significantly reduce the useful blade work. The effect of unsteady pressure fluctuation was not a structural concern in this case as the high amplitude frequency was predicted away from the potential damage modes. With the coupling of temperature and rotational loads, the dynamic response of the blade showed the thermal loads to cause localised stress regions as the blade wall temperature approached an adiabatic condition.

NOMENCLATURE

IBPA	interblade phase angle
EO	engine order
F	frequency
ω	blade natural frequency
K	blade stiffness matrix
Fr	rotational forcing term]
Ft	Thermal (flux) forcing term
X	displacement
m	blade mass
A	area
$\vec{v}$	velocity vector
p	surface pressure
da	change in area
dt	unit time
ψ	modal matrix term
M	mass matrix
k	reduced frequency

ACKNOWLEDGMENTS

This work was supported by the Australian International Postgraduate Research Scholarship (IPRS) scheme. Part funding was also provided by the Total E&P Nigeria Limited, Port Harcourt District.

REFERENCES

[1] A.V. Srinivasan, "Flutter and Resonant Vibration Characteristics of Engine Blades," in *The 1997 IGTI Scholar Paper. Journal of Engineering for Gas Turbine and Power, Volume 119, page 741-775*, Orlando, Florida, 1997.

[2] Anubhav Datt and Wayne Johnson, "An Assessment of the State of the art in Multidisciplinary Aeromechanical Analyses," Report ID: 0704-0188, USA, 2008.

[3] S. Moffat and L. He, "On Decoupled and Fully-Coupled Methods for Blade forced response Prediction," *Journals of Fluids and Structures,* vol. 20, pp. 217-234, 2005.

[4] Max. F. Platzer and Franklin O. Carta, "AGARD MANUAL ON AEROELASTICITY IN AXIAL-FLOW TURBOMACHINES VOLUME 2 STRUCTURAL DYNAMICS AND AEROELASTICITY," NORTH ATLANTIC TREATY ORGANIZATION- AGARD, ESSEX, 1988.

[5] M. Schmitz, O. Schafer, J. Szwedowicz and T. Secall-Wimmel, "Axial Turbine Blade Vibrations Induced by Stator Flow," in *Unsteady Aerodynamics, Aeroacoustics and Aeroelasticity of Turbomachines*, Duke, North Carolina, USA, Springer, Dordrecht, The Netherlands, 1996, pp. 107-118.

[6] A. R. Collar, "The Expanding Domain of Aeroelasticity," *Journal of Royal Aeronautical Society,* vol. 50, pp. 613-636, 1946.

[7] J. G. Marshall and A. Imregun, "A Review of Aeroelasticity Methods with Emphasis on Turbomachinery Applications," *Journals of Fluids and Structures,* vol. 10, pp. 237-265, 1996.

[8] D.S. Whitehead, "Bending Flutter of Unstalled Cascade Blades at Finite Deflection," Publication of Aeronautical Research Council, UK Ministry of Aviation; R&M No: 3386, UK, 1965.

[9] Y.L.Iau, R.C.K Leung and R.M.C. So, "Vortex-Induced Vibration Effects on Fatigue Life Estimation of Turbine Blades," *Journal of Sound and Vibration,* vol. 307, pp. 698-719, 2007.

[10] J. S Green, M. Imregun, A.I.Sayma and M. Vahdati, "Whole-Assembly Flutter Analysis of Low Pressure Turbine Blade," Publications of Imperial College, London, Undated.

[11] Xiuquan Huang, "Three-Dimensional Unsteady Flow in Oscillating Turbine Blade Row," Thesis of Durham University, Durham, UK, 2006.

[12] Roque Corral, Nelida Cerezal and Carlos Vasco, "Flutter Boundaries of Pairs of Low Pressure Turbine Blades," in *Unsteady Aerodynamics, Aeroacoustic and Aeroelasticity of Turbomachines*, Dordretch, Netherlands, Springer, 2006, pp. 3-16.

[13] R.J Zwaan and B. B Prananta, "Fluid-Structure Interaction in Numerical Aeroelastic Simulation," *International Journal of Non-Linear Mechanics,* vol. 37, pp. 987-1002, 2002.

[14] A. Jameson, "The Evolution of ComputationalMethods in Aerodynamics," *Journal of Applied Mechanics,* vol. 50, pp. 1052-1060, 1983.

[15] J. Denton, "The Calculation of three dimensional viscous flow through multi-stage turbomachines," *35th International Gas Turbine and Aeroeengine Congress and Exhibition,* pp. 90-GT-19, 1990.

[16] Staurt Moffat, "Forced Response Prediction for Industrial Gas Turbine Blades," PhD Thesis of Durham University, Durham, UK, 2006.

[17] Li He, "Unsteady flow and Aeroelasticity," Publication of University of Durham, UK, Undated.

[18] Nicola Donini, "Aeroelasticity of Turbomachines Linearised Flutter Analysis," Thesis of Politecnico Di Milano, Milan, 2012.

[19] Gabriel Saiz, "Turbomachinery Aeroelasticity Using a Time-Linearised Multi Blade-Row Approach," A PhD Thesis of University of London, Imperial College, London, UK, 2008.

[20] M.G.Dunn, W. A. Bennett, R.A. Delaney and K. V. Rao, "Investigation of Unsteady Flow Througha Transonic Turbine Stage: Data/Prediction Comparison for Time-Averaged and Phase-Resolved Pressure Data," *Journal of Turbomachinery,* vol. 114, pp. 91-99, 1992.

[21] K.V.Rao, R.A.Delaney and M.G.Dunn, "Vane-Blade Interaction in a Transonic Turbine , Part 1: Aerodynamics," *Journal of Propulsion and Power,* vol. 10, no. 3, pp. 305-311 , 1994.

[22] K. Rao, R. Delaney and M. G. Dunn, "Vane blade interaction in a transonic turbine. Part II: heat transfer," *Journal of Propulsion and Power,* pp. 312-317, 1994.

[23] Roache P. J, K Ghia and F. White, "Editorial Policy Statement on the Control of Numerical Accuracy," *ASME Journal of Fluids Engineering,* vol. 108, no. 1, p. 2, 1986.

[24] Vitaly Gnesin, Lyubov Kolodyazhnaya and Romauld Rzadkowski, "Coupled Aeroelastic oscillations of a Turbine Blade row in a 3D Transonic Flow," *Journal of Thermal Science,* vol. 10, no. 4, pp. 319-326, 2000.

[25] Davy-Allegret-Bourdon, "Experimental Study of Fluid-Structure Interactions on a Generic Model," A Licentiate of Engineering Thesis, Royal Institute of Technology (KTH), Sweden, 2004.

[26] X.Q. Huang, L.He and D.L.Bell, "Influence of Upstream Stator on Rotor Flutter Stability in Low Pressure Steam Turbine Stage," *Journal of Power and Energy,* vol. 220, no. A1, pp. 25-35, 2006.

[27] Joshua J Waite and Robert E Kielb, "Physical Understanding and Sensitivities of Low Pressure Turbine Flutter," *Journal of Engineering for Gas Turbine and Power,* vol. 137, no. 012502, pp. 1-9, 2015.

[28] Joshua J. Waites, Robert E. Kielb and Simon L. Bittner, "The Influence of Steady Loading Parameters on Low Pressure Turbine Flutter," in *56th AIAA/ASCE/AHS/ASC Structures, Structural Dynamics and Materials Conference, AIAA 2015-1417*, Kissimmee, Florida, January 2015.

[29] V. Gnesin and R. Rzadkowski, "A coupled fluid-structure analysis for a 3-D Inviscid flutter of IV standard Configuration," *Journal of*

Sound and Vibration, vol. 251, no. 2, pp. 315-327, 2002.

[30] Hirofumi Doi, "Fluid/Structure Coupled Aeroelastic Computations for Transonic flows in Turbomachine," Standford University, USA, 2002.

[31] L.He, "An Euler Solution for Unsteady Flows Around Oscillating Blades," in *The American Society of Mechanical Engineers 89-GT-279*, New York, 1989.

[32] Lei Qi and Zhengping Zou, "Unsteady Flows in Turbines, Noise Control, Reduction and Cancellation Solutions in Engineering," ISBN: 978-953-307-918-9, 2012.

[33] Andrew Halfpenny, "A frequency Domain Approach for Fatigue Life Estimation from Finite Element Analysis," in *A Paper presented at the International Conference on Damage Assessment of Structures*, Dublin, 1999.

[34] N.W.M.Bishop, L.W. Lack, T. Li and S.C. Kerr, "Analytical Fatigue Estimation of Vibration Induced Fatigue Damage," Publication of University College London, nCode International, London, Undated.

Proceedings of the 14th International Symposium on Unsteady Aerodynamics,
Aeroacoustics & Aeroelasticity of Turbomachines
ISUAAAT14
8-11 September 2015, Stockholm, Sweden

I14-S7-5

TIP CLEARANCE INFLUENCE ON AEROELASTIC STABILITY OF AN AXIAL TURBINE BLADE ROW

Mateus Augusto Malta Teixeira
Royal Institute of Technology
Department of Energy Technology
S – 10044 Stockholm, Sweden
mamt2@kth.se

Paul Petrie-Repar
Royal Institute of Technology
Department of Energy Technology
S – 10044 Stockholm, Sweden
paul.petrie-repar@energy.kth.se

Shogo Kameyama
University of Tokyo
Department of Aeronautics and Astronautics
Tokyo, Japan
kameyama@aero.t.u-tokyo.ac.jp

ABSTRACT

The present study investigates the influence of tip clearance on the aeroelastic behaviour of an oscillating low pressure turbine (LPT) blade row. The investigation was performed through full-scale time-marching RANS CFD simulations employing models with and without tip clearance. The unsteady blade row results were obtained in ANSYS CFX® v.15 using a time domain solution approach with periodic phase-shifted boundaries using Fourier coefficients allocation. Three tip clearance cases were investigated: no gap and tip clearances of 1% and 3% span. The influence of the mesh resolution in the tip clearance was investigated.

The effect of increasing the size of the tip clearance on the steady flow was found to be non-linear particularly on the suction side. There was a significant drop in the pressure at 95% span on the suction side near mid-chord for 3% tip clearance but not the 1% tip clearance. The stronger tip clearance flow for the 3% gap creates a region of high Mach number on the suction side that is not present for the 1% gap.

The tip clearance flow does not appear to have a significant influence on the aerodynamic damping for the AETR rig geometry at the prescribed flow condition. The least stable mode was IBPA = -45° for all tip clearance cases investigated. The unsteady work done on the blade was investigated in more detail by examining plots of aerodynamic damping versus span and the local work coefficient on the blade at various blade heights. Differences in the local work were observed, particular near the tip. No trend could be observed in the differences on the suction side, however, a trend was observed for the pressure side. There was a significant stabilizing shift in the work coefficient on the pressure side for 3% tip clearance near the tip.

INTRODUCTION

Aerodynamic loading in turbomachinery blades has continuously increased with the development of modern advanced high performance aero-engines. Concurrently, with the drive towards thinner and lighter blades increases the engine's susceptibility to flow induced vibrations such as flutter. Flutter is defined as an unstable and self-excited aeroelastic phenomenon resulting from the coupled action between unsteady aerodynamic forces and blade motion: once the blade starts to vibrate, it induces unsteady aerodynamic forces, which in turn will act back on the blade surface. Such fluid-structure interaction leads to a net energy exchange per period between the vibrating blade and the working fluid and might result in structural failure within a short period of time. Turbomachinery flutter investigation still represents a challenging research field.

Typically, unsteady CFD calculations are performed as part of the flutter analysis and the user must decide the level of the fidelity of the CFD calculation. This is always a trade-off between accuracy and computational effort. It is typical nowadays to perform 3D URANS calculations for a flutter analysis but it is not standard to include tip clearance flows. The influence of the tip clearance flow on the flutter analysis can be critical for some turbomachinery applications. For example, if it is assumed that the deflection of the critical flutter mode is proportional to the square of the span height and that the aerodynamic damping (normalized by local deflection) is constant as a function of span, then half of the unsteady flutter work is done in the top 12.9% of the blade. If the flow near the tip is transonic then the tip clearance flow will strongly influence the flow near the tip which will also influence the unsteady work near the tip and the flutter stability of the whole blade.

Tip leakage flow originates from the pressure side over to the suction side of the blade, creating the so called tip clearance vortex, which in turn interacts with the main flow field. Characterized by strong viscous forces and high velocity gradients, this region of the blade is related to energy losses and directly affects engine efficiency. The influence of tip clearance flow on flutter analysis has been investigated previously.

Fu et. al. [1] investigated the tip clearance effects on the aeroelastic stability of transonic axial compressor blades using two independent three-dimensional flutter prediction approaches (energy method and aeroelastic eigenvalue analysis). Tip clearances in the range from 0.17 mm to 2.70 mm were investigated and it was concluded that the aeroelastic stability was not monotonic with the size of the tip clearance. The least stable tip clearance predicted by both methods was 1.69mm

Glodic et. al. [2] studied the influence of tip clearance modeling in the predictions of the unsteady flow response in an oscillating low pressure turbine (LPT) cascade with low subsonic outflow Mach number ($M_2 = 0.4$). The study validated different numerical models (full-scale time-marching RANS CFD simulations) against experimental data (experimental sector cascade with one oscillating blade in axial bending mode). The authors had obtained an equivalent result for the imaginary part of the complex unsteady force up to 70% span on their model without tip clearance when compared to the nominal tip clearance case (with size of 1% span, or 0.97mm). Differences relative to the nominal tip clearance model at 90% span section amounted to about 25%. In addition, it was concluded that the resolution of the tip clearance mesh in the spanwise direction of the gap might have a considerable impact on prediction accuracy. Nonetheless, the results have not shown significant variations in aeroelastic stability when small variations in the tip clearance size were imposed (0.7%, 1% and 3% span gap sizes, 0.68mm, 0.97mm and 2.91mm respectively).

In contrast, experiments by Huang et. al. [3] found that the tip gap size had a significant influence on the unsteady flow response for an oscillating linear turbine cascade. The authors observed that stabilizing contributions could be observed near the tip for small tip clearances (1.25% chord or 1.79mm), while a bigger gap size (5% chord or 7.15mm) led to prevalent destabilizing contributions at the aft suction surface associated with a well-developed vortex.

The effects of tip clearance flow on turbomachinery blade flutter is probably highly case dependent and more work is required to validate CFD methods for the prediction of aerodynamic damping (flutter stability) when tip clearance flows are important.

The aim of the current work is to investigate whether the Aeroelastic Turbine Rig (AETR) at KTH can be used to investigate the influence of tip clearance flows on turbine flutter. Steady and unsteady CFD simulations were performed using the standard turbine geometry from the AETR. The flow conditions examined correspond to the maximum flow rate (5kg/s) currently possible in the facility. Two tip clearance sizes: 1 and 3 mm were investigated. The aim is to determine whether the different tip clearance sizes would result in significant differences in the measured unsteady pressure. If a significant difference in the unsteady pressure is predicted, the AETR could be used to obtain useful experimental data for the validation of CFD methods for flutter prediction when tip clearance flows are important.

NOMENCLATURE

A	Fourier coefficient
b	blade span
c	blade chord
c_p	static pressure coefficient
	$c_p = \dfrac{p - p_{ref}}{p_{0,ref} - p_{ref}}$
f	flow quantity
i	imaginary unit
k	blade index
M	Mach number
N	number of blades
$\vec{n}$	normal vector to surface element
p	pressure
S	surface
t	time
T	period of oscillation
$\vec{v}$	velocity vector
W	work per cycle

Greek symbols

α	nondimensional blade's maximum displacement
σ	interblade phase angle
ω^*	reduced frequency
ϕ	phase angle

Abbreviations

AETR	AeroElastic Turbine Rig
CFD	computational fluid dynamics
IBPA	interblade phase angle
LPT	low pressure turbine
PS	pressure side
SS	suction side
RANS	Reynolds Averaged Navier Stokes
RMS	root mean square
TBR-FT	Transient Blade Row Fourier Transformation Method

TWM	traveling wave mode
URANS	Unsteady Reynolds Averaged Navier Stokes

Subscripts

0	total quantity
1	quantity at the inlet
2	quantity at the outlet
aero	aerodynamic
ref	reference

METHODOLOGY

The present investigation is carried out numerically using a full-scale time-marching 3D viscous model to obtain the solution of the URANS equations in the time domain. All the flow field features, including nonlinearities, can be included in such unsteady flow solution. With the concept of traveling wave mode and IBPA, a fixed phase difference in the unsteady flow field between two consecutive blade passages can be established. If this phase is known *a priori*, periodic boundary conditions can be employed in conjunction with the use of a limited number of passages to model an entire blade row, which significantly reduce the computation time of the simulation. Efforts in the development of such an idea are attributed to Erdos et al. [4] (who proposed the Direct Store method), Giles [5] (with the Time-inclined method) and He [6], who established the Shape Correction method.

The cascade is modeled numerically by employing the commercial computational fluid dynamics code ANSYS CFX® v.15 with the Transient Blade Row Fourier Transformation (TBR-FT) [7] method. The potential of using the TBR-FT method for unsteady aerodynamic investigations was shown by Camara [8]. Elder et. al. [9] also points out that the TBR-FT method provides solutions 3.5 times faster than average periodic reference cases.

In order to assess the blade row flutter stability the energy method is employed. It assumes that flutter occurs with the blade natural mode shape (or some prescribed oscillation motion) and makes a flutter prediction by calculating the energy exchanged between the blade and the flow field. When mechanical damping is considered, the sum of mechanical and aerodynamic damping work determines the system stability. If the total damping work is negative, the aeroelastic system is unstable and flutter will occur.

The flutter stability of each traveling wave mode is assessed by calculating the work done by the unsteady pressure on the blade motion. This can be expressed as the aerodynamic damping which is equal to

$$\Xi = \frac{W_{aero}}{\pi\, b\, h_{max}^2 p_{ref}}$$

where W_{aero} is the aerodynamic work per cycle, b is the height of the blade, h_{max} is the maximum blade displacement and p_{ref} is the reference pressure which is set to the total pressure at the inlet minus the static pressure at the inlet.

While the aerodynamic damping is a measure of the overall stability, it does not show any detail about the underlying physics. This can be examined by looking at the local work coefficient on the blade. This can be calculated for simple mode shapes as,

$$w = \frac{-c\; p_{imag}\, \mathbf{h} \cdot \hat{\mathbf{n}}}{p_{ref}\, {h^2}_{max}}$$

where c is the chord length, p_{imag} is the imaginary (out of phase) component of the unsteady pressure, $\mathbf{h}$ is the vector of the local displacement on the blade and $\hat{\mathbf{n}}$ is the local normal.

CASE BACKGROUND

The geometry considered in this study was taken from the Aeroelastic Turbine Rig at the Royal Institute of Technology (KTH), Stockholm, Sweden. This experimental setup is characterized by a high subsonic low pressure axial turbine cascade. Figure 1 shows three spanwise profile sections of the blades in the cascade. The blade features a real chord of 54.34 mm at midspan and an aspect ratio of 1.78 (span of 97mm). The annular cascade is composed by 80 blades, with pitch to chord ratio of 0.62 at midspan. The nominal tip clearance is 1% of span (2.15% of axial chord or 0.97 mm), with hub and shroud radii of 383 mm and 480 mm, respectively.

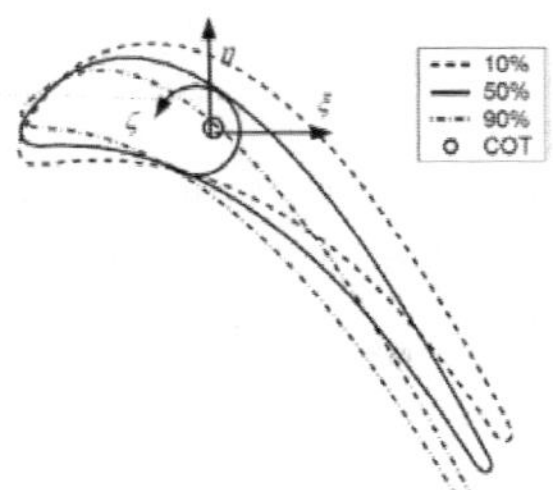

Figure 1 – AETR profile sections at different span positions

The investigated blade oscillation motion was a simplified rigid body bending mode perpendicular to the chord line. The displacement linearly increases from 5% span until the blade tip. The oscillation's maximum nondimensional displacement corresponds to that given by the blade's first mode shape, α = 0.00997 with h_{max} = 0.5422mm or approximately 1% of the midspan's chord, extracted from a pre-stressed modal analysis in ANSYS Mechanical APDL® v.15. A reduced frequency of ω^* = 0.3 based on full chord was used, corresponding to a blade frequency of 213.87 Hz.

STEADY STATE ANALYSIS

The AETR blade row was modeled in ANSYS CFX® v.15 using a mesh of finite volumes. The Reynolds stresses terms of the RANS equations were computed using a standard k-ε turbulence model with wall functions. High resolution scheme is used to deal with the advection term and second order backward Euler scheme is used for the transient term. The convergence target value was set to

1×10^{-4} for the RMS residuals of momentum and mass variables.

The flowing flow conditions were set at the inlet boundary: total pressure profile from hub to shroud, constant total temperature, flow direction and turbulence intensity. The average total pressure at the inlet was approximately 160 kPa. At the outlet a constant static pressure is prescribed. Table 1 summarizes the prescribed steady-state boundary conditions.

Table 1 - Boundary conditions

Parameter	Value	Units
p_{01}	Radial profile	kPa
T_{01}	303	K
Inlet flow direction	26	deg
Turbulence intensity	5	%
p_2	109.1	kPa

The single-passage hexahedral fluid mesh was generated with ANSYS CFX-TurboGrid® v.15 using the Automated Topology and Mesh (ATM) method. Three tip clearance sizes were investigated: no tip clearance, 1% of span (0.97mm) and 3% of span (2.91mm). The tip gap was discretized with three different spanwise resolutions: 6, 12 and 18 cells for the 1% gap size and 18, 27 and 36 cells for the 3% gap size. Table 2 summarizes the three tip clearance cases investigated.

Table 2 - Steady state case studies

Parameter	Description	Spanwise resolution
Case 1	No tip clearance	-
Case 2	1% span tip clearance	6, 12 and 18 cells
Case 3	3% span tip clearance	18, 27 and 36 cells

The multi-block grid with and O-mesh around the profile was used (Figure 2). In order to assess the influence of the mesh density in the main flow, three different meshes with no tip clearance with approximately 128k (coarse), 256k (medium) and 512k (fine) nodes per passage have been calculated. The table below shows the mesh statistics.

Table 3 – Mesh Statistics

Mesh resolution	Coarse	Medium	Fine
Nodes	131328	254606	508590
Cells on hub	2304	3586	5651
Cells in radial direction	56	70	89
Cells inside boundary layer	9	11	10

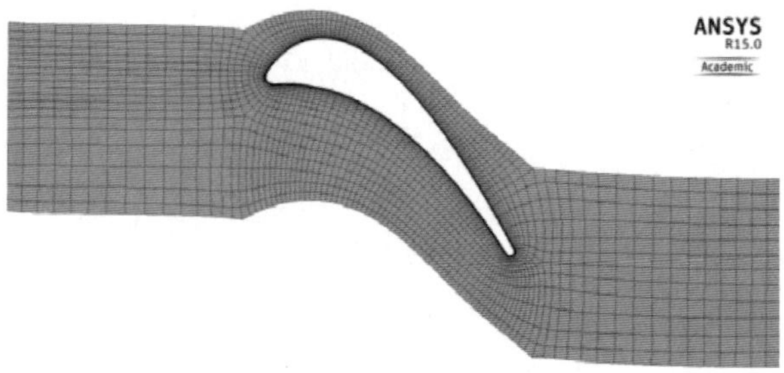

Figure 2 – Fluid mesh at 50% span (medium resolution – no tip clearance, one passage shown)

Blade loading curves for the three meshes (coarse, medium and fine resolutions) at 90% normalized spanwise position are shown in Figure 3. The result was similar for other span heights (not shown). As grid convergence for the medium resolution mesh was demonstrated, it was decided to continue all steady and unsteady simulations with the medium resolution mesh.

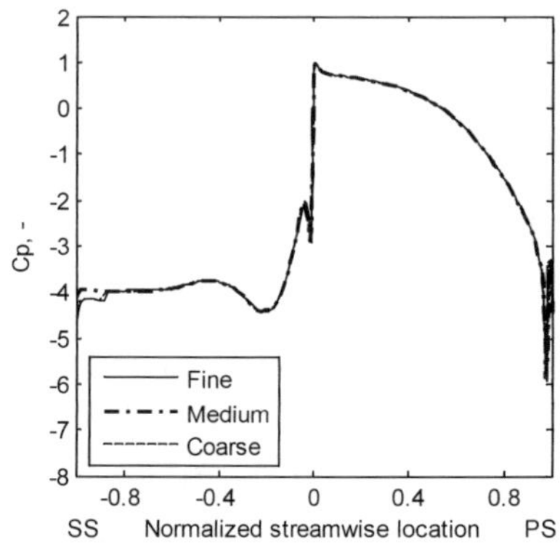

Figure 3 – Steady blade loading at 90% span (coarse, medium and fine mesh resolution – no tip clearance)

The contour plot of the Mach number distribution obtained from the medium density mesh at 90% span with no tip clearance is shown in Figure 4. The averaged mass flow Mach number at the inlet and outlet was approximately 0.32 and 0.74, respectively. In addition, the maximum Mach number at midspan was 0.8723, achieved at the aft suction side.

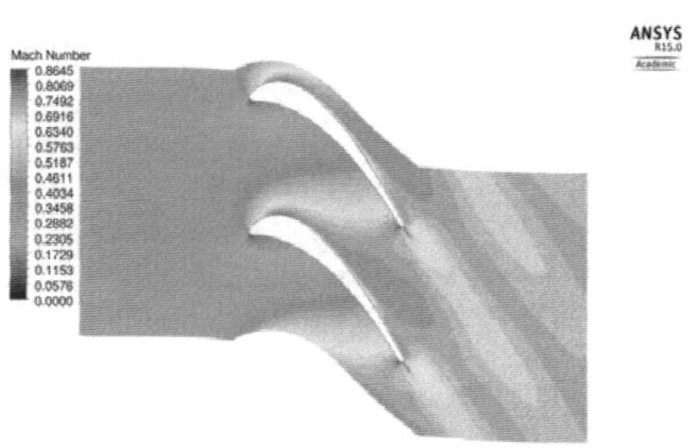

Figure 4 – Mach number contour plot at 90% span (medium resolution – no tip clearance)

The influence of the spanwise mesh resolution on the steady-state blade loading for the case with 1% tip clearance is shown in Figure 6. The mesh statistics are shown in Table 4. Figure 5 shows how the tip clearance mesh cells were simply added above the tip of the blade, in the spanwise direction. The blade loading curve for the 6 gap cells mesh differs from the 12 and the 18 gap cells meshes at around 70% normalized streamwise location on the suction side of the blade. This region is affected by the tip clearance flow, thus a finer mesh resolution is required to estimate the surface pressure accurately. The unsteady simulations for this case were performed on the mesh with 12 cells in the tip clearance.

Table 4 - Mesh statistics (medium resolution – 1% span tip clearance)

Cells at clearance	6 cells	12 cells	18 cells
Nodes	297178	388770	466170
Cells on hub	3586	3586	3586
Cells in radial direction	70+6	70+12	70+18

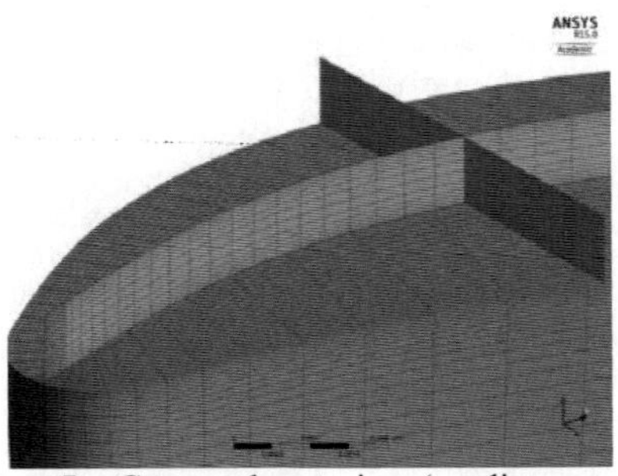

Figure 5 – Gap mesh cut view (medium resolution – 1% span tip clearance, 12 cells)

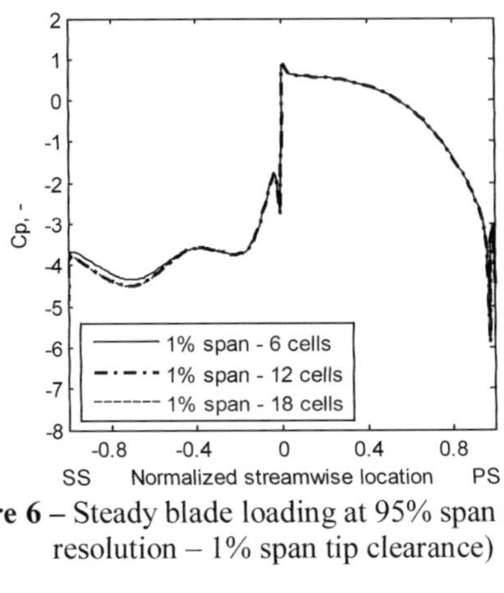

Figure 6 – Steady blade loading at 95% span (medium resolution – 1% span tip clearance)

The influence of the spanwise mesh resolution for the case with 3% tip clearance was also investigated. The mesh statistics for this case are shown in Table 5. The blade loadings at 95% span calculated with the three tip clearance mesh resolutions are shown in Figure 7. The mesh with 18 cells in the gap slightly underestimated the loading at approximately 50% streamwise location on the blade's suction side. The mesh with 36 cells in the tip gap was used for the unsteady flow simulations.

Table 5 - Mesh statistics (medium resolution – 3% span tip clearance)

Cells at clearance	18 cells	27 cells	36 cells
Nodes	376306	435652	494998
Cells on hub	3586	3586	3586
Cells in radial direction	70+18	70+27	70+36

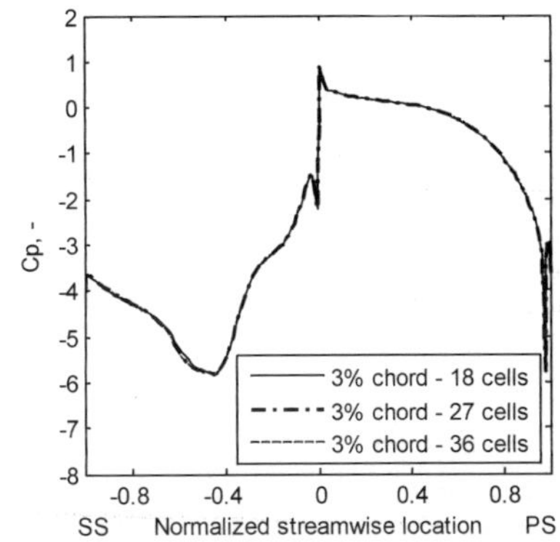

Figure 7 – Steady blade loading at 95% span (medium resolution – 3% span tip clearance)

The blade loading curves comparing the solutions calculated with different tip clearances at 50%, 90% and 95% span are shown in Figures 8 to 10 respectively. At 50% span, the differences between the loading curves are small because the influence of the tip flow is limited to the vicinity of the tip. At 90% span, differences in the blade loading can be seen particularly on the suction side. At 95% span, there are significant differences in the blade loadings particularly for the 3% tip clearance case where there is a significant drop in the pressure on the suction side near mid-chord. This pressure drop is due to a region of high Mach number flow on the suction side of the blade near mid chord. This can be seen in the Mach number contour plot for the 3% tip clearance at 95% span (Figure 11). There is a region on the suction surface near mid-chord where the flow is near sonic ($M = 0.9893$). Compare this with the contour plot of Mach number at 90% span for the no tip clearance case (Figure 4) where the maximum Mach number is 0.8723. There is also a significant change in the pressure on the pressure side at 95% span for the 3% tip gap. The pressure is lower in the front half and slightly higher in the back half of the blade.

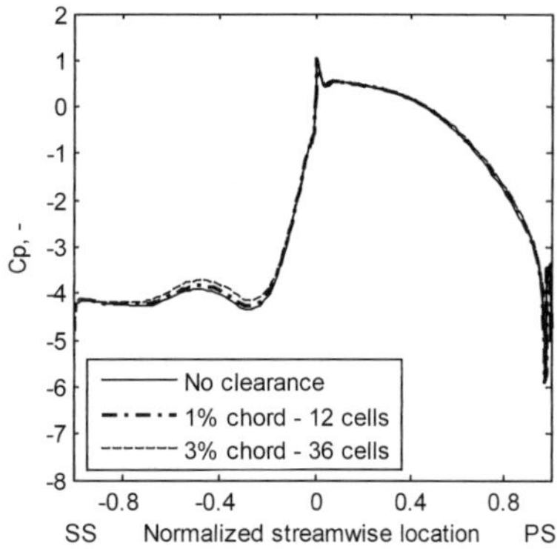

Figure 8 – Steady blade loading at 50% span (medium resolution mesh)

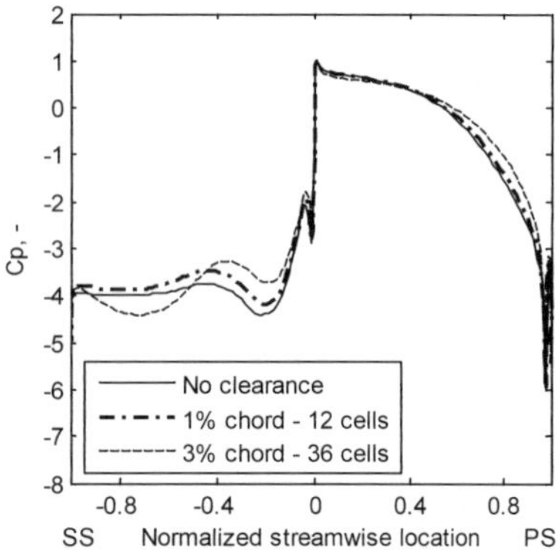

Figure 9 – Steady blade loading at 90% span (medium resolution mesh)

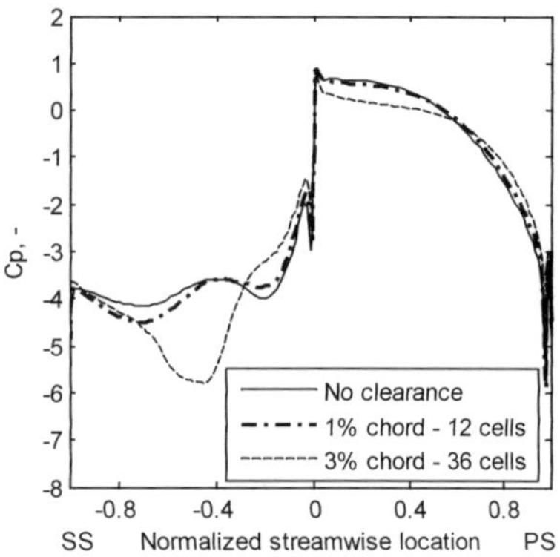

Figure 10 – Steady blade loading at 95% span (medium resolution mesh)

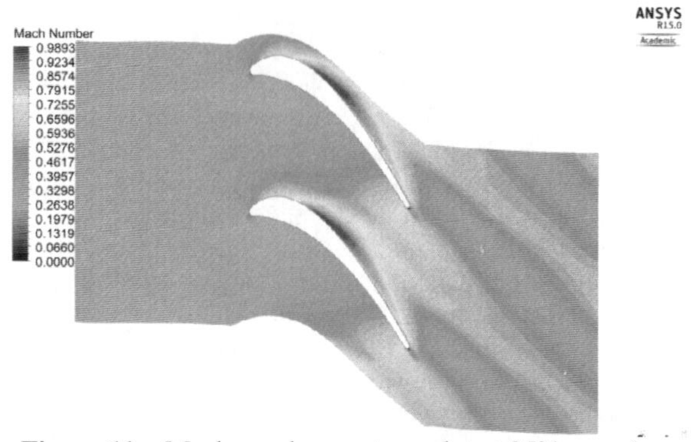

Figure 11 – Mach number contour plot at 95% span (medium resolution – 3% span tip clearance, 36 cells)

The Mach number increase (depicted in Figure 11) near the blade's tip can be related to a strong tip leakage flow from the pressure side to the suction side. Contour plots of the flow Mach number at axial slices (constant axial position) at mid chord near the tip region for the three tip clearance cases are shown in Figures 12 to 14. The changes in the flow-field due to the tip clearance flow are non-linear as the size of the tip clearance is increased. The tip flow for the 1 mm gap causes a reduction in the Mach number of the flow on the suction side just below the tip while there is an increase in the flow Mach number in the same region when the tip gap is 3 mm.

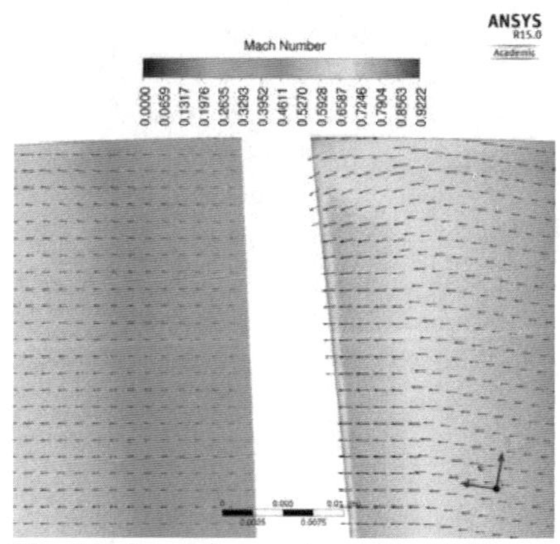

Figure 12 – Mach number contour plot at 50% normalized streamwise location (medium resolution – no tip clearance)

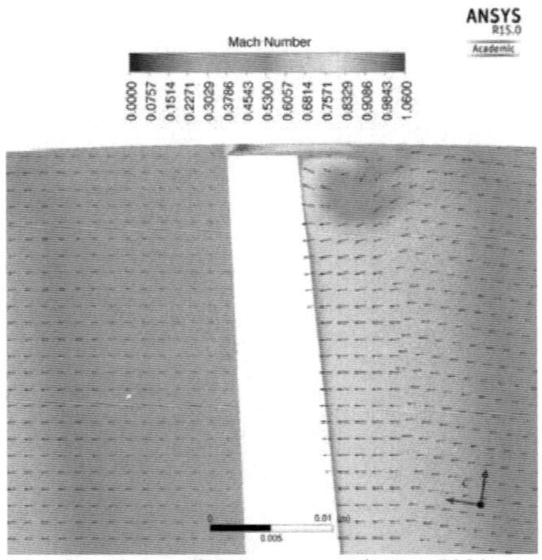

Figure 13 – Mach number contour plot at 50% normalized streamwise location (medium resolution – 1% span tip clearance, 12 cells)

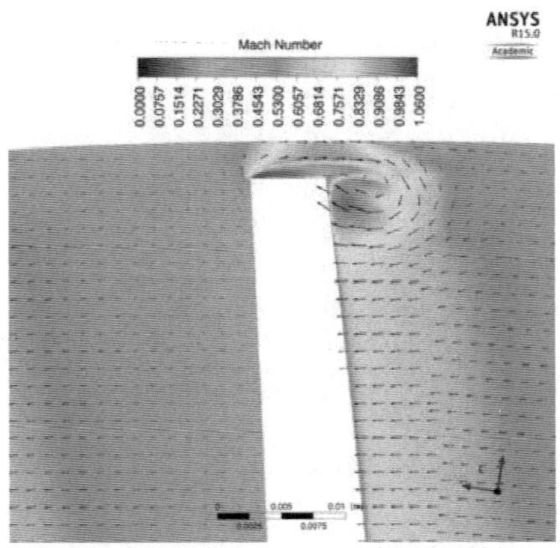

Figure 14 – Mach number contour plot at 50% normalized streamwise location (medium resolution – 3% span tip clearance, 36 cells)

UNSTEADY STATE ANALYSIS

Table 6 shows the cases selected for the aeroelastic investigation. The unsteady flow analyses were conducted using the TBR-FT approach in ANSYS CFX® v.15. For each case shown in Table 6, unsteady flow solutions were calculated for the following IBPAs: -180°, -135°, -90°, -45°, 0°, 45°, 90°, 135° and 180°. The unsteady flow simulations in the time domain ran for five blade oscillation periods and there were 50 timesteps per period for each tested IBPA. The unsteady CFD setup parameters are shown in Table 7.

Table 6 - Unsteady state case studies

Parameter	Description	Gap spanwise mesh resolution
Case 1	No tip clearance	-
Case 2	1% span tip clearance	12 cells
Case 3	3% span tip clearance	36 cells

Table 7 – Unsteady state setup parameters

Parameter	Value	Units
Blade oscillation frequency	213.87	Hz
Maximum displacement	0.5422	mm
Timesteps per period	50	-
Periods per run	5	-
RMS residual target	1×10^{-4}	-
α	0.00997	-

RESULTS AND DISCUSSION

Figure 15 shows the aerodynamic damping aeroelastic stability parameter for various IBPA's for different tip gap sizes. The tip clearance flow does not appear to have a significant influence on the aerodynamic damping for the AETR rig geometry at the prescribed flow condition. The least stable mode can be identified at σ = -45°, for all tip clearance cases investigated.

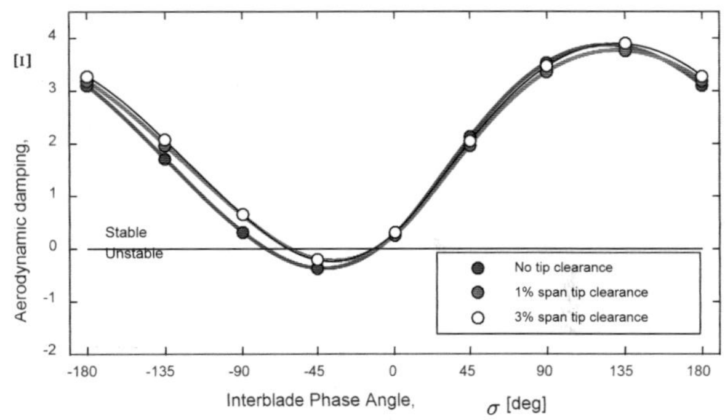

Figure 15 – Nondimensional aerodynamic damping stability parameter vs σ

In order to examine the aeroelastic behavior in greater detail, plots of aerodynamic damping as a function of span height for the three least stable IBPAs are shown in Figures 16 to 18. Note that the aerodynamic damping values are calculated from the local work at each span height but normalized by the maximum blade amplitude of the whole blade. This was done to directly compare the work done at various span heights. For all plots it can be seen that the difference become larger near the tip and this is where the magnitude of the aerodynamic damping is greatest. At IBPA = -90° (Figure 16), the plots for tip clearance 1% and 3% are similar and the plot for no tip clearance is less stable. This can also be seen in Figure 15. At IBPA = -45° (Figure 17), the three plots are different near the tip but the area under the plots for 1% and 3% tip clearance are similar. At IBPA = 0° (Figure 18), there are small differences between the plots but the overall dampings are similar. Overall, there appears to be no trend and the influence of the tip clearance flow on the aerodynamic damping appears to be dependent on the interblade phase angle.

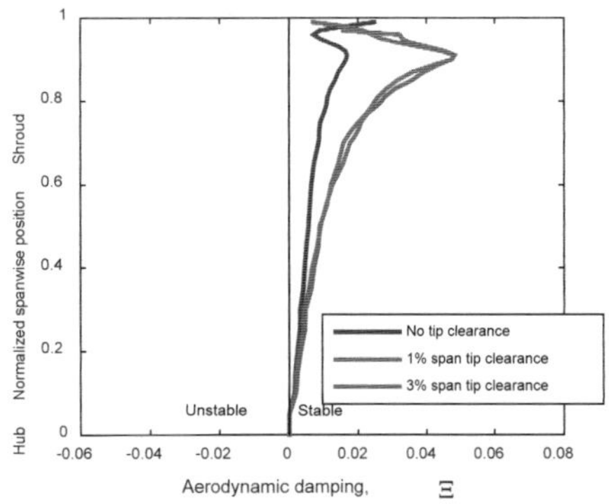

Figure 16 – Nondimensional aerodynamic damping stability parameter vs span position, σ = -90°

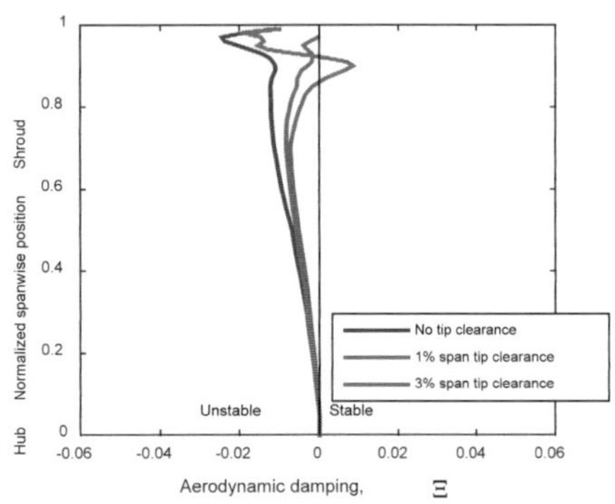

Figure 17 – Nondimensional aerodynamic damping stability parameter vs span position, σ = -45°

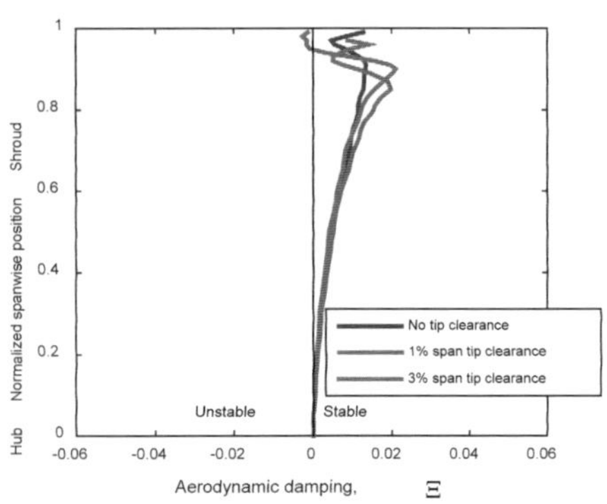

Figure 18 – Nondimensional aerodynamic damping stability parameter vs span position, σ = 0°

Figures 19 to 21 show the local work coefficient (positive unstable) on the blade surface for the least stable IBPA for all three tip clearance cases. These plots do show that the tip clearance flow is influencing the unsteady work done on the blade but the influence on the overall damping is small as the tip clearance flow appears to reduce the work done in one area but increase it in other areas.

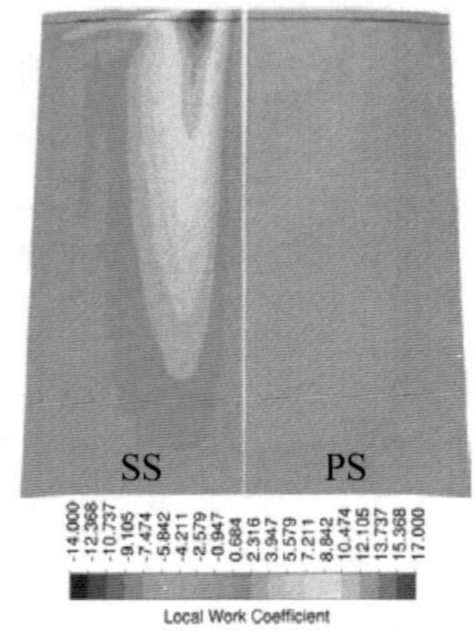

Figure 19 – Local work coefficient distribution on the blade surface, σ = -45° (no tip clearance)

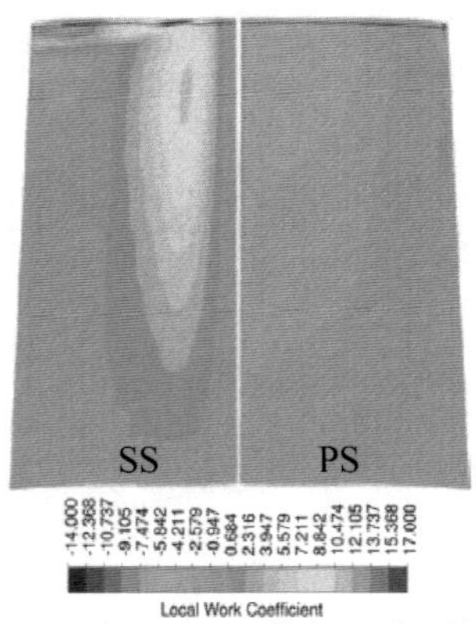

Figure 20 – Local work coefficient distribution on the blade surface, σ = -45° (1% span tip clearance)

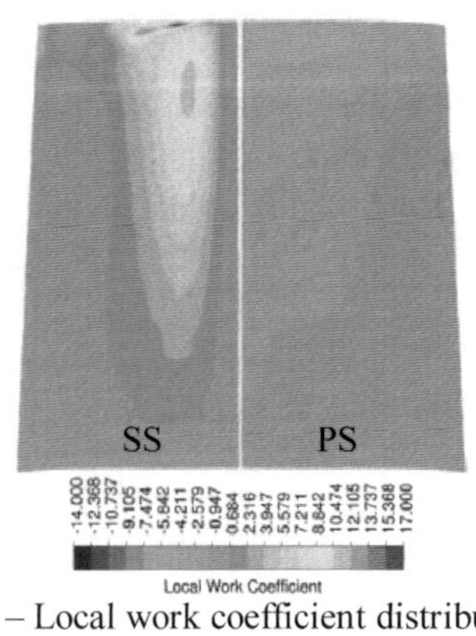

Figure 21 – Local work coefficient distribution on the blade surface, σ = -45° (3% span tip clearance)

The local work coefficient for the three tip clearance case for the least stable mode (IBPA = -45°) at 50%, 80%, 85%, 90%, 93% and 95% span heights are shown in Figures 22 to 27 respectively. The local work coefficient curves for the three tip clearance cases do not differ significantly from 50% and 80% span. Differences can be seen at 85% where the work coefficient on the suction side is more unsteady for the 1% and 3% tip clearance and the work coefficient on the pressure side is more stable for the 3% tip clearance. For 90%, 93% and 95% span there are differences between all three curves on the suction side and these differences become large nearer the tip and there is no clear trend. However, on the pressure side for 90%, 93% and 95% span there is no difference in the work coefficient for no tip clearance and 1% tip clearance but there is a significant stabilizing shift in the work coefficient on the pressure side for 3% tip clearance.

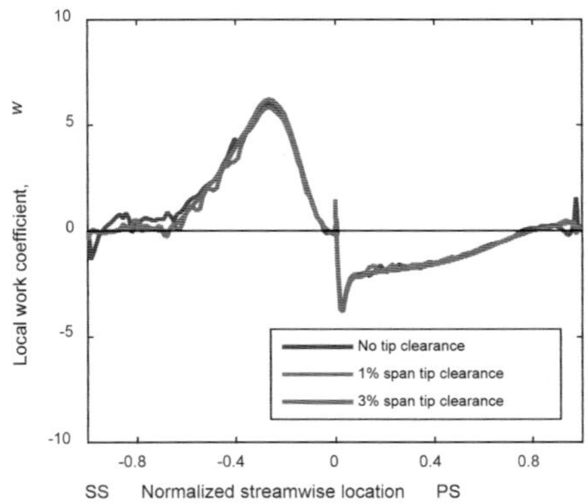

Figure 22 – Local work coefficient distribution along the blade's chord, σ = -45°, 50% span

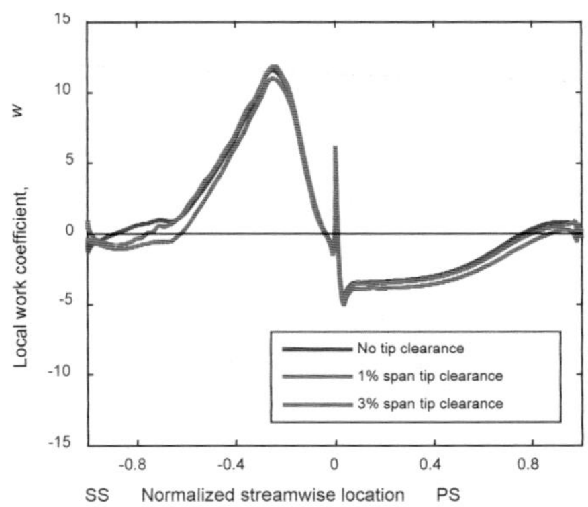

Figure 23 – Local work coefficient distribution along the blade's chord, σ = -45°, 80% span

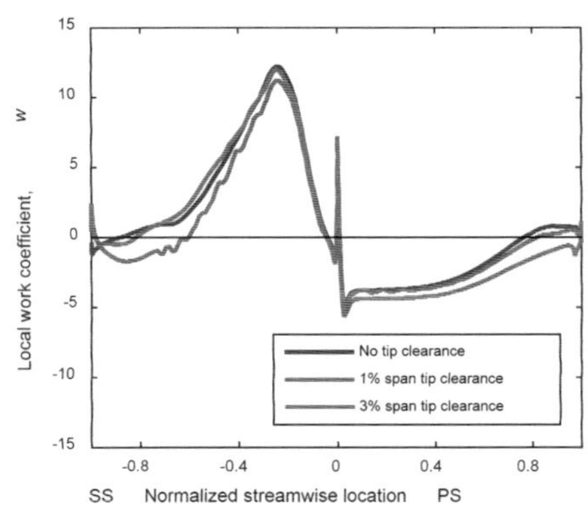

Figure 24 – Local work coefficient distribution along the blade's chord, σ = -45°, 85% span

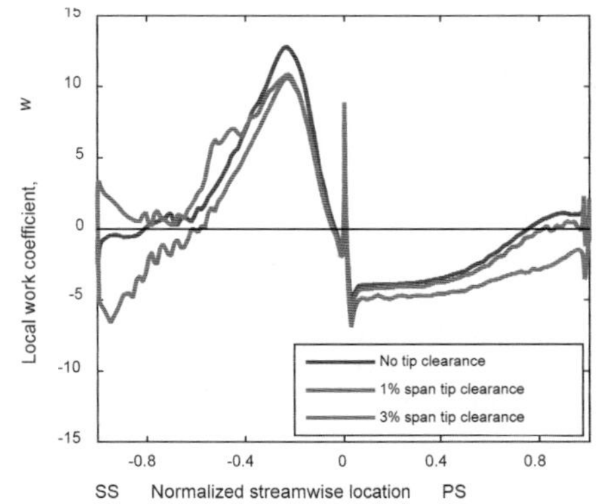

Figure 25 – Local work coefficient distribution along the blade's chord, σ = -45°, 90% span

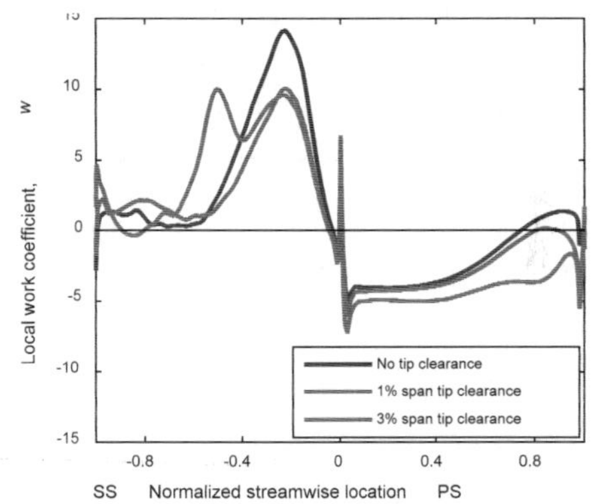

Figure 26 – Local work coefficient distribution along the blade's chord, σ = -45°, 93% span

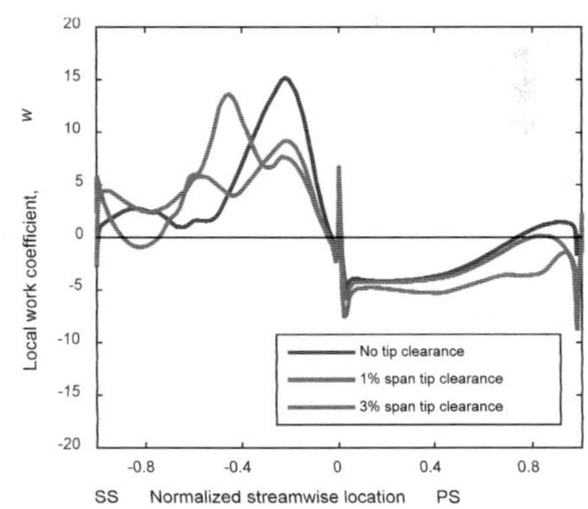

Figure 27 – Local work coefficient distribution along the blade's chord, σ = -45°, 95% span

CONCLUSIONS

The influence of the size of the tip clearance on the aeroelastic behaviour of an oscillating low pressure turbine (LPT) blade row has been investigated through full-scale time-marching URANS CFD simulations. Three tip clearance cases were investigated: no gap and tip clearances of 1% and 3% span. The effect of increasing the size of the tip clearance on the steady flow was found to be non-linear, particularly on the suction side. There was a significant drop in the pressure at 95% span on the suction side near mid-chord for the 3% tip clearance but not the 1% tip clearance. The stronger tip clearance flow for the 3% gap creates a region of high Mach number on the suction side that is not present for the 1% gap.

The tip clearance flow does not appear to have a significant influence on the overall aerodynamic damping for the AETR rig geometry at the prescribed flow condition. One factor affecting this is that the local blade amplitude for the AETR is proportional to span height. This causes more unsteady aerodynamic work to be done on the blade further away from the tip compared with real turbine blades where the local blade amplitude is proportional to the span height squared.

The least stable mode was IBPA = -45° for all tip clearance cases investigated. The unsteady work done on the blade was investigated in more detail by examining plots of aerodynamic damping versus span and the local work coefficient on the blade at various blade heights. Differences in the local work were observed, particularly near the tip. No trend could be observed in the differences on the suction side, however, a trend was observed for the pressure side. There was a significant stabilizing shift in the work coefficient on the pressure side for the 3% tip clearance near the tip.

REFERENCES

[1] Z. Z. Fu, Y. R. Wang, X. H. Jiang and D. S. Wei, "Tip Clearance Effects on Aeroelastic Stability of Axial Compressor Blades," *ASME Journal of Engineering for Gas Turbines and Power,* vol. 137, 2015.

[2] N. Glodic, D. Vogt and T. Fransson, "Influence of Tip Clearance Modelling in Predictions of Aeroelastic Response in an Oscillating LPT Cascade," in *Proceedings of the 13th International Symposium on Unsteady Aerodynamics, Aeronautics & Aeroelasticity in Turbomachines, ISUAAAT 13*, Tokyo, Japan, 2012.

[3] L. Huang, L. He and D. Bell, "An Experimental Investigation into Turbine Flutter Characteristics at Dif-ferent Tip-Clearances," in *Proceedings of GT2006, ASME Turbo Expo 2006: Power for Land, Sea and Air*, Barcelona, Spain, 2006.

[4] J. I. Erdos, E. Alzner and W. McNally, "Numerical Solution of Periodic Transonic Flow through a Fan Stage," *AIAA Journal,* vol. 15, pp. 1559-1568, 1977.

[5] M. Giles, "UNSFLO: A numerical method for the calculation of unsteady flow in turbomachinery," Gas Turbine Laboratory, Massachusetts Institute of Technology, 1991.

[6] L. He, "An Euler solution for unsteady flows around oscillating blades," *Journal of Turbomachinery,* vol. 112, pp. 714-722, 1990.

[7] Patange, S.; © 2014 ANSYS, Inc., "ANSYS Turbomachinery CFD System - 15.0 Update," 20 June 2014. [Online]. Available: http://www.ansys.com. [Accessed 2 August 2015].

[8] E. Camara, "Validation of Time Domain Flutter Prediction Tool with Experimental Results," Master of Science Thesis EGI-2015-011MSC EKV1081 , KTH - Royal Institute of Technology, Stockholm, Sweden, 2015.

[9] R. Elder, W. I., P. S., W. Holmes, S. R. and H. B., "Investigation of Efficient CFD Methods for the Prediction of Blade Damping," in *ASME Proceedings | Aerodynamic Excitation and Damping (With Turbomachinery)*, 2013.

Proceedings of the 14th International Symposium on Unsteady Aerodynamics, Aeroacoustics & Aeroelasticity of Turbomachines
ISUAAAT14
8-11 September 2015, Stockholm, Sweden

I14-S8-1

DETAILED INVESTIGATION OF THE PREDICTION OF COMPRESSOR CASCADE PERFORMANCE AT OFF DESIGN INCIDENCES WITH FREE STREAM DISTURBANCES

John Leggett
University of Southampton
Southampton, UK, j.leggett@soton.ac.uk

Stephan Priebe
GE Global Research
Niskayuna, NY, USA

Richard Sandberg
University of Southampton
Southampton, UK

Vittorio Michelassi
GE Global Research
Munich, Germany

Aamir Shabbir
GE Aviation
Cincinnati, OH USA

ABSTRACT

The need for improved loss prediction techniques of turbo machinery is ever present. The most common tool currently in use is Reynolds Averaged Navier-Stokes modelling which provides good estimation at design conditions but can struggle with off design conditions. With this in mind an alternative method such as Large Eddy Simulation is needed to accurately assess off design performance.

Large Eddy Simulation may provide this alternative and the initial assessment and comparison of RANS and LES at predicting axial compressor cascade losses are detailed here. The simulations are based on a NACA 65 linear cascade run over a range of incidences with incoming free stream turbulence. The comparison looks to explain difference seen in the loss prediction between the two methodologies by reviewing the statistical results and examining areas where these methodologies break down.

INTRODUCTION

Driven by rising demands on natural resources and growing populations that want to travel by air, the need for accurate and efficient methods for predicting axial compressor performance in turbo jet engines is ever increasing together with the need to further improve design efficiency and operability range.

In order to reduce weight and installation effects, modern compressors become smaller with fewer blades per stage and fewer stages, thereby pushing the limits of stage load and operability and increasing the demand for accurate performance predictions from the early design phase.

Historically the losses in a compressor are separated into "profile loss", "end wall loss", "leakage loss" and "shock loss", though this is now known as a strong simplification. The relative importance of these losses may change along the speed-line, from design to off-design. Each of these areas in their idealised form is governed by certain physical processes and such simplification can often be modelled more easily.

Here the focus is on computing and modelling the profile losses in a linear cascade at both design and off-design conditions. In this case the losses are governed by only the boundary layer with no effects from the end wall or tip leakage flows considered. While this idealised model of a compressor stage only accounts for about a third of the losses in a real stage [2], the processes are not independent from each other and improvements in one may well show improvements with others.

Today the design of axial compressors is largely performed by using correlations and further verified by Computational Fluid Dynamics (CFD) methods, most notably Reynolds Averaged Navier-Stokes (RANS). The use of CFD has allowed much faster iteration times and greater insight into potential compressor performance, reducing the need for experimental testing. With CFD being such a powerful design tool it is unlikely to be replaced in design processes in the future. However, there are limitations to current methodologies, most notably the loss of information and accuracy introduced by the modelling assumptions inherent in RANS models, which are not easy to overcome without great computational expense and time. The assumptions made in RANS models can break down in complex flow regions leading to inaccurate predictions. In axial compressors, complex flow phenomena of interest include upstream wake blade interactions, free-stream turbulence and the limit of off-design conditions in the presence of strong adverse pressure gradients and flow reversal. Events such as stall and surge can render an engine inoperable or lead to mechanical failure. These conditions are often found at the limit of the operational envelope, and their accurate prediction is of paramount importance.

While the move to more accurate CFD models or even Direct Numerical Simulation (DNS) to overcome the failings of RANS is certainly a desired one, for most CFD applications these are infeasible. Even with recent developments in high performance computing, many of the high fidelity CFD methods, most notably DNS are still confined to Reynolds numbers lower than those present in aircraft engine compressors, and simplified geometries. However Large Eddy Simulation (LES) is feasible for limited numbers of runs at low engine Reynolds numbers.

The prediction of loss over a range of incidences is the focus here. It has been seen that RANS produces good results for near design conditions and provide well resolved mean flow properties. However as the off design performance can be the limiting factor for engine operability, the main interest here is in improving these predictions using LES modelling and to study the viability and performance of LES compared with current RANS methods.

There has been previous work on compressors at off design conditions looking at the effects of incidence on blade performance, [3, 4]. This has provided good insight into the physics and complexities involved but such studies are largely limited to academic research. Making this information available to designers testing new designs is paramount to new advancements in working compressors.

With this in mind there has been notable increase in interest of LES application to turbo machinery with work on secondary flows [5, 6], heat transfer [7, 8] and acoustics [9] all performed recently.

The comparison of LES and RANS in recent work by Gourdain [10] shows the performance of the two CFD methods at design conditions of rotor stator interactions. The work highlights the benefits of LES though emphasises the added complexity and lack of a considered good practice making LES a "Pandora's box".

More relevant to the current study are works on off-design conditions [11, 12], highlighting the importance of capturing unsteady phenomena such as stream wise structures that can effect loss prediction at off design conditions, and accurate shock boundary layer interactions in transonic compressors, both improving statistical results.

The remainder of this paper outlines the cascade set up and running conditions, followed by the numerical set up and CFD methods employed is presented. A brief validation of the LES methodology applied to compressors cascades is discussed and finally the comparison of the LES and RANS results is shown.

CASE SET-UP

The focus of this paper is on assessing the performance of LES and RANS in simulating compressor cascade flows over a range of incidences.

The simulations have been performed with a fixed amplitude of free stream turbulence to compare with experimental data from Leipold et al. [1], which also

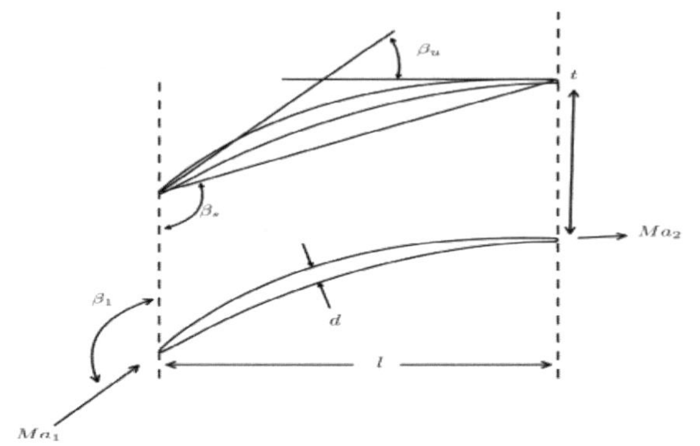

Figure 1 Schematic representation of linear compressor cascade NACA65 profile, based on outline in Leipold et al. [1]

Table 1 Nominal Parameters

Ma = 0.67	β_1 = 132°
RE = 300,000	t/l = 0.59
d/l = 0.059	β_s = 112.5°

provides the nominal design incidence working conditions.

The investigation focuses on the simulation of a NACA 65 profile at an axial chord Reynolds number of 300K based on the inflow free-stream parameters, and an inflow Mach number of 0.67. The set-up is shown in Figure 1 with the nominal set up parameters in Table 1. For the investigation a set of three incidence angles are studied, $\alpha = 37^0$, 40^0 and 44^0 all with a nominal free stream turbulence level of ~3% measured at the blade inlet plane. The angles are defined here as the incidence of flow with reference to the horizontal or with reference to Figure 1 $\alpha = \beta_1 - 90^o$.

The mesh structure for the LES simulations is an H-grid with an embedded O-grid type mesh set-up as shown by dashed lines in Figure 2.

For all the LES simulations the domain extends to 0.5 chord up stream of the leading edge and 1 chord length downstream of the trailing edge, with a span-wise extension of 0.2 chord length. The mesh contained 603,136 points in the plane with 128 modes in the span wise extent, totaling 155,609,088 points for the full domain, giving a wall resolution of Δx^+~20, Δy^+~1.5 and Δz^+~25.

The boundary conditions, elaborated on in the numerics section are shown in Figure 2. Periodic boundaries are applied in both the pitch wise and span wise directions, with a non-reflective zonal characteristic boundary for the exit. At the inflow, synthetic turbulence is superimposed onto a fixed mean velocity profile prescribed with constant density, the pressure being extrapolated from within the domain. The application of a digital filter turbulence generation method [13] is used to provide the synthetic turbulence at a prescribed intensity at the inlet. Finally a non-slip adiabatic boundary condition is applied to the blade wall.

The synthetic turbulence length scales are set to 10%

chord for the pitch-wise and stream-wise directions and 5% chord for the span-wise direction. The length scales are limited by the domain extent and filter width. These lengths scales were chosen based on previous simulations to avoid excessive decay of the turbulence upstream of the blade, as was seen to happen with initial LES simulations. Furthermore there is limited information on the turbulent length scale found in the experimental reference data leaving this a somewhat open question.

NUMERICS

The 2D RANS computations were run in ANSYS CFX using the SST turbulence model. For transition modelling, the two-equation γ-θ model was used. In order to match the LES setup, Sutherland's law, which is used in the LES, was also specified for the dynamic viscosity in the RANS.

The LES simulations were performed using an in-house compressible Navier-Stokes multi-block structured code, HiPSTAR, originally developed as a DNS solver and with the recent addition of LES capabilities. The code has previously been used to perform in-depth studies on compressible pipe flows [14], supersonic wakes [15] and turbulent jets in coflow [16]. More recently work by Sandberg et al. [17]used the code to simulate low pressure turbines, this work also details the development of the code for use in turbo machinery in greater detail.

The code employs a fourth-order accurate low-storage Runga-Kutta [18] method with a fourth-order accurate finite difference scheme for the in-plan discretisation and Fourier spectral discretisation in the span-wise direction. Additionally a skew-symmetric splitting [19] is employed to stabilise the convective terms.

The recent addition of the Wall Adaptive Local Eddy viscosity (WALE) [20] LES model to the code has enabled the move to higher Reynolds number flows such as this study.

The LES implementation assumes a filtering of the Navier-Stokes equations and the calculation and addition of the sub-gird stress contributions are added in the form of a sub-grid-scale viscosity.

In order to reduce error introduced by communication across discontinuous grid metrics at the block interfaces the code employs characteristic interfaces conditions [21]. Non-reflecting zonal characteristic boundary conditions [22] are used to reduce acoustic reflections. The exit boundary condition, enforcing only static pressure, is a simple characteristic correction of the incoming flux terms of the form $K(P-P_\infty)$, where K is a constant to improve stability.

CODE PERFORMANCE

A short overview of the cost and performance of the simulations is given here. For each of the simulations the time step was set to 1×10^{-5} and approximately 3 blade passing time periods were required for convergence of basic statistical data. The simulations were performed on the UK national computer Archer and required 30 hours on 7,344 cores to run the 300,000 iterations needed. A further 2 blade passing time periods were required to establish the flow, which was initialised with a mean flow domain based on the RANS results.

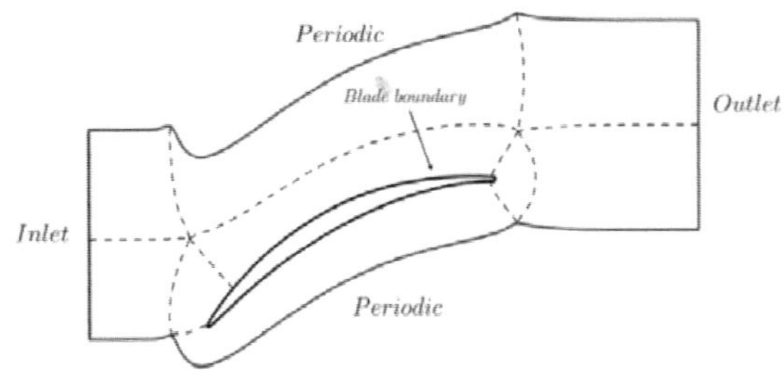

Figure 2 Boundary conditions and block layout.

VALIDATION

As the code has not previously been used for diffusive compressor flows a validation study is presented here. The study focuses on replicating the DNS results produced by [23], chosen based on the reduced Reynolds number making the case less computationally expensive. Only the statistically converged results are presented here comparing blade loading and skin friction, and touching briefly on wall normal profiles. The case uses the same NACA 65 profile described above, and the LES are performed at an axial chord Reynolds number of 138K based on inflow parameters and an inlet Mach 0.2 to minimise compressibility effects when compared with the incompressible reference. The linear cascade set-up is the same as that shown in Figure 1 run at an incidence of $\beta = 132^0$. For the validation study the case was run with a nominal level of inflow turbulence, Tu ~3.8% measured at a reference plane 0.4 chord upstream of the leading edge, slightly higher than the prescribed 3.25% at the same plane in the reference DNS case. The slightly higher level of turbulence is a result of the difficulty in estimating the dissipation of the turbulence prescribed at the inlet plane.

A reference case with a clean inflow is available, however it was deemed unfeasible to match a compressible simulation with the presence of acoustic fluctuations to a reference case without. Supported by an initial simulation with a clean inflow where it was found that acoustic waves caused turbulent break down of the boundary layer regardless of the care taken with non-reflecting boundary conditions. Therefore in order to minimize differences the reference case with a prescribed level of inflow turbulence was chosen.

The computational mesh used for the validation set-up employed the same H-type with embedded O-type as that in Figure 2, with a reduced number of grid points, using ~500,000 points in-plane and 66 collocation points in the span. This results in wall resolutions of $\Delta x^+\sim15$, $\Delta y^+\sim1.5$ and $\Delta z^+\sim10$

RESULTS

The comparison between the current LES with the incompressible reference DNS is detailed next.

The interest here is in matching the DNS reference results as accurately as possible. Figure 3 and Figure 4 show the comparison of blade loading and skin friction between the LES and DNS reference case. In both plots it is shown that the LES matches the DNS results well apart from discrepancies seen close to the separation point and transition. It is seen that the LES predicts a smaller separation bubble with a faster transition and reattachment likely attributed to the slightly higher turbulence levels and compressibility, as the separations point seems well matched. The LES does not appear to capture a recirculation region in the separation bubble as seen in the DNS, most likely due to the faster transition and reattachment. No other notable difference between the LES and DNS can be seen in the blade loading and skin friction plots.

The mean wall-tangential velocity profiles show good comparison for the three positions x/l = [0.1, 0.3, 0.4] shown in Figure 5. These profiles show the LES has little problem in capturing the development of the boundary layer up to transition. There is a slight difference in the profile at x/l = 0.4 where the DNS has a fuller boundary layer. It is noted in the reference paper that the effect of increased free stream turbulence reduces the boundary layer thickness as expected through energizing perturbations. The thinner boundary layer of the LES results are attributed to the higher free stream turbulence in the LES case.

The trends for increased free stream turbulence shown in [23] also support the variation seen in the pressure and skin friction plots, notably the shorter separation bubble. However the variation seen in the DNS results of the paper are not as sensitive to turbulence intensity as those seen here. It is suggested that the further discrepancies between the results are due to the presences of acoustic waves and compressible instabilities not present in the incompressible reference case.

LES RANS COMPARISON

Initial simulations of the case by Leipold et al. [1] raised some questions as to the correct nominal incidence as first runs at the experimentally specified 42° incidence did not match the blade loading.

In order to overcome this the RANS was first run using the operating conditions provided in the experimental paper by Leipold et al.[1], which are an inlet Reynolds number of Re = 300k, an inlet Mach number of M = 0.67, and an inlet flow angle of α = 42°. Since the loading from RANS did not match the experimental loading at these operating conditions (see Figure 6), a series of RANS were performed to investigate the dependence of the loading on the operating conditions. A parametric study was performed, varying the following parameters in the RANS: the Reynolds number Re, the Mach number M, the inlet turbulence intensity and length scale, and the flow angle α. Three-dimensional effects and Axial Velocity Density Ratio could also be relevant.

However, no information on these effects is available in the experimental paper and they were therefore not investigated in the RANS study. As expected, the loading

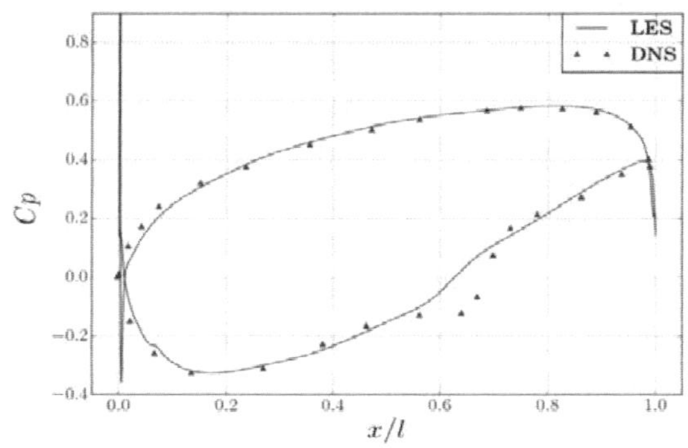

Figure 3 Pressure coefficient for LES and incompressible DNS reference [23].

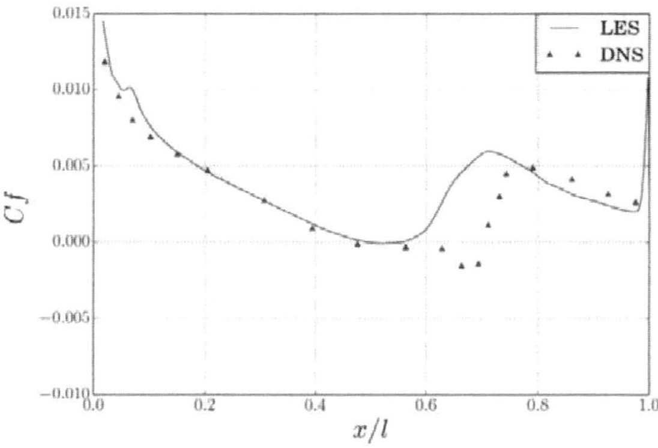

Figure 4 Friction coefficient for LES and incompressible DNS reference [23].

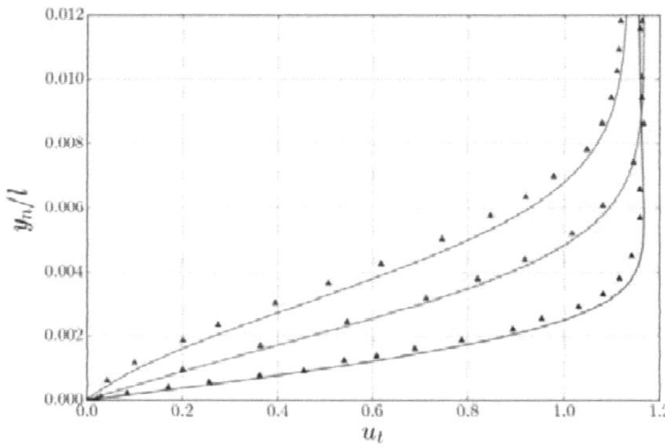

Figure 5 Mean tangential velocity profiles for x/l = 0.1 (—,▲) x/l = 0.3 (—,▲) and x/l = 0.4 (—,▲). Symbols are incompressible reference data [23].

is strongly affected by the inlet flow angle α (see Figure 6). A close match between the RANS loading and the experimental loading can be obtained by lowering the flow angle by 2°, from α = 42°, which is the value provided in Leipold et al., to α = 40°.

The RANS study thus suggests that the appropriate design operating conditions for this case are: Re = 300k, M = 0.67, and α = 40°, and these operating conditions were specified as baseline conditions for the LES. Note that performing the parametric study using LES would

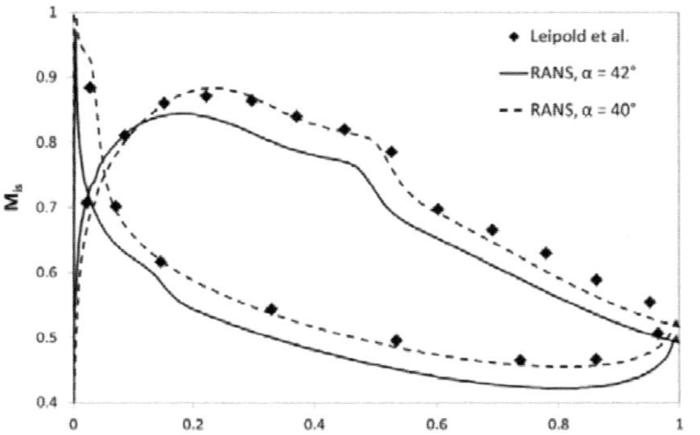

Figure 7 RANS loading at different inlet flow angles and comparison to the experimental loading of Leipold et al. [1]

have been prohibitively expensive and the operating conditions were therefore verified using RANS.

As an introduction to the results analysis instantaneous contours of the span-wise vorticity of two of the LES cases are plotted in Figure 7 showing the turbulent inflow and wake paths for the 37° and 40° cases.

It can be seen in this figure there is no large separation for either case on the suction surface, though there is substantial turbulent shedding on the pressure surface for the 37° case. Over the full incidence range chosen the pressure side undergoes transition early on either via a small leading edge separation bubble or via bypass transition at 44°, resulting in a turbulent boundary layer for the majority of the pressure surface. The suction surface over the incidence range shows laminar development up to the peak suction after which transition takes place either via a separation bubble or via bypass transition again seen in the 44° incidence case.

With the focus being the accurate estimation of compressor performance an appropriate unit of measure is needed. The stagnation pressure loss coefficient is used here and is defined as $\omega = (P_{t1}-P_{t2})/(P_{t1}-P_1)$. This metric is chosen as it is a convenient and easily calculated quantity.

In calculating the pressure loss coefficient it is common practice to take mean values of two planes, one up steam and one downstream of the blade to calculate these losses. It is important to note, though, that this method loses all information about the origins of these losses, as well as assuming a certain level of homogeneity of the wake at the point where the mean quantities are calculated. Often the errors involved in this process are negligible but as compressors become more compact and blade to blade spacing reduces this may not always be the case. The area averaged mean values are used here for the pressure loss coefficient.

Loss.

The comparison of loss prediction between the LES and RANS over the incidence sweep is shown in Figure 8. It is seen that both the LES and RANS predict a rise in loss as the incidence is increased, though the losses predicted by the LES show a smaller rise over the incidence range compared with the RANS.

It was found when attempting to run higher incidences

(a)

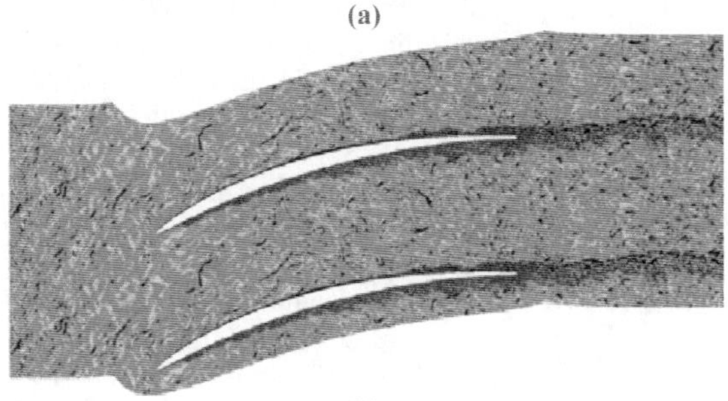

(b)

Figure 6 Instantaneous LES contours of the span-wise vorticity for 37° (a) and 40° (b) incidences.

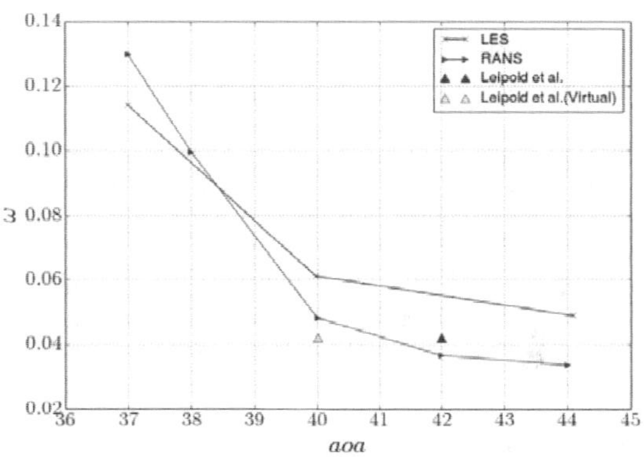

Figure 8 Losses for LES and RANS over incidence sweep 37° to 44° including virtual point for reference data to highlight the 2° change in incidence to match loading.

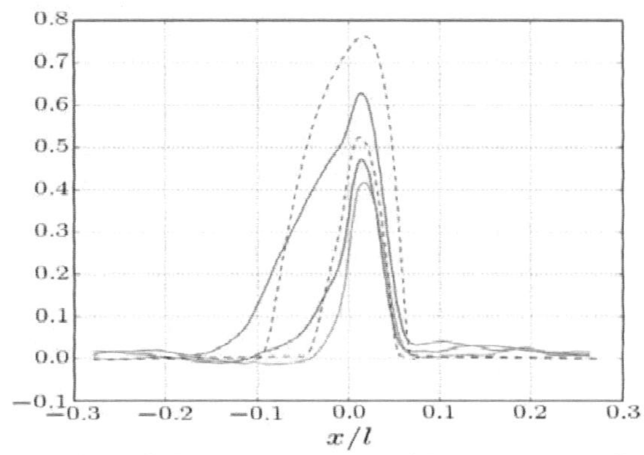

Figure 9 Total pressure $(P-P_{t1})/(P_1-P_{t1})$ wake profile 0.1 chord downstream of trailing edge, LES (—) and RANS (--) solutions at incidences 37° (●), 40° (●) and 44° (●). Pressure surface is negative.

that the loss prediction for the RANS was limited to a maximum incidence of 37° as the simulations could not be converged below this. It is seen in Figure 7 (a) that large turbulent structures are shed from the pressure surface leading edge and it is likely these are the cause of the RANS instabilities at higher incidences.

Upon decreasing the incidence above the nominal case of 40° to 44°, the differences in loss prediction between the two methods is more consistent. It is seen that the LES predicts a larger loss coefficient than the RANS for both the 40° and 44° compared with the lower losses at 37°.

The trend in loss prediction seen in Figure 8 shows the LES loss coefficient increases at a slower rate compared with the RANS as the incidence is increased. It is likely that much of this difference in loss is attributed to the increased unsteadiness of the pressure surface, which may not be accurately captured in the RANS. It is hard to confirm this without further investigation as the breakdown of loss production is not known from the pressure loss coefficient. However a qualitative overview of the suction surface gained from Figure 7 does not show large variation in its behavior and is less likely to be responsible for the large differences in the loss predictions. This is further supported by Figure 11 (a) and (b) showing the suction surface and pressure surface wall shear, respectively.

The comparison of the LES and RANS results with the experimental prediction is difficult to do with certainty. As mentioned previously the experimental data is performed at a reported incidence of 42°; however it was found that in order to match the loading profile an incidence of 40° is required for the LES and RANS simulations. The exact cause of the difference is uncertain as information regarding the full parameter set up of the experiments is lacking. Comparing the data that is available it is seen that a RANS simulation at the reported incidence under predicts the loss while an extrapolated prediction from the LES over predicts the loss. However looking at the loss prediction at 40° where the loading profiles match it is seen that both the RANS and LES over predict the loss compared with the experimental data which has been shifted by -2° and is represented by the virtual point in Figure 3. This suggests that while the loading (Figure 10) has been matched with a change of incidences it is likely that other parameters are still different between the experimental and simulation set-ups. Unfortunately further simulations have not been performed at the reported incidence for the LES due to computational cost.

Profile loading.

Looking next at the blade loading profiles given in Figure 10 in more detail the initial interpretation shows reasonable comparison for 44° and 40° cases while at 37° there is more substantial variation between the profiles.

Starting with the comparison of the loading at 40° with the experimental data, both the LES and RANS show reasonable correlation with the LES matching the suction surface downstream of x/l = 0.6 better than the RANS, though upstream of transition the opposite is true with the LES predicting a higher loading than the RANS and experimental data. The pressure surface comparison shows good correlation for both the RANS and LES downstream of x/l ≈ 0.5 though upstream of this the LES shows a slight increase in the loading compared with the RANS and experimental results which match well.

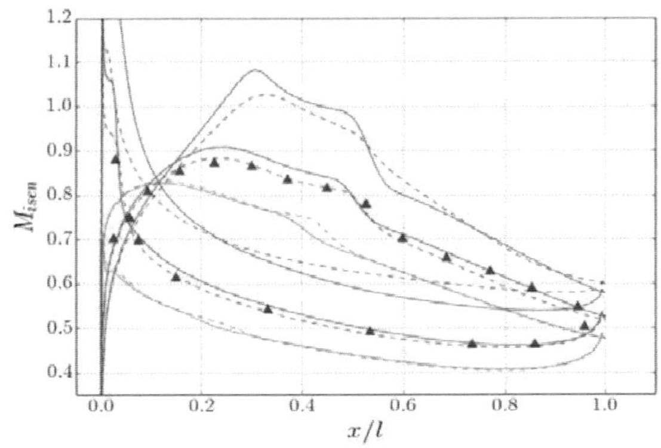

Figure 10 Isentropic Mach number loading of LES (—) and RANS (▪▪) solutions at incidences 37° (●), 40° (●) and 44° (●) with reference data from Leipold et al. (▲)[1].

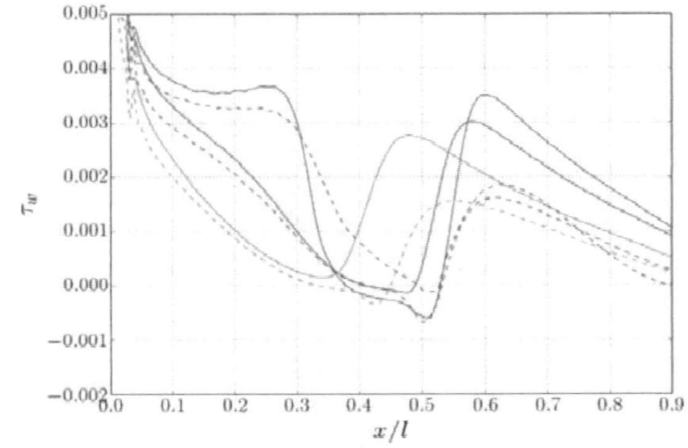

(a)

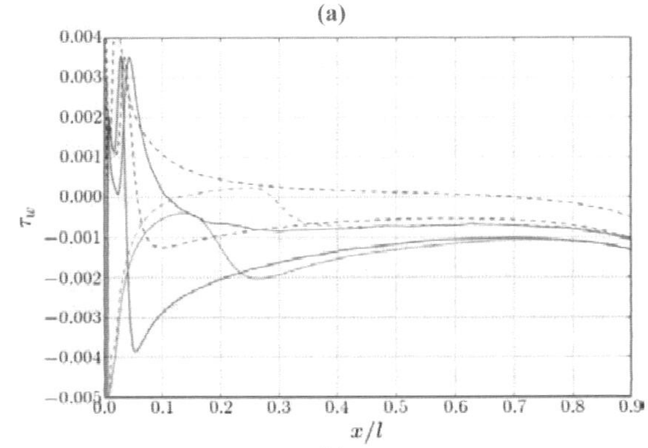

(b)

Figure 11 Wall shear stress profiles of LES (—) and RANS (▪▪) at incidence 37° (●), 40° (●) and 44° (●). Suction surface (a), Pressure surface (b) (Note: for the pressure surface a negative wall shear is attached flow.)

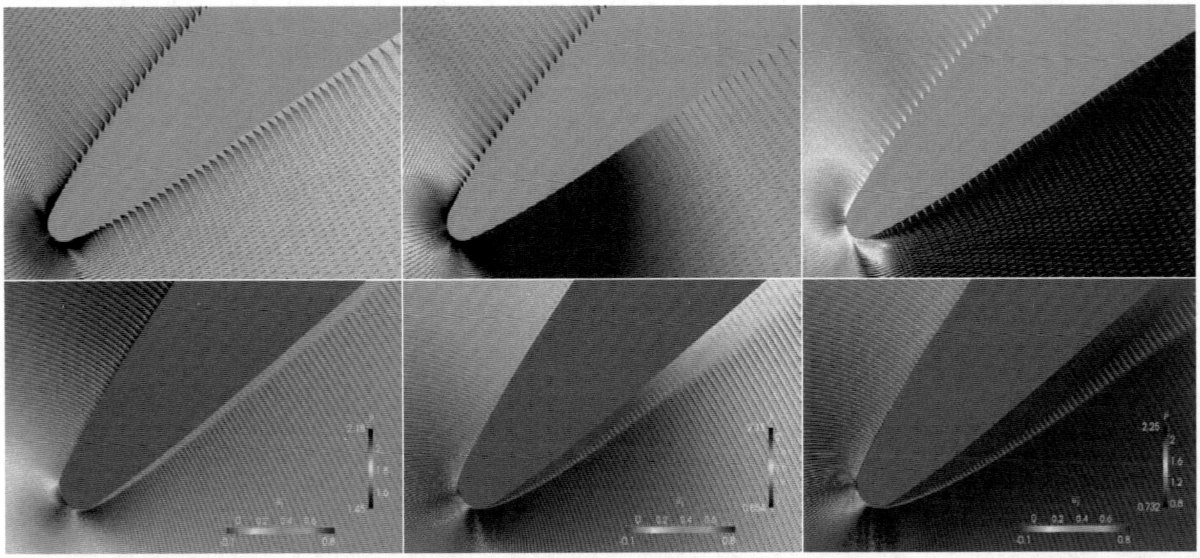

Figure 12 Pressure contour with superimposed velocity vector field for RANS (top) and LES (bottom) at incidences 44°, 40° and 37° from left to right respectively

This increased loading seen in the LES likely contributes to the faster transition (seen in Figure 11) in the same case due to a stronger deceleration of the flow at separation destabilising the shear layer further.

The performance of the off design incidence at 37° shows the most variation as expected, as these are the running conditions RANS finds most challenging. The first notable difference when examining the loading profiles in Figure 10 is the higher peak suction loading seen in the LES. It is clear here that the LES predicts a higher acceleration of the flow up to peak suction, possible attributed to the larger pressure surface separation as seen in the vorticity contours (Figure 7). This increased separation increases the effective blockage and forces a slightly higher peak blade passage velocity.

It is noted that for the 37° case it was found that the Mach number could not be increased to the desired 0.67 at inflow without chocking the flow. Instead the Mach number was increased as far as possible while maintaining an un-chocked condition and run at these conditions. As the RANS struggled most with converging the flow the final conditions were set by the RANS which settled on an inlet Mach number of 0.65. The final running conditions of the LES for the range of incidences are given in Table 2, measured at a reference plane 0.4 chord lengths upstream of the leading edge. It can be seen in the loading profile plots that the flow is locally supersonic over parts of the suctions surface. For the LES case this covered a region from $x/l \sim 0.25$ to 0.45 while the RANS predicted a shorter supersonic region from $x/l \sim 0.25$ to 0.4.

The pressure surface loading shows quite marked variation, notably the LES predicts a lower trailing edge isentropic Mach number which extends upstream to $x/l \approx 0.2$. This is likely attributed to a more substantial blockage in the LES, supported by the larger supersonic region on suction surface, and the resulting deceleration following this blockage.

Wall shear.

The LES suction surface wall shear shows reasonable comparison with the RANS at all but the highest incidence, though the magnitude of the wall shear is comparable and the trends are captured clearly in both LES and RANS. The pressure surface on the other hand shows a marked increase in the time-averaged wall shear for all the LES cases suggesting a more turbulent boundary layer though the trends are again captured by both methods.

Looking back at the 40° loading shown in Figure 10 the close match between the LES and RANS does not help much to explain the lower loss predicted by the RANS. However looking in greater detail at the wall shear trends plotted in Figure 11(a) it shows that the LES undergoes transition faster following separation, with earlier reattachment. It also shows a higher post transition wall shear maximum compared with the RANS suggesting a more turbulent boundary layer adding to the higher loss seen. On the pressure surface the wall shear trends and development match however the LES prediction is on average a factor of 2 higher than the RANS which likely plays a more significant role in the higher loses seen.

The loss prediction at incidence 44° shows a similar higher prediction for the LES compared with RANS. The higher loss predicted by the LES can in part be attributed to the earlier transition seen on the suction surface in Figure 11 (a), transition that most likely occurs via bypass transition in the LES case while a clear separation and turbulent reattachment is seen in the RANS case. Furthermore the pressure surface shear stress, seen in Figure 11 (b), shows notable difference in the two cases. For the LES case transition occurs without separation at $x/l \approx 0.18$ while the RANS predicts a separation from $x/l = 0.15$ to $x/l = 0.3$ followed by turbulent reattachment. The trends downstream of transition for both cases are similar though the LES again shows a wall shear an order of approximately 2 higher. These differences are noticeable in the loading plot in Figure 10 where the pressure surface

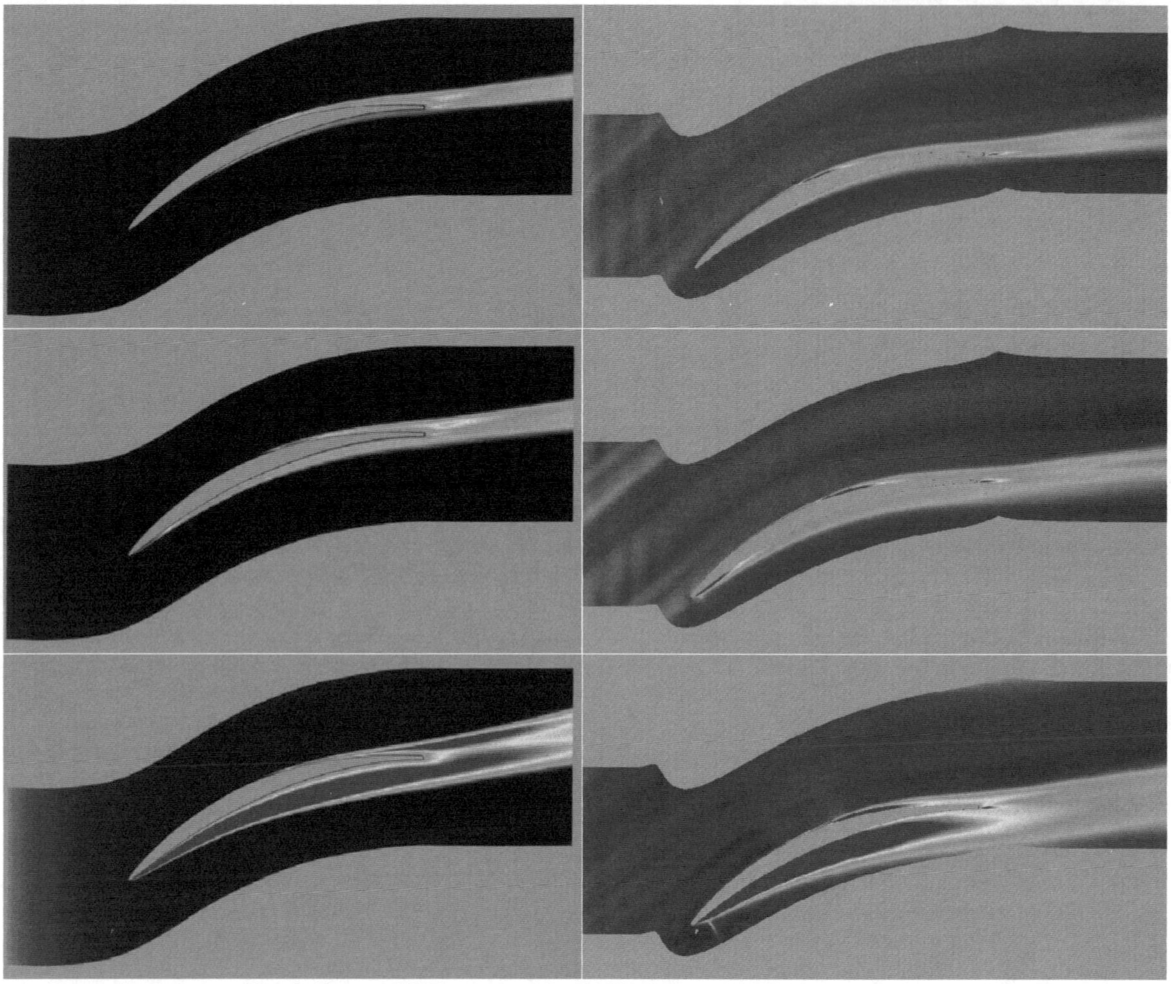

Figure 13 Turbulent Kinetic Energy (TKE) contours for RANS (left) and LES (right) at incidences 44°, 40° and 37° from top to bottom respectively. Highlighting TKE productions areas. Contour ranged [0, 0.02]

separation is clearly seen at x/l ~ 0.25 while the early transition of the LES on the suction surface is seen at x/l ~ 0.35.

For the highest incidence case, $\alpha = 37^o$, the wall shear plots show quite substantial variations as expected from the loading plots. The higher Mach number seen in the LES translates to a notably different wall shear plot as seen in Figure 11(a) where the LES predicts a higher wall shear up to x/l = 0.3 followed by a sharp fall and separation due to the flow locally going supersonic and the presences of a weak shock. The RANS on the other hand does not predict such a sharp reduction in wall shear though there is suggestion that it too captures the weak shock, seen in the initial sharp drop in wall shear at x/l = 0.3 but does not predict the rapid separation seen with the LES. The pressure surface plots for the 37° case suggest a more substantial slowdown of the flow in the LES case though the pressure wall shear Figure 11 (b) shows quite a marked increase in wall shear for the LES compared with the RANS. More interestingly the RANS predicts a notably longer separation compared with the LES which is due to the RANS failure to capturing the large turbulent shedding seen in vorticity contour plots which once temporally averaged in the LES cases result in a substantially shorter separation.

The earlier transition seen in the wall shear plots of the LES cases suggest the increased resolution of the LES captures the breakdown of the boundary layer more accurately than the RANS. Though there is still uncertainty as to whether it is the capturing of free-stream turbulent forcing, boundary layer instabilities or the interaction of these physical processes that lead to the earlier break down.

Wake profiles.

Figure 9 shows the normalised total pressure for a wake profile 0.1 chord lengths downstream of the trailing edge. The LES wake for the 40° case shows a slightly wider wake in comparison with the RANS, somewhat expected due to presence of the turbulence in the LES wake increasing dissipation. The profile for 37° case shows a substantially wider wake for the LES with greater wake dissipation at this position compared with the RANS. Again an expected outcome with the large turbulent structures captured in the LES increasing the wake dissipations and spreading. This is supported in Figure 9 by noting that the majority of the increase in wake width for the LES results is attributed to the pressure side. Furthermore it is seen in the comparison for the 37° case that the peak wake loss of the profile is substantially reduced in comparison with the RANS. Suggesting that the wake dissipation and mixing out in the LES cases is more rapid compared with the RANS simulations

Flow field contours.

To help further explain the trends seen in the loss prediction between the two methodologies the velocity vectors and pressure contours of the leading edge are plotted in Figure 12. These figures are included to qualify the separation and boundary layer variation seen at higher incidences. It is clear with the 37° case that a notable separation originating at the pressure surface leading edge extends far down stream supporting the positive wall shear seen in Figure 11 (b). The peak pressure on the suction surface for the 37° case is also clearly seen with a region of high pressure located near the leading edge supporting the high cross over points in the loading plot. The variation over the incidence range is also highlighted as the clear growth of the pressure surface leading edge separation is easily visible.

The turbulent kinetic energy (TKE) contour plots seen in Figure 13 also highlight this variation in the pressure surface flow. It is clear in the 44° case that little loss is generated by the leading edge and that turbulent break down occurs further along the blade. This is captured in both the RANS and LES though the LES predicts the turbulent break down sooner, producing a slightly wider wake at the trailing edge. However for the 37° and 40° incidences the pressure surface boundary layer is turbulent from the leading edge. It is also noted that the LES captures a weak shock at the pressure surface leading edge separation for the 37° case. This shock is not seen in the RANS case.

Boundary layer.

The displacement thickness, momentum thickness and shape factor for the LES suctions surface over the incidence range is plotted in Figure 14. It is seen in these three figures the development of the suctions surface boundary layer follows a clear laminar growth till separation or transition followed by the growth of a turbulent boundary layer in all cases.

Wall normal profiles.

The tangential mean velocity profiles seen in Figure 15 show the development of the boundary layer at positions x/l = [0.54, 0.64, 0.76,. 0.99] for incidences 37°, 40° and 44°. The tangential velocity profiles show the clear trend in increased flow rate for the three incidence cases, also detailing the slight increase in boundary layer thickness as the blade loading is reduced form 37° to 44°. For all the cases a fully developed turbulent boundary layer is established down stream of x/l = 0.64, while for the 37° case the boundary layer appears slightly transitional at x/l = 0.54 while the 40° and 44° cases are already turbulent. Comparing with the wall shear plot in Figure 11 this is near the reattachment point for the 37° case. The comparison between the RANS and LES at 40°

Table 2 Final running conditions for LES simulations

Angle	36.99°	39.97°	44.09°
Re	302,566	302,518	298,084
Mach	0.654	0.674	0.666
% Tu	0.029	0.033	0.032

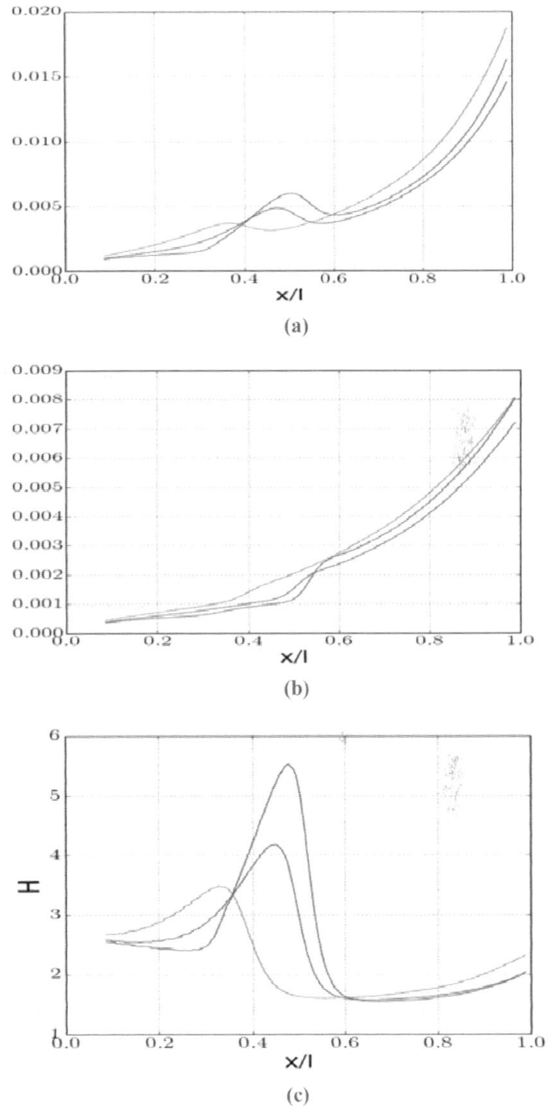

Figure 14 Momentum thickness (a), displacement thickness (b) and Shape factor (c) for LES suctions surface at incidences 37° (●), 40° (●) and 44° (●).

shows good agreement with boundary layer thickness though the RANS predicts a more constant velocity profile with a lower velocity gradient nearer the wall.

The trends seen in Figure 16 for the wall normal turbulence intensity are supportive of those seen in the tangential velocity plots. All three incidence shows high levels of turbulence in the boundary layer indicative of a turbulent boundary layer. The 37° case at x/l = 0.54 shows the most variation with a clear peak in turbulence intensity near the wall suggesting the turbulent sear layer of the separation bubble is reenergising the boundary layer and reattaching. This turbulent boundary layer is seen to establish itself and grows steadily for the remainder of the wall normal plots, with the peak intensity reducing and moving upward away from the wall. The 40° case also shows a similar trend in turbulence intensity progression. Showing a highly energetic think boundary layer at x/l = 0.54 following the reattachment of the separation and the establishment of a turbulent boundary layer. The progression downstream shows the TKE profile steadily progress to mirror that seen in the 37° case.

The 44° case does not have a separation in the LES simulations and the trend in wall normal turbulence intensity and tangential velocity shows this accordingly. It is seen at x/l = 0.54 that the boundary layer already has a turbulent TKE profile, with the peak turbulence intensity found further from the wall and a thicker distribution across the boundary layer. Again the progression downstream sees the boundary layer thicken and the peak intensity move further from the wall following the trends seen in the two higher incidences. The comparison of the RANS and LES at 40° sees the RANS predict a sharper TKE profile with slower grow up till x/l =0.99. Interestingly at x/l = 0.99 the TKE profile width matches that of the LES though the profile shape suggests that a shear layer maybe developing away from the blade surface as the peak intensity is found toward the outside of the boundary layer.

CONCLUSIONS

The investigation performed here compares the loss prediction performance of LES and RANS applied to an axial compressor. This initial investigation shows promise in overcoming short falls in current methodologies though further analysis in understanding and interpreting the differences in the results is needed.

It is seen in the statistical comparisons that the LES in this investigation predicts a higher loss than the RANS at the nominal and low incidences while it predicts lower losses at the high incidence. The reasoning behind the increased loss prediction at lower incidences is still unclear without further work though initial investigations suggest the LES predicts earlier breakdown and subsequently larger wakes than the RANS due to the improved resolution of the boundary layer and the capturing of free stream turbulence.

The loss prediction at the nominal incidence is still being investigated to explain the large variation between the RANS, LES and experimental data at design condition. It is likely though, as noted earlier, that what has been run as nominal parameters here may not be comparable when compared with the experimental data as the loading could not be matched at the specified incidence. This highlights the importance in matching all parameters when comparing to experimental data as values such as the turbulent length scales where not known in this investigation.

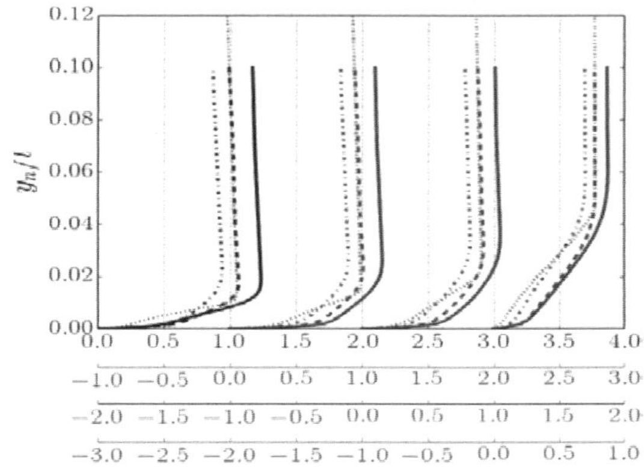

Figure 15 Comparison of LES and RANS (▪▪▪, 40° only) suctions surface wall normal tangential mean velocity profiles for 37° (—), 40°(— ▪) and 44°(— —) incidences at x/l = 0.56 (●), x/l = 0.64 (●), x/l = 0.76 (●) and x/l = 0.99 (●).

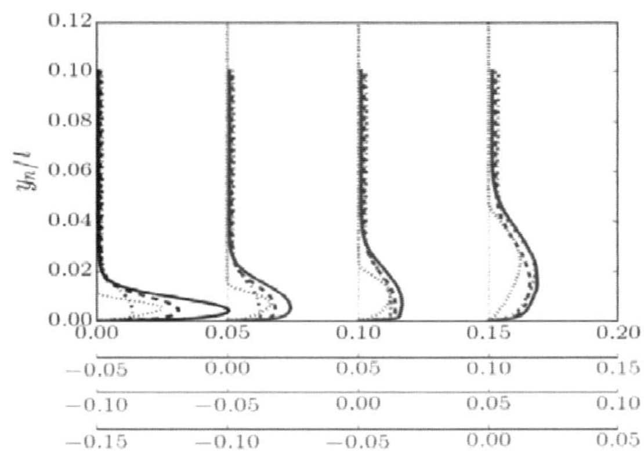

Figure 16 Comparison of LES and RANS (▪▪▪, 40° only) wall normal turbulent kinetic energy profiles for 37° (—), 40° (— ▪) and 44°(— —) incidence case at x/l = 0.56 (●), x/l = 0.64 (●), x/l = 0.76 (●) and x/l = 0.99 (●).

The lower loss predicted in the higher incidence case by the LES is attributed here to the improved prediction of the turbulent separation of the pressure surface. It has been shown that substantial variation exists between the RANS and LES in predicting the pressure surface behaviour, most notably when a leading edge separation and early transition are present.

It is seen in the wall shear plots that the LES in these cases predict a higher average value for the pressure surface than the RANS, the difference increasing with increasing incidence. Furthermore the prediction of separation and transition between the two methodologies for the suction surface boundary layer shows the LES capturing greater variation with change of incidence, while the RANS predicts quite similar behaviour in each case. The RANS shows a clear separation and turbulent reattachment in every case while the LES varies from bypass transition to separation due to a weak shock followed by turbulent reattachment. This can again be attributed to the better resolution of the LES and the ability to capture free-stream fluctuations that are likely to force the boundary layer, though further analysis of this data is required to explain fully the difference seen here.

Overall this initial investigation supports the notion that LES improves the capture of the flow of axial compressors at off design conditions, though some further study is required to explain the differences between RANS and LES at design conditions, which remains somewhat unclear.

ACKNOWLEDGMENTS

Acknowledgment is given to the UK turbulence consortium for the provision of computational time on the UK national computer Archer under grant number EP/L000261/1. The first author would like to acknowledge the financial support from the EPSRC and General Electric.

NOMENCLATURE

l	axial chord length
x	axial distance
y_n	wall normal distance
ω	pressure loss coefficient $(P_{01}-P_{02})/(P_{01}-P_1)$
P	static pressure
P_t	total pressure
α	inflow incidence
β	cascade parameter angle
C_p	coefficient of pressure
C_f	coefficient of friction
u_t/Tu	turbulence intensity
M_{isen}	isentropic Mach number
aoa	angle of incidence
τ_w	normalised wall shear
δ_0	displacement thickness
δ_1	momentum thickness
H	shape factor δ_0/δ_1

SUBSCRIPTS

1,2	inlet, outlet plane

REFERENCES

[1] Leipold, R., Boese, M., and Fottner, L., 2000, "The Influence of technical surface roughness caused by precision forging on the flow around a highly loaded compressor cascade," Journal of Turbomachinery, 122, pp. 416-425.

[2] Denton, J. D., 1993, "The 1993 IGTI Scholar Lecture: Loss Mechanisms in Turbomachines," Journal of Turbomachinery, 115(4), pp. 621-621.

[3] Steinert, W., and Starken, H., 1996, "Off-Design Transition and Separation Behaviour of a CDA Cascade," ASME, 118, pp. 204-209.

[4] Evans, S., Hodson, H., Hynes, T., and Wakelam, C., 2010, "Flow Control in a Compressor Cascade at High Incidence," Journal of Propulsion and Power, 26(4), pp. 828-836.

[5] You, D., Wang, M., Moin, P., and Mittal, R., 2007, "Large-eddy simulation analysis of mechanisms for viscous losses in a turbomachinery tip-clearance flow," Journal of Fluid Mechanics, 586, pp. 177-177.

[6] Boudet, J., Caro, J., Shao, L., and Lévêque, E., 2007, "Numerical studies towards practical large-eddy simulation," Journal of Thermal Science, 16(4), pp. 328-336.

[7] Duchaine, F., Corpron, a., Pons, L., Moureau, V., Nicoud, F., and Poinsot, T., 2009, "Development and assessment of a coupled strategy for conjugate heat transfer with Large Eddy Simulation: Application to a cooled turbine blade," International Journal of Heat and Fluid Flow, 30(6), pp. 1129-1141.

[8] Bhaskaran, R., and Lele, S. K., 2010, "Large eddy simulation of free-stream turbulence effects on heat transfer to a high-pressure turbine cascade," Journal of Turbulence, 11(July 2015), pp. N6-N6.

[9] Greschner, B., and Thiele, F., 2011, "Wall Modeled LES simulation of Rotor-Stator-Cascade Broadband Noise," AIAA Conference(June), pp. 05-08.

[10] Gourdain, N., 2015, "Prediction of the unsteady turbulent flow in an axial compressor stage. Part 1: Comparison of unsteady RANS and LES with experiments," Computers & Fluids, 106, pp. 119-129.

[11] McMullan, W. A., and Page, G. J., 2012, "Towards Large Eddy Simulation of gas turbine compressors," Progress in Aerospace Sciences, 52, pp. 30-47.

[12] Hah, C., 2009, "Large Eddy Simulation Of Transonic Flow Field In NASA," (January), pp. 1-13.

[13] Klein, M., Sadiki, a., and Janicka, J., 2003, "A digital filter based generation of inflow data for spatially developing direct numerical or large eddy simulations," Journal of Computational Physics, 186(2), pp. 652-665.

[14] Sandberg, R. D., Sandham, N. D., and Suponitsky, V., 2012, "DNS of compressible pipe flow exiting into a coflow," International Journal of Heat and Fluid Flow, 35, pp. 33-44.

[15] Sandberg, R. D., 2012, "Numerical investigation of turbulent supersonic axisymmetric wakes," Journal of Fluid Mechanics, 702, pp. 488-520.

[16] Sandberg, R., Suponitsky, V., and Sandham, N., 2011, "DNS of a canonical compressible nozzle flow," H. Kuerten, B. Geurts, V. Armenio, and J. Fröhlich, eds., Springer Netherlands, pp. 291-296.

[17] Sandberg, R. D., Michelassi, V., Pichler, R., Chen, L., and Johnstone, R., 2015, "Compressible direct numerical simulation of low-pressure turbines: part I - methodology," Journal of Turbomachinery, 137(5), pp. 51011-51011.

[18] Kennedy, C. a., Carpenter, M. H., and Lewis, R. M., 2000, "Low-storage, explicit Runge–Kutta schemes for the compressible Navier–Stokes equations," Applied Numerical Mathematics, 35(3), pp. 177-219.

[19] Kennedy, C. A., and Gruber, A., 2008, "Reduced aliasing formulations of the convective terms within the Navier-Stokes equations for a compressible fluid," Journal of Computational Physics, 227(3), pp. 1676-1700.

[20] Nicoud, F., and Ducros, F., 1999, "Subgrid-Scale Stress Modelling Based on the Square of the Velocity Gradient Tensor," Flow, Turbulence and Combustion, pp. 183-200.

[21] Kim, J. W. J. W., and Lee, D. J. D. J., 2003, "Characteristic Interface Conditions for Multiblock High-

Order Computation on Singular Structured Grid," AIAA journal, 41(12).
[22] Sandberg, R. D., and Sandham, N. D., 2006, "Nonreflecting zonal characteristic boundary condition for direct numerical simulation of aerodynamic sound," AIAA journal, 44(2), pp. 3-6.
[23] Zaki, T. a., Wissink, J. G., Rodi, W., and Durbin, P. a., 2010, "Direct numerical simulations of transition in a compressor cascade: the influence of free-stream turbulence," Journal of Fluid Mechanics, 665, pp. 57-98.

Proceedings of the 14th International Symposium on Unsteady Aerodynamics, Aeroacoustics & Aeroelaticity of Turbomachines
ISUAAAT14
8-11 September, 2015, Stockholm, Sweden

I14-S8-2

DETACHED-EDDY SIMULATION OF INTERMITTENT CORNER SEPARATION IN A LINEAR COMPRESSOR CASCADE

W. MA

School of Aeronautics and Astronautics,
Shanghai Jiao Tong University,
Shanghai 200240, P. R. China
mawei@sjtu.edu.cn

A.J. Wang

School of Mechanical Engineering,
Shanghai Jiao Tong University,
Shanghai 200240, P. R. China
aj.wang@sjtu.edu.cn

ABSTRACT

Nowadays it's widely accepted that in order to further investigate the mechanisms of corner separation we need to investigate deeply the instantaneous flow field, because corner separation is fully unsteady and highly intermittent. Recently bimodal phenomenon in corner separation has been found by Ma et al. (Experiments in Fluids, 2013, doi:10.1007/s00348-013-1546-y). Through detailed and accurate experimental results of the velocity flow field in a linear compressor cascade, they discovered two modes exist in the corner separation of a compressor cascade. In addition, these two modes are aperiodic. However the configuration and evolution of each mode in this bimodal phenomenon are still unclear, thus the bimodal phenomenon needs to be investigated further. The main objective of this paper is to further investigate the bimodal phenomenon in corner separation using detached-eddy simulation (DES), which is already widely used in foundational researches. DES can simulate successfully the intermittent flow in the corner separation within certain accuracy. Bimodal points exist in the DES results. Additional DES results can successfully capture the evolution of bimodal points along a station normal to blade surface, as well as their locations in this linear compressor cascade that has found in experiment. Through proper orthogonal decomposition (POD) analysis of the DES instantaneous flow field, no POD mode dominates the flow field. At the same time, there is not strong connection between modes. Meanwhile each POD mode is aperiodic, consistent with the bimodal phenomenon found in experiment.

1 INTRODUCTION

Corner separation, the three-dimensional separation in the junction of the endwall and blade corner region, is an inherent flow feature in compressors and relates to the efficiency and stability of compressors [1]. Generally speaking, corner separation is caused mainly by the combined effects of strong streamwise pressure gradient, presence of secondary flow and merging of the wall and blade boundary layers. Numbers of experimental and numerical studies have been carried out to investigate the impacts, the flow topology, the influencing factors, as well as global consequence of corner separation [2, 3], in which most of the studies are based on the steady flow fields. However until now an effective control of these effects is still very difficult to achieve, because the mechanisms of corner separation are not clearly understood [2]. Because corner separation is fully unsteady and highly intermittent, thus we need to investigate deeply the instantaneous flow field in order to further investigate the mechanisms of corner separation.

In some highly intermittent flow, a remarkable phenomenon is the existence of bimodal histograms of velocity (referred as "bimodal histograms" later), which have two obvious peaks. In the literature, this phenomenon has been found and discussed in different flow fields [4–7].

The most famous flow field is the junction flow firstly discussed by Devenport and Simpson [4], in which bimodal histograms are mainly found in the vicinity of the wing nose. They claimed that the flow switched aperiodically between two different modes. Following there are numerous numerical simulations to understand further the physics of such flow. Recently Gand et al. [5] reviewed comprehensively both the experiments and simulations. Now it is widely accepted that the bimodal histograms in the junction flow are consistent with large-scale low-frequency unsteadiness of the instantaneous flow structure associated with the horseshoe vortex.

In subsonic turbomachinery flows, which are more complex than the junction flow mentioned above, bimodal histograms usually exist in two situations. One is in the blade wakes that are similar to the Karman vortex street, due to the vortex shedding. The other one is in tip-clearance flows [6, 8], due to the interaction between the tip-leakage vortex and tip-leakage jet. These interactions are periodic, and the frequencies of the vortex shedding and interaction can be fixed in the spectra of instantaneous velocities. This is the obvious difference with the bimodal phenomenon in the junction flow.

Linear compressor cascade is a simplified model for investigating the flow field in the compressor. In the passage of a linear subsonic compressor cascade without tip-clearance, Hobson et al. [7] found bimodal histograms in the flow field at mid-span measured by one-component laser Doppler anemometry (LDA). However they only associated the bimodal histograms with the backflow coefficients, and did not interpret further its physics.

Recently Ma et al. [9] discovered firstly in the literature that bimodal phenomenon exists in the flow field of corner separation without leakage by experiments. Through detailed and accurate experimental results of the velocity flow field in the linear compressor cascade, they discovered that the points with bimodal histograms of velocity exist and mainly appear around the mean interface of separated flow and non-separated flow. Then they decomposed successfully the bimodal histogram into two Gaussian distributions, and thus claimed that two modes exist in the flow field of corner separation. Because obvious peak does not exist in the spectra of the velocity components at the points with bimodal histograms of velocity, they came to the conclusion that two modes in the flow field are aperiodic. However the configuration and evolution of each mode in this bimodal phenomenon are still unclear. In order to improve the knowledge of corner stall and then find more effective methods to control corner stall, the bimodal phenomenon needs to be investigated further.

Computational fluid dynamics (CFD) has been a more and more powerful tool to investigate the flow field in turbomachinery with the rapid developments of computing abilities. The experimental data of Ma et al. have been used in some numerical researches [10–12]. Gao et al. [10] carried out an unsteady simulation based on Reynolds-averaged Navier-Stokes equations (URANS) to investigate the influence of a fluctuation inflow. Their results show that the fluctuation does not show much influence on the mean performance of the compressor cascade. Wang et al. [11] carried out an unsteady simulation based on hybrid LES/RANS to investigate the unsteady separation structure and the sources of unsteadiness. Gao et al. [12] carried out large eddy simulation (LES) to study further the flow field in corner separation. In these numerical investigations, the main objectives dedicate to investigating the unsteady mechanisms of corner separation, instead of further investigating the bimodal phenomenon in corner separation.

The first objective of this paper is to further investigate the bimodal phenomenon in corner separation using DES, which is already widely used in designs and researches [13]. DES is carried out firstly in the same configuration and with the same inlet flow conditions as the experiment of Ma et al. [9]. The numerical results are then compared with the experimental results including the static pressure on the blade, and the velocity flow fields in the passage. As a consequence, unsteady flow field are presented and used to find bimodal points and further investigate the mechanisms of bimodal phenomenon. At last POD analysis is used to find the modes for cross-section in spanwise direction.

2 EXPERIMENTAL CONFIGURATION

The experimental investigation was carried out in a low-speed linear compressor cascade wind tunnel at a Reynolds number of 3.82×10^5 based on the blade chord and the inlet free stream velocity [9]. Figure 1 shows the schematic of the experimental arrangement. The parameters of the cascade are summarized in Table 1.

The experimental data include (i) the inlet flow conditions measured by one-dimensional hot-wire anemometry, (ii) the static pressure on the blade and the endwall measured by pressure taps, (iii) the outlet flow measured by a five-hole pressure probe, and (iv) the velocity field in the passage measured by two-component LDA.

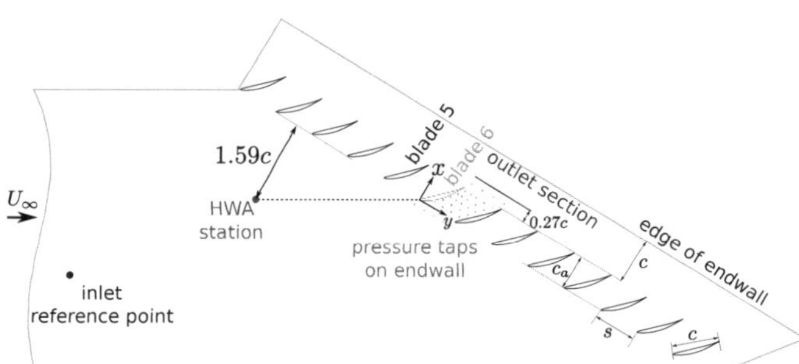

FIGURE 1. Schematic of the experimental arrangement.

TABLE 1. Parameters of the cascade.

Blade profile	NACA 65
Blade number	13
Chord	c = 150 mm
Camber angle	23.22°
Stagger angle	42.70°
Axial chord	c_a = 110 mm
Pitch	s = 134 mm
Solidity	1.12
Span	h = 370 mm
Aspect ratio	2.47
Optimum incidence	0.18°
Design upstream flow angle	54.31°
Incidence angle	4°
Inlet velocity	U_∞=40.0 m/s
Reynolds number	$Re = U_\infty c/\nu$=3.82×10^5

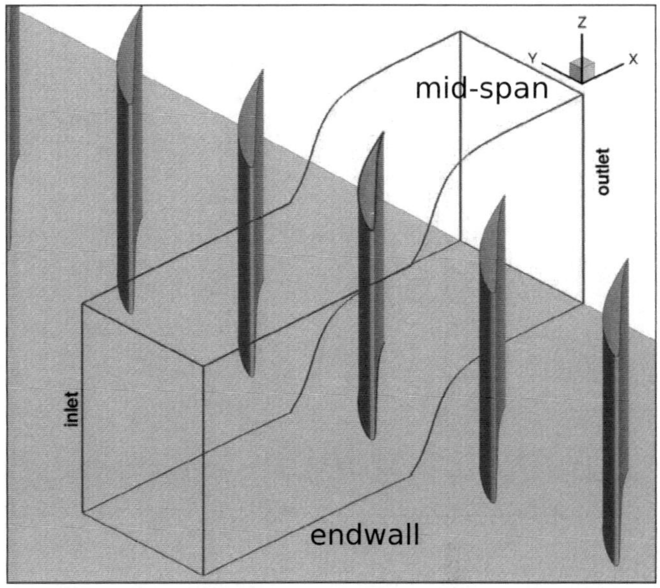

FIGURE 2. Grid distribution.

3 NUMERICAL METHODS

3.1 Computational grids

The computational mesh used in the current investigation is generated by AutoGrid5, a commercial software package.

The distribution of computational passage is shown in Fig. 2, including it's global view and detailed grids at the leading edge and trailing edge. To be able to compare these results with the experimental results, the inlet and the exit of the computational domain are placed at 1.59 chords upstream of the leading edge and 1.00 chord downstream of the trailing edge of the blade, respectively.

A O4H multi-block method of about 3.82×10^6 cells is used to ensure the grid quality. The cell height at the wall is *a priori* set to about Δy=1×10^{-5}, yielding Δy^+ <2 both for the blade and the endwall.

3.2 Boundary conditions

In the computation the flow is assumed to be unsteady and fully turbulent.

The inlet velocity profile is obtained from the hotwire in the experiment and the incidence angle is specified for the inlet boundary. The turbulent intensity of 0.8% is specified for inlet boundary, in accordance with the present experimental values.

A uniform atmospheric static pressure is employed for the outlet boundary. Furthermore, nonslip and adiabatic conditions are adopted for all of the solid walls. A symmetry condition is employed at mid-span because just one half of the blade span is simulated. Periodic conditions are imposed along the pitchwise boundaries.

3.3 Numerical scheme and turbulence model

The commercial computational fluid dynamics software packages ANSYS FLUENT 15.6 is used.

For the solver, the pressure-based Navier-Stokes solution algorithm is enabled, because in this simulation the inlet velocity is just about 0.12 and the flow field is incompressible. At the same time, absolute velocity formulation, and time-dependent solution are enabled. Detached-eddy model is used.

The pressure-velocity coupling is handled by the SIMPLE algorithm. A first order upwind scheme and a central-differencing scheme are used for the convection terms and the viscous terms of each governing equation, respectively.

A first order implicit method is used for time marching with a global constant time step of 5.0×10^{-5}s.

In this work, 2600 time steps are performed. And total simulation time of 0.13s is used to determine statistical quantities. About 35 periods that the flow passes through the blade passage (c/U_∞) has been achieved.

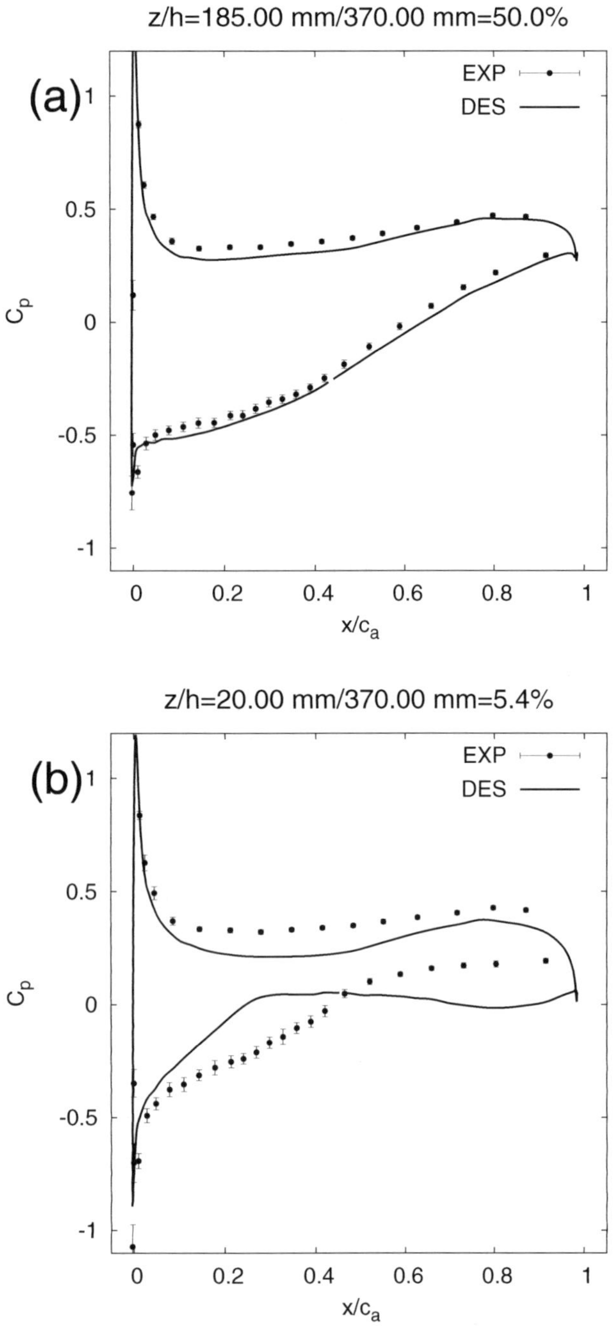

FIGURE 3. Mean static pressure coefficient C_p on blade surface. (a) Mid-span; (b) near endwall.

4 MEAN STATIC PRESSURE ON BLADE SURFACE

The experimental and time-averaged numerical results of static pressure distributions on the blade in the spanwise sections are compared in Fig. 3,in which the static pressure coefficient is defined as

$$C_p = \frac{p_s - p_{s\infty}}{p_{t\infty} - p_{s\infty}} \tag{1}$$

where $p_{t\infty}$, $p_{s\infty}$ are the reference inlet mean total and static pressures, respectively; p_s is the mean static pressure at one point. To make this paper more concise, only two representative sections are chosen here. The first one is at mid-span (z/h=50.0%); and the other one is near the endwall and in the region of corner stall (z/h=5.4%). DES reproduces acceptably the pressure distributions on the blade at mid-span; however, fail to reproduce the pressure distributions near the endwall where the three-dimensional separations occur. The agreements at mid-span partly reflect that the experimental set-up is reliable. On the other hand, the disagreements in the corner region reflect the necessity of such experimental data, which could be used to calibrate the advanced CFD.

5 MEAN VELOCITY IN PASSAGE

Experimental and numerical of mean velocity fields in the plane z=5.00 mm and 20.00 mm are shown in Fig. 4, in which $s^* = s/L$, where s is the length of arc from the leading edge to the measurement location, L is the length of arc of the blade suction side. In this figure, separation region is shown more clearly by the contour of the backflow coefficient η of u_s (i.e. the portion of the acquired velocity data per measurement point that includes negative velocities of the streamwise velocity).

DES results are shown in Fig. 5. In general DES numerical results can capture the overall pattern of corner separation, but not the flow details.

At z=5.00 mm, the streamlines of LDA experiment have a jump between s^*=0.50∼0.60. From the view of bimodal phenomenon in the corner separation, this jump shows that the separation increases extremely in this zone, and thus indicates that two configurations of two modes interact highly in this zone. DES result at plane (in Fig. 5) do not obtain this feature

For the separation points of DES results locate at s^*=0.31 at z=5.00mm and at s^*=0.41 at z=20.00 mm, while the experimental separation points locate at s^*=0.35 at z=5.00 mm and s^*=0.50 at z=20.00 mm. Thus DES result separates earlier compared with the LDA experiment.

6 BIMODAL PHENOMENON

Through detailed and accurate experimental results of the velocity flow field in the linear compressor cascade, Ma et al. [9]

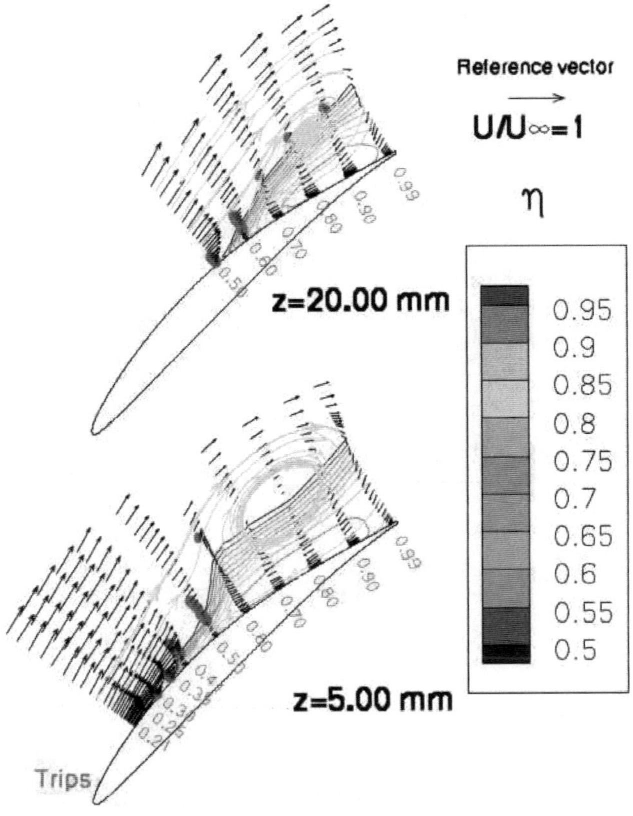

FIGURE 4. Experimental mean velocity at z=5.00 mm and 20.00 mm.

discovered that the points with bimodal histograms of velocity exist in the flow field. The pink points in Fig.4 indicate the positions of the bimodal points at the spanwise sections z=5.00 mm and 20.00 mm. They mainly appear around the mean interface of separated flow and non-separated flow.

In DES results, we indeed found bimodal points where two obviously peaks exist in its PDF of velocity. The locations of these bimodal points in the LDA measurement stations are shown by pink lines in Fig. 5. Consistent with the experimental data, these bimodal points also mainly appear around the mean interface of separated flow and non-separated flow.

Fig. 6 shows the PDFs of u_x at 10 points at s^*=0.50 and z=20.00 mm. This measurement station (see in Fig. 4) is located near the endwall and just upstream of the mean separation point in its spanwise cross-section. The main feature characterising these PDFs is the presence of one or two peaks, depending on the locations of the measurement points. Only one peak exists near the endwall as the measurement point where n=6.5 mm and 8.0 mm. Getting further from the wall, a second peak appears

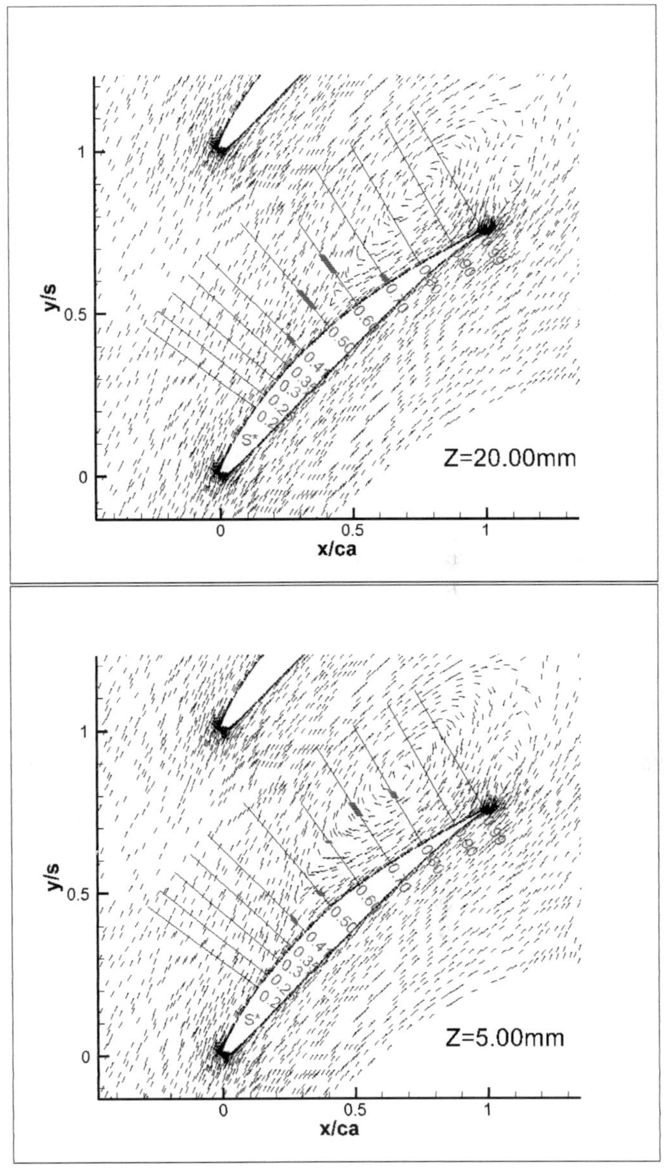

FIGURE 5. DES mean velocity at z=5.00 mm and 20.00 mm.

and becomes more and more important compared with the first peak, as the measurement points where n=9.5 mm $\sim$ 18.5 mm. At last, only the second peak exists at the measurement point where n=20.00 mm.

We then try to decompose the bimodal histogram into two Gaussian distributions.

$$P(u_x) = G_1(u_x; a_1, U_{x1}, \sigma^2_{u_{x1}}) + G_2(u_x; a_2, U_{x2}, \sigma^2_{u_{x2}}) \qquad \text{(2a)}$$

where

$$G_1(u_x;a_1,U_{x1},\sigma^2_{u_{x1}}) = \frac{a_1}{\sqrt{2\pi}\cdot\sigma_{u_{x1}}}\exp\left[-\frac{(u_x-U_{x1})^2}{2\sigma^2_{u_{x1}}}\right] \quad (2b)$$

$$G_2(u_x;a_2,U_{x2},\sigma^2_{u_{x2}}) = \frac{a_2}{\sqrt{2\pi}\cdot\sigma_{u_{x2}}}\exp\left[-\frac{(u_x-U_{x2})^2}{2\sigma^2_{u_{x2}}}\right] \quad (2c)$$

with the requirements

$$\begin{cases} a_1, U_{x1}, \sigma_{u_{x1}}, a_2, U_{x2}, \sigma_{u_{x2}} \in \mathbb{R}, \\ 0 < a_1, a_2 < 1.0, \\ a_1 + a_2 = 1.0, \\ u_{xmin} < U_{x1} < U_{x2} < u_{xmax}, \\ \sigma_{u_{x1}}, \sigma_{u_{x2}} > 0. \end{cases} \quad (2d)$$

where u_{xmin} and u_{xmax} are the minimum and maximum values of u_x, respectively. The parameter a_1 is the probability of the first Gaussian distribution. U_{x1} and U_{x2} are the mean velocities of the first and second Gaussian distribution, respectively. After the convergence of the fitting process, the correlation coefficients for the sum of two Gaussian distributions of the original histograms are above 0.97.

This means that DES can successfully capture the bimodal phenomenon in this linear compressor cascade that has found in experiment. And thus DES results could be used to further investigate the mechanisms of bimodal phenomenon, such as velocity configuration of each mode.

7 POD ANALYSIS

Proper orthogonal decomposition (POD) is a useful technique to analyse the complex flow phenomenon [14]. This technique has been proved to be an effective method for identifying dominant features and events both in experimental and numerical data [15–18], such as flow downstream of a half cylinder in terms of passive control, flow past a backward-facing step, annular jet, internal combustion engine flow, in-cylinder engine flow, natural convective flow above an isothermal horizontal plate, turbulent jet in cross flow, and flow around a circular cylinder. The idea of POD analysis is assumed that spatiotemporal velocity field $\boldsymbol{u}(\boldsymbol{x},t)$ can be decomposed into two individual parts, the spatial characteristic $\boldsymbol{\phi}_n(\boldsymbol{x})$ and temporal characteritic $\boldsymbol{a}_n(t)$.

$$\boldsymbol{u}(\boldsymbol{x},t) = \sum_{n=1}^{\infty} \boldsymbol{a}_n(t)\boldsymbol{\phi}_n(\boldsymbol{x}) \quad (3)$$

The modes obtained are simply the eigenfunctions of the velocity autocorrelation operator and form an orthogonal set. For a

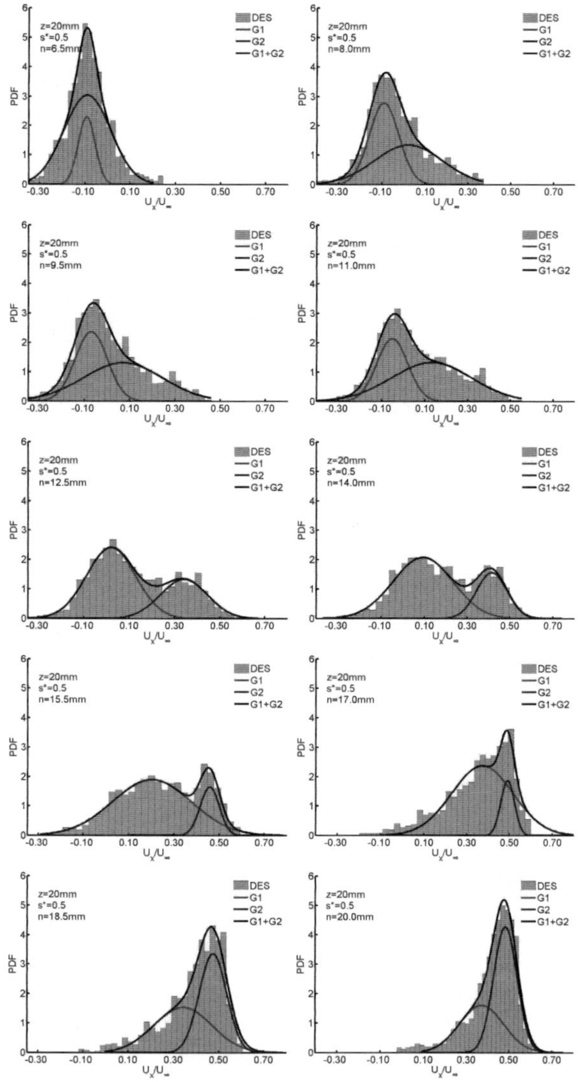

FIGURE 6. Bimodal points at s^*=0.50 and z=20.00 mm, n=6.5 mm $\sim$ 20.0 mm.

spatiotemporal velocity field $\boldsymbol{u}(\boldsymbol{x},t)$, the eigenfuction ϕ needs to be determined in order to make the following fuction maximized

$$\frac{\langle(\boldsymbol{u},\boldsymbol{\phi})(\boldsymbol{u},\boldsymbol{\phi})\rangle}{(\boldsymbol{\phi},\boldsymbol{\phi})} = \lambda \geqslant 0 \quad (4)$$

where λ is the eigenvalue that is also called as mode energy, $\langle\rangle$ is the ensemble-averaging operation, and $(,)$ is the standard inner produce. maximization of the above quantity in Eq. 4 is a classical problem of eigenvalue and leads to

$$\int \langle \boldsymbol{u}(\boldsymbol{x})\boldsymbol{u}^*(\boldsymbol{x}')\rangle \boldsymbol{\phi}(\boldsymbol{x})d\boldsymbol{x}' = \lambda \boldsymbol{\phi}(\boldsymbol{x}) \tag{5}$$

where the superscript $*$ denotes the comple conjugate.

In this paper the method of Meyer et al. [15] is used. This method is introduced in the following. Supporting flow field velocity fluctuations have form $\boldsymbol{u} = (u, v)$, all fluctuation velocity components from the N snapshots can be arranged in a matrix

$$\boldsymbol{U} = [\boldsymbol{u}_1 \boldsymbol{u}_2 \cdots \boldsymbol{u}_N] \tag{6}$$

The autocovariance matrix is then created by

$$\boldsymbol{M} = \boldsymbol{U}^T \boldsymbol{U} \tag{7}$$

where the superscript ^{T} denotes the transpose operator. The above eigenvalue problem can thus be resoved by

$$\boldsymbol{M}\boldsymbol{A}_i = \lambda_i \boldsymbol{A}_i \tag{8}$$

where λ is the eigenvalue and $\boldsymbol{A}$ is the eigenvector. The calculated eigenvalues are ordered from big to small in order to ensure that the most important modes in terms of energy are the first few modes. After that, the eigenvectors make up a basis for constructing the POD modes,

$$\boldsymbol{\phi}_i = \frac{\sum_{n=1}^{N} A_{i,n}\boldsymbol{u}_n}{\|\sum_{n=1}^{N} A_{i,n}\boldsymbol{u}_n\|}, i = 1, 2, \cdots, N \tag{9}$$

where $A_{i,n}$ is the nth component of the eigenvector corresponding to the ith eigenvalue λ_i from Eq. 7 and the discrete 2-norm is defined as

$$\|y\| = \sqrt{y_1^2 + y_2^2 + \cdots + y_M^2} \tag{10}$$

The POD coefficients are determined by projecting the fluctuating part of the velocity field onto the POD modes,

$$\boldsymbol{a}_n = \boldsymbol{\Psi}^T \boldsymbol{u}_n \tag{11}$$

where $\boldsymbol{\Psi} = [\boldsymbol{\phi}_1 \boldsymbol{\phi}_2 \cdots \boldsymbol{\phi}_N]$. Original data can then bereconstructed using the first Nth POD modes

$$\boldsymbol{u}_n = \boldsymbol{\Psi} \boldsymbol{a}_n \tag{12}$$

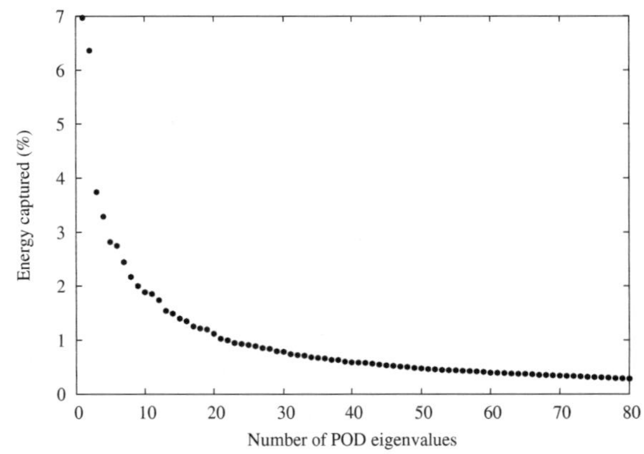

FIGURE 7. Percentage of the total kinetic energy of velocity fluctuations associated with POD modes

Since it required a large amount of computer resources while POD calculating, in this paper POD analysis is only used to find the modes for some cross-sections in spanwise direction. The data from each cross-section is based on 2600 snapshots. To investigate the dependence on the number of snapshots, the analysis has been repeated with only 1200 snapshots. The results were qualitatively identical to the results based on 2600 snapshots, even there are some slight variations. The accuracy of the POD based on 2600 snapshots is therefore found to be satisfactory for the present analysis. The POD results at different spanwise sections are similar.

Fig. 7 shows the percentage of the total kinetic energy of velocity y fluctuations associated with POD modes. The first ten modes contain together 34.4% of the total kinetic energy. The first 127 modes contain together 90.0% of the total kinetic energy. The percentages of the first mode to the fourth mode are 6.97%, 6.36%, 3.74% and 3.29%, respectively. This means that no mode from POD analysis dominates the flow field.

Here the POD analysis at z=5.00 mm are presented and discussed. Fig. 8 shows the flow fields of the first four modes at this spanwise plane. The flow fields of the first mode and the second mode are very similar. There is a relatively large vortex in the passage. These two configurations are also very similar to the mean velocity flow field shown in Fig. 5. The flow fields of the third mode and the fourth mode are very similar. There are lot small vortex in the passage compared with that of the first mode and the second mode.

Fig. 9 indicates the time history of the first four POD coefficients at z=5.00 mm spanwise plane. The scatter plot of the two coefficients found for all snapshots introduced in Meyer et al. [15] are used here to find the relation between two modes. However they are not show circular distributions. This means that there is not strong connection between two modes.

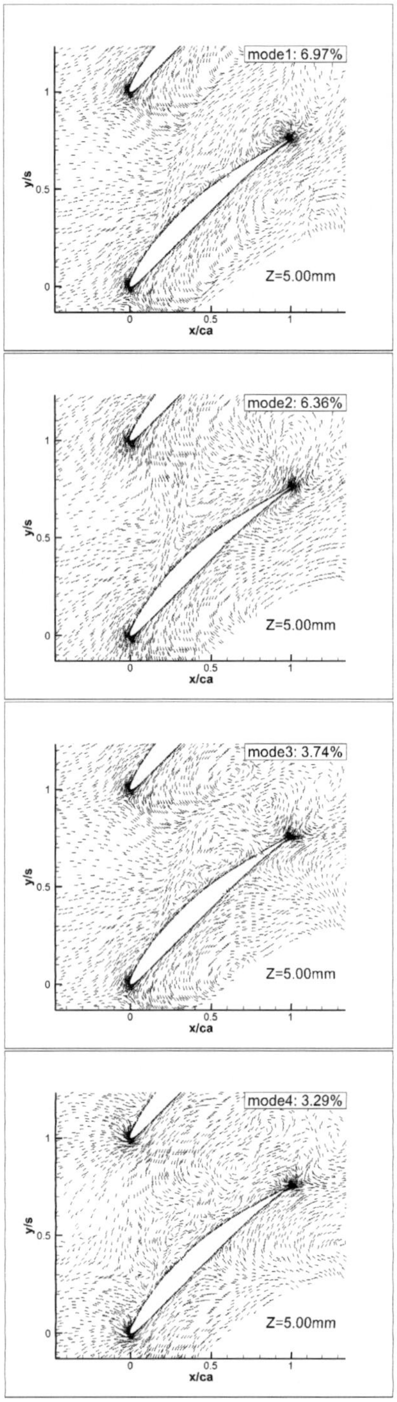

FIGURE 8. Flow fields of the first four POD modes at z=5.00 mm plane.

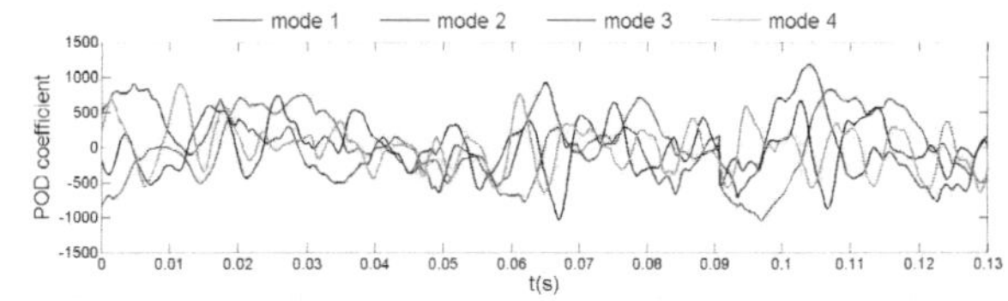

FIGURE 9. Time history of the first four POD coefficients at z=5.00 mm plane.

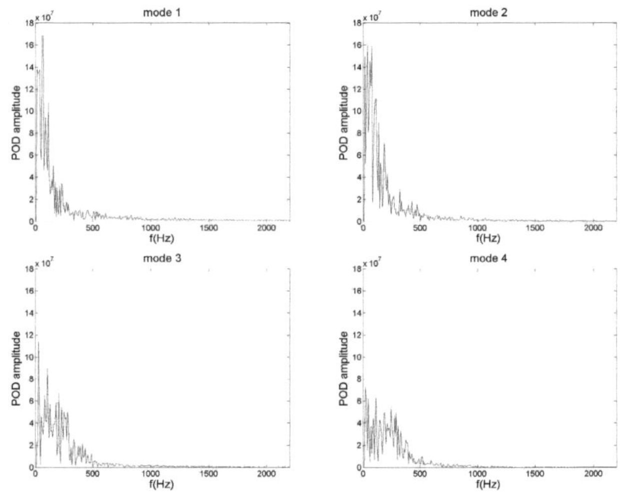

FIGURE 10. Spectra of the first four POD coefficients at z=5.00 mm plane.

Fig. 10 shows the spectra of the first four POD coefficients at z=5.00 mm spanwise plane. An obvious peak does not exist in each subfigure. Therefore, each POD mode is aperiodic. This is consistent with the bimodal phenomenon found in this linear compressor cascade.

In the future these POD modes will be used to investigate the two modes in bimodal phenomenon.

8 CONCLUSIONS

In this paper, DES has been used to investigate the intermittent corner separation in a linear compressor cascade.

DES can simulate successfully the intermittent flow in the corner separation within certain accuracy. Bimodal points exist in the DES results. Additional DES results can successfully capture the evolution of bimodal points along a station normal to blade surface, as well as their locations in this linear compressor cascade that has found in experiment.

Through POD analysis of the DES instantaneous flow field, no POD mode dominates the flow field, and there is not strong connection between modes Each POD mode is aperiodic, consistent with the bimodal phenomenon found in this linear compressor cascade. In the future these POD modes will be used to investigate the two modes in bimodal phenomenon.

NOMENCLATURE

c	chord [m]
c_a	axial chord [m]
C_p	coefficient of static pressure [–]
h	blade span [m]
L	length of arc from leading edge to trailing edge [m]
n	distance to blade surface [m]
P_s	mean static pressure [Pa]
P_t	mean total pressure [Pa]
$P_{s\infty}$	reference mean static pressure [Pa]
$P_{t\infty}$	reference mean total pressure [Pa]
Re	Reynolds number based on blade chord [–]
s	pitch of cascade [m]
s^*	position of LDA measurement station [–]
U_∞	reference velocity [m/s]
η	backflow coefficient [–]
CFD	computational fluid dynamics
DES	detached-eddy simulation
LDA	laser Doppler anemometry
LES	large eddy simulation
POD	proper orthogonal decomposition
RANS	Reynolds-averaged Navier-Stokes

ACKNOWLEDGMENT

Wei Ma wishes to express the sincere gratitude to the support from the Natural Science Foundation of Shanghai (Grant No. 15ZR1422500), and the Initiative Projects of Shanghai Jiao Tong University Young Teachers (Grant No. AF4130024). This work is sponsored by "2011 Aero-Engine Collaboration Innovation Plan" of P. R. China.

REFERENCES

[1] Cumpsty, N. A., 2004. *Compressor Aerodynamics*. Krieger Publishing Company.

[2] Gbadebo, S. A., Cumpsty, N. A., and Hynes, T. P., 2005. "Three-dimensional separations in axial compressors". *Journal of Turbomachinery,* ***127***(2), pp. 331–339.

[3] Lei, V.-M., Spakovszky, Z. S., and Greitzer, E. M., 2008. "A criterion for axial compressor hub-corner stall". *Journal of Turbomachinery,* ***130***(3), p. 031006.

[4] Devenport, W. J., and Simpson, R. L., 1990. "Time-dependent and time-averaged turbulence structure near the nose of a wing-body junction". *Journal of Fluid Mechanics,* ***210***, pp. 23–55.

[5] Gand, F., Deck, S., Brunet, V., and Sagaut, P., 2010. "Flow dynamics past a simplified wing body junction". *Physics of Fluids,* ***22***, p. 115111.

[6] Tian, Q., and Simpson, R. L., 2007. "Experimental study of tip leakage flow in the linear compressor cascade: Part I - stationary wall". *AIAA 2007-269*.

[7] Hobson, G. V., Williams, A. J. H., and Ganaim Rickel, H. J., 1998. "Laser-Doppler-velocimetry measurements in a cascade of compressor blades at stall". *Journal of Turbomachinery,* ***120***, pp. 170–178.

[8] Gourdain, N., Wlassow, F., and Ottavy, X., 2012. "Effect of tip clearance dimensions and control of unsteady flows in a multi-stage high-pressure compressor". *Journal of Turbomachinery,* ***134***(5), p. 051005.

[9] Ma, W., Ottavy, X., Lu, L., and Leboeuf, F., 2013. "Intermittent corner separation in a linear compressor cascade". *Experiments in Fluids,* ***54:1546***(6), pp. 1–17.

[10] Gao, F., Ma, W., Boudet, J., Ottavy, X., Lu, L., and Leboeuf, F., 2013. "Numerical analysis of three-dimensional separated flow in a linear compressor cascade". *ASME GT2013-95581*.

[11] Wang, Z., and Yuan, X., 2013. "Unsteady mechanisms of compressor corner separation over a range of incidences based on hybrid LES/RANS". *ASME GT2013-95300*.

[12] Gao, F., Ma, W., Zambonin, G., Boudet, J., Ottavy, X., Lu, L., and Shao, L., 2015. "Large-eddy simulation of 3-D corner separation in a linear compressor cascade". *Physics of Fluids,* ***27***, p. 085105.

[13] Spalart, P. R., 2009. "Detached-eddy simulation". *Annual Review of Fluid Mechanics,* ***41***, pp. 181–202.

[14] Berkooz, G., Holmes, P., and Lumley, J. L., 1993. "The proper orthogonal decomposition in the analysis of turbulent flows". *Annual Review of Fluid Mechanics,* ***25***(1), pp. 539–575.

[15] Meyer, K. E., Pedersen, J. M., and Özcan, O., 2007. "A turbulent jet in crossflow analysed with proper orthogonal decomposition". *Journal of Fluid Mechanics,* ***583***, pp. 199–227.

[16] Shah, M. K., and Tachie, M. F., 2009. "Proper orthogonal decomposition analysis of separated and reattached pressure gradient turbulent flows". *AIAA Journal*.

[17] Feng, L., Wang, J., and Pan, C., 2011. "Proper orthogonal decomposition analysis of vortex dynamics of a circular cylinder under synthetic jet control". *Physics of Fluids,* ***23***, p. 014106.

[18] Oberleithner, K., sieber, M., Nayer, C. N., Paschereit, C. O., Petz, C., Hege, H.-C., Noack, B. R., and Wygnanski, I., 2011. "Three-dimensional coherent structures in a swirling jet undergoing vortex breakdown: stability analysis and empirical mode construction". *Journal of Fluid Mechanics,* ***679***, pp. 383–414.

Proceedings of the 14th International Symposium on Unsteady Aerodynamics, Aeroacoustics & Aeroelasticity of Turbomachines
ISUAAAT14
8-11 September 2015, Stockholm, Sweden

I14-S8-3

THE UNSTEADY DISCRETE ADJOINT METHOD FOR TURBOMACHINERY APPLICATIONS

Georgios Ntanakas
Rolls-Royce Deutschland
Georgios.Ntanakas@Rolls-Royce.com

Marcus Meyer
Rolls-Royce Deutschland
Marcus.Meyer@Rolls-Royce.com

ABSTRACT

The unsteady discrete adjoint method for the calculation of objective function gradients for transient turbomachinery compressible viscous flows is presented. The URANS state equations are used along with the one-equation Spalart-Allmaras model. The implementation relies on an existing unsteady three-dimensional unstructured grid solver and discrete adjoint capabilities previously developed for steady flows. The formulation of the adjoint equations is introduced within a generic time-dependent CFD optimal design problem. An adjoint sliding plane functionality is also presented in order to enable multi-row time-dependent adjoint calculations, exploiting the rotor-stator interface. Algorithmic Differentiation (AD) is used to compute differential terms. Two testcases, a turbine vane and a turbine stage, are selected to demonstrate the feasibility of the method.

1 INTRODUCTION

The use of the adjoint method (both discrete and continuous) [1, 2, 3, 4, 5] is well established in the field of Computational Fluid Dynamics (CFD) for optimal design problems. In optimization problems, the sensitivity derivatives of a number of objective functions with respect to a number of inputs (design variables) is needed for gradient-based optimization algorithms. Conventional finite difference methods [6] suffer from step-size limitations and involve a computational cost that grows linearly with the number of design variables. Forward differentiation methods [7] overcome the step-size obstacle but still exhibit the high computational cost.
The adjoint method allows obtaining these sensitivities with a cost which is only proportional to the number of objectives and essentially independent of the number of design variables. The fact that in many CFD-based optimization tasks the number of objectives and constraints is significantly smaller (quite often a few orders of magnitude) compared to the number of design variables makes the adjoint method quite popular in this field.
Despite the fact that the method has been used heavily for time-independent (steady) problems, its usage for time-dependent (unsteady) flows is still quite limited, see for example [9, 11]. In the field of turbomachinery unsteady flows the authors are only aware of the work of He [10] where the adjoint version of a nonlinear harmonic balance solver is employed to optimize the aeroelastic behaviour of rotor blades.
Due to the continuous advances in High Performance Computing (HPC), time-accurate simulation of the inherently unsteady turbomachinery flows is now becoming feasible for industrial problem sizes in project-relevant time scales. Therefore the extension of the adjoint method from steady to unsteady turbomachinery CFD models is needed to extend its benefits into the unsteady domain.

In this paper, a mathematical overview of the unsteady adjoint equations and some details of its implementation into a CFD solver are provided. Note that in contrast to [10], in this paper, the adjoint method is formulated and solved in the time domain. In addition, a method is proposed for handling rotor-stator interaction, thus allowing multi-row calculations. Finally, the method is applied to two turbomachinery testcases.

2 FORMULATING THE UNSTEADY ADJOINT SYSTEM

The state equations of the primal (non-linear) problem consist of:

- 1 equation for mass conservation
- 3 equations (x,y,z) for momentum conservation
- 1 equation for energy conservation
- 1 equation for turbulence modeling (Spalart-Allmaras [20] including wall functions is used in this work).

The aforementioned equations form the well-known Reynolds-Averaged Navier-Stokes (RANS) system and can be written as

$$R(Q, X) = 0, \tag{1}$$

where Q is the flow vector,

$$Q = [\rho \quad u \quad v \quad w \quad p \quad \tilde{\nu}]^T, \tag{2}$$

which contains density, the three components of velocity, pressure and the Spalart variable respectively. X denotes the mesh points coordinate matrix. The mesh is generated using the initial geometry, which is defined by the design variables α. This process can be represented by the mesh residual equation

$$R_m(\alpha, X) = 0. \tag{3}$$

In order to expand eq. (1) for unsteady flows (URANS), we need to add the volume-weighted time derivative of the flow, thus getting the unsteady flow residual

$$G(Q, X, t) = V\frac{\partial Q}{\partial t} + R(Q, X) = 0. \tag{4}$$

It should be noted that in this paper the mesh will not be treated as deforming, so X is a constant matrix, apart from the rigid mesh movement due to rotation (rotor-stator interaction), whose details and treatment can be found in section 4.

In order to solve eq. (4), the discretization of both space and time domain are needed. The spacial discretization is represented by the generated mesh X together with a 2nd order accurate method, see [17] for details. For the time discretization, we employ a backwards-difference formula (BDF) of 2nd order accuracy. The time step will be denoted by n. Thus the equation (4) can be written as

$$\begin{aligned} &G(Q^n, Q^{n-1}, Q^{n-2}, X) = \\ = \ & V\tfrac{3Q^n - 4Q^{n-1} + Q^{n-2}}{2\Delta t} + R(Q^n, X) = 0 \end{aligned} \tag{5}$$

The equation system (5) is solved iteratively for Q^n using

$$Q^n_{m+1} = Q^n_m + \Delta Q, \tag{6}$$

where m denotes the pseudo-time-stepping towards convergence of the steady problem per time step. By substituting eq. (6) to eq. (5) and using a Taylor expansion we get

$$\begin{aligned} & G(Q^n_{m+1}, Q^{n-1}, Q^{n-2}, X) = 0 \\ \Leftrightarrow \quad & G(Q^n_m + \Delta Q, Q^{n-1}, Q^{n-2}, X) = 0 \\ \Leftrightarrow \quad & G(Q^n_m, Q^{n-1}, Q^{n-2}, X) + \tfrac{\partial G}{\partial Q^n_m}\Delta Q = 0 \\ \Leftrightarrow \quad & \tfrac{\partial G}{\partial Q^n_m}\Delta Q = -G(Q^n_m, Q^{n-1}, Q^{n-2}, X) \\ \Leftrightarrow \quad & \left(\tfrac{3V}{2\Delta t} + \tfrac{\partial R}{\partial Q^n_m}\right)\Delta Q = \\ & = -R(Q^n_m, X) \\ & -\tfrac{3V}{2\Delta t}Q^n + \tfrac{2V}{\Delta t}Q^{n-1} - \tfrac{1V}{2\Delta t}Q^{n-2} \end{aligned} \tag{7}$$

System (7) can for example be solved using an explicit 5-stage Runge-Kutta scheme, where the matrix sum $\left(\frac{3V}{2\Delta t} + \frac{\partial R}{\partial Q^n_m}\right)$ is substituted by a block-Jacobi preconditioning matrix.

During or after the time-dependent solution of the system of equations, i.e. the computation of the flow vector at each time step n, $(Q^1, Q^2, \ldots, Q^{n-1}, Q^n, \ldots, Q^N)$, an objective or constraint function of interest, such as lift and drag of an airfoil, the efficiency or capacity of a turbine or efficiency and pressure ratio of a compressor stage, can be calculated. This scalar function J, describing the objective or constraint of interest, can thus be formulated mathematically as

$$\begin{aligned} J = \ & \tfrac{1}{T_p}\int_0^{T_p} \tilde{J}\left(Q\left(t\right), X, a\right) dt \\ \simeq \ & \tfrac{1}{T_p}\sum_{n=1}^{N} \tilde{J}^n(Q^n, X, a). \end{aligned} \tag{8}$$

This means we are interested in the mean value J of $\tilde{J}$ over a certain period of time T_p. In gradient-based optimization we need the gradient of J w.r.t. the design variables α in order to minimize (or maximize) J. Developing the gradient in a straightforward way, we get

$$\frac{dJ}{d\alpha} = \frac{\partial J}{\partial Q}\frac{\partial Q}{\partial \alpha} + \frac{\partial J}{\partial X}\frac{\partial X}{\partial \alpha} + \frac{\partial J}{\partial \alpha}. \tag{9}$$

Whether we choose to calculate directly $\frac{dJ}{d\alpha}$ or use the expansion of equation 9 (which means calculating $\frac{\partial Q}{\partial \alpha}$ and $\frac{\partial X}{\partial \alpha}$ first) we cannot avoid perturbing the design space for every single component of the matrix to calculate the gradient. The cost of that is the product of a primal calculation times the dimension of the design space. The fact that the cost of the primal calculation for large scale unsteady problems can be in the range of days and the dimension of a realistic design space can be of order 10^2 and more makes the forward calculation of gradients prohibitive.

Therefore we will expand the objective function of eq. (8) as follows:

$$J_{aug} = J + \Lambda^T G + \Psi^T R_m. \tag{10}$$

Eq. (10) is the so called Lagrangian. Λ denotes the flow adjoint vector and Ψ denotes the mesh adjoint vector. Since we have added terms (R_m and G) which are (close to) zero after the successful solution of the system of equations, $J = J_{aug}$, and the gradients will also be equal. Taking the derivative, we obtain:

$$\begin{aligned}
\frac{dJ}{d\alpha} = & \frac{dJ_{aug}}{d\alpha} \\
= & \frac{\partial J}{\partial Q}\frac{\partial Q}{\partial \alpha} + \frac{\partial J}{\partial X}\frac{\partial X}{\partial \alpha} + \frac{\partial J}{\partial \alpha} \\
+ & \Lambda^T \left(\frac{\partial G}{\partial Q}\frac{\partial Q}{\partial \alpha} + \frac{\partial G}{\partial X}\frac{\partial X}{\partial \alpha} \right) \\
+ & \Psi^T \left(\frac{\partial R_m}{\partial \alpha} + \frac{\partial R_m}{\partial X}\frac{\partial X}{\partial \alpha} \right) \\
= & \frac{\partial J}{\partial \alpha} + \Psi^T \frac{\partial R_m}{\partial \alpha} \\
+ & \frac{\partial Q}{\partial \alpha} \left(\frac{\partial J}{\partial Q} + \Lambda^T \frac{\partial G}{\partial Q} \right) \\
+ & \frac{\partial X}{\partial \alpha} \left(\frac{\partial J}{\partial X} + \Lambda^T \frac{\partial G}{\partial X} + \Psi^T \frac{\partial R_m}{\partial Q} \right)
\end{aligned} \quad (11)$$

As mentioned before, we want to avoid calculating the expensive terms $\frac{\partial Q}{\partial \alpha}$ and $\frac{\partial X}{\partial \alpha}$. Thus, we form the following equations:

$$\frac{\partial J}{\partial Q} + \Lambda^T \frac{\partial G}{\partial Q} = 0 \quad (12)$$

$$\frac{\partial J}{\partial X} + \Lambda^T \frac{\partial G}{\partial X} + \Psi^T \frac{\partial R_m}{\partial Q} = 0 \quad (13)$$

Eq. (12) is the flow adjoint equation and eq. (13) is the mesh adjoint equation. One can also calculate the useful surface sensitivities $\frac{\partial J}{\partial X_s}$ (where X_s is the surface mesh) before calculating the full derivative $\frac{\partial J}{\partial \alpha}$ by using the adjoint method to the (morphing) technique that is used to generate the volume mesh X starting from the surface mesh X_s.

In this paper, we will focus on the flow adjoint equation, which introduces the unsteadiness. It can be rewritten as

$$\left[\frac{\partial G}{\partial Q}\right]^T \Lambda = -\left[\frac{\partial J}{\partial Q}\right]^T . \quad (14)$$

Using the discretization of eq. (5), $\left[\frac{\partial G}{\partial Q}\right]^T$ can be written in an upper tridiagonal matrix form as:

$$\left[\frac{\partial G}{\partial Q}\right]^T = \quad (15)$$

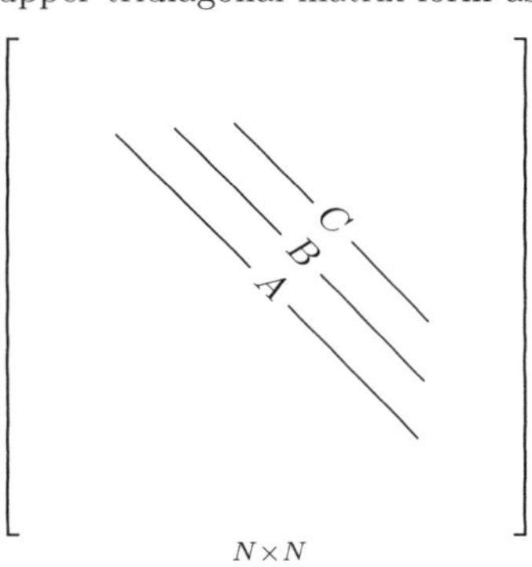

where:

- A has elements of value: $\frac{\partial G^n}{\partial Q^n} = \frac{3V}{2\Delta t} + \frac{\partial R}{\partial Q^n}$,
- B has elements of value: $\frac{\partial G^n}{\partial Q^{n-1}} = -\frac{2V}{\Delta t}$ and
- C has elements of value: $\frac{\partial G^n}{\partial Q^{n-2}} = \frac{1V}{2\Delta t}$.

We thus can solve system of eq. [14) by backward substitution starting at the last time step and marching backwards in time. The flow adjoint equation which has to be solved per time-step is given as

$$\begin{aligned}
&\left(\tfrac{3V}{2\Delta t}+\tfrac{\partial R}{\partial Q^n}\right)\Lambda^n-\tfrac{2V}{\Delta t}\Lambda^{n+1}+\tfrac{1V}{2\Delta t}\Lambda^{n+2}-\tfrac{\partial J}{\partial Q^n}=0 \\
&\Leftrightarrow \overline{\mathrm{G}}(\Lambda^n, \Lambda^{n+1}, \Lambda^{n+2}, Q^n) = 0
\end{aligned} \quad (16)$$

In order to solve eq. (16), we use the adjoint analog to the solution procedure in the primal solver. Assuming that

$$\Lambda^n_{m+1} = \Lambda^n_m + \Delta\Lambda \quad (17)$$

we can use the Taylor expansion for the flow adjoint residual:

$$\begin{aligned}
& \overline{\mathrm{G}}(\Lambda^n_{m+1}, \Lambda^{n+1}, \Lambda^{n+2}) = 0 \\
\Leftrightarrow \quad & \overline{\mathrm{G}}(\Lambda^n_m + \Delta\Lambda, \Lambda^{n+1}, \Lambda^{n+2}) = 0 \\
\Leftrightarrow \quad & \overline{\mathrm{G}}(\Lambda^n_m, \Lambda^{n+1}, \Lambda^{n+2}) + \tfrac{\partial \overline{\mathrm{G}}}{\partial \Lambda^n_m}\Delta\Lambda = 0 \\
\Leftrightarrow \quad & \tfrac{\partial \overline{\mathrm{G}}}{\partial \Lambda^n_m}\Delta\Lambda = -\overline{\mathrm{G}}(\Lambda^n_m, \Lambda^{n+1}, \Lambda^{n+2}) \\
\Leftrightarrow \quad & \left(\tfrac{3V}{2\Delta t} + \tfrac{\partial R}{\partial Q^n_m}\right)\Delta\Lambda = \\
& = \underbrace{\frac{\partial J}{\partial Q^n} - \frac{\partial R}{\partial Q^n}\Lambda^n_m}_{\text{steady adjoint residual}} \\
& -\tfrac{3V}{2\Delta t}\Lambda^n_m + \tfrac{2V}{\Delta t}\Lambda^{n+1} - \tfrac{1V}{2\Delta t}\Lambda^{n+2}
\end{aligned} \quad (18)$$

Eq. (18) is very similar to eq. (7) and confirms once more the duality between the primal and the adjoint system.

Note that for the adjoint calculations, the primal flow of the corresponding time-step is needed (Q^n). Since the primal problem is solved forward, but the adjoint problem is solved backwards in time, and the adjoint solver needs the primal solution at each time step to calculate its linearization, it means that we have to solve and store the primal solution for every time-step first and then continue with the adjoint calculations. In the present work we use the brute force approach and store each time step on the disk. Future work will address options for increasing the solution speed of the adjoint solver by either checkpointing, reduced order approximations, or parallel I/O and pre-fetching of solutions.

3 CODE IMPLEMENTATION

This section will give a short description of the terms of the derivatives of eq. (18) which are constructed

throughout the code. It is not intended to give a detailed view of the code but to set the ground for the next section (4).
In the primal code, the flux-contribution to the residual $R(Q_m^n, X)$ of eq. (7) are calculated through a series of operations. For the adjoint code, Tapenade [21], a tool for generating source code by Algorithmic Differentiation (AD), is used in order to obtain $\frac{\partial R}{\partial Q^n}\Lambda^n$ for eq. (18) and a short explanation of the procedure to obtain this term is given below:
Assume that there is a (primal) piece of code with I as input and O as output. Tapenade (when using the reverse mode, see [21]) is capable of producing a new piece of code where $\left[\frac{\partial O}{\partial I}\right]^T \kappa$ is also computed (κ is a user chosen vector). Assume now that we have a series of operations:

$$W \xrightarrow{a} X \xrightarrow{b} Y \xrightarrow{c} Z$$

or i.e. $Z = Z(Y(X(W)))$. Assume also that we want to calculate $\left[\frac{\partial Z}{\partial W}\right]^T \kappa$ (where κ is a vector). Since we do not want to apply Tapenade to the entire sequence to reduce the complexity of the generated code, we expand as follows:

$$\begin{aligned} & \left[\frac{\partial Z}{\partial W}\right]^T \kappa \\ = & \left[\frac{\partial Z}{\partial Y}\frac{\partial Y}{\partial X}\frac{\partial X}{\partial W}\right]^T \kappa \\ = & \underbrace{\left[\frac{\partial X}{\partial W}\right]^T \underbrace{\left[\frac{\partial Y}{\partial X}\right]^T \underbrace{\left[\frac{\partial Z}{\partial Y}\right]^T \kappa}_{c}}_{b}}_{a} \end{aligned}$$

We have shown that in order to obtain the desired gradient, the sequence of the operations should be reversed. Similarly, in the flow solver, a sequence of operations is needed to obtain the residual R using the flow Q as an input. In order to obtain $\left[\frac{\partial R}{\partial Q}\right]^T \Lambda$, these operations should be reversed after AD is applied to every single one of them. $\left[\frac{\partial J}{\partial Q}\right]^T$ is calculated in a similar way, by applying AD to the bits of code that calculate the objective function using the flow vector as input.

4 ROTOR-STATOR INTERFACE

For turbomachinery applications, an interface that connects the stationary stator with the rotating rotor mesh is required for multi-row calculations, which is typically called a sliding plane. Firstly, the formulation of this sliding plane method that is implemented in the primal will be reviewed and then its adjoint counterpart will be formulated.
Meshes for the stator and rotor rows are generated in such a way that they exhibit a one-cell overlap between two adjacent rows (Fig. 1). Thus, two contact surfaces are created, on which the solution variables between stationary and rotating zones are interpolated. The inner/donor surface of one row is attached to the

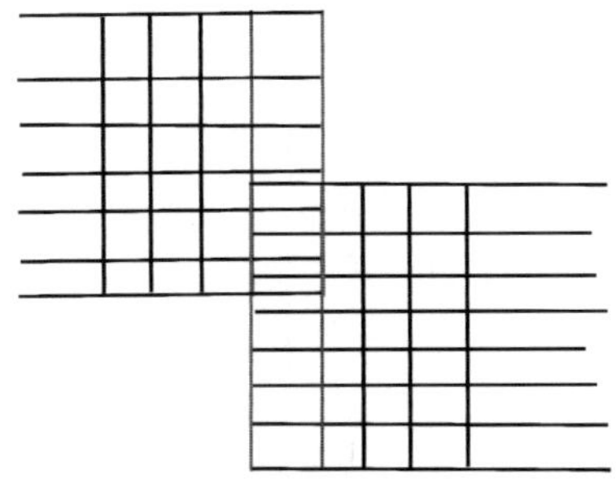

Figure 1: One-cell overlap of adjacent rows' meshes on radial cut. Red: inner/donor surface. Green: outer/receiver surface.

outer/receiver surface of the other row and vice versa. Flow variables are interpolated from the nodes of the donor face to the ones of the receiver face. Since the code is working on each row independently in a relative frame of reference, we need to perform the interpolation in a universal absolute frame. The following three steps are necessary:

1. Transformation to the absolute frame of reference

$$Q^0 T^{abs} = Q^1 \tag{19}$$

where T^{abs} is the transformation matrix to the absolute frame of reference.

2. Interpolation of values from donor to receiver nodes
A search is performed to locate receiver nodes that lie inside donor faces. Then, weight coefficients (w_i) are calculated based on the distance between the receiver and the donors (Fig. 2).

$$w_i = \frac{d_i}{\sum\limits_{j=1}^{\sigma} d_j} \tag{20}$$

where σ is the number of nodes of the corresponding face and d_i is the distance between receiver and donors.

We then can calculate the interpolated values using

$$\sum_{j=1}^{\sigma} w_j Q_j^1 = Q^2 \tag{21}$$

3. Transformation back to the relative frame of reference

$$Q^2 T^{rel} = Q^3 \tag{22}$$

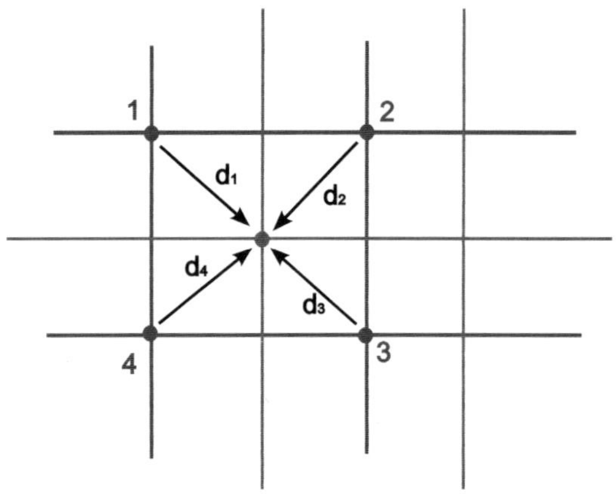

Figure 2: Node interpolation on axial cut. Red: inner/donor nodes. Green: outer/receiver nodes.

The method can be viewed as a sequence of operations on the flow variables ($Q^0 \rightarrow Q^1 \rightarrow Q^2 \rightarrow Q^3$) which is similar to the one introduced in section (3). In order to use the method in the adjoint code, we need to apply AD to the different steps and reverse them. Thus we perform the following calculations:

$$\left[\frac{\partial R}{\partial Q^0}\right]^T \Lambda =$$

$$= \underbrace{\left[\frac{\partial Q^1}{\partial Q^0}\right]^T \underbrace{\left[\frac{\partial Q^2}{\partial Q^1}\right]^T \underbrace{\left[\frac{\partial Q^3}{\partial Q^2}\right]^T \left[\frac{\partial R^1}{\partial Q^3}\right]^T \Lambda}_{3}}_{2}}_{1}. \quad (23)$$

5 EXAMPLE TESTCASES

Two representative turbomachinery testcases are chosen to demonstrate the unsteady adjoint solver and the adjoint sliding plane functionality in this paragraph. Obtaining the unsteady adjoint flow is the main target of those applications. Full optimization cycles and evaluation of the optimized results is the objective of future work.

5.1 Turbine Nozzle Guide Vane

The first testcase involves the first stator of a High Pressure Turbine (HPT - Fig. 3a). The 3D mesh of a single passage consists of ~ 1 million nodes (Fig. 3b, 3c). The following boundary conditions are employed:

- inflow: prescribed total pressure, total temperature, flow angles and turbulence profile
- outflow: prescribed static pressure profile
- viscous wall treatment for all gas-washed surfaces
- periodicity treatment for periodic boundary conditions

(a) Full annulus

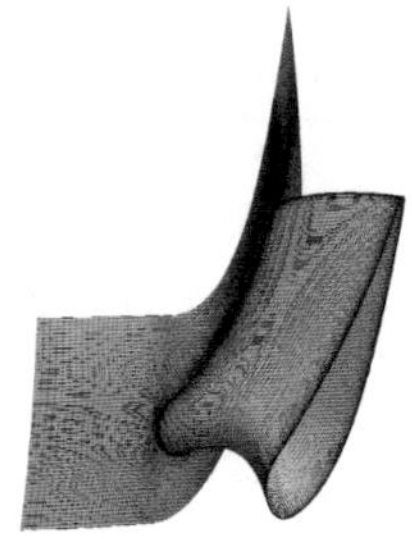

(b) Surface mesh

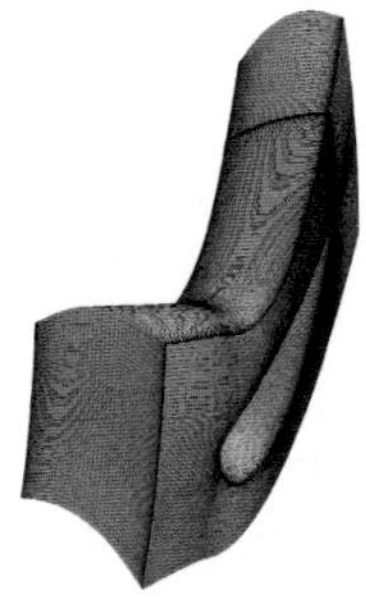

(c) Volume mesh

Figure 3: Geometry and mesh

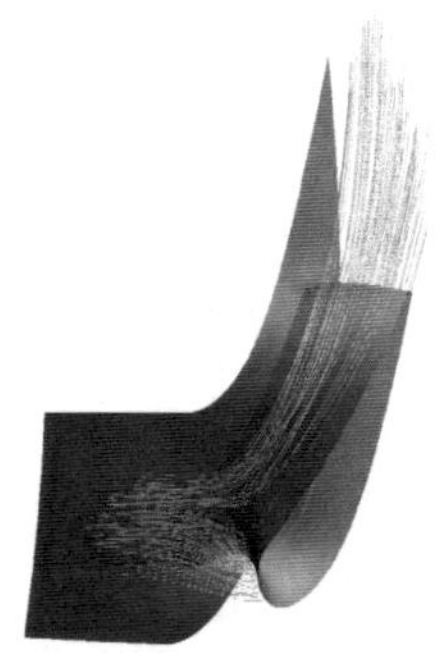

Figure 4: Static pressure on surface and velocity magnitude streamlines

For the primal calculations a time period of duration $T_p = 1.8\dot{1}0^{-3}$ s is selected consisting of 128 time-steps. The simulation loops over the period until periodicity of the flow is observed. Once converged, a vortex-shedding appears downstream of the trailing edge. Even though the unsteadiness is not obvious in Fig. 5.1, it can be seen in Fig. 5a, 5b, where relative velocity is shown at two different time-steps or Fig. 6a, 6b,where one mach number isosurface is plotted.

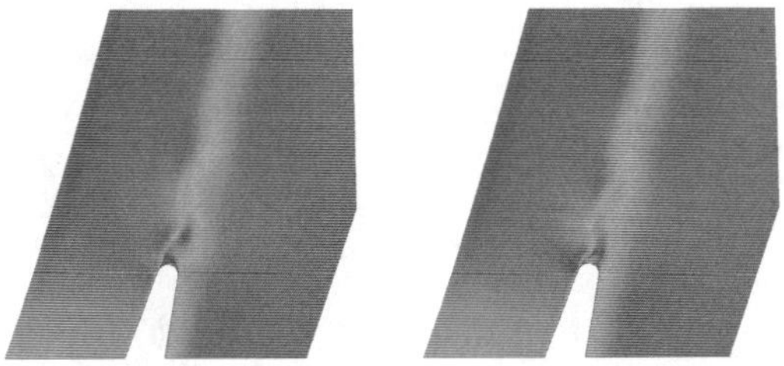

(a) (b)

Figure 5: Relative velocity vector near trailing edge at two different time-steps on radial cut at mid height

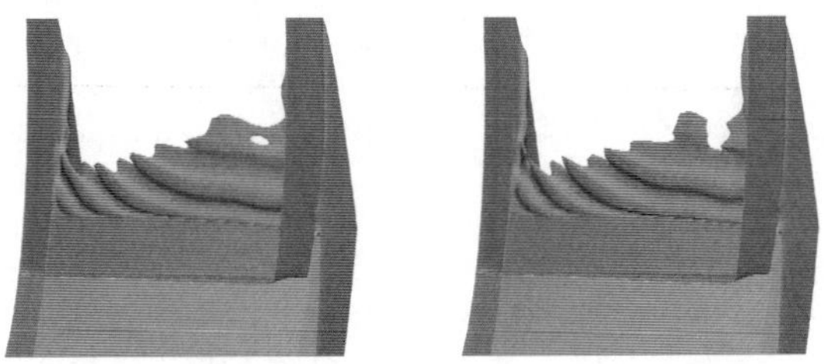

(a) (b)

Figure 6: Mach number isosurface near trailing edge at two different time-steps

The unsteady adjoint solver can now be executed based on what has been described in section 2. The objective function that is used for demonstration purposes is reduced mass flow at the inflow boundary. This quantity is very important for turbine nozzle guide vanes, since it sets the operating point of the whole engine. Firstly, the flow adjoint vector Λ is computed. Different variables of the adjoint vector are shown in Figs. 7 and 8 at different time-steps.

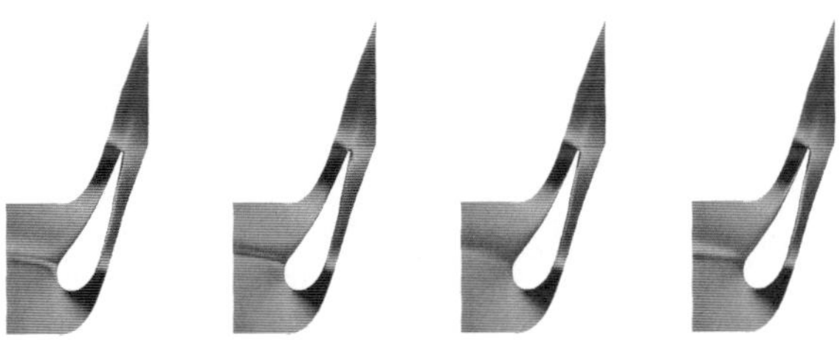

(a) (b) (c) (d)

Figure 7: Ajoint velocity magnitude at four different time-steps on radial cut

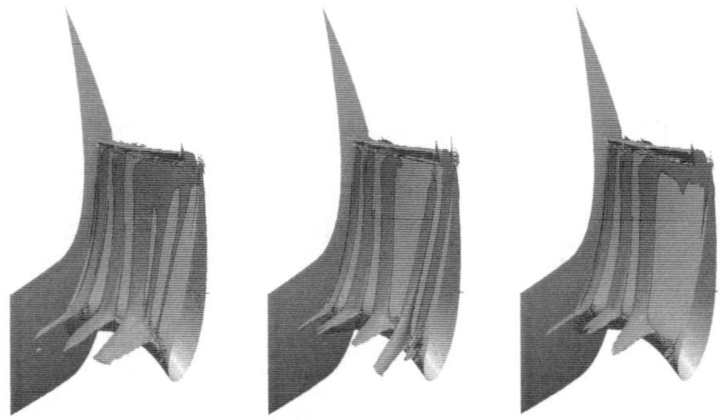

(a) (b) (c)

Figure 8: Adjoint density isosurface at three different time-steps

We can also compute the more visually intuitive surface sensitivities $\frac{\partial J}{\partial X_s}$. They indicate sensitive areas on the gas-washed surfaces and provide the direction of surface modification (inwards/outwards) in order to increase the objective function. The differences between different time-steps are shown in Fig. 9. Note that the differences are small since the unsteadiness is restricted to the trailing edge domain.

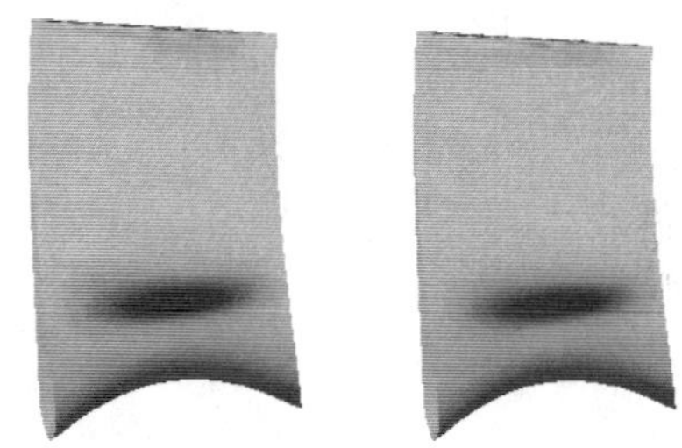

(a) Suction side (b) Suction side

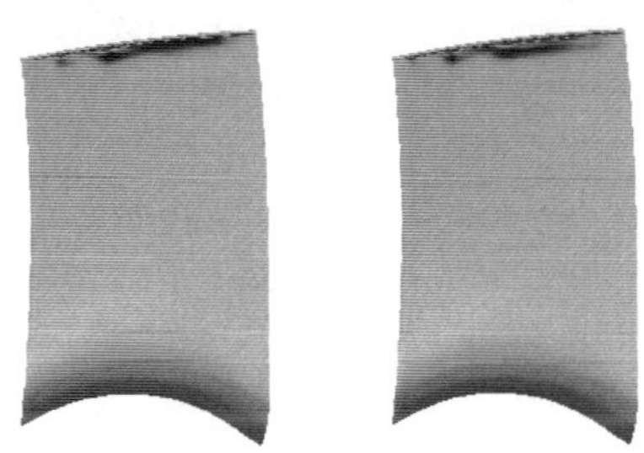

(c) Pressure side (d) Pressure side

Figure 9: Surface sensitivities at different time-steps. Blue: push inwards, red: pull outwards to increase objective

5.2 High Pressure Turbine Stage

The second testcase involves a turbine stage in order to show the sliding plane capabilities developed and presented in this paper. To limit the computational effort, we restrict ourselves to a small slice in radial direction at mid-height of the turbine stage as shown in fig. 10. Nevertheless, the method described in this paper can also be applied to large 3D meshes, which is the topic of future work.

(a) (b)

Figure 10: Geometry and mesh of the turbine stage test case

The mesh consists of $\sim 10^5$ nodes. The time period is $T_p = 1.9\,10^{-3}$ s and 64 time-steps are computed per period.

For visualization, we use repeats of the computational domain: In Fig. 11 the primal flow is shown (note the different position of the rotor row). The wakes are passing through the rotor-stator interface smoothly, which indicates the sliding plane functionality is properly working.

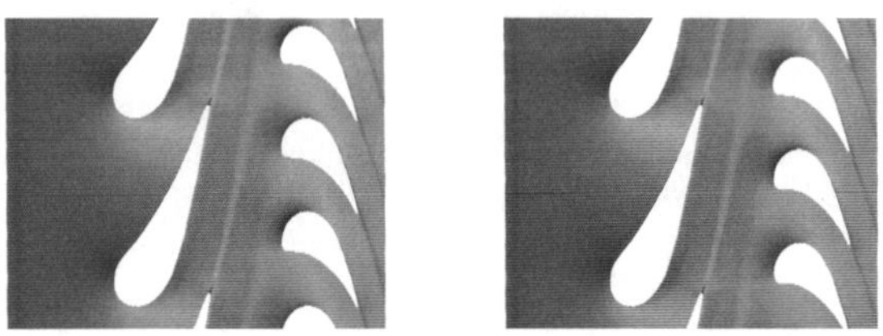

(a) (b)

Figure 11: Absolute mach number at two different time-steps on radial cut

Similarly, the adjoint sliding plane functionality is shown in Fig. 12. The objective function in this case is the mass flow at the outflow boundary. Note that the line that appears between the two rows is due to the backward flow of information between receivers and donors, i.e. purely a visualization artefact, and does not affect the adjoint gradient. One can see that the adjoint variables are also continuous across the sliding plane interface, thus indicating visually the proper implementation of the adjoint sliding plane. Additional numerical validation of the implementation is currently under way and will be presented in subsequent work.

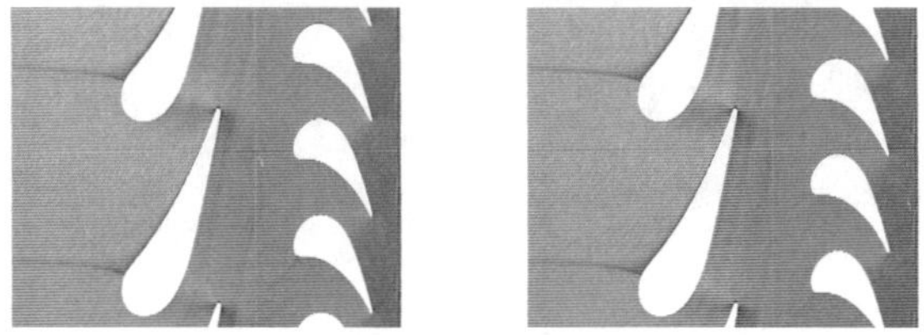

(a) (b)

Figure 12: Adjoint density at two different time-steps on radial cut

6 CONCLUSIONS

A time-dependent discrete adjoint capability has been developed and implemented in a 3D URANS solver in order to compute unsteady gradients of chosen objective functions for further use in unsteady optimization. An adjoint sliding plane capability in order to handle multi-row turbomachinery applications has been developed and implemented. The developed code has been demonstrated for two testcases, a high pressure turbine nozzle guide vane and a high pressure turbine stage. The results for those rather simple testcases already indicate the advantages of the adjoint approach for unsteady CFD models, i.e. providing insight into the *unsteady* surface sensitivities can guide the analyst in search for improvements of the shape of the model as well as the use of adjoint-calculated gradients of unsteady functionals for optimization algorithms to for example reduce forcing of turbomachinery blades.

ACKNOWLEDGMENTS

The authors gratefully acknowledge the permission of Rolls-Royce Deutschland to publish this paper.
This work has been conducted within the *AboutFlow* project (http://aboutflow.sems.qmul.ac.uk), funded by the European Union FP7 Programme for research, technological development and demonstration under Grant Agreement No. 317006.

REFERENCES

[1] Giles, M. Pierce, N. An introduction to the adjoint approach to design. *Flow Turbulence and Combustion.* 65 (3), pp. 393-415, 2000.

[2] Jameson, A. Aerodynamic shape optimization using the adjoint method. VKI Lecture Series on Aerodynamic Drag Prediction and Reduction. von Karman Institute of Fluid Dynamics, Rhode St Genese, Belgium, 2003.

[3] Jameson, A. Sriram, S. and Martinelli, L. A Continuous Adjoint Method for Unstructured Grids. AIAA Paper 2003- 3955, 16th AIAA CFD Conference, Orlando, Florida, 2003.

[4] Giles, M.B. and Pierce, N.A. Adjoint equations in CFD: duality, boundary conditions and solution behavior. AIAA 97-1850, 1997.

[5] Giles, M.B. Adjoint methods for aeronautical design. ECCOMAS CFD conference, 2001.

[6] Vanderplaats, G.N., Hicks, R.N., and Murman, E.M., "Application of Numerical Optimization Techniques to Airfoil Design," NASA Conference on Aerodynamic Analysis Requiring Advanced Computers," NASA SP-347, Part II, 1975.

[7] Baysal, O., and Eleshaky, M.E., Aerodynamic Sensitivity Analysis Methods for the Compressible Euler Equations, Journal of Fluids Engineering, Vol. 113, pp. 681-688, 1991.

[8] Mavriplis, D. Discrete adjoint-based approach for optimization problems on three- dimensional unstructured meshes. AIAA journal 45.4, pp 741-750, 2007.

[9] Nielsen, E.J. Diskin B. Discrete Adjoint-Based Design for Unsteady Turbulent Flows on Dynamic Overset Unstructured Grids. *AIAA Journal.* Vol. 51, No. 6, pp. 1355-1373, 2013..

[10] He, L. Wang, D. X. Concurrent Blade Aerodynamic-Aero-elastic Design Optimization Using Adjoint Method J. Turbomach. 133(1), 2010.

[11] Mani, K. and Mavriplis, D. J., Unsteady Discrete Adjoint Formulation for Two-Dimensional Flow Problems with Deforming Meshes, AIAA Journal, Vol. 46-6, pp. 13511364, 2008.

[12] Nadarajah, S. and Jameson, A. A Comparison of the Continuous and Discrete Adjoint Approach to Automatic Aerodynamic Optimization. *Proceedings of the 38th Aerospace Sciences Meeting and Exhibit.* Reno NV, AIAA Paper 2000-0667, 2000.

[13] Nadarajah, S. Kim, S. Jameson, A. Alonso, J. Sonic boom reduction using an adjoint method for supersonic transport aircraft configuration. Symposium transsonicum IV, International Union of Theoretical and Applied Mechanics, September 2-6, DLR Gottingen, Germany, 2002.

[14] Nadarajah, S. Jameson, A. Studies of the continuous and discrete adjoint approaches to viscous automatic aerodynamic optimization, AIAA p. 2001-2530. AIAA 15th computational fluid dynamics conference, Anaheim, CA, 2001.

[15] Campobasso, M.S. Duta, M. Giles M.B. Adjoint methods for turbomachinery design. ISABE conference, Bangalore, 2001.

[16] Campobasso, M.S. Duta, M. Giles M.B. Adjoint calculation of sensitivities of turbomachinery objective functions AIAA J Propulsion Power, 19 (4), 2003.

[17] Moinier, P. Algorithm developments for an unstructured viscous flow solver. PhD Thesis, Oxford University, 1999.

[18] Denton, J.D. and Singh, U.K. Time-marching methods for turbomachinery flow calculations. VKI-LEC-SER-1979-9, von Karman Inst. for Fluid Dynamics, Belgium, 1979.

[19] Mohammadi, B. Pironneau, O. Shape optimization in fluid mechanics. Annu Rev Fluid Mech, 36, pp. 225-279, 2004.

[20] Spalart, P.R. Allmaras, S.R. A one-equation turbulence model for aerodynamic flows. La Recherche Aerospatiale, pp. 5-21, 1994.

[21] Hascoët, L. Tapenade: a tool for automatic differentiation of programs. In Proceedings of 4th European Congress on Computational Methods, ECCOMAS 2004, Jyvaskyla, Finland, 2004.

Proceedings of the 14th International Symposium on Unsteady Aerodynamics, Aeroacoustics & Aeroelasticity of Turbomachines
ISUAAAT14
8-11 September 2015, Stockholm, Sweden

I14-S8-4

UNSTEADY BLADE FORCE COMPUTATION SENSITIVITY IN A TRANSONIC TURBINE TO ROTOR TIP GAP, HUB AND SHROUD CAVITY MODAL DETAIL

Tobias Gezork
KTH, Heat and Power Technology
Stockholm, Sweden

Mauricio Gutierrez
KTH, Heat and Power Technology
Stockholm, Sweden

Pieter Groth
GKN Aerospace Sweden AB
Trollhättan, Sweden

Damian M. Vogt
University of Stuttgart
Stuttgart, Germany

Torsten Fransson
KTH, Heat and Power Technology
Stockholm, Sweden

ABSTRACT

The influence of including geometric detailing features on blade forcing predicted by CFD calculations is investigated. Various features such as rotor tip gap, rotor tip shroud cavity with substantial leakage flow and large rotor upstream hub rim cavity are investigated. The test case is based on a single stage transonic test turbine rig in which the unsteady aerodynamics and blade forcing have previously been investigated numerically.

Including the tip gap and the tip shroud leakage in the computational model resulted in an overall increase in unsteady forcing. The change in forcing is mainly due to a change in secondary flow structure, but also due to a change of the stage flow condition. Conversely, it is shown that including the hub cavity in the computational model, even without the presence of purge-flow, reduces the unsteady force. Lastly, even though there are changes in unsteady blade loading, the blade response amplitude is only changed marginally (maximum 7%) due to forcing variations alone.

NOMENCLATURE

BPF	Blade passing frequency	(Hz)
CFD	Computational fluid dynamics	
EO	Engine order	
F	Force vector	(N)
$F_{g(s/u)}$	Generalized force (shrouded/ unshrouded)	(N)
HCF	High cycle fatigue	
LE	Leading edge	
T	Torque and Period time	(Nm and s)
TE	Trailing edge	
PS	Pressure side	
SS	Suction side	
c_p	Pressure coefficient	(-)
p	Pressure	(Pa)
p_0	Total pressure	(Pa)
β	Incidence angle	(°)
ϕ	Mode shape vector	(m)
Δ	Difference	

INTRODUCTION

Blade failure in turbomachinery is mainly related to high cycle fatigue (HCF) when vibratory stresses exceed fatigue limits (see [1] through [10]). Cyclic vibration can occur when blades flutter at blade eigenfrequency or due to aerodynamic forcing. Aerodynamic forcing in turbomachinery can be non-synchronous or synchronous to the shaft speed. Synchronous forcing is due to upstream or downstream obstructions or non-uniformities in the flow path, such as adjacent blade rows, struts or distorted flow and occurs at engine orders (EO). Critical HCF failure occurs when the force acts at a resonance frequency (resonance crossing). In some cases, however, not all of these resonance crossing can be avoided within the operating range. To be able to predict the severity of a crossing, both aerodynamic and structural calculations are performed. Aerodynamic excitation calculations customarily exclude detailing features and are computed on clean flow path geometries to reduce complexity and computational cost, especially in the design phase which may require several iterations.

Several studies have shown that the flow in the clean main annulus flow path is significantly affected by geometric detailing features. Paniagua et. al [11] investigated the interaction between hub rim purge flow and mainstream flow for a transonic blow-down turbine stage for various purge flow rates both numerically (steady computations) and experimentally (−0.5% to +1.5% main passage mass flow hub purge and 3% vane coolant). Even very small purge flow was shown to interact with the rotor hub vortex and increase rothalpy in the rotor exit plane. An increase in purge flow showed increased blockage corresponding to a reduction in vane exit Mach number and a reduction in unsteady rotor force. The authors conclude that not modeling the cavity leads to a mismatch in simulated rotor leading edge Mach number predictions compared with experimental data. Increasing the purge flow rate showed a decrease in rotor torque. However increasing purge flow also showed a decrease in total pressure loss at the high span sections.

Ella et.al [12] experimentally investigated the influence of a smooth upstream hub cavity on flow structures in a transonic blow-down linear turbine cascade. The authors found that the secondary flow strength and structure significantly changed below 20% blade span comparing the cavity with a flat end-wall configuration and an increase in overturning due to cavity momentum loss. Changes in the flow field were measured with seven-hole pressure probes at 40% chord downstream of the blade trailing edge.

For subsonic turbine geometries, previous work has shown that the geometric detail of cavities can have a significant influence on the unsteady flow. Engelmann et. al [13] investigated the influence of a shroud tip cavity and casing steam injection slot using steady mixing-plane simulations showing non-decaying end-wall vortex structures and upstream and downstream changes in the local flow. Jenny et. al [14] studied the influence of hub rim seal purge flow both numerically with unsteady simulations and experimentally showing the influence of purge on the passage vortex core and non-deterministic unsteadiness. Schrewe et. al. [15] investigated the influence of rim seal purge flow experimentally showing a change in blade loading. Rosic and Xu [16] studied the influence of shroud cavity leakage flow showing an increase in passage vortex strength compared to a clean geometry and Chilla et. al. [17] investigated the influence of hub rim seal geometry using transient simulations for a high-speed turbine showing an effect on the main passage flow and vortex shedding non-synchronous to the vane passing.

The change of forcing due to tip shroud cavity leakage has been investigated previously for a subsonic model turbine geometry [18]. Blade forcing was significantly sensitive to leakage flow through a tip shroud cavity and it was concluded that omitting the tip shroud cavity leakage flow in the calculations significantly changes the unsteady force on the rotor blade. The present investigation extends this to a transonic test case.

The objective of the work is to investigate how the blade forcing is affected by changes in the main passage flow due to including the geometric detail of the cavity and leakage features for a transonic case to show if cavity features are similarly dominant to transonic flow.

METHOD

TEST CASE

The investigated test case is based on the transonic experimental turbine stage MT1 at the Pyestock Light Piston Facitlity which had been used for aerodynamic and heat transfer measurements [19]. The test setup consists of a one stage turbine operated in blow-down mode at constant speed for a short time. The original setup is shown in Fig. 1 and available hub and shroud curves as well as stator and rotor blade surfaces were used to set up the test case.

The original experimental test geometry is shown in Fig. 2 and the turbine stage parameters are listed in Tab. 1.

CASE SETUP

Six configurations based on a test geometry with clean end walls are studied. Different levels of geometric detail are included in a purely numerical investigation to analyze the influence of hub cavities, the tip gap and a tip shroud leakage cavity on the prediction sensitivity of forcing calculations.

Table 1: Design stage parameters (from [20] and [21])

Inlet total pressure	461.5	*kPa*
Inlet total temperature	444.4	*K*
Vane inlet Mach number	1.034	
Stator inflow direction	aligned axial	
Rotor exit static pressure	142.5	*kPa*
Number of stator vanes	32	*(scaled: 30)*
Number of rotor blades	60	
Shaft speed	9500	*rpm*

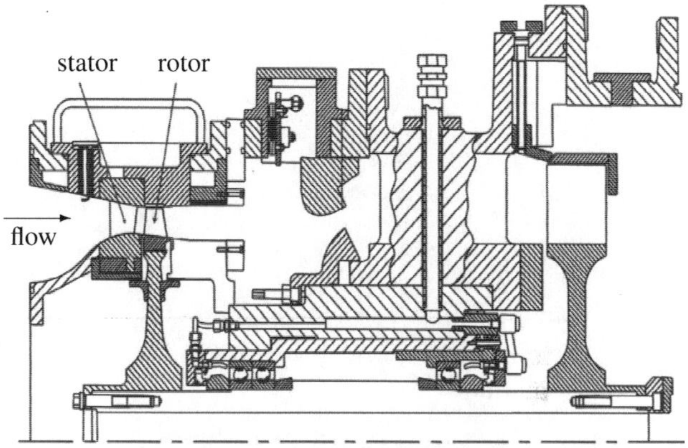

Figure 1: Test case geometry (from [20])

Figure 2: Picture of instrumented stage [22]

Table 2: Configuration cases

Feature	Label	ΔT(%)	$\Delta\beta$ (°)
no feature/base	B	-	-
tip gap	T	−3.0	1.1
tip shroud	S	−5.1	4.2
hub cavity	HB	−0.7	−0.7
hub cav. + tip gap	HT	−4.2	0.1
hub cav. + tip shroud	HS	−5.7	2.9

The base case (B) features clean end walls (no gap). A second case includes a rotor tip gap (T) and a third case features a tip shrouded rotor blade (S) with a leakage flow around the shroud. Another set of three cases with an added large hub cavity in between the stator and the rotor (HB, HT and HS respectively) are based on these three cases. The cases with hub cavity were run with 0.002% purge flow. Without this flow rate, the low

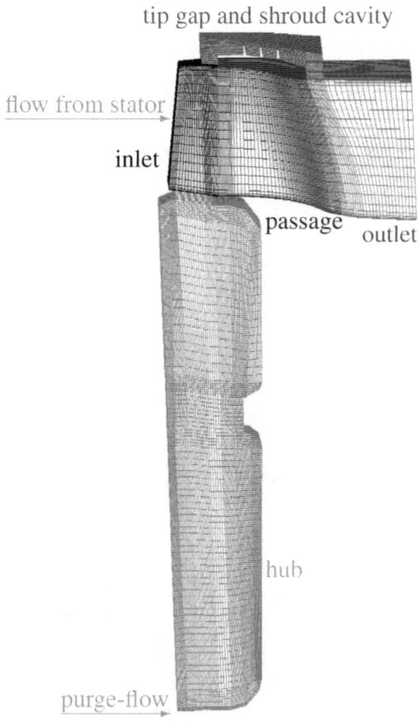

Figure 3: Modular mesh setup (coarse mesh shown)

momentum flow in the hub cavity was considered unstable. All cases are summarized in Tab. 2. The feature geometry of the hub and tip shroud geometry together with the rotor blade mesh is shown in Fig. 3.

The base case and the shrouded blade configuration (B and S) are set up featuring a rotating casing. The tip clearance configuration (T) has a stationary casing. Having the casing stationary in the base case shows a negligible increase in pressure on the suction side blade tip, which is due to the wall shear, as the only observable difference compared to the base case with rotating casing. Therefore the base case with rotating casing can also be used as a reference for the tip gap configuration.

The tip clearance is an offset to the shroud contour and amounts to 1.2% of the blade span. The clearance of the fins in the tip shroud cavity is 2.5% blade span. The calculated leakage mass flow across the tip gap is approximately 1.2% of the blade passage mass flow. The leakage flow in the tip shroud cavity is approximately 4.5% of the passage mass flow.

The addition of the mentioned design features (hub/shroud cavity, tip gap) does affect the stage performance and the stator or rotor blade loading. This could be corrected by adjusting the stage boundary conditions, however, for this investigation the boundary conditions are kept constant following a design approach. Typically the unsteady blade forcing is assessed with given stage boundary conditions on a clean geometry without a high level of detailing and based on predetermined inflow and outflow conditions (e.g. based on a through flow or meridional analysis). The change in calculated blade loading in terms of time averaged blade torque difference (ΔT) and rotor mid-span incidence $\Delta\beta$ (positive incidence impinging on the pressure side) is also shown in Tab. 2. The calculated mass flow difference between the cases is less than 0.8% (mass flow is mainly constrained by the stator vane choking).

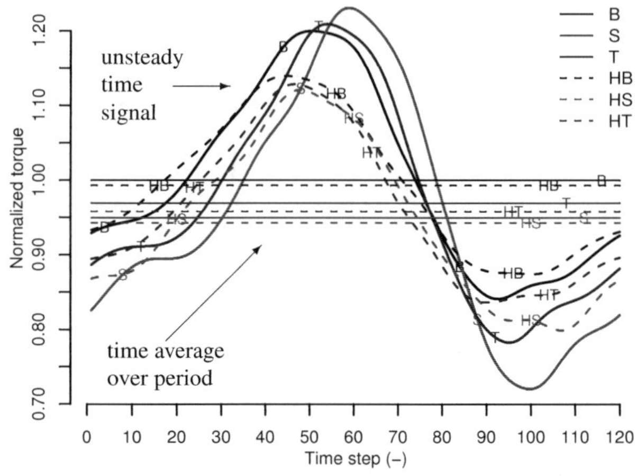

Figure 4: Comparison of unsteady blade moment

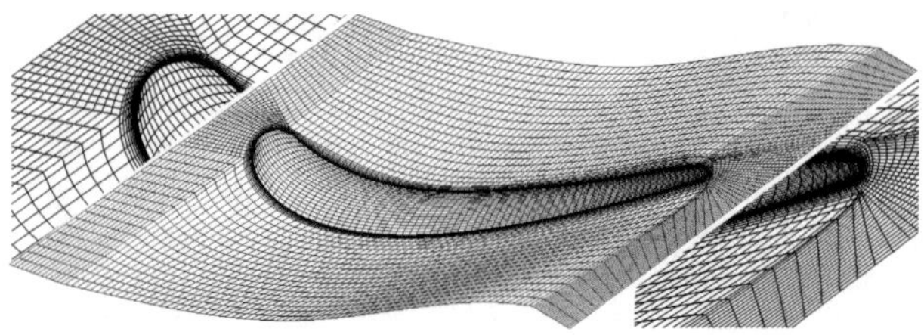

Figure 5: Rotor blade to blade mesh in tip gap plane with magnified leading and trailing edge

CFD MESH AND SOLVER SETUP

Hexahedral meshes with matching grid interfaces between the different detailing features were generated using Ansys ICEM CFD. This was necessary to be able to use in-house tools that require structured multi-block grids. A modular mesh setup is used where the same core blocking is used for the main passage in all configurations, adding the different features with node-to-node match at interfaces. This is done to eliminate possible influences of mesh resolution and interpolation errors at the interfaces when comparing the flow in the main passage between the different cases. This comes with a trade-off of losing flexibility in the blocking (visible in the mesh constrained at the LE and TE).

The mesh consists of individual blocks for *i*) the passage base mesh with a separate inlet and outlet block; *ii*) the hub cavity; *iii*) the tip shroud cavity and *iv*) the tip gap fluid block. The computational domain extends 70% chord upstream of the stator blade and 120% chord downstream of the rotor blade. The rotor stator interface is placed close to the stator trailing edge (TE) to facilitate the setup of the hub cavity case. The modular mesh setup is shown in Fig. 3. The rotor blade-to-blade mesh at blade tip is shown in Fig. 5.

The CFD computations were performed with the commercial solver Ansys CFX 15.0. The RANS SST turbulence model with a second order backward Euler scheme was applied. Transient results were obtained using a sliding mesh approach using 120 time steps for the passing of the rotor blades across one stator vane pitch. This is sufficient to resolve the first harmonic of the unsteady force signal corresponding to the stator engine order. Five inner iteration loops for each position when moving the transient rotor/stator sliding mesh interface were used. The

RMS residual for the whole calculations was approximately 10^{-4} at the last coefficient loop step.

The $y+$ value on the rotor blade for the finest mesh is less than 6.3 (average $y+$ 1.9) and at least 10 cells in the boundary layer were used. Uniform total pressure and total temperature were prescribed at the inlet and average static outlet pressure was specified at the outlet. Inflow is assumed as aligned with the axial direction.

A sector based approach using rotational periodicity is used to reduce the computational domain. The number of the stator vanes was scaled from 32 to 30 blades resulting in an integer blade count ratio of two, hence one stator vane and two rotor blades form the computational domain. This corresponds to a scaling factor of about 6%. It has been shown that scaling is a valid option for the present case [23]. The hub and tip shroud leakage cavities were also modeled as a sector. This approach will not capture large rotational flow structures with low nodal diameters because their circumferential wave lengths do not coincide with the sector. It is assumed that the forcing pattern is dominated by structures related to the blade passing frequency (BPF) and the passage flow across the airfoil is driving blade forcing. This could be confirmed by means of a full annulus simulation, which however has not been part of this investigation.

MESH CONVERGENCE AND TIME PERIODICITY

Three different mesh configurations were used to demonstrate mesh convergence for the rotor blade. Only results obtained with the finest mesh are presented. The mesh configurations are listed in Tab. 3. For the stator domain the upstream

Table 3: Mesh setup

	Mesh nodes per passage (-)		
Blade row	coarse	medium	fine
S1	568 512	568 512	568 512
R1	148 420	503 494	1 134 486
Tip gap	4 560	14 535	32 775
Shroud cavity	41 608	137 686	309 265
Hub cavity	188 020	657 943	1 459 095

unsteadiness is low with no significant difference between the meshes. Mesh convergence for the EO excitation signal was demonstrated previously for the stator blade mesh [24]. The finest mesh for the rotor blade has 19 nodes in the o-grid, 58 nodes across the span and 184 cells around the blade.

The unsteady forcing magnitude along the span at the first and second blade passing frequencies for the most complex case which proved most difficult to convergence (HS) are shown in Fig. 6. For this case grid convergence is achieved along most of the blade height, except locally at low span, to an average of 7% (for the other cases to less than 1.5% on average). The differences between the cases are larger than the differences due to different mesh resolutions (compare Fig. 21). Obtaining unsteady results on an even finer mesh for all configurations was not feasible with the available computational resources and the finest setup in total comprised of 6.2M nodes[1].

[1]this number is smaller than the sum of nodes in Tab. 3 since interface nodes in the two blade rotor domain are counted once only

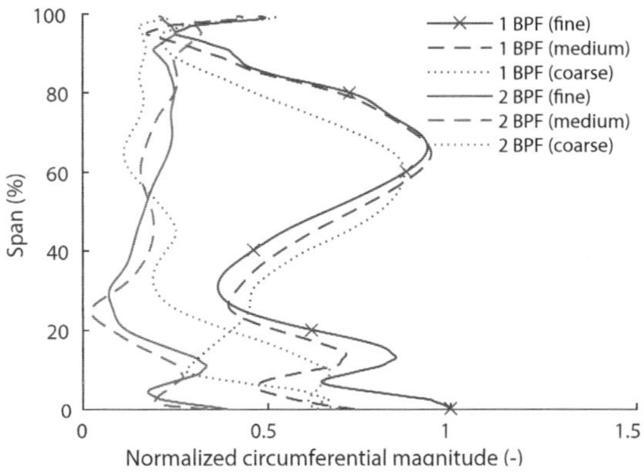

Figure 6: Unsteady force at BPF and twice BPF

A forced response calculation was performed using the unsteady aerodynamic forcing for disk mode shapes for the shrouded and unshrouded configurations at the mode eigenfrequency. The maximum amplitude of a damped structural response was calculated and the relative maximum amplitude compared with the most refined mesh for all cases shows good convergence for CFD mesh refinement (Tab. 4). This integral amplitude combining all the significant influences in the setup is (together with the vibratory stresses) the relevant quantity for blade forcing.

Table 4: Relative critical blade amplitude difference from forced response calculation with respect to value obtained with the respective fine mesh

	Mesh configuration		
Cases	coarse	medium	fine
B	6.7%	−1.3%	–
T	−3.9%	−0.6%	–
S	9.6%	−3.4%	–
HB	−0.6%	−3.0%	–
HT	−3.4%	−0.0%	–
HS	52.8%	6.0%	–

Time-periodic convergence is judged based on the pressure signal of several points in the domain and the overall blade torque. Once these quantities are converged to a time-periodic state (after at least 10 excitation signal pitches), parameters of the full flow field are extracted and stored for further post-processing. Quantitatively the correlation of the integrated time force signal for the blade passage force for two subsequent model pitches converged to a time periodic value. Correlation for the axial force and circumferential blade torque was above 99.9% for all the cases. Correlation for 36 equally spaced pressure monitor points on four planes up and downstream of both stator and rotor row with a stimulus of more than 5% was above 99.8% for all cases, except for the case with hub and shroud cavity with a value of 98.4%. This case does not reach a perfectly time periodic state. For the clean case (which reaches a time periodic state) a correlation of greater than 99.9% of these 36 points corresponds to a periodic state of the surface pressure signal in all the blade points and thus also the blade force. This was also shown for generalized force. Also here, for the cases with hub detailing calculations a perfectly time periodic state cannot be reached.

REFLECTIVITY OF INLET AND OUTLET

No significant reflections were observable at the sliding mesh interface or the upstream domain inlet. On the pressure outlet boundary with specified static pressure, an upstream propagating reflection of the unsteady flow for the first signal harmonic was observable. The reflections propagated up to the rotor blade trailing edge but were low in magnitude compared to the unsteadiness arising from the stator-rotor interaction. A good match to the experimental data was achieved also with this simulation artifact. The detailed investigation of the reflections was left for future investigation.

RESULTS

TIME AVERAGED BLADE LOADING

The time averaged stator blade loading at mid-span is shown in Fig. 7. The stator vane is almost entirely chocked near the trailing edge. Including the tip gap or the tip shroud cavity on the rotor blade (reduced back pressure) pushes the shock location further downstream and slightly increases the shock strength on the stator blade, also slightly reducing the inlet pressure to the rotor. Including the hub cavity downstream of the stator has an opposing effect on shock location.

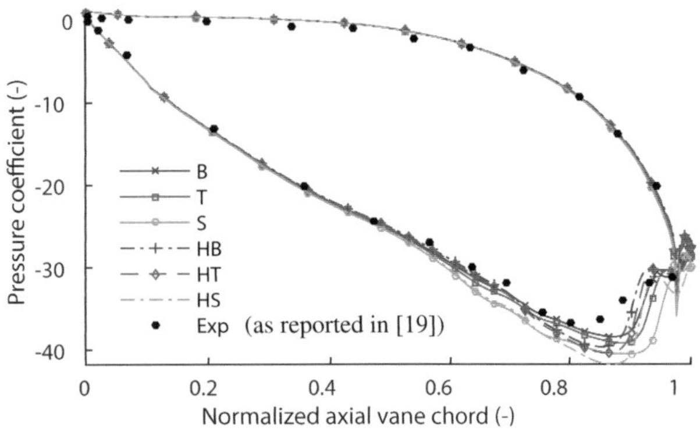

Figure 7: Stator mid-span time averaged blade loading

The time averaged rotor blade loading at mid-span is shown in Fig. 8. It appears to be more sensitive to the shroud detailing features than the hub cavity feature. The inclusion of the tip gap and tip shroud cavity reduces the blade loading. This is already visible in the blade torque comparison (Fig. 4). On the pressure side the flow is accelerated from about 50% chord towards the TE. The cases with tip or shroud cavity show a uniform reduction in pressure on the PS. On the suction side the flow is accelerated up to about 20% chord. The rate of acceleration is higher for the tip gap case and even higher (with different inflow incidence) for the tip shroud cavity case. The flow decelerates from about 75% chord to the trailing edge. The rate of deceleration varies from case to case. Changes can be due to a change in rotor incidence, especially for the cases with shroud cavity (compare Fig. 2). There is a weak passage shock detached from the blade surface and the blade passage is almost choked. The shock is strongest for the base case. Reference pressure is the

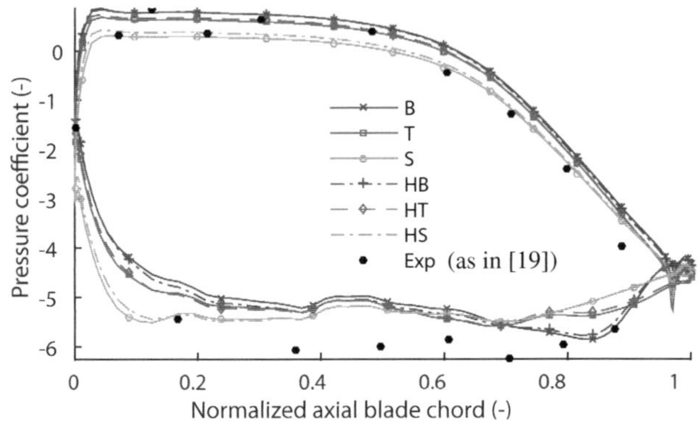

Figure 8: Rotor mid-span time averaged blade loading

relative row inlet pressure and the pressure coefficient defined as in Eqn. 1 and Eqn. 2.

$$c_p = \frac{p - p_{ref}}{p_{0,ref} - p_{ref}} \tag{1}$$

$$\hat{c}_{p,unsteady} = \frac{\hat{p}}{p_{0,ref} - p_{ref}} \tag{2}$$

For the tip gap and shroud cavity case, blade loading is shifted from the trailing edge towards the leading edge of the blade. Adding the hub cavity does not significantly change the time averaged blade loading for all cases.

UNSTEADY BLADE LOADING

From the blade torque and the time averaged mean pressures it is evident that the blades see different steady loading. The magnitude content at stator BPF of the unsteady pressure also changes between the cases as can be compared in Fig. 9. Compared are the base case (clean end-walls), the tip gap and the tip shroud cavity case at mid-span. On the pressure side the unsteady pressure magnitude is overall slightly increased including the tip gap and tip shroud cavity. On the suction side the same trend is visible, however the magnitude increase is larger towards the leading edge and the magnitude pattern is changed from 75% chord to the TE. In this region the maximum peak is reduced for both the case with tip gap and shroud cavity and an additional peak is present for the tip gap case at about 70% chord.

In Fig. 10, 11 and 13 the unsteady blade loading for all cases at 10%, 50% and 90% span is shown, compared also to

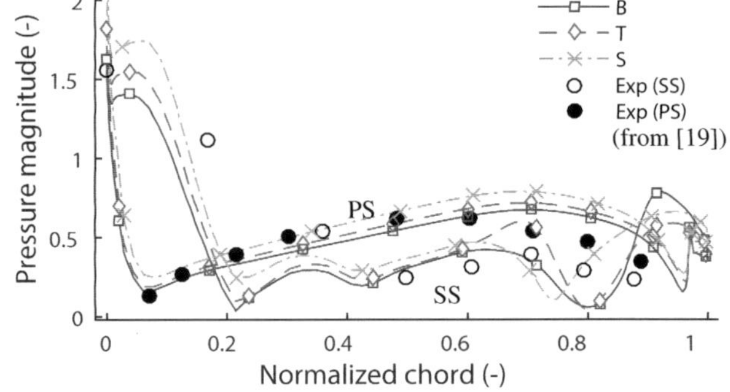

Figure 9: Unsteady R1 blade loading, 30EO, 50% span

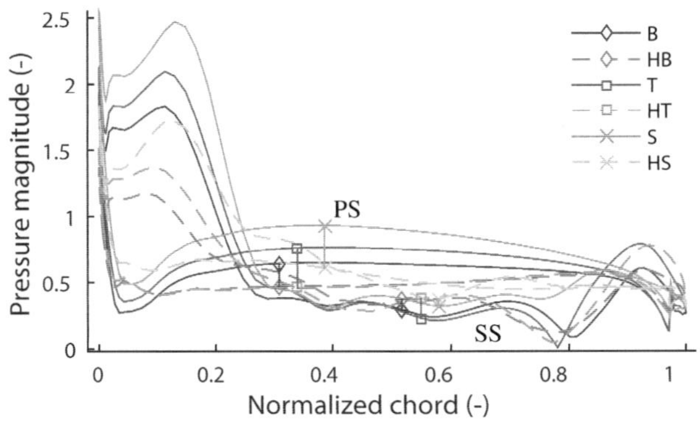

Figure 10: Unsteady R1 blade loading, 30EO, 10% span

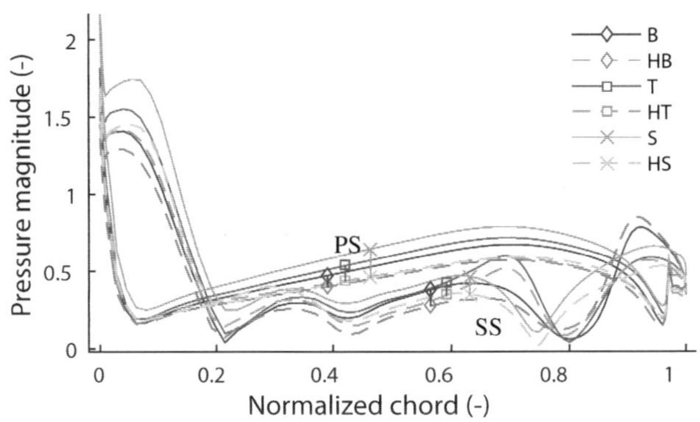

Figure 11: Unsteady R1 blade loading, 30EO, 50% span

the cases with added hub cavity. At 10% span (Fig. 10) adding the hub cavity (dashed lines) shows a decrease in unsteady load on the pressure side. The suction side leading edge peak is amplified and no distinct trend can be identified after 25% chord. At mid-span (Fig. 11), adding the hub cavity uniformly reduces the unsteady pressure magnitude, especially on the pressure side from 25% chord and on the suction side between 40% and 80% chord and close to the leading edge. At 90% span up until 50% chord, unsteady loading is increased when adding the hub cavity.

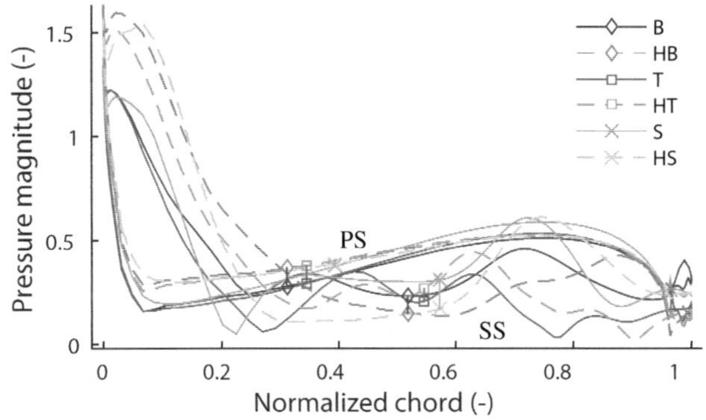

Figure 13: Unsteady R1 blade loading, 30EO, 90% span

For the rest of the blade, unsteady loading is reduced, except for the tip gap case and the shroud gap case locally towards the trailing edge. For 10% and 90% span, especially on the suction side the unsteadiness is mostly influenced by the local secondary flow. At the hub (compare 10% span) relative unsteady load over the blade chord is highest at the LE. At the shroud (compare 90% span) unsteady load is distributed more towards the trailing edge.

COMPARISON TO AVAILABLE TEST DATA

The case without shroud and hub cavity can be compared to experimental results obtained on the actual test rig on which the derived cases are based. The steady and unsteady test data is shown in Fig. 7, 8 and 9. The computations show a reasonably good match with the experimental data both for the steady and the unsteady pressure magnitude.

The simulated mean pressure is slightly over-predicted, especially on the suction side, compared to the measurements. This is already observed in calculations done by [19], [25] and [26] who all show the same trend when comparing to the experimental data.

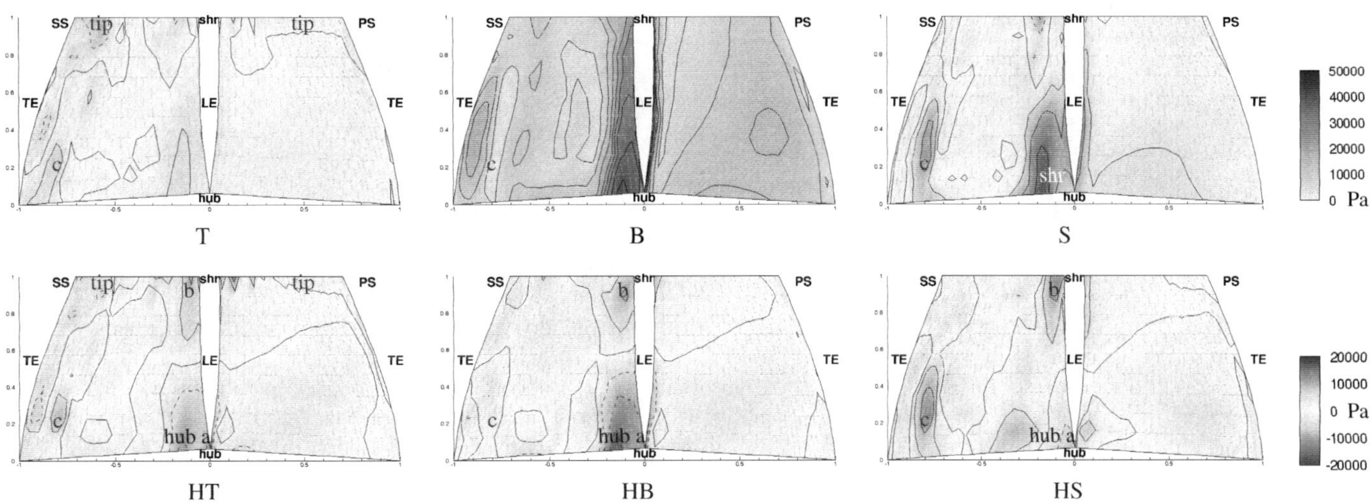

Figure 12: Unsteady pressure magnitude at BPF on rotor blade surface (top: clean base case (middle), tip gap (left), shroud cavity (right); bottom: with added hub cavity; cases with detailing features are shown as difference to the base case)

UNSTEADY BLADE SURFACE LOADING

The unsteady pressure magnitude on the unwrapped blade surfaces at BPF is shown in Fig. 12. For the tip gap case (T) the unsteady pressure magnitude is reduced on the suction side close to the trailing edge (tip). For the case with shroud cavity (S) the unsteady pressure on the pressure side at low span is slightly increased as well as on the suction side leading edge (shr).

Comparing the cases without hub cavity (B, T, S) to the cases with added hub cavity (HB, HT, HS) two trends can be observed on the leading edge: First, the unsteady pressure magnitude at the hub casing is reduced (a). Second, however, the unsteadiness at high span increases (b), leading to a span-wise more even distribution. On the suction side at high chord and low span the unsteadiness is increased for all cases (c).

FLOW FIELD

Looking at the tip features and comparing the case with the tip gap and the tip shroud cavity with the base case it can be observed, that with the small tip clearance, the tip leakage flow increases the vorticity and mixing after the rotor blade, does however not significantly influence the flow in the blade row (shown in Fig. 14). The leakage flow (t_1) seems to however strengthen the tip vortex (t_2).

For the tip shroud leakage case, there is little interaction between the passage and leakage flow in the shroud cavity (s_2), likely due to the large pressure drop across the cavity and large mass flow. For this case the flow in the blade to blade passage is also more aligned with the blade (s_1). Fluid exiting the cavity after the blade locally increases the back pressure at higher span for the rotor blade (s_3). Accordingly, the local rotor inlet pressure is reduced in the vicinity of the tip shroud cavity (s_2).

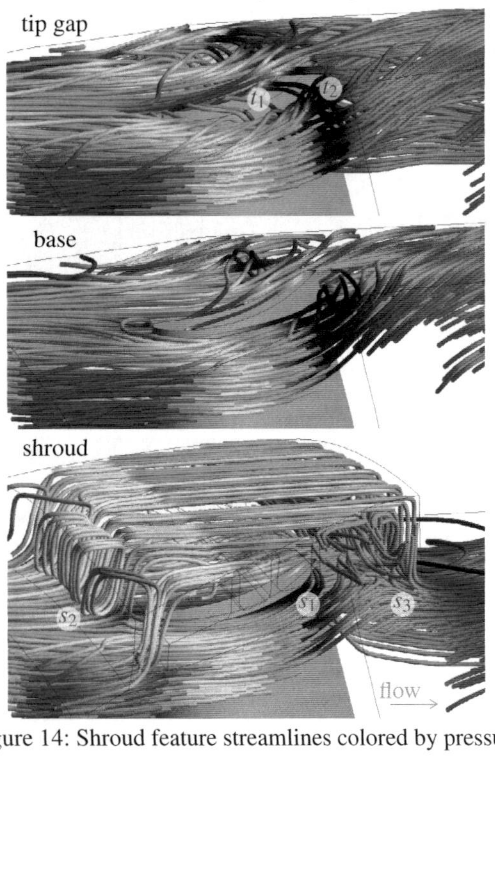

Figure 14: Shroud feature streamlines colored by pressure

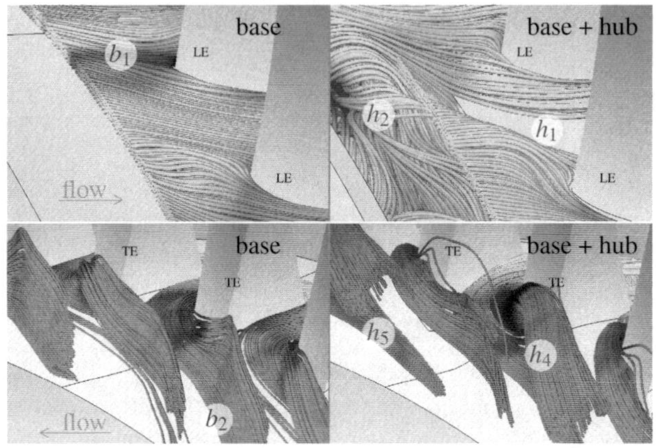

Figure 15: Hub feature streamlines colored by pressure

For the hub cavity (Fig. 15), there is some interaction of the flow upstream of the rotor blade. The flow sheet of aligned flow is distorted by the cavity interaction (b_1, h_1, h_2). The potential field passing with the rotor blade along the cavity circumferentially leads to an interaction with the main flow, leading to a stronger horseshoe vortex (h_4, h_5 compared to b_2). The reduction in unsteady pressure at the leading edge of the blade can directly be related to the increased circumferentially averaged pressure distribution due to the cavity.

BLADE TO BLADE FLOW FIELD

The unsteady flow for the base case (clean end-walls, B) is shown in Fig. 16. Shown are streamlines as well as Mach number contours. The slice is colored by instantaneous unsteady pressure. To the left the periodic sector is shown at $t_0 = 0$, to the right after half the period time $t_{T/2}$. Two neighboring blade passages are 180° out of phase due to the blade count ratio. Thus time steps $t_0 = 0$, $t_1 = \frac{T}{4}$, $t_2 = \frac{T}{2}$, $t_3 = \frac{3T}{4}$ are represented. In the first time step (t_0) the passage is fully choked with a high pressure gradient across the blade throat. With the upstream

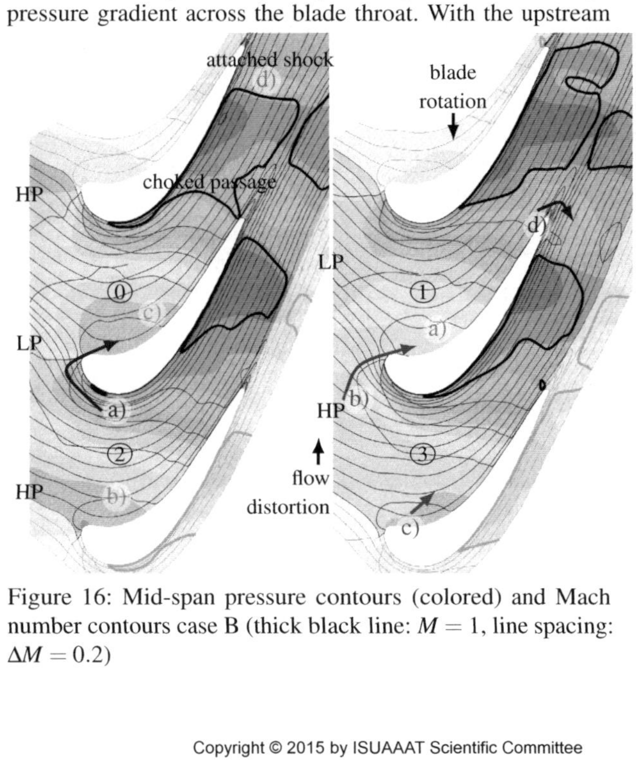

Figure 16: Mid-span pressure contours (colored) and Mach number contours case B (thick black line: $M = 1$, line spacing: $\Delta M = 0.2$)

flow disturbance traveling circumferentially, the low pressure field propagates from the blade LE to mid-passage and the high pressure region on the pressure side at the passage inlet cancels out (t_1, a). This leads to a slow break-down of the shock and the supersonic flow regime. Next the high pressure inlet disturbance travels downstream on the suction side (t_2, t_3, b, c). With a higher pressure gradient the flow in the passage then becomes fully choked again.

This highly unsteady transonic flow is compared in terms of Mach number comparing all cases in Fig. 17 to 19. Shown is the averaged Mach number and contours at sonic condition as time average, the standard deviation of the time average and the minimum and maximum envelope.

UNSTEADY BLADE TORQUE AND GENERALIZED FORCE

Unsteady forcing between the cases is compared in terms of the BPF signal of the blade torque (compare Fig. 4) and in terms of generalized force for the first 32EO BPF (computed as scaled 30EO) crossing with the 28 nodal diameter mode shape. Relative differences with respect to the base case are shown in Tab. 5.

Table 5: Comparison of unsteady blade torque and generalized force at BPF between the cases with respect to reference case B

Cases	Unsteady torque	F_{gu}	F_{gs}
B	-	-	-
T	10.4%	0.3%	8.8%
S	29.8%	18.8%	23.4%
HB	−20.8%	3.2%	−7.7%
HT	−16.2%	−3.2%	−5.8%
HS	−8.2%	−8.1%	−12.2%

The shrouded and unshrouded configuration feature different mode shapes, however for comparison, the generalized force for both modes (F_{gu}, F_{gs}) is presented with the obtained forcing. Generalized force as a measure of the unsteady aerodynamic work done is defined as shown in Eqn. 3 with the complex force F for the mode shape ϕ, here depicted in terms of cosine and sine part.

$$F_g = \sqrt{\sum_{a=cos,sin} \sum_{b=cos,sin} \sum_{i=x,y,z} \left(\phi_{a,i}^T F_{b,i}\right)^2} \tag{3}$$

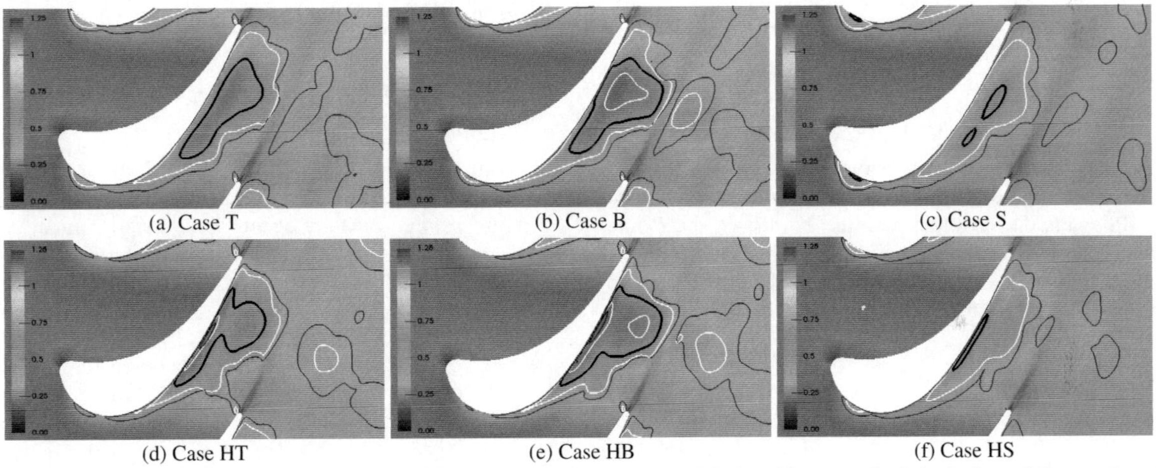

(a) Case T (b) Case B (c) Case S
(d) Case HT (e) Case HB (f) Case HS

Figure 17: Rotor Mach number at 10% span (contours: black: time average M=1, white: standard deviation of fluctuating signal, red/blue: sonic mach number envelope at maximum/minimum condition)

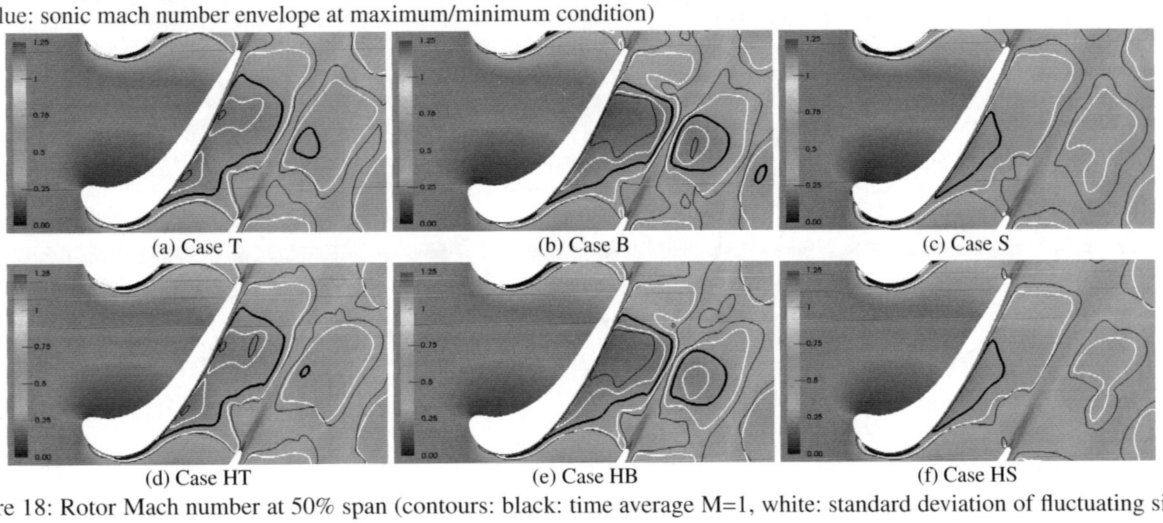

(a) Case T (b) Case B (c) Case S
(d) Case HT (e) Case HB (f) Case HS

Figure 18: Rotor Mach number at 50% span (contours: black: time average M=1, white: standard deviation of fluctuating signal, red/blue: sonic mach number envelope at maximum/minimum condition)

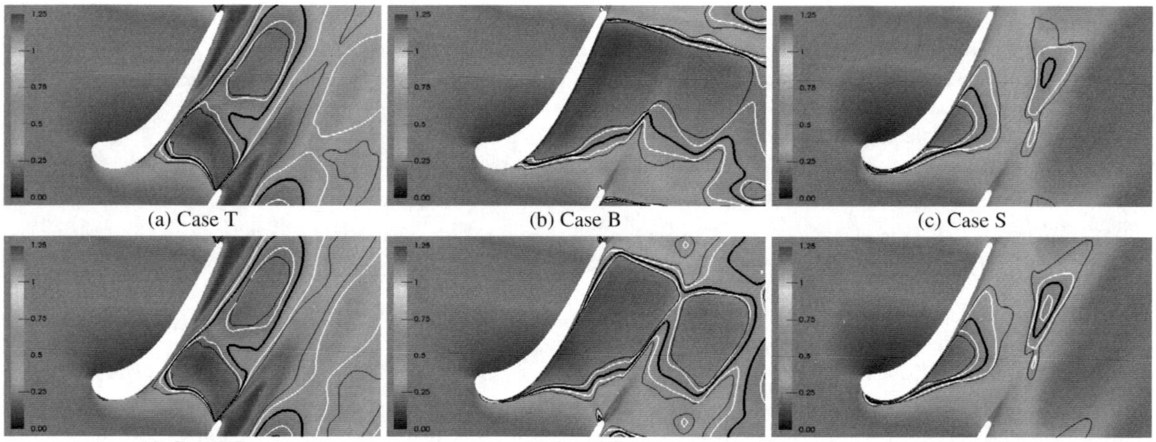

(a) Case T (b) Case B (c) Case S

(d) Case HT (e) Case HB (f) Case HS

Figure 19: Rotor Mach number at 90% span (contours: black: time average M=1, white: standard deviation of fluctuating signal, red/blue: sonic mach number envelope at maximum/minimum condition)

It can be seen, that the change in generalized force is an order of magnitude less sensitive to the change in unsteady blade torque and that the hub cavity seems to reduce flow unsteadiness. Further it can be seen, that the case with tip shroud shows the biggest changes in terms of generalized force and unsteady torque.

DISTRIBUTED FORCE AGAINST SPAN

The harmonic axial force from hub to shroud extracted from various sampled span-lines is shown in Fig. 20. Shown is the BPF excitation together with the 2nd and 3rd harmonic component both in terms of the absolute value as well as the absolute difference compared to the reference case (B, clean end-walls). The values for the 2nd as well as the 3rd harmonic components are small with respect to the first harmonic (except for case S).

The cases with tip gap and shroud cavity (T, S) show increase in force magnitude for the first harmonic both below 80% span as well as a local decrease at the blade tip. The cases with hub cavity show a decrease in 1st harmonic force at mid-span. Close to the hub at 10% span these cases show a localized increase in force magnitude.

The harmonic circumferential force from hub to shroud is shown in Fig. 21. The 2nd and 3rd harmonic component of the circumferential force are again mostly small compared to the first harmonic, except for case S. The cases with tip gap and shroud cavity (T, S) show a decreased localized forcing at the blade tip, however a local increase around 85% span related to

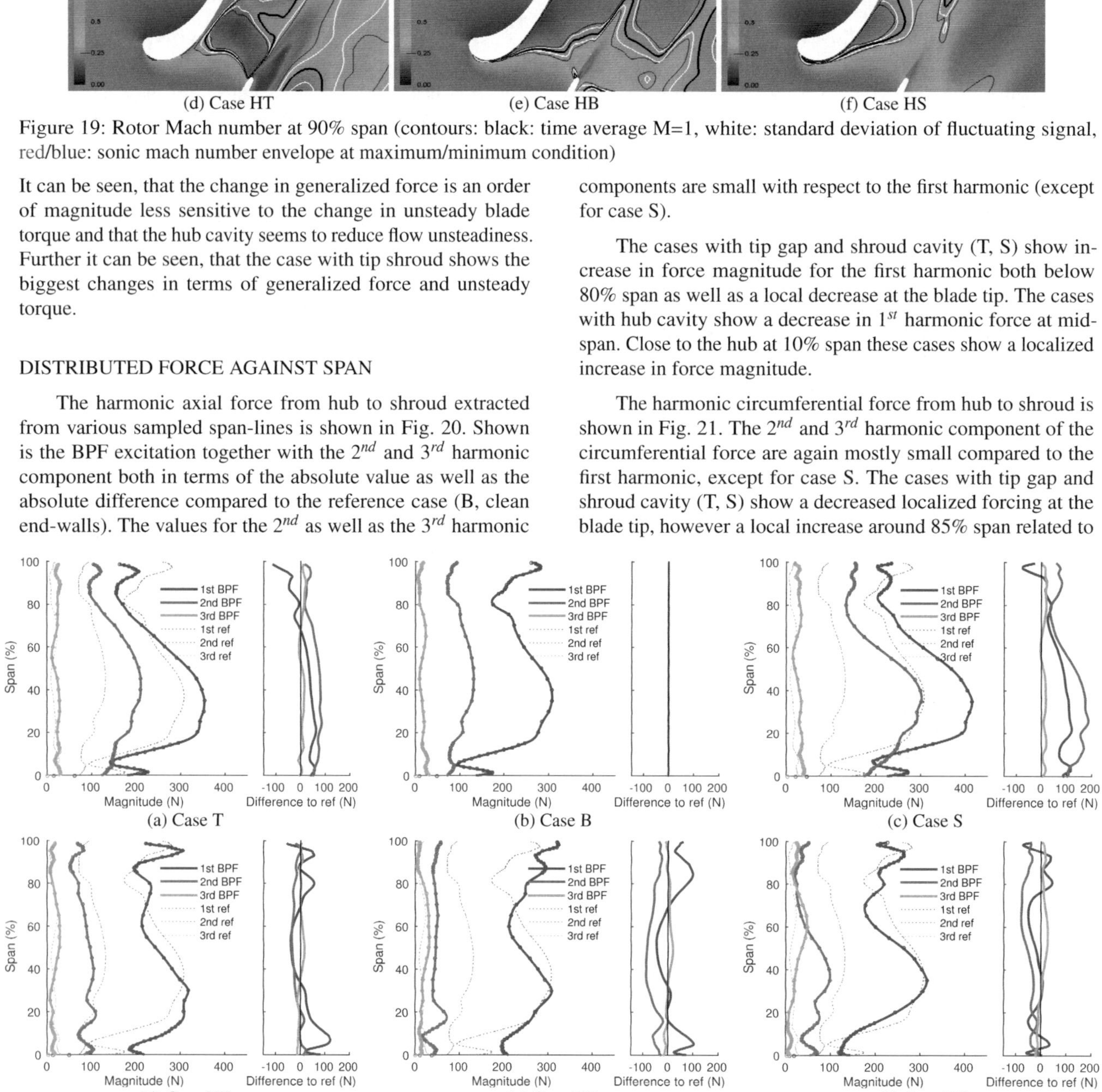

(a) Case T (b) Case B (c) Case S

(d) Case HT (e) Case HB (f) Case HS

Figure 20: Axial force magnitude at BPF for different blade span

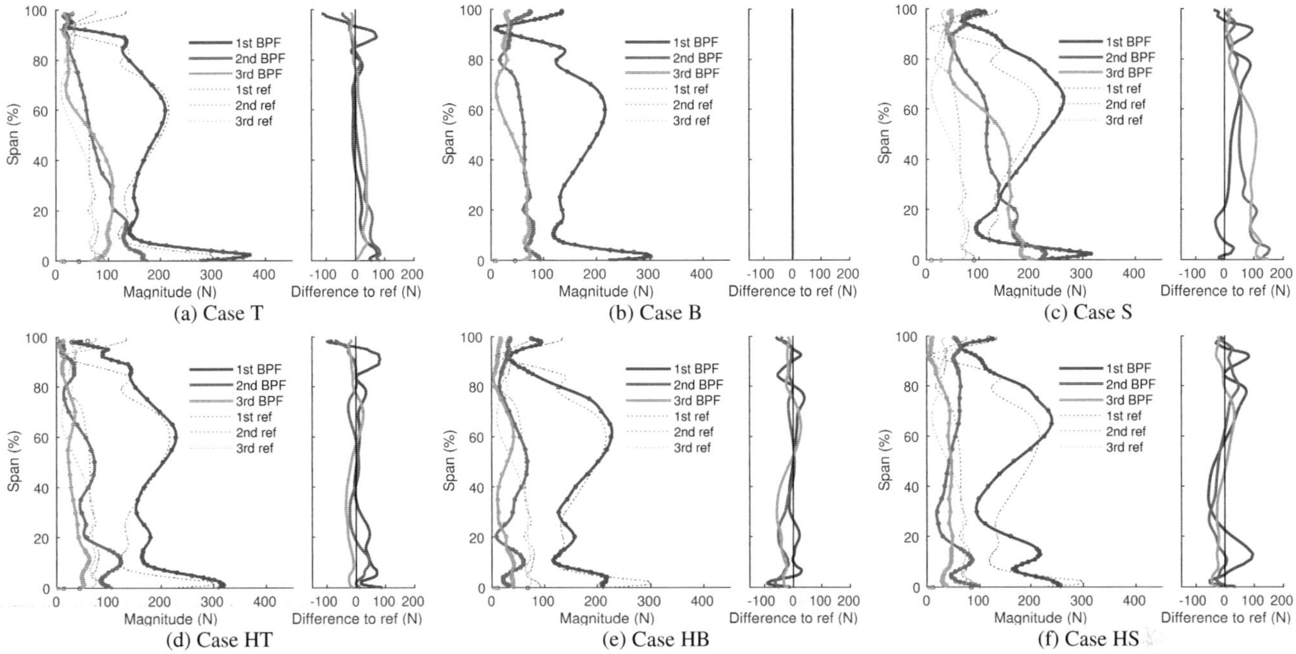

(a) Case T (b) Case B (c) Case S

(d) Case HT (e) Case HB (f) Case HS

Figure 21: Circumferential force at BPF for different blade span

the shroud vortex. Both cases also show an increase in forcing towards the hub compared to the base case. Adding the hub cavity does not significantly alter the unsteady circumferential forcing pattern.

The span-wise comparison of unsteady forcing to the base case shows that changes in the unsteady excitation force magnitude at BPF are very localized and represent small changes. Further, the very localized detailing features show an effect not strictly bound to the directly affected or adjacent flow regions. Comparing the unsteady distributed loading presented above, it is shown, that the investigated features do not significantly alter the unsteady forcing nature. A forced response calculation with the disk modes using the obtained forcing pattern did not show variations larger than 7% in the calculated forcing amplitude.

CONCLUSION

In the current investigation, rotor tip gap, rotor tip shroud cavity and a hub cavity upstream of the rotor blade were investigated as detailing features for a transonic turbine case. This was done to estimate the influence of these features on rotor blade forcing. The hub cavity shows a reduction in force unsteadiness, where the tip gap and shroud cavity seem to increase the unsteadiness of the forcing signal. The tip gap does not change the flow pattern significantly whereas the shroud cavity clearly has an effect on the flow in terms of a by-pass changing the blade pressure ratio towards to tip. For the hub cavity a slight interaction between the passage flow and hub fluid is apparent together with vortex shedding not at blade passing frequency at the hub cavity.

In terms of the forcing signal at blade passing frequency, only small changes between the cases could be observed, up to a maximum difference of 25% for the case with shroud cavity alone. At the same time, using a coarse design mesh, especially in the most complex case with hub and shroud cavity, produced a very different result showing that for these flow configurations mesh resolution is especially important.

The variation in calculated forced response amplitude for the several cases was limited to 7%, even given the larger variability in generalized force.

This leads to the conclusion that for aeromechanical blade failure the detailing features for the present case could be omitted to obtain the excitation signal alone.

FUTURE WORK

An analysis investigating the effect of the various features on aerodamping is pending. The analysis was limited to the single stage test case, effects of the features on downstream flow and blade excitation, could thus not be shown. A sector model approach was chosen, which comes with limitations and convergence towards a time periodic solution is not achieved for the case with hub and shroud cavity. For these cases a full annulus simulation is pending.

ACKNOWLEDGMENT

The work has been funded by the Swedish National Space Technology Research Program (NRFP), the support of which is gratefully acknowledged. The simulations were partly performed on resources at PDC-HPC (Centre for High Performance Computing).

REFERENCES

[1] Srinivasan, A. V., 1997. "Flutter and Resonant Vibration Characteristics of Engine Blades". *ASME. J. Eng. Gas Turbines Power,* **119**, pp. 742–775.

[2] Mazur, Z., Luna-Ramírez, A., Juárez-Islas, J., and Campos-Amezcua, A., 2005. "Failure analysis of a gas turbine

blade made of inconel 738lc alloy". *Engineering Failure Analysis,* **12**(3), pp. 474–486.

[3] Mazur, Z., Hernández-Rossette, A., and García-Illesoas, R., 2006. "L-0 Blades Failure Investigation of a 110 MW Geothermal Turbine". *ASME Paper No. POWER2006-88024*, pp. 281–289.

[4] Mazur, Z., García-Illescas, R., Aguirre-Romano, J., and Pérez-Rodríguez, N., 2007. "Last-Stage Blade Failure Evaluation". *ASME Paper No. POWER2007-22010*, pp. 83–90.

[5] Song, K.-S., Kim, S.-G., Jung, D., and Hwang, Y.-H., 2007. "Analysis of the fracture of a turbine blade on a turbojet engine". *Engineering Failure Analysis,* **14**(5), pp. 877–883.

[6] Vardar, N., and Ekerim, A., 2007. "Failure analysis of gas turbine blades in a thermal power plant". *Engineering Failure Analysis,* **14**(4), pp. 743–749.

[7] Australian Transport Safety Bureau, 2009. Investigation: AO-2009-061 - Engine failure - VH-SBA, Wagga Wagga Aerodrome, NSW, 4 October 2009.

[8] Barella, S., Boniardi, M., Cincera, S., Pellin, P., Degive, X., and Gijbels, S., 2011. "Failure analysis of a third stage gas turbine blade". *Engineering Failure Analysis,* **18**(1), pp. 386–393.

[9] Kim, H., 2011. "Crack evaluation of the fourth stage blade in a low-pressure steam turbine". *Engineering Failure Analysis,* **18**(3), pp. 907–913.

[10] Qu, S., Fu, C., Dong, C., Tian, J., and Zhang, Z., 2013. "Failure analysis of the 1st stage blades in gas turbine engine". *Engineering Failure Analysis,* **32**(0), pp. 292 – 303.

[11] Paniagua, G., Dénos, R., and Almeida, S., 2004. "Effect of the hub endwall cavity flow on the flow-field of a transonic high-pressure turbine". *ASME Paper No. GT2004-53458.*

[12] Ella, H. A. E., Sjolander, S., and Praisner, T., 2010. "Effects of an upstream cavity on the secondary flow in a transonic turbine cascade". *ASME Paper No. GT2010-22843.*

[13] Engelmann, D., Kalkkuhl, T. J., Polklas, T., and Mailach, R., 2012. "Influence of Shroud Cavity Jet and Steam Admission Through a Circumferential Slot on the Flow Field in a Steam Turbine". *ASME Paper No. GT2012-68465.*

[14] Jenny, P., 2013. "Unsteady Rotor Hub Passage Vortex Behavior in the Presence of Purge Flow in an Axial Low Pressure Turbine". *Journal of Turbomachinery,* **135**(5), June, p. 051022.

[15] Schrewe, S., 2013. "Experimental Analysis of the Interaction Between Rim Seal and Main Annulus Flow in a Low Pressure Two Stage Axial Turbine". *Journal of Turbomachinery,* **135**(5), June, p. 051003.

[16] Rosic, B., and Xu, L., 2012. "Blade Lean and Shroud Leakage Flows in Low Aspect Ratio Turbines". *Journal of Turbomachinery,* **134**(3), May, p. 031003.

[17] Chilla, M., 2013. "Unsteady Interaction Between Annulus and Turbine Rim Seal Flows". *Journal of Turbomachinery,* **135**(5), June, p. 051024.

[18] Gezork, T., Mayorca, M. A., Groth, P., Vogt, D. M., and Fransson, T., 2014. "Influence of Tip Shroud Cavity Detailing on Turbine Blade Forcing Calculations". *ASME Paper No. GT2014-26724.*

[19] Laumert, B., 2002. "Numerical Investigation of Aerodynamic Blade Excitation Mechanisms in Transonic Turbine Stages". Ph.D. Thesis, Royal Institute of Technology, Stockholm. ISBN: 9172833548.

[20] Beard, P. F., Povey, T., and Ireland, P. T., 2008. "Mass flow rate measurement in a transonic turbine test facility with temperature distortion and swirl". *Flow Measurement and Instrumentation,* **19**(5), pp. 315–324.

[21] Hilditch, M. A., Smith, G. C., and Chana, K. S., 2001. "Measurements and Predictions of Endwall Heat Transfer in Two High Pressure Turbines". *RTO-MP-069(I),* **069**(May), pp. 7–11.

[22] European Commission, 2012. "TATEF 2 Turbine Aero-Thermal External Flows 2". `http://ec.europa.eu/research/transport/projects/items/tatef_2_en.htm`.

[23] Hosseini, S. M., Fruth, F., Vogt, D. M., and Fransson, T. H., 2011. "Effect of Scaling of Blade Row Sectors on the Prediction of Aerodynamic Forcing in a Highly-Loaded Transonic Turbine Stage". *ASME Paper No. GT2011-45813.*

[24] Yang, S., 2014. "CFD investigation of aeromechanic forcing sensitivity for a generic transonic turbine". Thesis No. EGI-2014-115 MSC EKV1073, Royal Institute of Technology, Stockholm.

[25] Gourdain, N., Wang, G., Duchaine, F., and Gicquel, L., 2013. "Application of Large-Eddy Simulation to rotor / stator configurations". *21ème Congrès Français de Mécanique Bordeaux.*

[26] Wang, G., Papadogiannis, D., Duchaine, F., Gourdain, N., and Gicquel, L. Y. M., 2013. "Towards Massively Parallel Large Eddy Simulation of Turbine Stages". *ASME Paper No. GT2013-94852*, June.

Proceedings of the 14th International Symposium of Unsteady Aerodynamics, Aeroacoustics & Aeroelasticity of Turbomachines
ISUAAAT 14
8-11 September 2015, Stockholm, Sweden

I14-S9-1

ANALYSIS OF BLADE VIBRATIONS IN A 1.5-STAGE TRANSONIC COMPRESSOR

Maximilian Jüngst
Felix Holzinger
Fabian Wartzek
Christoph Brandstetter
Daniel Möller
Heinz-Peter Schiffer
Technische Universität Darmstadt
Institute of Gas Turbines and Aerospace Propulsion
Darmstadt, Germany
Email: juengst@glr.tu-darmstadt.de

ABSTRACT

The experimental investigation presented in this paper included tip gap variations of the casing at the axial position of the rotor. Blade vibrations occurred, which limited compressor operation. A transient measurement close to the stability limit of the compressor was investigated using unsteady instrumentation and different methods that analyse blade vibrations. Strain gauges, blade tip timing and unsteady pressure transducers were used. The analysed data can be compared to the phenomenon "rotating instabilities" or "rotating disturbances", as well as non-synchronous vibrations that display similarities to classic "flutter" near stall. The presented measurement is unique, since both phenomena are observed in one transient manoeuvre of approximately 8 seconds duration close to the stability limit. The observed phenomena were compared to previous reports and further insight into the unsteady flow in the tip region prior to and during the occurrence of critical blade vibrations is presented.

NOMENCLATURE

AR	Amplitude Rise
BTT/BTC	Blade Tip Timing/Blade Tip Clearance
BPF	Blade Passing Frequency
CPSD	Cross-Power Spectral Density
FTW/BTW	Forwards / Backwards Travelling Wave
IBPA	Inter-Blade Phase Angle
LE/TE	Leading Edge/Trailing Edge
LPR/HPR	Low Pressure Region/High Pressure Region
ND	Nodal Diameter
RI	Rotating Instability
SG	Strain Gauges
TO	Throttle Opening
WPT	Wall Pressure Transducer
$A_{max}(\Delta f)$	Max. Amplitude in a specified Frequency Range
EO^{rot}/EO^{stat}	EO in rotating/stationary frame of reference
f	Frequency
$f_{aliased}$	Aliased Frequency
f_{rev}	Rotor speed expressed as rotational frequency
f^{rot}/f^{stat}	Frequency in rotating/stationary frame of reference
f_{sample}	Sampling Frequency
m	Integer Value: 0,1,2,...
N_{blades}	Number of blades
p	Static Pressure
p_{26}	Static Wall Pressure of Position 26 in axial array
$\overline{p}$	Mean Static Pressure
$p_{t,inlet}$	Total Pressure in the settling chamber
r_{shroud}	Radius at Shroud
T_{period}	Periodic Time
U_{tip}	Tip Velocity
$\Delta\theta$	Angular separation of two sensors
$\Delta\Psi$	Angular Resolution of Wall Pressure Data
Δt	Time lag between two sensors
Δx	Spatial Resolution of Wall Pressure Data
Π_{mean}	Mean Pressure related to $p_{t,inlet}$
ϕ	Phase angle of two signals
Ψ	Static-to-Total-Pressure-Rise Coefficient
ρ_{inlet}	Density in the settling chamber
σ_{rel}	Relative Standard Deviation

INTRODUCTION

In recent years Technische Universität Darmstadt Institute of Gas Turbines and Aerospace Propulsion Publications have focused on blade vibrations in transonic compressor rotors. Research on the transonic compressor focused on both, aerodynamic and aeroelastic issues. Synchronous vibrations, flutter and general non-synchronous vibrations are investigated experimentally and numerically using state-of-the-art BLISK rotors that represent the front stage of modern high pressure compressors [1–6]. The investigations demonstrated the influence of the blade tip vortex on rotor-induced blade vibrations. For the measurements presented in this paper, the strength of the blade tip vortex was influenced by enlarging the tip gap in the rotor section of the compressor, achieved by increasing the shroud diameter (illustrated in Figure 1).

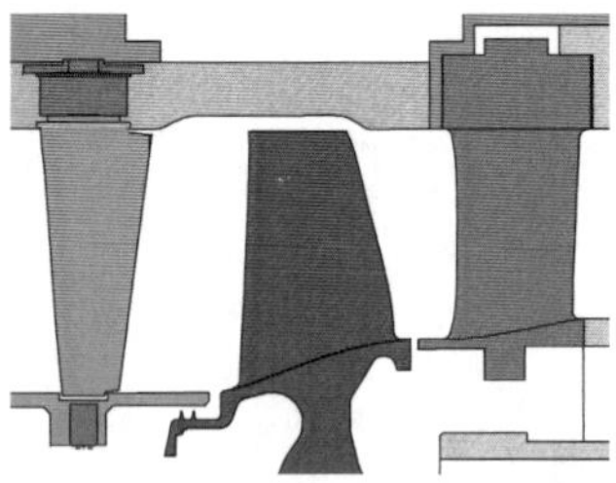

Figure 1. COMPRESSOR STAGE WITH VARIABLE INLET GUIDE VANES AND ENLARGED TIP GAP

Variations in the tip gap in axial compressors during transient operation of aero-engines can be enormous. The chosen tip gap of 2.2% of mean blade height is representative of transient manoeuvres [7].

Several publications have shown that compressors with large tip gaps tend to show different pre-stall behaviour to rotors with smaller tip gaps. Rotors with enlarged tip gaps are exposed to high pressure fluctuations that rotate in the the duct, which are called "rotating instability" (in short: RI) or "rotating disturbances" [8–14].

In [15] non-synchronous vibrations near stall have been studied for several speed lines. A comparison to earlier publications is given [8, 11, 16]. Besides some similarities a global picture cannot be drawn.

The Investigation by [3] showed that a variation of the tip gap can extensively change the aerodynamic and aeroelastic phenomena of axial compressors. RI-induced vibrations and non-synchronous vibrations that show similarities to classic *"flutter"* near stall were observed. Also, a detailed categorization of the observed phenomena and flow induced vibrations is given [3, 5]. The current paper will contribute to this, analysing a single measurement in detail.

The presented measurement is unique, since these phenomena were captured in one transient manoeuvre of the Darmstadt Transonic Compressor of approximately 8 seconds duration. The compressor was back-pressured into its stability limit, starting at the last stable operating point close to the stability limit at part speed (2).

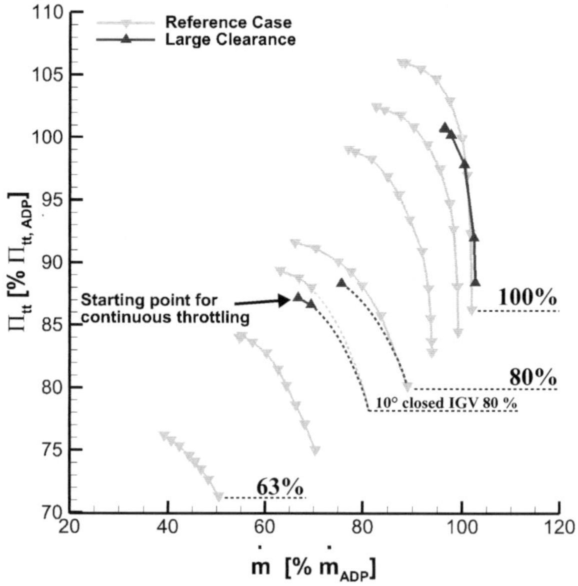

Figure 2. PERFORMANCE MAP AND ILLUSTRATION OF THE TRANSIENT THROTTLING PROCESS OF THE ANALYSED MEASUREMENT IN THE COMPRESSOR MAP

The point marked as *"Starting point for continuous throttling"* is represented by Revolution 0 in Figures 9,11 and 13. The phenomena that occurred during the 8 seconds of the measurement (1500 revolutions respectively) will be analysed in this paper using different measurement techniques:

- blade-mounted strain gauges (SG): Analysis of the blade vibration (eigenmode)
- casing-mounted blade tip timing (BTT): Analysis of the BLISK vibration (nodal diameter ND)
- casing mounted wall pressure transducer (WPT): Analysis of the static wall pressure in the rotor section (flow phenomena, propagation of pressure waves, reconstruction of the flow field)

DARMSTADT COMPRESSOR RESEARCH FACILITY

Figure 3 illustrates the research facility. The test rig is driven by an 800 kW D/C drive (6). A gearbox (5) enables the test rig to run at up to 21.000 rpm shaft speed and 350 Nm shaft torque, which is measured with a torquemeter (4). The incoming flow enters the test rig through a variable inlet throttle (0), a settling chamber (1) and the intake (2), which includes the mass flow measurement. The transonic compressor (3) is run as a 1.5-stage compressor with variable inlet guide vanes, as introduced in Figure 1.

The flow leaves is guided through a radial diffuser back to ambience. Of particular importance for the measurements close to the stability limit is the emergency bleed of the rig. It enables transient throttling near stall.

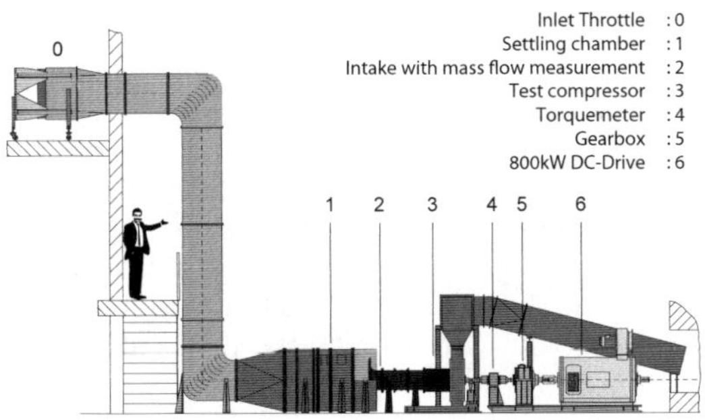

Figure 3. SKETCH OF THE TEST LAB

UNSTEADY INSTRUMENTATION AND ANALYSIS TECHNIQUES

The following section provides an overview of the unsteady instrumentation. Blade vibrations are measured in the rotating frame of reference by applying strain gauges (SG) to the blades, while capacitive tip timing sensors (BTT) measure vibrations in the stationary frame of reference. Unsteady pressure transducers in the casing wall are used to analyse flow phenomena in the rotor section.

Instrumentation in the Rotating Frame of Reference

Figure 4 illustrates the blade-mounted strain gauges. Strain measurement is mode optimized at several blade positions; different strain gauges are used to monitor different modes. During transient measurements the strain is monitored online to keep blade stress in acceptable ranges with moderate loss of durability. Since the SG measure in the rotating frame of reference, the actual blade vibration frequency is measured independently of the rotor speed. However, vibration of only a single blade is captured.

Instrumentation in the Stationary Frame of Reference

The test rig is instrumented with casing-mounted capacitive sensors (Figure 4). A big advantage of this measuring system is contactless blade vibration monitoring, plus the supervision of all blades. On this account, BTT/BTC data are always documented during critical operation. Every time a blade passes a sensor the subsequent capacitance change is measured by the sensor. Consequently, the gap between blade and sensor (/shroud) can be detected (BTC). By comparing the nominal passing times of the blades with the actual passing times, blade deflection can be derived (BTT). By doing so, blade vibrations can be measured.
The disadvantage of the mounted capacitive sensors is their frequency resolution. There is only a signal impulse when a

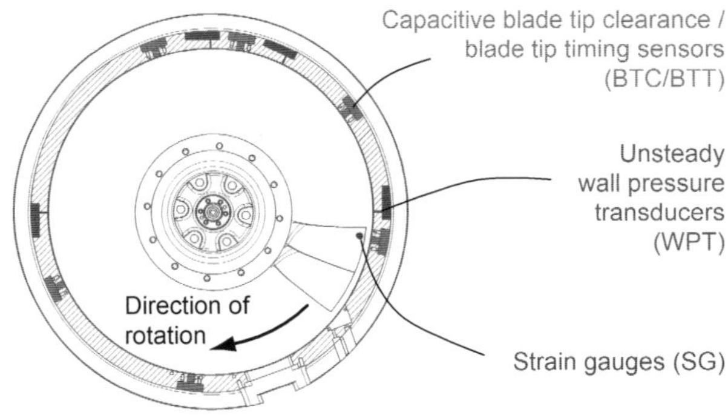

Figure 4. UNSTEADY CIRCUMFERENTIAL DISTRIBUTED INSTRUMENTATION IN THE ROTOR SECTION OF THE COMPRESSOR [FROM [3]]

blade passes a sensor, therefore the sampling frequency equals the blade passing frequency (BPF). Using the reasoning of the Nyquist-Shannon sampling theorem, every frequency component above half of the BPF is aliased to lower frequencies. Conclusions can be drawn from the aliased frequency $f^{\text{stat}}_{\text{aliased}}$ and the actual frequency f^{stat} can be calculated. Spectra that use every blade for sampling are called All-Blade-Spectra. Using just the signal of one specific blade, the sampling frequency is reduced further to once per revolution. These spectra are called Single-Blade-Spectra. The detailed theoretical background is given by [17, 18].
This equation can be used to calculate aliased frequencies:

$$f^{\text{stat}}_{\text{aliased}} = |N_{\text{blades}} f_{\text{rev}}\, m - f^{\text{stat}}|. \tag{1}$$

As a result, the aliased frequency in the stationary frame of reference $f^{\text{stat}}_{\text{aliased}}$ depends on the actual value of the frequency f^{stat} compared to the sampling frequency. It is mirrored to lower frequencies when it is larger. Due to the fact that the measuring system gives only positive frequencies, the absolute value of $f^{\text{stat}}_{\text{aliased}}$ is shown in the spectrum. As well as aliasing effects, the frequency shift, known as Doppler shift between the rotating and the stationary frame of reference, has to be considered. The vibration frequency of the blade is shifted to higher frequencies for forwards travelling waves and lowered for backwards travelling waves, depending on the nodal diameter ND. This following equation considers both effects:

$$f^{\text{stat}}_{\text{aliased}} = |\pm(f^{\text{rot}} + \text{ND}\cdot f_{\text{rev}}) + m N_{\text{blades}} f_{\text{rev}}| \tag{2}$$

Figure 5 illustrates sub-sampling of an arbitrary first blade mode, up to an approximate rotor speed of 4000 rpm. The red line indicates the Nyquist frequency, which is equal to half the BPF (All-Blade-Spectrum). Below the range of 4000 rpm the actual blade vibration frequency (green) is sub-sampled and therefore mirrored to frequencies below the red line. Above 4000 rpm, the sampling frequency is high enough to measure the actual blade

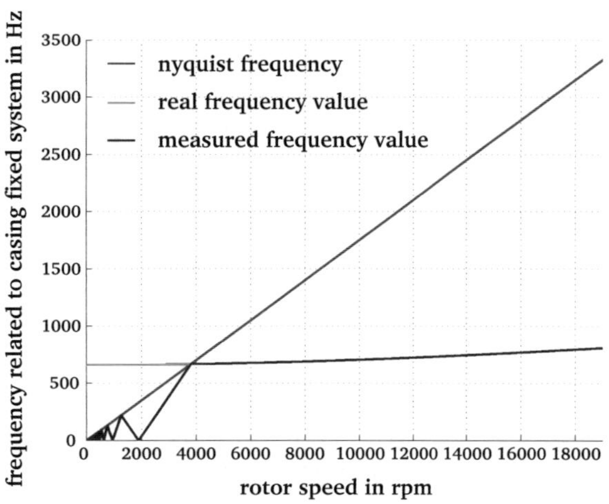

Figure 5. BTT INSTRUMENTATION: PARTLY SUBSAMPLED BLADE FREQUENCY

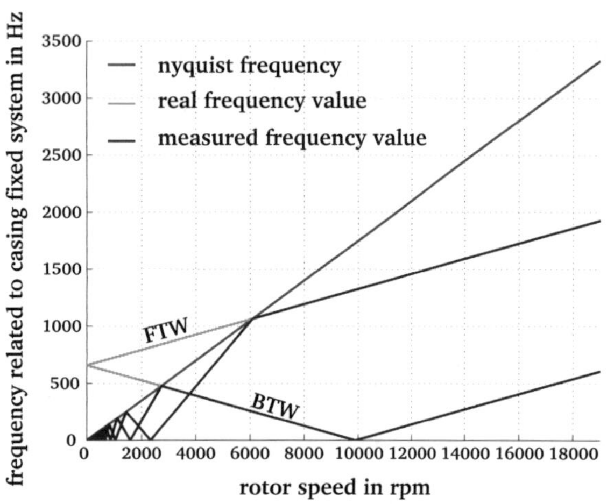

Figure 6. BTT INSTRUMENTATION: EIGENFREQUENCY SHIFTED BY TRAVELLING WAVE MODES

frequency. Stiffening effects of mode 1 are considered in this example, which results in the frequency rise with increased rotor speed. The second effect that occurs when measuring with the BTT system is the frequency shift due to travelling waves (which is a result of the inter-blade phase angle IBPA or ND), moving relative to the stationary frame of reference. The effect is known as the Doppler-shift. Figure 6 presents Mode 1, this time not as a ND 0 vibration (all blades in phase). Shown is a forwards (FTW) and backwards travelling wave (BTW) of ND 1. The forwards travelling wave is shifted to higher frequencies; the backwards travelling wave is shifted to lower frequencies when measured in the stationary reference frame according to Equation 2. Both are sub-sampled for low rotor speeds.

In addition to the BTT/BTC sensors unsteady pressure transducers are used in the stationary frame of reference to analyse flow phenomena. These WPT sensors measure the static pressure difference between the static wall pressure of the compressor section and the atmospheric pressure in the test laboratory. Sampling is performed with a frequency of 500 kHz, however the eigenfrequency of these pressure transducers lies at about 150 kHz. Low pass filtering is therefore applied.
The pressure transducers are clustered in axial and circumferentially distributed arrays. The circumferential array is shown in Figure 4. This array is used for the analysis of circumferentially moving pressure waves. It can also be used to analyse stall inception [3]. The ND can be calculated by comparing the angular separation of two sensors $\Delta\theta$ to the phase difference of the analysed wave ϕ:

$$\mathrm{ND} = \frac{\Delta\theta}{\phi} \tag{3}$$

A description is given by [19, 20]. The phase difference ϕ can be evaluated in the frequency domain by computing a cross-power spectrum of two sensors or in the time domain by analysing the time lag between the measured signals. The time domain method is used here (Figure 17).
The axially distributed array of the pressure transducers is shown in Figure 7. This array can be used to resolve the unsteady pressure field at the casing wall. Since the blades pass the stationary axial array during their rotation around the annulus, the spatial resolution is a function of the rotor speed f_{rev}. One rotor cycle of 360° is divided by the number of samples taken:

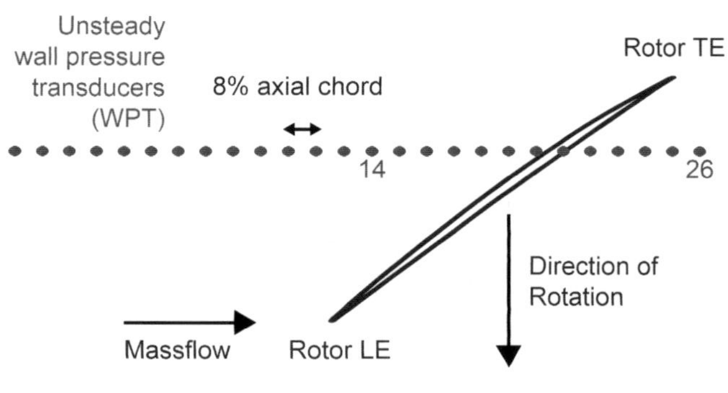

Figure 7. AXIAL DISTRIBUTION OF UNSTEADY WALL PRESSURE TRANSDUCERS

$$\Delta\psi = \frac{360^\circ}{f_{\mathrm{sample}}/f_{\mathrm{rev}}} \tag{4}$$

The spatial resolution can be computed as

$$\Delta x = \frac{2\pi r_{\text{shroud}}}{f_{\text{sample}}/f_{\text{rev}}}, \tag{5}$$

as the WPT sensors are mounted in the casing wall at radius r_{shroud}. At the investigated rotor speed, one passage is resolved with approximately 90 measurements. The axial resolution is given by the distance between two WPT sensors in the casing. For the investigated setup, the distance is 8% of the axial blade chord length.

By using a 1/rev-signal the WPT data is structured which allows ensemble averages of several revolutions. The left picture in Figure 8 illustrates the unstructured, time-continuous WPT data. This dataset can be split into separate revolutions using the trigger signal. The result is shown in the middle illustration of Figure 8. The identified revolutions can be averaged, calculating the mean value at every grid point of the revolutions.

This can be taken further by dividing the revolutions into blade pitches (illustrated on the right hand side of Figure 8) to analyse the data using samples of one distinct position in a blade pitch (marked as a red dot). This spatial point of a blade pitch can be used for further data analysis. The frequency resolution is reduced as only one sample per pitch is taken into account. Note that the flow field is still resolved at every point of the measured grid.

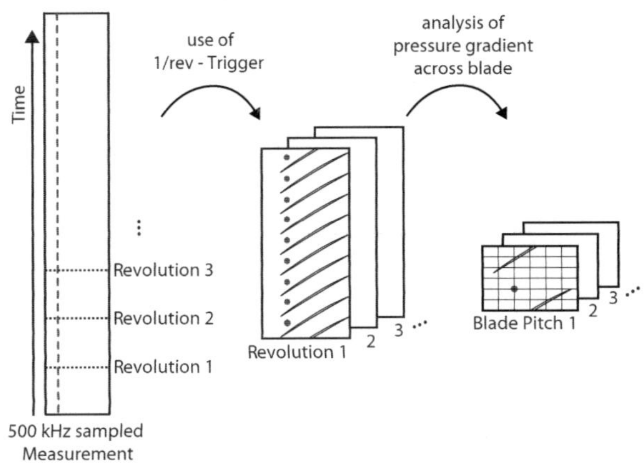

Figure 8. STORAGE AND ANALYSIS OF THE KULITE DATA

STRUCTURAL VIBRATION ANALYSIS - COMBINED ANALYSIS METHOD OF STRAIN GAUGES AND BLADE TIP TIMING

This section analyses the structural vibration of the transient measurement, introduced in Figure 2. First, blade-mounted SG are used, to analyse the active blade modes during the performed measurements. Second, the SG and BTT sensors are compared to determine the complete BLISK vibration pattern.

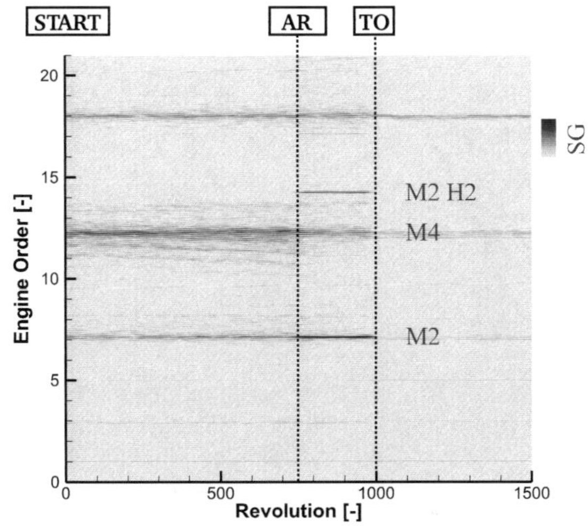

Figure 9. SPECTROGRAM OF THE STRAIN GAUGE SIGNAL DURING TRANSIENT OPERATION

Blade Mode Analysis

In Figure 9 the strain gauge amplitude is plotted against revolution and engine order. The SG with the highest response is shown, it is also the relevant one for the monitoring of the modes. A short-time Fourier transformation was used to compute the spectrogram. The revolution axis represents the time domain of the measurement. Revolution 0 marks the starting point of the measurement, introduced in Figure 2. From left to right, the throttle is continuously closed and the rotor is back-pressured to its stability limit. The engine order represents the frequency domain of the measurement. Engine order is the frequency divided by the current rotor speed. Eigenmodes of the blades can be identified by horizontal lines, as the rotor speed is approximately constant for the whole measurement. The measurement is divided into three parts for the analysis. The first part (Revolution 0 to 750) of the measurement is dominated by a broadband vibration of mode 4 (M4) and a moderate vibration of mode 2 (M2 / first torsion). The broadband excitation shows a pattern that consists of parallel lines of constant distance, which changes the frequency over the measured time. The second part of the measurement, starting at line AR (amplitude rise), is dominated by a rapid amplitude rise of mode 2 and its harmonics (M2H2). This section is initialized promptly and immediate vanishing of the broadband excitation of mode 4 is visible. At approximately Revolution 1000 the back-pressure is reduced rapidly by the throttle opening (TO). The high vibration amplitudes triggered before then calm to moderate ranges.

In summary,

• a broadband vibration of blade mode 4 occurs, followed by a rapid rise of blade mode 2 amplitudes and its harmonics. The broadband vibration vanishes when the harsh amplitudes of mode 2 rise. For further illustration of the identified blade vibration see Figure 10.

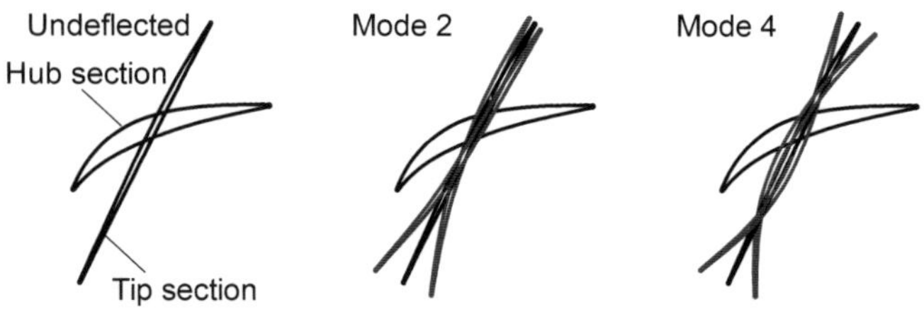

Figure 10. IDENTIFIED BLADE MODES OF THE INVESTIGATED BLISK ROTOR [ILLUSTRATION FROM [5]]

BLISK Vibration Analysis

To picture the vibration pattern of the entire BLISK, the strain gauge and BTT instrumentation are combined. By comparing a SG spectrum with a BTT spectrum, measurements in the stationary and rotating frame of reference are matched. As a result of the frequency shift between both reference frames the nodal diameter can be determined (simple version of Equation 2 / only for positive ND):

$$EO^{stat} = EO^{rot} + ND \tag{6}$$

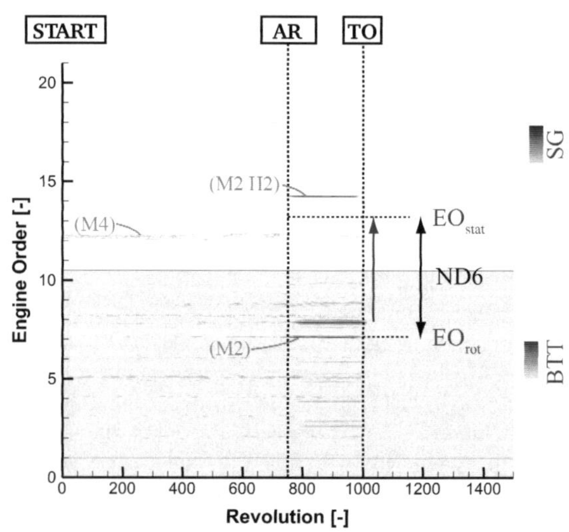

Figure 11. SPECTROGRAM OF THE STRAIN GAUGE AND BLADE TIP TIMING SIGNAL DURING TRANSIENT OPERATION

Only integer values can be used for the ND to match the frequencies in both frames. This only works if sub-sampling effects of the BTT system are taken into consideration during measurement. Due to this effect the BTT frequency is mirrored at engine order 10.5. The green arrow in Figure 11 illustrates the frequency shift based on aliasing effects.

Similar to the spectrogram of the strain gauge (Figure 9), the measurement can be divided into three parts. Again, up to Revolution 750 the low vibration amplitude of mode 4 is detectable. Once again, the vanishing of the mode 4 vibration is conspicuous at line AR, and a sharp rise in the BLISK vibration (mode 2 and ND 6) occurs. The high amplitudes are lowered by the final throttle opening.

In short:

• Combining strain gauge and BTT instrumentation, the high amplitude BLISK vibration is identified as a blade mode 2 vibration with a dominant circumferential wave pattern of 6 nodal diameter. This means that a forwards travelling wave is rotating around the annulus with a sine wave of 6 periods, which results from the phase angle between the distinct blade vibrations. Further illustration is given in Figure 12.

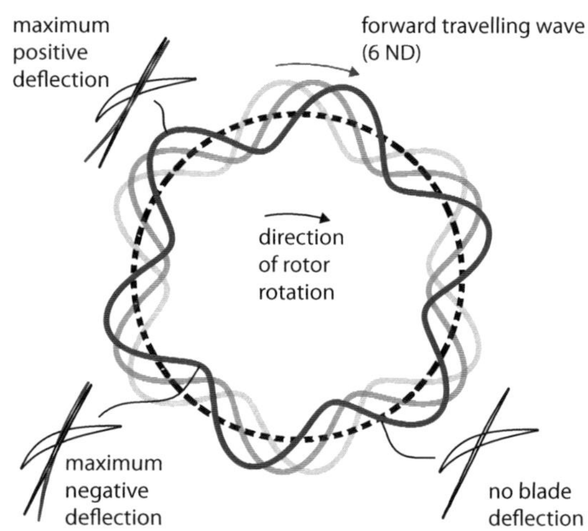

Figure 12. IDENTIFIED BLISK VIBRATION PATTERN OF THE INVESTIGATED ROTOR

BLADE VIBRATION AND UNSTEADY PRESSURE - GENERAL MEASUREMENT OVERVIEW

Figure 13 is a general picture of the analysed measurement. Strain gauge instrumentation is combined with the unsteady pressure transducers to analyse the aeroelastic behaviour of the rotor section during measurement. The whole measurement is divided into three parts as before.

The upper diagram in Figure 13 shows the current blade loading of the rotor, both aerodynamically and structurally. The aerodynamic loading of the rotor is estimated by the total-to-static pressure rise coefficient:

$$\Psi = \frac{p_{26} - p_{t,\text{inlet}}}{0.5\,\rho_{\text{inlet}} U_{\text{tip}}^2} \qquad (7)$$

Static pressure p_{26} is measured downstream of the rotor (Figure 7 / highlighted as WPT 26). The total pressure $p_{t,\text{inlet}}$ and density ρ_{inlet} are measured in the settling chamber. The aerodynamic loading has reached its maximum point for the speed-line. This is in direct contrast to measurements that were investigated in [6], where the aerodynamic loading could be continuously raised during measurement.

The emergency throttle opening at Revolution 1000 is visible by a rapid drop in Ψ (marked with "throttle opening" TO).

The structural blade loading is presented as a percentage of the high cycle fatigue limit of the eigenmode. This limit is critical for rig safety, since the rotor BLISK must be checked for cracks if the limit is exceeded. A warning is given when the 50% limit is

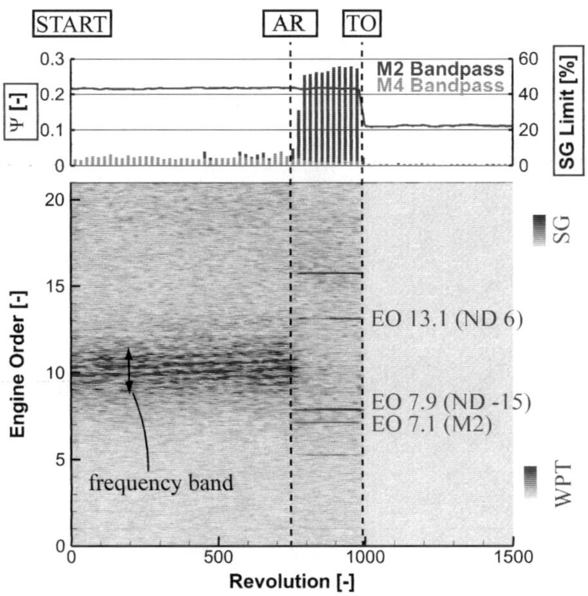

Figure 13. OVERVIEW OF THE ANALYSED MEASUREMENT (STRAIN GAUGE AND STATIC WALL PRESSURE DATA) - UPPER PLOT: BLADE LOADING - LOWER PLOT: SPECTROGRAM

reached, which was the case for the presented measurement. As taken from the SG data, the amplitudes of mode 2 rose sharply during Revolution 750 (marked with "amplitude rise" AR) and Revolution 1000. Compared to the high cycle fatigue limit, the 50% limit is exceeded.

In the lower diagram, a spectrogram of the SG and unsteady WPT instrumentation is given, similar to the ones given for SG and BTT in Figures 9 and 11. The pressure probe is located 8% axial chord downstream of the LE (Figure 7, highlighted as probe 14). At this location, the highest amplitudes were observed for the axial pressure probe array.

From Revolution 0 to 750 (Start to AR), a broadband frequency band is detectable in the spectrum of the pressure transducer. The high amplitude pattern consists of parallel lines. This pattern vanishes rapidly at Revolution 750, which is marked with AR. Pprevious investigations show, that these broadband excitations can lead to coupling with eigenmode 4 [6], which is not the case for the presented measurement. Between AR and TO, high pressure amplitudes rise - together with the amplitudes of the blades (mode 2) - at definite frequencies. These frequencies can be matched to pressure waves in the annulus. By comparing the frequencies to the one measured in the rotating frame of reference ND 6 and ND -15 can be determined (similar to Equation 2). The method of using circumferential distributed WPT is presented in the following section. This certifies the value. The negative value indicates that the wave is rotating in the opposite direction to the rotor turning direction (backwards) in the annulus. The connection and origin of these pressure waves are described in [5].

Both completely distinct parts of the measurement (START to AR and AR to TO) will now be studied in detail, using several data analysis methods.

CHARACTERIZATION OF THE PHENOMENA - DETAILED STUDY OF THE FLOW FIELD USING CIRCUMFERENTIAL AND AXIALLY DISTRIBUTED WALL PRESSURE ARRAYS

An overview of the analysed measurement is given in Figure 13. Distinct frequency bands and revolutions are of special interest. Between part START and AR, revolutions 500 to 600 analysed, for AR to TO, revolutions 750 to 1000 are the focus.

Study of measurement part START to AR

Figure 14 presents a study of the frequencies, using revolutions 500 to 600 for an FFT with a high frequency resolution of 2 Hz. The maximum peak in the spectrum represents the blade passing frequency. Since the highest amplitude occurs at this frequency, all amplitudes are normalized with the BPF amplitude (scale shown until 0.5 times BPF amplitude). At around half the BPF the frequency band with higher amplitudes is visible, which was illustrated previously in the spectrogram at about EO 10 (Figure 13). The distance of the peaks in this frequency band equates to half the rotor speed at this operation point. A similar frequency band at half the BPF, which consists of peaks with a discrete distance, was reported in [8], which provides a theoretical model to explain the pattern. According to the model, a fluctuating pressure pattern with distinct circumferential modes rotates relative to the rotor at half the rotor speed. To verify the determined value of the FFT, a study of the phase is shown in Figure 15, using revolutions 500 to 700 for a cross-power spectral density calculation (CPSD). The phase between two circumfer-

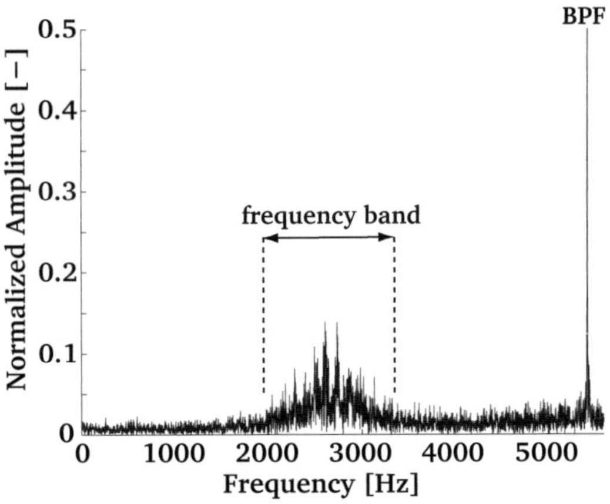

Figure 14. FAST FOURIER TRANSFORMATION OF AN UNSTEADY PRESSURE TRANSDUCER 8% DOWNSTREAM OF ROTOR LE; REVOLUTION 500 TO 600 USED FOR FFT

entially distributed sensors is calculated using the CPSD. Calcu-

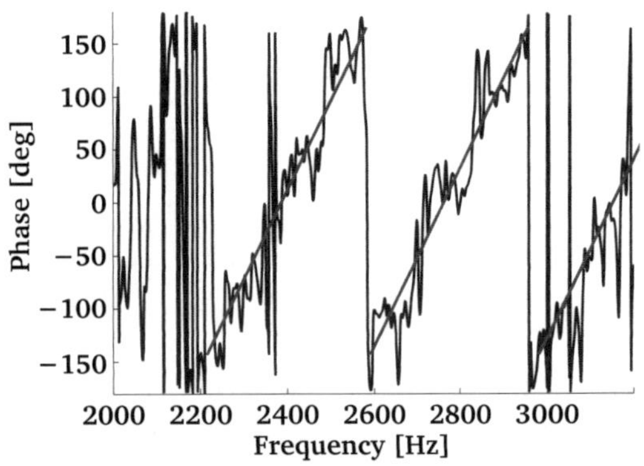

Figure 15. CALCULATED PHASE ANGLE OF TWO CIRCUMFERENTIALLY DISTRIBUTED UNSTEADY PRESSURE TRANSDUCER

lation of the phase velocity is performed by [9] using the gradient of the phase angle φ in the frequency domain. The flow structure in the presented measurement is now analysed the same way, using

$$f_{\text{rev}}^{\text{stat}} = \Delta\theta / (\partial\varphi / \partial f). \tag{8}$$

A linear polynomial with a slope of 0.84°/Hz is used to fit the gradient (the red line in Figure 15). The angle difference Δθ between the sensors is known. Consequently, the mean rotational speed of the flow structure related to the stationary frame of reference can be computed using Equation 8. The rotational frequency of the flow structure is shown to be half the rotor speed. Hence, the frequency band excitation is verified as rotating around the annulus at half the rotor speed. The flow structure, that belongs to the frequency band excitation, is similar to ones described as *'rotating instability'* (in short: RI) or *'rotating disturbances'* [8–14].

In Figure 16 the axial distribution of the pressure amplitudes is illustrated. Therefore, 26 FFTs are calculated from the axial pressure transducer array illustrated in Figure 7.

From LE to 50% axial blade chord, higher amplitudes are visible for the whole range of EO 1 to 21. This indicates that random fluctuations in a broad frequency band occur in the region of the LE when throttling the compressor close to the stability limit. The highest amplitude frequency band, similarly to RI and identified previously, is also strictly bound to the rotor section of the compressor. It spans from the LE of the blades to 75% of the axial chord length. Close to the LE, the frequencies span from EO 8 to 12. The larger the distance from LE, the smaller the range of frequencies in the spectrum. The connection between the identified region to distinct flow phenomena is determined by studying the flow field (Figure 19).

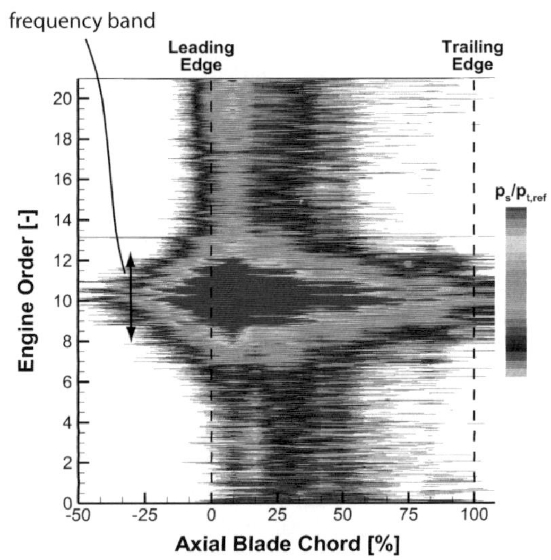

Figure 16. ANALYSIS OF THE AXIAL DISTRIBUTION - part 1: START to AR - FAST FOURIER TRANSFORMATION OF THE PRESSURE SIGNAL RECORDED BY THE 26 UNSTEADY PRESSURE TRANSDUCERS (SENSORDISTRIBUTION: FIGURE 7 / DATA STORAGE TYPE: FIGURE 8 LEFT)

Study of measurement part AR to TO

By comparing the signals of casing-mounted, unsteady WPT in the circumferential array (Figure 4), an analysis of the rotating pressure waves in the compressor duct between revolutions 750 and 1000 can be performed. The chosen peak lies at EO 13.1 in the spectrogram (Figure 13). Since the vibration does not occur at multiples of the rotor speed, it is non-synchronous.

A transient analysis is performed in the time domain. Therefore, the pressure time signal is band-pass filtered in a 200 Hz range around the analysed frequency (compare EO 13.1) using a direct-form FIR equiripple filter. The time difference of the pressure wave between two circumferentially distributed sensors is analysed. To compute the phase difference from the time lag Δt between two sensors, this equation is used:

$$\Delta\varphi = 360^\circ / T_{\text{period}} \cdot \Delta t, \qquad (9)$$

Dividing the calculated phase by the sensor angle difference, the ND of the frequency peak can be determined. This analysis method was described by [19], however, the phase difference was calculated in the frequency domain. The result of the analysis is shown in Figure 17. No clear determination of an ND is possible for several revolutions, but at approximately Revolution 750 the values jump to ND 6, which matches the result of the strain gauge and BTT analysis (Figure 11). The benefit of this analysis is that only the pressure data is used to determine the wave pattern of the pressure field in the duct. The same analysis could be performed with the BTT data.

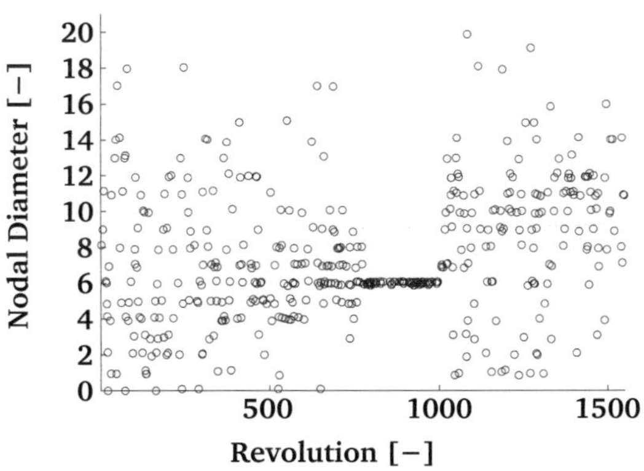

Figure 17. TRANSIENT CALCULATED NODAL DIAMETER OF CIRCUMFERENTIALLY DISTRIBUTED UNSTEADY PRESSURE TRANSDUCERS

Since the vibration frequency is strictly bound to the blade eigenmode (by 4% of the first blade mode)[1] the non-synchronous vibration observed here has similarities to *"flutter"*, seen over the last couple of decades [19,21,22].

The strict bound of the vibration frequency to the blade eigenmode contrasts observations of [6], where the excitation of blade mode 4 by RI occurred in a frequency range of 16% of the first blade mode. Figure 9 illustrates the difference.

Figure 18 illustrates the axial distribution of the pressure waves in the duct that result from non-synchronous vibration. Similarly to Figure 16 (measurement part START to AR), broadband (random) fluctuations span from LE to 50% of axial blade chord. Note that the same scale is used for both Figures. The high amplitude frequency band (identified as RI) vanished in the whole rotor section, however the distinct frequencies at EO 7.9 and 13.1 (in Figure 13, ND -15 and 6, respectively) span the 26 pressure transducers as a result of the high amplitudes of the blade vibration.

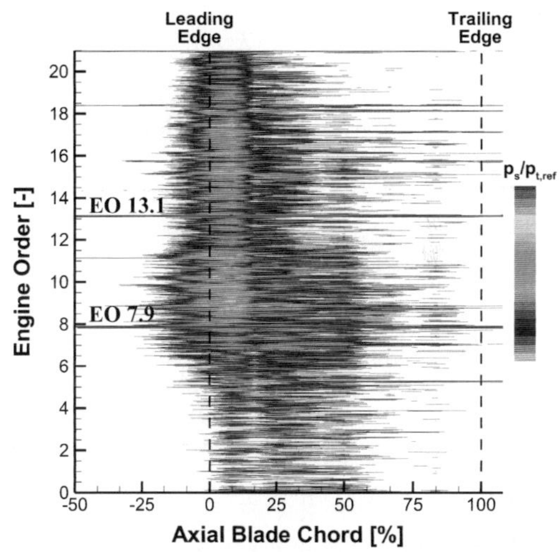

Figure 18. ANALYSIS OF THE AXIAL DISTRIBUTION - part 2: AR to TO - FAST FOURIER TRANSFORMATION OF THE PRESSURE SIGNAL RECORDED BY THE 26 UNSTEADY PRESSURE TRANSDUCERS

Summary of the detailed measurement analysis using circumferential and axial distributed pressure arrays

The casing wall pressure analysis leads to the understanding that two different phenomena occurred during the analysed measurement. Up to Revolution 750 a wide frequency band excitation at about half the BPF is detectable; it rotates around the annulus at approximately half the rotor speed.

- START to AR: The phenomenon is similar to *"rotating instability"* and reported by other authors [8–12, 14].
- The "rotating instability" randomly excite blade mode 4

It vanishes immediately when the flow is dominated by a wave pattern with distinct ND.

- AR to TO: The phenomenon is a non-synchronous vibration, which shows similarities to classic *"flutter"* near stall of the first torsion blade mode [19,21,22].

[1] frequency is resolved with 0.4% of the first blade mode

Note that both phenomena occur at a similar operation point, with a minimum difference in throttling. The non-synchronous vibration of the blades ends the occurrence of RI.

ANALYSIS OF THE FLOW AT THE BLADES TIP AND IDENTIFICATION OF THE AEROELASTIC MECHANISM

The axial unsteady pressure transducer array is used in this section (Figure 7). The pressure field at the casing wall is resolved to capture the flow features at the casing wall and link them to the flow field at the blade tip.

Pressure field in the rotor section - mean pressure, standard deviation, spectral map

Three different types of analysis are used.
Firstly, the ensemble average of the pressure data is used to analyse the mean flow field. The mean pressure is calculated at every point of the measured grid using equation

$$\overline{p} = \frac{1}{M}\sum_{k=1}^{M} p_k, \quad (10)$$

where M is the number of samples taken at every grid point (number of revolutions). It is related to the total pressure of the settling chamber $p_{t,inlet}$, and plotted as pressure ratio Π_{mean}. The revolution-wise data sorting is used for this (compare to Figure 8).
Secondly, unsteady flow and pressure fluctuations can be estimated by computing the relative standard deviation at every grid point (same data storage format):

$$\sigma_{rel} = \frac{1}{\overline{p}}\sqrt{\frac{1}{M-1}\sum_{k=1}^{M}(p_k - \overline{p})^2} \quad (11)$$

Thirdly, to investigate the associated frequencies of the estimated pressure fluctuations in the blade pitch, the WPT data are further structured. It is necessary to identify the pitches from a given set of revolutions. The structure of the analysed WPT data is illustrated in Figure 8 on the right. An analysis in the frequency domain can then be performed at every grid point of a blade pitch (one of the grid points marked with a red dot in Figure 8). A spectrogram is calculated using the same grid point in adjacent pitches. The amplitude A_{max} of a given frequency band Δf can now be plotted as a function of axial and circumferential coordinates.
In summary:

1. Π_{mean} represents the mean flow field.
2. σ_{rel} estimates the pressure fluctuations in the passage.
3. $A_{max}(\Delta f)$ connects the intensity of a specified frequency range to the estimated fluctuations.

These three methods of data analysis are also used in [6] and [3]. Figure 19 illustrates the distribution of measurement part 1 (START to AR). The mean pressure field is dominated by the pressure rise across the rotor in axial direction. The high pressure almost reached the LE at the PS of the blades due to throttle closing. The low pressure region at the LE is a small expansion

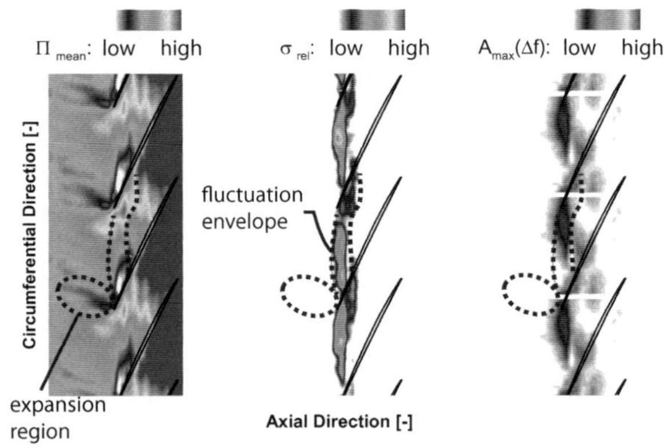

Figure 19. PRESSURE FIELD IN THE ROTOR SECTION FOR ROTATING INSTABILITY (part 1: START to AR) - MEAN PRESSURE FIELD, RELATIVE STANDARD DEVIATION (ESTIMATION OF THE FLUCTUTATION), SPECTRAL MAP

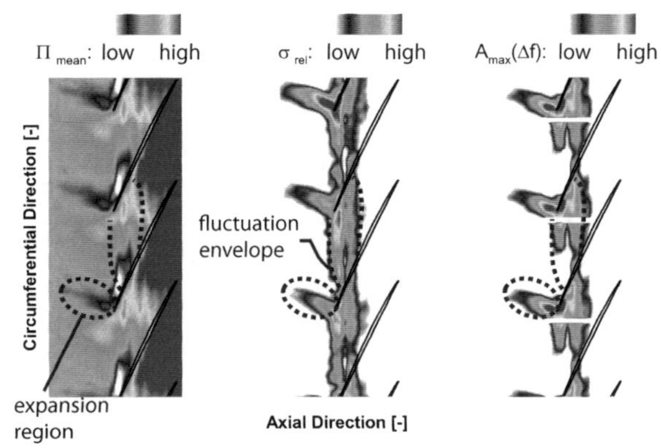

Figure 20. PRESSURE FIELD IN THE ROTOR SECTION FOR FLUTTER (part 2: AR to TO) - MEAN PRESSURE FIELD, RELATIVE STANDARD DEVIATION (ESTIMATION OF THE FLUCTUTATION), SPECTRAL MAP

region, as the rotor is driven under part speed conditions.
The highest pressure fluctuations (σ_{rel}) are visualized right next to the mean values. The unsteadiness of the pressure in the tip region is at the LE of the blades and the PS of the adjacent blade

at the LE. This is an indicator for operation points near stall, and the region represents the fluctuation envelope of the tip leakage flow and vortex [23, 24].
When adding the information of the spectral map, the frequency band identified in Figures 13, 14 and 16 can be matched to the same region of the blade tip vortex. This observation was also made by [10]. The identified region can be connected to Figure 16: it explains the distribution of the frequency band from 0 to 75% of the axial chord and the strict binding to the rotor section.
During measurement part 2 (AR to TO), the mean pressure field is similar to the one observed previously (Figure 20). This emphasises the small change in overall flow conditions and minimal difference in throttling and operating points. The fluctuations are extremely high at the LE of the blades, upstream of the LE, and the pressure is oscillating in the blade pitch between 0% and 25% of the axial chord length with maximum values close to the LE.
In contrast to Figure 19, the expansion region on the suction side of the blades is fluctuating. This can possibly be explained with the vibration of the blade: a vibrating blade in first torsion (mode 2) leads to a variation in incidence. A varying incidence will then lead to a fluctuation of the expansion region.
The fluctuation envelope of the blade tip vortex is broader than the one observed in Figure 19, though the involvement of the blade tip vortex is indicated. Using the spectral map, the dominant frequencies (EO 7.9 and 13.1) can be matched to the high pressure fluctuations. This is supported by the numerical and experimental investigations of [1, 2, 5, 25].

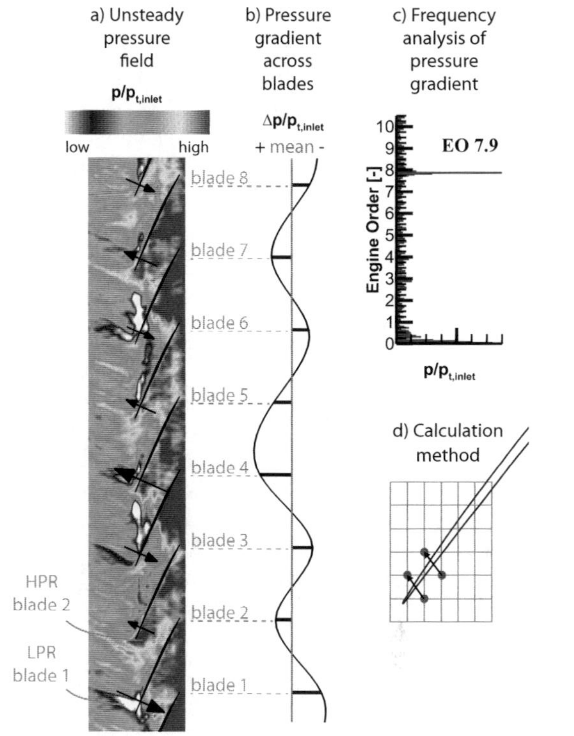

Figure 21. ANALYSIS OF THE UNSTEADY PRESSURE FIELD - AEROELASTIC COUPLING: PRESSURE GRADIENT ACROSS BLADE TIP (REVOLUTION 900)

Process of aeroelastic coupling: pressure gradient across blade tip

The final investigation is used to connect the analysis of the flow at the rotor tip to the mechanical blade vibration.
The aeroelastic connection between a blade vibration and a pressure fluctuation in the blade passage is the resulting static pressure on the blade surface.
The pressure difference across the blade tip results in a pressure force on the blade surface. A vibrating blade will lead to pressure waves that propagate in the flow field.
Thus, the pressure gradient across the blade tip is analysed it is the main indicator of the work input or exchange between blade vibration and the flow at the blade tip.
A snapshot of the unsteady pressure field in measurement part AR to TO is given in Figure 21a. The rotor is rotating from top to bottom and flow direction is from left to right.
In general, the flow field is characterized by the pressure rise in axial direction, and by a variation in circumferential direction. At the LE of blades 1, 3 and 6, large regions of low pressure span the PS and SS (Figure 21: LPR, blade 1). In contrast, the static pressure on the PS of blades 2, 4 and 7 is very high (Figure 21: HPR, blade 2). Consequently, the unsteadiness of the pressure field leads to variation of the pressure gradient across the blade tip in circumferential direction.
The black arrows in Figure 21a represent the calculated pressure gradient across the blade at LE. As a result of the higher pressure on the PS, the mean value of the pressure gradient always results in a force directed from PS to SS. It is thus feasible to subtract the mean value of the pressure gradient. Consequently, the black arrows illustrate fluctuation of the pressure gradient.
The length of the arrow is used as an indicator of the intensity of the pressure gradient across the blade. The unsteadiness of the gradient is obvious when comparing the blades.
Figure 21b shows the difference in the pressure gradient between the blades. The deviation from the mean pressure gradient is given, as described previously.
The large low pressure regions, identified in Figure 21a at the LE of blades 1, 3 and 6, result in variation of the pressure gradient in the negative direction. This meets expectations. The high pressure on the pressure side of blades 2, 4 and 7 results in a higher pressure gradient across the blade.
Figure 21c presents the result of the fourier transformed variation of the pressure gradient in circumferential direction. The pressure gradient across the blade at LE fluctuates with the frequency identified in Figure 13 and 18. Since the pressure gradient across the blades is the source of the tip leakage flow, the strength of the tip leakage flow will vary with the frequency of the blade vibration.
The pressure gradient is calculated orthogonal at the LE between 8% and 16% of the axial chord of the blades (Figure 21d). The analysis of the gradient consequently locates the region of excitation at the LE of the blades and matches the numerical investigations of [1, 25].

CONCLUSION

This paper presents the analysis of non-synchronous vibrations in a BLISK rotor tested in the Darmstadt Transonic Compressor. One transient manoeuvre of the compressor with an enlarged tip gap in the rotor section was investigated. Two frequently reported phenomena appear during the transient manoeuvre:

1. Close to the stability limit fluctuations occur at half the BPF, which is similar to the reported *"rotating instabilities"* or *"rotating disturbances"* that occur prior to stall events.
2. Non-synchronous vibrations of the first torsion mode occur. The vibration is strictly bound to the eigenfrequency of the blade as are the fluctuations in the flow. The phenomena is comparable to *"flutter"* events near stall in the literature.

• Investigation of the pressure field leads to the conclusion that the main driver for both identified phenomena is the tip leakage flow and the resulting vortex.

• Aeroelastic coupling between the pressure fluctuations in the flow and the blade vibration of mode 2 (first torsion) is mainly driven by the pressure gradient across the blades at the LE of the blades. Thus, the strength of the tip leakage flow will vary in the same manner as the vibration of the blades. The experimentally identified region of excitation (8% to 16% of the axial blade chord length) matches numerical investigations.

ACKNOWLEDGMENTS

This research was co-funded by the Bundesministerium für Wirtschaft und Energie (BMWi) and Rolls-Royce Deutschland Ltd and Co KG within AG Turbo 2020 (AG Turbo 2020 1.2.3 "Aeroelastische Vorgänge im Blattspitzenbereich hochbelasteter Verdichter", Förderkennzeichen 0327719D). The authors gratefully acknowledge this financial support. The authors would like to thank Rolls-Royce Deutschland Ltd and Co KG for permission to publish this work.

REFERENCES

[1] Leichtfuss, S., Holzinger, F., Brandstetter, C., Wartzek, F., and Schiffer, H. P., 2013. "Aeroelastic investigation of a transonic research compressor". In Proceedings of the ASME Turbo Expo 2013: Power for Land, Sea and Air.

[2] Leichtfuss, S., 2015. "Zum einfluss des spaltwirbels auf das aeroelastische verhalten transsonischer verdichter". PhD thesis, Technische Universität Darmstadt, Darmstadt.

[3] Holzinger, F., Wartzek, F., Jüngst, M., Leichtfuss, S., and Schiffer, H. P., 2015. "Self-excited blade vibration experimentally investigated in transonic compressors - rotating instabilities and flutter". In Proceedings of ASME Turbo Expo 2015: Turbine Technical Conference and Exposition.

[4] Holzinger, F., Wartzek, F., Nestle, M., Leichtfuss, S., and Schiffer, H. P., 2015. "Self-excited blade vibration experimentally investigated in transonic compressors - acoustic resonance". In Proceedings of ASME Turbo Expo 2015: Turbine Technical Conference and Exposition.

[5] Holzinger, F., 2015. "Tip clearance vortex coupled blade vibration in transonic compressors". PhD thesis, Technische Universität Darmstadt, Darmstadt.

[6] Jüngst, M., Holzinger, F., Leichtfuss, S., and Schiffer, H. P., 2015. "Analysing non-synchronous blade vibrations in a transonic compressor rotor". In 11th European Turbomachinery Conference (ETC).

[7] Freeman, C., 1985. "Effect of tip clearance flow on compressor stability and engine performance". *van Karman Institute for Fluid Dynamics, LectureSeries*.

[8] Baumgartner, M., Kameier, F., and Hourmouziadis, J., 1995. "Non-engine order blade vibration in a high pressure compressor". In Proceedings of 12th International Symposium on Airbreathing Engines (ISABE).

[9] Kameier, F., and Neise, W., 1997. "Rotating blade flow instability as a source of noise in axial turbomachines". *Journal of Sound and Vibration*, ***203***(5), pp. 833–853.

[10] Mailach, R., Lehmann, I., and Vogeler, K., 2001. "Rotating instabilities in an axial compressor originating from the fluctuating blade tip vortex". *Journal of Turbomachinery*, ***123***(3), pp. 453–460.

[11] Mailach, R., Sauer, H., and Vogeler, K., 2001. "The periodical interaction of the tip clearance flow in the blade rows of axial compressors". In Proceedings of the 46th ASME Turbo Expo.

[12] März, J., Hah, C., and Neise, W., 2002. "An experimental and numerical investigation into the mechanisms of rotating instability". *Journal of Turbomachinery*, ***124***(3), pp. 367–374.

[13] Young, A., Day, I., and Pullan, G., 2011. "Stall warning by blade pressure signature analysis". In Proceedings of ASMETurbo Expo 2011.

[14] Pardowitz, B., Tapken, U., Sorge, R., Thamsen, P. U., and Enghardt, L., 2014. "Rotating instability in an annular cascade: Detailed analysis of the instationary flow phenomena". *Journal of Turbomachinery*, ***136***(6), p. 061017.

[15] Martensson, H., Andersson, C., and Edin, N., 2012. "Aeroelastic instability of antransonic compressor near stall". In 13th International Symposium on Unsteady Aerodynamics, Aeroacoustics and Aeroelasticity of Turbomachines (ISUAAAT) 2012.

[16] Thomassin, J., Vo, H. D., and Mureithi, N. W., 2008. "Blade tip clearance flow and compressor nonsynchronous vibrations: The jet core feedback theory as the coupling mechanism". *Journal of Turbomachinery*, ***131***(1), p. 011013.

[17] Zielinski, M., and Ziller, G., 2000. "Noncontact vibration measurements on compressor rotor blades". *Measurement Science and Technology*, ***11***(7), pp. 847–856.

[18] Zielinski, M., and Ziller, G., 2005. "Noncontact blade vibration measurement system for aero engine application". In 17th International Symposium on Airbreathing Engines (ISABE).

[19] Gill, J. D., and Capace, V. R., 2004. "Experimental investigation of flutter in a single stage unshrouded axial-flow fan". In 42nd AIAA Aerospace Sciences Meeting and Exhibit.

[20] Kurkov, A. P., 1984. "Formulation of blade-flutter spectral analyses in stationary reference frame". In *NASA Technical Paper 2296*.

[21] Sisto, F., 1988. "Introduction and overview". In *AGARD Manual on Aeroelasticity 1988*, Vol. AGARD-AG-298-VOL-2. pp. 12–1/12–2.

[22] Srinivasan,A.V., 1997. "Flutter and resonant vibration characteristics of engine blades". In Proceedings of the 1997 International Gas Turbine & Aeroengine Congress & Exposition.

[23] Biela, C., 2012. "Experimental investigation of the aerodynamic influence of inlet-guide-vanes on the flow features of a one-and-a-half stage axial transonic compressor". PhD thesis, Technische Universität Darmstadt, Germany.

[24] Brandstetter, C., Kegalj, M., Wartzek, F., Heinichen, F., and Schiffer, H. P., 2014. "Stereo piv measurement of flow structures underneath an axial-slot casing treatment on a one and a half stage transonic compressor". In Proceedings of the 17th International Symposium on Applications of Laser Techniques to Fluid Mechanics.

[25] Möller, D., Jüngst, M., Holzinger, F., Brandstetter, C., and Heinz-PeterSchiffer, 2015. "Investigation of flutter in a transonic axial compressor". In Proceedings of the 14th International Symposium of Unsteady Aerodynamics,Aeroacoustics & Aeroelasticity of Turbomachines.

Proceedings of the 14th International Symposium of Unsteady Aerodynamics, Aeroacoustics & Aeroelasticity of Turbomachines
ISUAAAT14
8-11 September 2015, Stockholm, Sweden

I14-S9-2

INVESTIGATION OF FLUTTER IN A TRANSONIC AXIAL COMPRESSOR

Daniel Möller*
Maximilian Jüngst
Felix Holzinger
Christoph Brandstetter
Heinz-Peter Schiffer
Technische Universität Darmstadt
Gas Turbines and Aerospace Propulsion
Darmstadt, Germany

Sebastian Leichtfuß
TurboScience GmbH
Darmstadt, Germany

Email: moeller@glr.tu-darmstadt.de

ABSTRACT

This paper presents a numerical study on flutter for the transonic compressor rig at the Technische Universität Darmstadt. Flutter was experimentally observed for the second eigenmode of the rotor blades and is simulated for two rotational speeds using a time-linearized approach. The numerical simulation results are in close agreement with the experiment in both cases. The simulations and comparison with the experimental observations showed that flutter occurs near the compressor stability limit due to excitation by a pressure fluctuation pattern originating from the tip clearance flow. The tip clearance flow pattern is travelling in the backward direction, seen from the rotating frame of reference, and causes a forward travelling structural vibration pattern with the same phase difference between blades. When decreasing the rotor tip gap size or the eigenfrequency of the second eigenmode, stable conditions are predicted in the numerical simulations. Therefore, the flutter phenomenon can probably be avoided if the compressor design is modified. Variation of the eigenfrequency also shows that the velocity with which the tip clearance flow pattern sweeps around the circumference near the blade tip must be 40% to 50% of the rotational blade tip velocity for flutter to occur.

NOMENCLATURE

BLISK	blade integrated disk
TUD	Technische Universität Darmstadt
ND	nodal diameter
M2	2nd eigenmode of the rotor blade
(V)IGV	(variable) inlet guide vane
SG	strain gauge
BTC/BTT	blade tip clearance / blade tip timing
WPT	wall pressure transducer
RANS	Reynolds-Averaged-Navier-Stokes
DLR	German Aerospace Center
N1/N2	rotational speed 1/2
IA1/IA2	IGV angle 1/2
ADP	aerodynamic design point
LE	leading edge
TE	trailing edge
PS	pressure side
SS	suction side
EO	engine order
Λ	aerodynamic damping
K	aerodynamic stiffness
$\bar{C}$	complex coefficient
T_{trans}	convection time between blades
f	eigenfrequency
N_b	total number of rotor blades
r	radius
V	velocity

*Address all correspondence to this author

1 INTRODUCTION

The design of modern compressor rotors tends towards blade integrated disks (BLISKs). This design reduces the weight of the compressor rotors but increases susceptibility to blade vibration because of reduced structural damping.

One special kind of blade vibration is a phenomenon known as flutter, which occurs at non-synchronous frequencies (i.e. frequencies which are not an integer multiple of compressor rotational frequency) related to an eigenmode of the BLISK and results from interaction between fluid and structural forces.

Many flutter events in compressors and fans have been reported in the literature in recent years. Holzinger et al. [1] present experimental observations for the transonic compressor test rig at the Technische Universität Darmstadt (TUD), which is also investigated in this paper. Experimental results from tests with nominal rotor tip clearance, enlarged clearance, and nominal clearance with a non-axisymmetric casing treatment are shown. In the case of nominal clearance, flutter for the 2nd eigenmode of the rotor blade (M2) occurred, which results from the interaction between the blade vibration and the tip clearance flow. For the enlarged clearance, the stable operating range is reduced, compared to the nominal clearance and limited by the occurrence of rotating instabilities, that lock on to the 4th eigenmode. Extension of the operating range for the nominal gap example could be achieved by applying a non-axisymmetric casing treatment and M2 flutter could be avoided.

In Leichtfuß et al. [2], Leichtfuß et al. [3], and Leichtfuß [4], the M2 flutter of the TUD transonic compressor is investigated numerically for nominal speed. The numerical results agree well with the experimental observations in [1]. The flutter phenomenon could be attributed to the tip clearance flow impinging on the suction side of the adjacent blade. It was shown that flutter occurence depends strongly on the tip clearance size and that, for small tip clearance, flutter can be avoided.

Johann et al. [5] present subsonic stall flutter in a 4.5-stage compressor at part speeds, which was initiated by malscheduling the variable vanes. A numerical study confirmed the experimental findings, with the only difference being the unstable nodal diameter (ND). Another flutter event for a multistage compressor was reported by Andersson et al. [6], who observed it experimentally in a 3.5-stage compressor at different rotational speeds. Flutter occurred at 70% speed in the first stage rotor and at 95.6% speed in the third stage rotor.

Gill and Capece [7] experimentally observed flutter for a fan BLISK rotor at part speed, initiated by back-pressuring the rotor into stall. Numerical investigations of fan flutter can be found in Vahdati et al. [8] and Vahdati and Cumpsty [9]. Vahdati et al. [8] specifically investigated intake duct effects on fan flutter and distinguished between stall flutter caused by suction side separation, and acoustic flutter caused by reflected pressure waves in the intake duct. Vahdati and Cumpsty [9] show that flutter occurs in a fan at speeds with tip Mach number close to unity when two conditions are fulfilled: first, a separation region at the blade suction side; and second, for the aerodynamic pressure field caused by the blade vibration, cut-on conditions upstream and cut-off conditions downstream of the rotor.

This paper develops the numerical investigations of the TUD transonic compressor test rig in [2, 3, 4] further by simulating the flutter at part speed conditions and comparison with the results at nominal speed. A parameter study of the rotor tip clearance size and the M2 eigenfrequency is carried out. Special attention is paid to flow features in the rotor tip region and the interaction between individual blades of the BLISK.

2 TEST RIG

The TUD transonic compressor rig was first commissioned in 1994 [10]. Initially, research focused on aerodynamics and later also on aeroelastic investigations of different compressor configurations. The layout of the test rig can be seen in Figure 1.

In 2010, a new 1.5-stage configuration with a titanium BLISK rotor was installed, as shown in [11] and [12]. This configuration features a variable inlet guide vane (VIGV). A cross-sectional view of the configuration is shown in Figure 2. The VIGV is shown in green, the rotor in red, and the stator in blue.

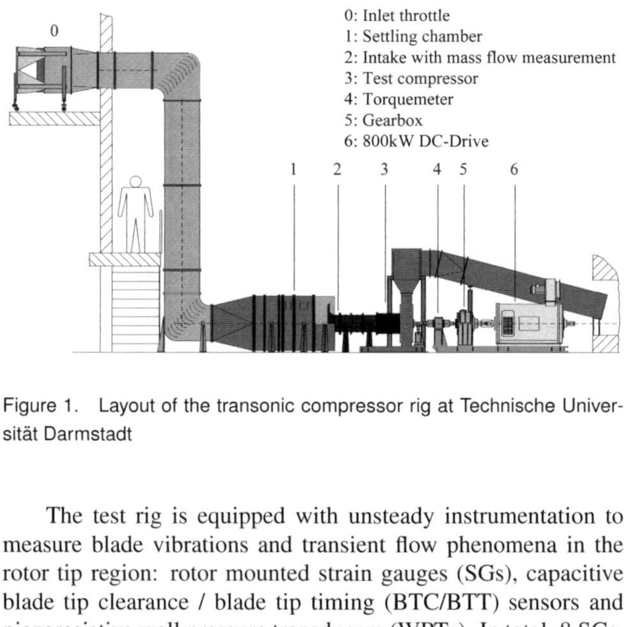

Figure 1. Layout of the transonic compressor rig at Technische Universität Darmstadt

The test rig is equipped with unsteady instrumentation to measure blade vibrations and transient flow phenomena in the rotor tip region: rotor mounted strain gauges (SGs), capacitive blade tip clearance / blade tip timing (BTC/BTT) sensors and piezoresistive wall pressure transducers (WPTs). In total, 8 SGs, 6 BTC/BTT sensors and up to 72 WPTs are used. SGs and the BTC/BTT sensors measure the structural response of the blades; the WPTs resolve the unsteady wall pressure above the rotor. The SGs are sampled with a frequency of 100 kHz, the BTC/BTT sensors have a sampling rate of 2500 kHz, and the WPTs record data with a frequency of 500 kHz. The signals of the WPTs are lowpass filtered at 100 kHz. The WPTs are mounted in the casing above the rotor to form an axial array resolving the unsteady wall pressure field. This field can later be analyzed using time and frequency domain techniques.

To characterize blade vibrations, the SG signals are used to determine the active eigenmode during a vibration event. The ND can be determined by comparing the frequency of the signal

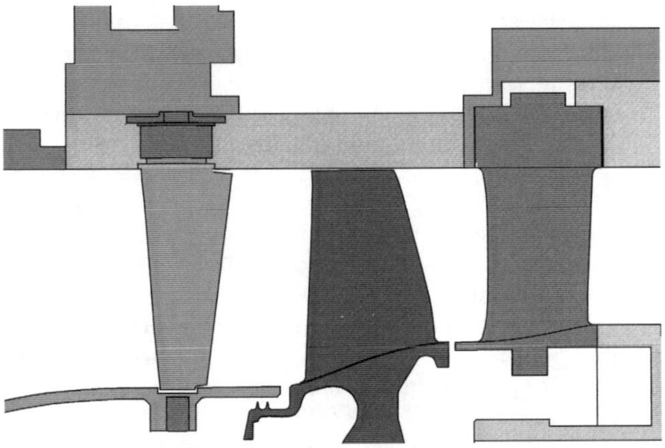

Figure 2. 1.5-stage compressor configuration with VIGV (green), rotor (red), and stator (blue)

from the SGs in the rotating frame of reference and the signal from the BTC/BTT sensors or the WPTs in the stationary frame of reference. Further information on how signals from the measurements are processed is presented by Jüngst et al. [13].

3 NUMERICAL SETUP

The Reynolds-Averaged-Navier-Stokes (RANS) solver TRACE developed by the German Aerospace Center (DLR) [14] is used for all numerical simulations in this paper. A block-structured single-passage grid is created for the VIGV, rotor, and stator. The total number of grid nodes is 1.60M (the VIGV grid consists of 585k nodes, the rotor grid of 866k nodes, and the stator grid of 144k nodes). A constant gap is applied at the VIGV tip to model the effects of the penny slot [15]. All walls are resolved using wall functions, except the walls at the VIGV and rotor blade tip and shroud. For these walls, the grid fully resolves the turbulent boundary layer up to the viscous sublayer. As a result, a sufficiently large number of radial layers can be used to resolve the tip gap. A schematic view of the generated mesh can be seen in Figure 3 (only every second grid line is shown in the plot).

First, steady state, single-passage simulations are carried out for the complete 1.5-stage compressor. A mixing plane approach is used at the interface between adjacent blade rows. Turbulence is modeled using Wilcox's two-equation model [16] with corrections for stagnation points [17] and streamline curvature [18]. With this setup, the compressor characteristics were simulated for nominal speed (N1) and one part speed (N2). For N1, the VIGV was set to the nominal IGV angle (IA1); for N2, the VIGV was closed to IA2.

The steady state results are used as a starting point for the unsteady aeroelastic simulations. A time-linearized approach has been used as it has proven to be an efficient tool for flutter prediction ([19] and [20]). For the time-linearized computations, only the rotor is simulated. Therefore, interactions between blade rows are not captured. A mesh deformation is applied at the rotor surface according to M2. The eigenmodes and eigenfrequencies of the rotor blades are obtained using the open-source finite element solver CalculiX. Modeshapes for the first two eigenmodes can be seen in Figure 4. The first eigenmode is a flap eigenmode and the second eigenmode is a torsional eigenmode.

Phase-lag boundary conditions are used at periodic boundaries to simulate different ND. 2D non-reflecting boundary conditions are applied at inlet and outlet boundaries to avoid spurious wave reflections [21]. Since no linearized equations for the turbulence quantities are available, a constant eddy viscosity assumption is applied.

An energy method is used to determine the aerodynamic damping, i.e. the blade is moved at a fixed amplitude according to the eigenmode and the sum of work over a complete vibration cycle is calculated using the resulting aerodynamic forces on the blade. This approach requires linearity of the flutter phenomenon under investigation. In the experiment, a linear relation between the measured pressure fluctuations from the WPTs and the deflection of the rotor blade recorded by the SGs was observed during the onset of flutter. Therefore, the assumption of linearity of the flutter phenomenon is valid for the simulations presented in this paper.

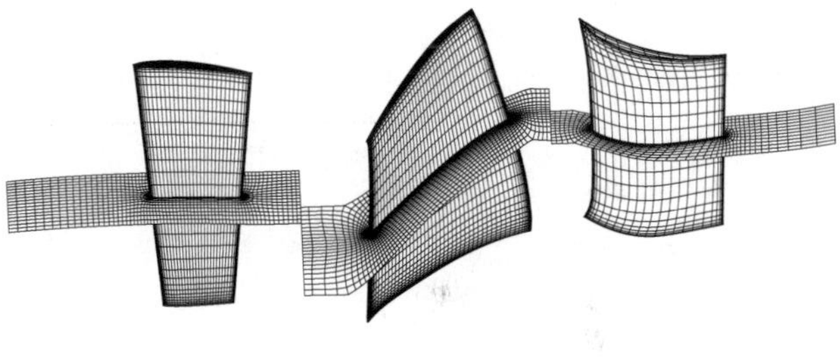

Figure 3. Single passage mesh for the 1.5-stage compressor (every second grid line is shown)

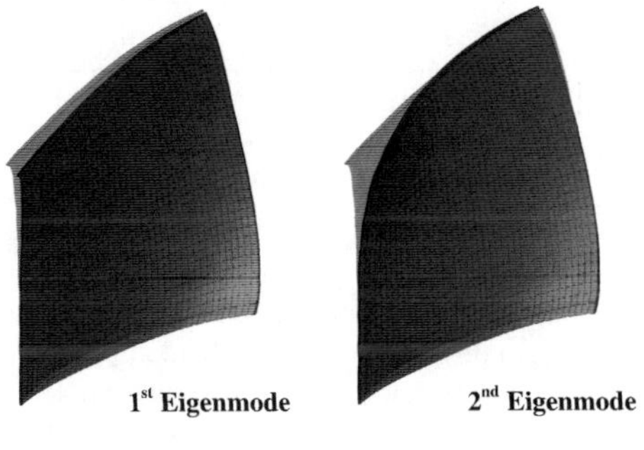

Figure 4. First two eigenmodes of the rotor blades

4 RESULTS

This section presents the results of the steady-state simulation of the complete 1.5-stage compressor, which are then compared to experimental data (Section 4.1). Section 4.2 presents the experimentally observed flutter phenomenon near the compressor stability limit, which is compared to the time-linearized aeroelastic simulation predictions. Further investigation of the numerical simulations provides insight into the excitation mechanism for blade vibration (Section 4.3). Finally, the influence of rotor tip clearance and M2 eigenfrequency on the flutter phenomenon is presented in Section 4.4 and 4.5.

4.1 Compressor speed lines and aerodynamics

A comparison between the measured and simulated speed lines can be seen in Figure 5. Note that the values for reduced mass flow and total pressure ratio are referenced to the aerodynamic design point (ADP) on the nominal operating line and that only three operating points before the stability limit are available from the experiment for N2. The characteristics from experiment and simulation are in good agreement, though the total pressure ratio is slightly higher in the simulations for all reduced mass flows and for both speed lines. The compressor stability limit also agrees well. In the experiment, the measurement was stopped at the stability limit when blade vibrations limited further throttling of the compressor rig; in the simulations, the stability limit was identified when an aeroelastic instability was predicted by the time-linearized aeroelastic simulations.

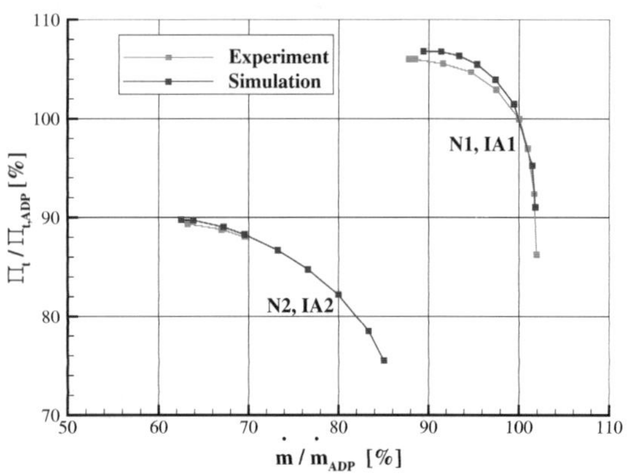

Figure 5. Compressor characteristics from experiment and simulation

Figure 6 shows relative Mach number contours of the flow at a constant radial height, near the blade tip, for the last operating point near the stability limit, for rotational speeds N1 and N2. Additionally, isolines of entropy further visualize the location of the tip clearance vortex. The levels for the contour plot and the isolines are kept constant in both cases.

In case of speed N1, the flow is supersonic near the blade tip. At the stability limit, the shock is detached from the blade leading edge (LE) by about 20% of the chord length. The shock shows a dent near the suction side (SS), which is caused by the tip clearance vortex interaction. This interaction contributes to an increased growth of the vortex, resulting from reduced flow velocity after the vortex passes the shock. The tip clearance vortex is causing a large blockage region in the passage. The direction of the tip clearance vortex is almost circumferential and it impinges on the adjacent blade near the LE on the pressure side (PS), which is shown later in more detail (Figure 12).

In case of speed N2, the flow is mostly subsonic, except for a small supersonic patch near the LE at the SS of the blade. The tip clearance vortex can also be identified in the plot and a large blockage zone can be seen in the passage. The strength of the tip clearance vortex and therefore the extend of the resulting blockage zone is lower in this case, which is a result of the reduced pressure difference between PS and SS. Also for speed N2, the direction of the tip clearance vortex is almost circumferential and it impinges on the PS of the adjacent blade near the LE.

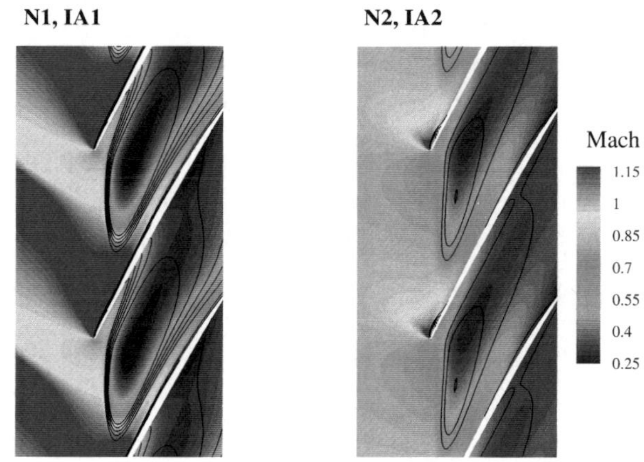

Figure 6. Relative Mach number contour and isolines of entropy near the blade tip at compressor stability limit

4.2 Observations at compressor stability limit

Experiment: In the experiment, transient measurements were carried out at the stability limit by continuously throttling the compressor, while the unsteady instrumentation (SG, WPT, and BTC/BTT sensors) was used to record blade vibrations and unsteady flow phenomena at the casing above the rotor. The measurement is stopped by opening the emergency valve when stall is encountered or strong vibration of the rotor blades is detected. Stall and blade vibrations were both encountered during different transient measurements on the same speedline (cf. [1]).

Figures 7 and 8 show spectrograms of an SG and a WPT signal during a transient measurement for the N1 and N2 speed lines. The frequency is normalized by rotational speed, yielding the engine order (EO) shown on the y-axis and the time is shown on the x-axis as the number of rotor revolutions. From left to right, the back-pressure of the rotor is continuously increased until the measurement is stopped by opening the emergency valve.

For the transient measurement at nominal speed N1 (Figure 7), activity in M2 can be detected from the SG signal in the rotating frame of reference. The WPT shows peaks at EO 7.2 and 13.8, which switches to EO 8.2 and 12.8 with further throttling the compressor. Frequency comparison with the SG signal (cf. [1] and [13]) identifies 8 and 7 ND pressure waves (peaks for EO 13.8 and 12.8), which travel in the direction of rotor rotation seen from the rotating frame of reference. These waves are caused by the structural vibration pattern of the BLISK (cf. [1]). The peaks at EO 7.2 and 8.2 correspond to pressure waves travelling backwards, seen from the rotating frame of reference, with NDs -13 and -14 (the minus sign indicates pressure waves travelling against the direction of rotor rotation). Analysis of the WPTs signals by Holzinger [22] showed, that the backward travelling pressure wave signal is not just a reflection of the signal from the forward travelling pressure wave, it has its own distinctive characteristics. Since the forward and the backward travelling wave have the same phase difference between adjacent blades, it is possible that both pressure waves can be observed at the same time during one vibration event.

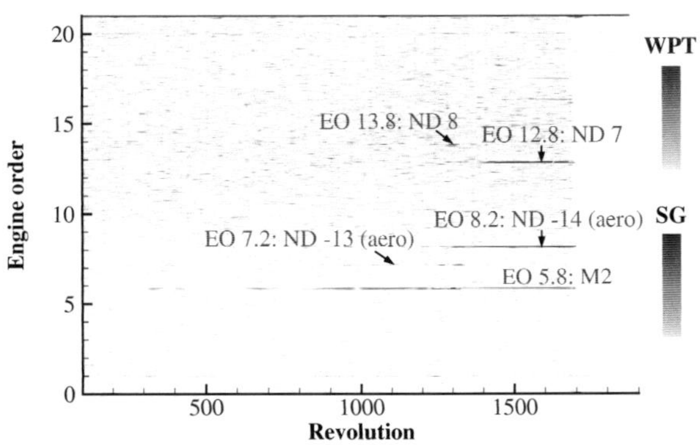

Figure 7. Spectrogram of SG and WPT signal during transient measurement at stability limit for N1, IA1

In the case of the part speed N2 shown in Figure 8, the SG also shows a peak at the M2 eigenfrequency and the second harmonic of the M2 blade vibration. The WPT shows the strongest peaks at EO 14.2 and 6.8, which correspond to a 7 ND structural vibration pattern and the corresponding -14 ND backward travelling pressure wave. Minor peaks are also present in the WPT signal, belonging to structural vibration patterns with higher NDs and their corresponding backward travelling pressure waves. This could be due to a vibration pattern which is not just 7 ND, but consists of a superimposed pattern with minor contributions from other NDs as well.

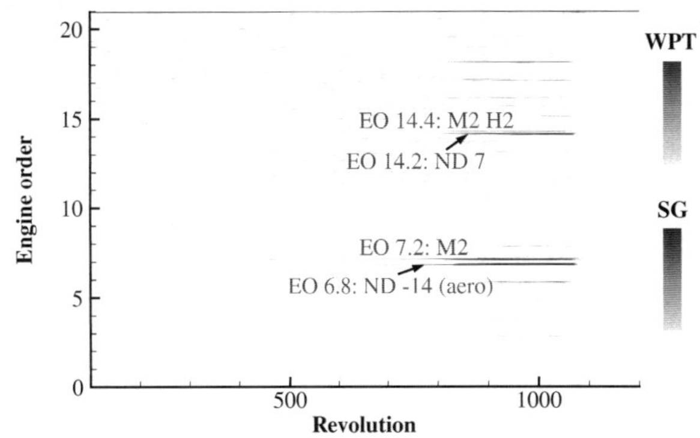

Figure 8. Spectrogram of SG and WPT signal during transient measurement at stability limit for N2, IA2

Simulation: The vibration observed in the experiments is numerically investigated using a time-linearized approach, as presented in Section 3. For the last operation point before the compressor stability limit in the simulations (Figure 5), aerodynamic damping is presented in Figure 9 for both rotational speeds. In this paper, aerodynamic damping is shown as the logarithmic decrement Λ of the blade vibration. In both cases, negative aerodynamic damping and therefore a self-excited blade vibration is predicted for one ND. Considering both the numerical and experimental observations, the vibration phenomenon can be classified as flutter since the vibration is connected to M2 and an unstable condition is encountered from a small displacement of the blade, which causes exponentially growing amplitudes (cf. [23]).

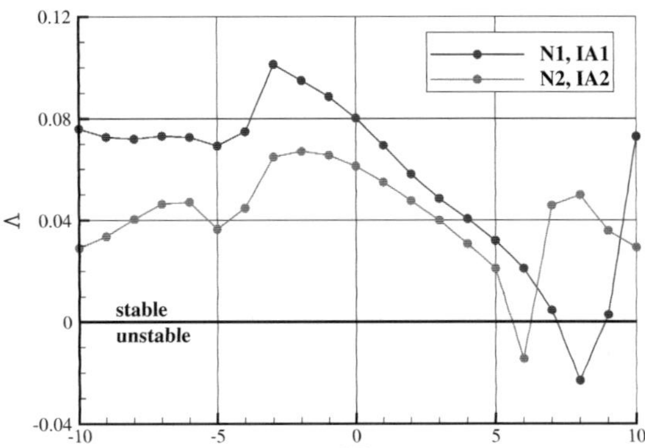

Figure 9. Aerodynamic damping versus ND for speed lines N1, IA1 and N2, IA2 at stability limit

Negative aerodynamic damping occurs for the nominal speed case at ND 8. This agrees well with the experimental observation (Figure 7): a small peak corresponding to an 8 ND vibration pattern can be seen at the onset of the flutter phenomenon, which matches exactly the predicted unstable ND. However, the strongest peak in the experiment is observed for a 7 ND vibration pattern. In that case, the predicted ND differs only by one compared to the measured ND. For the part speed case, the predicted unstable ND is 6. This also agrees well with the experiment (cf. Figure 8), since the unstable ND differs only by one between numerical simulation and experiment.

4.3 Mechanism for flutter

The numerical simulations provide further insight into the underlying mechanism for blade excitation leading to flutter by investigating local aerodynamic work input into the M2 blade vibration on the blade surface. For the case N1, IA1, this distribution is shown on blades PS and SS for the unstable ND 8 in Figure 10. A large region near the LE on the blade PS has aerodynamic work input into blade vibration, i.e. excitation of the vibration. Near the LE at the blade SS, there is a region where the damping is high.

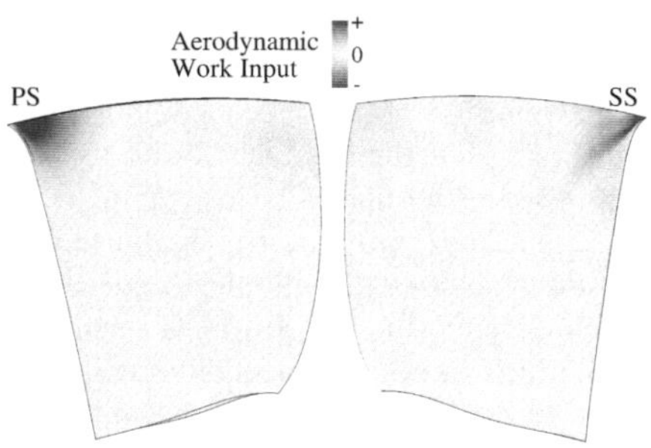

Figure 10. Local aerodynamic work input for nominal speed N1 and unstable ND 8

To further interpret these regions, closer examination of the flow at the rotor tip region in the experiment and the simulation can be made by visualization of the pressure fluctuation at the casing wall above the rotor. Figure 11 shows pressure fluctuations corresponding to the M2 frequency from the experiment and the simulation. The experimental data was obtained by bandpass filtering the unsteady pressure signal from the WPTs with the M2 eigenfrequency. For the time-linearized numerical simulations, only fluctuations with the M2 eigenfrequency are computed. Pressure fluctuations from the experimental data and the numerical simulations are not on the same scale because the simulation is valid for small perturbations only. Therefore, only qualitative comparison between the two is possible.

Figure 11 shows that the local excitation at the PS near the LE observed in Figure 10 corresponds to a region of large pressure fluctuation. This region is caused by the impinging tip clearance flow from the adjacent blades (Figure 6 and 12). For the local damping near the LE on the SS of the blade, no associated strong pressure fluctuation can be found. It is most likely

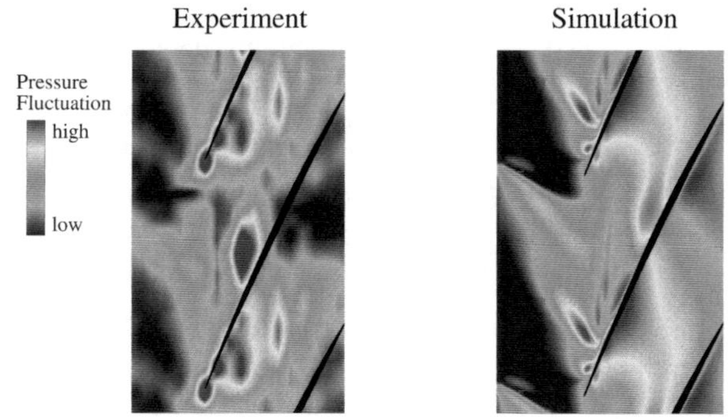

Figure 11. Pressure fluctuation corresponding to M2 frequency at casing above rotor from experiment and simulation for nominal speed N1

the result of the tip clearance vortex originating from the blade at this location. A region with high pressure fluctuations can be seen at midchord on the blade SS in Figure 11. No significant damping or excitation can be observed in the local work input plot, since the amplitude of M2 is small at this location. Figure 12 shows streamlines that impinge at the location on the reference blade L0, where excitation occurs. Many streamlines pass the adjacent blade L+1 at midchord. Therefore, the high pressure fluctuation at midchord on the SS can be attributed to the tip clearance flow. Another interesting observation from Figure 12 is that the streamlines are able to pass through multiple passages before impinging on the reference blade. This is possible because of negative axial velocity observed in the blockage region (Figure 6) for operating points near the compressor stability limit.

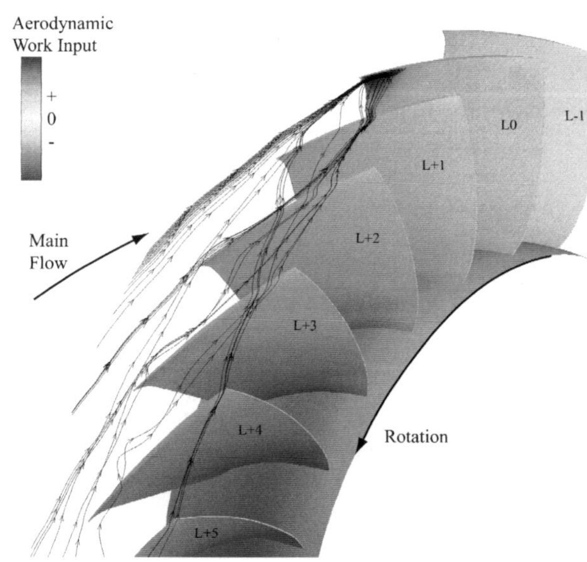

Figure 12. Streamlines impinging on excitation region at blade PS near the LE for nominal speed N1

In the case of flutter at part speed N2, the contour of local work input for the unstable ND 6 shown in Figure 13 is very similar to the example with nominal speed N1. The reason for excitation of the M2 blade vibration is a region near the LE at the PS of the blade again, which is caused by impingement of the tip clearance flow from the adjacent blades. This can also be seen in the plot of pressure fluctuation for the experiment and the simulation in Figure 14. A region of high pressure fluctuation exists at this location. A damping region near the LE on the blade SS can also be seen for part speed. This is likely to be due to the tip clearance vortex originating at this location; an incidence driven flow separation was also observed near the LE at about 80% radial height and is shown in Figure 15. This separation contributes to the damping effect. In the simulations, pressure fluctuations can also be observed near the LE on the blade SS. These regions are a result of the fluctuating tip clearance flow and fluctuation of the supersonic patch (Figure 6). Increased fluctuation is observed in this region in the experiment as well.

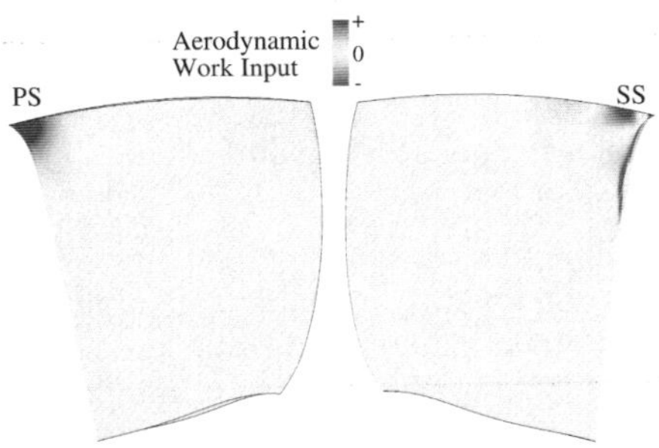

Figure 13. Local aerodynamic work input for part speed N2 and unstable ND 6

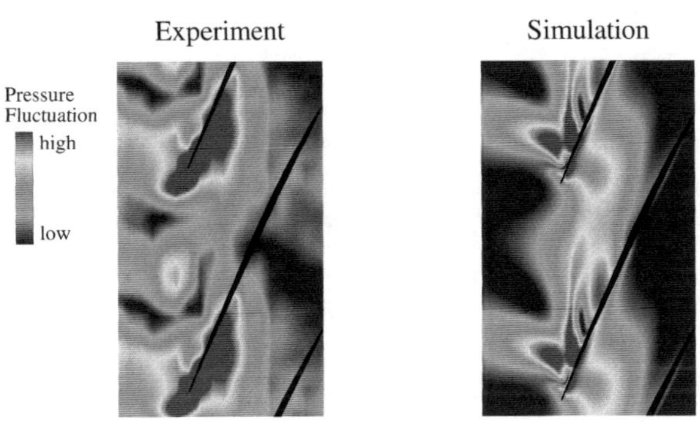

Figure 14. Pressure fluctuation corresponding to M2 frequency at casing above rotor from experiment and simulation for part speed N2

The method of aerodynamic influence coefficients is applied reverse to derive the damping contribution of all blades in the rotor row to the total aerodynamic damping, as shown by Leichtfuß [4]. This is done for the unstable NDs 8 (nominal speed case) and 6 (part speed case). The total aerodynamic damping Λ_{ND} for one ND is obtained as the sum of the contributions from all blades in the assembly N_b

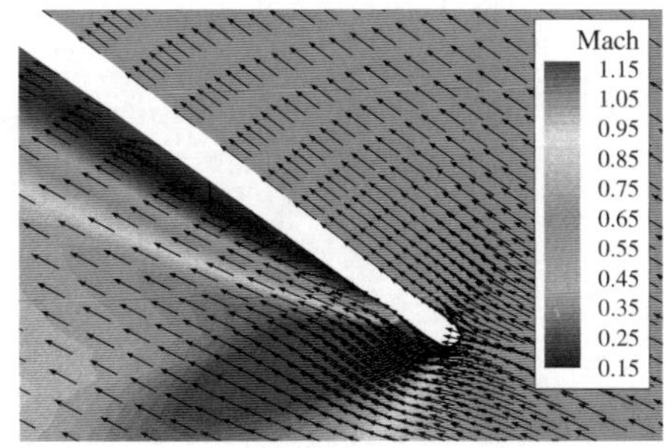

Figure 15. Incidence driven flow separation at blade SS near the LE at around 80% radial height (velocity vectors are scaled to uniform length)

$$\Lambda_{ND} + iK_{ND} = \sum_{n=1}^{N_b} \bar{C} e^{-iT_{trans,n} 2\pi f_E} e^{i2\pi \frac{(N_b - ND)(N_b - n)}{N_b}}.$$

K_{ND} on the left hand side denotes the aerodynamic stiffness for one ND. The part $\bar{C}e^{-iT_{trans,n}2\pi f_E}$ describes the aerodynamic influence of one blade in the row on the reference blade, with $\bar{C}$ being the complex coefficient, $T_{trans,n}$ being the convection time from the blade to the reference blade, and f_E being the blade eigenfrequency. The other part, $e^{i2\pi \frac{(N_b-ND)(N_b-n)}{N_b}}$, describes the geometric phase difference between one blade and the reference blade for one ND. Since the left hand side of the equation is known from the aeroelastic simulations, the damping contributions of each blade in the rotor row to the total aerodynamic damping for every ND can be calculated.

Figure 16 shows the damping contributions of neighboring blades to the total aerodynamic damping for the unstable NDs 8 (nominal speed case) and 6 (part speed case). Blades that are located ahead of the reference blade L0, as viewed in the direction of rotor rotation, are given ascending positive position numbers, and blades located behind the reference blade are given descending negative position numbers.

Figure 16 shows that the blade influence on itself is damping for both cases and that excitation results from the influence of the neighboring blades. Even blades that are located far ahead of the reference blade have a significant influence. This agrees well with observations of the streamlines in Figure 12 and is probably a result of the large blockage zone in the passage at blade tip height (Figure 6). This makes it possible for fluid to travel across multiple passages. For rotational speed N1, the largest excitation results from the blade L1 located directly ahead of the reference blade; for rotational speed N2, blade L4 has the largest excitation effect.

The flutter mechanism is very similar for both, N1 and N2, and needs the tip clearance flow to impinge on the PS of the

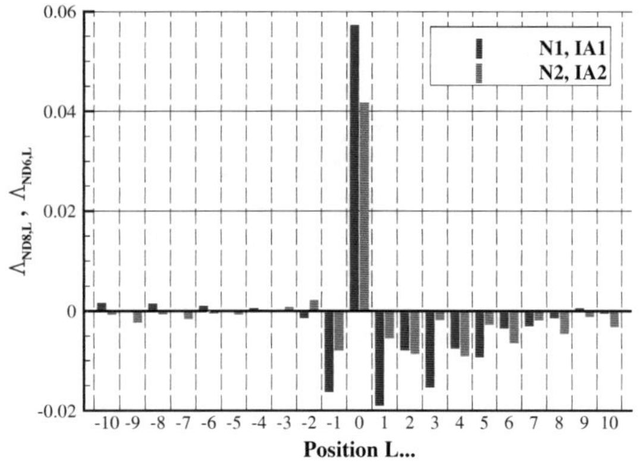

Figure 16. Damping contribution of each rotor blade towards aerodynamic damping for ND 8 (case N1) and ND 6 (case N2)

adjacent blade near the LE, as well as the formation of a large blockage zone in the passage near the blade tip to enable fluid to travel through several passages. Comparison of both rotational speeds shows that the occurrence of flutter is independent of the presence of a shock in the passage and its interaction with the tip clearance flow for the presented phenomenon in this paper. Flutter can occur as long as the formation of a large blockage zone is possible, causing negative axial velocity in the passage near the tip at operating points close to the compressor stability limit. Compared with the experimental observations shown in Section 4.2, the backward travelling wave is accounted for by the convection of the tip clearance flow that excites the blade vibration. As a result, a forward travelling structural vibration pattern is excited. This structural vibration causes a forward travelling pressure wave which is also measured by the WPTs.

4.4 Variation of rotor tip gap

Since the previous investigations have shown that the tip clearance vortex is the main mechanism for the occurrence of flutter, a parameter study of the rotor tip gap size was carried out. Two additional setups were created - increased and a decreased tip gap size - and then simulated for rotational speed N1.

Figure 17 shows the resulting speed lines together with the simulation of the nominal tip gap and the experimental data. Aeroelastic simulations were also carried out using the previously described time-linearized method. Figure 18 shows the aerodynamic damping across all NDs for the three tip gap sizes at the respective leftmost operating point on the speedline (Figure 17). For the large gap, flutter occurrs at a higher reduced mass flow. This is due to the increased tip clearance flow resulting from the enlarged clearance, which is exciting the M2 blade vibration. The unstable ND shifts to 10. In contrast, if the tip gap size is reduced, flutter can be avoided. Even for the lowest reduced mass flow, the aerodynamic damping stays positive across all NDs. The leftmost operating point for the small tip gap in Figure 17 is also the last operating point for which a converged steady solution can be obtained in the simulations. This indicates that the aerodynamic compressor operation limit is reached at this point, rather than the aeroelastic limit.

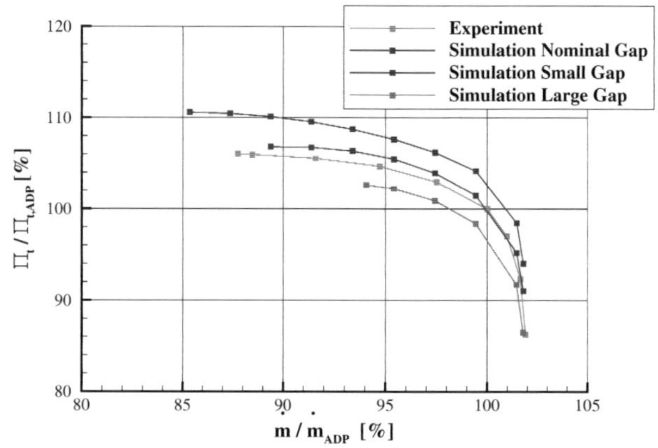

Figure 17. Compressor characteristics for three tip gap sizes at nominal speed N1

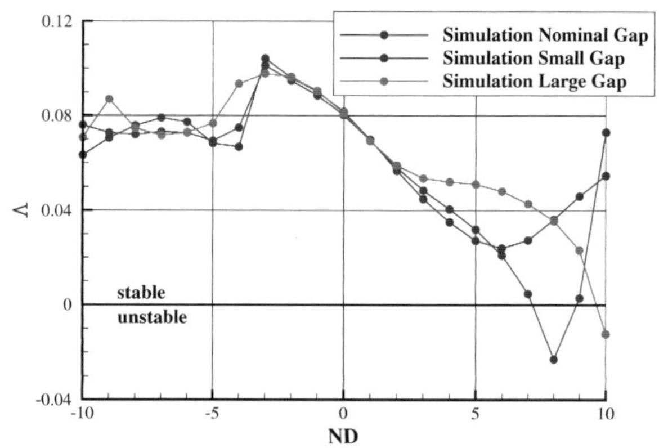

Figure 18. Aerodynamic damping versus ND at the leftmost operating point on the speedline for three tip gap sizes

In the aeroelastic simulations with different tip gap sizes, the damping contribution of each blade in the rotor row was determined using the procedure described above and is shown in Figure 19. The damping contribution is always shown for the respective least stable ND in the simulations of the three gap sizes. The damping contribution of blades located far ahead of the reference blade L0 is largest for the large clearance. As the clearance is reduced, the damping contribution of these blades decreases, since the tip clearance flow is reduced for this case and the flow in the tip region is blocked from passing over multiple rotor passages.

This effect contributes to decreased excitation of the M2 blade vibration, therefore no flutter is predicted for the case with small tip clearance.

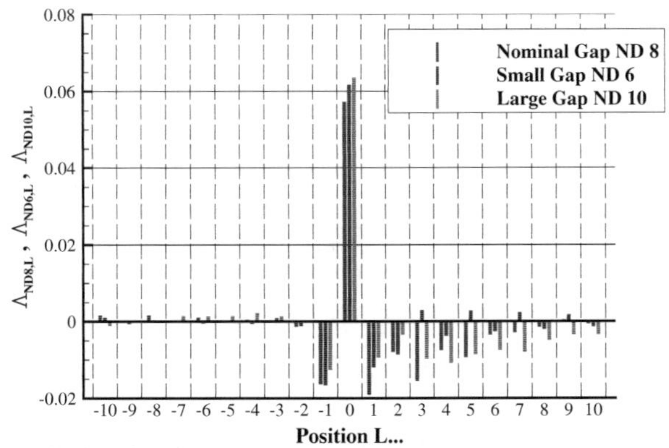

Figure 19. Damping contribution of each rotor blade towards aerodynamic damping for the least stable ND of different tip gap sizes

4.5 Variation of blade eigenfrequency

Section 4.4 demonstrated that the flutter phenomenon can be avoided by reducing the rotor tip gap. However, reduction in rotor tip gap size might not always be possible due to safety restrictions during compressor operation. Another way to prevent flutter is to tune the M2 eigenfrequency of the rotor blade. For this reason, aeroelastic simulations with varying M2 eigenfrequency were carried out. The modeshape was kept constant during the simulations.

Figure 20 shows aerodynamic damping over the ND for five M2 eigenfrequencies at the last operating point at the stability limit for rotational speed N1. Two main observations can be made:

1. The unstable ND shifts towards smaller values as the eigenfrequency increases.
2. For a small enough frequency, the aerodynamic damping remains positive for all NDs, i.e. flutter can be avoided.

From the rotating frame of reference, a forward travelling pressure wave, corresponding to the structural vibration of the BLISK, and a backward travelling pressure wave can be observed during one vibration event (Section 4.2). Section 4.3 showed that the backward travelling pressure wave results from the aerodynamic excitation mechanism of the BLISK vibration. The velocity with which this wave is travelling around the circumference at the blade tip radius r_{tip} can be calculated with the M2 eigenfrequency f_{M2} and the ND of the backward travelling pressure

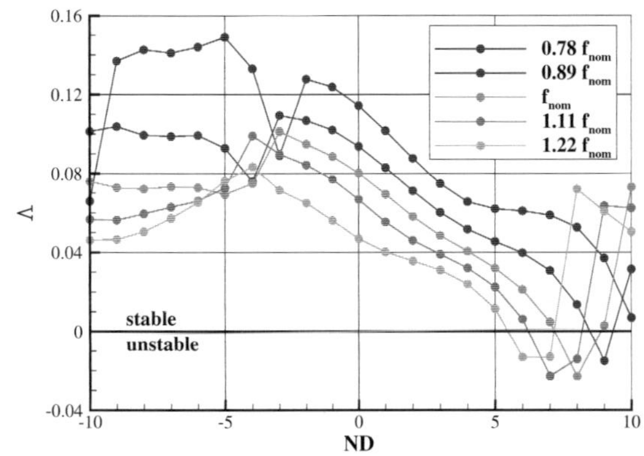

Figure 20. Aerodynamic damping versus ND at the leftmost operating point for nominal speed N1 and five M2 eigenfrequencies

wave using this expression:

$$V_{\text{pressure wave}} = \frac{2\pi r_{\text{tip}} f_{\text{M2}}}{|ND_{\text{pressure wave}}|}.$$

As the M2 eigenfrequency varies in the aeroelastic simulations for one fixed ND, the travelling wave velocity also varies. Figure 21 shows the aerodynamic damping from the simulations against the velocity of the backward travelling pressure wave normalized with the rotational blade tip velocity. Different curves are from simulations with different prescribed NDs. Negative aerodynamic damping values are obtained for all curves and the minimum is always near 0.4 to 0.5 normalized velocity. This means that flutter is always encountered if the velocity of the backward travelling pressure wave (seen from the rotating frame of reference) is about 40% to 50% of the rotational blade tip velocity. Therefore, as the M2 eigenfrequency is varied, the unstable ND shifts so that this condition is fulfilled. The travelling velocity in the stationary frame is clearly below the speed of sound; as a result, this excitation mechanism is not of acoustic origin but arises from convection of the tip clearance flow.

The velocity regime of the pressure wave causing the flutter is close to the velocity regime with which rotating stall cells would travel around the circumference. This indicates that the condition required for the occurrence of the observed flutter is very close to the condition under which rotating stall would be initiated. Therefore, this could explain why both rotating stall and flutter were encountered in the experimental campaign during different transient measurements near the compressor stability limit (Section 4.2).

As was described earlier, no flutter is predicted when the eigenfrequency of M2 is sufficiently small. The decrease required to avoid flutter is predicted here as being about 20% of the nominal M2 eigenfrequency, which is a significant amount and might only be achieved by a change in blade material or large changes to the rotor blade geometry. However, this might also

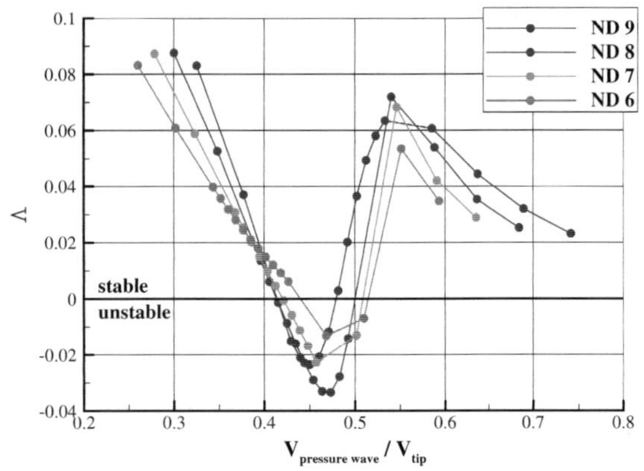

Figure 21. Aerodynamic damping versus the normalized velocity of the backward travelling pressure wave exciting the blade vibration

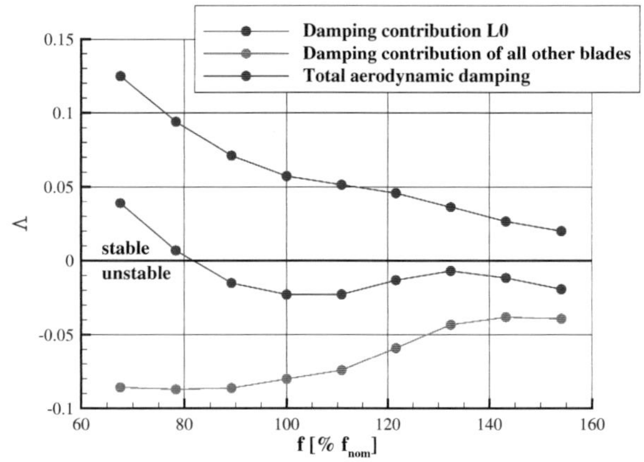

Figure 22. Damping contribution of blade on itself (L0) and from all other blades in the row towards total aerodynamic damping as the blade eigenfrequency varies (the damping for the least stable ND is shown)

markedly influence the shape of M2.

To understand the mechanism for stabilization in this case, the damping contribution of a reference blade on itself and the accumulated contributions from all other blades in the row during variation of the M2 eigenfrequency are shown in Figure 22. Summation of the two yields the total aerodynamic damping in Figure 22. The curve for the damping contribution of the reference blade on itself is always positive. This indicates a damping influence. In contrast to this, the curve for the accumulated contributions from all other blades in the row is always negative (excitation of the blade vibration). When the eigenfrequency is increased, the damping of the blade on itself and the excitation by neighboring blades both become smaller. The total aerodynamic damping remains negative in this case.

The damping of the blade on itself grows in relation to the excitation by neighboring blades if the eigenfrequency is decreased. As a result, the total aerodynamic damping has positive values if the eigenfrequency is decreased by at least about 20%. Stable conditions are predicted even though the excitation from the neighboring blades is still present because the damping of the blade on itself is increased in relation to the excitation. An explanation is that, by reducing the eigenfrequency of the blade, the blade motion and the fluid flow surrounding the blade (with its inherent frequency content) are decoupled. The entire system is stabilized as a result.

5 CONCLUSION

This paper presented numerical investigations of an M2 flutter phenomenon that has been experimentally observed in the TUD transonic compressor rig. Aeroelastic simulations using a time-linearized approach were carried out for two different rotational speeds and the numerical results agreed well with the experimental observations.

The simulations showed that the mechanism for flutter is similar in both cases. The tip clearance flow impinging on the PS of the adjacent blade close to the LE was observed. This excites the blade vibration. The formation of a large blockage zone in the passage at rotor tip height enables fluid to pass through several blade passages before impinging on the blade. Therefore, even blades located far ahead of one reference blade have a significant excitation effect on the reference blade vibration. A backward travelling pressure wave (seen from the rotor relative frame of reference and recorded by the WPTs in the experiment), could be accounted for by the tip clearance flow, which excites a forward travelling vibration pattern of the BLISK with the same phase difference between blades as the backward travelling wave.

In the simulations, as the rotor tip gap size was reduced, stable conditions were predicted. This was due decreased excitation contribution to the aerodynamic damping of a reference blade from blades located far ahead of the reference blade. The simulations also showed that reduction of the M2 eigenfrequency by about 20% would also lead to stable conditions. This is due to increased aerodynamic damping when a blade moves in the surrounding fluid in relation to the excitation by the other blades in the row. By tuning the M2 eigenfrequency, a shift of the unstable ND was observed. Evaluation of the velocity with which the backward travelling pressure wave propagates around the annulus near the blade tip showed that, for flutter to occur, this velocity must be 40% to 50% of the rotational velocity of the blades near the tip. This region is close to the regime for which rotating stall would be expected. The proximity of the two effects could the reason why aerodynamic and aeroelastic stability limits are encountered during different transient measurements for the same speed line.

Decreasing the rotor tip gap size or the M2 eigenfrequency would prevent flutter. However, it remains questionable whether the required changes are possible due to other constraints during compressor design. An alternative approach to preventing flutter could be the use of casing treatments above the rotor (as experimentally shown in [1] and numerically for a generic casing treatment in [4]), as they prevent the emergence of the excitation mechanism that was observed in the rotor tip region.

ACKNOWLEDGMENTS

Thanks go to Deutsche Forschungsgemeinschaft and Rolls-Royce Deutschland for providing a PhD scholarship (DFG-GRK1344) and Rolls-Royce Deutschland for permission to publish this work.

The measurement campaign was funded by the Bundesministerium für Wirtschaft und Energie and Rolls-Royce Deutschland and was carried out under AG Turbo 2020 (AG Turbo 2020 1.2.3 "Aeroelastische Vorgänge im Blattspitzenbereich hochbelasteter Verdichter", Förderkennzeichen 0327719D), the financial support for which is gratefully acknowledged.

REFERENCES

[1] F. Holzinger, F. Wartzek, M. Jüngst, H.-P. Schiffer, and S. Leichtfuß. Self-Excited Blade Vibrations Experimentally Investigated in Transonic Compressors - Rotating Instabilities and Flutter. In *Proceedings of the ASME Turbo Expo 2015, Montréal, Canada*, 2015. GT2015-43628.

[2] S. Leichtfuß, F. Holzinger, C. Brandstetter, F. Wartzek, J. Ratz, and H.-P. Schiffer. Aerodynamic and Aeroelastic Investigation of a Transonic Compressor Rig. In *Proceedings of the ISUAAAT13, Tokyo, Japan*, 2012. ISUAAAT13-DE-7.

[3] S. Leichtfuß, F. Holzinger, C. Brandstetter, F. Wartzek, and H.-P. Schiffer. Aeroelastic Investigation of a Transonic Research Compressor. In *Proceedings of the ASME Turbo Expo 2013, San Antonio, Texas, US*, 2013. GT2013-94730.

[4] S. Leichtfuß. *Zum Einfluss des Spaltwirbels auf das aeroelastische Verhalten transsonischer Verdichter*. PhD thesis, Technische Universität Darmstadt, 2014.

[5] E. Johann, B. Mück, and J. Nipkau. Experimental and Numerical Flutter Investigation of the 1st Stage Rotor in 4-Stage High Speed Compressor. In *Proceedings of the ASME Turbo Expo 2008, Berlin, Germany*, 2008. GT2008-50689.

[6] C. Andersson, H. Mårtensson, and N. Edin. Non-Synchronous Vibrations in the 3½ Stage Transonic Test Compressor Blenda. In *Proceedings of the ISABE 2013, Busan, South Korea*, 2013. ISABE-2013-1135.

[7] J. D. Gill and V. R. Capece. Experimental Investigation of Flutter in a Single Stage Unshrouded Axial-Flow Fan. In *Proceedings of the 42nd AIAA Aerospace Sciences Meeting and Exhibit, Reno, Nevada, US*, 2004. AIAA 2004-686.

[8] M. Vahdati, G. Simpson, and M. Imregun. Mechanisms for Wide-Chord Fan Blade Flutter. In *Journal of Turbomachinery*, volume 133, 2011.

[9] M. Vahdati and N. A. Cumpsty. The Mechanism of Aeroelastic Instability in Transonic Fans. In *Proceedings of the ISUAAAT13, Tokyo, Japan*, 2012. ISUAAAT13-I-5.

[10] G. Schulze, C. Blaha, D. K. Hennecke, and J. M. Henne. The Performance of a New Axial Single Stage Transonic Compressor. In *Proceedings of the ISABE 1995, Melbourne, Australia*, 1995. ISBN 1-56347-177-9.

[11] C. Biela, C. Brandstetter, F. Holzinger, and H.-P. Schiffer. Influence of Inlet Guide Vane Wakes on Performance and Stability of a Transonic Compressor. In *Proceedings of the ISABE 2011, Gothenburg, Sweden*, 2011. ISABE-2011-1209.

[12] C. Brandstetter, M. Kegalj, C. Biela, and H.-P. Schiffer. PIV-Measurements in a Transonic Compressor Test Rig with Variable Inlet Guide Vanes. In *Proceedings of the ISABE 2011, Gothenburg, Sweden*, 2011. ISABE-2011-1222.

[13] M. Jüngst, F. Holzinger, D. Möller, F. Wartzek, and H.-P. Schiffer. The Analysis of Blade Vibrations of a 1.5-Stage Transonic Compressor. In *Proceedings of the ISUAAAT14, Stockholm, Sweden*, 2015. I14-S9-1.

[14] D. Nürnberger, F. Eulitz, S. Schmitt, and A. Zachcial. Recent Progress in the Numerical Simulation of Unsteady Viscous Multistage Turbomachinery Flow. In *Proceedings of the ISABE 2001, Bangalore, India*, 2001. ISABE-2001-1081.

[15] S. Leichtfuß, C. Biela, H.-P. Schiffer, and F. Heinichen. Influence of Inlet Guide Vane Wakes on the Passage Flow in a Transonic Axial Compressor. In *Proceedings of the ASME Turbo Expo 2012, Copenhagen, Denmark*, 2012. GT2012-69485.

[16] D. C. Wilcox. *Turbulence Modeling for CFD*. DCW Industries Inc., La Cañada, California, US, 2004.

[17] M. Kato and B. E. Launder. The Modelling of Turbulent Flow Around Stationary and Vibrating Square Cylinders. In *Proceedings of the 9th Symposium on Turbulent Shear Flows, Kyoto, Japan*, 1993.

[18] J. Bardina, J. H. Ferziger, and R. S. Rogallo. Effect of rotation on isotropic turbulence: computation and modelling. In *Journal of Fluid Mechanics*, volume 154, 1985.

[19] H.-P. Kersken, C. Frey, C. Voigt, and G. Ashcroft. Time-Linearized and Time-Accurate 3D RANS Methods for Aeroelastic Analysis in Turbomachinery. In *Proceedings of the ASME Turbo Expo 2010, Glasgow, UK*, 2010. GT2010-22940.

[20] H.-P. Kersken, G. Ashcroft, C. Frey, O. Pütz, H. Stüer, and S. Schmitt. Validation of a Linearized Navier-Stokes Based Flutter Prediction Tool Part 1: Numerical Methods. In *Proceedings of the ASME Turbo Expo 2012, Copenhagen, Denmark*, 2012. GT2012-68018.

[21] M. B. Giles. Nonreflecting Boundary Conditions for Euler Equation Calculations. In *AIAA Journal*, volume 28, number 12, 1990.

[22] F. Holzinger. *Tip clearance vortex coupled blade vibration in transonic compressors*. PhD thesis, Technische Universität Darmstadt, 2015.

[23] F. Holzinger, F. Wartzek, M. Nestle, H.-P. Schiffer, and S. Leichtfuß. Self-Excited Blade Vibrations Experimentally Investigated in Transonic Compressors - Acoustic Resonance. In *Proceedings of the ASME Turbo Expo 2015, Montréal, Canada*, 2015. GT2015-43618.

Proceedings of the 14th International Symposium on Unsteady Aerodynamics, Aeroacoustics & Aeroelasticity of Turbomachines
ISUAAAT14
8-11 September 2015, Stockholm, Sweden

I14-S9-3

CONCURRENT MULTI-BLADEROW AERODYNAMIC-AEROELASTIC DESIGN OPTIMIZATION USING ADJOINT APPROACH

O. Valero
Department of Engineering Science,
University of Oxford
Park Road, Oxford, OX1 3PJ, UK

L. He
Department of Engineering Science,
University of Oxford
Park Road, Oxford, OX1 3PJ, UK

D. X. Wang
Siemens Industrial Turbomachinery (SIT) Ltd,
Lincoln, LN5 7FD, UK

Y. S. Li
Siemens Industrial Turbomachinery (SIT) Ltd,
Lincoln, LN5 7FD, UK

R. G. Wells
Siemens Industrial Turbomachinery (SIT) Ltd,
Lincoln, LN5 7FD, UK

ABSTRACT

Given the increasing demands on turbomachinery, there is a need to look at features like blade flutter stability with more detail.

This paper presents the development and validation of an adjoint method to concurrent aerodynamic and aeroelastic blade design optimization in a multi-bladerow environment. The system is formulated consistently following the same treatment as the flow equations [1]. The fact that the equations are solved in the frequency domain and that the adjoint method is cost-independent of the number of variables makes the optimization system computationally efficient.

In order to effectively capture the aeroelastic sensitivities due to reflecting boundary conditions an interface treatment is used to pass the adjoint unsteady disturbances across bladerows. This interface treatment is consistent with the adjoint characteristics and validated numerically by comparing the results against a direct finite difference method

A set of case studies are presented which will demonstrate the multi-component and multi-disciplinary capabilities of this method.

INTRODUCTION

The development of turbomachinery blades is driven towards highly loaded and compact configurations, which inevitably leads to enhanced interactions between the surrounding working fluid and the blades, even at design conditions. This challenges the widely used design rules on aero-mechanical stability margins where blade aeroelasticity is typically evaluated after a successful aerodynamic design, and if required, the design is modified to meet the margins. Modern turbomachinery blade designs increasingly point to a need to look at aerodynamic and aeroelasticity in a more simultaneous (or concurrent) way in order to avoid long time-scales and expensive iterations in a design process.

With the progress of computer technology and numerical methods for fluid dynamics, there has been a large development of the so called optimization algorithms. These algorithms try to maximize the performance of a design by modifying its parameters, with minimal intervention from the designer. However, there has been little or none development for this type of concurrent optimization.

An important issue regarding the development of optimization methods for aeroelasticity is the computational cost of the unsteady flow equations. A full time-domain unsteady three-dimensional method [3] is still expensive and prohibitive for turbomachinery design optimization. This has motivated the development of efficient methods that can evaluate aero-mechanical performance under realistic computational times [1,4-7].

The nonlinear harmonic phase solution method [1] splits the unsteady flow equation into one time-averaged equation and one-time harmonic equation with the unsteady perturbation represented in a set of Fourier series. The nonlinear harmonic method effectively translates a problem of solving an unsteady flow to equivalently solving two or three 'steady-like' flows. This is highly efficient as the unsteady nonlinear harmonic flow solver provides the time-averaged aerothermal and aeroelastic performance at the same time. Particularly for the study of blade flutter and forced response, a harmonic solution retaining only one harmonic would be sufficiently accurate for the unsteady flow of interest.

Another costly factor in a blade shape optimization is the number of design variables. The adjoint method has been proven as a cost effective approach to calculate gradients of an objective function to a large number of variables. The first application to aerodynamic design optimization was pioneered by Jameson [8,9]. However, it has not been until the last decade that it gained attention in the CFD community [11,12] while multi-bladerow aerodynamic design optimization has only been researched for the past 5 years [13-16].

On the aeroelastic side, some efforts have been done to formulate a system of adjoint equations that could efficiently calculate the aeroelastic sensitivities of turbomachinery blades. Duta et al. [17] was one the first to evaluate this with a discrete adjoint system based on the linear harmonic flow equations. Other discrete developments followed them [18,19]. The most relevant work for this thesis is the continuous discretization performed by He and Wang [20], which captures the aerodynamic damping and forcing sensitivities to blade geometry changes in a single-bladerow environment. However, this can be restrictive as the multi-bladerow interaction effects (which may be considerable) are not included. Li and He [21] showed that up to 100% change in the rotor aeroelastic damping can result from the stator reflection for a transonic compressor, which was in fact supported recently by Rahmati et al. [2].

It is clear that in order to perform efficient aerodynamic and aeroelastic optimization, turbomachinery blades must be optimised in a multi-bladerow environment, where blade-row interactions can be taken into account. As an attempt to address the issues above, a new system of adjoint equations is re-derived based upon the Nonlinear Harmonic Phase Solution Method and the work done by He and Wang [20]. The harmonic multi-bladerow treatment [2] used in the flow solution system is adopted as a basis to establish a new unsteady adjoint interface method to pass the adjoint disturbances across the bladerows. The new adjoint system that is presented is able to perform concurrent multi-disciplinary and multi-component design optimization of turbomachinery blades.

NOMENCLATURE

α	Design variable
δ	Logarithmic-decrement (aero damping)
λ	Adjoint variables
η	Stage isentropic efficiency
σ	Inter-blade phase angle (IBPA)
π	Pressure ratio
ω	Angular frequency of unsteadiness
Δs	Entropy generation rate
A, B	Harmonic coefficients
E	Energy
I	Objective function
$\dot{m}$	Mass flow rate
u_g, v_g, w_g	Mesh vibration velocities
U	Flow variables
V_i	Viscosity terms of flow equation
W	Work sum

FLOW SOLUTION METHOD

The Nonlinear Harmonic Phase Solution Method developed by [1] discretizes the URANS equations in the frequency domain at three distinctive phase angles of a period of unsteadiness and calculates the flow field by solving a system of three coupled "steady-like" equations. This solution method is able to predict blade flutter efficiently and it sets the framework for the adjoint design optimization method. The unsteady-flow solver has been validated extensively in previous literature [1,2] and used for design optimization [20,21].

The URANS equations are written in cylindrical coordinates as follows:

$$\frac{\partial U}{\partial t} + R(U) = 0 \tag{1}$$

where the flow residual term $R(U)$ collects the spatial derivatives:

$$R(U) = \frac{\partial(F - Uu_g - V_x)}{\partial x} + \frac{\partial(G - Uv_g - V_\theta)}{r\partial\theta} + \frac{\partial r(H - Uw_g - V_r)}{\partial r} - S = 0 \tag{2}$$

The Nonlinear Harmonic Phase Solution Method assumes that the flow variable can be decomposed in a time-averaged part and a harmonic part:

$$U(x,t) = \overline{U}(x) + U_A(x)\sin\omega t + U_B(x)\cos\omega t \tag{3}$$

The flow variable U, is a function of space and time but $\overline{U}$,U_A and U_B are functions of space only. With this assumption, Eq. 3 is added to Eq. 1, thus eliminating the time derivative term and yielding the following:

$$\omega(U_A\cos\omega t - U_B\sin\omega t) + R(\overline{U} + U_A\cos\omega t + U_B\sin\omega t) = 0 \tag{4}$$

Eq. 4 replaces the time derivative term for an extra term that can be treated as a source term. However, the expression is still time-dependent and has three unknowns: $\overline{U}$, U_A and U_B. At three particular phase angles ωt=-π/2, 0, π/2, the flow variable $U(x,t)$ can be written as follows:

$$U\left(\omega t = -\frac{\pi}{2}\right) = U_{-1} = \overline{U} - U_A \tag{5a}$$

$$U(\omega t = 0) = U_0 = \overline{U} + U_B \tag{5b}$$

$$U\left(\omega t = +\frac{\pi}{2}\right) = U_1 = \overline{U} + U_A \tag{5c}$$

Consequently, the three unknowns: $\overline{U}$,U_A and U_B can also be expressed in terms of the flow solutions at these three phase angles:

$$\overline{U} = \frac{U_1 + U_{-1}}{2} \tag{6a}$$

$$U_A = \frac{U_1 - U_{-1}}{2} \tag{6b}$$

$$U_B = U_0 - \frac{U_1 + U_{-1}}{2} \tag{6c}$$

Substituting this back into Eq. 4 leads to the final system of three "steady-like" equations:

$$\omega\left(U_0 - \frac{U_1+U_{-1}}{2}\right) + R(U_{-1}) = 0 \qquad (7a)$$
$$\omega\left(\frac{U_1-U_{-1}}{2}\right) + R(U_0) = 0 \qquad (7b)$$
$$-\omega\left(U_0 - \frac{U_1+U_{-1}}{2}\right) + R(U_1) = 0 \qquad (7c)$$

The three equations above are simultaneously solved in a similar way to that of the steady RANS equations with the extra term being treated as a source term. This effectively reduces the computational cost of the unsteady problem to the same order of magnitude as a steady problem for blade flutter applications.

The spatial discretization is done with a second order cell-centered finite volume scheme with second and fourth order artificial viscosity. In order to march the solutions to a steady state, a pseudo time derivative term is added to the three equations, so that the Runge-Kutta stepping technique can be applied. The convergence is accelerated with time-stepping and multigrid techniques.

UNSTEADY ADJOINT SYSTEM OF EQUATIONS

The objective of a gradient-based optimization method is to evaluate the gradient of the objective function with respect to the design variable in order to determine the best search direction to find the optimum. This function I, is a scalar that is a function of the design variable and the flow variable vector as follows:

$$I = I(\alpha, U()) \qquad (8)$$

where α and U denote the design variable and flow variable respectively. The relation between the design and flow variable is determined by the nonlinear flow equations:

$$R[\alpha, U(\alpha)] = 0 \qquad (9)$$

The equality above is true for any design variable. Using the chain rule, the total derivative of the objective function I with respect to the design variable α can be written as:

$$\frac{dI}{d\alpha} = \frac{\partial I}{\partial \alpha} + \frac{\partial I}{\partial U}\frac{\partial U}{\partial \alpha} \qquad (10)$$

For a large number of design variables, a direct evaluation of the derivative $\partial U/\partial \alpha$ becomes very expensive, since it requires a solution of the flow equations for each design variable. The main feature of the adjoint method is that it eliminates the flow variable sensitivity from the gradient equation. Consequently, the cost of calculating the gradient expression above (Eq. 10) is no longer proportional to the number of variables.

The derivation of the adjoint system of equations starts with the flow equation from the Nonlinear Harmonic Phase Solution Method (Eq. 4). A Lagrange multiplier, called the adjoint variable vector λ^T is added multiplying the flow equations as follows:

$$\lambda^T[(U_A \;\; \cos\omega t - U_B \;\; \sin\omega t) + R(U)] = 0 \qquad (11)$$

This product holds true, independently of the value of λ^T. The above equation is a function that must be integrated over time:

$$\int_0^T \lambda^T[(U_A \;\; \cos\omega t - U_B \;\; \sin\omega t) + R(U)]dt = 0 \qquad (12)$$

Due to the periodicity of the flow, the equality above remains intact and it still holds if it is assumed that the adjoint variable has the same harmonic form as the flow variable as follows:

$$\lambda = \bar{\lambda} + \lambda_A \sin\omega t + \lambda_B \cos\omega t \qquad (13)$$

The next step is to linearize the flow equations. The source terms containing the unknowns U_A and U_B are rearranged using the properties of Fourier series for time-periodic functions:

$$\frac{T}{2}\omega(\lambda_B^T U_A - \lambda_A^T U_B) + \int_0^T \lambda^T R(U)dt = 0 \qquad (14)$$

where

$$\lambda_A^T = \frac{2}{T}\int_0^T \lambda^T \sin\omega t\, dt \qquad (15a)$$
$$\lambda_B^T = \frac{2}{T}\int_0^T \lambda^T \cos\omega t\, dt \qquad (15b)$$
$$U_A = \frac{2}{T}\int_0^T U \sin\omega t\, dt \qquad (15c)$$
$$U_B = \frac{2}{T}\int_0^T U \cos\omega t\, dt \qquad (15d)$$

Substituting U_A and U_B for Eqs. 15c and 15d in Eq. 14 yields:

$$\int_0^T [\omega(\lambda_B^T \sin\omega t - \lambda_A^T \cos\omega t)U + \lambda^T R(U)]dt = 0 \qquad (16)$$

After this rearrangement, Eq. 16 is linearized as follows:

$$\int_0^T \left[\omega(\lambda_B^T \sin\omega t - \lambda_A^T \cos\omega t) + \lambda^T \frac{\partial R}{\partial U}\right]\frac{\partial U}{\partial \alpha}dt + \int_0^T \frac{\partial R}{\partial \alpha}dt = 0 \qquad (17)$$

The equality holds again after linearization if the adjoint variable is proven to be periodic. The objective function is also a function of time $I = I(\alpha, U)$, and so the gradient expression (Eq. 10) is written down as follows:

$$\frac{dI}{d\alpha} = \int_0^T \left(\frac{\partial I}{\partial \alpha} + \frac{\partial I}{\partial U}\frac{\partial U}{\partial \alpha}\right)dt \qquad (18)$$

Subtracting Eq. 17 from the gradient expression (Eq. 18) and rearranging around the flow variable sensitivity gives:

$$\frac{dI}{d\alpha} = \int_0^T \left[\frac{\partial I}{\partial U} - \omega(\lambda_B^T \sin\omega t - \lambda_A^T \cos\omega t) - \lambda^T \frac{\partial R}{\partial U}\right] \frac{\partial U}{\partial \alpha} dt + \int_0^T \left(\frac{\partial I}{\partial \alpha} - \lambda^T \frac{\partial R}{\partial \alpha}\right) dt \quad (19)$$

The adjoint equation is defined so that the flow variable sensitivity term $\partial U/\partial\alpha$, can be eliminated from the gradient expression. In order to do so, the adjoint variable is chosen in a way that the following equation is satisfied:

$$\frac{\partial I}{\partial U} - \omega(\lambda_B^T \sin\omega t - \lambda_A^T \cos\omega t) - \lambda^T \frac{\partial R}{\partial U} = 0 \quad (20)$$

The above can be rewritten as follows:

$$\frac{\partial I}{\partial U} - \omega(\lambda_A^T \cos\omega t - \lambda_B^T \sin\omega t) + \lambda^T \frac{\partial R}{\partial U} \quad (21)$$

The resulting unsteady adjoint equation has a source term with the same form as the prime flow equation but with a negative sign. This is in line to the "Anti-physics" adjoint information path which was first mentioned by Giles [18] and systematically shown by Wang and He [19,20].

Because the adjoint variable has the same harmonic form as the flow variable, the adjoint equation can be solved following the same procedure as the Nonlinear Harmonic Phase Solution Method. This leads to an equivalent system of three steady-like adjoint equations as shown below:

$$-\omega\left(\lambda_0 - \frac{\lambda_1+\lambda_{-1}}{2}\right) + R(\lambda_{-1}) = 0 \quad (22a)$$

$$\omega\left(\frac{\lambda_1-\lambda_{-1}}{2}\right) + R(\lambda_0) = 0 \quad (22b)$$

$$\omega\left(\lambda_0 - \frac{\lambda_1+\lambda_{-1}}{2}\right) + R(\lambda_1) = 0 \quad (22c)$$

After the elimination of the flow variable sensitivity, the new gradient expression becomes:

$$\frac{dI}{d\alpha} = \int_0^T \left(\frac{\partial I}{\partial \alpha} - \lambda^T \frac{\partial R}{\partial \alpha}\right) dt \quad (23)$$

This expression is still linked to the design variable α, but the computational cost of calculating $\partial R/\partial\alpha$ for a large number of design variables does not exceed the computational time of solving one flow solution. Using the adjoint method, there are three main equations to be solved: the nonlinear system of flow equations, the linear system of adjoint equations and the gradient equation. The computational cost, however, is roughly two times the cost of solving the baseline unsteady flow equations, independently of the number of design variables.

The same spatial and temporal discretization schemes used for the flow equations can be used to solve the system of adjoint equations. Regarding the boundary conditions for the adjoint equation, there is a corresponding adjoint flow characteristic travelling at a speed of the same magnitude but in opposite direction of each flow characteristic. This "Anti-physics" adjoint information path forces the number of inlet and exit adjoint boundary conditions to be the opposite as for the flow inlet and exit boundary conditions. Their implementation in this work is the same as the steady and unsteady work of Wang and He [14,19,20]. However the continuous periodic boundary condition is unique and it is explained next.

PERIODIC BOUNDARY CONDITION

Each blade is subject to unsteadiness of a circumferential traveling wave pattern. Hence, there is a phase difference that has to be modelled at the periodic boundaries of the blade passage (Fig. 1).

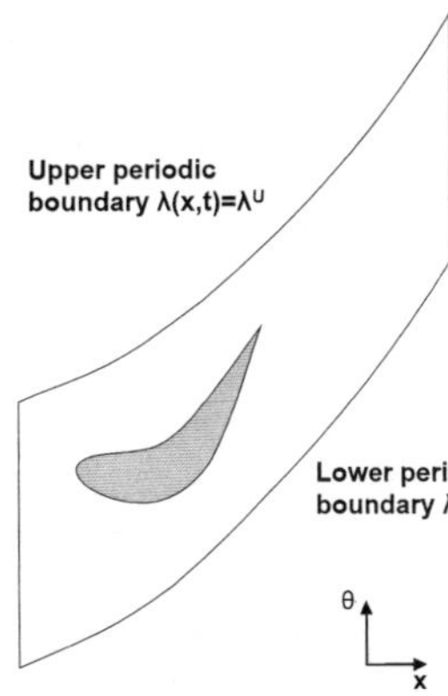

Figure 1. Single-passage domain with adjoint periodic boundaries

This phase difference is called inter-blade phase angle σ and it is defined as:

$$\sigma = \frac{2\pi n}{N} \quad (24)$$

where N is the number of blades of the bladerow of interest and n the number of nodal diameters of the unsteady disturbance.

In a physical time domain, the flow field at the initial time is defined by the boundary conditions. The opposite is true with respect to the adjoint field: it is null at the final time (as imposed by the adjoint equation and boundary conditions) while it is unknown at the initial time. In the physical time domain t_p, the adjoint variable can be written as:

$$\lambda = \bar{\lambda} + \lambda_A \sin\omega t_p + \lambda_B \cos\omega t_p \quad (25)$$

An adjoint time domain can be defined as being the opposite of the physical time domain $(t_p = -t_a)$. In this time domain, the adjoint variable has the following form:

$$\lambda = \bar{\lambda} - \lambda_A \sin\omega t_a + \lambda_B \cos\omega t_a \quad (26)$$

The switch between the physical time and the adjoint time changes the sign of the harmonic terms. The harmonic form of the adjoint variable allows to assume that the upper and lower periodic boundaries are phase-shifted in a similar way as the flow field. Hence in a physical time domain the upper boundary condition has a phase lead over the lower boundary as follows:

$$\lambda^U = \overline{\lambda^L} + \lambda_A^L \sin(\omega t_p + \sigma) + \lambda_B^L \cos(\omega t_p + \sigma) \quad (27)$$

In the adjoint time domain, the relation between the upper and lower periodic boundary is completely reversed. Changing to the adjoint time domain, the above equation at the upper boundary can be rewritten as:

$$\lambda^U = \overline{\lambda^L} + \lambda_A^L \sin(-\omega t_a + \sigma) + \lambda_B^L \cos(-\omega t_a + \sigma) \quad (28)$$

This leads to:

$$\lambda^U = \overline{\lambda^L} - \lambda_A^L \sin(\omega t_a - \sigma) + \lambda_B^L \cos(\omega t_a - \sigma) \quad (29)$$

In the adjoint time domain, the upper boundary is lagged respect to the lower boundary. Consequently, the lower boundary is now the boundary that has the phase lead:

$$\lambda^L = \overline{\lambda^U} - \lambda_A^U \sin(\omega t_a + \sigma) + \lambda_B^U \cos(\omega t_a + \sigma) \quad (30)$$

ADJOINT MULTI-BLADEROW TREATMENT

The time-averaged and the harmonic components of the adjoint variable have to be exchanged successfully among bladerows. The mixing-plane treatment used in this work for the adjoint time-averaged part is the one developed by Wang and He [19,20].

The new feature of this work is the adjoint interface treatment that is developed as an extension of the technique applied to the flow solution method [2] and it follows the same two-steps strategy:

1) A time-trace signal of the adjoint variable is built at both sides of the interface
2) The different harmonic terms of the adjoint variable are separated and only those corresponding to the first harmonic are retained.

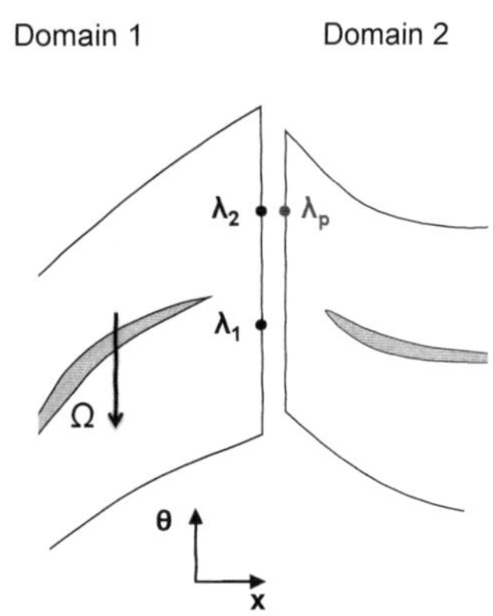

Figure 2. Relative position between domains (physical time domain)

A time trace of the adjoint variable has to be created. First, the relative position of a point p from domain 2 at the interface with respect to domain 1 is calculated at different time instances based on its initial position, rotational speed, relative motion and time instant (Fig. 2). The value of the adjoint variable at the point p, at different time instances can be given as follows:

$$\lambda_p\left(t = t_{p_i}\right) = \lambda_i(i = 0,1,2,\dots,n) \quad (31)$$

where λ_i is the local value of the adjoint variable in domain 2 located at the point of the interface that is coincident with p at the physical time domain instance i. If the same period of unsteadiness is considered, a time-trace of the adjoint variable for the same point in the adjoint time domain is given by:

$$\lambda_p(t = t_{a_i}) = \lambda_{n-i}(i = 0,1,2,\dots,n) \quad (32)$$

Both time traces will show a mirrored shape over a period of unsteadiness, due to the periodicity of the adjoint variable (Fig. 3)

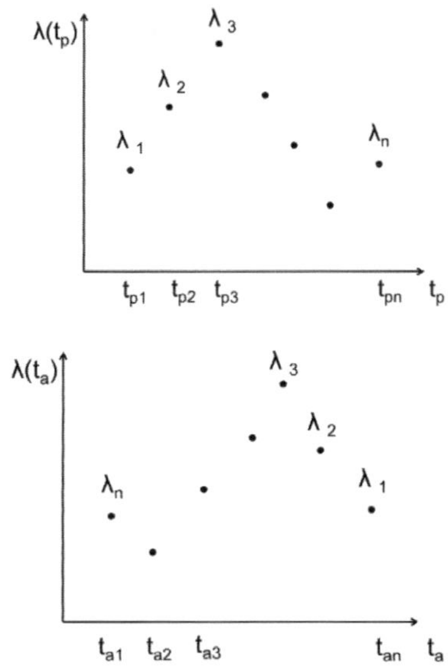

Figure 3. Time-trace of the adjoint variable over a period of unsteadiness. Physical (top) and adjoint (bottom) time domains

Once a time trace is created for all the points at the interface of each domain, the components of the adjoint variable that correspond to the first harmonic are extracted while the rest of the terms are filtered out as follows (physical time domain):

$$\lambda_A = \frac{\omega}{\pi}\sum_{i=1}^{n} \lambda_i \sin\left(\omega t_{p_i}\right) \Delta t_{p_i} \quad (33a)$$

$$\lambda_A = \frac{\omega}{\pi}\sum_{i=1}^{n} \lambda_i \cos\left(\omega t_{p_i}\right) \Delta t_{p_i} \quad (33b)$$

$$\bar{\lambda} = \frac{\omega}{2\pi}\sum_{i=1}^{n} \lambda_i \Delta t_{p_i} \quad (33c)$$

The harmonic form of the adjoint variable leads to a very simple interface treatment that is essentially an extension of the unsteady flow harmonic interface method. The procedure steps shown through Eqs. 33a-c are directly and equally applicable to a physical harmonic flow variable as well as to an adjoint harmonic variable in both time domains.

HARMONIC TREATMENT VALIDATION

The aeroelastic log-dec damping of a vibrating blade is calculated as follows

$$\delta = -\frac{W}{E_s} \tag{34}$$

where E_s is the total strain energy of the vibrating solid blade and W is defined as the sum of the work done by the fluid to the vibrating blade in one vibration cycle.

The work sum is a function of space and time and it is calculated by the integration of the unsteady pressure weighted by mesh vibration velocities on the blade surface.

Positive work sum leads to reduced aeroelastic stability in the blade while a negative work sum means that energy is being dissipated from the blade into the working fluid, increasing aero-damping, thus aeroelastic stability.

The effectivity of the harmonic interface treatment is very important as adjacent bladerows can cause an unsteady reflection (Fig. 4) that will affect the work sum of the vibrating blade [2]. Strong reflections are expected from the downstream stators for a vibrating compressor rotor blade and from upstream stator blades for a turbine rotor blade as discussed by [26]

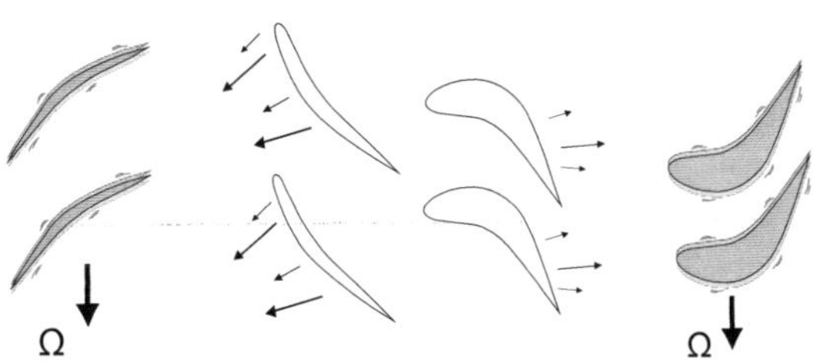

Figure 4. Unsteady reflections from adjacent stators. Compressor (left) and turbine (right) stages.

These reflections are captured and exchanged between adjacent bladerows by the flow solver through the flow harmonic interface treatment. The adjoint harmonic interface treatment, on the other side, needs to capture the work sum sensitivity of the vibrating blade to changes in the geometry of adjacent bladerows so correct shape optimization of adjacent rows can be performed.

In order to validate the adjoint harmonic interface treatment, the work sum gradient at the surface of a vibrating blade is calculated using the adjoint and the direct finite difference method. Three validation cases are performed for the two different geometries that are studied in the results section: The DLR transonic compressor stage [27] and a low-pressure industrial turbine stage. For the first validation cases, a small shape perturbation is only applied to the rotor to show the overall accuracy of the adjoint method predicting the work sum gradient to rotor geometry changes. For the second and third set of cases, a small shape perturbation is only applied to the stator. The adjoint harmonic interface treatment is disabled on the second case. This means that there are no unsteady reflections going through the stator-rotor interface on the second case and that any change in the stator geometry that could affect the vibrating blade is only due to the changes in average flow field through the steady mixing plane. Table 1 and 2 shows the work sum gradients obtained by the adjoint and direct finite difference for the three cases. The relative errors between these two values are also given.

By looking at both tables, it can be observed that the adjoint and finite difference methods are in good agreement, as their differences are always less than 10%. Another important point is that there is a disparity of 1-2 orders of magnitude between the gradients of the second and third case. The smallest of both values corresponds to the case when no harmonic interface is enabled. This demonstrates that the adjoint harmonic interface is capturing the sensitivities from the aeroelastic reflections. Otherwise, these values would have at least the same order of magnitude.

	Adjoint	F. Difference	%error
Rotor	$-2.12 \cdot 10^{-3}$	$-2.27 \cdot 10^{-3}$	-6.6%
Stator (no interface)	$-1.91 \cdot 10^{-5}$	$-1.98 \cdot 10^{-5}$	-3.53%
Stator	$-3.41 \cdot 10^{-4}$	$-3.68 \cdot 10^{-4}$	-7.34%

Table 1. Work sum gradients. DLR compressor stage

	Adjoint	F. Difference	%error
Rotor	$-5.12 \cdot 10^{-3}$	$-5.27 \cdot 10^{-3}$	-2.84%
Stator (no interface)	$-3.01 \cdot 10^{-5}$	$-3.1 \cdot 10^{-5}$	-2.91%
Stator	$-2.21 \cdot 10^{-3}$	$-2.38 \cdot 10^{-3}$	-7.14%

Table 2. Work sum gradients. Low-pressure turbine stage

DESIGN OPTIMIZATION SYSTEM

FE-CFD interactions

For every change to the shape of the blade, the modeshape and frequency of vibration of the blade corresponding to the first mode have to be updated in the CFD model through FE modal analysis using ABAQUS

Objective function & Worksum

The objective function is a weighted sum of performance parameters of interest. In this work, it is defined as:

$$I = \frac{W_0}{W} + \sigma_1 \left(\frac{\dot{m}}{\dot{m}_0} - 1\right)^2 + \sigma_2 \left(\frac{\pi}{\pi_0} - 1\right)^2 + \sigma_3 \max[\text{sign}(1, \Delta s - \Delta s_0), 0] \cdot \left(\frac{\Delta s}{\Delta s_0} - 1\right)^2 \tag{35}$$

where W_0 is the worksum value of the baseline design, $\dot{m}$ is the mass flow rate and π the stage pressure ratio. σ_1 and σ_2 are the weighting factors of the aero-thermal constraints. If concurrency is enabled ($\sigma_3 > 0$) the aeroelastic objective function will increase aero-damping while avoiding the entropy generation rate Δs, to be increased more than its original ratio

Aeroelastic optimization procedure

Fig. 5 shows the overall structure of a gradient-based design optimization system using the unsteady adjoint method. The overall process is summarised in the next steps:

1. The baseline design geometry and flow field are initialised.
2. The flow solution at the three phase angles are calculated for the baseline design
3. The adjoint flow field is initialised and the unsteady adjoint equations at the three phases are solved
4. The gradient is calculated and a step is taken in the negative direction
5. A new blade geometry is generated
6. The mode shape and vibration frequency of the new blade geometry are obtained by FE analysis.
7. The flow solution at the three phases is calculated for the new geometry. Its objective function (performance) is calculated.

Steps 3-7 form an adjoint optimization cycle, and this process is repeated until a maximum number of iterations, specified by the user, are reached. The computational cost of each cycle rarely exceeds three times the cost of solving the unsteady flow equations.

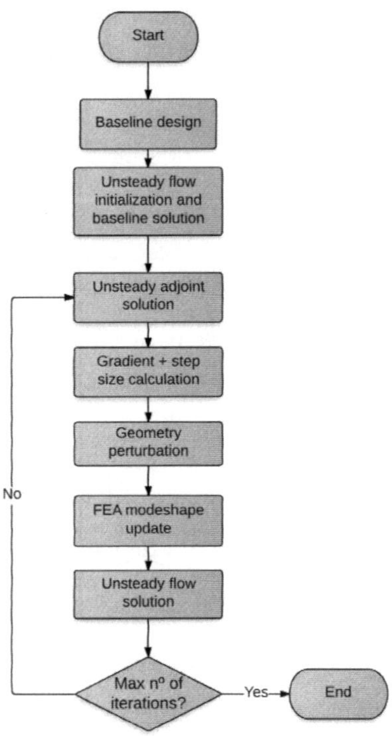

Figure 5. Flowchart of the unsteady adjoint optimization system

Optimization settings

For each blade whose shape is optimised, 81 hump functions (or variables) are used. The design optimization is carried out with only 1 CPU and the average computational cost per design cycle for the DLR compressor case is 4.5 hours and for the LP turbine stage is 5.5 hours, which is very low compared to other gradient-based and non-gradient based methods [22] considering the large number of variables that are used (162 for a full stage optimization).

MULTI-BLADEROW DESIGN OPTIMIZATION: DLR COMPRESSOR STAGE

The objective function for the first two cases is defined as "aeroelastic-only" as no entropy generation rate constraint is enabled ($\sigma_3 = 0$). The first case is intended to evaluate the performance of the adjoint harmonic interface treatment and see the effect that adjacent blades can have in the aeroelastic performance of a vibrating rotor blade. The second case will show the margin of improvement that can be obtained with a full stage (rotor and stator) design.

1) Stator "aeroelastic-only" optimization

The design optimization converges after 30 design cycles as shown in (Fig. 6). The final values of the objective function parameters are shown in Table 3. The rotor blade aeroelastic log-dec is slightly increased from 2.81 to 2.92 (i.e 4% increase). Despite the margin is small and not as large as what can be achieved with a rotor re-design, this is expected as the rotor work sum gradient with respect to a change in stator geometry is two orders of magnitude lower than the rotor's gradient (Table 1). The aerodynamic performance is not decreased by the new stator shape and the entropy of the stage barely changes.

The new stator blade's influence on the local work sum contours at the rotor blade surface for the first vibration mode is as shown in Figs. 8-9. It can be seen how the two stable areas located before the shock wave on both the pressure and suction side are increased.

The changes to the shape of the stator blade are shown in Fig. 13. The leading edge is slightly humped up while the rest of the profile is more significantly humped down. The midspan area shows the biggest changes.

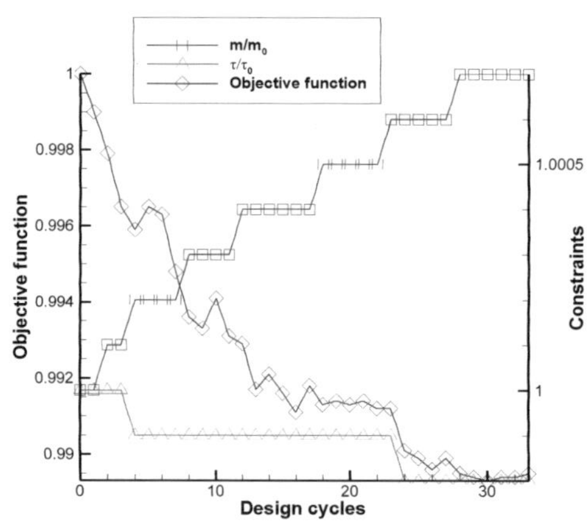

Figure 6. Convergence of objective function and constraints

	Isentropic Efficiency (%)	Log-dec (%)	Mass flow rate (kg/s)	Pressure ratio
Original	86.79	2.81	16.95	1.5939
Optimized	86.8	2.92	16.96	1.5935
Change	+0.01%	+0.11%	-0.07%	-0.07%

Table 3. Final performance values after convergence

2) Stage "aeroelastic-only" optimization

Table 4 shows the final values of the design optimization, where convergence is reached nearly after 35 design cycles (Fig. 7). The objective function drops its value below 0.7, showing an improvement of 52% for the rotor blade log-dec, which increases its value up to 4.28%. Despite, the entropy generation rate is not controlled by the objective function; the improvement in aeroelasticity is also increasing the isentropic efficiency by 0.48%, up to 87.21. The mass flow and pressure ratio changes are kept at reasonable levels by the constraints.

The local work contours at the rotor blade surface for the first mode are shown in Figs. 8-9. The log-dec improvement comes from the stable gradients being significantly increased on both sides, while the unstable area on the suction side is also decreased.

The local Mach number contours at midspan are shown in Fig. 10. The redesign strategy shown by the adjoint method is to weaken the shock-wave and push it downstream. The pressure coefficient distribution at midspan (Fig. 11) shows a new blade loading distribution with a better diffusion rate.

The geometry changes to both stator and rotor blades are shown in Figs. 12-13. The rotor is the blade that shows the largest changes. The optimised rotor profile is humped up after midchord and slightly humped down before midchord. The largest changes are located around 25% span and 75% chord. The stator blade only shows a small negative (down) humping spanwise, near the trailing edge. This change is much smaller compared to the stator-only case.

	Isentropic Efficiency (%)	Log-dec (%)	Mass flow rate (kg/s)	Pressure ratio
Original	86.79	2.81	16.95	1.5939
Optimized	87.21	4.28	16.91	1.592
Change	+0.42%	+1.47%	-0.23%	-0.12%

Table 4. Final performance values after convergence

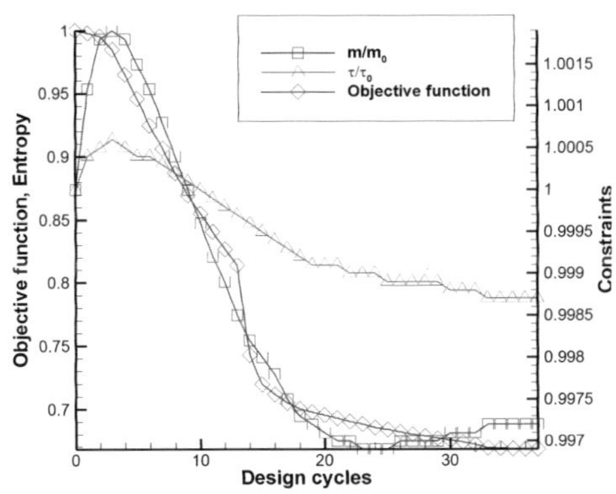

Figure 7. Convergence of objective function and constraints

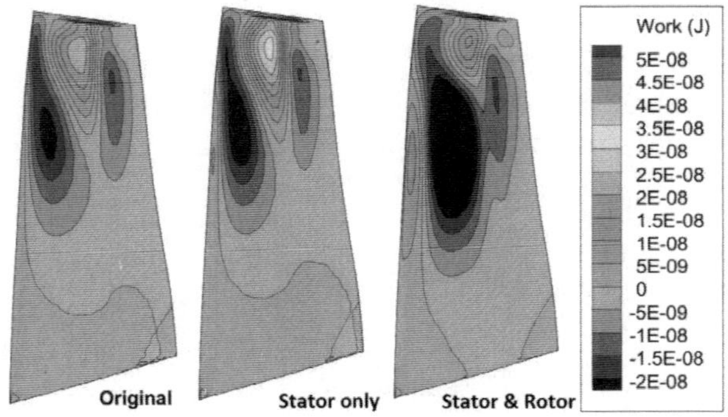

Figure 8. Local work sum contours on rotor blade surface (first mode). Pressure side

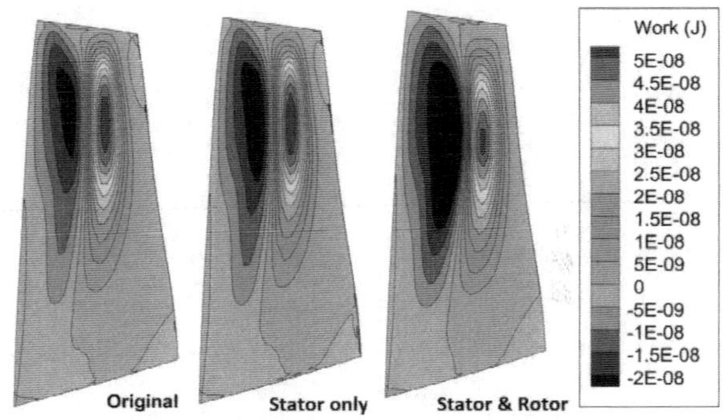

Figure 9. Local work sum contours on rotor blade surface (first mode). Suction side

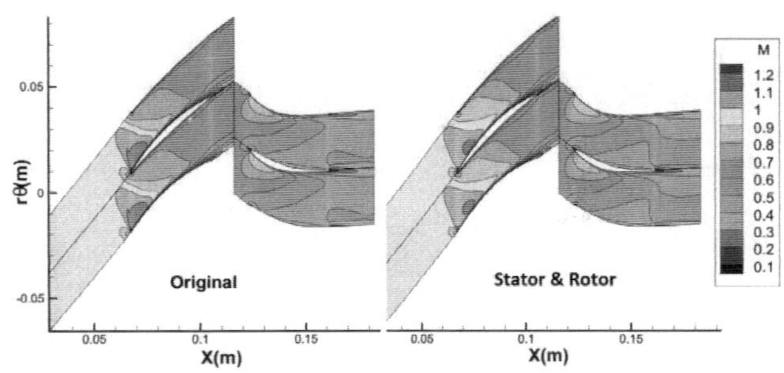

Figure 10. Local Mach number contours at 50% span

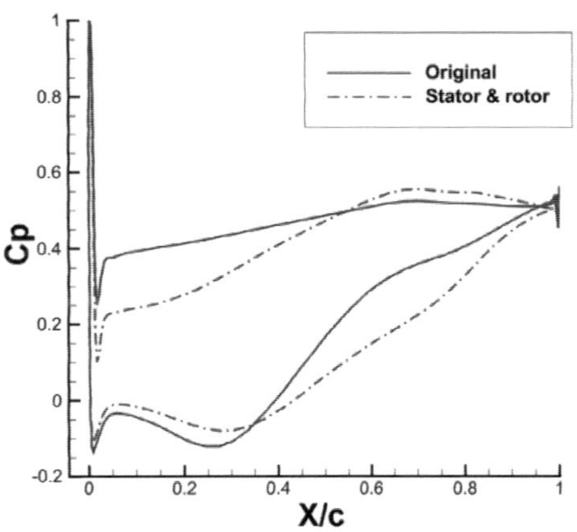

Figure 11. Rotor blade surface pressure coefficient at 50% span

Suction side

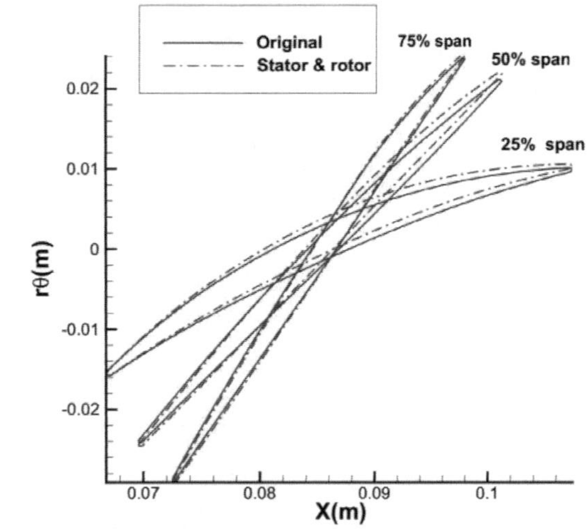

Figure 12. Blade profile geometry comparison. Rotor blade

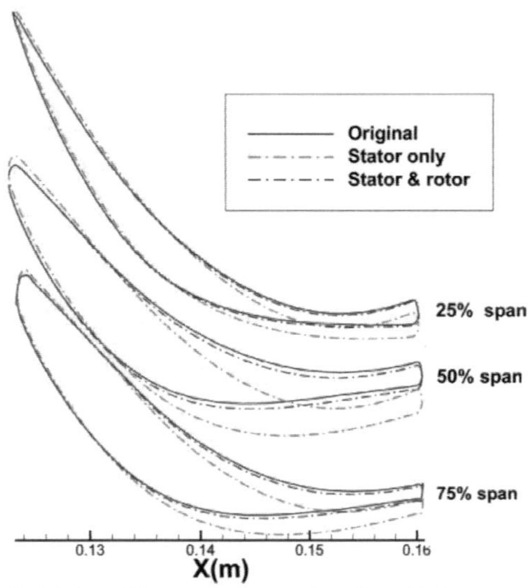

Figure 13. Blade profile geometry comparison. Stator blade

CONCURRENT DESIGN OPTIMIZATION: LOW-PRESSURE TURBINE STAGE

An "aeroelastic-only" and a concurrent stage design optimization of a low-pressure turbine stage are presented next and compared to show the strengths of a concurrent objective function.

3) Stage "aeroelastic-only" optimization

The aeroelastic-only design optimization converges after 35 design cycles as shown in Fig. 14 and the objective function decreases at a similar rate as the rotor-only case. Table 5 shows a final value of rotor blade log-dec 30% larger than the baseline design value. The entropy generation, however, is now increased due to the lack of entropy generation constraints, as opposed to the DLR compressor case. The mass flow is now kept constant but the pressure ratio increases to a maximum change of 0.22%.

	Isentropic Efficiency (%)	Log-dec (%)	Mass flow rate (kg/s)	Pressure ratio
Change	-0.23%	+1.15%	-0.05%	+0.22%

Table 5. Final performance values after convergence

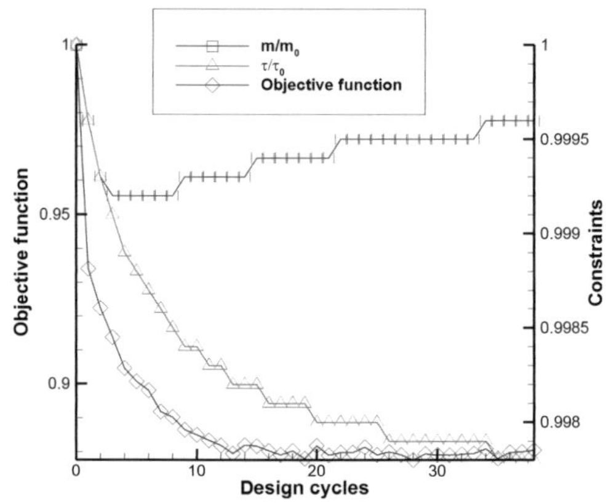

Figure 14. Convergence of objective function and constraints

4) Stage concurrent aeroelastic optimization

The last design optimization case is performed with a fully concurrent aeroelastic objective function. Hence, the entropy generation rate constraint is enabled ($\sigma_3 > 0$). The concurrent design optimization takes a similar time to converge as the aeroelastic-only case (Fig. 15). The final values are given in Table 6. The rotor blade log-dec change leads to 8% less improvement compared to the aeroelastic-only case. However, the entropy constraint is now able to control the entropy generation rate and as a result, the isentropic efficiency is almost unchanged compared to the original design. The other two constraints keep both the mass flow rate and pressure ratio at similar levels as the original design.

	Isentropic Efficiency (%)	Log-dec (%)	Mass flow rate (kg/s)	Pressure ratio
Change	-0.02%	+0.84%	-0.11%	+0.09%

Table 6. Final performance values after convergence

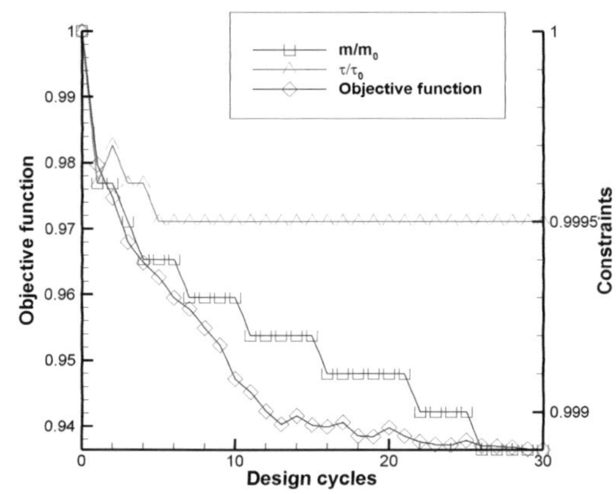

Figure 15. Convergence of objective function and constraints

Comparison

A concurrent design optimization improves performance in rotor blade log-dec with a much better control of the entropy generation rate (Fig. 16) and its impact in the stage efficiency. The local work sum contours for the first mode on both blade surfaces are shown in Figs. 17-18. On the pressure surface the stable gradient for both cases seems to be weaker with respect to the original contour. However on the suction surface it can be seen how the unstable gradient for the two new designs is weakened and the stable gradient increases its strength immediately after the shock. The margin of aeroelastic improvement is greater for the aeroelastic-only design.

In terms of the aerodynamic performance, the aeroelastic-only case increases the local peak Mach number at the stator passage, which has a negative effect on the passage flow losses as shown in Fig. 19. However, the concurrent case keeps the local Mach numbers to almost the same level at the stator passage. Regarding the rotor passage, the pressure coefficient distribution at midspan is shown in Fig. 20. Near the throat, both the aeroelastic-only and concurrent designs show a higher loading and higher peak velocity with a much steeper deceleration until the trailing edge; which increases the flow losses in the passage. However, the concurrent case reduces the loading at the leading edge, thus reducing the secondary flows and incidence losses and compensating for the increased peak velocity at the throat.

The changes to the blade geometry are compared in Figs. 21-22. The magnitude of the shape perturbation is twice as large for the aeroelastic-only design as for the concurrent design, with a smaller upwards humps at the leading edge of the rotor. The changes in the stator, however, are very similar for both cases and smaller in magnitude when compared to the rotor blade shape changes.

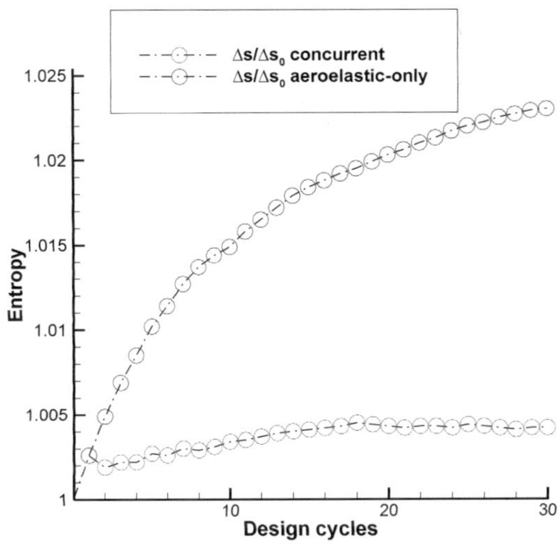

Figure 16. Entropy generation rate over design cycles

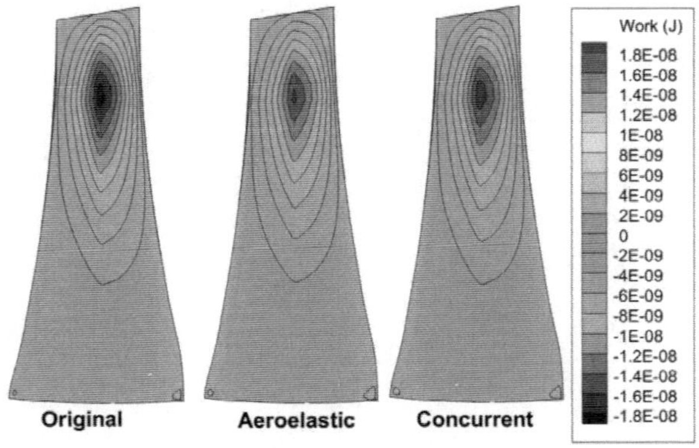

Figure 17. Local work sum contours on rotor blade surface (first mode). Pressure side

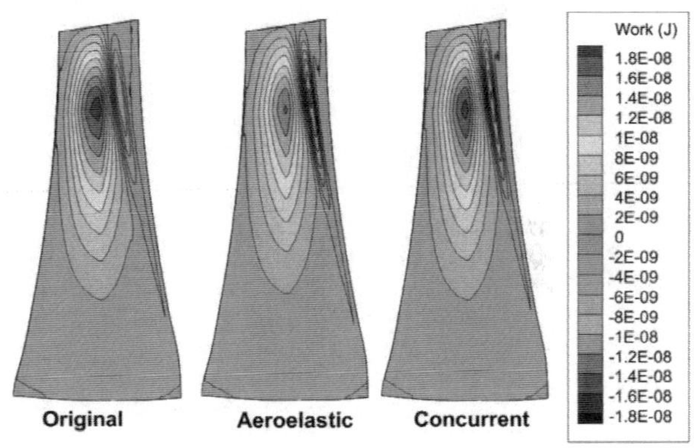

Figure 18. Local work sum contours on rotor blade surface (first mode). Suction side

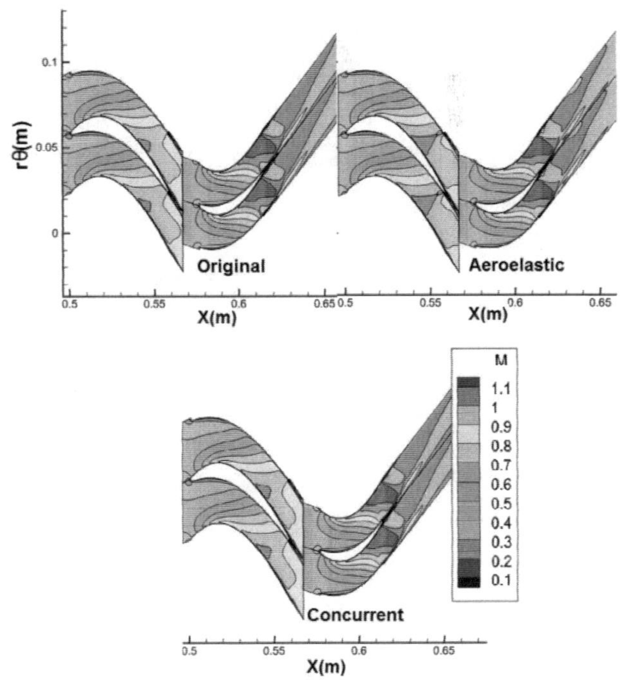

Figure 19. Local Mach number contours at 50% span

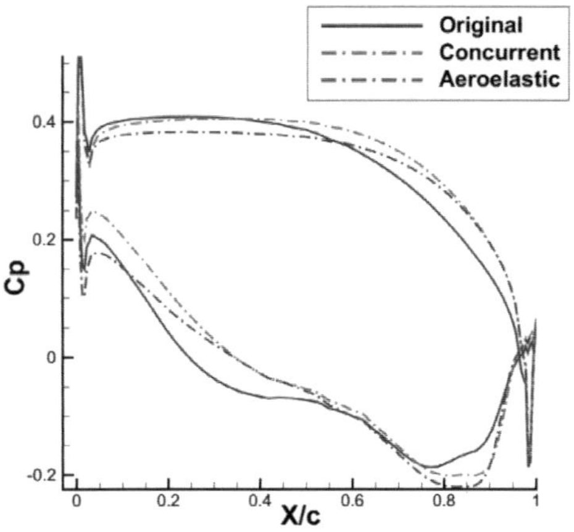

Figure 20. Rotor blade surface pressure coefficient at 50% span

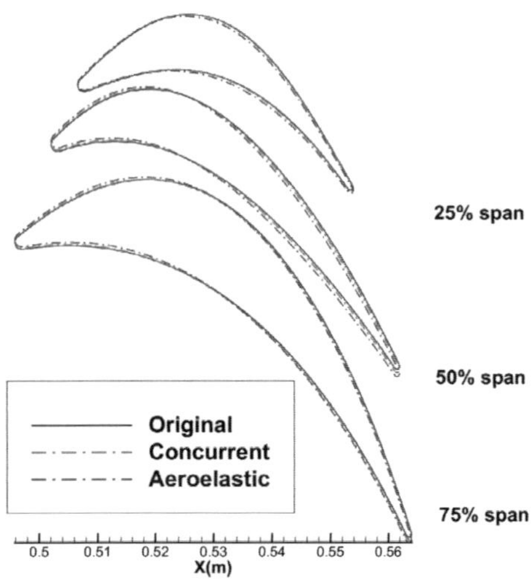

Figure 21. Blade profile geometry comparison. Stator blade

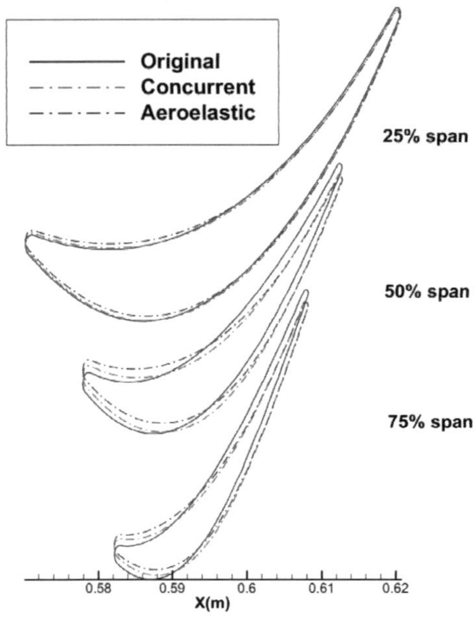

Figure 22. Blade profile geometry comparison. Rotor blade

CONCLUSIONS

A new system of unsteady adjoint equations is formulated based upon the Nonlinear Harmonic Phase Solution Method and the continuous approach. This system shows better the relation between the flow field and adjoint field in the source terms and periodic boundary conditions and extends the use of a harmonic interface treatment to perform multi-disciplinary and multi-component adjoint optimization.

Different case studies are presented which has demonstrated the effectiveness of the new adjoint harmonic interface treatment and the potential of the stator to influence the aeroelastic performance of adjacent rotor blades. The effect of the aero-elastic improvement on the aerothermal efficiency is also different from case to case. The strengths of a concurrent design optimization have been demonstrated controlling the negative impact on aerodynamics while increasing flutter stability.

ACKNOWLEDGMENTS

The authors would like to thank Siemens Industrial Turbomachinery Ltd for their feedback and support, funding this work. Special thanks are also due to M. Rahmati.

REFERENCES

[1]. L. He, Harmonic solution of unsteady flow around blades with separation. *AIAA Journal,* 46, pp 1299-1307, 2008

[2]. Rahmati, M. T. and He, L. and Wang, D. X. and Li, Y. S. and Wells, R. G. and Krishnababu, S. K., "Nonlinear time and frequency domain methods for multirow aeromechanical analysis", *ASME Journal of Turbomachinery* 136, 4 (2013), pp. 041010.

[3]. He, L., "An Euler solution for unsteady flows around oscillating blades", *ASME Journal of Turbomachinery* 112 (1989), pp. 714-722.

[4]. Hall, K. and Lorence, C., "Calculation of three-dimensional unsteady flows in turbomachinery using the linear harmonic Euler equations", *ASME Journal of Turbomachinery* 115, 4 (1993), pp. 800-809.

[5]. He, L. and Ning, W., "An efficient approach for analysis of unsteady viscous flows in turbomachines", *AIAA Journal* 36, 11 (1998), pp. 2005-2012.

[6]. Chen, T. and Vasanthakumar, P. and He, L., "Analysis of unsteady blade row interaction using nonlinear harmonic approach", *Journal of Propulsion and Power* 17, 3 (2001).

[7]. Hall, K. and Thomas, J. and Clark, W., "Computation of unsteady nonlinear flows in cascade using a harmonic balance technique", *AIAA Journal* 40, 5 (2002), pp. 879-886.

[8]. Jameson, A., "Aerodynamic design via control theory", *Journal of Scientific Computing* 3, 3 (1988), pp. 233-260.

[9]. Jameson, A., "Computational aerodynamics for aircraft design", *Journal of Science* 245 (1989), pp. 361-371.

[10]. Mohammadi, B. and Pironneau, O., "Shape optimization in fluid mechanics", *Annual Review of Fluid Mechanics* 36 (2004), pp. 255-279.
[11]. Yang, S. and Wu, H. and Liu, F. and Tsai, H., "Aerodynamic design of cascade by using an adjoint equation method", *AIAA Paper* 03-1068 (2003).
[12]. Wu, H. and Yang, S. and Liu, F., "Comparison of three geometric representation of airfoils for aerodynamic optimization", *AIAA Paper* 2003-4095 (2003).
[13]. Wang, D. X. and He, L., "Adjoint aerodynamic design optimization for blades in multi-stage turbomachines: Part I - methodology and verification", *ASME Journal of Turbomachinery* 132, 2 (2010), pp. 021011.
[14]. Wang, D. X. and He, L., "Adjoint aerodynamic design optimization for blades in multi-stage turbomachines: Part II - validation and application", *ASME Journal of Turbomachinery* 132, 2 (2010), pp. 021012.
[15]. Walther, B. and Nadarajah, S, "Optimum shape design for multirow turbomachinery configurations using a discrete adjoint approach and an efficient radial bas...", *ASME Journal of Turbomachinery* 137, 8 (2015), pp. 081006.
[16]. Walther, B. and Nadarajah, S, "Constrained adjoint-based aerodynamic shape optimization of a single-stage transonic compressor", *ASME Journal of Turbomachinery* 135, 2 (2012), pp. 021017.
[17]. Duta, M. C. and Giles, M. B. and Campobasso, M. S., "The harmonic adjoint approach to unsteady turbomachinery design", *International Journal for Numerical Methods in Fluids* (2002), pp. 323-332.
[18]. Thomas, J. and Hall, K. and Dowell, E., "A discrete adjoint approach for modelling unsteady aerodynamic design sensitivities", *AIAA Paper* 03-0041 (2003).
[19]. Huang, H. and Ekici, K., "A discrete adjoint harmonic balance method for turbomachinery shape optimization", *Aerospace Science and Technology* 39 (2014), pp. 481-490.
[20]. He, L. and D. X. Wang, "Concurrent blade aerodynamic-aero-elastic design optimization using adjoint method", *ASME Journal of Turbomachinery* 133, 1 (2011), pp. 011021-011021-10.
[21]. Li, H. D. and He, L., "Blade aerodynamic damping variation with rotor-stator gap: a computational study using single-passage approach", *ASME Journal of Turbomachinery* 127, 3 (2005), pp. 573-579.
[22]. Valero, O. and He, L. and Y. S. Li, "Concurrent blade aerodynamic-aeroelastic design optimization with re-scaled response surface", *ASME Paper* GT2014-25854 (2014).
[23]. Vatsa, V., "Computation of sensitivity derivatives of Navier-Stokes equations using complex variables", *NASA Technical Report* (2004).
[24]. Giles, M. B. and Pierce, N., "Adjoint equations in CFD: duality, boundary conditions and solution behaviour", *Flow, Turbulence and Combustion* 65, 3/4 (2000), pp. 393-415.
[25]. Moffat, S. and He, L., "Blade forced response prediction for industrial gas turbine Part I: Methodologies", *ASME Paper* (2003).
[26]. He, L., "Unsteady Flow and Aeroelasticity", *Cambridge Turbomachinery Course Notes* (2012).
[27]. Dunker, R. and Rechter, H. and Starken, H. and Weyer, H., "Redesign and performance analysis of a transonic axial compressor stator and...", *Journal of Engineering for Gas Turbines and Power* (1984), pp. 279-287

Proceedings of the 14th International Symposium on Unsteady Aerodynamics, Aeroacoustics & Aeroelasticity of Turbomachines
ISUAAAT14
8-11 September 2015, Stockholm, Sweden

I14-S9-4

AEROELASTIC BEHAVIOUR OF THREE AND HALF STAGE COMPRESSOR ROTOR BLADES WITH FORIEGN OBJECT IN ENGINE INLET

Romuald RZADKOWSKI
Institute of Fluid-Flow Machinery
Polish Academy of Sciences
Fiszera 14, PL 80-231 , Gdańsk, Poland
z3@imp.gda.pl

Vitally GNESIN, Luba KOLODYAZHNAYA
Department of Unsteady Gasdynamics and Aeroelasticity
Institute of Mechanical Engineering Problems
National Academy of Sciences of Ukraine
2/10 Pozharsky st., Kharkov,61046 Ukraine
gnesin@ipmach.kharkov.

ABSTRACT

A bird strike can cause damage to stationary and rotating aircraft engine parts, especially the engine fan. This paper presents a bird strike simulated by blocking four stator blade passages in the engine inlet. It includes the numerical results of the unsteady low-frequency aerodynamic forces and the aeroelastic behaviour, caused by a non-symmetric upstream flow, affecting the first three rotor blade stages in the axial-compressor of a jet engine.

An in-house code was used to calculate the unsteady non-viscous flow of ideal gas through a 3D model of a 3.5 stage compressor. The forces were analyzed in operating conditions, using a partially blocked inlet.

INTRODUCTION

The foreign object (e.g. a bird) was characterized in tests as a water-like hydrodynamic response by Wilbeck and Rand [1]. Storace et al. [2] developed a computer program to predict structural response caused by soft body impact. Commercial codes are now available for such impact studies; Rao and Srinivas [3] used LS Dyna for a bird impact on fan blades, whereas Bianchi [4] used RADIOSS for a bird strike on a helicopter blade and rotor control chain. Heidari, Carlson and Yantis [5] used a nonlinear transient analysis for a propulsion system during bird strike induced fan blade loss. Experiments were carried out on SO-3 compressor first stage rotor blades [6]. A crack was initiated by placing rectangular blocks on the stator blades, simulating birds engulfed in the engine.

The Fluent code was used to analyze the low frequency excitation caused by inlet blockage of 1.5 stages of an SO-3 engine, but it did not take into account rotor blade vibration [7,8].

The aeroeleastic behaviour of rotor blades in 1.5- and 2.5-stage compressors was analyzed by Rzadkowski et al. [9-10].

In this paper the unsteady forces acting on rotor blades of a 3.5 stage compressor were calculated using the 3D non-viscous flow and an in-house code, taking into account rotor blade vibration..

NOMENCLATURE

- p pressure
- T temperature
- ν natural mode frequency
- F_{1i},F_{2i} Fourier coefficients
- $\overline{L}$ dimensionless blade length
- F_y F_z circumferential and axial aerodynamic forces
- M aerodynamic moment
- h_y h_z circumferential and axial blade section displacements
- φ rotating angle
- S0, S1, S2, S3 stator blade rows
- R1, R2, R3 rotor blade rows

AERODYNAMICAL MODEL

Blade failures in the first stage of SO-3 compressor were reported in the years 1975-1991 [6]. Experiments were carried out on a first stage rotor blade in the compressor of an SO-3 engine at the Air Force Institute of

Technology in Warsaw to initiate a crack by placing rectangular blocks on the stator blades, which in real life could be caused by birds engulfed in the engine (see Fig. 1).

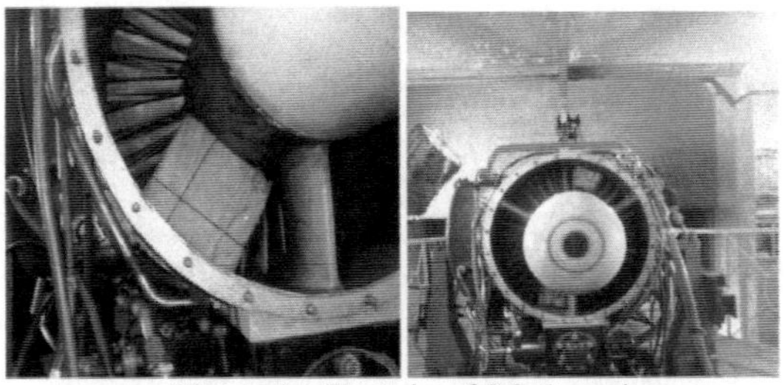

Figure 1a: Test rig of SO-3 engine

The 3D transonic flow of an inviscid non-heat conductive gas through the first compressor stages of an SO-3 aircraft engine (Fig. 1a). The physical domain, including the nozzle cascade (S0), the rotor wheel (R1) rotating with constant angular velocity the nozzle cascade (S1).

A 3D whole-annulus model of the first stage of a SO-3 jet engine compressor is shown in Fig. 1b. The reference rotor blade in R1 cascade is divided into 10 cross-sections.

In order to model the block in the inlet of the engine (see Fig. 1), four stator blade passages was blocked..

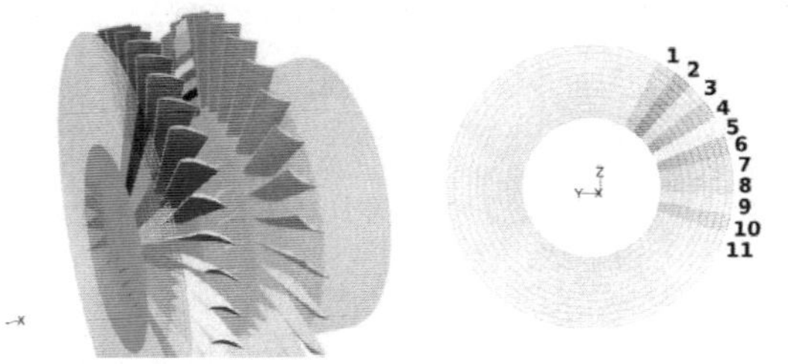

Figure 1b: Computational model of the first stage (stator-rotor-stator)

The 3D transonic flow of an non-viscous, non-heat conductive gas through the 3.5-stage compressor of an aircraft engine was simulated (Fig. 1c). A 3D whole-annulus model of the 3.5-stage jet engine compressor is shown in Fig. 1d. The model consists of : 44 blades in the inlet stator cascade (S0), 28 in the rotor cascade (R1), 33 in the stator cascade of the first stage (S1), 42 (R2), 44 (S2), 42(R3) and 66(S3). The numerical calculations were carried out using computational H-grids: 10*40*84 grid points for the S0 passage, 10*60*78 for the R1 passage, 10*48*68 for the S1 passage, 10*44*64 for the R2 passage, 10*42*62 for the S2 passage, 10*44*62 for the R3 passage and 10*28*60 for the S3 passage. In order to model the engine inlet partial admission, four stator blade passages were blocked (Fig. 1b). Figures 2a, 2b show the tangential grids in the peripheral sections of rotor R1 and stator S1.

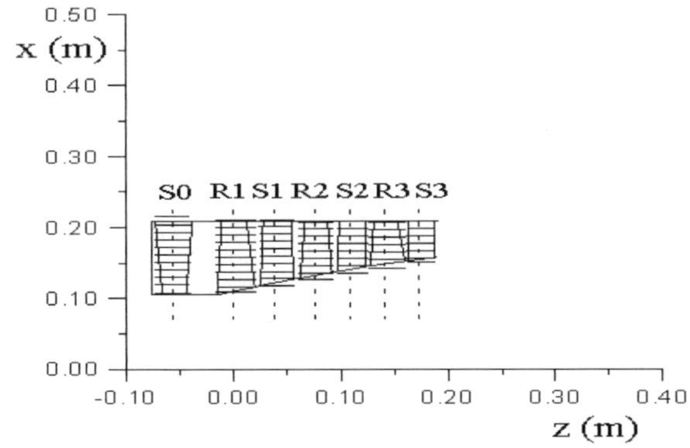

Figure 1c. The meridional section of a compressor

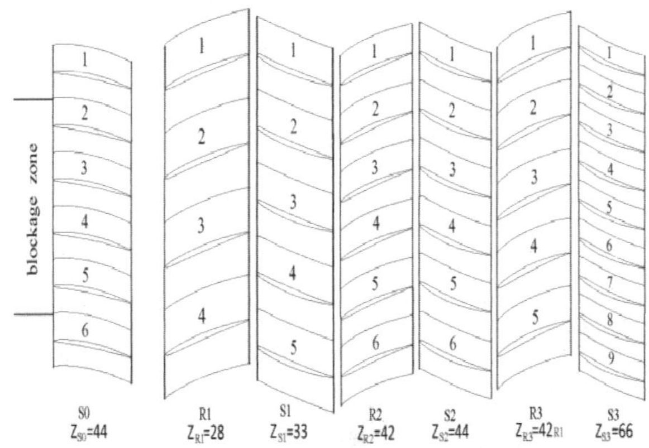

Figure 1d. 3.5 stage compressor

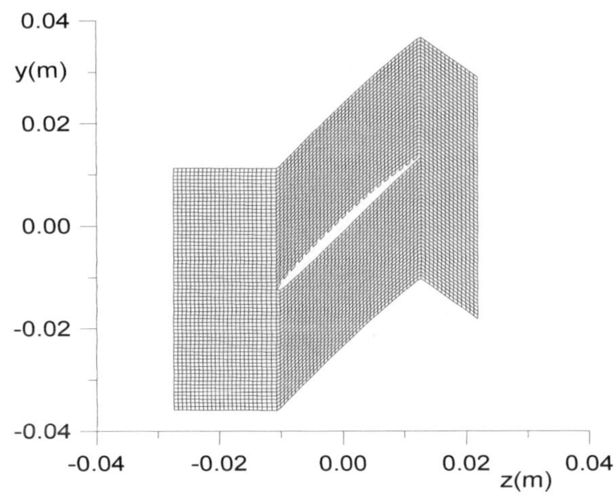

Figure 2a. Difference grid in the peripheral sections of rotor and stator blades R1

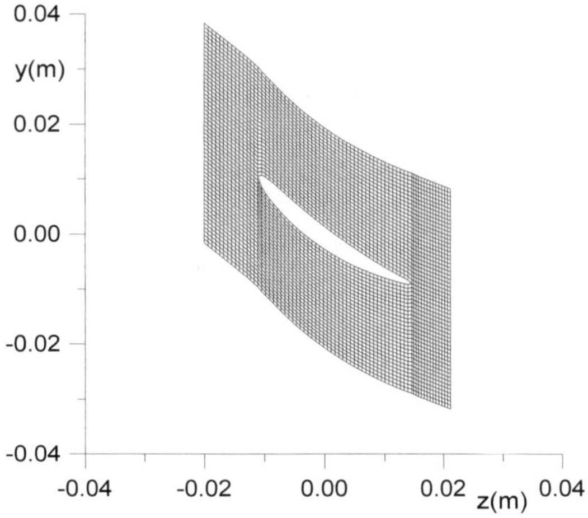

Figure 2b. Difference grid in the peripheral sections of stator blades S1

The spatial transonic flow (including, in the general case, strong discontinuities in the form of shock waves and wakes behind the exit edges of blades) is written in a relative Cartesian coordinate system rotating with the constant angular velocity ω in accordance with full non-stationary Euler equations, and presented in the form of integral conservation laws of mass, impulse and energy [11]. A discrete form of Euler equations is obtained from a arbitrary deformable grid using the Godunov-Kolgan difference scheme with 2nd order approximation of coordinates and time [11]. Each of the passages is meshed, using an H-type grid for the stator domain and a hybrid H-H grid for the rotor domain ([11]). Here the outer H-grid of the rotor remains stationary during the calculation, while the inner H-grid is rebuilt in each iteration by a given algorithm, so that the external points of the inner grid remain stationary, but the internal points (on the blade surface) move according to the blade motion. The blade vibration formulation is based on a modal approach of the fluid structure coupled problem. The blade displacement and velocity obtained from the dynamic problem are used as boundary conditions in an aerodynamic problem at each iteration.

In the general case, when axial velocity is subsonic at the inlet boundary, the initial values of total pressure, total temperature and flow angles are used, while at the outlet boundary only static pressure is imposed. Non-reflecting boundary conditions, i.e., incoming waves (three at inlet, one at the outlet) must be used and suppressed, which is accomplished by setting their time derivative to zero.

On the blade's surface, because the grid moves with the blade, the normal relative velocity is set to zero

$$(v - w)\, n=0. \quad (1)$$

STRUCTURAL MODEL

The blade vibration formulation is based on a modal approach of the coupled problem ([12], [13]). The dynamic model of the oscillating blade in linearized formulation is governed by matrix equation:

$$M\ddot{u}+C\dot{u}+Ku=F, \quad (2)$$

where *M*, *C*, *K* are the mass, mechanical damping and stiffness matrices of the blade respectively; *u(t)* is the blade displacement; *F* is the unsteady aerodynamic forces vector, which is a function of blade displacement.

The first step of the modal approach consists of solving the problem of the natural mode shapes and eigenvalues without damping and in a vacuum. Then the displacement of each blade can be written as a linear combination of the first *N* modes shapes with the modal coefficients depending on time:

$$u=Uq=\sum_{i=1}^{N} U_i q_i, \quad (3)$$

where U_i is the displacement vector corresponding to i- th mode shape; $q_i(t)$ is the modal coefficient of i- th mode. Taking into account the equation (3) and the orthogonality property of the mode shapes the equation (2) can be written in form of:

$$I\ddot{q}+H\dot{q}+\Omega q=\lambda(t), \quad (4)$$

where I=diag (1, 1, ... 1), H=diag (2 h_1, 2 h_2, ... 2 h_n), Ω =diag (ω_1^2, ω_2^2, ... , ω_n^2) are diagonal matrices; ω_i is i-th natural blade frequency; $\{\lambda(t)\}$ is the modal forces vector corresponding to the mode shapes, $h_i = \omega_i \xi_i$, where ξ_i is the i-th modal damping coefficient ([12]). Thus the dynamic problem (2) reduces to the set of independent differential equations relatively to modal coefficients of natural modes ([10], [11]):

$$\ddot{q}_i+2h_i\dot{q}_i+\omega_i^2 q_i=\lambda_i, \quad (5)$$

The equations of motion (5) can be solved using any standard integration method.

The modal forces λi are calculated at each iteration with the use of the instantaneous pressure field in the following way ([11]):

$$\lambda_i=\frac{\iint_{\sigma} p\overline{U}_i \cdot \overline{n}^{\circ}\, d\sigma}{\iiint_{v} \rho\overline{U}_i^2\, dv}. \quad (6)$$

where: p is the pressure along the blade surface.

NUMERICAL RESULTS

The numerical analysis was performed in a nominal regime under the parameters:

- number of rotor revolutions per minute n=15400 rpm;
- full pressure at the inlet P_o =101000 Pa;
- full temperature at the inlet T_o=288 K;

- static pressure at the outlet p_2=140000-149000 Pa.

The blade vibration was defined by taking into account the first five natural mode shapes of rotating R1, R2 and R3 blades. The values of natural frequencies are given in Table 1. Mechanical damping was not considered. Nevertheless, aerodynamic damping was part of the calculation process [11].

The calculations were performed for two versions: with full and partial admission.

Table 1. Natural frequencies of rotating compressor rotor blades R1, R2, R3

Natural mode	1	2	3	4	5
ν, Hz Rotor 1	540	1620	2160	3240	4320
ν, Hz Rotor 2	670	2010	2680	4020	6030
ν, Hz Rotor 3	1194	3580	4780	7165	8360

The investigation of unsteady characteristics was performed using the Fourier analysis

$$F(t) = F_0 + \sum_{i=1}^{\infty} F_{1i} \cos(2\pi\nu i t) + F_{2i} \sin(2\pi\nu i t),$$

where: F(t) is a physical unsteady characteristic; F_o is the averaged value; F_{1i}, F_{2i} are the Fourier coefficients; *i* is the harmonic number; ν is the 1st harmonic frequency.

Presented below is a comparative analysis of unsteady aerodynamic loads and blade oscillation amplitudes for rotor R1 with partial and full admission.

The average values of unsteady loads (circumferential and axial forces and aerodynamic moment) acting on a rotor R1 blade's root, mid and peripheral sections are presented in Tables 2a and b. The average values of circumferential forces were the highest in the root, whereas the axial forces and moments were the highest in the peripheral blade cross-section.

Table 2a. The averaged values of unsteady loads acting on R1 rotor blade root, mid and peripheral sections (partial admission)

$\bar{L}$	Fy, N	Fz, N	M, N*m
0.05	-5.79	-0.673	0.0091
0.5	-5.47	-2.364	0.0244
0.95	-3.20	-2.772	0.0338

The amplitude-frequency characteristics of the unsteady aerodynamic circumferential forces acting on the peripheral layer of a R1 rotor blade with partial and full admission are presented in (Fig. 6a, b). In the case of partial admission, the 6th low-frequency harmonic of the circumferential R1 force was the highest in the peripheral cross-section.

Table 2b. The averaged values of unsteady loads acting on R1 rotor blade root, mid and peripheral sections (full admission)

$\bar{L}$	Fy, N	Fz, N	M, N*m
0.05	-5.83	-0.646	0.0097
0.5	-5.39	-2.33	0.0232
0.95	-3.050	-2.631	0.0312

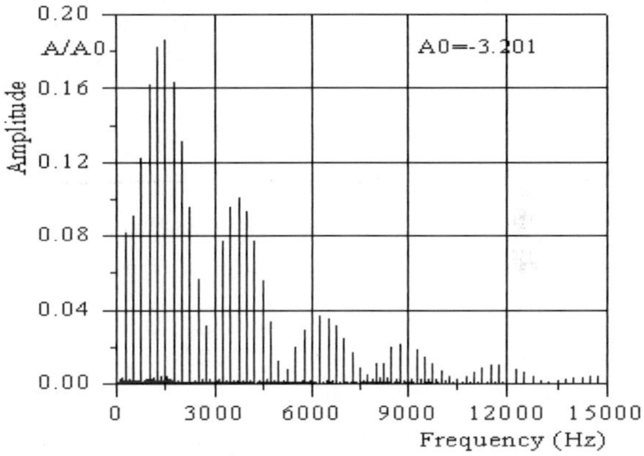

Figure 6a. Amplitude-frequency spectrum **of** the circumferential force acting on the R1 rotor blade with partial admission (peripheral layer)

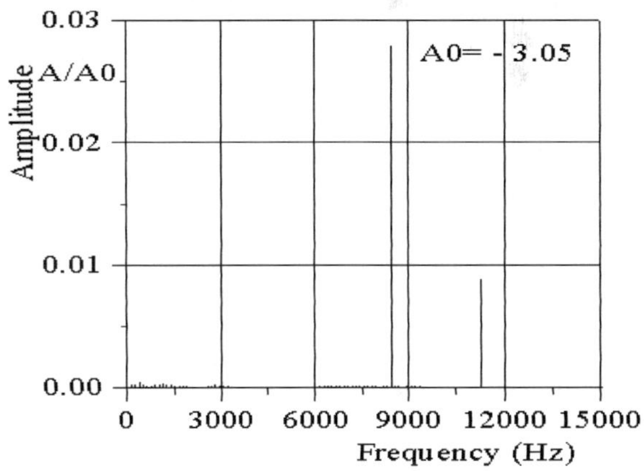

Figure 6b. Amplitude-frequency spectrum **of** the circumferential force acting on the R1 rotor blade with full admission (peripheral layer)

Fig. 6a shows that the main components of unsteady aerodynamic forces acting on R1 rotor blades with partial admission are the harmonics of the multiples of 256 Hz, namely, 1536 Hz, 4096 Hz and 6144 Hz. Harmonic amplitudes corresponding to the travelling frequency of stator S0 ($\nu_{rev} \times z_{s0} = 256 \times 44 = 11264\ Hz$) and stator S1 ($\nu_{rev} \times z_{s1} = 256 \times 33 = 8448\ Hz$) are negligible in relation to the low frequency components. The average values of aerodynamic loads under full admission were nearly the same as under partial admission, but pulsation amplitudes were five times as large, and the main contribution came from harmonics with frequencies equal to the travelling frequencies of stators S0 and S1.

Figure 7 shows unsteady circumferential forces acting on the peripheral layer of R2 rotor blades. The 3rd low-frequency harmonic of circumferential R2 force was the highest. The amplitudes of low frequency harmonics caused by partial admission were smaller than the loads acting on the R1 rotor blades. However, it is important to note that in the R2 blades low frequency excitation also appeared. The maximal harmonic amplitudes under partial admission were four times larger than under full admission.

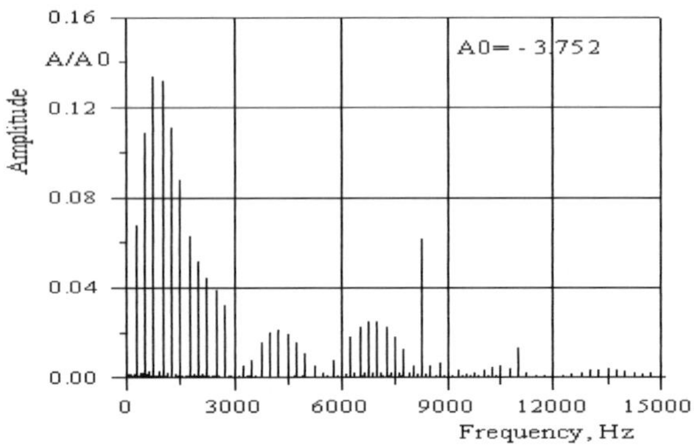

Figure 7. Amplitude-frequency spectrum of the circumferential force acting on the R2 rotor blade with partial admission (peripheral layer)

Figure 8 shows unsteady circumferential forces acting on the peripheral layer of R3 blades. The 4th low-frequency harmonic of the circumferential R3 force was the highest in the peripheral layer. The amplitudes of low frequency harmonics caused by partial admission were smaller than the loads acting on the R1 blades. As in the case of R2 blades, low frequency excitation also appeared in the R3 blades, which is a very significant finding with regard to blade life. The maximal harmonic amplitudes under partial admission were four times larger than under full admission.

The blade oscillations caused by unsteady aerodynamic loads are presented in form of blade peripheral section displacements for rotor R1, R2 and rotor R3 in Fig. 9 -12. Figure 9 presents the peripheral bending oscillations of rotor R1 under partial admission as well as their amplitude-frequency characteristics. The main contribution to the bending oscillations was brought about by the 1st natural mode with a frequency closed to the 1st natural frequency. The high frequency forced oscillations were negligible. The frequency in the flow was 450 Hz. The unsteady bending of hy (circumferential) components was 82% of the average displacements. The amplitudes of torsion vibration were 4000% at 1500 Hz (Fig. 10). Therefore, blocking the inlet caused a greater decrease of rotor blade vibration frequency in the flow than when the inlet was unblocked. In both cases the form of vibration was bending-torsion and different to the 1st mode shape.

The average values of blade deformation under full and partial admission were practically the same, but partial admission caused an up to fivefold blade amplitude increase and included torsional components.

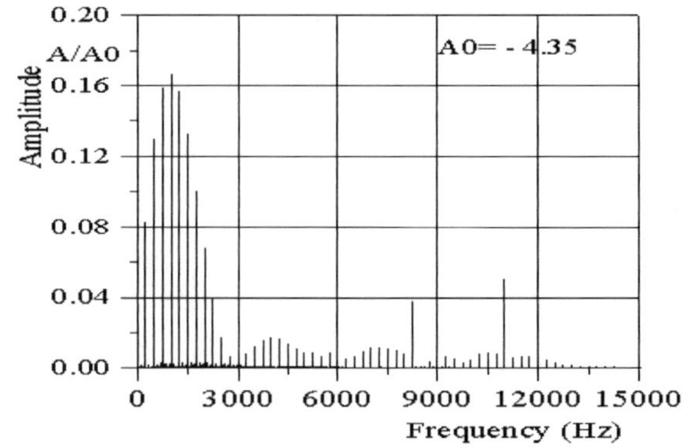

Figure 8. Amplitude-frequency spectrum of the circumferential force acting on the R3 rotor blade with partial admission (peripheral layer)

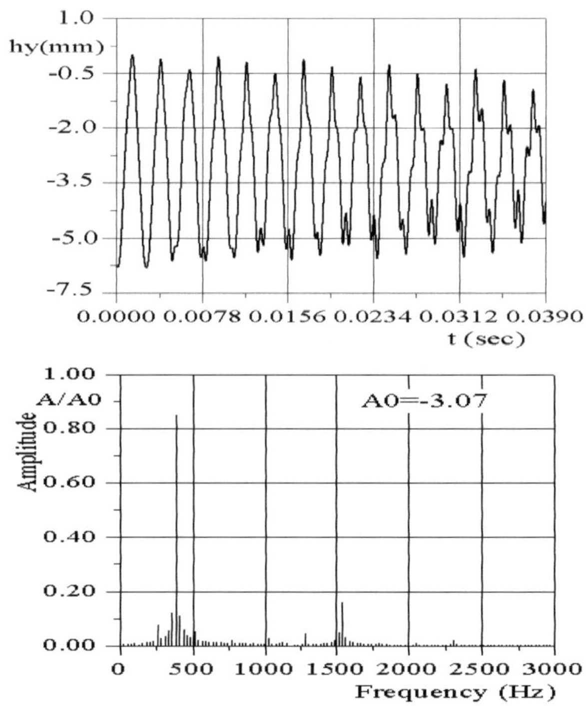

Figure 9. Peripheral blade section displacement in a circumferential direction (rotor R1), a – displacement under partial admission; b - amplitude frequency spectrum under partial admission

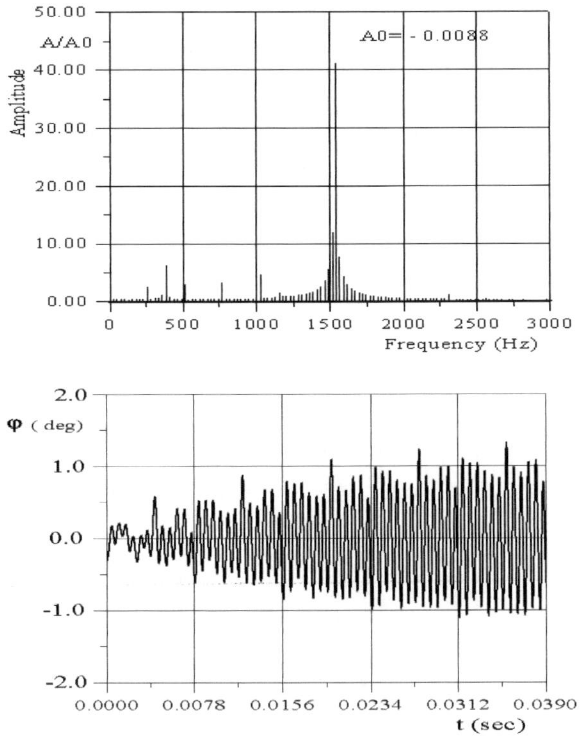

Figure 10. Peripheral blade section torsion (rotor R1), a – angle under partial admission; b - amplitude frequency spectrum under partial admission;

The bending oscillation of blade peripheral sections in rotor R2 are shown in Figures 11 and 12. The averaged values of blade deformation under partial admission and full admission were practically the same, but partial admission increased the blade amplitudes almost threefold and included torsional components.

The first natural frequency of R2 blades was 670 Hz (Tab. 1), and the frequency in the flow decreased to 500 Hz. The unsteady bending of hy (circumferential) components was 66% of the average displacements. The amplitudes of torsion vibration were 75% of the average twist angle at 450 Hz and 50% at 1500 Hz. R2 blades with a blocked inlet vibrated with a complex bending–twisting form, bending components at 500 Hz and twisting components at 250 Hz, 450 Hz, 750 Hz, 1100 Hz and 1500Hz. This form of vibration was different to the 1st mode shape of the R2 blade.

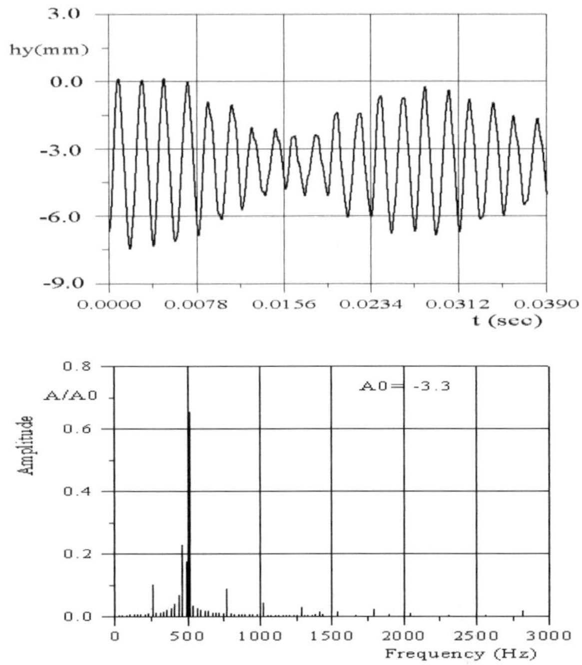

Figure 11. Peripheral blade section displacement in a circumferential direction (rotor R2), a – displacement under partial admission; b - amplitude frequency spectrum under partial admission

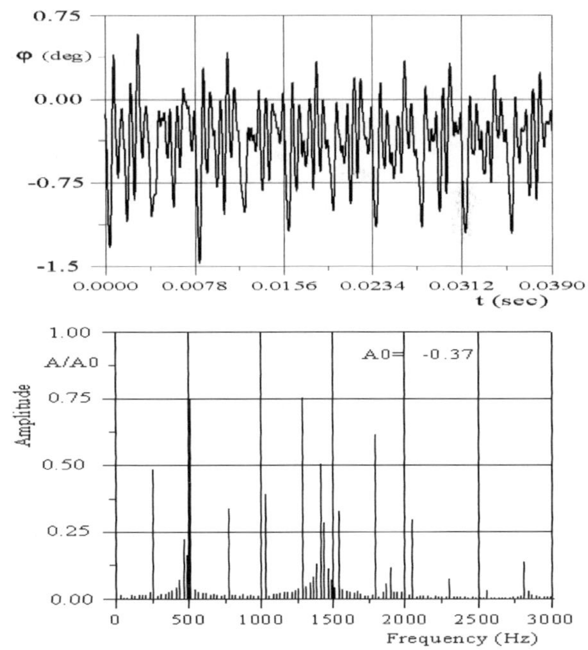

Figure 12. Peripheral blade section torsion (rotor R2), a – angle under partial admission; b - amplitude frequency spectrum under partial admission;

The bending oscillation of blade peripheral sections in rotor R3 are shown in Figures 13 and 14. The averaged values of blade deformation under partial admission and full admission were practically the same, but partial admission increased the blade amplitudes almost threefold and included torsional components.

The first natural frequency of R3 blades was 1194 Hz (Tab. 1), and the frequency in the flow decreased to 700 Hz. The unsteady bending of hy components was 66% of the average displacements. The amplitudes of torsion vibration were 129% of the average twist angle at 250 Hz, 90% at 500 Hz, 140% at 750 Hz and 90% at 2500 Hz. R3 blades with an blocked inlet vibrated with a complex bending–twisting form. This form of vibration was different to the 1st mode shape of the R3 blade.

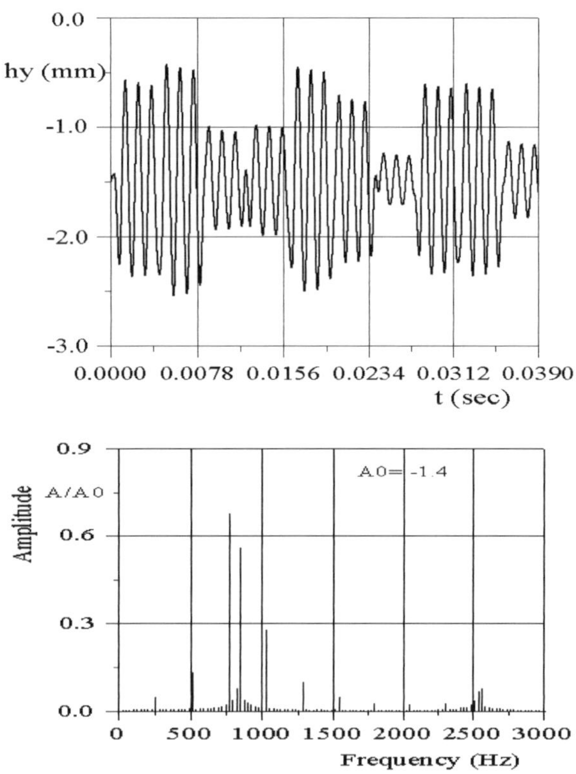

Figure 13. Peripheral blade section displacement in a circumferential direction (rotor R2), a – displacement under partial admission; b - amplitude frequency spectrum under partial admission

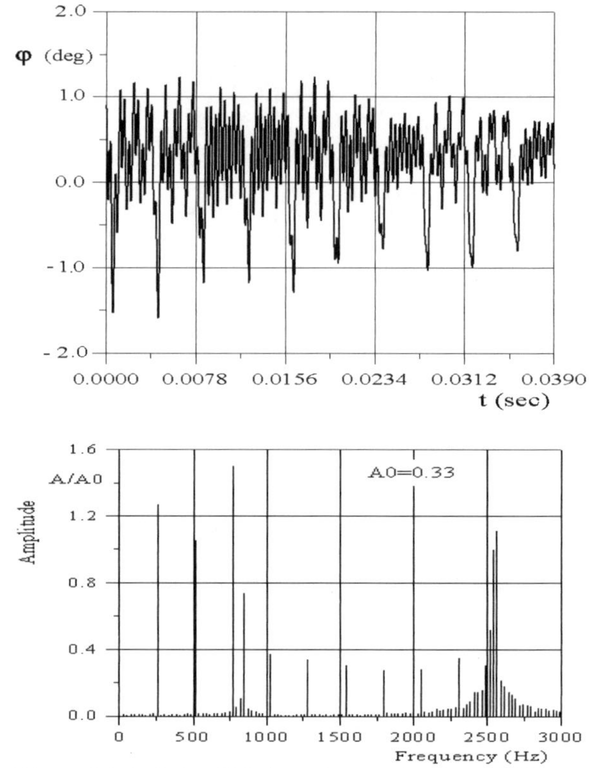

Figure 14. Peripheral blade section torsion (rotor R2), a – angle under partial admission; b - amplitude frequency spectrum under partial admission;

CONCLUSIONS

This paper has presented the calculation results for unsteady low- and high-frequency aerodynamic forces acting on R1, R2 and R3 blades in a 3.5-stage jet engine compressor, taking into account blade vibration, in the case of a bird strike. The forces were analyzed in two operating conditions: with an unblocked and partially blocked engine inlet. The obtained results show that disturbances in the engine inlet strongly influence the level of unsteady forces acting on the rotor blades (in the 1st, 2nd and 3rd stages). With a partially blocked inlet a whole spectrum of low-frequency harmonics was observed, including the natural frequencies of blades.

With the inlet partially blocked, R1, R2 and R3 blade vibrations differed from their first mode shape and included torsional components.

This paper has for the first time shown that partial admission causes low-frequency harmonics to affect not only the 1st and 2nd rotor blade stages, but also the 3rd stage, with only slightly smaller amplitude values. This indicates that under partial admission low-frequency harmonics will also affect subsequent stages.

ACKNOWLEDGMENTS

The authors wish to acknowledge NCBiR (project PBS1/B4/5/2012) for the financial support of this work. All numerical calculations were made at the Academic Computer Centre TASK (Gdansk, Poland).

REFERENCES

[1] Wilbeck, J. S. and Rand, J. L., The Development of a Substitute Bird Model, Journal of Engineering for Power, ASME, vol. 103, p. 725, 1981.

[2] Storace, A. F., Nimmer R. P. and Ravenhall, R., Analytical and Experimental Investigation of Bird Impact on Fan and Compressor Blading. Journal of Aircraft, Vol. 21, July 1984.

[3] Rao, J. S. and Srinivas, S., Bird Impact on Rotating Fan Blades, 2nd Indian LS-DYNA Users Conference, Chennai, Feb 20-21, 2003.

[4] Bianchi, F., Analysis with RADIOSS of a bird strike onto a helicopter blade and onto a rotor control chain, Altair HTC09, Ludwigsburg 2nd-4th November 2009.

[5] Heidari, M. A., Carlson, D. L. and Yantis, T., Rotor-dynamics Analysis Process, p.3601, Worldwide Aerospace Conference & Technology Showcase, April 8 - 10, Toulouse, 2002.

[6] Szczepanik R., Experimental Analysis of Rotor Blades of Aircraft Engines in Various Operating Conditions, DSc Thesis, ITWL Warsaw, (in Polish) 2010.

[7] Rządkowski R., Soliński M., Szczepanik R., The Unsteady Low-Frequency Aerodynamic Forces Acting on the Rotor Blade in the First Stage of an Jet Engine Axial Compressor, Advances in Vibration Engineering, 11(2), 193-204, 2012.

[8] Soliński M., Rzadkowski R., Pająk A., Szczepanik R., The Unsteady Low-Frequency Forces Acting on the Rotor Blade in the First Stage on an Axial Compressor of SO-3 Jet Engine, Proceedings of the 8th International Conference on Vibration Engineering and Technology of Machinery, VETOMAC –VIII, p. 277-278, Gdansk, Sept 3-6, 2012.

[9] Rzadkowski R, Gnesin V., Kolodyazhnaya L., Szczepanik R.: Aeroelastic Behaviour of First Stage Compressor Rotor Blades with Foreign Object in Engine Inlet, *ISUAAAT'2012, 13TH International Symposium on Unsteady Aerodynamics, Aeroacustics and Aeroelasticity of Turbomachines*, Tokyo, September 11-14, S4-3, 2012.

[10] Rzadkowski R., Gnesin V., Kolodyazhnaya L., Szczepanik R.: The Unsteady Low-Frequency Aerodynamic Forces Acting on Rotor Blades in the First Two Stages of a Jet Engine Axial Compressor in the Case of a Bird Strike, 10th European Conference on Turbomachinery, Fluid Dynamics and Thermodynamics, 15-19 April 2013 Lappeenranta, Finland, Editor J. Backman, G. Bois, O. Leonard, Conference Proceedings, 222-234, 2013.

[11] Gnesin V., Kolodyazhnaya L, Rzadkowski R., A Numerical Model of Stator-Rotor Interaction in a Turbine Stage with Oscillating Blades, Journal of Fluids and Structures, 19(8), 1141-1153, 2004.

[12] Bathe K., Wilson E, Numerical Methods in Finite Element Analysis, Prentice-Hall, Inc Englewood Cliffs, New Jersey, 1976.

[13] Rządkowski, R., Dynamics of Steam Turbine Blading, Part Two Bladed Discs, Ossolineum, Wrocław-Warszawa, 1998.

Proceedings of the 14th International Symposium on Unsteady Aerodynamics, Aeroacoustics & Aeroelasticity of Turbomachines
ISUAAAT14
September 8-11, 2015, Stockholm, Sweden

ISUAAAT14-S11-1

THEORETICAL ANALYSIS OF THE UNSTEADY AERODYNAMICS OF VIBRATING AIRFOILS AT LOW REDUCED FREQUENCY

Roque Corral*
Advanced Engineering Direction
Industria de TurboPropulsores S.A.
28108 Madrid, Spain
e-mail: roque.corral@itp.es

Almudena Vega
School of Aeronautics
Universidad Politécnica de Madrid
28040 Madrid, Spain
e-mail: almudena.vega@upm.es

ABSTRACT

This paper studies the unsteady aerodynamics of vibrating airfoils in the low reduced frequency regime with special emphasis on its impact on the scaling of the work per cycle curves using an asymptotic approach. It is found that both the loading of the airfoil and the symmetry of the cascade play an essential role in the trends of the phase and modulus of the unsteady pressure caused by the vibration of the airfoil. For lightly loaded airfoils, the unsteady pressure and the influence coefficients scale linearly with the reduced frequency whereas the phase departs from $\pi/2$ and changes linearly with the reduced frequency. As a consequence, the work-per-cycle curves scale linearly with the reduced frequency and then their stability does not change with the reduced frequency. For highly loaded airfoils, the unsteady pressure modulus is fairly constant exhibiting only a small correction with the reduced frequency, while the phase varies linearly with it and then the mean value of the work-per-cycle scales linearly with the reduced frequency while the first harmonic is constant. On the other hand for symmetric cascade configurations, independently of whether they are made up of loaded or unloaded airfoils, it is found that the work-per-cycle curves follow exactly the same trend than for an unloaded case.

*Also associate professor at the Department of Fluid Dynamics and Engine Propulsion of the School of Aeronautics, UPM

Introduction

The unsteady pressure field caused by airfoil vibration plays a major role in the flutter and forced response of turbomachinery blade rows. Flutter is a self excited aeromechanic instability by which energy is extracted from the main stream and transferred to the structure. When the energy dissipated by the structure is smaller than that taken out from the flow, the vibration amplitude grows, leading eventually to either an instantaneous or a high cycle fatigue (HCF) failure of the structure, depending on the mechanical damping level. On the other hand, blades and vanes undergo periodic excitations due to the relative motion of neighboring blades. The vibration caused by these excitations may also provoke an HCF failure if the resulting vibration amplitude is large enough. It is well known that the largest response of the structure takes place at resonant conditions, when the frequency of the excitation matches one of the natural frequencies of the system. Under these conditions the vibration amplitude is directly proportional to the excitation force and inversely proportional to the overall damping of which a significant part may be due to the air stream in which the airfoil is immersed. In both phenomena, the unsteady pressure field induced by the airfoil vibration and its subsequent interaction with the structure are essential in determining the aerodynamic damping and, therefore, the life of the component.

Aerodynamics of vibrating airfoils is equally important for both flutter and forced response. However the range of reduced frequencies, k, i.e.:, the ratio between the flow-through or residence time of the fluid, $t_r = c/v_c$ and the angular frequency of the vibration, ω, may be quite different. While flutter typically occurs at the lowest modes of the assembly, forced response is usually of interest at higher modes since resonant crossings with the first modes are avoided whenever possible by design, an exception of this case being the analysis of crossings at large off-design conditions. As a consequence the reduced frequency of a flutter analysis is, generally speaking, much lower than that of a forced response study.

Aeronautical low pressure turbines (LPTs) are made of very slender and thin airfoils because their weight and cost have a large impact on the engine (about 20% of the total weight and 15% of the total cost). Natural frequencies of LPT rotor blades are very low exhibiting very low reduced frequencies ($k \sim 0.1 - 0.2$) as well and, as a consequence, LPT bladed-discs are prone to flutter. Nowadays flutter may become a dominant constraint on the design of modern LPTs precluding the use of more efficient aerodynamic configurations [1, 2]. Compressor airfoils have a much lower aspect ratio than that of the LPTs and usually the reduced frequency of their lowest modes is of order unity ($k \sim 1.0$). However, it is still interesting to study their behavior in the $k \ll 1$ limit as a means to increase the overall understanding of vibrating airfoils at low frequency, and in particular the sensitivity to the airfoil shape.

The unsteady aerodynamics associated with the vibration of turbine and compressor bladed-discs and stator vanes is nowadays routinely analyzed within the design loop of the aeroengine companies and it has been the subject of dedicated experiments in the past [3, 4, 5]. The aim of these simulations is the derivation of the aerodynamic stability of the rotor blades and the quantification of the aerodynamic damping, which is the result of the application of the unsteady pressures on the airfoil displacements.

With a few exceptions in the early days of aeromechanics [6], little attention has been paid to the understanding of vibrating airfoil aerodynamics since this is not a figure of merit in itself for the aeroelastic analyst. Designers obtain and use aerodynamic data from numerical tools, but they do not usually examine critically the unsteady flow-field since they are really just interested in the global aerodynamic damping. As a direct consequence of using aerodynamic codes as black boxes, little understanding has been gained in recent years about the physics of these type of flows and its impact on the aerodynamic damping. Although there are several trends that are well known, such as the stabilization with the reduced frequency, the physics which is behind this behavior is not really well understood and the statement lacks of a sound theoretical support. Another example is the damping curves as a function of the inter-blade phase angle (IBPA) of LPTs and compressors, that look quite different without a clear rational of which is the underlying reason for this. On the other hand, vibrating cascade experiments are scarce, and performed well apart in time in different institutions [7] with different techniques and, therefore, are not well suited for conceptual studies. Recently, however, there is a renewed interest in increasing the understanding of the problem fundamentals [8, 9], especially in the context of turbine flutter which is considered the simplest flutter instability of all.

This paper first reviews the mathematical dependence of the unsteady aerodynamics with the reduced frequency and conducts an asymptotic expansion of the governing equations in the low reduced frequency limit. In the analysis, the displacements of the airfoil are given and are supposed in practice to correspond to a structural mode-shape or a combination of several. Therefore, mode-interaction of any kind has to be taken into account at a different level since here we just study the unsteady aerodynamic field due to a given arbitrary motion of an airfoil. Finally practical conclusions derived purely from the theoretical analysis are drawn. The numerical verification of the theory is carried out in [10].

NOMENCLATURE

c	Axial chord
h.o.t.	higher order terms
IBPA	Inter Blade Phase Angle
IC	Influence Coefficient
k	Reduced frequency
LPT	Low Pressure Turbine
M	Mach number
Re	Reynolds number
s	Pitch
t	Time
TW	Traveling-Wave
$\mathbf{U}$	Vector of conservative variables
v_c	Characteristic velocity
$\tilde{\mathbf{W}}$	Non-dimensional velocity vector
δ	Vector of the airfoil vibration amplitude
ϕ	Unsteady pressure phase
θ	Stagger angle
ω	Angular frequency
S	Blade surface
p'	Unsteady pressure
W_{cycle}	Work-per-Cycle per span unit

GOVERNING EQUATIONS

The time-dependent Navier-Stokes equations of an ideal gas in conservative and compact form can be written as:

$$\frac{\partial \mathbf{U}}{\partial t} + \nabla.\mathbf{F}_c(\mathbf{U}) = \nabla.\mathbf{F}_v(\mathbf{U}) \quad (1)$$

where $\mathbf{U} = \{\rho, \rho\mathbf{v}, \rho E\}^T$ is the vector of conservative variables and $\mathbf{F_c}$ and $\mathbf{F}_v$ are the convective and viscous fluxes respectively. Without loss of generality the simplest boundary conditions for a two-dimensional airfoil cascade could be written as: $P_t = P_0$, $T_t = T_0$ and $\alpha = \alpha_{in}$ at the inlet; $p = p_{exit}$ at the exit, null velocity, $v = 0$, and heat flux, $\partial T/\partial n = 0$, are prescribed at the airfoil's solid walls while periodic boundary conditions are enforced at the lateral boundaries of the airfoil passage.

The above equations may be non-dimensionalised using $p_c = P_0$, $T_c = T_0$ and $v_c^2 = 2c_p T_0(1-(p_{exit}/P_0)^{(\gamma-1)/\gamma})$ as the characteristic pressure, temperature and velocity respectively. The characteristic velocity used here is the isentropic exit velocity that may be readily computed from the boundary conditions. The airfoil chord, c, is used as the characteristic length, L_c. Then, choosing the inverse of ω as the characteristic time, $t_c = \omega^{-1}$, Eq. 1 can be written in non-dimensional form as:

$$k\frac{\partial \tilde{\mathbf{U}}}{\partial \tau} = -\tilde{\nabla}.\mathbf{F}_c(\tilde{\mathbf{U}}) + \frac{1}{Re}\tilde{\nabla}.\mathbf{F}_v(\tilde{\mathbf{U}}) = \tilde{\mathbf{R}}(\tilde{\mathbf{U}}) \quad (2)$$

where $\tau = \omega t$ is the non-dimensional time, $\tilde{x} = x/c$ the non-dimensional distance and $\tilde{\mathbf{U}} = \{\tilde{\rho}, \tilde{\rho}\tilde{\mathbf{W}}, \tilde{\rho}\tilde{E}\}^T$ the vector of non-dimensional conservative variables, with $\tilde{\rho} = \rho/\rho_0$, $\tilde{p} = p/P_0$ $\tilde{\mathbf{W}} = \mathbf{v}/v_c$ and $E = \frac{1}{2\gamma}\tilde{T}/(1-\Pi_R^{-(\gamma-1)/\gamma}) + \frac{1}{2}\tilde{\mathbf{W}}^2$. The non-dimensional boundary conditions become $\tilde{P}_t = 1$, $\tilde{T}_t = 1$ and $\alpha = \alpha_{in}$ at the inlet and $\tilde{p} = \Pi_R^{-1}$ at the exit where $\Pi_R = P_0/p_{exit}$ is the pressure ratio of the row or the cascade. The non-dimensional solution of the problem depends on:

$$\tilde{\mathbf{U}} = f(\tilde{\mathbf{x}}, \tau, \Pi_r \text{or } M_{exit} Re.k, \gamma,) \quad (3)$$

where Re is the Reynolds number based on the characteristic velocity and the chord and k, is the so-called reduced frequency:

$$k = \frac{\omega c}{v_c} \quad (4)$$

which represents the ratio between the flow-through time and the characteristic time of the motion, or the inverse of the frequency if the motion is harmonic.

For a non-vibrating airfoil the no-slip boundary condition at the solid wall reduces to $\tilde{\mathbf{W}} = 0$ (or $\tilde{\mathbf{W}}.\mathbf{n} = 0$ if the flow is inviscid). However for moving walls this boundary condition has to be generalized to specify that the velocity of the fluid and the airfoil are the same at the airfoil surface, $\mathbf{x_c}(\xi, t)$. In non-dimensional form the non-slip boundary condition is:

$$\tilde{\mathbf{W}}(\tilde{\mathbf{x}}_c, \tau) = k\frac{d\tilde{\delta}}{d\tau} \quad (5)$$

where $\tilde{\delta} = \tilde{\mathbf{x}}_\mathbf{c} - \tilde{\mathbf{x}}_{c0}$ is the displacement of the airfoil with respect a reference position, $\mathbf{x}_{c0}(\xi)$, typically the initial or the mean position of the airfoil if the motion is harmonic (see Fig. 1). The reduced frequency appears both in the governing equations and in the boundary condition. It is important to note that (i) if $k \ll 1$ then at the wall $\tilde{\mathbf{W}} \simeq 0$, exactly as in a non-vibrating airfoil, and that the time derivatives may be neglected in first approximation in the governing equations, (ii) the no-slip boundary condition (Eq. 5) is a Lagrangian description of the problem since it is specified at the moving boundary, $\mathbf{x}_c$.

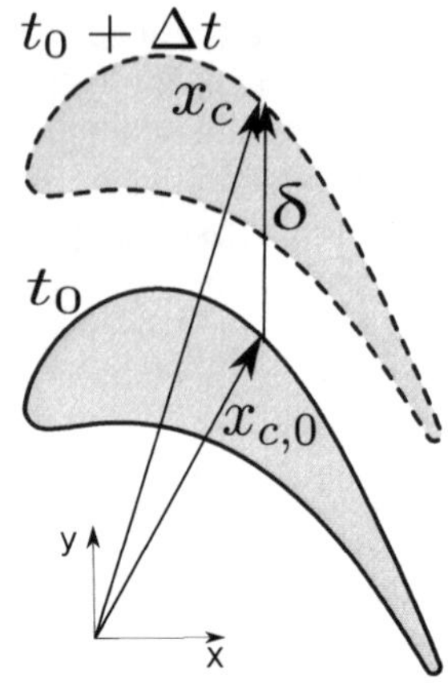

Figure 1. POSITION OF THE AIRFOIL IN A SPECIFIC TIME INSTANT

ASYMPTOTIC ANALYSIS FOR LOW K

Linearization of the Governing Equations

The non-dimensional governing equations (Eq.2) and their boundary conditions can be simplified if it assumed that the solution may be decomposed as the sum of a mean baseline flow, $\widetilde{\mathbf{U}}_0(\tilde{\mathbf{x}})$, plus a small unsteady perturbation, $\widetilde{\mathbf{u}}(\tilde{\mathbf{x}}, \tau)$:

$$\widetilde{\mathbf{U}}(\tilde{\mathbf{x}}, \tau) = \widetilde{\mathbf{U}}_0(\tilde{\mathbf{x}}) + \widetilde{\mathbf{u}}(\tilde{\mathbf{x}}, \tau) \quad (6)$$

such that $\widetilde{\mathbf{u}} \ll \widetilde{\mathbf{U}}_0$. It may be shown that this approximation is valid if the vibration amplitude, $\|\delta\|$, is much smaller than the airfoil's chord, i.e.: $\|\delta\| \ll c$, and highly non-linear phenomena, such as shock-waves and separation bubbles, are not present in the mean flow-field. Under these hypotheses the Navier-Stokes equations may be linearized about the mean flow to yield:

$$k\frac{\partial \widetilde{\mathbf{u}}}{\partial \tau} + A_0(\widetilde{\mathbf{U}}_0)\frac{\partial \widetilde{\mathbf{u}}}{\partial \mathbf{x}} = 0 \quad (7)$$

where $A_0 = (\partial \mathbf{R}/\partial \mathbf{U})_0$ is the Jacobian matrix and $\tilde{\mathbf{U}}_0$ satisfies $\tilde{\mathbf{R}}(\widetilde{\mathbf{U}}_0) = 0$. The linearization of the wall boundary conditions deserves special attention and it is addressed in the following sub-section.

Linearization of the non-slip boundary condition

It has been previously mentioned than the airfoil's non-slip boundary condition (Eq. 5) is expressed in Lagrangian form because it is imposed in a moving boundary. Linearized Navier-Stokes codes designed to deal with vibrating airfoils usually implement the no-slip boundary condition in a moving grid exactly as specified in Eq. 5 (see for instance [11, 12]).

However for analytical studies this form hides relevant information. When $k \ll 1$ the condition reduces to $\tilde{\mathbf{W}} \simeq 0$ at $\tilde{\mathbf{x}} = \tilde{\mathbf{x}}_c(\tau)$ concealing the fact that there is a quasi-steady displacement of the airfoil that makes the airfoil loading change with time. This trivial observation is not easy to explicitly state from a mathematical point of view within this formulation.

Alternatively, the non-slip boundary condition may be linearized and expressed in Eulerian form in the same way that was done by Verdon [13] in the resolution of the transonic unsteady potential equation. The non-dimensional fluid velocity field may be expressed as:

$$\tilde{\mathbf{W}}(\tilde{\mathbf{x}}, \tau) = \tilde{\mathbf{W}}_0(\tilde{\mathbf{x}}) + \tilde{\mathbf{w}}(\tilde{\mathbf{x}}, \tau) \tag{8}$$

where $\tilde{\mathbf{W}}_0(\tilde{x})$ represents the mean velocity and $\tilde{\mathbf{w}}(\tilde{x}, \tau)$ the perturbed velocity, with $\tilde{\mathbf{w}} \ll \tilde{\mathbf{W}}_0$. The no-slip boundary condition (Eq. 5) can then be written as:

$$\tilde{\mathbf{W}}_0(\tilde{\mathbf{x}}_{c0} + \tilde{\delta}) + \tilde{\mathbf{w}}(\tilde{\mathbf{x}}_{c0} + \tilde{\delta}, \tau) = k \cdot \frac{d\tilde{\delta}}{d\tau} \tag{9}$$

The boundary condition in the actual position of the airfoil may be referred to its mean or initial position, $\tilde{\mathbf{x}}_{c0}$, if $\left\|\tilde{\delta}\right\| \ll 1$, expanding $\tilde{\mathbf{W}}$ in Taylor series, neglecting second-order terms and recalling that $\tilde{\mathbf{W}}_0(\tilde{\mathbf{x}}_{c0}) = 0$. In this case the previous equation reduces to:

$$\nabla\tilde{\mathbf{W}}_0\Big|\Big|_{x=x_{c0}} \cdot \tilde{\delta} + \tilde{\mathbf{w}}(\tilde{\mathbf{x}}_{c0}) = k \cdot \frac{d\tilde{\delta}}{d\tau}$$

and the boundary condition for the linearised non-dimensional problem is simply:

$$\tilde{\mathbf{w}}(\tilde{\mathbf{x}}_{c0}) = k \cdot \frac{d\tilde{\delta}}{d\tau} - \nabla\tilde{\mathbf{W}}_0\Big|\Big|_{x=x_{c0}} \cdot \tilde{\delta} \tag{10}$$

The important fact is that this boundary condition is imposed in the original undeformed airfoil and physically represents a blowing velocity in the wall boundary. Numerical codes that implement this type of boundary condition do not need to resort to the use of moving grids because they implicitly assume that the airfoil displacements are small.

Perturbation Analysis of the Linearized Equations

Different levels of approximation of the linearized Navier-Stokes equations may be obtained taking advantage of the smallness of k and assuming that the displacement of the airfoil and the flow are periodic in time, i.e.: $\delta = \hat{\delta} e^{i\omega t}$ and $\mathbf{u} = \hat{\mathbf{u}} e^{i\omega t}$.

0th Order Approximation If we assume that $k = 0$ the linearized system of equations reduces to:

$$A_0(\tilde{\mathbf{U}}_0)\frac{\partial \tilde{\mathbf{u}}_0}{\partial \mathbf{x}} = 0, \text{ with } \tilde{\mathbf{w}}_0(\tilde{\mathbf{x}}_{c0}) = -\nabla\tilde{\mathbf{W}}_0\Big|\Big|_{x=x_{c0}} \cdot \tilde{\delta} \tag{11}$$

where the characteristic amplitude of $\tilde{\mathbf{u}}_0$ is of order $\tilde{\mathbf{u}}_c \sim \left\|\tilde{\delta}\right\|$, which is small if $\nabla\tilde{\mathbf{W}}_0\Big|\Big|_{x=x_{c0}} \sim o(1)$. The pressure and the rest of the variables are in-phase with the displacement since physically are due to the perturbation of the base flow associated to the static displacement of the airfoil. In this approximation the time plays the role of a parameter. For a harmonic motion whose solution may be expressed in the complex domain the solution is real and the phase of the solution with the airfoil displacement, $\phi_0 = 0$.

1st Order Approximation The first order correction of the solution with k may be obtained if we express:

$$\tilde{\mathbf{u}}(\tilde{x}, \tau) = \tilde{\mathbf{u}}_0(\tilde{x}, \tau) + \tilde{\mathbf{u}}_1(\tilde{x}, \tau) \tag{12}$$

introducing the above expression in Eqs 7 and 10, we obtain in the frequency domain:

$$ik\hat{\mathbf{u}}_0 + A_0(\tilde{\mathbf{U}}_0)\frac{\partial \hat{\mathbf{u}}_1}{\partial \mathbf{x}} = 0, \text{ with } \hat{\mathbf{w}}_1(\tilde{x}_{c0}) = ik\hat{\delta} \tag{13}$$

where for the sake of simplicity we have omitted the tilde to indicate non-dimensional variables and assumed that the motion is harmonic in time and the displacement has only a real component, $\hat{\delta} = \hat{\delta}_R$, i.e.: it is the origin of the phases and we are dealing with a real mode. In this case $\hat{\mathbf{u}}_1$ is purely imaginary, $\hat{\mathbf{u}}_1 = \mathbf{u_{1I}} + \text{h.o.t.}$ The characteristic value of the correction $\mathbf{u}_{1c} \sim k\|\delta\|$ is much smaller than the characteristic value of the zero order approximation $\mathbf{u}_{0c} \sim \nabla\tilde{\mathbf{W}}_{0c} \cdot \|\delta\|$since $k \ll 1$. The 1st order correction of the phase ϕ_1 may be estimated as:

$$\tan\phi_1 \sim \frac{\mathbf{u}_{1Ic}}{\mathbf{u}_{0c}} \sim \frac{k(1 + \alpha\nabla\tilde{\mathbf{W}}_{0c})}{\nabla\tilde{\mathbf{W}}_{0c}} \tag{14}$$

which is small as well since $k \ll 1$, actually $\phi_1 \propto k$ if $\nabla\tilde{\mathbf{W}}_{0c} \sim o(1)$, exactly as it was shown by the authors in [8] for a LPT airfoil and it will be shown in [10] for other airfoils as well.

2nd Order Approximation The next approximation of the solution provides the correction with the frequency of the real part of the solution. As it will become apparent in the following section this correction provides further insight of the solution for lightly loaded airfoils of which the flat plate is the limit case. When the gradient of the velocity of the mean flow vanishes the quasi-steady motion of the airfoil does not cause any perturbation and the next approximation of the flow is needed.

Following a consistent perturbation analysis, the 2nd order approximation is given by:

$$-k\,\mathbf{u}_{1I}+A_0(\widetilde{\mathbf{U}}_0)\frac{\partial \mathbf{u}_{1R}}{\partial \mathbf{x}}=0, \text{ with } \hat{\mathbf{w}}_{1R}(\widetilde{\mathbf{x}}_{c0})=0 \qquad (15)$$

where it has been assumed that $\hat{\mathbf{u}}_1=\mathbf{u}_{1R}+i\mathbf{u}_{1I}$ and that $\mathbf{u}_{1I}$ is already known from the previous expansion. It can be observed that the time derivative is responsible of coupling the real and imaginary parts of the solution. The characteristic value of $\mathbf{u}_{1R}$ is, according with the previous equation, $u_{1Rc}\sim ku_{1Ic}\sim k^2\|\delta\|$. The 2nd order approximation of the phase, ϕ_2, is then:

$$\tan\phi_2\sim\frac{u_{1Ic}}{u_{0c}+u_{1Rc}}\sim\frac{k(1+\alpha\nabla\tilde{\mathbf{W}}_{0c})}{a\nabla\tilde{\mathbf{W}}_{0c}+bk^2} \qquad (16)$$

where a and b are real constants of order unity. It may be readily observed that for highly loaded airfoils (i.e.: $\nabla\tilde{\mathbf{W}}_{0c}\sim o(1)$) if $k\ll 1$ then $\phi\sim k$, which is one of the first conclusions of the analysis. However, for lightly loaded airfoils (i.e.: $\nabla\tilde{\mathbf{W}}_{0c}\simeq 0$) $\phi\simeq\pi/2-dk$ even if $k\ll 1$ (d is as well a real constant of order unity). This is a second major conclusion.

Estimate of the Unsteady Pressure

For highly loaded airfoils the leading term of the previous expansion is $\tilde{\mathbf{u}}_0$, which is real and in phase with the displacement, being its characteristic value $\tilde{\mathbf{u}}_{c0}\sim\|\delta\|\nabla\tilde{\mathbf{W}}_{0c}$. Since $\tilde{p}=p/P_0$ and $\tilde{p}$ varies from 1 at the inlet to Π_R^{-1} at the exit, the characteristic unsteady pressure is:

$$p'_{c0}\sim\frac{1}{2}\rho_c v_c^2\left(\frac{\delta_c}{c}\right)\nabla\tilde{\mathbf{W}}_{0c} \qquad (17)$$

since the perturbation of the flow is purely static and therefore the unsteady pressure is proportional to the dynamic head and the displacement.

Its first order correction with k is due to the airfoil vibration velocity and is associated to $\tilde{\mathbf{u}}_1$, whose characteristic value is $\tilde{u}_{1c}\sim k\|\delta\|$ and therefore the correction of the pressure perturbation due to the unsteady terms is of the order of

$$p'_{c0}\sim\frac{1}{2}\rho_c v_c^2\left(\frac{\delta_c}{c}\right)\left(\frac{\omega c}{v_c}\right)\sim\rho_c v_c\omega\delta_c. \qquad (18)$$

This second scaling is only relevant if either the steady aerodynamic loading is low or vanishes. Strictly speaking this is the order of the unsteady pressure when $\tilde{\nabla}\tilde{\mathbf{W}}_c\ll k$. This means that even for low reduced frequencies, i.e.: $k\ll 1$, if the pressure gradient, $\tilde{\nabla}\tilde{\mathbf{W}}_c$, is small enough, the value of the characteristic pressure is given by 18. This scaling is also valid for reduced frequencies of order unity (i.e.: $k\sim 1$), since in this case Eqs. 17 and 18 are equivalent.

Conclusions

According with the previous perturbation analysis, some conclusions may be drawn for vibrating airfoils at low reduced frequency. In this context low means much smaller than one. From a mathematical point of view is enough to have $k\sim 0.1$ to formally support the analysis although, as it will be shown in [10], the conclusions hold as well for higher values of k, being the limit case dependent. Flutter boundaries for cantilever LPTs are in the range of $k\sim 0.1-0.2$ [1] and as a consequence the following conclusions may be applied to realistic aeroelastic configurations:

1. The phase of highly loaded airfoils where the velocity changes of the order of itself in distances of the order of the chord, i.e. $\tilde{\nabla}\tilde{\mathbf{W}}_c\sim o(1)$, changes linearly with k, i,e,: $\phi\propto k\ll 1$, and therefore is small. The unsteady pressure induced by the vibration is given by Eq. 17 and therefore the influence coefficients are in first approximation independent of k and proportional to the dynamic head. The influence coefficient of the j-th airfoil, l_j, can be written as:

$$l_j=\frac{1}{2}\rho_c v_c^2\left(\frac{\delta_c}{c}\right)\nabla\tilde{\mathbf{W}}_{0c}c\left(\tilde{l}_{0j}+\tilde{l}_{1j}\frac{k}{\nabla\tilde{\mathbf{W}}_{0c}}\right) \qquad (19)$$

   if we retain the dependence with k. This means that in second approximation the influence coefficient varies linearly with k. This fact was numerically shown in [8].
2. The phase of lightly loaded airfoils, i.e. when $\tilde{\nabla}\tilde{\mathbf{W}}_c\simeq 0$ or more precisely $\tilde{\nabla}\tilde{\mathbf{W}}_c\ll k$, departs from $\pi/2$ and varies linearly with k, $\phi-\pi/2\propto k$. This fact has a deep impact on the work-per-cycle curves as we will see later on. The unsteady pressure induced by the vibration is given in this case by Eq. 18 and hence the influence coefficients scale with k for lightly loaded airfoils and can be written as:

$$l_j=\rho_c v_c\omega\delta_c c\,\tilde{l}_{0j} \qquad (20)$$

   that can be seen as the limit of Eq. 19 when $\tilde{\nabla}\tilde{\mathbf{W}}_c\to 0$.

Assessment of Work-per-Cycle Curves

The previous assertions have direct implications on the shape of the work-per-cycle curves. It has been previously reported [5, 8], that for LPT airfoils, and in general for highly loaded subsonic airfoils, vibrating in flap and edge-wise modes, only the vibrating airfoil, $j=0$, and the adjacent airfoils, $j=\pm 1$, contribute significantly to the work-per-cycle, W_{cycle}. The same assumption has been adopted here. It may be shown that the work-per-cycle can be expressed as a function of the inter-blade phase angle, σ, and the influence coefficients in the following form:

$$W_{cycle}=\pi p'_c\delta_c c\left[\tilde{l}_0\sin\phi_0+\tilde{l}_{+1}\sin(\phi_1-\sigma)+\tilde{l}_{-1}\sin(\phi_{-1}+\sigma)\right] \qquad (21)$$

where $\widetilde{l}_j$ and ϕ_j are respectively the modulus and phase of the non-dimensional ICs and δ_c the vibration amplitude.

Loaded Airfoils For loaded airfoils vibrating at $k \ll 1$ the scaling of the W_{cycle} is as follows:

$$W_{cycle} = \mathbb{C}\left[\widetilde{l}_0\phi_0 + \widetilde{l}_{+1}\sin(\phi_1 - \sigma) + \widetilde{l}_{-1}\sin(\phi_{-1} + \sigma)\right] \quad (22)$$

where $\mathbb{C} = \frac{1}{2}\pi\rho_c v_c^2 \delta_c^2 \nabla\tilde{\mathbf{W}}_{0c}$, and since $\phi_0 = k(1 + \alpha_0\nabla\tilde{\mathbf{W}}_{0c})/\nabla\tilde{\mathbf{W}}_{0c} = \alpha_0' k$, $\phi_{\pm1} = k(1+\alpha_{\pm1}\nabla\tilde{\mathbf{W}}_{0c})/\nabla\tilde{\mathbf{W}}_{0c} = \alpha_{\pm1}' k$, where α_0' and $\alpha_{\pm1}'$, where α_0' and $\alpha_{\pm1}'$ are constants of order unity, and the non-dimensional IC moduli, $\widetilde{l}_0$ and $\widetilde{l}_{\pm1}$, are constant with a small correction proportional to k, then the mean value of the work-per-cycle, $\bar{W}_{cycle-mean}$, scales with k whereas the variation with σ is constant in first approximation, i.e.:

$$W_{cycle} = \bar{W}_{cycle-mean} + W_{cycle}'(\sigma, k) = k\, W_{cycle}^0 + W_{cycle}'(\sigma, k)$$

where $\max_{k,\sigma} W_{cycle}$ is bounded with k in first approximation, exactly as it was shown by Vega and Corral [8] for a LPT airfoil vibrating in flap, edge and torsion modes. Numerical experiments have been carried out in [10] to confirm that this is a generic result independent of the geometry. If more ICs were needed to be retained to accurately represent the solution, as it is sometimes the case for some torsion modes, the conclusions would be the same but more harmonics would be needed to be retained in Eq. 22.

It is interesting to remark that the mean value of the work-per-cycle, $\bar{W}_{cycle}$ increases with the airfoil loading, since $W_{cycle} \sim \mathbb{C}\phi \sim \frac{1}{2}\pi\rho_c v_c^2 \delta_c^2 \nabla\tilde{\mathbf{W}}_{0c} \cdot k(1 + +\alpha\nabla\tilde{\mathbf{W}}_{0c})/\nabla\tilde{\mathbf{W}}_{0c} \sim \frac{1}{2}\pi\rho_c v_c^2 \delta_c^2 k \cdot (1 + \alpha\nabla\tilde{\mathbf{W}}_{0c})$, and the variation of W_{cycle} with σ, W_{cycle}', scales directly with this parameter.

Lightly Loaded or Unloaded Airfoils For unloaded airfoils vibrating at $k \ll 1$ the scaling of the W_{cycle} is as follows:

$$\begin{aligned} W_{cycle} &= \tfrac{1}{2}\pi\rho_c v_c^2 \delta_c^2\, k\left[\widetilde{l}_0 + \widetilde{l}_{+1}\sin(\tfrac{\pi}{2} - \sigma) + \widetilde{l}_{-1}\sin(\tfrac{\pi}{2} + \sigma)\right] \\ &= \tfrac{1}{2}\pi\rho_c v_c^2 \delta_c^2\, k\left[\widetilde{l}_0 + (\widetilde{l}_{+1} + \widetilde{l}_{-1})\cos\sigma\right] \end{aligned} \quad (23)$$

since all the phases are approximately $\pi/2$. The first observation is that the work-per-cycle scales linearly with the reduced frequency and, what is more striking, that the damping curves are symmetric with respect $\sigma = 0^\circ$, i.e.:

$$W_{cycle} = k\, W_{cycle}^0(\sigma) = k W_{cycle}^0(-\sigma)$$

It is interesting to observe that (i) this symmetry is possible even if $\widetilde{l}_{+1} \neq \widetilde{l}_{-1}$ and (ii) increasing or decreasing the reduced frequency does not produce any noticeable effect on the stability, i.e.: the sign of W_{cyc} does not change with k. This behaviour will be confirmed in [10] using numerical experiments. It is important to highlight that for lightly loaded airfoils it may well happen that more ICs are needed to represent the mode. It is enough to retain the complete expansion of the W_{cycle} as a function of the ICs in Eq. 23 to conclude that in any case the W_{cycle} scales linearly with k, whereas the symmetry with σ may be spoiled if the phase of these ICs is different from $\pi/2$.

Symmetric Cascades The general form of the W_{cycle}, Eq. 21, can be particularized for symmetric cascades, where $l_{-1} = -l_{+1}$ and $\phi_{-1} = \pi + \phi_{+1}$, in which case Eq. 21 reduces to:

$$W_{cycle} = \pi p_c' \delta_c c\left[\widetilde{l}_0\sin\phi_0 + 2\widetilde{l}_{+1}\sin\phi_{+1}\cos\sigma\right] \quad (24)$$

It has been shown that for loaded airfoils vibrating at $k \ll 1$, $\phi \propto k$, while the influence coefficients are fairly independent of k, (see Eq. 19) and then:

$$W_{cycle} = \frac{1}{2}\pi\rho_c v_c^2 \delta_c^2 \nabla\tilde{\mathbf{W}}_{0c}\, k\left[\alpha_0\widetilde{l}_{00} + 2\alpha_{+1}\widetilde{l}_{+1}\cos\sigma\right] \quad (25)$$

Two important conclusions can be drawn at this point. The first is that for a symmetric case the W_{cycle} curves are symmetric, i.e. $W_{cycle}(\sigma) = W_{cycle}(-\sigma)$, as it could be expected because of the symmetry of the problem. The second conclusion is that the work-per-cycle scales linearly with the reduced frequency for all the inter-blade phase angles, i.e.:

$$W_{cycle}(\sigma) \propto k \quad (26)$$

Since the W_{cycle} curve of all the unloaded airfoils are symmetric and scale with k, as well, it can be concluded that all the symmetric cases, for different reasons, fulfill these two properties.

CONCLUDING REMARKS

The unsteady aerodynamics of vibrating airfoils and its impact in the work-per-cycle curves in the low reduced frequency regime has been analytically studied by means of a perturbation analysis that takes advantage of the smallness of k. The non-dimensional airfoil loading and the symmetry or asymmetry of the cascade have been identified as two key parameters to understand the different trends of the influence coefficient modulus and phase, as well of the work-per-cycle curves with the reduced frequency.

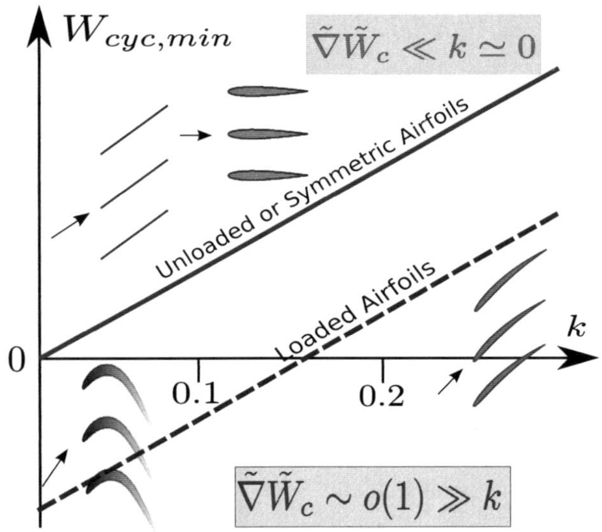

Figure 2. SKETCH OF THE MAIN CONCLUSIONS

The main conclusions for airfoils vibrating at low reduced frequency are the following (see Fig, 2 for a brief graphical interpretation):

1. The phase of the unsteady pressure with respect the airfoil displacement is proportional to the reduced frequency for highly loaded airfoils whereas for lightly loaded airfoils is around $\pi/2$.
2. The work-per-cycle of symmetric cascades scales linearly with the reduced frequency independently of whether the airfoil is loaded or not (see Fig. 2).
3. The work-per-cycle of lightly loaded airfoils scales linearly as well with the reduced frequency (see Fig. 2) and exhibits a symmetry with the IBPA.
4. A corollary of the two previous conclusions is that the stability of this type of airfoils does not change with the reduced frequency, i.e., stable configurations remain stable whatever the reduced frequency is.
5. The mean value of the work-per-cycle of highly loaded airfoils (irrespectively of whether they correspond to a compressor or turbine geometry) scales linearly with k (see Fig. 2) in the same way than the phase of the influence coefficients, nevertheless the influence coefficients modulus are fairly insensitive to the reduced frequency.
6. For highly loaded airfoils the variation of the work-per-cycle with the IBPA is bounded for any value of the reduced frequency.
7. The minimum damping of highly loaded airfoils increases linearly with k but does not scale directly with it.
8. Aerodynamic loading and symmetry are two key parameters to determine the correct scaling of the W_{cyc}
9. A new aeroelastic non-dimensional parameter, $\nabla\tilde{\mathbf{W}}_c.\mathbf{n}$, that measures the velocity gradient in the wall normal direction has been identified.
10. These conclusions are in first approximation independent of the type of mode.

The predictions of the analytical theory for reduced frequencies of order unity still provide certain insight on the behavior observed in the simulations. In particular it is concluded that the phase of the ICs for high values of the reduced frequency tends to $\pi/2$ and hence the work-per-cycle curves have a cosine-like shape, which matches the experience obtained running simulations for compressor blades that actually operate at much higher reduced frequencies than LPT cantilever blades. It is important to highlight that this shape of W_{cycle} curves is due to reduced frequency effect and it is not associated to the actual shape of the airfoil.

The perturbation analysis conducted in this work provides a general framework to greatly improve the theoretical understanding of the aerodynamic damping of vibrating airfoils. The scaling laws derived here may be used to predict the sensitivity of a computed damping curve with the frequency without resorting to new simulations, at least in the preliminary design phase of the turbomachine. Moreover the methodology and the conclusions derived here may be used at the academic level in future courses of turbomachine aeroelasticity.

ACKNOWLEDGMENTS

RC and AV wish to thank Industria de Turbopropulsores S.A. for allowing the publication of this paper and for its support during the project. This work has been partially funded under grant DPI2013-46403-R by the Spanish Ministry of Economy and Competitiveness.

References

[1] Corral, R., Gallardo, J. M., and Vasco, C., "Aeroelastic Stability of Welded-in-Pair Low Pressure Turbine Rotor Blades: A Comparative Study Using Linear Methods," *Journal of Turbomachinery*, Vol. 129, January 2007, pp. 72–83.

[2] Corral, R., Gallardo, J. M., and Martel, C., "A Conceptual Flutter Analysis of a Packet of Vanes Using a Mass-Spring Model," *Journal of Turbomachinery*, Vol. 131, April 2009, pp. 021016–1–7.

[3] Buffum, D. H. and Fleeter, S., "Aerodynamics of a Linear Oscillating Cascade," TM 103250, NASA, 1990.

[4] He, L., "Unsteady flow in oscillating turbine cascades : Part 2-Computational study," *Journal of Turbomachinery*, Vol. 120, No. 2, 1998, pp. 269–275.

[5] Nowinski, M. and Panovsky, J., "Flutter Mechanisms in Low Pressure Turbine Blades," *Journal of Engineering for Gas Turbines and Power*, Vol. 122, 2000, pp. 82–88.

[6] Theodorsen, T., "General Theory of Aerodynamic Instability and the Mechanism of Flutter," Tech. Rep. 496, NACA, 1935.

[7] Bölcs, A. and Fransson, T. H., "Aeroelasticity in Turbomachines: Comparision of Theoretical and Experimental Cascade Results," Tech. rep., Laboratoire de Thermique Appliquee et de Turbomachines, EPFL, 1986.

[8] Vega, A. and Corral, R., "Physics of Vibrating Airfoils at Low Reduced Frequency," *ASME paper GT2013-94906*, Proceedings of the 58th ASME Gas Turbine and Aero engine Congress, Exposition and Users Symposium, San Antonio, Texas, USA, June 2013.

[9] Waite, J. and Kielb, R., "Physical Understanding and Sensitivities of Low Pressure Turbine Flutter," *Journal of Turbomachinery*, Vol. 137, No. 1, 2014, pp. 012502–1–9.

[10] Vega, A. and Corral, R., "Numerical Analysis of the Unsteady Aerodynamics of Vibrating Airfoils at Low Reduced Frequency," *ISUAAAT14-S7-4*, Proceedings of the 14th International Symposium on Unsteady Aerodynamics, Aeroacoustics & Aeroelasticity of Turbomachines, Stockholm, Sweden, September 2015.

[11] Sbardella, L. and Imregun, M., "Linearized Unsteady Viscous Turbomachinery Flows Using Hybrid Grids," *Journal of Turbomachinery*, Vol. 123, 2001, pp. 568–582.

[12] Corral, R., Escribano, A., Gisbert, F., Serrano, A., and Vasco, C., "Validation of a Linear Multigrid Accelerated Unstructured Navier-Stokes Solver for the Computation of Turbine Blades on Hybrid Grids," *AIAA Paper 2003-3326*, 9th AIAA/CEAS Aeroacoustics Conference, Hilton Mead, South Carolina, May 2003.

[13] Verdon, J. and Caspar, J., "A linearized unsteady aerodynamic analysis for transonic cascades," *Journal of Fluid Mechanics*, Vol. 149, 1984, pp. 403–429.

Proceedings of the 14th International Symposium on Unsteady Aerodynamics, Aeroacoustics & Aeroelasticity of Turbomachines
ISUAAAT14
September 8-11, 2015, Stockholm, Sweden

ISUAAAT14-S11-2

NUMERICAL ANALYSIS OF THE UNSTEADY AERODYNAMICS OF VIBRATING AIRFOILS AT LOW REDUCED FREQUENCY

Almudena Vega
School of Aeronautics
Universidad Politécnica de Madrid
28040 Madrid, Spain
e-mail: almudena.vega@upm.es

Roque Corral*
Advanced Engineering Direction
Industria de TurboPropulsores S.A.
28108 Madrid, Spain
e-mail: roque.corral@itp.es

ABSTRACT

This paper studies the unsteady aerodynamics of vibrating airfoils in the low reduced frequency regime with special emphasis on its impact on the scaling of the work per cycle curves by means of numerical experiments. Simulations using a frequency domain linearized Navier-Stokes solver have been carried out on rows of a low-pressure turbine airfoil section, the NACA0012 and NACA65 profiles, and a flat plate cascade operating at different flow conditions to show the generality and correctness of the conclusions. Both, the traveling-wave and influence coefficient formulations of the problem are used in combination to investigate the nature of the unsteady pressure perturbations. It is shown that the behavior of the unsteady pressure modulus and phase as well as the work per cycle curves with the reduced frequency is independent of the geometry of the airfoil and the modeshape in first order approximation in the reduced frequency, and that the airfoil loading and the symmetry of the cascade play an essential role on the final shape of the work per cycle curves. Simulations at higher values of reduced frequency, revealed that in this regimen the influence coefficients modulus varies linearly with the reduced frequency, while the phase is always $\pi/2$, and then, the classical sinusoidal shape of the work per cycles curves when operating at low values of reduced frequency turns into a cosinusoidal shape. It is then concluded that the classical cosinusoidal shape of compressor airfoils is not either a geometry effect nor a flow effect effect but a direct consequence of the fact that the natural frequencies of their lowest modes are higher than for a turbine airfoil.

*Also associate professor at the Department of Fluid Dynamics and Engine Propulsion of the School of Aeronautics, UPM

Introduction

Aeroengine low pressure turbines (LPTs) are made of very slender and thin airfoils because their weight and cost have a large impact on the engine (about 20% of the total weight and 15% of the total cost). The natural frequencies of the blades are very low and, as a consequence, LPT assemblies are prone to flutter. Nowadays flutter may become a dominant constraint on the design of modern LPTs precluding the use of more efficient aerodynamic configurations [1, 2]. Compressor airfoils have a much lower aspect ratio than that of the LPTs and usually the reduced frequency of their lowest modes is of order unity ($k \sim 1.0$). However, it is still interesting to study their behavior in the $k \ll 1$ limit as a means to increase the overall understanding of vibrating airfoils at low frequency, and in particular the sensitivity to the airfoil shape.

The unsteady aerodynamics of low-pressure turbine rotor blades and stator vanes associated with the airfoil vibration is nowadays routinely analyzed within the design loop of the aeroengine companies and it has been the subject of dedicated experiments [3, 4, 5]. The standard application of all these efforts is the derivation of the aerodynamic stability of the rotor blades and the quantification of the aerodynamic damping, which is the result of the application of the unsteady pressures on the airfoil displacements.

However, little attention has been paid to the understanding of vibrating airfoil aerodynamics since this is not a figure of merit in itself for the aeroelastic analyst. Designers obtain and use aerodynamic data from numerical tools, but they do not even inspect the unsteady flow field once they have produced it, since they are really just interested in the aerodynamic damping. As a direct consequence of using aerodynamic codes as black boxes, little understanding has been gained in recent years about the physics of these type of flows and although there are several trends that are well known, such as the stabilization with the reduced frequency, the physics which is behind this behavior is not really well understood. On the other hand, vibrating cascade experiments are scarce, and performed well apart in time in different institutions [6] with different techniques and, therefore, are not well suited for conceptual studies. Recently, however, there is a renewed interest in increasing the understanding of the problem fundamentals [7, 8], especially in the context of turbine flutter which is considered the simplest flutter instability of all.

In [9] a perturbation analysis of the linearized Navier-Stokes equations at low reduced frequency was conducted. It was concluded that the non-dimensional airfoil loading and the symmetry or asymmetry of the cascade are two key parameters to understand the different trends of the influence coefficient modulus and phase, as well of the work-per-cycle curves with the reduced frequency. One of the main conclusions derived for airfoils vibrating at low reduced frequency was that for an unloaded case the influence coefficients phase departs from $\pi/2$ and varies linearly with k, $\phi - \pi/2 \propto k$, while the influence coefficients modulus scales with k , and then the work-per-cycle curves are cosinusoidal and scale linearly with the reduced frequency. On the other hand, for highly loaded airfoils the influence coefficients phase is proportional to the reduced frequency while its modulus is constant and therefore the mean value of the work per cycle scales linearly with the reduced frequency but it is not the case for its harmonics. The perturbation analysis also showed that for symmetric cascades, the work per cycle curves scale linearly with the reduced frequency and have a cosinusoidal shape.

The predictions of the analytical theory for reduced frequencies of order unity still provide certain insight on the behavior observed in the simulations. In particular it is concluded that the phase of the ICs for high values of the reduced frequency tends to $\pi/2$ and hence the work-per-cycle curves have a cosine-like shape, which matches the experience obtained running simulations for compressor blades that actually operate at much higher reduced frequencies than LPT cantilever blades. It is important to highlight that this shape of W_{cycle} curves is due to reduced frequency effect and it is not associated to the actual shape of the airfoil.

This paper numerically verifies the analytical conclusions derived in [9] using a set of two-dimensional linear cascades including flat plates, the NACA65 compressor, a turbine airfoil and a cascade made up of NACA0012 airfoils operating at different flow conditions. The Mu^2s^2T suite of codes has been used in the present work. This set of numerical tools has been well validated for vibrating airfoils in the past [10] and more recently in [11].

NOMENCLATURE

c	Axial chord
IBPA	Inter Blade Phase Angle
IC	Influence Coefficient
k	Reduced frequency
LPT	Low Pressure Turbine
M	Mach number
s	Pitch
t	Time
TW	Traveling-Wave
$\mathbf{U}$	Vector of conservative variables
v_c	Characteristic velocity
δ	Vector of the airfoil vibration amplitude
ϕ	Unsteady pressure phase
θ	Stagger angle
S	Blade surface
p'	Unsteady pressure
W_{cycle}	Work-per-Cycle per span unit

NUMERICAL FORMULATION

Linearized Navier-Stokes Equations

The three-dimensional Navier-Stokes equations in conservative form for an arbitrary control volume can be written as:

$$\frac{d}{dt}\int_{\Omega}\mathbf{U}d\Omega + \int_{\Sigma}(\mathbf{f}-\mathbf{U}v_x, \mathbf{g}-\mathbf{U}v_y, \mathbf{h}-\mathbf{U}v_z)dA = 0 \quad (1)$$

where $\mathbf{U}$ is the vector of conservative variables, $\mathbf{f}$, $\mathbf{g}$, and $\mathbf{h}$ the sum of the inviscid and viscous fluxes, Ω the flow domain, Σ its boundary, $d\mathbf{A}$ the differential area pointing outward to the boundary and v_x, v_y and v_z the three components of the boundary velocity. Now we can decompose the flow into two parts: a steady or mean flow, plus a small but periodic unsteady perturbation, which in turn can be expressed as a Fourier series in time. If we retain just the first harmonic any variable can be expressed as:

$$\mathbf{U}(\mathbf{x},t) = \mathbf{U}_0(\mathbf{x}) + Re(\hat{\mathbf{u}}(\mathbf{x})e^{i\omega t}) \quad (2)$$

where $\mathbf{U}_0$ represents the mean flow and $\hat{\mathbf{u}}$ is the complex perturbation. The Navier-Stokes equations can then be linearized about the mean flow to obtain:

$$(\frac{d}{d\tau}+i\omega)\int_{\Omega}\hat{\mathbf{u}}d\Omega+\int_{\Sigma}(\frac{\partial\mathbf{f}}{\partial\mathbf{U}}\hat{\mathbf{u}},\frac{\partial\mathbf{g}}{\partial\mathbf{U}}\hat{\mathbf{u}},\frac{\partial\mathbf{h}}{\partial\mathbf{U}}\hat{\mathbf{u}})dA+ \quad (3)$$

$$+i\omega\int_{\Omega}\mathbf{U}_0 d\Omega'+\int_{\Sigma}(\mathbf{f}_0,\mathbf{g}_0,\mathbf{h}_0)dA'=i\omega\int_{\Sigma}\mathbf{U}_0\mathbf{x}'dA$$

which is a linear equation of complex coefficients and where the first term is an additional time derivative added to solve the equations marching in the pseudo-time τ. The main advantage of this approach is that the governing equations do not need to be marched in physical time until a periodic state is reached and therefore it is much more computationally efficient than a standard non-linear method. Moreover the formulation of the problem in the Fourier space allows the implementation of more accurate boundary conditions in the inlet and outlet boundaries.

Spatial Discretization

The code used in the present study, Mu^2s^2T-L, solves the three-dimensional Navier-Stokes equations in conservative form [12]. The spatial discretization is obtained linearizing the discretized equations of the non-linear version of the code Mu^2s^2T [13], from which the background solution is obtained. The spatial domain is discretized using hybrid unstructured grids that can contain cells with an arbitrary number of faces and the solution vector is stored at the vertex of the cells. The code uses an edge-based data structure, a typical grid is discretized by connecting the median dual of the cells surrounding an internal node. For the node i the semi-discrete form of Eq.3 can be written as

$$\frac{d(\Omega_i\hat{\mathbf{u}_i})}{d\tau}+\Sigma_{j=1}^{n_{edges}}\frac{1}{2}S_{ij}(\hat{\mathbf{F}}_i+\hat{\mathbf{F}}_j)-\hat{\mathbf{D}}_{ij}=\hat{S}(\hat{\mathbf{u}}_i) \quad (4)$$

where S_{ij} is the area associated with the edge ij, and n_{edges} the number of edges that surround node j. The resulting numerical scheme is cell centered in the dual mesh and second-order accurate. It can be shown that for triangular grids the scheme is equivalent to a cell vertex finite volume scheme. A blend of second and fourth order artificial dissipation terms, $\hat{\mathbf{D}}_{ij}$, is added to capture shock waves and prevent the appearance of high frequency modes in smooth flow regions, respectively. The second-order terms are activated in the vicinity of shock waves by means of a pressure-based sensor and locally the scheme reverts to first order in these regions. The artificial dissipation terms can be written as

$$\hat{\mathbf{D}}_{ij}=|A_{ij}|S_{ij}[\mu_{ij}^{(2)}(\hat{\mathbf{u}}_j-\hat{\mathbf{u}}_i)-\mu_{ij}^{(4)}(L_j-L_i)] \quad (5)$$

where $\mu_{ij}^{(2)}$ and $\mu_{ij}^{(4)}$ are the average of the artificial viscosity coefficients in the nodes i and j. $|A_{ij}|$ is a 5x5 matrix that plays the role of a scaling factor. If $|A_{ij}|=(|u|+c)_{ij}I$, where I is the identity matrix, the standard scalar formulation of the numerical dissipation terms [14] is recovered. When $|A_{ij}|$ is chosen as the Roe matrix [15] the matricial form of the artificial viscosity [16] is obtained. The scalar version of the numerical diffusion terms has been used in this work since the difference between both approaches is negligible for the Mach numbers of interest [17] and is more robust.

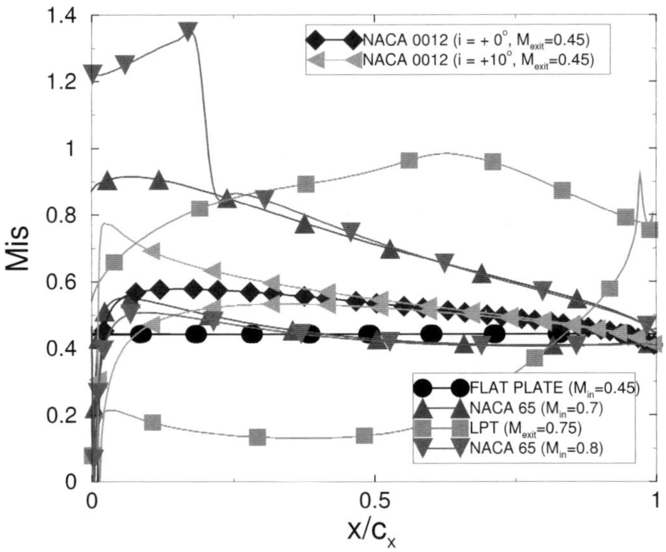

Figure 1. ISENTROPIC MACH NUMBER DISTRIBUTIONS ALONG THE AIRFOIL CHORD

The viscous fluxes can be evaluated in a multiple number of ways. The gradients of the flow variables are approximated at the nodes using the divergence theorem in the same way that the convective fluxes are computed. The approximation of the gradients at the midpoint of the edge is obtained by a simple average,

$$\overline{\nabla\mathbf{U}}_{ij}=\frac{1}{2}(\nabla\mathbf{U}_i+\nabla\mathbf{U}_j) \quad (6)$$

For the present study, no viscous effects are considered.

Temporal Discretization

Equation 4 can be expressed in compact form as

$$\frac{d(\Omega_i\hat{\mathbf{u}_i})}{d\tau}=\mathbf{R}(\hat{\mathbf{u}})=\mathbf{C}(\hat{\mathbf{u}})+\mathbf{H}(\hat{\mathbf{u}})+\mathbf{S}(\hat{\mathbf{u}}) \quad (7)$$

where the residual can be split in the convective, $\mathbf{C}$, numerical diffusion and viscous, $\mathbf{H}$, parts and a source term $\mathbf{S}$. The integration in time is performed using an explicit five stage Runge-Kutta scheme, where the artificial viscosity terms are evaluated only in three stages of the Runge-Kutta.

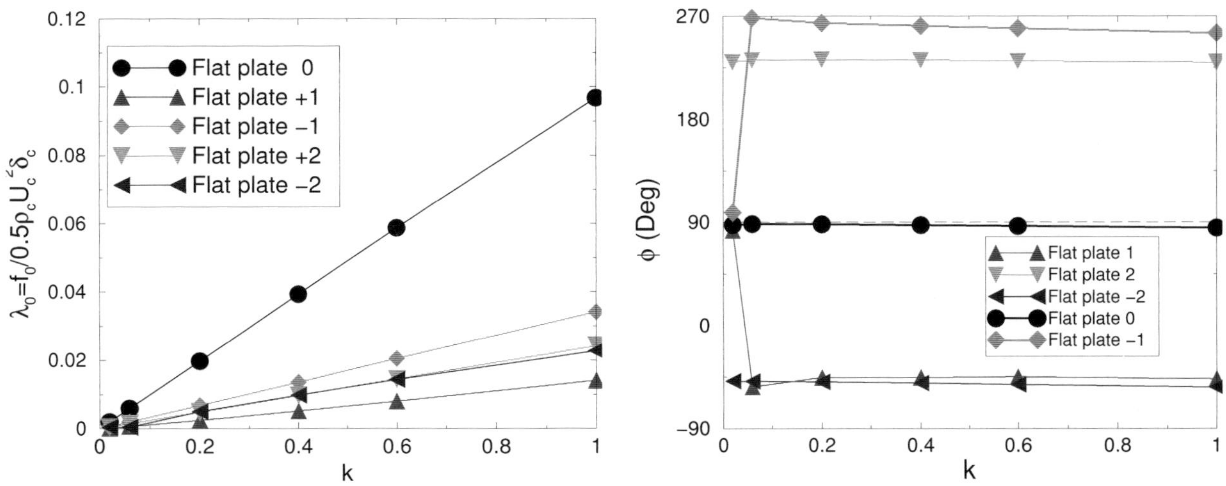

Figure 2. NON-DIMENSIONAL INFLUENCE COEFFICIENT MODULUS (LEFT) AND PHASE (RIGHT) AS A FUNCTION OF THE REDUCED FREQUENCY FOR AN ASYMMETRIC CASCADE OF FLAT PLATES VIBRATING IN BENDING ($M_{in} = 0.45, \theta = 45°$, $s/c = 0.5$)

The scheme is second-order accurate in time. Implicit residual smoothing and full coarsening multigrid can be used to increase the stability limit of the scheme and speed up it convergence. A more detailed description of the numerical scheme and details of the validation process can be found in [12].

Boundary Conditions

The exact 2D unsteady non-reflecting boundary conditions [18] have been used at the inlet and outlet in a plane by plane basis. At the solid walls half volumes associated with the walls nodes are used to march in pseudo time. For viscous cases the velocity is imposed at the boundary nodes. The code uses phase-shifted boundary conditions in the periodic boundaries to simulate annular cascades of airfoils vibrating sinusoidally with a common angular frequency, ω, and common Inter Blade Phase Angle, IBPA.

Mesh Generation

Two-dimensional hybrid unstructured inviscid grids are generated using an in-house grid generator (G2D) . Moving grids are computed in a pre-processing step. The information of the wall displacements is used to derive the boundary velocities which are transferred to the inner nodes in a smoothing step. When the blade motion is harmonic the grid displacements are filtered out using a Fourier decomposition to avoid the potential presence of non-harmonic components associated to the grid displacements.

NUMERICAL VALIDATION OF THE THEORY

In order to verify the conclusions derived in [9] and to study the flutter characteristics of aerodynamic sections, four different airfoil geometries have been investigated namely: a flat plate, a low-pressure turbine airfoil and the NACA 0012 and NACA 65 sections. The airfoil's motion has been defined for all the cases as a rigid body flap mode in a direction orthogonal to the engine axis and the stability of the airfoils has been investigated in a wide range of reduced frequencies $k \in [0.01 - 0.5]$. Figure 1 shows the isentropic Mach number distributions along the airfoil chord for all the sections object of study. It can be seen that they are quite different among them, in terms of both geometry and static pressure distribution, and are intended to cover a wide spectrum of flow conditions including off-design and transonic cases. While there are aspects that are not verified in this battery of cases for the sake of brevity, such as the three-dimensionality or the flexibility of the modes, these factors are of secondary importance from a conceptual point of view and are rarely investigated even in laboratory experiments designed for code validation. The effect of the mode-shape has not been investigated here but this problem was partially addressed by the authors in [7]. It is thus expected that the conclusions derived in the present work are general and not restricted to a particular type of geometry since the theory developed in [9] does not make any hypothesis in this regard.

Asymmetric Flat Plate Linear Cascade

The limit case of an unloaded airfoil is a flat plate, for this reason a linear cascade of inviscid vibrating flat plates has been chosen

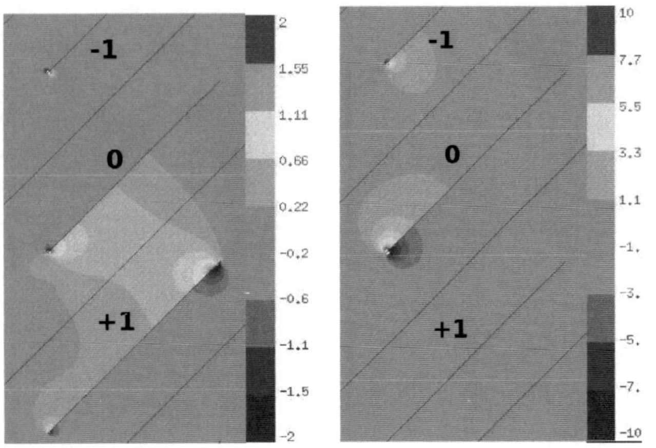

Figure 3. REAL (LEFT) AND IMAGINARY (RIGHT) COMPONENTS OF THE UNSTEADY PRESSURE FOR AN ASYMMETRIC CASCADE OF FLAT PLATES VIBRATING IN BENDING ($M_{in} = 0.45, \theta = 45^\circ$, $s/c = 0.5, k = 0.1$)

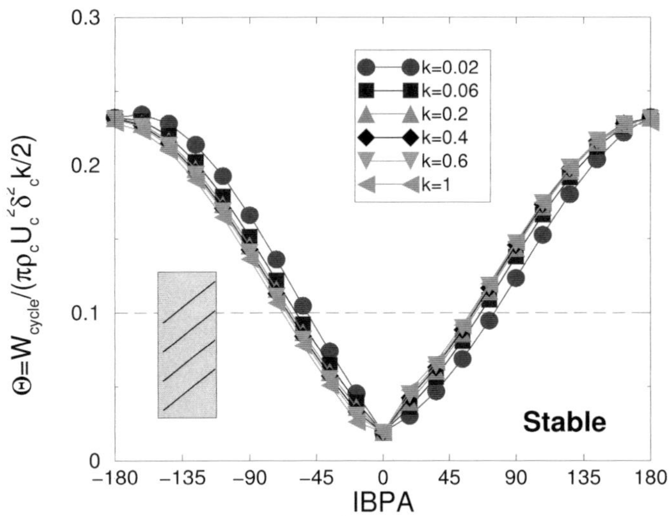

Figure 4. NON-DIMENSIONAL DAMPING AS A FUNCTION OF THE IBPA AND THE REDUCED FREQUENCY FOR AN ASYMMETRIC CASCADE OF FLAT PLATES VIBRATING IN BENDING ($M = 0.45, \theta = 45^\circ$, $s/c = 0.5$)

as the vehicle to verify the conclusions derived in [9] for lightly loaded airfoils. The chosen cascade consists of a flat plate vibrating in flap mode with a stagger angle $\theta = 45^o$, $M_{exit} = 0.45$, and a pitch of $s/c = 0.5$ (see Fig. 3). The cascade constructed to run the simulations using the IC formulation is made up of five plates.

Figure 2 (left) represents the influence coefficient modulus for the vibrating and neighboring plates of the cascade. It can be seen that the largest contributions to the W_{cycle} come from the 0th and -1^{st}airfoils and that all the ICs vary linearly with the reduced frequency and are null for $k = 0$, exactly as described by the asymptotic method. Moreover, it can be seen that the range of validity of the theory extends well beyond the $k \ll 1$ limit. The phases of the influence coefficient are shown in Fig.2 (right) and it can be appreciated that they remain fairly constant in the frequency range of interest for all the airfoils in the cascade, showing only a small correction with the reduced frequency. The phase of the largest ICs, those corresponding to the 0th and -1st airfoils, is either $\pi/2$ or $-\pi/2$, exactly as predicted by the theory as well. The iso-contours of the real and imaginary parts of the unsteady pressure of the flat plate vibrating at $k = 0.1$ in flap mode are displayed in Fig. 3. It may be appreciated firstly how the real component is much smaller than the imaginary one, exactly as predicted by the theory, which is consistent with the fact that the phase is $\pm\pi/2$. Secondly, it may be seen how most of the work is generated in the leading-edge region of the plate, in the vicinity of the singularity.

A direct consequence of the dependence of the ICs and phases with the reduced frequency is that the work-per-cycle scales linearly with the reduced frequency, $W_{cycle}(\sigma) \propto k$, and that the curves are symmetric, $W_{cycle}(\sigma) = W_{cycle}(-\sigma)$, despite of the fact that the cascade is asymmetric ($\theta = 45^\circ$). These two conclusions derived theoretically are corroborated by Fig. 4 that plots the non-dimensional work-per-cycle, Θ_1, defined as

$$\Theta_1 = \frac{W_{cycle}}{\frac{1}{2}\pi\rho_c v_c^2 \delta_c^2\, k}. \tag{8}$$

as a function of the IBPA computed using the travelling-wave approach. It may be observed that the damping curves corresponding to a wide range of reduced frequencies nicely collapse into a single universal curve as predicted by the analysis described in [9], showing surprisingly that it is not possible to destabilize the cascade decreasing the reduced frequency. It may be seen that the theoretical predictions are valid well beyond the $k \ll 1$, since even the $k = 1$ case exhibits a good fit with the rest curves.

The contribution in this case of cascade airfoils different from the +0 and -1 is not small, however in this case it may be observed that the phases of the ±2 airfoils are roughly $\phi_{\pm2} = \pi/2 \pm 3\pi/4$ and hence do not contribute to spoil the symmetry of the damping curve.

NACA65 Configuration

The NACA65 compressor airfoil [6, 19] has been chosen as the vehicle to verify the theoretical predictions of the effect of the airfoil loading on the ICs modulus and phases and on the work-per-cycle. This case has been used as well to understand the

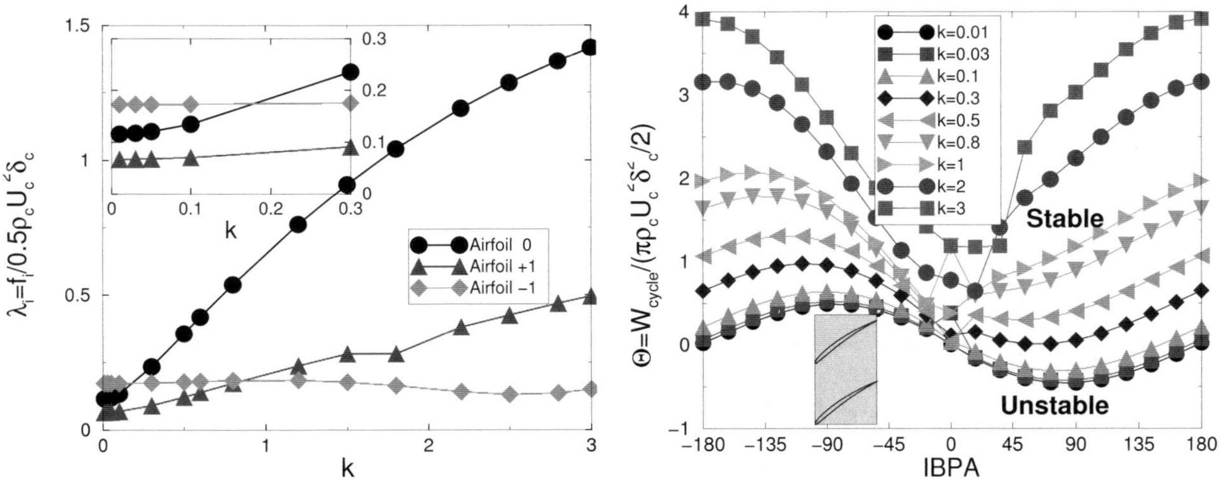

Figure 5. NON-DIMENSIONAL IC MODULUS OF THE 0th AND $\pm 1^{st}$ AIRFOILS AS A FUNCTION OF THE REDUCED FREQUENCY (LEFT) AND NON-DIMENSIONAL DAMPING AS A FUNCTION OF THE IBPA AND THE REDUCED FREQUENCY FOR THE NACA65 COMPRESSOR ($M_{in} = 0.7$)

trends for $k \sim o(1)$. The flat plate case of the previous subsection was chosen to have the same conditions and stagger angle than the exit of the NACA 65 airfoil whose operating point is defined by an inlet angle $\alpha_1 = 55^o$, exit angle $\alpha_2 = 40^o$, inlet and exit isentropic Mach number, $M_{in} = 0.7$ and $M_{exit} = 0.45$, respectively, and airfoil spacing, $s/c = 1$. Linearised inviscid simulations were carried out using a grid containing about 16,000 nodes per passage and the simulations for the IC formulation were carried out with five airfoils. The structure of the analysis is the same than for the flat plate case.

Figure 5 (left) represents the modulus of the ICs as a function of the reduced frequency. It may be observed that reduced frequency range of the plot extends well beyond the low reduced frequency regime. When $k \leq 0.1-0.2$ the ICs are fairly constant, (see the inset of Fig. 5 (left)) however when $k \sim$o(1) the ICs grow proportionally with k, exactly as it predicts the theory derived for $k \ll 1$.It is important to recall that the unsteady pressure associated solely to the vibration velocity of the airfoil, not to the loading, is proportional to k. It can be seen as well that only the 0th and -1st airfoils contribute significantly to the W_{cycle} at low reduced frequency.

Figure 5 (right) shows the non-dimensional work per cycle, Θ_2 defined as

$$\Theta_2 = \frac{W_{cycle}}{\frac{1}{2}\pi\rho_c v_c^2 \delta_c^2} \tag{9}$$

as a function of the IBPA and the reduced frequency. It can be clearly seen that for $k \ll 1$ all curves collapse into a single one, as it was previously reported in [7], and have a sinusoidal shape. It can also be seen that for $k \sim o(1)$ the curves have a co-sinusoidal shape, what is a direct consequence of the fact that when k is increased the phase of the ICs tends to $\pi/2$, as can be seen in Fig, 6, that displays the phase of the ICs as a function of k.

For loaded airfoils the mean value of the damping increases with k. At low reduced frequency, i.e $k \ll 1$, it is due to the fact that $\phi \propto k$, whereas at higher values of k (for $k \sim o(1)$) the phase is $\phi \simeq \pi/2$ but $p_c' \propto k$. Simultaneously the sinusoidal shape of the W_{cycle} curve at $k \ll 1$ becomes a cosine-like curve for $k \sim o(1)$ because of $\phi \to \pi/2$. This behavior was already described in [7] for a low-pressure turbine and it is consistent with the model for the work-per-cycle presented in [20]. This actually means that this is a generic behavior independent of the airfoil shape. Actually it becomes apparent that the cosine-like shape typical of the damping curves of compressor airfoils is not due to the airfoil's shape or flow conditions but it is a direct consequence of the fact that their reduced frequency is close to one, since at $k \ll 1$ compressors have as well damping curves with a sine-likeshape.

The stabilization effect due to the reduced frequency increase, on the contrary that it happens for unloaded profiles, is not absorbed by the non-dimensionalization because, as reported in [7], it is necessary to scale differently the mean damping and its variation with the IBPA. This scaling is properly predicted as well by the theory.

Figure 7 displays the real and imaginary part of the unsteady pressure. It may be observed that the real part is much larger than the imaginary part, as predicted by the theory. The real part

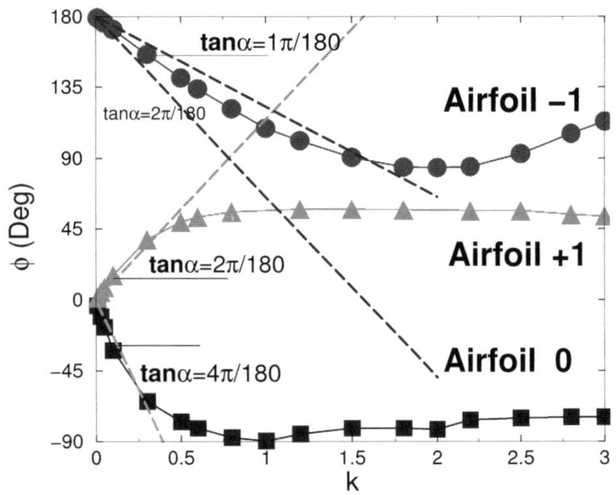

Figure 6. PHASES OF THE 0TH AND ±1 AIRFOILS AS A FUNCTION OF THE REDUCED FREQUENCY FOR THE NACA65 COMPRESSOR ($M_{in} = 0.7$)

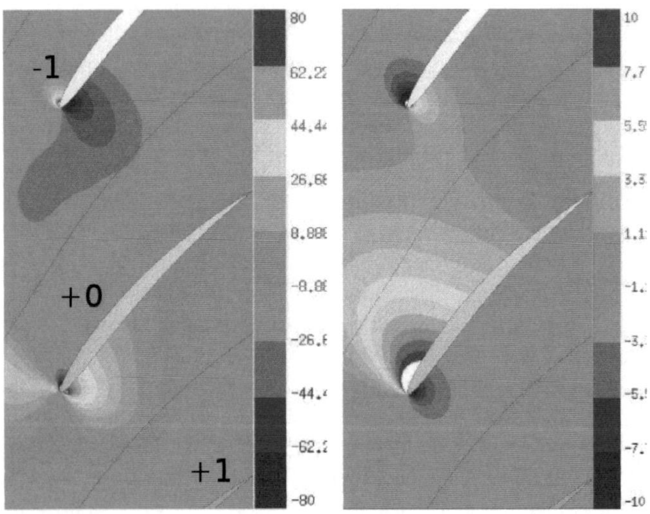

Figure 7. REAL (LEFT) AND IMAGINARY (RIGHT) PART OF THE UNSTEADY PRESSURE FOR THE SUBSONIC CASE OF THE NACA65 COMPRESSOR ($k = 0.1$)

represents the flow-field modification due to the static displacement of the airfoil and covers the whole passage. The imaginary part that is associated to the vibration velocity indicates that the damping is caused by a change in the apparent angle of attack of the airfoil.

NACA65 Configuration. Transonic Case In order to verify these conclusions are sensitive less to the presence of shock waves, the inlet angle and Mach number were raise to $\alpha = 58°$ and $M_{in} = 0.8$, respectively, to obtain a base flow with a shock-wave (Fig. 1 displays the isentropic Mach number distribution of this case along the chord, where it may be appreciated a strong shock-wave in the front part of the airfoils).

Aerodynamic damping curves and influence coefficients were obtained without any qualitative change when compared with the purely subsonic case presented above. The only noticeable difference is that, as expected, the airfoil is more unstable. All these data are not included here for the sake of brevity but the aerodynamic damping curves are shown at Fig. 8and the mean and minimum damping of this case are included in Fig. 17 at the last sub-section where all the numerical results of the paper are summarized. Figure 9 displays the real and imaginary part of the unsteady pressure. It may be observed that the real part is much larger than the imaginary part, exactly as it occurs for the subsonic case. It can also be seen that the unsteady flow field is very different to that observed for the subsonic case (Fig. 7), as could be expected since the presence of the shock wave alters significantly the steady flow field, but nevertheless it does not finally affect to the work per cycle as could be seen in Fig. 8, which represents the non-dimensional work per cycle curves as a function of the IBPA and the reduced frequency for the transonic case.

High Lift Low-Pressure Turbine Airfoil

For the sake of completeness a few results previously reported by the authors [7] to investigate the influence of the reduced frequency, Mach number and mode-shape on the unsteady aerodynamics of a highly loaded LPT airfoil are revisited here. The conclusions obtained there may be seen as a particular case of the theoretical support presented in this work. The operating point is defined by an inlet and exit angle of , $\alpha_1 = 44°$ and $\alpha_2 = 60°$ respectively, $M_{exit} = 0.75$ and an airfoil spacing, $s/c = 0.96$. The linearized NS equations are discretized using an hybrid grid made up of triangles and quadrilaterals, containing about 30,000 nodes per passage.Turbulence was accounted for using the Baldwin-Lomax model with imposed transition.

The unsteady pressure phases scale linearly with the reduced frequency as can be seen at Figure 10, which shows the dependence of the influence coefficient phase of the central, ϕ_0, and $+1^{st}$, ϕ_{+1}, airfoils as a function of k and M_{exit}. The trends described for highly loaded airfoils, and confirmed by the NACA 65 airfoil simulations, are reconfirmed here again, namely the proportionality of the phase with the reduced frequency and the second order impact of the dynamic head or Mach number on the phase that only takes place through a slight modification of the airfoil loading shape.

Figure 11 describes the variation of the non-dimensional influence coefficient modulus of the central airfoil with k and

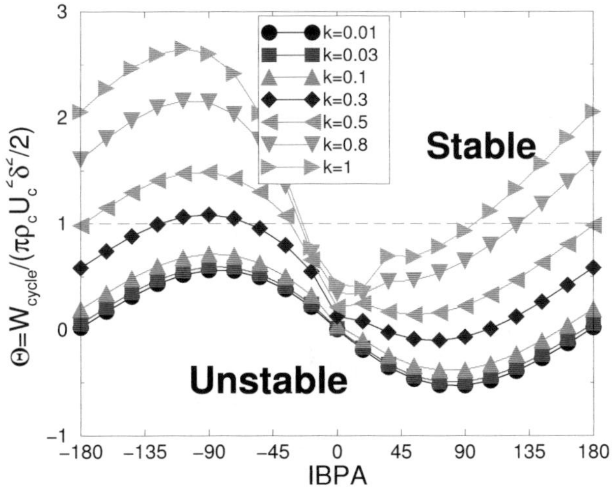

Figure 8. NON-DIMENSIONAL DAMPING AS A FUNCTION OF THE IBPA AND THE REDUCED FREQUENCY FOR THE NACA65 COMPRESSOR ($M_{in} = 0.8$)

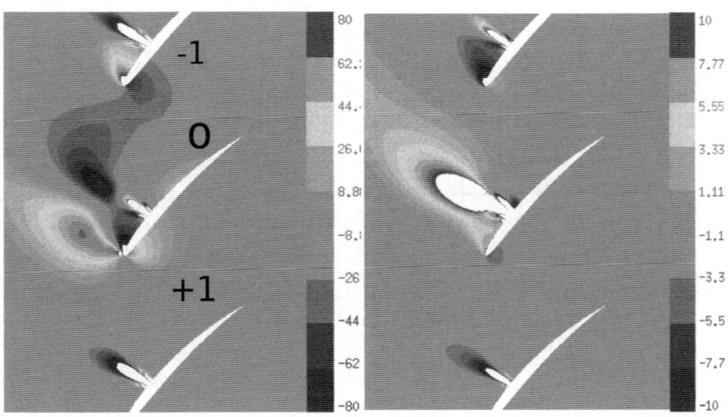

Figure 9. REAL (LEFT) AND IMAGINARY (RIGHT) PART OF THE UNSTEADY PRESSURE FOR THE TRANSONIC CASE OF THE NACA65 COMPRESSOR ($k = 0.1$)

M. It may be seen that the variation with the Mach number squared, or the dynamic head, is properly absorbed by the non-dimensionalization since the variation of the IC with M_{exit}^2 in the range $M_{exit} \in [0.4 - 0.7]$ is of second order compared with the variation of k, that is small as well, of the order of 10% in the range $k \in [0 - 0.2]$. This is a clear indication that the unsteady pressure scales proportionally to the dynamic head, exactly as it was predicted by the theoretical analysis.

The non-dimensional W_{cycle} curves as a function of k are presented in Figure 12, where it can be seen that they follow exactly the same behavior than that described for the NACA 65 airfoil.

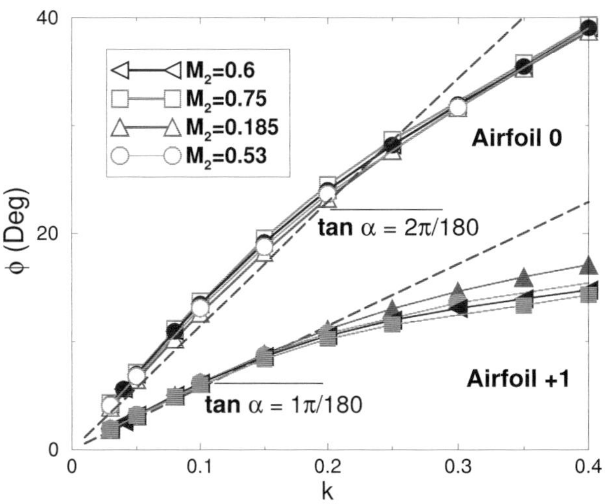

Figure 10. PHASE OF THE INFLUENCE COEFFICIENTS OF THE 0TH AND +1ST AIRFOILS AS A FUNCTION OF THE REDUCED FREQUENCY AND MACH NUMBERS FOR A LPT AIRFOIL.

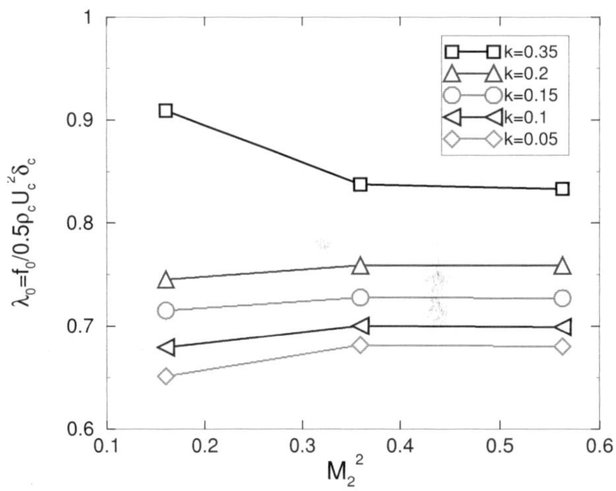

Figure 11. VARIATION OF THE NON-DIMENSIONAL INFLUENCE COEFFICIENT MODULUS OF THE CENTRAL AIRFOIL WITH THE SQUARE OF THE MACH NUMBER AND THE REDUCED FREQUENCY

The study of the LPT case confirms that all the theoretical results are valid for quite different airfoil geometries and that the trends that were derived from an individual case in [7] are general and fairly independent of the airfoil shape.

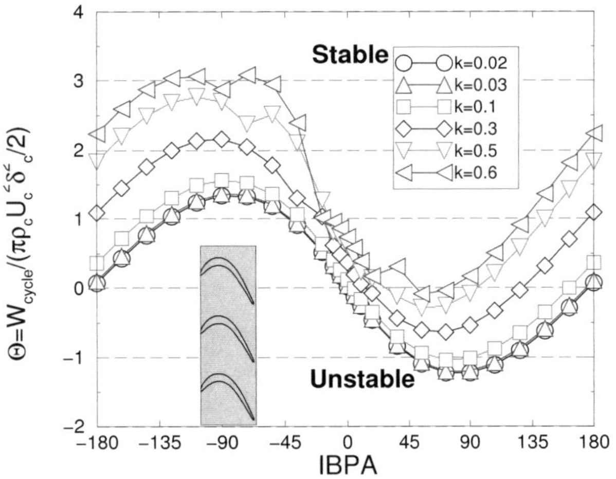

Figure 12. NON-DIMENSIONAL DAMPING AS A FUNCTION OF THE IBPA AND THE REDUCED FREQUENCY FOR THE LPT CASE ($M_{exit} = 0.75$)

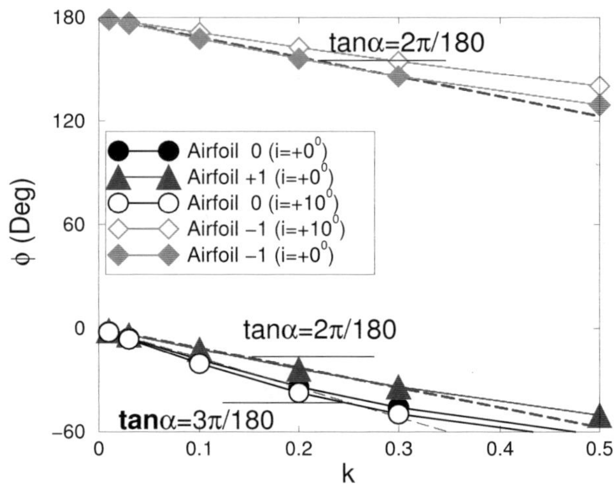

Figure 13. INFLUENCE COEFFICIENT PHASES OF THE 0TH AND ±1ST AIRFOILS AS A FUNCTION OF THE REDUCED FREQUENCY FOR THE NACA 0012 SYMMETRIC (FILLED SYMBOLS) AND ASYMMETRIC (OPEN SYMBOLS) CASES.

Symmetric NACA 0012 Configuration

A NACA 0012 section has been chosen to build a linear symmetric cascade and verify the theoretical conclusions derived for symmetric loaded configurations. Both the inlet and outlet flow angles are $\alpha_{1,2} = 0^o$, the exit isentropic Mach number is $M_{exit} = 0.45$ and the pitch to chord ratio $s/c = 1$. For the asymmetric case constructed with the same airfoil the boundary conditions are the same except for the fact that an inlet angle of 10º is considered. Linearized inviscid simulations have been carried out using grids containing about 45,000 nodes per passage.

Symmetric cascades with symmetric airfoils are a particular case which implies that with a general character, $\widetilde{l}_{+1} = -\widetilde{l}_{-1}$ and $\phi_{-1} = \pi + \phi_{+1}$. As it has been already outlined in [9][9], $W_{cycle}(\sigma) \propto k$ and the W_{cycle} curves are symmetric, i.e $W_{cycle}(\sigma) = W_{cycle}(-\sigma)$. Figure 13 shows the dependence of the influence coefficient phase of the central, ϕ_0, and ± 1, airfoils, $\phi_{\pm 1}$, as a function of k. It may be seen that all of them, in a large range of reduced frequencies, scale linearly with the reduced frequency and that $\phi_{-1} = \pi + \phi_{+1}$, as expected for a symmetric case.

The moduli of the ICs , represented in Fig. 14, follow two different trends because the NACA 0012 is a lightly loaded section compared with the LPT or NACA 65 sections. Figure 1 shows how the Mach number or the velocity gradient along the airfoils chord, which is actually only a part of the contribution to the gradient, is very small for the NACA 0012, being this case an intermediate case between a highly loaded case and a flat plate with strictly zero loading. This means that the range of reduced frequencies at which the ICs remain constant is much narrower than in the NACA 65 or LPT cases. As a consequence, the ICs exhibit a small constant region for $k < 0.1$, in the same way than highly loaded profiles and then a linear increase with k, exactly as in the flat plate (unloaded) case.

For a loaded symmetric profile the work-per cycle scales with the reduced frequency as follows:

$$\Theta_2 = \frac{W_{cycle}}{\frac{1}{2}\pi\rho_c v_c^2 \delta_c^2\, k} \tag{10}$$

The non-dimensional W_{cycle} curves for different values of k are displayed in Fig. 15, where it can be clearly seen that all curves nicely collapse into a single cosine-like shape in a wide range of reduced frequencies, which tightly follows the theoretical predictions well beyond the low reduced frequency limit.

The region $-45^\circ < \sigma < 45^\circ$ exhibit a series of peaks whose amplitude is determined by its separation to the resonant conditions. Since resonances change with k, while the matrix of computed IBPAs is constant, the same IBPA is affected in a different extent depending on the actual value of k. Moreover the resonance peaks are shifted with k and therefore curve collapsing is not possible in the resonances.

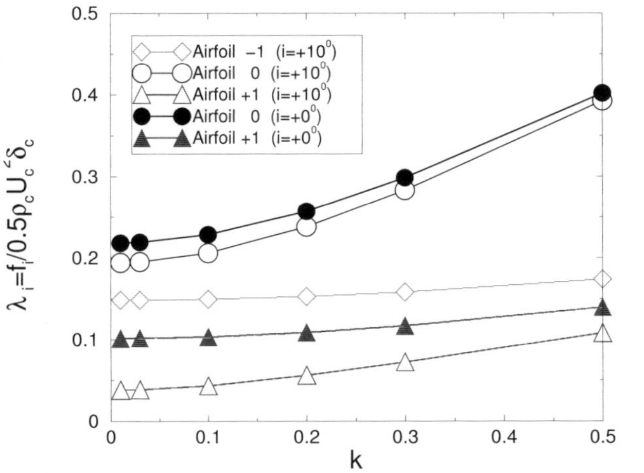

Figure 14. NON-DIMENSIONAL IC MODULUS OF THE 0th AND $+1^{st}$ AIRFOILS AS A FUNCTION OF THE REDUCED FREQUENCY FOR THE NACA 0012 SYMMETRIC (FILLED SYMBOLS) AND ASYMMETRIC (OPEN SYMBOLS) CASES.

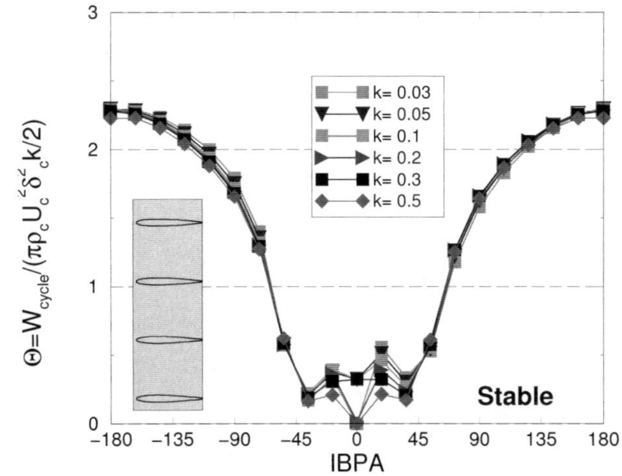

Figure 15. NON-DIMENSIONAL DAMPING AS A FUNCTION OF THE IBPA AND THE REDUCED FREQUENCY FOR THE NACA 0012 SYMMETRIC CASE

Asymmetric NACA 0012 Configuration (off-Design Case)

Figure 16 shows the non-dimensional work-per-cycle curves for a symmetric linear cascade of NACA 0012 airfoils for different values of k, operating at an incidence of 10°, which is not symmetric anymore. Most of the conclusions that were obtained for the LPT and NACA 65 cases are re-confirmed once again as it could be expected since we are dealing with a loaded profile. The damping curves have a sinusoidal shape which is a direct indication that only the airfoils adjacent to the central one significantly contribute to the work-per-cycle and that the phases of the ICs are small. For the configuration under study it is the -1st airfoil the one that has the largest contribution while the influence of the $+1^{st}$ airfoil is much smaller (see Fig. 14) in a similar manner that it is observed in the NACA 65 airfoil and therefore the curve is a +sin, instead of a -sin as it occurs in the LPT case where the $+1^{st}$ airfoil is the dominant adjacent airfoil. It may be appreciated how the effect of the off-design is small in the central airfoil, but significantly increases and decreases the unsteady pressure fluctuations in the upper (-1st) and lower ($+1^{st}$) airfoils respectively, being the mean IC of both airfoils similar to that of the purely symmetric case.

It may be appreciated that the phases of all the airfoils are weakly effected by the off-design conditions. The increase in loading of the suction side is nearly balanced by a similar decrease in the pressure side (see Fig. 1), only some effect may be appreciated in the -1 airfoil that detects the increase in loading of the suction side of the $+0^{th}$ airfoil giving rise to a smaller slope of the phase with k. It should be remember that the larger the loading is the smaller is the slope of the phase angle of the IC.

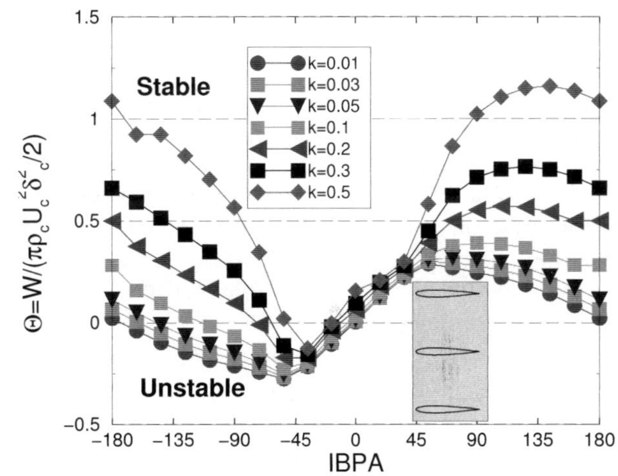

Figure 16. NON-DIMENSIONAL DAMPING AS A FUNCTION OF THE IBPA AND THE REDUCED FREQUENCY FOR THE NACA 0012 ASYMMETRIC CASE

CASE COMPARISON

A in depth comparison of all the cases presented here is out of the scope of the present work, but it is worth to attempt a brief read across and a summary of the basic information provided by the different configurations. Such summary is graphically sketched in Fig. 17. The mean value of all the damping curves vanish for zero reduced frequency and is proportional to k. The minimum damping value of the flat plate (open circles) and the symmetric NACA 0012 cascade (open diamonds) depart from zero as well, exactly as predicted by the theory.

The slope of the mean damping curves as a function of k is proportional to the loading being the highest and the lowest the corresponding to the LPT airfoil and the flat plate respectively. The slopes of NACA 65 and NACA 0012 airfoils are very similar and lie in between the two previous, being slightly smaller that corresponding to the NACA 0012 airfoils that has a lower loading.

The mean damping of the NACA 0012 cascade for the design and off-design conditions is nearly the same since at it has been outlined the mean damping does not depend on the actual loading in first approximation for a given airfoil. A similar behaviour may be seen when the mean damping of the NACA 65 airfoils is compared for two different M_{in}, one subsonic and another transonic. It has been shown [7] that for subsonic airfoils the effect of the Mach number is absorbed by the non-dimensionalization, therefore the difference between both cases could only be attributed to a change of the loading distribution, that actually is very large as it may be appreciated in Fig. 1, however, exactly as in the case of the NACA 0012 airfoil, the difference in the mean damping is negligible.

The most unstable configuration is the LPT airfoil, because it is the one with the largest loading and influence of the central airfoil in the adjacent airfoils. The NACA 65 airfoil is more unstable operating at $M_{in} = 0.8$ than at $M_{in} = 0.7$ because the presence of shock waves increases the influence of the 0th airfoil in the -1st airfoil. The NACA 0012 with zero angle of attack ($i = 0°$) is always stable and the minimum damping is proportional to k (Fig. 17, open diamonds).

The difference between the mean and minimum damping of the LPT and NACA 65 airfoils (highly loaded airfoils) is fairly constant, which is one more indication that the ICs are in first approximation independent of k. It may be concluded that Fig. 17 allows to see at a glance the main conclusions of the work.

MODE SHAPE EFFECT

The validity of the analytical model presented in the work for the three fundamental modes (i.e.:, flap, edge and torsion), was already shown in [7], but it could not be fully appreciated because space limitations prevented us from displaying the results for all the modes. Figure 18 displays the mean work-per-cycle, $\bar{W}_{cycle-mean}$ and its perturbation with σ, $\max_\sigma W'_{cycle}(k)$, as a function of k, for the highly loaded LPT airfoil. It may be seen that the mean damping grows linearly with k, while the variation with σ is fairly constant in a large range of reduced frequencies for the three modes. The behavior of the three modes is qualitatively the same exactly as expected, since the theory developed in **(author?)** [9] does not made any hypothesis about the

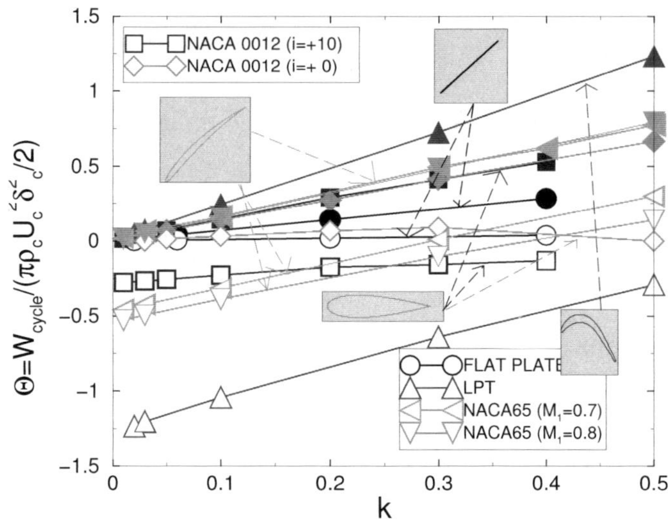

Figure 17. MEAN (FILLED SYMBOLS) AND MINIMUM (OPEN SYMBOLS) DAMPING AS A FUNCTION OF THE REDUCED FREQUENCY OF ALL THE CONFIGURATIONS.

airfoil's motion. It is worth mentioning that, while the flap and edge modes may be well represented by using only $\widetilde{I}_0$, $\widetilde{I}_{+1}$ and $\widetilde{I}_{-1}$, the torsion mode requires a large number of ICs for its representation, however this does not modify the trends previously outlined by the theory. The ultimate objective of this figure is to dissipate doubts about the generality of the theory when applied to different mode-shapes

CONCLUDING REMARKS

All the theoretically derived conclusions in [9] have been numerically verified using a well validated linearized Navier-Stokes solver on a multiplicity of two-dimensional airfoils operating at design/off-design and subsonic/transonic conditions. Both the traveling-wave and the influence coefficient techniques have been used. The numerical study has been focused on the behavior of the bending mode but a few results for the three fundamental modes are also shown to prove that the conclusions obtained in the paper are independent of the mode shape.

The main conclusions for airfoils vibrating at low reduced frequency are the following:

1. The work-per-cycle of symmetric cascades is proportional to the reduced frequency and symmetric with respect IBPA=0 independently of whether the airfoil is loaded or not. This behaviour is the same for cascades made up of lightly loaded airfoils (symmetric or asymmetric). A direct consequence of

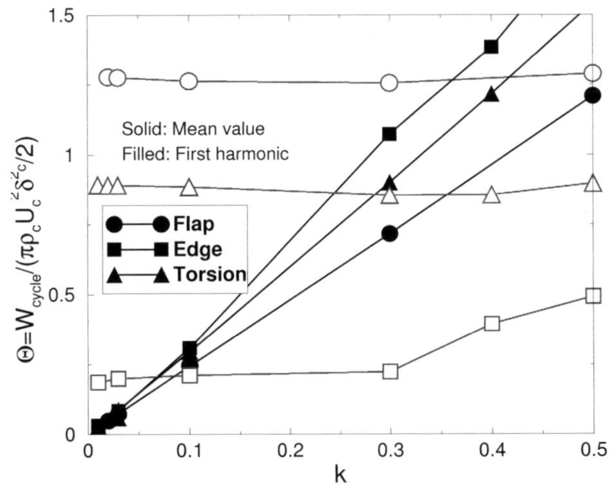

Figure 18. MEAN VALUE, $\bar{W}_{cycle-mean}$, (SOLID SYMBOLS) AND FIRST HARMONIC, $\max_\sigma W'_{cycle}(k)$, (FILLED SYMBOLS) OF THE NON-DIMENSIONAL AERODYNAMIC DAMPING FOR THE FLAP, EDGE AND TORSION MODES OF A LPT AIRFOIL AS A FUNCTION OF THE REDUCED FREQUENCY (AFTER [7])

this first conclusion is that for this two type of cases the stability does not change with the reduced frequency.The mean value of the work-per-cycle is always proportional to k independently of whether it is a loaded or an unloaded airfoil and it is being considered a symmetric or an asymmetric cascade.

2. For highly loaded airfoils (irrespectively of whether they correspond to a compressor or turbine geometry) the phase of the influence coefficients is proportional to k nevertheless the influence coefficients modulus are fairly insensitive to the reduced frequency and the Mach number. On the contrary, for lightly loaded airfoils the phase of the influence coefficients is always $\pi/2$ and the influence coefficients modulus is proportional to k.
3. For highly loaded airfoils the variation of the work-per-cycle with the IBPA is bounded for any value of the reduced frequency and the difference between the mean and minimum dampings is fairly constant.
4. All the conclusions are independent of the airfoil geometry and the mode shape in first order approximation of the reduced frequency.

Simulations at higher values of reduced frequencies, of order unity,show that the phase of the ICs for large values of the reduced frequency tends to $\pi/2$ while the modulus of the ICs varies linearly with k and hence the work-per-cycle curves exhibit a cosine-like shape, which matches the experience obtained running simulations for compressor blades that actually operate at much higher reduced frequencies than LPT cantilever blades. It is important to highlight that this shape of W_{cycle} curves is due to a reduced frequency effect and it is not associated to the actual shape of the airfoil. The second conclusion (i.e the mean value is proportional to k) can be extended to higher values of reduced frequency.

ACKNOWLEDGMENTS

AV and RC wish to thank Industria de Turbopropulsores S.A. for allowing the publication of this paper and for its support during the project. This work has been partially funded under grant DPI2013-46403-R by the Spanish Ministry of Economy and Competitiveness.

References

[1] Corral, R., Gallardo, J. M., and Vasco, C., "Aeroelastic Stability of Welded-in-Pair Low Pressure Turbine Rotor Blades: A Comparative Study Using Linear Methods," *Journal of Turbomachinery*, Vol. 129, January 2007, pp. 72–83.

[2] Corral, R., Gallardo, J. M., and Martel, C., "A Conceptual Flutter Analysis of a Packet of Vanes Using a Mass-Spring Model," *Journal of Turbomachinery*, Vol. 131, April 2009, pp. 021016–1–7.

[3] Buffum, D. H. and Fleeter, S., "Aerodynamics of a Linear Oscillating Cascade," TM 103250, NASA, 1990.

[4] He, L., "Unsteady flow in oscillating turbine cascades : Part 2-Computational study," *Journal of Turbomachinery*, Vol. 120, No. 2, 1998, pp. 269–275.

[5] Nowinski, M. and Panovsky, J., "Flutter Mechanisms in Low Pressure Turbine Blades," *Journal of Engineering for Gas Turbines and Power*, Vol. 122, 2000, pp. 82–88.

[6] Bölcs, A. and Fransson, T. H., "Aeroelasticity in Turbomachines: Comparision of Theoretical and Experimental Cascade Results," Tech. rep., Laboratoire de Thermique Appliquee et de Turbomachines, EPFL, 1986.

[7] Vega, A. and Corral, R., "Physics of Vibrating Airfoils at Low Reduced Frequency," *ASME paper GT2013-94906*, Proceedings of the 58th ASME Gas Turbine and Aero engine Congress, Exposition and Users Symposium, San Antonio, Texas, USA, June 2013.

[8] Waite, J. and Kielb, R., "Physical Understanding and Sensitivities of Low Pressure Turbine Flutter," *ASME Paper GT2014-25133*, Proceedings of the 59st ASME Gas Turbine and Aero engine Congress, Exposition and Users Symposium, Dusseldorf, Germany, June 2014.

[9] Corral, R. and Vega, A., "The Low Reduced Frequency Limit of Vibrating Airfoils. Part I: Theoretical Analysis," *ASME Paper GT2015-42440*, Proceedings of the 60th

ASME Gas Turbine and Aero engine Congress, Exposition and Users Symposium, Montreal, Canada, June 2015.

[10] Corral, R., Escribano, A., Gisbert, F., Serrano, A., and Vasco, C., "Validation of a Linear Multigrid Accelerated Unstructured Navier-Stokes Solver for the Computation of Turbine Blades on Hybrid Grids," *AIAA Paper 2003-3326*, 9th AIAA/CEAS Aeroacoustics Conference, Hilton Mead, South Carolina, May 2003.

[11] Vega, A., Corral., R., Zanker, A., and Ott, P., "Experimental and Numerical Assessment of the Aeroelastic Stability of Blade Pair Packages," *ASME Paper GT2014-25607*, Proceedings of the 59st ASME Gas Turbine and Aero engine Congress, Exposition and Users Symposium, Dusseldorf, Germany, June 2014.

[12] Corral, R., Escribano, A., Gisbert, F., Serrano, A., and Vasco, C., "Validation of a Linear Multigrid Accelerated Unstructured Navier-Stokes Solver for the Computation of Turbine Blades on Hybrid Grids," *AIAA Paper 2003-3326*, 2003.

[13] Burgos, M., Corral, R., and Contreras, J., "Efficient Edge Based Rotor/Stator Interaction Method," *AIAA Journal*, Vol. 41, No. 1, January 2011, pp. 19–31.

[14] Jameson, A., Schmidt, W., and Turkel, E., "Numerical Solution of the Euler Equations by Finite Volume Techniques using Runge-Kutta Time Stepping Schemes," *AIAA Paper 81-1259*, 1981.

[15] Roe, P., "Approximate Riemman Solvers, Parameters, Vectors and Difference Schemes," *Journal of Computational Physics*, Vol. 43, 1981, pp. 357–372.

[16] Swanson, R. C. and Turkel, E., "On Central-Difference and Upwinding Schemes," *Journal of Computational Physics*, Vol. 101, 1992, pp. 292–306.

[17] Corral, R., Burgos, M. A., and Garcia, A., "Influence of the Artificial Dissipation Model on the propagation of Acoustic and Entropy Waves," *ASME Paper 2000-GT-563*, 2000.

[18] Giles, M. B., "Non-reflecting Boundary Conditions for Euler Equation Calculations," *AIAA Journal*, Vol. 28, 1990, pp. 2050–2057.

[19] Fransson, T. H. and Verdon, J. M., "Updated Report on Standard Configurations for Unsteady Flow," Tech. rep., KTH, 1992.

[20] Vega, A. and Corral, R., "The Low Reduced Frequency Limit of Vibrating Airfoils. Part II: Numerical Experiments," *ASME Paper GT2015-42439*, Proceedings of the 60th ASME Gas Turbine and Aero engine Congress, Exposition and Users Symposium, Montreal, Canada, June 2015.

Proceedings of the 14th International Symposium on Unsteady Aerodynamics,
Aeroacoustics & Aeroelasticity of Turbomachines
ISUAAAT14
8-11 September 2015, Stockholm, Sweden

I14-S11-3

PREDICTION OF FLUTTER STABILITY OF AN AEROELASTICALLY UNSTABLE LINEAR LOW-PRESSURE TURBINE CASCADE BY APPLYING THE INFLUENCE COEFFICENTS APPROACH ON NUMERICAL DATA

Christopher Fuhrer, Damian M. Vogt

ITSM Institute of Thermal Turbomachinery and Machinery Laboratory, University of Stuttgart
Stuttgart, Germany

ABSTRACT

The flutter behavior of an aeroelastically unstable linear Low-Pressure Turbine cascade is investigated experimentally and numerically. The objective is to predict the critical reduced flutter frequency by applying the aerodynamic influence coefficients approach on numerical data.

The Remote Flutter Lab (RFL) is located at KTH's Division of Heat and Power Technology (HPT) in Stockholm, Sweden. Each of the eleven central blades of the linear cascade is suspended elastically around a fixed torsion axis, therefore rigid body motion with defined suspension stiffness is possible. With this setup operating points with self-excited and self-sustained flutter were reached by changing the reduced frequency, either by variation of the flow velocity or the stiffness of the blade suspension. At five central blades data acquisition of unsteady blade pressure at midspan and torsional blade motion is available. The measurements indicate a critical reduced frequency of $k_{\text{crit}} \approx 0.15$ for different stiffness setups.

On the numerical side CFD calculations with a commercial solver are performed, using a 3D-model of the cascade blade row. To reproduce the measurements measured total pressure profiles are used as inlet boundary conditions. By applying the influence coefficients method a prediction of aeroelastic stability for the given blade geometry is possible.

INTRODUCTION

Blade flutter phenomena are self-excited and self-sustained and can lead to destruction of turbomachinery blading within a short period of time. Therefore it is of major importance to be capable of reliably predicting flutter stability for turbomachinery setups with appropriate tools. To assess flutter stability the ratio of flow velocity u, oscillation frequency f and blade chord c is used, known as the reduced frequency [1]

$$k = \frac{2\pi f c}{u}. \tag{1}$$

On the experimental side performing free flutter tests is one of the most intuitive ways to investigate flutter stability of turbomachinery blades and therefore determine the critical reduced frequency. The current setup at the RFL is a simplified validation experiment with well-controlled physics and easy handling compared to a rotating blade row.

With numerical tools a fast prediction of aerodynamic stability for a specific blade setup and different operating points is possible. For the actual investigation the influence coefficient approach is employed to determine aerodynamic damping. One blade is made oscillating and the different blade influences are superimposed with an interblade phase angle assuming traveling wave mode behavior [2,3,4]. By adding mechanical damping to this aerodynamic damping the overall system stability for the cascade can be identified. The present work compares numerical influence coefficient results to experimental data gained during free flutter tests at the RFL with focus on critical reduced frequency.

NOMENCLATURE

A	blade oscillation amplitude (deg)
c	blade chord (m)
C_p	static pressure coefficient (-), $C_p = (p - p_{\mathrm{s,ref}}) / (p_{0,\mathrm{ref}} - p_{\mathrm{s,ref}})$
$\hat{C}_{\mathrm{p,A}}$	normalized unsteady pressure coefficient (-)
δ	logarithmic decrement (-)
f	frequency (Hz)
J	moment of inertia (kgm^2)
k	reduced frequency (-)
K_E	kinetic energy (J)
n,m	blade indices (-)
N	number of blades (-)
p	pressure (Pa)
σ	interblade phase angle (deg)
S	spring free length (mm)
T	static friction (Nm)
u	flow velocity (m/s)
W_cycle	work per oscillation cycle (J)
x,y,z	cartesian coordinates (m)
Ξ	stability parameter (-)
ζ	damping ratio (-)
y^+	dimensionless wall distance (-)

Subscripts

0	total
ax	axial
crit	critical
dyn	dynamic
IC	influence coefficients
in	inlet, -35 % c_ax
ref	reference
s	static
TWM	travelling wave mode

Acronyms

CFD	Computational Fluid Dynamics
HPT	Division of Heat and Power Technology, KTH
IBPA	interblade phase angle
IC	influence coefficient
KTH	Kungliga Tekniska Högskolan - Royal Institute of Technology, Sweden
PS	pressure side
LPT	Low Pressure Turbine
RFL	Remote Flutter Lab
SS	suction side

EVALUATION METHODOLOGY

Flutter phenomena are commonly described as traveling wave mode (TWM), where all blades are oscillating at the same frequency and in the same mode with different interblade phase angles (IBPA) [2, 3]. In a blade row of N blades oscillating in travelling wave mode each blade has contributions to the pressure on a single reference blade m which superimpose linearly with a certain IBPA σ

$$\hat{C}^{m,\sigma}_{p\mathrm{A,TWM}}(x,y,z) = \sum_{n=-(N/2)}^{n=+(N/2)} \hat{C}^{n,m}_{p\mathrm{A,IC}}(x,y,z)e^{-i\sigma n}. \tag{2}$$

The complex pressure coefficient $\hat{C}^{m,\sigma}_{p\mathrm{A,TWM}}(x,y,z)$ at point x,y,z acting on blade m, with the cascade travelling at a phase lag of σ, contains contributions of each complex pressure coefficient $\hat{C}^{n,m}_{p\mathrm{A,IC}}(x,y,z)$ of the vibrating blade n. Blade indices are descending in the direction of the pressure side.

Complex pressure values for each blade surface are generated numerically from time-resolved data, with phase in respect to blade motion. This complex pressure is normalized by the oscillation amplitude A and a dynamic reference pressure $p_{\mathrm{dyn,ref}}$ calculated from the total pressure at the inlet and static outlet pressure [4]. Since only torsional movement is present in the current investigation, the infinitesimal normalized torque component dm per surface element ds around the torsion axis ζ is given by

$$dm_\zeta = \left(r \times \hat{C}_{\mathrm{p,A}}\right) \cdot \mathrm{e}_{\zeta_\perp} \cdot \mathrm{d}s. \tag{3}$$

According to [5] the work per oscillation cycle for pure torsion yields from integration around the whole blade to

$$W_\mathrm{cycle} = \pi \cdot A \cdot \mathrm{Im}\left(m_\zeta\right). \tag{4}$$

Whereas stability is determined by the phase angle between torque and angular displacement. If the response is lagging the excitation the flow absorbs work from the structure and therefore has a stabilizing character. In this case the reported stability parameter Ξ is positive

$$\Xi = \frac{-W_\mathrm{cycle}}{\pi \cdot A}. \tag{5}$$

To compare aerodynamic and mechanical damping, work per cycle needs to be converted into a damping ratio ζ. Referring to [6] logarithmic decrement δ yields from the ratio of W_cycle

and the average kinetic energy of the blade $\overline{K}_{\mathrm{E}}$ to

$$\delta = \frac{-W_{\mathrm{cycle}}}{4 \cdot \overline{K}_{\mathrm{E}}}. \tag{6}$$

NUMERICAL MODEL

For the numerical calculations the commercial solver ANSYS CFX 15 is used. To account for computational effort only a 20 % span section of the cascade is modeled. The adjacent boundary walls are set as free slip walls. For all simulations the SST-turbulence model is applied and the fluid is assumed as ideal gas. A structured multiblock mesh with 146493 nodes per channel, and an average $y^+ < 1$ is used. The mesh for one passage is shown in Figure 1.

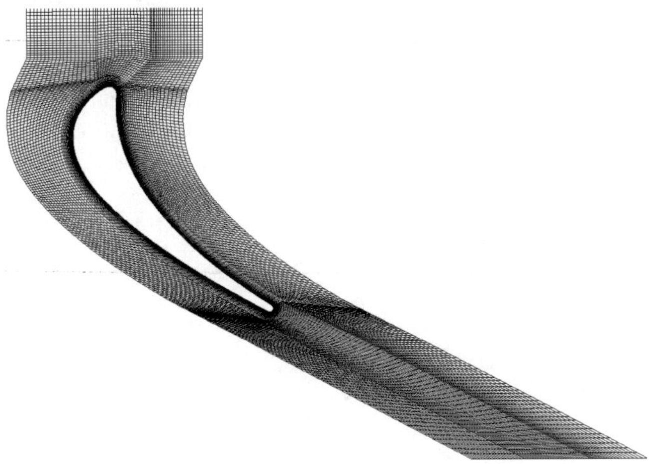

FIGURE 1: CFD multiblock mesh for a single blade passage

For the steady calculations one passage with periodic boundaries in pitchwise direction is employed. The inlet of the numerical model is located at the axial position of the upstream probe traverse of the test rig. Pitchwise averaged total pressure profiles from measurements are applied as inlet boundary conditions. At the outlet measured averaged static pressure is used. To perform time-marching calculations for the influence coefficients approach, seven blades are modelled. The central blade is moved in torsional mode around the pivot point with an oscillation amplitude $A = 0.25°$ which corresponds to a deflection of about 0.3 % chord length at trailing edge.

To judge the simulation parameters the least stable point, which is the minimum of the particular stability S-curve resulting from the IC-method, is taken. Figure 2 shows exemplarily the characteristics for a specific reduced frequency with a variation of timesteps per period.

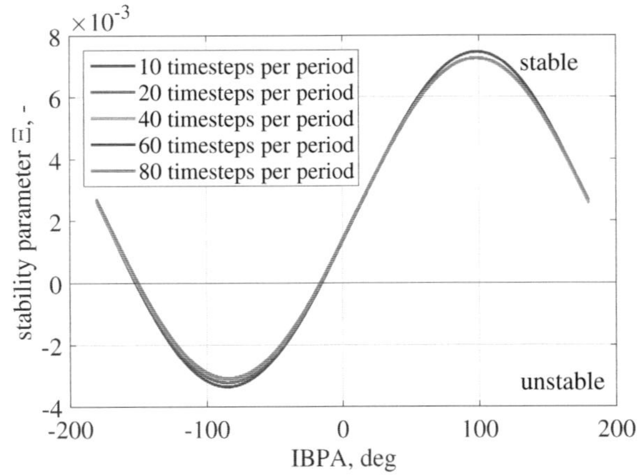

FIGURE 2: S-curve for a variation of timesteps per period with a reduced frequency of $k = 0.3$

The overall relative difference for the stability parameter at the least stable point shown in Figure 3 is about 10 %. A timestep independent solution seems to be reached with 60 timesteps per period, the absolute difference between 80 and 60 steps per period at the least stable point is about 0.001 %.

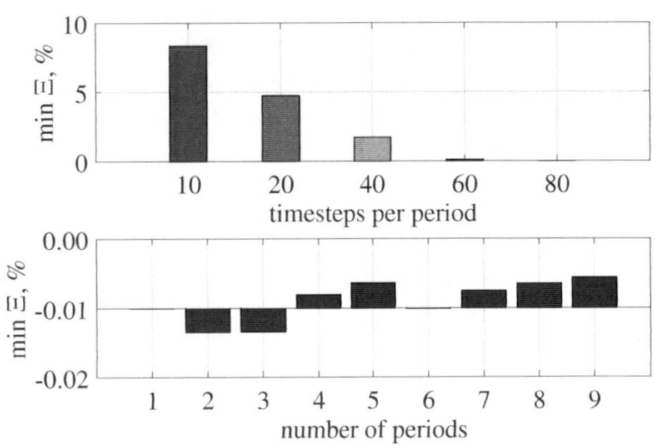

FIGURE 3: Convergence of the least stable point with timestep variation and number of calculated periods for a reduced frequency of k =0.3 and 40 timesteps per period

Although convergence was mostly reached after one oscillation period, all calculations have been running about four periods before unsteady pressure data was exported. Figure 3 shows calculation results at k =0.3 over nine oscillation periods, the relative difference compared to the first period is less than 0.03 % for the following periods.

EXPERIMENTAL SETUP

The Remote Flutter Lab (RFL) is located at KTH's Division of Heat and Power Technology (HPT) in Sweden. Figure 4 shows the test rig with a schematic representation of the blade suspension. The RFL is a low pressure turbine (LPT) linear cascade with the possibility to run free-flutter tests in torsion mode at variable blade stiffness, inflow angles of $\pm 45°$ and flow velocities up to Mach numbers of 0.4 at the outlet. Steady flow characteristics can be measured by probe traverses upstream and downstream of the cascade. In addition measurements of unsteady blade loading and blade movement are possible. The outlet flow is discharged to ambient [7].

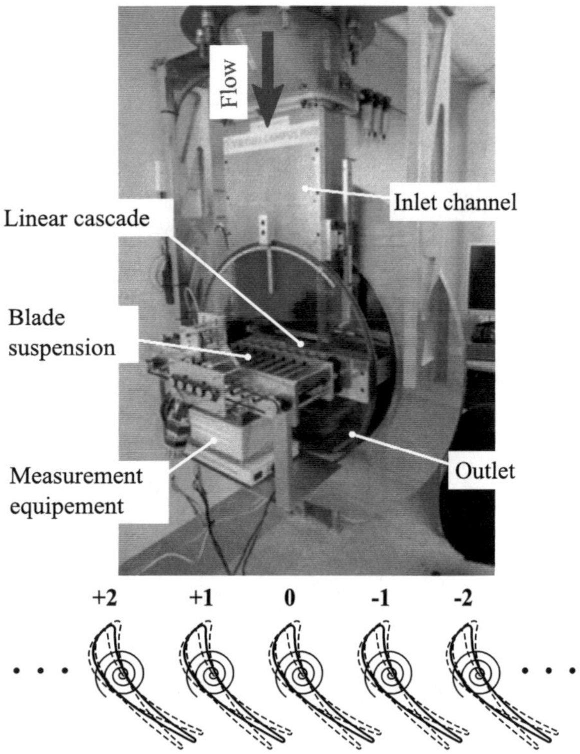

FIGURE 4: Linear cascade test rig RFL with flutter module and schematic blade suspension for the five instrumented blades

The tested cascade consists of 15 prismatic blades manufactured by stereolithography (SLA) 3D printing. The LPT profiles have an axial chord of 45 mm and a span of 100 mm. Two blades at each end are rigidly fixed, eleven blades in between are suspended elastically as rigid bodies around a specific torsion axis, allowing only pure torsional movement. Figure 5 shows the variable suspension for one blade.

The highest stiffness corresponds to the lowest free-length of the flat springs. The free length of the springs can be adapted

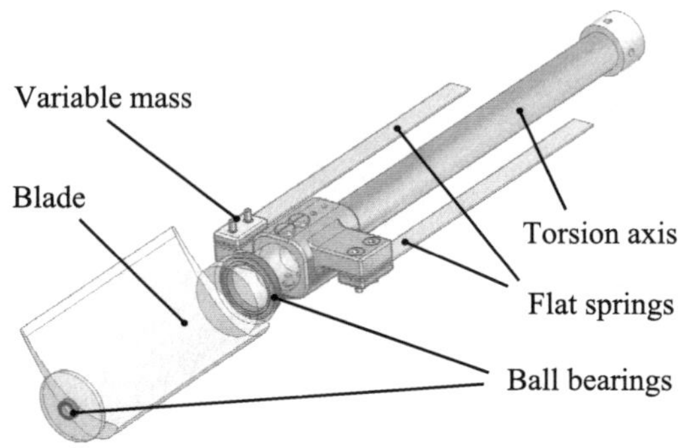

FIGURE 5: Elastically variable blade suspension [7]

for all blades by moving the clamped support of the springs. For the current investigation free-lengths from 50 mm to 90 mm were tested. To measure the unsteady blade surface pressure each of the five central blades is instrumented with 15 pressure taps at midspan and with low-torque potentiometers at the shaft to measure the angular displacement. The LPT blade profile with arcwise pressure tap positions and the pivot point for torsional movement is depicted in Figure 6.

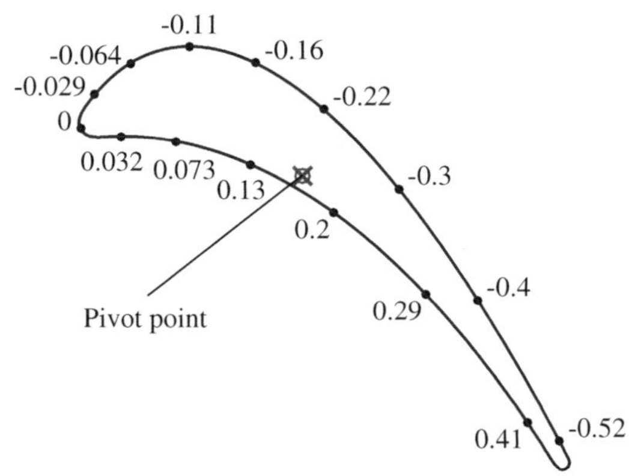

FIGURE 6: LPT blade profile section with pivot point and pressure tap distribution at midspan

The indicated positions are referenced to an arcwise coordinate normalized by the total arc length of the blade. The pressure side is in positive and suction side in negative direction. The pivot point is located at 46 % c_{ax} axial chord downstream of the leading edge. The test facility is equipped with pitchwise and spanwise probe traverses at 37 % c_{ax} upstream and at 36 % c_{ax} downstream of the blades. Upstream, a three-hole wedge probe and downstream a five-hole cone probe is used to measure pressure profiles. In pitchwise direction the flow field of about five blades can be covered. Accuracy of the pressure transducers for blade loading is $\pm 0.25\,\%$ and for the barometer $\pm 11.5\,\text{Pa}$.

EXPERIMENTAL RESULTS

Free flutter tests at different inlet velocities and suspension stiffness levels were performed. All results are for a horizontal cascade position, which corresponds to a incidence angle of $\alpha = 23^\circ$. The following results were produced by changing the reduced frequency via fan speed and therefore flow velocity. Figure 7 shows angular displacement at the start of blade flutter for the five instrumented blades for a spring free length of $S = 90\,\text{mm}$. In most cases cascade flutter started from a single blade in this example Blade +2. For the presented low stiffness level, variations in time after flutter had started, are very little compared to cases with higher blade stiffness, where non periodic overlaid oscillations were present. To account for these variations the time data sets of angular displacement and unsteady pressure were split into blocks and evaluated with a FFT-analysis for each block. The results of this short-time Fourier transformation are depicted in Figure 8 for different blade stiffness values. Error bars indicate variations over time for each blade.

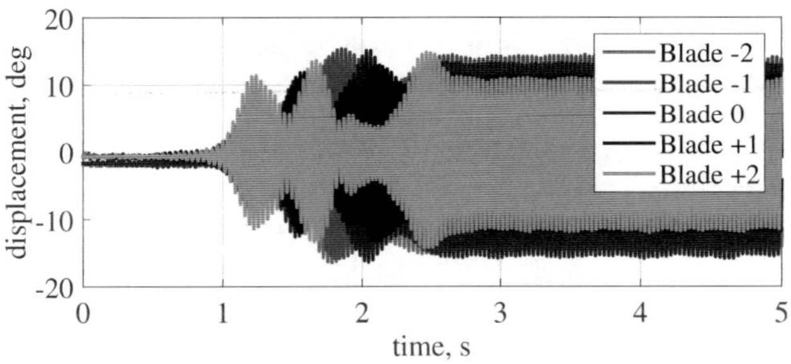

FIGURE 7: Time signal for flutter measurements with spring free length $S = 90\,\text{mm}$ and increasing fan speed

Obviously, the oscillation frequency decreases with increasing spring free length because the eigenfrequency of the blades is changed, just as the oscillation amplitude increases. As aforementioned the variation in time is less for increased spring free length. This becomes obvious especially for the IBPA, which is almost constant for a free length of $S = 90\,\text{mm}$. In agreement to literature, e. g. [8], the IBPA for this linear cascade is not constant amongst the blades. Nonetheless, an evaluation of the IBPA indicates a tendency for average IBPAs between σ = -180° to σ = -70°, with increasing tendency towards higher free length. The reduced frequency varies from $k \approx 0.13$ for the highest free length to $k \approx 0.16$ for the lowest free length of the springs.

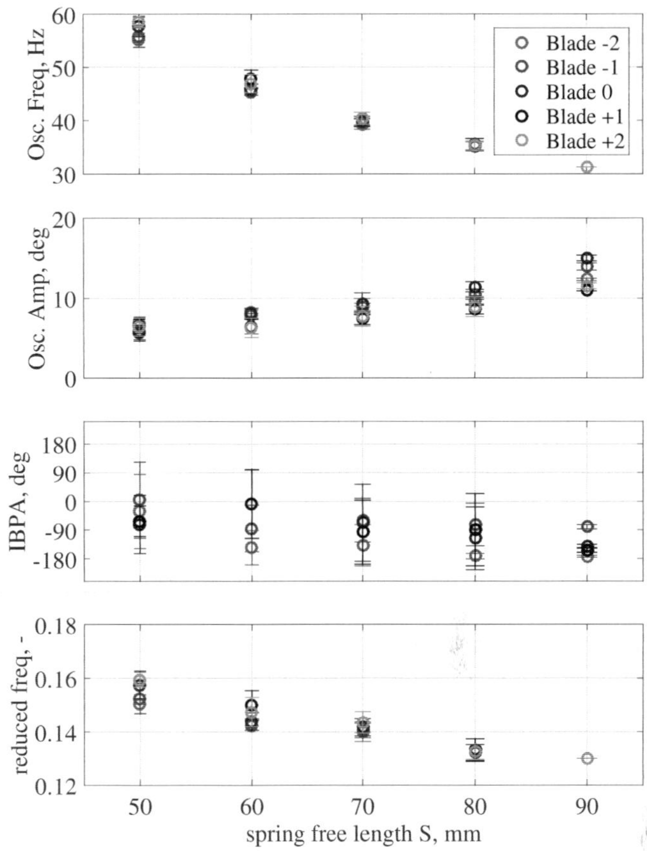

FIGURE 8: Measurement data during flutter for specific free lengths of the springs

In Figure 9 the unsteady pressure on midspan during flutter at a reduced frequency of $k = 0.13$ is depicted for the five instrumented blades. The highest amplitude is located near an arcwise distance of -0.1 on the suction side and is decreasing towards both sides. The pressure amplitude on the pressure side is less than 50 % of the suction side. For all blades the variation over time is insignificant for the shown operating point. The trend of the unsteady pressure amplitude is comparable for the different blades. Whereas the differences in the pressure amplitude are considerably high. Generally, the phase change from suction to pressure side is within the range of -180°. On the pressure side the phase in reference to the motion is in good agreement for all blades near the leading edge, but is diverging towards the trailing edge. On the suction side the differences among the blades is up to 60°. Likewise diverging from leading to trailing edge, the difference in standard deviation over time result from data clipping at 180°.

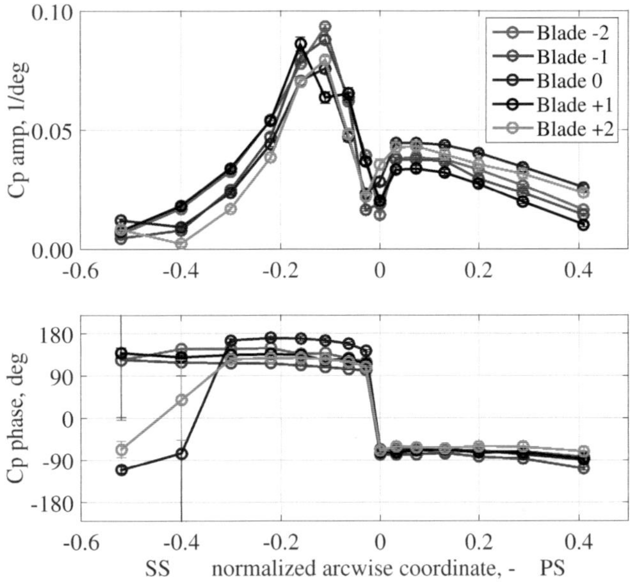

FIGURE 9: Unsteady blade surface pressure data at midspan during flutter with a spring free length of $S = 90$ mm, $k = 0.13$

PREDICTION OF FLUTTER STABILITY

In Figure 10 a comparison of steady-state results for simulations and measurements are depicted. Due to the mentioned incidence the stagnation point is not included in the measurement data for this operating point. The measurement results are represented by an average over all blades. The aerodynamic characteristics are captured very well on the suction side by the CFX-Solver. On the pressure side the differences are higher near the leading edge and decreasing towards the trailing edge. Generally the calculated pressure is within the margin of the differences between the single blades.

To assess flutter stability numerically, transient simulations for different reduced frequencies were performed by changing the frequency f of the oscillating blade. With the influence coefficient approach a stability curve comparable to Figure 2 was generated for the different reduced frequencies. In general a higher reduced frequency shifts the stability curves to a more stable condition. Figure 11 contains the determined least stable point for the variation of reduced frequency.

The trend of the least stable point is almost perfectly linear for the investigated pivot point. With this aerodynamic results the turning point from an unstable to a stable condition is predicted at a reduced frequency of $k = 0.7$ with a corresponding IBPA of $\sigma = -74°$.

To capture the stability of the whole aeroelastic system mechanical damping is needed additionally to the aerodynamic damping. Those damping values are taken from ring-down measurements for each blade at the investigated stiffness levels

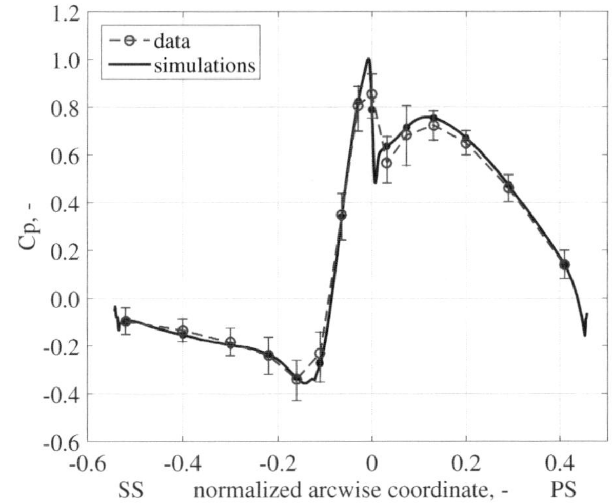

FIGURE 10: Comparison of steady blade loading at midspan between simulations and measurement data

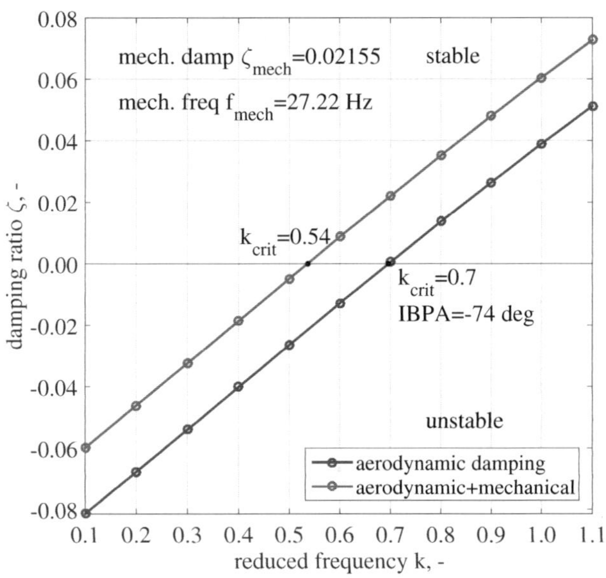

FIGURE 11: Damping ratio of the least stable point for different reduced frequencies predicted with the IC-approach and added mechanical damping

in terms of logarithmic decrement. The ring-down measurements were performed by deflecting the blades manually under no-flow conditions, therefore not aerodynamically loaded but not in a vacuum environment either. To convert W_{cycle} to a damping ratio ζ the eigenfrequency of the blades is needed, which is likewise taken from ring-down measurements. Considering this viscous damping a critical reduced frequency of $k_{crit} = 0.54$ can be identified by this approach. A discrepancy between measured and predicted critical reduced frequency is apparent.

Taking a closer look at a comparison of the unsteady pressure distribution depicted in Figure 12, the data reveals some differences in amplitude and phase. The numerical IC-results in Figure 11 are shown for the least stable point with σ = -74° and for an average IBPA taken from the measurements of σ = -150°. The included measurement data are averaged over time and blades taken from Figure 9.

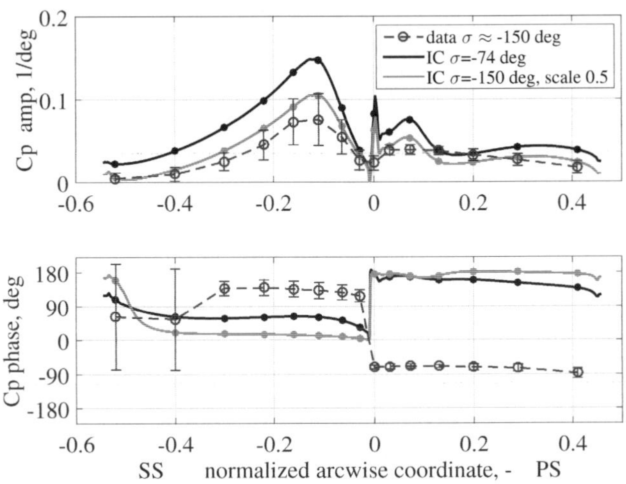

FIGURE 12: Comparison of unsteady pressure data during flutter at $S = 90$ mm and numerical prediction for the IBPA of the least stable point and an IBPA averaged from measurements

The results from the influence coefficients approach are superimposed on the reference blade for the aforementioned IBPAs . The amplitude of σ = -150° is scaled by 0.5 to match the data. On the suction side the trend of the pressure amplitude is similar to the measurements but on a higher level. The trend on the pressure side differs between measurements and numerical results. Especially near the leading edge the amplitude is overpredicted. The phase of the unsteady pressure on the suction side is about 70° higher for σ = -74° and approximately 120° higher for σ = -150°. Whereas the trend is similarly increasing towards trailing edge for some blades. The decreasing trend on the pressure side is captured by the numerical simulations as well. The difference between pressure and suction side is not captured correctly for the least stable point. But for the IC-results with σ = -150° there is almost the same difference compared to the measurement data.

PROBABLE REASONS FOR THE DIFFERENCES

The analysis above indicates significant differences between predicted and actual critical reduced flutter frequency. The cascade seems to be more stable in the experiments than predicted. Conceivable reasons for these discrepancies are:

- Uncertainties in manufacturing and therefore slightly different position of the torsional axis
- Additional damping because of e. g. bearing friction or prestress of the blades due to the plastic material
- Deformation of the blade during operation, leading to a different torsion axis location as well as added bearing friction
- Influence of the finite extent of the linear cascade
- Mistuning between the blades

To investigate deviations of the pivot point a tye-die plot for different torsion axis positions for the reference blade, according to [9], is depicted in Figure 13.

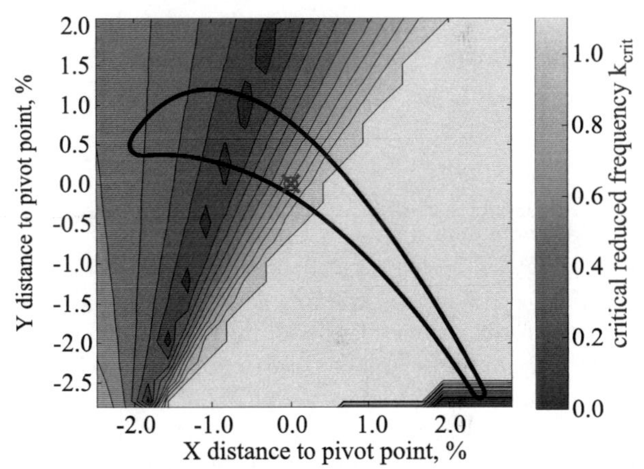

FIGURE 13: Critical reduced frequency from IC-approach for a variation of torsion axis position in simplified tye-die plot

To generate this plot, numerical results for the unsteady pressure distribution for two reduced frequencies were used. Since the changes of unsteady blade loading for slight pivot point variations are minor, only those two results were taken and the resulting torque was evaluated for different pivot point positions in post-processing. The stability plot reveals a major gradient of critical reduced frequency near the actual pivot point location, which could change stability behavior significantly.

For the evaluation of additional damping due to static friction, the aforementioned ring-down measurements were investigated. Figure 14 contains data compared to a model of the differential equation of motion with the displacement φ, the moment of inertia J, the viscous damping d, the stiffness c and additional static friction T.

$$J\ddot{\varphi} + d\dot{\varphi} + T \cdot sign(\dot{\varphi}) + c\varphi = F(t). \quad (7)$$

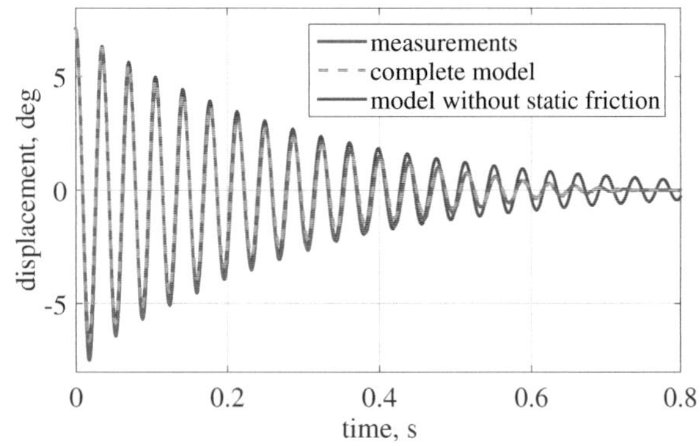

FIGURE 14: Modelling of ring-down measurements with viscous damping and additional static friction for $S = 90$ mm

The model with only viscous damping for $T = 0$ Nm fits the measurement data for the first periods, but then underpredicts the decay of the blade oscillation. With additional static friction $T = 0.001$ Nm the oscillation fades away like the measured blade displacement. This additional damping could result from pressure pipes at the instrumented blades. Comparison of the static friction to the aerodynamic blade torque plotted in Figure 15, where static torque from CFD calculations without blade motion is plotted over varied flow velocity, reveals a lower magnitude of static friction.

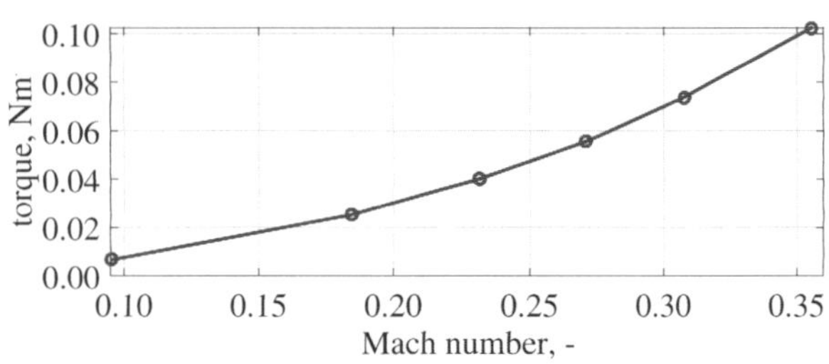

FIGURE 15: Steady CFD results of blade torque around the torsion axis for different flow velocities

This leads to the conclusion that static friction is easily overcome under flow conditions. Additional damping during flutter measurements could also occur due to blade deformation. This could lead to prestress within the bearing system and therefore to higher damping than measured in the ring-down measurements under no flow conditions.

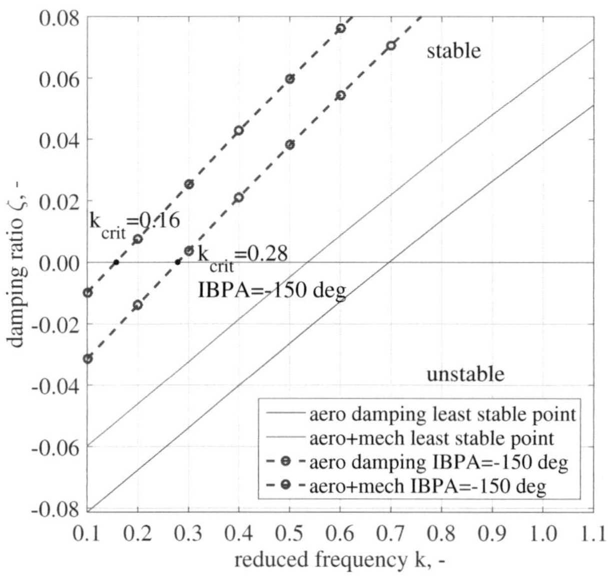

FIGURE 16: Stability characteristics from the IC-approach for $S = 90$ mm evaluated at the average measured IBPA

The abovementioned evaluation is valid for flutter at the least stable point of the stability characteristics. Looking at the IBPA during the measurements plotted in Figure 8 shows a different IBPA for $S = 90$ mm. Taking an average of $\sigma = -150°$ for the IC-evaluation, instead of the least stable point leads to the stability characteristics depicted in Figure 16. Compared to the least stable approach the stability is higher. The resulting critical reduced frequency with added mechanical damping is $k_{\text{crit}} = 0.16$ which corresponds very well to the measured $k_{\text{crit}} = 0.15$. Seen in this light, the IC approach is capable of capturing the critical reduced frequency with a measured IBPA. The reason for the cascade oscillating at an IBPA other than the minimum of the predicted aerodynamic damping curve needs further investigations.

It is believed that the finite extent of the linear cascade leads to a more stable characteristics. As reported by [10] for a five bladed supersonic linear cascade, the outer blades could be more stable due to their fixed neighboring blades and therefore missing excitation. Since the current cascade has 11 moving blades this fact would be reduced, but still present.

Another reason for a more stable setup is mistuning between the blades. Figure 17 shows the eigenfrequencies of the blades determined from the ring-down measurements. For the investigated free length $S = 90$ mm the differences are lowest, but still up to 2 Hz. This mistuning does lead to additionally lowering the critical reduced frequency.

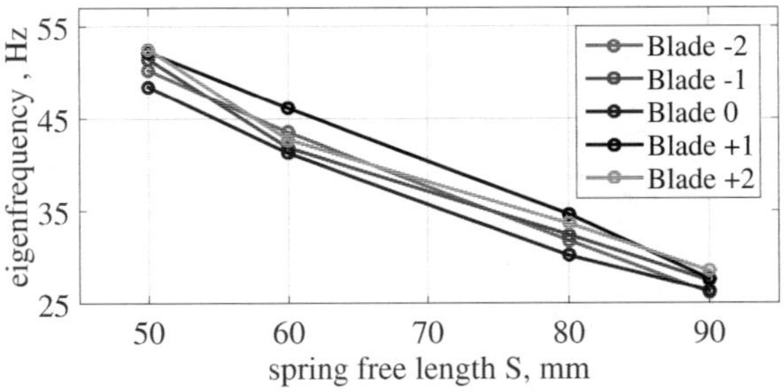

FIGURE 17: Eigenfrequencies of the blades from ring-down measurements for different stiffness levels

CONCLUSION

Free flutter tests at an aeroelastically unstable linear LPT cascade have been performed to determine the critical reduced frequency. The commercial flow solver ANSYS CFX 15 has been used to predict flutter stability with the influence coefficients approach. The reduced frequency was varied by variation of the oscillation frequency. A purely linear trend for the least stable point has been recognized for variations of reduced frequency at the actual pivot point. With this findings a critical reduced frequency of $k_{\text{crit}} \approx 0.7$ was determined. Adding mechanical damping from ring-down measurements lead to $k_{\text{crit}} \approx 0.5$.

For the measurements a critical reduced frequency of $k_{\text{crit}} \approx 0.15$ was identified. Several probable reasons for the discrepancy between measurements and CFD prediction with the IC-method were reviewed. Minor variations of the pivot point position, e. g. due to blade deformation or manufacturing, could lead to a lower k_{crit}. Likewise could additional mechanical damping during flutter occur resulting from deformations in the bearing system. In this setup, where aerodynamic forces and therefore aerodynamic damping is low this could lead to a more stable behavior.

Since the predicted IBPA of $\sigma = -74°$ for the IC-approach differs from the measured $\sigma = -150°$ the stability characteristics was evaluated at the IBPA from the measurements instead of the least stable point. This leads to $k_{\text{crit}} = 0.16$, which corresponds very well to the critical reduced frequency at the beginning of flutter. This in turn leads to the conclusion that the critical reduced frequency can be determined with the IC-approach, but it is currently unknown why the cascade does not oscillate at the predicted least stable point. As one possible reason mistuning between the blades was identified.

ACKNOWLEDGMENTS

The authors wish to thank the Division of Heat and Power Technology at KTH in Stockholm for the opportunity to perform measurements at the Remote Flutter Lab.

REFERENCES

[1] Meldahl, A., 1946. "Selbsterregte Flügelschwingungen mit einem Freiheitsgrad". *Brown Boveri Mitteilungen*, Dec., pp. 386–393.

[2] Hanamura, Y., Tanaka, H., and Yamaguchi, K., 1980. "A Simplified Method to Measure Unsteady Forces Acting on the Vibrating Blades in Cascade". *JSME Int. J.*, ***23***(180-12), June, pp. 880–887.

[3] Crawley, 1988. "Aeroelastic Formulations for Tuned and Mistuned Rotors". In AGARD Manual on Aeroelasticity in Axial-Flow Turbomachines – Structural Dynamics and Aeroelasticity, M. F. Platzer and F. O. Carta, eds., Vol. 2. AGARD-AG-298.

[4] Vogt, D., 2005. "Experimental Investigation of Three-Dimensional Mechanisms in Low-Pressure Turbine Flutter". PhD thesis, Royal Institute of Technology Stockholm, ISBN-91-7178-034-3.

[5] Verdon, J. M., 1987. "Linearized Unsteady Aerodynamic Theory". In Platzer and Carta [11], pp. 2/1–2/31. AGARD-AG-298.

[6] Carta, F. O., 1967. "Coupled Blade-Disk-Shroud Flutter Instabilities in Turbojet Engine Rotors". *Journal of Engineering for Power*, July, pp. 419–426.

[7] Monaco, L., Fransson, T. H., Vogt, D. M., and Bergmans, J., 2015. "A Remotely Operated Aeroelastically Unstable LPT Cascade for Turbomachinery Aeromechanics Education and Training - Remote Flutter Lab". *Trans. ASME, J. Eng. Gas Turb. Power,* ***137***(032507), Oct.

[8] Fleeter, S., and Jay, R. L., 1987. "Unsteady Aerodynamic Measurements in Flutter Research". In Platzer and Carta [11], pp. 8/1–8/18. AGARD-AG-298.

[9] Panovsky, J., and Kielb, R. E., 2000. "A Design Method to Prevent Low Pressure Turbine Blade Flutter". *Trans. ASME, J. Eng. Gas Turb. Power,* ***122***, Jan., pp. 89–98.

[10] Snyder, L. E., and Commerford, G. L., 1974. "Supersonic Unstalled Flutter in Fan Rotors; Analytical and Experimental Results". *Trans. ASME, J. Eng. Power*, Oct., pp. 379–386.

[11] Platzer, M. F., and Carta, F. O., eds., 1987. AGARD Manual on Aeroelasticity in Axial-Flow Turbomachines – Unsteady Turbomachinery Aerodynamics, Vol. 1. AGARD-AG-298.

Proceedings of the 14th International Symposium on Unsteady Aerodynamics, Aeroacoustics&Aeroelasticity of Turbomachines
ISUAAAT14
8-11 September 2015, Stockholm, Sweden

I14-S11-4

AN IMMERSED BOUNDARY METHOD FOR SIMULATING AN OSCILLATING CASCADE

Guotun Hu
Beihang University
Beijing, China, huguotun@126.com

Lin Du, Xiaofeng Sun
Beihang University
Beijing, China

ABSTRACT

Based on the immersed boundary method, a fast explicit numerical method is established to solve the unsteady flow with an oscillating cascade. In order to consider more practical problems, a turbulence model is adopted for turbulence closure. The detailed method is solving the Reynolds averages Navier-Stokes equations with low Reynolds number turbulence model Lam-Bremhorst model that is one of the $k \sim \varepsilon$ models. In order to validate the method, two cases, the laminar boundary layer and turbulent boundary layer, are performed and the results agree well with Blasius solution and the rate of wall. Then the method is extended to the high Reynolds number for an oscillating cascade. In the oscillating cascade simulation, it is found that the oscillation of the cascade is influenced greatly by the reduced velocity, which is just the same conclusion of the laminar results. It is worth noting that in the present scheme, the governing equations are discretized on simple Cartesian meshes using immersed boundary method and the coupling process is not necessary to be generating any body-fitting grid, which makes it much faster in computational process for such a fluid-cascade interaction problem.

INTRODUCTION

With the improvement of aircraft and engine performance, the turbomachinery development trends towards higher-power and higher-load. This requires the reduction of the blade and disk thickness and the increase of blade aspect ratio. Besides, with the application of new type of lightweight alloy steel, the blade stiffness is greatly reduced. Therefore, the flutter of the compressor/turbine blade rows becomes more and more serious, which has become one of the critical factors that need to be considered during the engine design period. Effective flutter prediction tools and the understanding of flutter mitigation strategies are critical in successful engine design. It is intended to find a way to improve the physical understanding of coupled instability in compressor/turbine blade rows, which not only involves the interaction between the fluid and blade but also involves the interaction between adjacent blades.

The interaction of aerodynamic, inertial and elastic forces that are acting on the blades may cause blades vibration. The understanding on the physics of mutual interaction between flow and oscillating cascade is essential for improving overall efficiency, durability and reliability. Flutter is a kind of typical fluid-structure interaction problem, which includes fluid, structure and the interaction between them. Aerodynamic force on the blade surface determines the displacement and velocity of the oscillating cascade, which will influence the flow field and the distribution of the aerodynamic force. Therefore, in order to solve the fluid-structure interaction problem, the coupling between fluid and structure must be considered, which can capture the details of the unsteady flow field.

Because of the difficulty of the fluid-structure interaction problem, the iterative coupled method[1-8] is the most popular method in the past few decades, which solves the fluid and structural equations separately but exchanges the information between them at each time step, so that the solution from one domain is used as boundary condition for the other domain. In other words, in every time step, the unsteady aerodynamic forces which are obtained from the previous time step will be used to calculate the new position of the blades. Then, the new position and the solid body velocity are used as the new boundary conditions for the next time step when the unsteady aerodynamic forces are computed. Since the fluid domain boundaries are time-dependent, it becomes necessary to perform the calculation of the fluid equations on a moving mesh. In the fluid domain, most studies use body-fitted grid and re-gridding architecture to accommodate changes in the structure deformation or displacement. As we know that this approach will cost a lot time, besides, it is very hard to generate high quality grid when the geometry is complex especially when multi-body interference appears. Therefore, the efficiency and reliability of the body-fitted grid become one of the important factors of restricting the coupling calculation. In order to improve the computing speed, the method which Pioerno[9] adopted is a staggered procedures with a structural predictor for fluid/structure interaction simulations. That means, before each fluid time step, grid point locations are simply linearly time-interpolated instead of re-gridding, which saves computational time because the re-gridding costs much more time than the computing time. Because of the structural predictor, it is not the actual motion trajectory. So it cannot ensure the accuracy of calculation. It just can get a more accurate calculation results only in certain cases, besides, this requires a lot of computing experience as a support, so it does not have universality.

In addition, there is a time delay when information transmits between fluid domain and structure domain. So it is not a real coupled way. Many researchers attempted to develop a full coupled method, which can discrete structure and flow equations in a unified Lagrange-Euler space and integrate in the same time layer. Gottfried[10] has developed a program named TAM-ALE3D based on this idea. And he adopted finite element method to solve the equations. However, as we know, using finite element method to solve viscous hydrodynamics problem will addthe problem's complexity. In 2010, the material point method(MPM) combined with adaptive mesh refinement technique was applied to investigate complicated fluid-structure interactions such as aircraft wing flutter and other aeroelastic problems by Patrick G.Hu et al.[11-12]. MPM discretized the material into material points. These points bring all material information and the movement represents the deformation and sports. It is very difficult to simulate fluid-structure in a unified coordinate system. Besides, the matching problem between fluid and structure interface is also a problem. Flow and the oscillating cascade impose boundary conditions on each other, so the idea of taking the fluid and structural dynamics as a single system may result in more convenient and timesaving methods. In the present work, we try to find another way avoid the errors caused by the moving boundary.Considering fluid-structure interaction, as we know, the IB method, which avoids re-gridding for moving boundary can be applied to such a moving boundary problem. The IB method was first proposed by Peskin[13] in the flow around heart.Using this method, the coupling processes are simulated on simple Cartesian meshes. The interfacial force is calculated using Virtual Boundary Formulation, as presented by Goldstein[14]. Zhong[15] has combined the immersed boundary method with the operator splitting technique to investigate the two-dimensional fluid-structure interaction problemsunder the condition of laminar flow.From his work, it is found that flow-cascade interaction may be a new application direction of the immersed boundary method. But the coordinate system in his research is two-dimensional, and the circumferential periodical condition is not well satisfied. In contrast with his work, we[16]has simulated the oscillating cascadeunder the condition of laminar flowbased on a quasi-three-dimensional coordinate system, which fits the real conditions better. The main advantage of this method is that it enables the calculation of this force field, even if the interface is moving or deforming.

In this paper, in order to consider more reality problems, a turbulence model is adopted for turbulence closure based on the quasi-three-dimensional numerical simulation platform (Ref. [16-17]). The detailed method is solving the Reynolds averages Navier-Stokes equations with low Reynolds number turbulence model Lam-Bremhorst[18-19] model that is one of the $k \sim \varepsilon$ models for a numerical simulation of the oscillating cascade under high Reynolds number. Lam-Bremhorst[18-19] model that is one of the $k \sim \varepsilon$ models for a numerical simulation of the oscillating cascade under high Reynolds number.

NOMENCLATURE

A	=	cylinder oscillation amplitude
c	=	chord
C_D	=	drag coefficient
C_L	=	lift coefficient
D	=	diameter of the cylinder
f	=	oscillating frequency
$\vec{f}(\vec{x}_k,t)$	=	boundary force density
F	=	boundary forces
F_h	=	lift in transverse direction
h	=	transverse movement displacement
$\dot{h}$	=	transverse movement velocity
$\ddot{h}$	=	transverse movement acceleration
I_θ	=	inertial moment around the blade elastic center
KC	=	Keulegan–Carpenter number
K_h	=	structural stiffness ratio in transverse direction
K_θ	=	structural stiffness ratio in rotational direction
m	=	mass of the blade
M_θ	=	moment around the airfoil elastic center
N_x	=	mesh number in x direction
N_y	=	mesh number in y direction
Re	=	Reynolds number
S_θ	=	static moment around the blade elastic center
θ	=	rotational angle
$\dot{\theta}$	=	rotational angular velocity
$\ddot{\theta}$	=	rotational angular acceleration
$U_{\max}$	=	maximum velocity of the cylinder motion
U_∞	=	free stream velocity
V	=	moving velocity of rotor blade
S_t	=	Strouhal number

ρ	=	density
U^*	=	the reduced velocity
Δt	=	time step
y^+	=	typical Reynolds number

TURBULENCE MODEL AND THE VALIDATION

Governing equations and the turbulence model

The incompressible Navier-Stokes equations are

$$\frac{\partial u_i}{\partial x_i}=0 \quad (1)$$

$$\frac{\partial u_i}{\partial t}+u_i\frac{\partial u_i}{\partial x_j}=-\frac{1}{\rho}\frac{\partial p}{\partial x_i}+v\nabla^2 u_i \quad (2)$$

where u is velocity, t is time, ρ is density, p is pressure, v is dynamic viscosity coefficient. Based on the Reynolds averages, instantaneous speed is decomposed into the average velocity and fluctuation velocity and instantaneous pressure is decomposed into mean and fluctuating value.

$$u_i=U_i+u_i^{'} \quad (3)$$

$$p=P+p^{'} \quad (4)$$

Then, put formula (3)-(4) into the governing equations will get the Reynolds equations.

$$\frac{\partial U_i}{\partial x_i}=0 \quad (5)$$

$$\frac{\partial U_i}{\partial t}+U_i\frac{\partial U_i}{\partial x_j}=-\frac{1}{\rho}\frac{\partial P}{\partial x_i}+v\nabla^2 U_i+\frac{1}{\rho}\frac{\partial(-\rho\overline{u_i^{'}u_j^{'}})}{\partial x_j} \quad (6)$$

Afterintroducing the Boussinesq eddy viscosity assu mption (1877), it will become as follows.

$$\frac{\partial U_i}{\partial x_i}=0 \quad (7)$$

$$\frac{\partial U_i}{\partial t}+U_i\frac{\partial U_i}{\partial x_j}=-\frac{1}{\rho}\frac{\partial P}{\partial x_i}+\frac{\partial}{\partial x_j}\left[(v+v_T)\left(\frac{\partial U_i}{\partial x_j}+\frac{\partial U_j}{\partial x_i}\right)\right] \quad (8)$$

The above equation is focused on solving eddy viscosity coefficient. At present, there are two kinds of method to calculate flow near the wall, namely wall function method and low Reynolds number model. The former method needs to put the first node in the turbulent region and then solves turbulent regiondirectly on the basis ofwall function. This method skips calculate the viscous sublayer and cannot get the numerical results in the viscous sublayer. Low Reynolds number method is a direct method to calculate the viscous sublayer, so it needs a very dense grid layoutin the near-wall region, and therefore require a large amount of computing. The purpose of the introduction of low Reynolds number model is to make numerical calculation from the high Reynolds number region to the solid wall, and not just for low Reynolds number flow.The low Reynolds number model is adopted in this paper. The turbulence model used in this paper is Lam-Bremhorst model that is one of the $k\sim\varepsilon$ models which ismost widely used.For the $k\sim\varepsilon$ model the eddy viscosity coefficient is as follows,

$$v_T=C_\mu f_\mu k^2/\tilde{\varepsilon} \quad (9)$$

here, k and $\tilde{\varepsilon}$ are decided by the following equations

$$U\frac{\partial k}{\partial x}+V\frac{\partial k}{\partial y}=v_T\left(\frac{\partial U}{\partial y}\right)^2-\varepsilon+\frac{\partial}{\partial y}\left[\left(v+\frac{v_T}{\sigma_k}\right)\frac{\partial k}{\partial y}\right] \quad (10)$$

$$U\frac{\partial\tilde{\varepsilon}}{\partial x}+v\frac{\partial\tilde{\varepsilon}}{\partial y}=f_1C_{\varepsilon1}\frac{\tilde{\varepsilon}}{k}v_T\left(\frac{\partial U}{\partial y}\right)^2-f_2C_{\varepsilon2}\frac{\tilde{\varepsilon}^2}{k}+E+\frac{\partial}{\partial y}\left[\left(v+\frac{v_T}{\sigma_\varepsilon}\right)\frac{\partial\tilde{\varepsilon}}{\partial y}\right] \quad (11)$$

Here v_T is eddy viscosity coefficient, k is turbulent kinetic energy and $\tilde{\varepsilon}$ is turbulent dissipation rate. The other parameters in Eqs. (10) and (11) are described as

$$f_\mu=\left(1-e^{-0.0165R_y}\right)^2\left(1+20.5/\mathrm{Re}_T\right)$$

$$f_1=1+\left(0.05/f_\mu\right)^3$$

$$f_2=1-e^{-\mathrm{Re}_T^2}\qquad E=0\qquad \varepsilon_0=0$$

$$\varepsilon=\tilde{\varepsilon}+\varepsilon_0$$

$$\mathrm{Re}_T=\frac{k^2}{\tilde{\varepsilon}v},\quad R_y=\frac{k^{1/2}y}{v},\quad y^+=\frac{u_\tau y}{v}$$

$$C_{\varepsilon1}=1.44, C_{\varepsilon2}=1.92, C_\mu=0.09, \sigma_k=1.0, \sigma_\varepsilon=1.3$$

By calculating Eqs. (9), (10) and (11) can be obtained turbulent viscosity, and thus can solve the Eqs. (7) and (8).

Validation and discussion of numerical results

In order to validate the program, two cases, the laminar boundary layer and turbulent boundary layer, are performed.

Firstly, it is about the laminar boundary layer. We know that similarity solution for a flat plate laminar boundary layer is the Blasius solution. Liepmann did experiments in 1943 and compared the theoretical result with the experimental value. The results were agreed well with each other, thus proved the validity of the Blasius solution which became one of the criterions to validatenumerical program. This paper also calculated the numerical results of a flat plate laminar boundary layer. The computational domain is shown in Fig. 1, which is divided into a 2048×512 mesh. The Reynolds number is 10000. The comparison with the Blasius solution is shown in Fig. 2. In the picture, u_e is a free flow velocity outside the boundary layer and y is the distance of the wall. From Fig. 2,good agreement is found between the present numerical results and the Blasius solution.

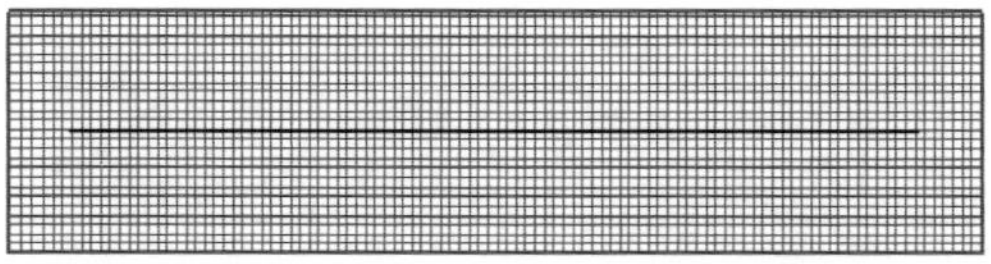

Fig.1 The computational domain and grid of the flat-plate

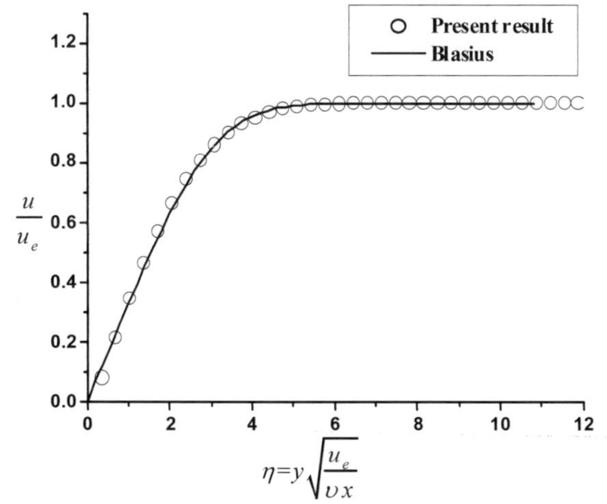

Fig.2 Comparison of the numerical result and Blasius result of the boundary layer of laminar flow

On the basis of the successful simulation about the flat plant laminarboundary and in order to examine the validity and correctness of turbulence models, the calculation of the turbulence boundary layer of a plate is simulated. Compared to the laminar boundary layer, the turbulent boundary layer is much more complex. It is divided into inner layer and outer layer, and the inner layer is divided into the viscous sublayer and logarithmic law layer, the outer layer is divided into upper and linear speed loss layer.The logarithmic law layer and speed loss layer are fully turbulent layers. In order to study the turbulent boundary layer, it needs to introduce several parameters[20].

$$u_\tau = \left(\frac{\tau_w}{\rho}\right)^{\frac{1}{2}} \tag{12}$$

Here $\tau_w = \mu\left(\frac{\partial u}{\partial y}\right)_{y=0}$, u_τ is Friction velocity. Due to the nearer the wall, the smaller the effective measure of vortex, so y can be seen as the length of the vortex scale, and the Reynolds number

$$y^+ = \frac{u_\tau y}{\nu} \tag{13}$$

can be seen as typical Reynolds number, the y reaction the influence of the viscosity changing with y. After the introduction of these relationships, the mean velocity profile of the inner layer can be expressed as:

$$u^+ = \phi_1\left(y^+\right) \tag{14}$$

Where,

$$u^+ = \frac{U}{u_\tau} \tag{15}$$

Eqs. (14) is called the law of the wall, which is proposed by Prandtl in 1925. Typically, at very small y, the velocity profile is linear which can be write as

$$u^+ = y^+ \tag{16}$$

This region near the wall is called the viscous sublayer and the velocity profile of the logarithmic layer can be write as

$$u^+ = \frac{1}{\kappa}\ln y^+ + C \tag{17}$$

where κ is Von karman constant which is usually taken 0.41. C is a constant, for smooth wall, C = 5.0 ~ 5.2. Computational domain and grid number are just the same as above and take the same Reynolds number 10000. Fig. 3 shows the numerical results and the results of above two equations. It is found that our numerical results are in relatively good agreement with the results of above two equations, which means the application of the turbulence model in this paper is correct. Also it is important to note that the use of the immersed boundary method in these problems does reveal obvious advantages, particularly in its simple mesh generation.

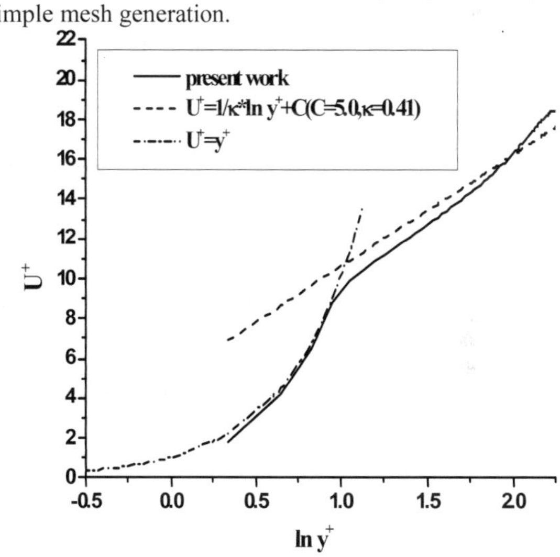

Fig.3 Comparison of the numerical result and the wall law of the boundary layer of turbulent flow

NUMERICAL ANALYSIS AND DISCUSSION OF AN OSCILLATING CASCADE

Base on the checks above, we will extend its application to the oscillating cascade in this section. Meanwhile, because of the application of turbulence model, the computational time cost has been increased. So the program parallelization is used to reduce the computing time. In previous work, we have established a quasi-three-dimensional computing platform and have been some preliminary and detailed study on the oscillatingcascades in laminar flow. That has proved the feasibility about solving fluid-structure interaction problem with immersed boundary method, and preliminary illustrates the immersed boundary method which is applied to the numerical simulation of complex turbomachinery is efficient and accurate.But the work just dealt withthe physical phenomena related to the low Reynolds number and it is far from the reality problems. So this paper will do some research about the oscillating cascade on the high Reynolds number.

Just as previous work, the oscillating cascade consists of five blades, just as shown in Fig. 4. Each blade has two degrees of freedom. The equations for each blade are given as follows:

$$m\ddot{h} + S_\theta \ddot{\theta} + c_h \dot{h} + k_h h = F_h \quad (18)$$

$$S_\theta \ddot{h} + I_\theta \ddot{\theta} + c_\theta \dot{\theta} + k_\theta \theta = M_\theta \quad (19)$$

all parameters in Eqs. (18) and (19) are described as

$$b = c/2 \quad \omega_h = \lambda\omega_\theta \quad k_h = m\omega_h^2 \quad c_h = 2\xi_h\omega_h m$$

$$S_\theta = mbx_\theta \quad I_\theta = mb^2\gamma_\theta^2 \quad k_\theta = I_\theta\omega_\theta^2 \quad c_\theta = 2\xi_\theta\omega_\theta I_\theta$$

We defined the reduced velocity $U^* = U_\infty / f_n D = 2\pi U_\infty / \omega_h c$. The $Newmark-\beta$ method is adopted to solveEqs. (18) and (19). The following parameters are considered: m=1.0 , c=1.0 , $a_h = -1$, $x_\theta = 1.8$, $\gamma_\theta = 1.865$, $\xi_h = 0$, $\xi_\theta = 0$ and $\lambda = 1$. For all ofthe following simulations, the initial angle of attack is set to 10. The system damping coefficients ξ_h and ξ_θ are set to zero in order toexcite high-amplitude oscillations.

With the reference length and velocity chosen to bethe chord length and inflow velocity, the Reynolds number is set to 10000 in this case, the lift coefficient is $C_L = \frac{F_L}{(1/2\rho U_\infty^2 c)}$. The blade profile is NACA0012. The mesh density is just as the mesh in the simulation of the turbulence boundary layer. The time step is $\Delta t = 0.8\times10^{-4}$. Before the calculation, we must verify that the independence of time step. As shown in Fig. 5, it is found that thecomputational time step is suitable.

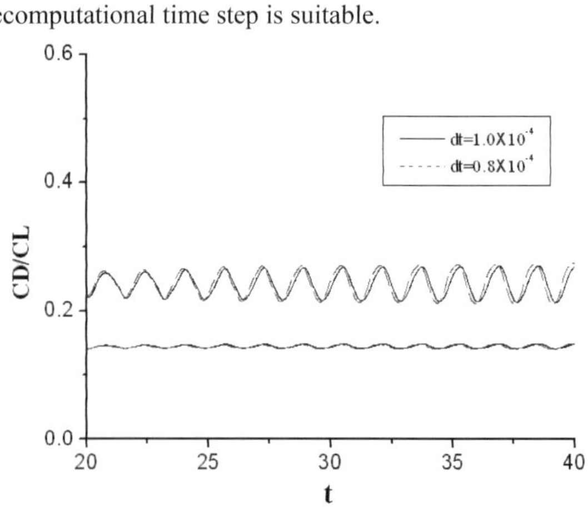

Fig.5 Comparison of results with different time steps

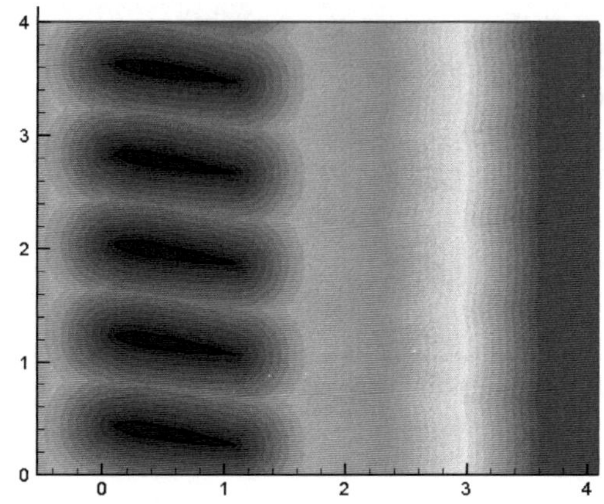

Fig.6 The picture of the distance from the wall y

We know that the distance from the wall is the basis of calculating the turbulence. Fig. 6 shows the results of the distance from the wall y and it is found that the result is correct and reasonable. Fig. 7-Fig. 12 shows time evolution of the transverse displacement and the rotation angle of the blades when the reduced velocity is different. From these pictures, we can find that when the reduced velocity is small, Fig. 13-Fig. 15 gives the variation of rotational angles versus rotational velocities. From Fig. 7-Fig. 12, we can find that the oscillations in two degrees of freedom are damped and after a certain time, the blade oscillation would eventually reach a fixed transverse displacement and angle of attack. The damping time is prolonged with the increase of the reduced velocity. When reduced velocity approaches to 3.0, the oscillation amplitudes in two degrees of freedom are not damped or amplified, the airfoil oscillates in a quasi-steady amplitude and period.When the reduced velocityequals 4.0, the rotational angle and transverse displacement become large rapidly and the system becomes unstable, the cascade will oscillate in high amplitude and the oscillation amplitude is amplified rapidly to infinity.

It is much clearer to find this phenomenon from Fig. 13-Fig. 15.The systems show different behaviors at different reduced velocity. For $U^* = 3.0$, the blade is attracted to an equilibrium state of oscillation, while for $U^* = 4.0$, the equilibrium state of oscillation seems not steady because all the parameters are amplified fast, which means nonlinear oscillation occurs. And for $U^* = 2.0$, the system eventually converges to a steady state.

Fig. 16 is the picture of the blade vibration. The principle of energy method is that in a period of vibration T, the positive or negative of the energy that exchanges between blades and the fluid determines whether the system is flutter. From this picture, we can calculate the aerodynamic work which is shown as Eqs. (20). When $W > 0$,the system is unstable, otherwise, the system is stable. But in this paper, because of the fluid and structure are coupled, there is no prescribed motion. So the period of vibration T is not known in advance. But we can present results are shown as functions of a temporal series and the trend of the aerodynamic work canbe applied to judge the system is stable or not. Fig. 17-Fig. 19 are time evolution of aerodynamic work at different reduced velocity. From the trend of the curve, we can also found that when $U^* = 2.0$, the system is stable, while for $U^* = 3.0$, the system is critical stable. And for $U^* = 4.0$, the system is unstable.

$$dW = -F_h \cdot dh + F_d \cdot ds + M_\theta \cdot d\theta$$
$$W = -\oint_T F_h \cdot dh + \oint_T F_d \cdot ds + \oint_T M_\theta \cdot d\theta \quad (20)$$

Fig. 20-Fig. 21 are the amplitude-frequency spectrum of blade oscillations at different reduced velocity. From these, we can find that the oscillation of the blade associated with the first harmonic oscillation of the lift. The frequency of oscillation is different from the reduced velocity. Based on the previous analysis, it isfound that the reduced velocity is the critical stable boundary at its value close to 3.0 in this case.

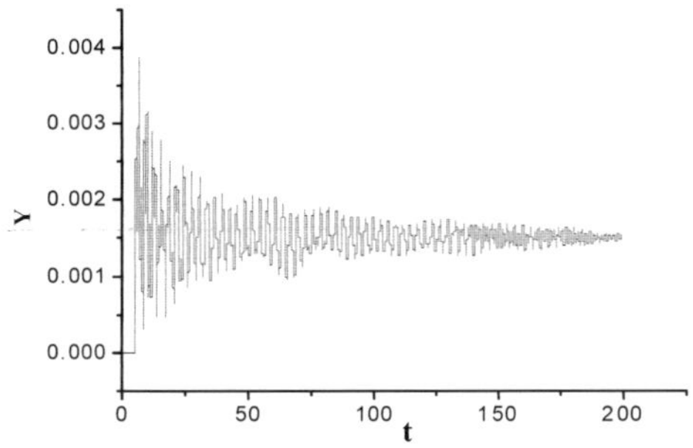

Fig.7 The time evolution of the transverse displacement, $U^* = 2.0$

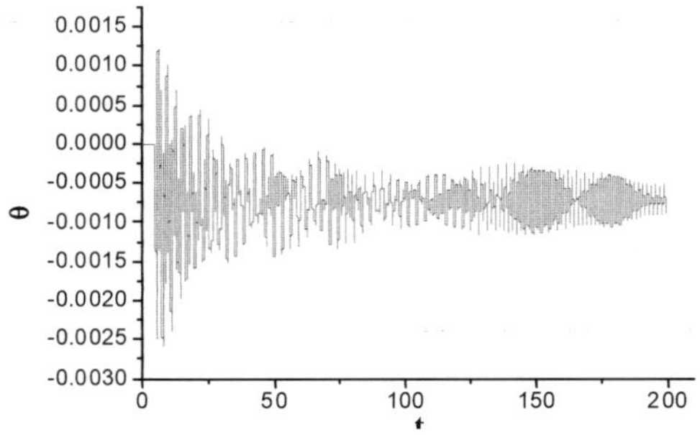

Fig.8 The time evolution of the rotational angle, $U^* = 2.0$

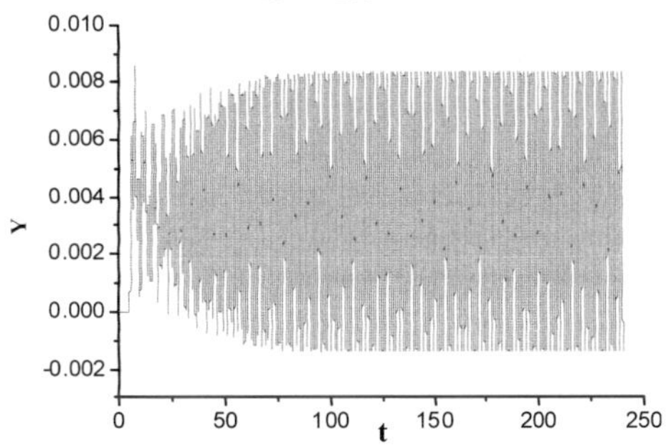

Fig.9 The time evolution of the transverse displacement, $U^* = 3.0$

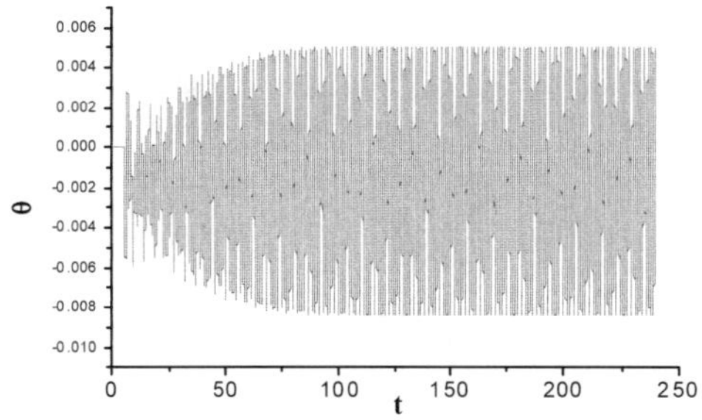

Fig.10 The time evolution of the rotational angle, $U^* = 3.0$

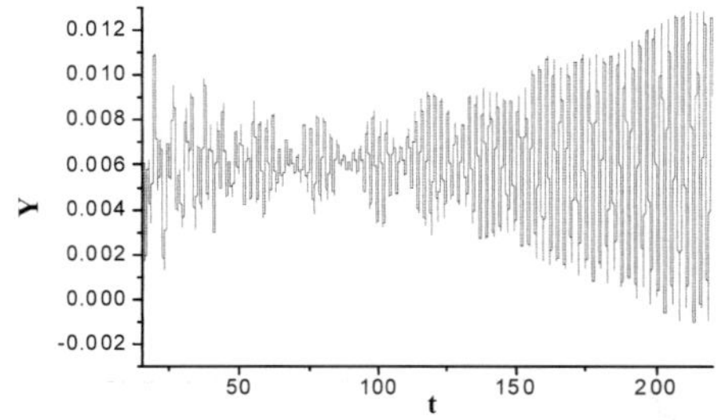

Fig.11The time evolution of the transverse displacement, $U^* = 4.0$

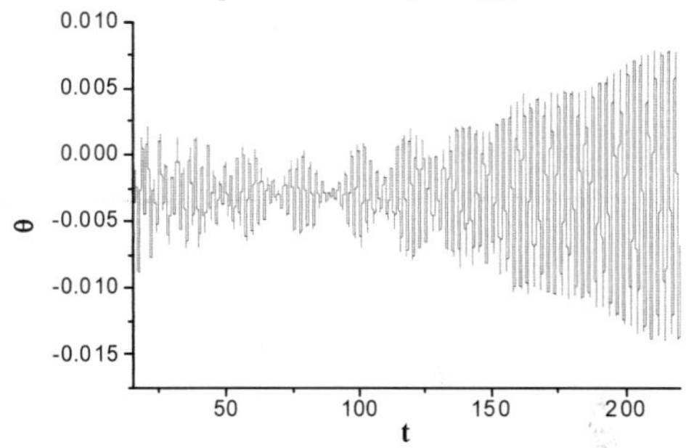

Fig.12 The time evolution of the rotational angle, $U^* = 4.0$

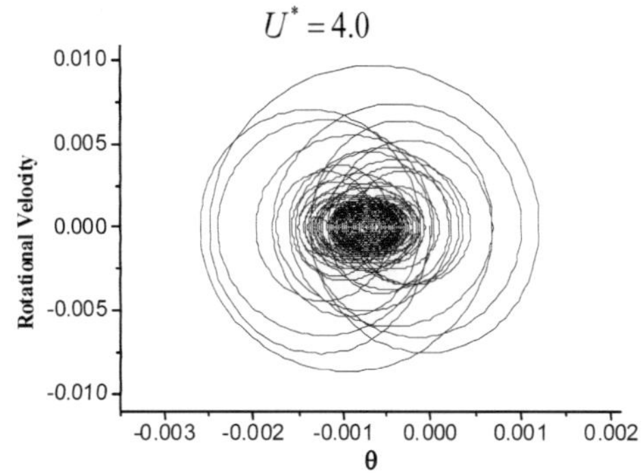

Fig.13 Variation of the rotational angle versus rotationalVelocity, $U^* = 2.0$

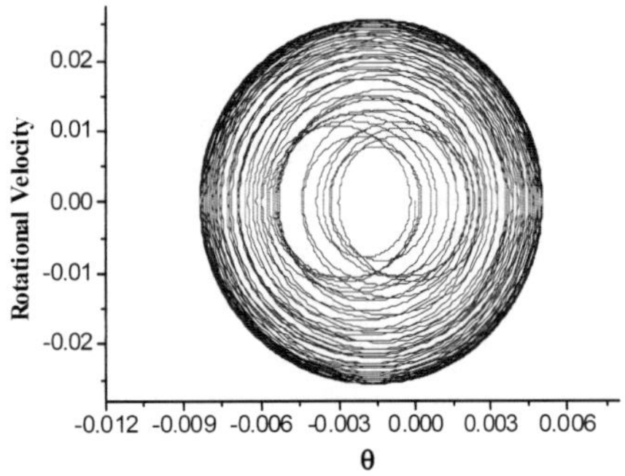

Fig.14 Variation of the rotational angle versus rotational Velocity, $U^* = 3.0$

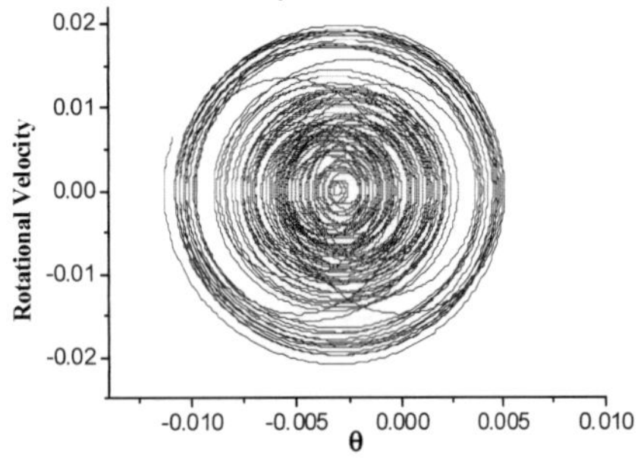

Fig.15 Variation of the rotational angle versus rotational Velocity, $U^* = 4.0$

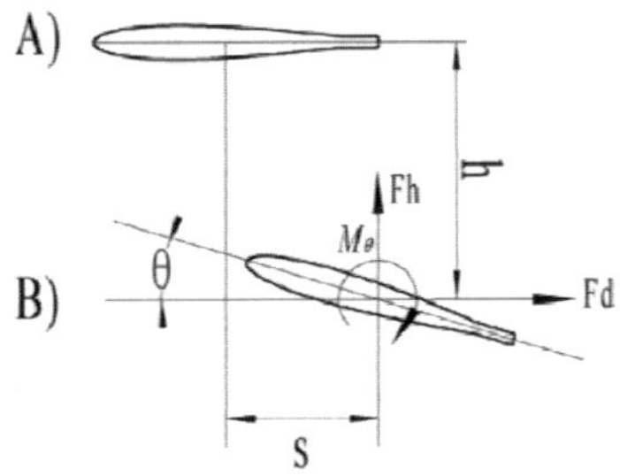

Fig.16 Blade vibration, A): Initial position; B): Instantaneous position

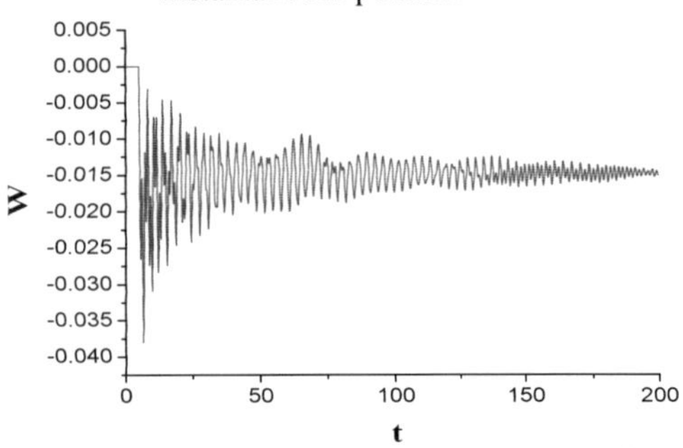

Fig.17 The time evolution of aerodynamic work, $U^* = 2.0$

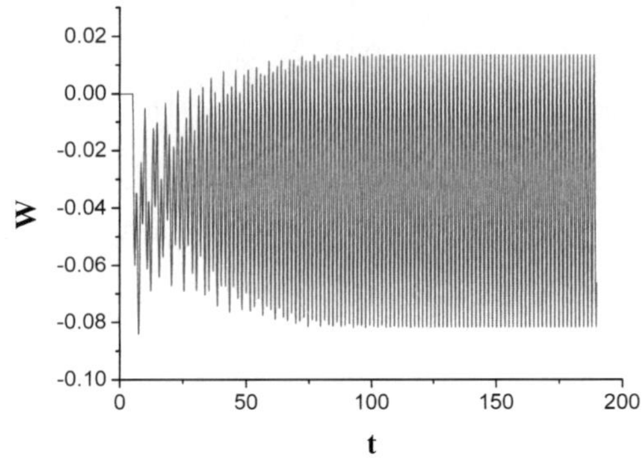

Fig.18 The time evolution of aerodynamic work , $U^* = 3.0$

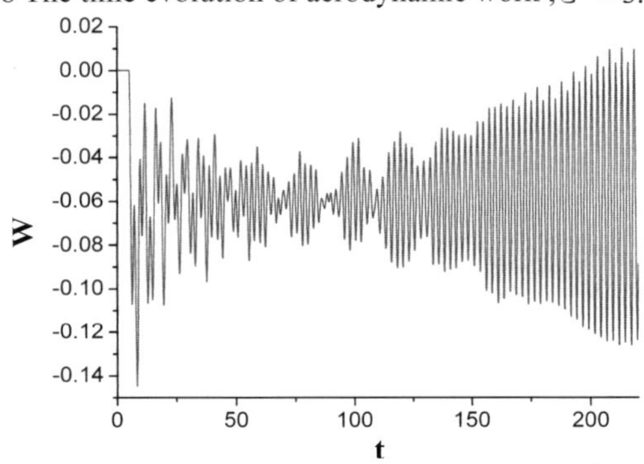

Fig.19 The time evolution of aerodynamic work, $U^* = 4.0$

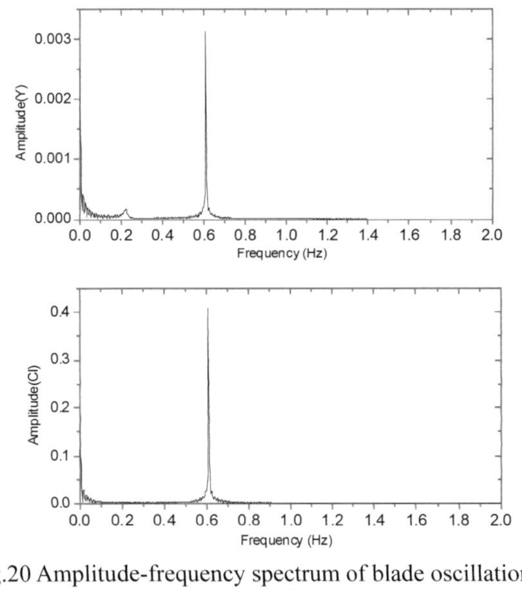

Fig.20 Amplitude-frequency spectrum of blade oscillations with time, $U^* = 3.0$

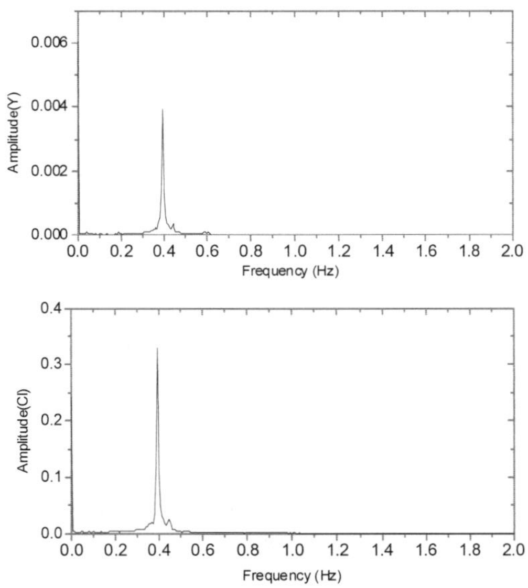

Fig.21 Amplitude-frequency spectrum of blade oscillations with time, $U^* = 4.0$

Fig. 22 shows the time evolution of lift coefficient of different blade when $U^* = 3.0$. For this picture, it is found that the blades vibrate in the same amplitude and in the same phases, that means the IBPA is 0 degree. Fig. 23 shows vorticity distributions when the system is stable. It is also found that the vortex sheds in a 0 degree. Also, Fig. 24 gives the time evolution of lift coefficient of different blade when $U^* = 4.0$ and the blades vibrate in different phases, which is 72 degrees. From the vorticity distributions as shown in Fig. 25, we can see that clearly. From the above various numerical results, we can find that when the reduced velocity is different, the oscillation of the blades in the cascade are different, which leads to different IBPA between adjacent blade. This conclusion is consistent with that in the laminar flow which have been studied.

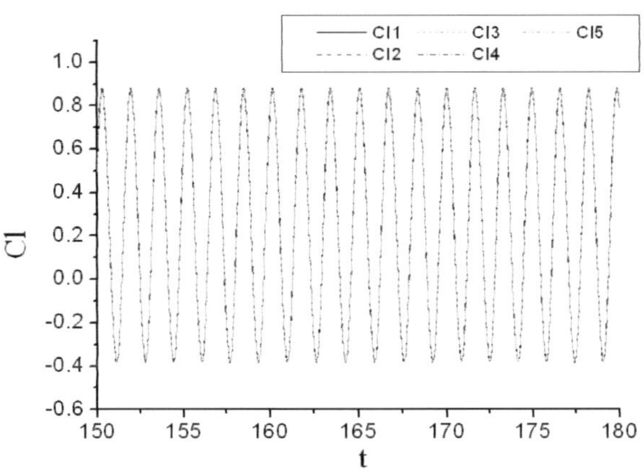

Fig.22 The life force coefficient of each blade with time when system is stable, $U^* = 3.0$

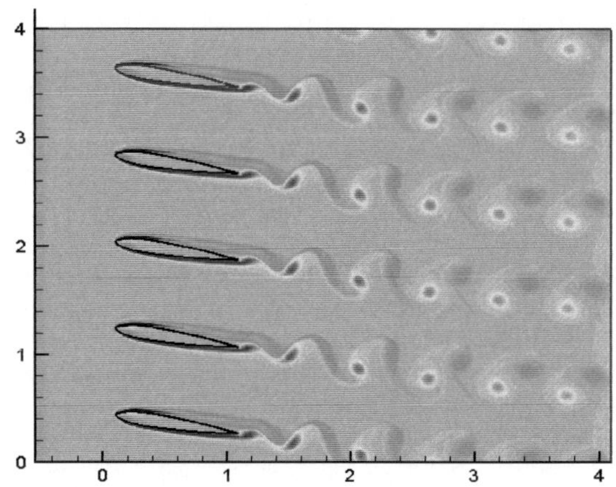

Fig.23 Vorticity distributions when system is stable, $U^* = 3.0$

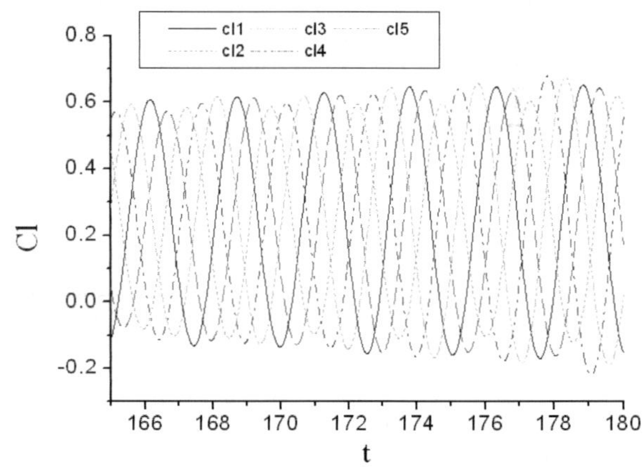

Fig.24 The life force coefficient of each blade with time, $U^* = 4.0$

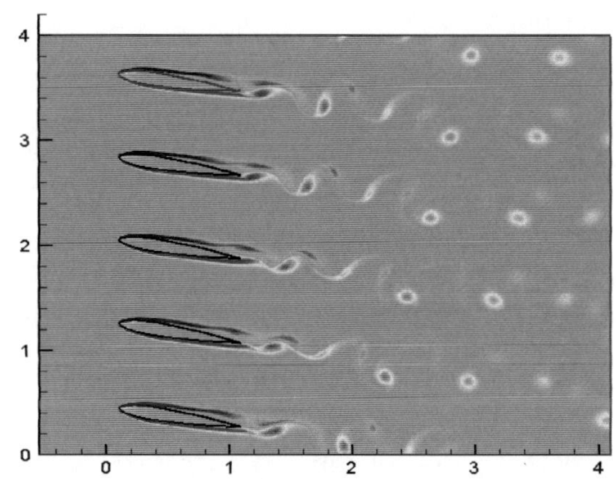

Fig.25 Vorticity distributions when system is stable, $U^* = 4.0$

CONCLUSION

In order to consider more reality problems, this paper established a fast explicit numerical method combines the turbulence model for the turbulence closure based on the immersed boundary method. The numerical simulation method presented in this paper is first validated by two cases: laminar boundary layer and turbulent boundary layer. With these checks, the present work is extended to investigate the cascade oscillating in high Reynolds number with an emphasis on the physical understanding of the fluid-structure interaction. We found that the reduced velocity is a sensitive parameter, which affects the oscillation of cascade and the IBPA, which is just the same conclusion as the laminar results. Also, present results are shown as functions of a temporal series, it is convenient to see when and how the system becomes unstable. Because of the introduction of the turbulence model, the program parallelization is used to reduce the computing time. Besides, It is worth noting that the coupling process is not necessary to be generating anybody-fitting grid, which makes it much faster in computational process for such a cascade problem. However, in this paper, the fluid influence rather than the disk and blade profile's oscillation modes are taken into account. So, further research work is still required to find the reduced velocity corresponding to a critical stable boundary for a real cascade aeroelasitic analysis.

ACKNOWLEDGMENTS

This research work has been supported by National Natural Science Foundation of China (Nos. 51236001, 51106154), and National Basic Research Program of China (No. 2012CB720201).

REFERENCES

[1] Ballhaus W F. Goorjina P M. Computation of unsteady transonic flows by the indicial method[J]. AIAA Journal, 1978, 6: 117-124.

[2] He L. An Euler solution for unsteady flows around oscillating blades[J]. Journal of Turbomachinery, 1990, 112(4): 714-722.

[3] He L. Denton J D. Three dimensional time-marching inviscid and viscous solutions for unsteady flows around vibrating blade[R]. ASME Paper 1993-GT-92.

[4] Sadeghi M. Liu F. Computation of mistuning effects on cascade flutter[J]. AIAA Journal, 2001, 39(1): 22-28.

[5] Sadeghi M. Liu F. Computation of cascade flutter by uncoupled and coupled methods[J]. International Journal of Computational Fluid Dynamics. 2005, 8: 559-569.

[6] Sadeghi M. Yang S. Liu F. et al. Parallel computation of wing flutter with a coupled Navier-Stokes/CSD method [J], AIAA-03-1347, 2003.

[7] Sadeghi M. Liu F. Investigation of mistuning effects on cascade flutter using a coupled method[J], Journal of propulsion and power, 2007 23(2):266-272.

[8] Kazawa J. Watanabe T. Numerical analysis toward active control of cascade flutter with smart structure[R]. AIAA-02-4079, 2002.

[9] Piperno S. Explicit/implicit fluid/structure staggered procedures with a structural predictor and fluid subcycling for 2D inviscid aeroelastic simulations[J]. International Journal for Numerical Methods in Fluids, 1997, 25:1207-1226.

[10] Gottfried D A. Simulation of fluid-structure interaction in turbomachinery[D]. West Lafayette, Engineering, Mechanical: Purdue University, 2000.5.

[11] Hu P G. et al. Material point method applied to fluid-structure interaction (FSI)/aeroelasticity problems[R], AIAA-2010-1464, 2010.

[12] Hu P G. et al. Material point method with least squares technique for nonlinear aeroelasticity and fluid-structure interactions (FSI) in ASTE-P Toolset[R]. AIAA-2010-8224, 2010.

[13] Peskin C S. Numerical analysis of blood flowin the heart, Journal of Computational Physics[J]. 1977, 25(3):220-252.

[14] Zhong G H, Sun X F. A simulation strategy for an oscillating cascade in the turbomachinery using immersed boundary method[J]. AIAA Journal of Propulsion and Power, 2009, 25(2): 312 -321.

[15] Goldstein D. Handler R. Sirovich L. Modeling a no-slip flow with an external force field[J]. Journal of Computational Physics, 1993, 105(2):354-366.

[16] Hu G T. Sun X F. A numerical modeling of the vortex-induced vibration of cascade in turbomachinery using immersed boundary method[J]. Journal of thermal science, 2011, 20(3):229-237.

[17] Hu G T. Du L. Sun XF. An immersed boundary method for simulating oscillating rotor blades[J]. Acta Aeronauticaet Astronautica Sinica, 2014, 35(8):2112-2125

[18] Lam C K. Bremborst K A. Modified form of the $k \sim \varepsilon$ model for predicting wall turbulence[J]. Journal of Fluids Engineering,1981,103:456-460.

[19] Wilcox D C. Comparison of two-equation turbulence models for boundary layers with pressure gradient[J]. AIAA Journal, 1993, 31(8):1414-1421.

[20] Maozhang Chen. Fundamentals of viscous Fluid Dynamics[M]. Higher Education Press, 2002.

Proceedings of the 14th International Symposium on Unsteady Aerodynamics, Aeroacoustics & Aeroelasticity of Turbomachines
ISUAAAT14
8-11 September 2015, Stockholm, Sweden

I14-S11-5

TIME-LINEARIZED AND HARMONIC BALANCE NAVIER-STOKES COMPUTATIONS OF A TRANSONIC FLOW OVER AN OSCILLATING BUMP

Q. Rendu
UMR CNRS 5509, LMFA,
Université de Lyon, France

M. Philit
UMR CNRS 5509, LMFA,
Université de Lyon, France

Y. Rozenberg
UMR CNRS 5509, LMFA,
Université de Lyon, France

S. Aubert
UMR CNRS 5509, LMFA,
Université de Lyon, France

P. Ferrand
UMR CNRS 5509, LMFA
Université de Lyon, France
Affiliate Professor at KTH Royal Institute of Technology, Sweden

S. Labit
Insititut Jean le Rond d'Alembert,
UPMC-CNRS UMR7190,
Univ Paris 6, 75005 Paris, France

J.-C. Chassaing
Insititut Jean le Rond d'Alembert,
UPMC-CNRS UMR7190,
Univ Paris 6, 75005 Paris, France

ABSTRACT

To predict onsets of aeroelastic instabilities in turbomachineries, as the unsteady flow is mainly periodic, linearized and harmonic-balance methods have been developed to reduce computational costs. Harmonic balance and time-linearized methods are evaluated in this work on shock-wave/boundary layer interaction over an oscillating bump in a transonic nozzle. The results are compared with experimental data obtained at Royal Institute of Technology (KTH, Sweden). Harmonic-balance calculations with Reynolds stress model and nonlinear unsteady RANS calculations with the revisited k-ω turbulence model of Wilcox show good agreement with experimental results. Time-linearized computations succeed in reproducing the URANS results when the turbulence model is fully derivated, whereas frozen turbulence assumption does not give satisfactory results to predict the aeroelastic stability.

INTRODUCTION

The prediction of oscillating loads on a structure is a key component of fluid-structure interaction analysis. This prediction can nowadays be achieved thanks to nonlinear Unsteady Reynolds-Averaged Navier-Stokes (URANS) methods. These methods are accurate enough to identify some starting instabilities, such as surge or flutter, and they can deal with complex geometries, but still at high restitution delays. Such computations are prohibitive for design or optimization purposes where a lot of configurations should be examined. For periodic flows, time-linearized, harmonic linearized and Time Harmonic Balance (THB) methods can predict the harmonic response of the flow without computing the transient regime between steady and periodic states. These methods are thus 10 to 100 times faster than standard nonlinear time-marching techniques.

Time-linearized methods have been widely used in turbomachinery for aeroelastic studies (1; 2; 3; 4). In time-linearized methods, the unsteady flow is assumed to be described as the sum of the steady solution and a harmonic perturbation at a prescribed frequency. Linearizing the Navier-Stokes equations around a steady state results in a system of linear equations which can be quickly solved using iterative Krylov subspace methods (5; 6). Through linearized methods, only a single harmonic of the periodic response of the flow is obtained. This may be enough to estimate the aerodynamic damping around a vibrating blade but not sufficient for a deeper analysis of the flow.

THB methods have been developed in the past decades to compute the nonlinear unsteady flow, amongst others for turbomachinery applications (7; 8; 9). They rely on the Fourier decomposition of the unsteady periodic flow. Such methods solve the nonlinear coupling between the time-averaged flow and a prescribed number of harmonics N_h of the unsteady periodic flow. To this end, $2N_h + 1$ quasi-steady computations at different times are needed to reconstruct the unsteady response of the flow. The number of steady computations may be important to recover a reasonable number of harmonics, but the use of classical techniques to improve convergence properties, such as local time stepping and implicit methods, greatly reduces the computational costs.

If these methods have been widely used for inviscid or viscous attached flows, the prediction of turbulence response is a great challenge, especially for dynamic separation of boundary layers. In the present work, time-linearized method with k-ω turbulence models of Wilcox (10; 11) and THB method with Reynolds Stress Model (RSM) closure are used to predict shock-wave/boundary-layer interaction over an oscillating bump. The results are compared to nonlinear RANS computations as well as experimental data from KTH-VM100 facility.

In the first part of this paper is presented the experimental setup used for validation. Then, numerical methods are presented, evaluated and compared to experimental results.

NOMENCLATURE

A	amplitude of bump oscillation	[m]
c	sound celerity	[m.s^{-1}]
c_{ax}	axial chord	[m]
Cp	pressure coefficient	[-]
E	fluid total-energy	[m^2.s^{-2}]
ε	turbulent dissipation	[m^2.s^{-3}]
f	frequency	[Hz]
ζ	damping coefficient	[-]
k_r	reduced frequency	[-]
k	turbulent kinetic energy (TKE)	[m^2.s^{-2}]
L	spanwise dimension of the bump	[m]
λ	wavelength	[m]
M	Mach number	[-]
N_h	prescribed number of harmonics for harmonic-balance computations	[-]
P	exchanged power	[J.s^{-1}]
Ps	static pressure	[Pa]
Pt	total pressure	[Pa]
ρ	density	[kg. m^{-3}]
t	physical time	[s]
Δt	physical time step	[s]
τ	pseudo time for dual time-stepping	[s]
$\Delta\tau$	pseudo time step	[s]
T	time period of a bump oscillation	[s]
Ts	static temperature	[K]
$\mathcal{U}$	maximal kinetic energy of vibration	[J]
$\boldsymbol{u}$	fluid velocity vector	[m.s^{-1}]
x	position in streamwise direction	[m]
y	position in spanwise direction	[m]
Δy	displacement of the bump	[m]
$\mathcal{W}$	exchanged work	[J]
ω_f	pulse ($\omega_f = 2\pi f$)	[s^{-1}]
ω	turbulent dissipation rate	[s^{-1}]
Ω	fluid-structure contact surface	[m^2]
$^0w(\boldsymbol{x})$	steady part of instantaneous complex variable $\widetilde{w}(\boldsymbol{x},t)$	[-]
$^i\widetilde{w}(\boldsymbol{x})$	complex amplitude of i-th harmonic of instantaneous complex variable $\widetilde{w}(\boldsymbol{x},t)$	[-]
j	complex number ($j^2 = -1$)	[-]
$\lvert z\rvert$	amplitude of complex number z	[-]
$\Phi(z)$	phase of complex number z	[degree]

KTH-VM100 FACILITY

The experimental setup used for validation consists in a generic model whose profile is similar to a typical compressor blade. It is located at Royal Institute of Technology (KTH) in the VM100 facility transonic nozzle. In the present work, the generic model considered is a bi-dimensional polyurethane bump, which can oscillate at different frequencies to reproduce the first bending mode of a compressor blade. The height of the nozzle is equal to the axial chord length of the bump $c_{ax} = 120mm$.

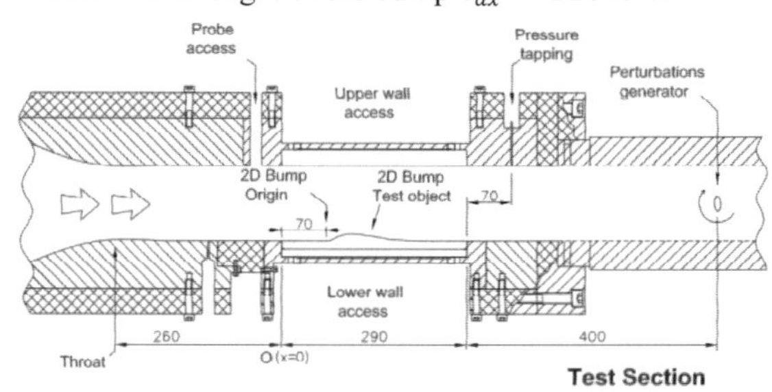

Figure 1 Bi-dimensional bump in KTH-VM100 facility from Bron (12)

The bi-dimensional geometry of the polyurethane bump, presented in Figure 1, is obtained from Bron (13). The bending mode is excited thanks to a rotating camshaft (see Figures 2 and 3). The prismatic shape of the camshaft allows an oscillation of the flexible bump up to a frequency of $500Hz$ (corresponding to a driven rotation speed of $10{,}000\,rpm$). The motion of the bump is sinusoidal with expected amplitude $\pm 0.5mm$.

Steady pressure coefficients were obtained along the bump by Ferria (14). These experimental results are used by Labit (15) to calibrate 2D steady RANS computations and are further presented in this paper to validate the shock-wave position in steady computations.

Unsteady pressure coefficients were obtained along the bump for different frequencies ranging from $10Hz$ to $300Hz$, corresponding to a range of reduced frequency $k_r \in [0.03,0.88]$. According to Ferria (14), the calculation of reduced frequency is based on the full axial chord:

$$k_r = \frac{\omega_f}{u_{exit}} * c_{ax}$$

As pointed out by Andrinopoulos (16), the polyurethane bump does not present the expected modes, especially at high frequencies. An aluminium bump has been designed to correct this issue, however the dynamic results are still to be obtained. Thus, the experimental results used for validation in this work are obtained with the polyurethane bump. The deformation measured at $f = 100Hz$ (corresponding to a reduced frequency $k_r = 0.29$) and presented in Figure 5 is used for unsteady computations.

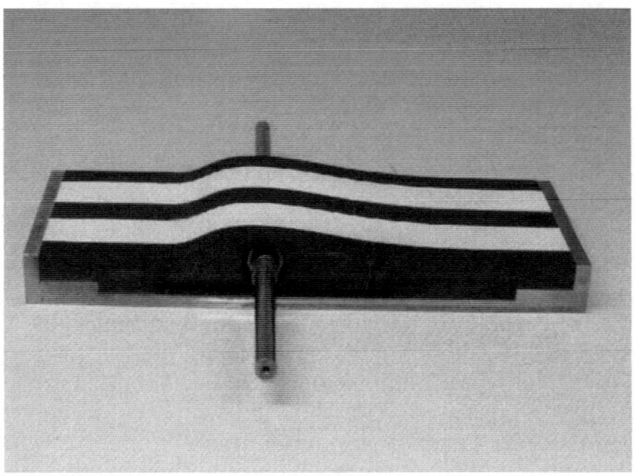

Figure 2 Polyurethane bump with camshaft from Ferria (14)

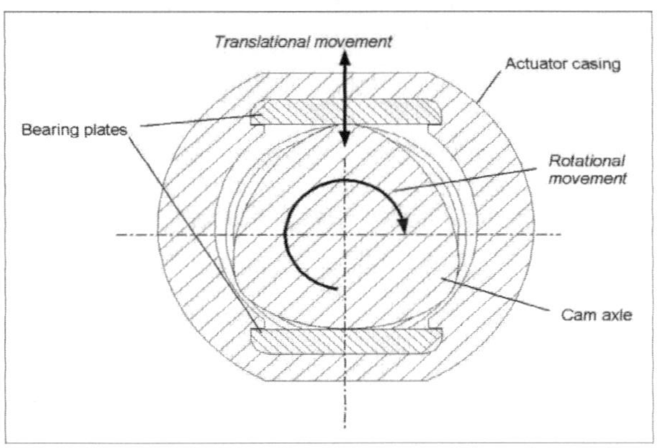

Figure 3 Camshaft of the oscillating polyurethane bump from Allegret-Bourdon (17)

AERODYNAMIC DAMPING COEFFICIENT

To obtain quantitative information on aerodynamic stability, one may want to compute the aerodynamic damping coefficient. If positive, this coefficient represents the damping of the vibrating structure by the surrounding flow. If negative, it quantifies the amount of energy given by the flow to the structure. In case of aerodynamic instability, the overall stability is given by the sum of the negative aerodynamic damping coefficient and the structural damping coefficient (computed from the type of embedding, the materials ...). If the global damping coefficient is positive in spite of an aerodynamic instability, the system succeeds in dissipating mechanically the energy extracted from the surrounding flow.

The aerodynamic damping coefficient ζ is defined as:

$$\zeta = \frac{1}{4\pi}\frac{\iint_{\Omega} \mathcal{W}\, d\Omega}{\mathcal{U}} \quad (1)$$

where $\mathcal{W} = \int_0^T P(\boldsymbol{x},t)\,dt = \Re\left(\int_0^T \tilde{P}(\boldsymbol{x},t)\,dt\right)$ is the exchanged work between the flow and the structure, $\tilde{P}$ is the instantaneous complex exchanged power and $\mathcal{U}$ is the maximal kinetic energy of the vibration. $\mathcal{U}$ is defined as :

$$\mathcal{U} = \frac{1}{2}\rho_{solid}\, SL\omega_f{}^2 \Delta y_{max}{}^2 \quad (2)$$

where $\rho_{solid}\, SL = 0.08\,kg$ is the mass of the oscillating part of the bump for a spanwise dimension $L = 0.1m$ ($S = 8.04\,cm^2$, $\rho_{solid} = 1030kg/m^3$). In the case of a pure bending mode, the exchanged power can be expressed as:

$$\tilde{P}(\boldsymbol{x},t) = -\widetilde{Ps}(\boldsymbol{x},t) * Ldx * \boldsymbol{n}(\boldsymbol{x}) \cdot \widetilde{\boldsymbol{V}}(\boldsymbol{x},t) \quad (3)$$

where $\widetilde{Ps}$ is the instantaneous complex pressure, $\widetilde{\boldsymbol{V}}$ the instantaneous complex velocity of the structure, Ldx the local steady surface of the bump (dx being an infinitesimal streamwise length) and $\boldsymbol{n}$ a normal vector to this surface, oriented outwards of the flow. Knowing the fundamental frequency $f = \omega_f/2\pi$ of the unsteady flow, the instantaneous variables can be written :

$$\widetilde{w}(\boldsymbol{x},t) = {}^0w(\boldsymbol{x}) + \sum_{i>0} {}^i\widetilde{w}(\boldsymbol{x})e^{j(\omega_f * i)t}$$

where ${}^0w(\boldsymbol{x})$ is the steady part of the variable $\widetilde{w}$ and ${}^i\widetilde{w}(\boldsymbol{x})$ the complex amplitude of the i-th harmonic. Keeping only the first harmonic ($i = 1$), we can write:

$$\int_0^T \tilde{P}(\boldsymbol{x},t)\,dt = \int_0^T -\widetilde{Ps}(\boldsymbol{x},t) * Ldx * \boldsymbol{n} \cdot \widetilde{\boldsymbol{V}}(\boldsymbol{x},t)\,dt$$
$$= T * {}^1\widetilde{Ps}(\boldsymbol{x}) * Ldx * \boldsymbol{n}(\boldsymbol{x}) \cdot {}^1\widetilde{\boldsymbol{V}}(\boldsymbol{x})$$

and we can finally obtain the expression of the aerodynamic damping coefficient:

$$\zeta = \frac{1}{4\pi}\frac{T.\Re\left(\int_{wall} {}^1\widetilde{Ps}(\boldsymbol{x}) * \boldsymbol{n}(\boldsymbol{x}) \cdot {}^1\widetilde{\boldsymbol{V}}(\boldsymbol{x})\,dx\right)}{\frac{1}{2}\rho_{solid}\, S\omega_f{}^2 \Delta y_{max}{}^2}$$

TIME LINEARIZED RANS EQUATIONS

By linearizing the Navier-Stokes equations around a steady solution, we assume that the unsteady flow can be expressed as the sum of a steady part and a harmonic perturbation. Formally, if $\boldsymbol{q}(\rho, \rho\mathbf{u}, \rho E, \rho k, \rho\omega)$ is the vector of conservative and turbulent variables, the finite volume formulation of Navier-Stokes equations can be written as:

$$\frac{d}{dt}[\boldsymbol{J}(\boldsymbol{p})\boldsymbol{q}] + \boldsymbol{F}(\boldsymbol{p},\boldsymbol{q}) = \mathbf{0} \qquad (4)$$

where $\boldsymbol{p}$ is the vector determining boundary conditions, $\boldsymbol{J}(\boldsymbol{p})$ the vector of mesh cells volume and $\boldsymbol{F}$ is a nonlinear operator representing the balance between convective and viscous fluxes. Let assume that the vectors $\boldsymbol{p}$ and $\boldsymbol{q}$ can be written as the sum of a steady part (only space dependent) and a harmonic unsteady part:

$$\boldsymbol{p}(\boldsymbol{x},t) = {}^{0}\boldsymbol{p}(\boldsymbol{x}) + {}^{1}\widetilde{\boldsymbol{p}}(\boldsymbol{x})e^{j\omega_f t},$$
$$\boldsymbol{q}(\boldsymbol{x},t) = {}^{0}\boldsymbol{q}(\boldsymbol{x}) + {}^{1}\widetilde{\boldsymbol{q}}(\boldsymbol{x})e^{j\omega_f t}$$

where ω_f is a prescribed pulse. Note that the complex amplitude of perturbation vectors ${}^{1}\widetilde{\boldsymbol{p}}$ and ${}^{1}\widetilde{\boldsymbol{q}}$ must be small in comparison to $\boldsymbol{p}$ and $\boldsymbol{q}$ to achieve linearization. Within this formalism, the space-dependent steady vectors $\boldsymbol{p}$ and $\boldsymbol{q}$ satisfy:

$$\boldsymbol{F}({}^{0}\boldsymbol{p}, {}^{0}\boldsymbol{q}) = \mathbf{0} \qquad (5)$$

Finally, we can write the Linearized RANS equations:

$$\left\{j\omega_f \boldsymbol{J}({}^{0}\boldsymbol{p}) + \frac{\partial \boldsymbol{F}}{\partial \boldsymbol{q}}({}^{0}\boldsymbol{p}, {}^{0}\boldsymbol{q})\right\} {}^{1}\widetilde{\boldsymbol{q}} = -\left\{j\omega_f \frac{\partial \boldsymbol{J}}{\partial \boldsymbol{p}}({}^{0}\boldsymbol{q}) + \frac{\partial \boldsymbol{F}}{\partial \boldsymbol{p}}({}^{0}\boldsymbol{p}, {}^{0}\boldsymbol{q})\right\} {}^{1}\widetilde{\boldsymbol{p}}$$

In order to solve this equation, one must compute the derivatives of the nonlinear operator $\boldsymbol{F}$ according to the vectors $\boldsymbol{p}$ and $\boldsymbol{q}$. To this end, the partial derivatives are exactly evaluated by applying an automatic derivation tool to the $\boldsymbol{F}$ function using the RANS solver Turb'Flow (18; 19). These partial derivatives are then read by the LRANS solver Turb'Lin (20) used in this work.

Concerning the turbulence modelisation, two different approaches are tested here. If the time for the turbulence to react to the unsteadiness is high in comparison to the characteristic time of the unsteadiness (typically a period of oscillation for flutter analysis), it can be assumed that the turbulence behaviour is frozen to its steady state. Usually, frozen turbulence is set through a constant value of the eddy viscosity, with respect to time. This approached is used in this paper, following the work of Philit (21). Other implementations of frozen turbulence are exposed by Labit (15) such as imposing a constant value for the turbulent Reynolds number or for the turbulent variables k and ε.

On the other hand, one can assume that the turbulence depends on the unsteady flow. Philit (21) showed that this approach, called Harmonic Turbulence Linearized RANS (HT LRANS), is mandatory to reproduce the unsteady behaviour of shock-wave/boundary-layer interaction. The turbulence model used in this work is the classical k-ω model of Wilcox (10). The underlying assumptions of such a model are turbulence isotropy and immediate response of the turbulence to the flow. These assumptions, if not well suited for shock-wave/boundary-layer interaction, are specific to the turbulence model and not to the linearized computations. The main objective of this study remains the comparison between LRANS and URANS computations, even if the physical mechanisms are not perfectly captured.

NUMERICAL METHODS

Two different codes are used and compared in this work. The solver Turb'Flow (18; 19), developed at Laboratory of Fluid Mechanics and Acoustics in Ecole Centrale de Lyon, is used for steady RANS, Unsteady RANS (URANS) and Linearized RANS (LRANS) computations. Turb'Flow solves the compressible 3D Navier-Stokes equations using finite volume formulation. Centered and upwind spatial scheme are implemented, as well as several turbulence models. Both explicit and implicit methods are available for temporal discretization to achieve faster computations.

In this work, spatial discretization of convective fluxes is achieved by an upwind Roe scheme with 3rd order MUSCL interpolation and HCUI (Harmonic Cubic Upwind Interpolation) flux limiter on conservative variables and with a second order MUSCL interpolation with SMARTER (Sharp Monotonic Algorithm for Realistic Transport Equation Revisited) flux limiter on turbulent variables. Center finite difference is used to discretize the viscous fluxes of both conservative and turbulent variables.

The temporal discretization for URANS computations is achieved thanks to a dual time-stepping method (22). This method consists in the estimation of the time derivative of conservative variables in equation (4) thanks to a pseudo time-step τ and can be written:

$$\frac{3\boldsymbol{J}^{n+1}\boldsymbol{q}^{k+1} - 4\boldsymbol{J}^{n}\boldsymbol{q}^{n} + \boldsymbol{J}^{n-1}\boldsymbol{q}^{n-1}}{2\Delta t} + \boldsymbol{J}^{n+1}\frac{\boldsymbol{q}^{k+1} - \boldsymbol{q}^{k}}{\Delta\tau} + \boldsymbol{F}(\boldsymbol{q}^{k+1}) = 0 \qquad (6)$$

where $\boldsymbol{q}$ is the vector of conservative variables known at instants k, n and $n-1$. Note that vector of mesh cells volume is time dependent to take into account the mesh deformation. In our computations, the temporal discretization of the pseudo time τ is achieved by an implicit method:

$$\boldsymbol{F}(\boldsymbol{q}^{k+1}) \approx \boldsymbol{F}(\boldsymbol{q}^{k}) + \frac{\partial \boldsymbol{F}}{\partial \boldsymbol{q}}(\boldsymbol{q}^{k+1} - \boldsymbol{q}^{k})$$

On the other hand, the steady computations with second order turbulence model are performed using a structured multi block cell-vertex finite volume solver developed by Vallet (23) with near-wall Reynolds-stress closure (24; 25). The governing equations are solved using a van Leer flux-vector-splitting scheme with 3rd order MUSCL interpolations and van Albada limiters (26). Local dual time stepping procedure is also employed to improve the convergence properties of the steady flow solution in the presence of separated flows (27). The Time Harmonic Balance (THB) computations are run using a solver developed by Labit (15).

For nonlinear unsteady RANS computations of the oscillating bump, the mesh must be deformed during the computation. As the mesh cells are thinner at the wall, a rigid body motion is imposed to the cells in the boundary layer. The mesh deformation is achieved at the center of the domain, where the cells are larger, as shown in Figure 4. With this technique, the volume variation of the most deformed cell is lower than 1%, guaranteeing almost the same quality as the steady mesh.

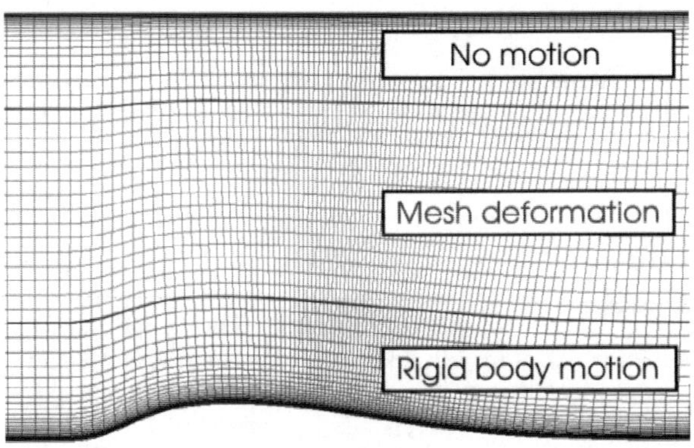

Figure 4 Mesh deformation due to bump displacement - rigid body motion is imposed close to the bump (small cells) and mesh deformation is achieved in freestream part of the flow where cells are larger, involving small relative volume variation (< 1%)

The numerical imposed deformation is computed thanks to the results of Andrinopoulos (16) on the polyurethane bump:

$$\Delta y = a_0 - a_2(x - x_0)^2 + a_3(x - x_0)^3$$

x	$[70.5\,, 191.0]\ [mm]$
x_0	$125.0\ [mm]$
a_0	$0.412\ [mm]$
a_2	$1.18 * 10^{-4}\ [mm^{-1}]$
a_3	$3.70 * 10^{-7}\ [mm^{-2}]$

The amplitude of the numerical deformation compared to experimental data is plotted along the bump in Figure 5 as well as the steady and extremes numerical positions of the bump in Figure 6. The origin of the abscises axis is chosen $70mm$ upstream of the bump (see Figure 1).

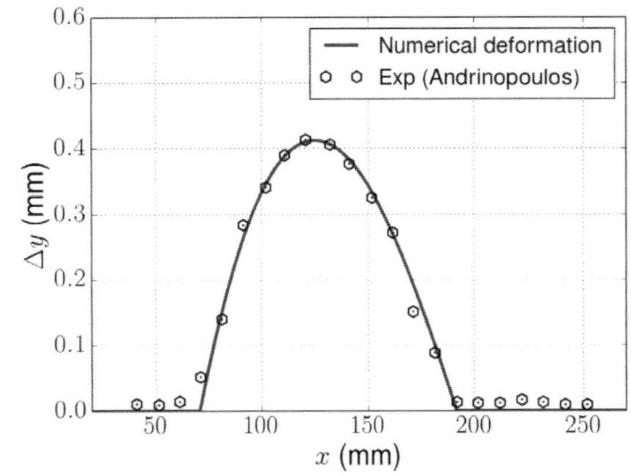

Figure 5 Amplitude of bump deformation along streamwise axis - experimental data from Andrinopoulos (16)

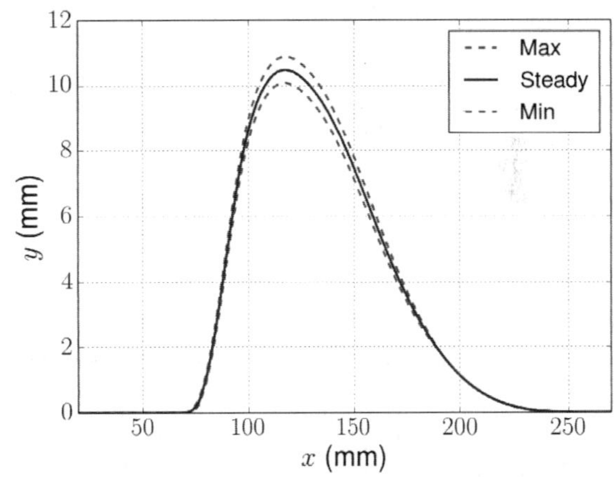

Figure 6 Steady and extremes positions of the bump in URANS computations (corresponding to a maximum section variation of 0.5%)

To compare experimental and numerical results, the steady pressure coefficient is computed for steady state as:

$$Cp = \frac{Ps}{Pt_{inlet} - Ps_{inlet}} - Cp_{inlet} \tag{7}$$

where Cp_{inlet} is defined as:

$$Cp_{inlet} = \frac{Ps_{inlet}}{Pt_{inlet} - Ps_{inlet}} \tag{8}$$

Note that to take into account the change in outlet static pressure in 2D computations, the experimental and numerical values of Cp_{inlet} are different.

For unsteady calculations, the unsteady pressure coefficient is computed, consistently with Ferria formulation (14) as:

$$^1\tilde{C}p = \frac{^1\tilde{P}s}{Pt_{inlet} - Ps_{inlet}} * \frac{c_{ax}}{A} \tag{9}$$

where $^1\tilde{P}s$ is the instantaneous static pressure, $c_{ax} = 120mm$ is the axial chord length and $A = 0.5mm$ is the theoretical amplitude of the bump motion.

STEADY STATE RESULTS

The experimental setup is composed of two converging ducts upstream of the bump which are not computed in this work. Philit (28) took into account these ducts in 3D computations to obtain an inlet boundary-layer profile suitable for 2D studies, as well as the corresponding mesh. In this study, we use the 2D structured 209*84 H mesh of Philit, presented in Figure 7.

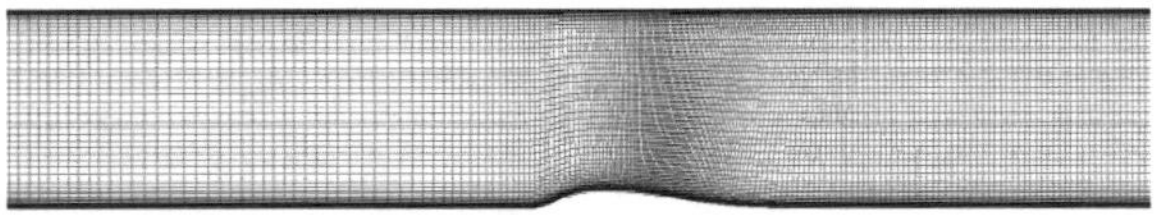

Figure 7 Structured H mesh of KTH-VM100 bi-dimensional bump from Philit (28)

Temporal discretization is achieved through implicit method with smoothed local time step (10% maximum variation between adjacent grid points) and maximum CFL of 50. The resulting system of linear equations is solved with a GMRES Krylov subspace method (29). Turbulence models of Wilcox (1988) k-ω (10) and Wilcox (2006) k-ω (11) are investigated.

In the studied configuration, inlet total pressure of $160kPa$ is imposed. The experimental results were obtained with an outlet static pressure of $104kPa$, corresponding to a strong shock wave with a separated boundary layer. To reproduce this phenomenon in 2D computations, the outlet static pressure was set to $108kPa$, as proposed by Labit (15).

Steady contours of Mach number, computed with Wilcox (1988) k-ω turbulence model, are plotted in Figure 8 in the bump vicinity. The transonic nozzle is choked, a strong shock-wave impacting the bump at the steady separation point S ($x = 145mm$) and the boundary-layer reattaching at steady reattachment point R ($x = 193mm$).

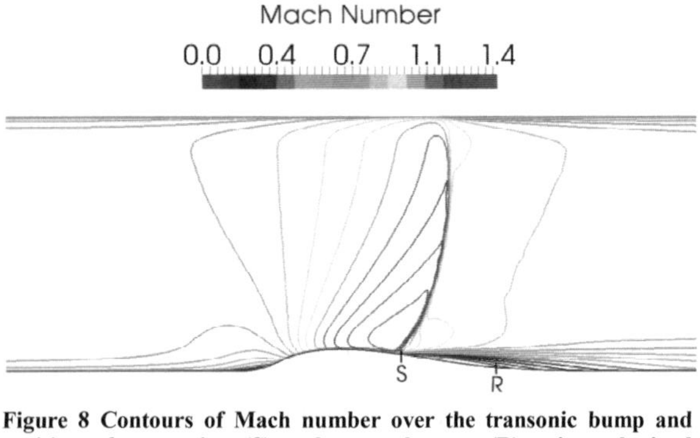

Figure 8 Contours of Mach number over the transonic bump and position of separation (S) and reattachement (R) points obtained from steady computation with Wilcox (1988) k-ω turbulence model

The steady pressure coefficients obtained with different turbulence models are compared in Figure 9 with experimental results of Ferria (14) and with numerical results of Labit (30) using a Reynold Stress Model (RSM) for turbulence closure (25). The experimental position of the shock-wave obtained by Bron (12) is also presented in Figure 9. Wilcox (2006) k-ω turbulence model and RSM turbulence closure show very good agreement with the experimental result, especially for the shock-wave position and the pressure recovery in the separated boundary layer. With the (1988) k-ω of Wilcox, the position of the shock-wave is predicted further downstream and the pressure recovery is overpredicted in the separated boundary layer.

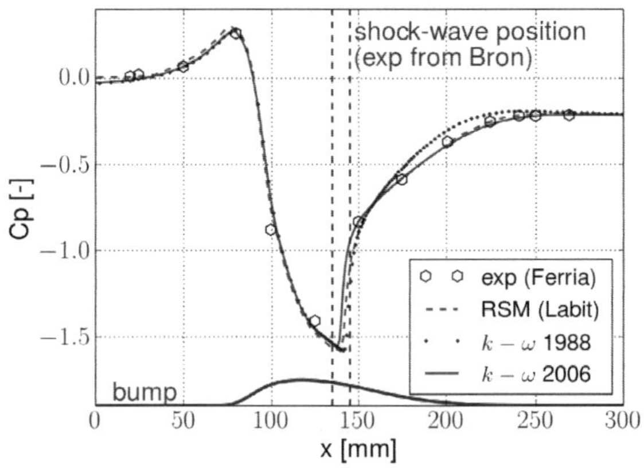

Figure 9 Steady pressure coefficient along the bump - numerical results with three different turbulence models compared with experimental results of Ferria (14) and Bron (12).

URANS RESULTS

The accuracy of the temporal discretization, presented in equation (6), is mainly driven by the number of physical iterations during one period of oscillation. The studied frequency of $100Hz$ corresponds to a period of $T = 0.01s$ for which we intend to investigate three different discretizations: 16, 64 and 256 physical time steps Δt. For this numerical study, the (1988) k-ω turbulence model of Wilcox is used.

In Figure 10 is plotted the time-dependent signal of wall pressure on the bump, at the steady separation point S (see Figure 8). This point was chosen as it shows great amplitude for the static pressure fluctuations. We can notice that the transient regime is very short (around one period of oscillation), leading quickly to a fully established periodic flow. For the further analysis of the periodic flow, we will not take into account the first four periods. In established periodic flow, numerical convergence seems to be obtained for 64 time steps per period of oscillation.

Focusing on the amplitude of the first harmonic of unsteady pressure coefficient along the bump, plotted in Figure 11 for different temporal discretization, we confirm that 64 time steps per period are enough to achieve numerical convergence. The same conclusion can be driven looking at the phase of the first harmonic along the bump (see Figure 11). With only 16 physical time steps per period, the amplitude is underpredicted in the separated region and the phase is globally shifted by 10 degrees. As we are mainly interested in the first harmonic, URANS results are from now on obtained with 64 physical time-steps per period of oscillation.

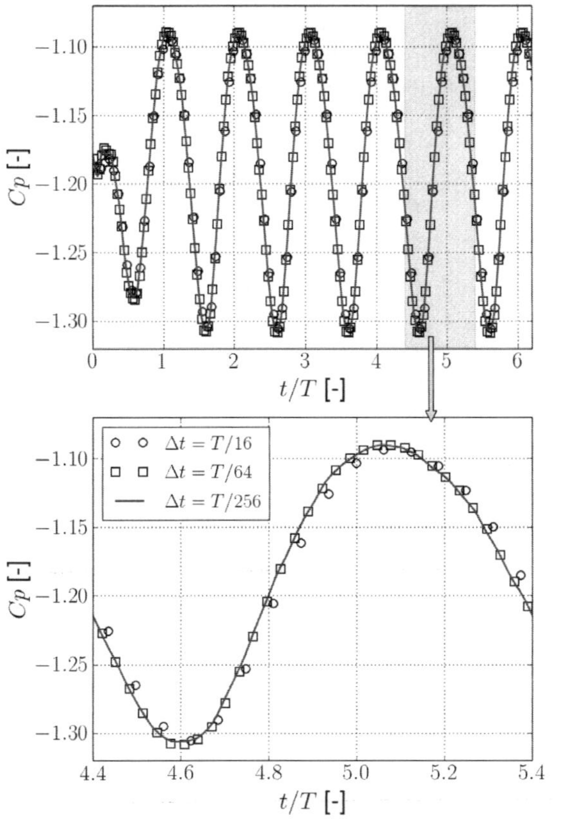

Figure 10 Signal of pressure coefficient on the bump at the root of the strong shock-wave (point S in Figure 8) as a function of dimensionless time for different temporal discretizations

In Figure 12 is plotted the ratio between the amplitudes of first and second harmonic of pressure coefficient:

$$|\,^2\tilde{C}p| / |\,^1\tilde{C}p|$$

Expecting from experimental data that the physical response of the flow is mainly driven by the first harmonic, this ratio should be small. It is inferior to 1% along the bump, except at steady separation point S ($x = 145mm$) where it grows up to 5%. This peak value is a consequence of the shock-wave moving over the pressure probes. These record the discontinuity induced by the shock as a strong nonlinearity, which does not exist in the frame of reference of the oscillating shock-wave.

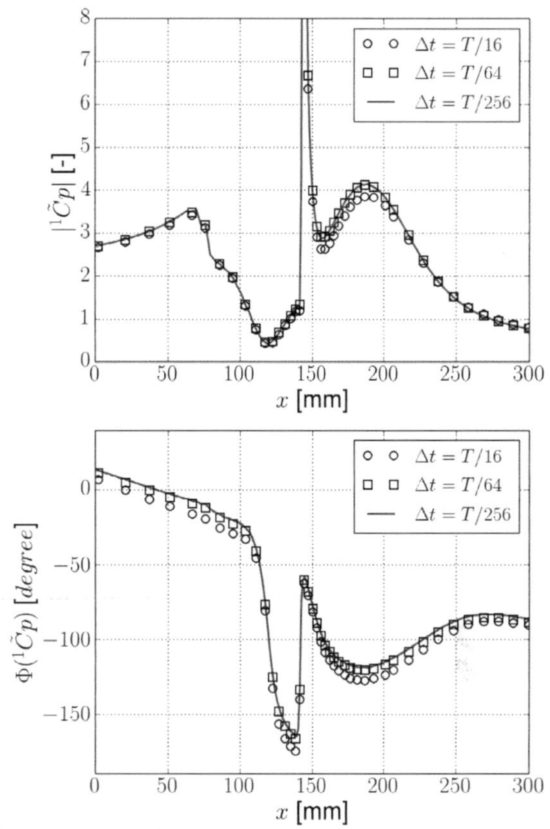

Figure 11 Amplitude (top) and phase (bottom) of the first harmonic of unsteady pressure coefficient along the bump for different temporal discretizations

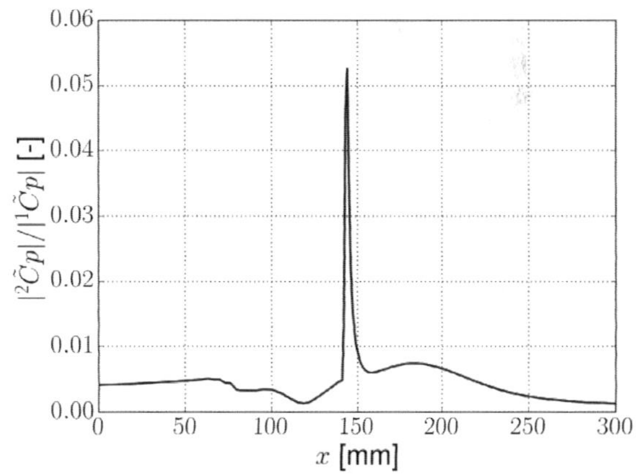

Figure 12 Ratio of amplitude of first and second harmonic of unsteady pressure coefficient plotted along the bump

TIME-LINEARIZED RESULTS

The amplitude and phase of the first harmonic of unsteady pressure coefficient $^1\tilde{C}p$ for LRANS computations (with both Frozen Turbulence and Harmonic Turbulence assumptions) are plotted along the bump in

Figure 13 and compared to the nonlinear URANS results discussed previously. Once again, only the (1988) k-ω turbulence model is used in this section.

The amplitude predicted with harmonic turbulence shows very good agreement with the nonlinear URANS results. The main discrepancy is a lower level of amplitude in the separated boundary layer and upstream of the bump. The frozen turbulence assumption shows overall good agreement on the amplitude but it overpredicts the drop behind the shock-wave as well as the rise in the separated boundary layer.

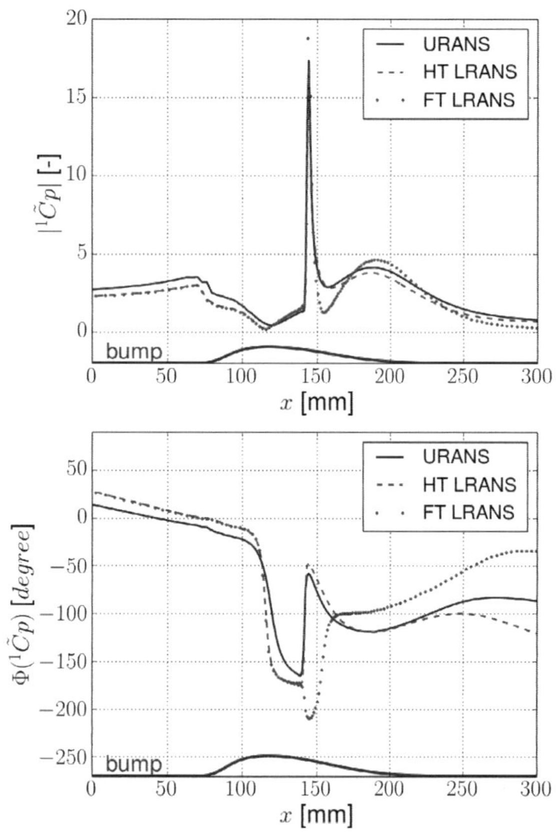

Figure 13 Amplitude (top) and phase (bottom) of the first harmonic of unsteady pressure coefficient along the bump for linearized RANS computations (HT: Harmonic Turbulence, FT: Frozen Turbulence) and comparison with nonlinear URANS computations

The amplitude of the pressure coefficient only represents the aerodynamic unsteady loads applied on the structure over one period of oscillation. The phase of the pressure coefficient, representing the temporal phase shift between pressure and imposed wall velocities, gives us insight on the stability of the system. Thus, the prediction of the phase difference between pressure and imposed wall velocities is the main challenge in fluid-structure interaction analysis. In Figure 13, we see that only the Harmonic Turbulence method succeeds in reproducing the phase computed by the nonlinear URANS method. The Frozen Turbulence fails to predict the phase behind the shock-wave, in the separated boundary layer where turbulence is likely to depend on the fluctuating aerodynamic flow. Such results were firstly exposed by Philit (21) and are confirmed by this work. According to the convention used by Ferria (14), the system should be stable for $\Phi \geq -180°$ and unstable for $\Phi \leq -180°$. With this convention, the nonlinear and Harmonic Turbulence computations predict absolute stability along the bump, whereas the Frozen Turbulence linearized calculation predicts aerodynamic instability in the vicinity of the shock-wave ($140mm < x < 150mm$).

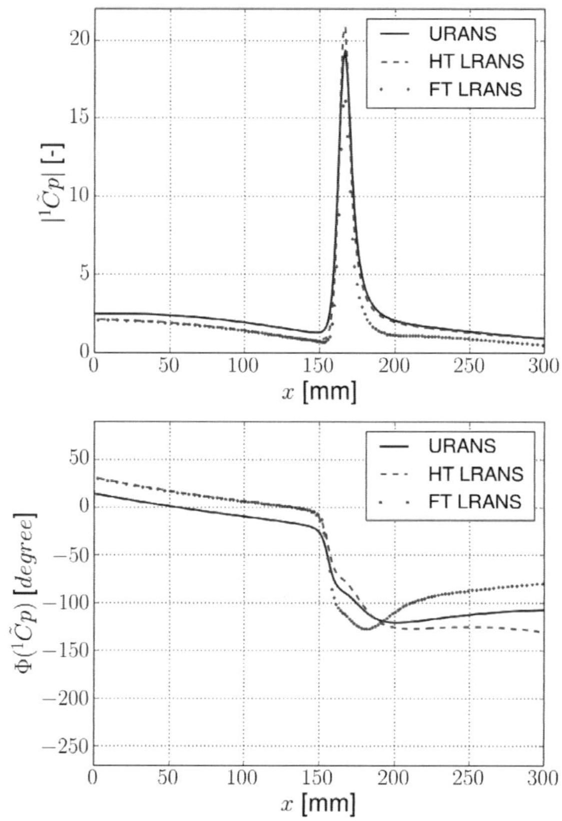

Figure 14 Amplitude (top) and phase (bottom) of the first harmonic of unsteady pressure coefficient along the top wall for linearized RANS computations (HT: Harmonic Turbulence, FT: Frozen Turbulence) and comparison with nonlinear URANS computations

In Figure 14 are plotted the amplitude and phase of the first harmonic of unsteady pressure coefficient along the top wall. Once again, time-linearized results using Harmonic Turbulence are in good agreement with nonlinear results whereas Frozen Turbulence assumption fails to reproduce the unsteady pressure amplitude and phase downstream of the shock-wave. This indicates an important influence of turbulence on the pressure fluctuations, even along the top wall. Indeed, the strong shock-wave impacts the top wall and thickens the boundary-layer there also.

In next section, to further investigate the influence of turbulence, we use different turbulence closures in URANS and THB computations.

INVESTIGATION OF TURBULENT MODELISATION

The unsteady experimental results of Ferria (14) at frequency $f = 100Hz$, corresponding to a reduced frequency $k_r = 0.29$, are chosen to assess the unsteady numerical results.

Three different turbulence closures are compared in this section to these experimental data. Labit (30) used Time Harmonic Balance (THB) method with Reynold Stress Model (RSM) to compute the unsteady aerodynamic flow. This method has been evaluated and validated over the transonic bump with downstream fluctuating static pressure (31). For the present work, THB results are obtained with $N_h = 2$ Fourier coefficients.

Two turbulence models of Wilcox are investigated using URANS computations: the (1988) k-ω and the (2006) k-ω turbulence models. In the (2006) k-ω model, the dependence on freestream values of turbulence has been corrected with a cross-diffusion term, as proposed by Kok (32), and a stress limiter has been added to correct the overprediction of turbulent kinetic energy in anisotropic parts of the flow (as proposed by Coakley (33), Menter (34) and Moore and Moore (35) amongst others).

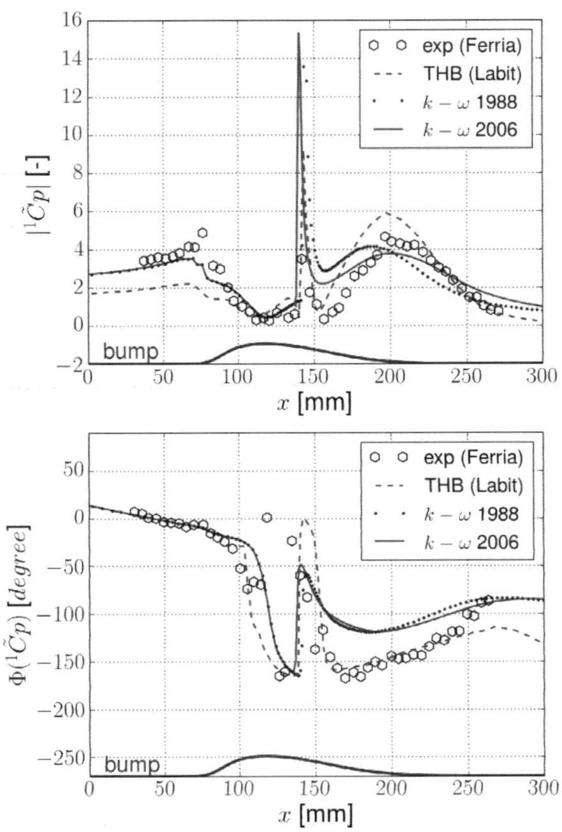

Figure 15 Amplitude and phase of the first harmonic of unsteady pressure coefficient along the bump for nonlinear URANS computations (with Wilcox (1988) k-ω and (2006) k-ω turbulence models), numerical results of Labit (30) using THB method with $N_h = 2$ Fourier coefficients, and the experimental results of Ferria (14)

The amplitude of the first harmonic of the unsteady pressure coefficient is presented in Figure 15 for both numerical and experimental results. Experimental data show three zones of high amplitude: upstream of the bump, in the vicinity of the strong shock-wave and at the reattachment of the separated boundary-layer ($x = 200mm$). Low values of amplitude are experimentally found in the supersonic zone (i.e. at the top of the bump, where the deformation of the bump is maximum) and downstream of the shock-wave root.

Results of Labit underestimate the amplitude upstream of the bump by 50%, and overestimate by 30% the amplitude in separated boundary-layer, while the position of the shock-wave and the drop of amplitude around $x = 155mm$ are in very good agreemens with experimental data.

URANS computations using two-equation turbulence models predict accurate amplitudes from inlet to the shock-wave. The drop of amplitude downstream of the shock-wave and the fluctuating pressure in separated boundary layer are underestimated. If both turbulence models give the same predictions up to the supersonic zone, the position of the shock-wave slightly differs, as well as the response in downstream boundary layer. In agreement with steady state results (Figure 9), the shock-wave predicted by the (2006) k-ω model is located upstream of the location predicted by the (1988) k-ω model. Moreover, (2006) k-ω model shows better agreement downstream of the shock-wave, especially for the position of maximum amplitude at the reattachment point of the boundary layer.

In Figure 15 is also plotted the phase of the first harmonic of the unsteady pressure coefficient. As noted earlier, the phase drives the stability of the bump motion. The configuration studied by Ferria at $f = 100Hz$ is stable ($\Phi \geq -180°$), what is correctly predicted by THB and URANS computations. THB method shows very good agreement downstream of the shock-wave for $160mm < x < 240mm$. A 180° phase shift is predicted at shock-wave root ($x = 140mm$) and another one in separated boundary-layer ($x = 150mm$). As the uncertainties on experimental data are important, we do not know the physical behaviour of the phase. From the authors' experience, we expect at least one 180° phase shift through the shock-wave.

One can notice the linear change of phase with the distance in the upstream zone of the nozzle ($x \leq 75mm$). This may correspond to the propagation of a plane wave. From Figure 15, we can compute the wave length of such a wave thanks to the inverse of the rate of change of the phase with respect to the distance:

$$|\Delta\Phi| = 2\pi\frac{|\Delta x|}{\lambda} \Rightarrow \lambda = 2\pi\frac{|\Delta x|}{|\Delta\Phi|} = 1.00\,m$$

Knowing the inlet Mach number $M_1 = 0.687$ and the inlet static temperature $Ts_1 = 279.6\,K$, we can also compute the wavelength of a backward moving plane wave of frequency $f = 100Hz$:

$$\lambda = \left|(u - c) * \frac{1}{f}\right| = 1.05\ m$$

Thus, the linear evolution of phase with distance upstream of the bump is probably related to a backward travelling plane wave generated by the bump oscillation.

URANS computations show good agreement in this region. The predictions of the phase at the top of the bump are quite similar to the THB results. However, there is a 30° to 50° phase shift between the experimental results and the URANS results in the separated boundary layer ($150mm < x < 200mm$). Phase predictions by the two k-ω turbulence models are almost identical.

The mechanisms of the shock-wave/boundary-layer interaction are thus quite well predicted with RSM and two-equation models turbulence closure. However, it is to keep in mind that the experimental flow is three-dimensional and exhibits large corner separations downstream of the shock-wave. The discrepancies observed in this work between computations and experimental results could be the result of such three-dimensional effects.

AEROELASTIC STABILITY ANALYSIS

If the phase shift between pressure fluctuations and imposed wall velocities gives qualitative insights in the aerodynamic stability of the system, it is generally not sufficient to conclude on the global stability. To go further, exchanged work $\mathcal{W}$ is plotted in Figure 16 along the bump for the nonlinear URANS computations, using only the first harmonic, as well as linearized results with both Frozen and Harmonic Turbulence assumptions. A stabilizing peak is observed in the vicinity of the shock-wave, where pressure fluctuations are strong, and the contribution of the separated boundary layer downstream of the shock-wave is also important. Notice that when the exchanged work is positive, the energy of the structure is dissipated by the surrounding fluid, leading to aerodynamic stability.

Once again, absolute aerodynamic stability is obtained for nonlinear and Harmonic Turbulence time-linearized computations, whereas Frozen Turbulence time-linearized computation predicts aerodynamic instability in the vicinity of the shock-wave, which is a direct consequence of the discrepancies seen on the phase prediction (see Figure 13). The damping coefficient is underestimated by 20% by Harmonic Turbulence linearized computation in comparison with nonlinear URANS results. This may be a consequence of the discrepancies in pressure fluctuations amplitude and phase shift upstream of the shock-wave prediction (see Figure 13).

As the agreement on damping coefficient is good between URANS and Harmonic Turbulence LRANS computations, we can now take advantage of time-linearized computations to study the aerodynamic stability over a range of frequencies.

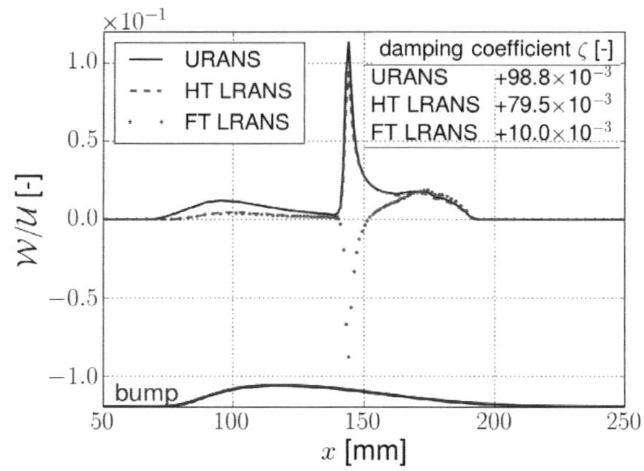

Figure 16 Local aerodynamic work along the bump at $f = 100Hz$ for linearized RANS computations (HT: Harmonic Turbulence, FT: Frozen Turbulence) and comparison with nonlinear URANS computations

The aerodynamic damping coefficient ζ predicted using the (1988) k-ω turbulence model of Wilcox is presented in Table 1 for $100Hz < f < 800Hz$ corresponding to a range of reduced frequency $0.29 < k_r < 2.34$. Up to $f = 300Hz$, the damping coefficient gradually decreases and reaches a minimum value $\zeta = -0.7 * 10^{-3}$. Then, the damping coefficient increases and aerodynamic stability is recovered for $f = 800Hz$. It is to notice that the modal shape has been kept identical, which may not be the case if it depends on the excitation frequency. Nevertheless, this methodology allows us to find the frequency at which the system is the most unstable. We can now run a full nonlinear unsteady RANS computation to further investigate the physical behaviour of the system at $f = 300Hz$, corresponding to $k_r = 0.88$.

Table 1 Aerodynamic damping coefficient predicted by Harmonic Turbulence LRANS computations for different frequencies of bump oscillation

$f\ [Hz]$	$k_r\ [-]$	$\zeta[-]$
100	0.29	$+\ 79.5 * 10^{-3}$
200	0.58	$+\ 2.0 * 10^{-3}$
300	0.88	$-\ 0.7 * 10^{-3}$
400	1.16	$-\ 0.3 * 10^{-3}$
500	1.46	$+\ 0.1 * 10^{-3}$
600	1.75	$+\ 0.6 * 10^{-3}$
700	2.04	$+\ 1.3 * 10^{-3}$
800	2.34	$+\ 2.3 * 10^{-3}$

In Figure 17 is plotted the exchanged work $\mathcal{W}$ along the bump for this frequency. As previously, the Frozen Turbulence assumption fails to reproduce nonlinear URANS results, showing once again a strong influence of the boundary layer on the aerodynamic stability. Results achieved with Harmonic Turbulence assumption show good agreement with nonlinear computations behind the shock-wave ($x > 150mm$). However, the exchanged work computed with linearized calculation is underestimated by

30% in the vicinity of the shock-wave and on the first half of the bump. This leads to a higher damping coefficient ($\zeta = +0.0 * 10^{-3}$) for URANS computation, thus predicting a system at its stability limit.

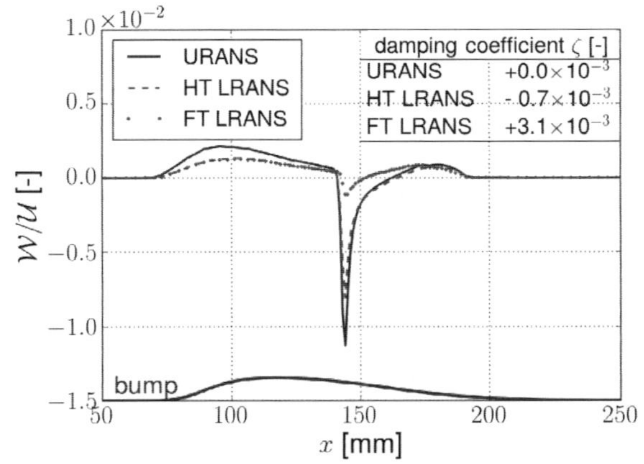

Figure 17 Local aerodynamic work along the bump at $f = 300Hz$ for linearized RANS computations (HT: Harmonic Turbulence, FT: Frozen Turbulence) and comparison with nonlinear URANS computations

We showed that LRANS computations are able to reproduce URANS predictions on shock-wave/boundary-layer interaction if the unsteady behaviour of turbulence is taken into account, which implies the derivation of the full turbulence model. Frozen Turbulence assumption is not able to predict the phase shift between pressure and wall velocity fluctuations, and it is thus not recommended for stability analysis in case of shock-wave/boundary-layer interaction.

CONCLUSION

In this work, the ability of time-linearized and harmonic-balance methods to reproduce nonlinear RANS results has been demonstrated in case of shock-wave/boundary-layer interaction over an oscillating bump.

Turbulence closure should be closely investigated when dealing with strong shock-wave/separated boundary-layer interaction. In our work, we show good agreement between experimental data, Reynolds stress model turbulence closure and k-ω turbulence models of Wilcox. The (2006) k-ω turbulence model of Wilcox improves both steady state results (better prediction of shock-position and pressure recovery in separated boundary layer) and unsteady ones (better position of maximum amplitude at reattachment).

Concerning time-linearized computations, the frozen turbulence assumption fails to correctly predict the phase of the unsteady pressure coefficient and thus should not be used for aeroelastic stability analysis in case of shock-wave/boundary-layer interaction. However, computations with harmonic turbulence succeed in reproducing the predictions of nonlinear computations. As the linearized computations are up to 100 times faster than URANS computations, they allow fast parametric studies of the unsteady flow. The influence of frequency on the aeroelastic stability has been quickly evaluated and full time-nonlinear calculation has been run on the most critical configuration for deeper analysis of the unsteady flow.

The promising results, obtained with 2D computations, should be extended to three-dimensional flows. Indeed, the bi-dimensional hypothesis made here on a flow with corner separations may have greater effects than turbulence modelisation and physical hypothesis (periodic flow assumed for time-linearized and harmonic balance computations). Thus, three-dimensional computations are mandatory to further evaluate these numerical methods in shock-wave/boundary-layer interaction.

REFERENCES

1. **Clark, W.S. and Hall, K.C.** *A time-linearized Navier-Stokes analysis of stall flutter.* in Journal of Turbomachinery, vol. 122 (3), 2000.
2. **Campobasso, M.S. and Giles, M.B.** *Effects of flow instabilities on the linear analysis of turbomachinery aeroelasticity.* in Journal of Propulsion and Power, vol.19 (2), 2003.
3. **Kersken, H.-P., et al.** *Time-linearized and time-accurate 3D RANS methods for aeroelastic analysis of turbomachinery.* in ASME Paper GT2010-22940, 2010.
4. **Chassaing, J.-C. and Gerolymos, G.A.** *Time-Linearized Time-Harmonic 3D Navier-Stokes Shock-Capturing Schemes.* in International Journal for Numerical Methods in Fluids, vol. 56 (3), 2008.
5. **Ipsen, I.C.F. and Meyer, C.D.** *The Idea Behind Krylov Methods.* , 1998.
6. **Chassaing, J.-C., Gerolymos, G. A. and Jérémiasz, J.-G.** *GMRES Solution of Compressible Linearized Navier-Stokes Equations without Pseudo-Time Marching.* 44th AIAA Aerospace Sciences Meeting, 2006.
7. **He, L.** *An Euler solution for unsteady flows around oscillating blade.* in Journal of Turbomachinery, vol. 112, 1990.
8. **Hall, K.C., Thomas, J.P. and Clark, W.S.** *Computation of Unsteady Nonlinear Flows in Cascades Using a Harmonic Balance Technique.* in 9th ISUAAAT, Lyon, France, 2000.
9. **Vasanthakumar, P., Chen, T. and He, L.** *Three-Dimensional Viscous Computation of Blade Flutter and Forced Response Using Nonlinear Harmonic Approach.* in 9th ISUAAAT, Lyon, France, 2000.
10. **Wilcox, D.C.** *Reassessment of the Scale Determining Equation for Advanced Turbulence Models.* in AIAA Journal, vol.26 (11), 1988.
11. **Wilcox, D.C.**. *Formulation of the k-w Turbulence Model Revisited.* in AIAA Journal, vol. 46 (11), 2008.
12. **Bron, O.** *Numerical and experimental study of the Shock-Boundary Layer Interaction in Transonic Unsteady Flow.* PhD thesis, Ecole Centrale de Lyon, France, 2004.

13. **Bron, O., Ferrand, P. and Fransson, T.H.** *Experimental and Numerical Study of Non-Linear Interactions in Two Dimensional Transonic Nozzle Flow.* in 11th ISUAAAT, Moscow, Russia, 2006.
14. **Ferria, H.** *Experimental Campaign on a Generic Model for Fluid Structure Interaction Studies.* Master thesis, KTH, Sweden, 2007.
15. **Labit, S.** *Formulations temporelle et fréquentielles pour la simulation numérique d'écoulements transsoniques en oscillations forcées.* PhD thesis, Université Pierre et Marie Curie, France, 2014.
16. **Andrinopoulos, N., et al.** *Design and Testing of a Vibrating Test Object for Investigating Fluid-Structure.* in ASME Paper GT2008-50740, 2008.
17. **Allegret-Bourdon, D.** *Experimental Study of Fluid-Structure Interctions on a Generic Model.* Lic. thesis, KTH, Sweden, 2004.
18. **Smati, L., Aubert, S. and Ferrand, P.** *Numerical study of unsteady shock motion to understand transonic flutter.* in Proceedings of Euromech 349 on Simulation of fluid-structure interaction in aeronautics, 1996.
19. **Smati, L., et al.** *Comparison of numerical schemes to investigate blade flutter.* in Proceedings of the 8th ISUAAAT, 1997.
20. **Philit, M., et al.** *Frequency parametrization to numerically predict flutter in trubomachinery.* in 4th International Conference on Computational Methods for Coupled Problmes in Science and Engineering, Kos, Greece, 2011.
21. **Philit, M., et al.** *Derivated turbulence model to predict harmonic loads in transonic separated flows over a bump.* in 28th ICAS, 2012.
22. **Jameson, A.** *Time Dependent Calculations Using Multigrid, with Applications to Unsteady Flows Past Airfoils and Wings.* in 10th AIAA Computational Fluid Dynamics Conference, Honolulu, Hawaï, 1991.
23. **Vallet, I.** *Aérodynamique Numérique 3D Instationnaire avec Fermetrue Bas Reynolds au Second Ordre.* in Ph.D Thesis, Université Pierre et Marie Curie, Paris, France, 1995.
24. **Gerolymos, G.A. and Vallet, I.** *Near-Wall Reynolds-Stress 3D Transonic Flows Computation.* in AIAA Journal, vol 35 (2), 1997.
25. **Gerolymos, G.A., Sauret, E. and Vallet, I.** *Contribution to the single-point-closure Reynolds-stress modelling of inhomogeneous flow.* in Theoretical and Computational Fluid Dynamics Journal, vol. 17, 2004.
26. **Gerolymos, G.A. and Vallet, I.** Aerodynamics package software. in http://sourceforge.net/projects/aerodynamics.
27. **Chassaing, J.-C., G.A., Gerolymos and Vallet, I.** *Efficient and Robust Computation of 3D Compressible Flows with Reynolds-Stress Closure.* in AIAA Journal, vol. 41 (5), 2003.
28. **Philit, M.** *Modélisation, simulation et analyse des instationnarités en écoulement transsonique décollé en vue d'application a l'aéroélasticité des turbomachines.* PhD thesis, Ecole Centrale de Lyon, France, 2013.
29. **Saad, Y. and Schultz, M. H.** *GMRES : a Generalized Minimal Residual Algorithm for Solving Nonsymmetric Linear Systems.* in Society for Industrial and Applied Mathematics, 1986.
30. **Labit, S., et al.** *Time-harmonic Navier-Stokes computations of forced shock-wave oscillations in a transonic nozzle.* in revison for Computers and Fluid, 2015.
31. **Labit, S., et al.** *Frequency-Domain Navier-Stokes Computations of Transonic Nozzle Flows.* in ECCOMAS, 2012.
32. **Kok, J.C.** *Resolving the Dependence on Freestream Values for the k-w Turbulence Model.* in AIAA Journal, vol. 38 (7), 2000.
33. **Coakley, T.J.** *Turbulence Modeling Methods for the Compressible Navier-Stokes Equations.* in AIAA 16th Fluid and Plasma Dynamics Conference, Danvers, Massachusetts, 1983.
34. **Menter, F. R.** *Zonal Two-Equation k-w Turbulence Models for Aerodynamic Flows.* in 24th AIAA Fluid Dynamics Conference, Orlando, Florida, 1993.
35. **Moore, J.G. and Moore, J.** *Realizability in Two-Equation Turbulence Models.* in 30th AIAA Fluid Dynamics Conference, Norfolk, Virginia, 1999.
36. **Hall, K.C., J.P., Thomas and W.S., Clark.** *Computation of Unsteady Nonlinear Flow in Cascades using a Harmonic Balance Technique.* in AIAA Journal, vol. 40 (5), 2002.
37. **Gerolymos, G.A. and Vallet, I.** *Wall-Normal-Free Near-Wall Reynolds-Stress Closure for 3D Compressible Separated Flows.* in AIAA Journal, vol. 29, 2001.

Proceedings of the 14th International Symposium on Unsteady Aerodynamics, Aeroacoustics&Aeroelasticity of Turbomachines
ISUAAAT14
8-11 September 2015, Stockholm, Sweden

I14-S11-6

ON NON-SYNCHRONOUS VIBRATIONS: LOCK-IN EFFECTS OF AN ECCENTRICALLY ROTATING CYLINDER

Sina Samarbakhsh
THRUST Program Graduate
Department of Energy Technology
KTH Royal Institute of Technology
Stockholm, Sweden

Christopher E. Meinzer
Institute of Turbomachinery and Fluid Dynamics
Leibniz Universität Hannover
Hannover, Germany

Björn Laumert
Department of Energy Technology
KTH Royal Institute of Technology
Stockholm, Sweden

Joerg R. Seume
Institute of Turbomachinery and Fluid Dynamics
Leibniz Universität Hannover
Hannover, Germany

ABSTRACT

The interaction of fluctuating vortex shedding with blade vibration can lead to a new class of aeromechanical instability referred to as non-synchronous vibrations (NSV). This phenomenon has been observed in the front stages of high-pressure compressors and fan blades, vanes and blisks (Kielb, Barter, Jeffrey, & Kenneth, 2003).By investigating a well-known case that shows comparable NSV features such as shedding flow over a circular cylinder an understanding of physics behind NSV can be developed (Kielb, Hall, Spiker, & Thomas, 2006).

An eccentrically rotating cylinder is introduced in this research as an alternative paradigm to study vortex dynamics and lift-force behavior in the range of the synchronization region. The main focus of this work is to understand to what extend a general feature of free vibration can be observed in the case of an eccentrically rotating cylinder. A quasi-3D computational model was established in order to explore the existence of a lock-in region and predict the response and the vorticity structure in the near-cylinder wake region over a specific parameter space of the introduced alternative case. The present work is a preliminary numerical study for the preparation of experimental work.

The existences of a synchronization region with a striking similarity in the lift-force coefficient and the wake mode has been observed in the current study. A larger range for the lock-in region and some new synchronized wake modes have been found. A simultaneous occurrence of cross-flow and in-line oscillation can justify the larger range of the synchronization region, since the lock-in happens at the frequency of twice the Strouhal frequency in the case of an in-line oscillation. Despite some very good similarities in the wake states of the eccentrically rotating and the freely moving cylinder, further work is required to determine how the case of rotation with eccentricity can be used to predict the vortex-induced motion.

INTRODUCTION

This work discusses the most important features of the NSV phenomena through studying the force characteristics and the vorticity pattern over a circular cylinder. The vortex shedding behind flexible bluff bodies occurs in many engineering situation. It is of interest in civil engineering structure like bridges and towers as well as in designing transmission lines, offshore platforms, heat exchangers, marine cables and other marine application, (Belivins, 1990) and (Williamson & Govardhan, 2004). It was also seen that in turbine engines, the interaction of fluctuating vortex shedding with the vibration of blade can lead to this form of aeroelastic instability. By investigating a well-known case that shows comparable NSV features such as shedding flow over a circular cylinder an understanding of physics behind NSV can be developed (Kielb, Hall, Spiker, & Thomas, 2006).

Fundamental study in this field shows that vortex wakes tends to be very similar regardless of the structure. The so-called lock-in phenomenon is observed in vortex-induced vibration, where the shedding process is being greatly affected by the motion of the elastic structure. When the frequency of the energy input to the system due to fluctuating vortex approaches enough to the natural vibration frequency of the structure, synchronization happens. This synchronization effect was first reported by Bishop and Hassan as a conclusion of their experimental study on the lift and drag forces on the oscillating cylinder in a flowing fluid (Bishop & Hassan, 1964). In the synchronization range the frequency of the vortex shedding locks at the structural motion frequency (forcing frequency) and the magnitude of the fluid force on the structure increases abruptly. This can lead to the destruction of the structure. In literature lock-in phenomena are sometimes referred to as "resonance", "synchronization", "wake capture" and "Strouhal excitation", and can be categorized as "self-excited" phenomena, as no external unsteadiness collaborates into the system, (Fransson, 2012). Due to the practical importance of vortex induced vibration, there is a wide literature on VIV.

A common approach to further investigate vortex-induced vibration and the corresponding NSV, is controlling the motion of the cylinder and allowing the response of the wake to the motion to be studied in isolation. It has been found very important to carefully match the experimental conditions between free and controlled vibration (Morse & Williamson, 2006). The majority of research in the field of vortex-induced vibration applies a rigid cylinder mounted horizontally and moving transversely to the flow stream, as a model for detailed understanding the physics behind this phenomenon. Where the free vibration follows a sinusoidal motion the controlled sinusoidal motion leads to the fluid forces conforming the free oscillating results.

This work develops a simple tool to predict force characteristics and wake mode of NSV excitation. Regarding the difficulties of the implementation of a vertically moving cylinder in an experimental study, an eccentrically rotating cylinder is introduced in this research as an alternative paradigm to study vortex dynamics and lift force behavior in the range of the synchronization region. The main focus of this work is to understand to what extend a general feature of free vibration can be observed in the case of an eccentrically rotating cylinder. The working hypothesis of this work is: if the present case captures the essential characteristics of a freely oscillating cylinder, e.g., the wake states, the transition between wake modes and the existence of the synchronization region, the results of a forced motion via an eccentrically rotating cylinder can be applied to predict the motion of an elastically mounted body. The relationship between an eccentrically rotating cylinder and an elastically mounted cylinder is very complex.

A quasi-3D computational model was established in order to explore the existence of a lock-in region and predict the response and the vorticity structure in the near-cylinder wake region over a specific parameter space of the introduced alternative case. The bounds of the lock-in have been determined through the study of the drag and lift force characteristics, by spectral analyses of the fluctuating lift force and by exploring the vortex synchronization in the frequency-amplitude plane. The manner in which the wake mode varies with the frequency of rotation (forcing frequency) was studied for a range of amplitude-diameter ratios (A/D) at a Reynolds number of Re 4000. This value for Re has been chosen in accordance to the experimental results for a transversely oscillating cylinder in a water channel found in literature. Besides that, over a very wide range of Re, namely $300<Re<3x10^5$, the wake structure is almost similar and the frequency of the vortex shedding and the fluctuating force coefficient remains constant. For this subcritical range, the boundary layer over the cylinder surface remains laminar and the wake is completely turbulent (Summer & Fredsoe, 1997). This fact can lead to extend the results of the current work to higher and more practical Re values.

NOMENCLATURE

A	Amplitude of vibration - Cylinder Eccentricity (m)
Cd	Drag force coefficient
Cl	Lift force coefficient
D	Cylinder diameter (m)
f_e	Forcing frequency Hz
fs	Vortex shedding frequency for stationary cylinder Hz
F	Frequency ratio f_e/f_s
u	Flow velocity at the wall (m/s)
y^+	Dimensionless distance from the cylinder wall
Δx	Mesh size in x-direction (m)
Δy	Mesh size in y-direction (m)
ν	Kinematic viscosity (m^2/s)

NUMERICAL STUDY

1 Stationary cylinder

1.1 Domain geometry

A rectangular domain was considered to simulate the flow over the cylinder. This domain can be considered as a wind tunnel or water channel in the experimental study. In order to simulate the available wind tunnel, dimensions of the rectangular domain was chosen as follow: height 300 mm, length 600 mm and widths 10 mm. A circular cylinder was placed in the mid-height of the channel such that its center coordinate is shown in Figure 1. A circular cylinder having a diameter D of 38.1 mm was applied in the simulation. The center of the cylinder is the origin of the cartesian coordinate system in the present CFD model.

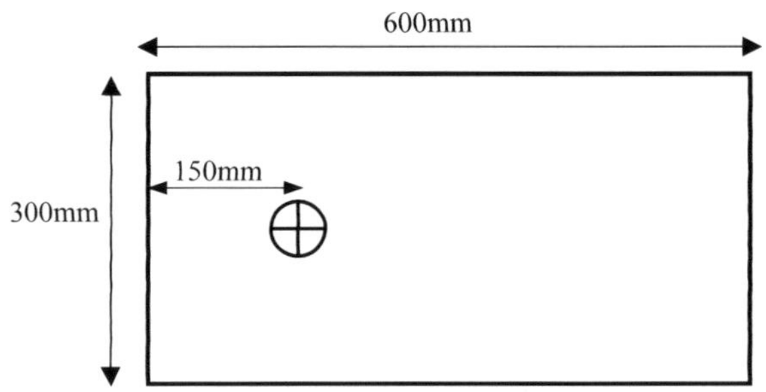

Figure 1 A rectangular CFD domain corresponding to wind tunnel/ water channel in experimental study

1.2 Grid generation

Y+ is the dimensionless distance from the wall. One of the important factors for the optimal performance of turbulence modes is the proper resolution of the boundary layer. The judgment about the near wall mesh spacing is based on the standard y+ definition. The definition of the y+ in terms of Δy, the mesh spacing between the wall and the first node away from wall, flow velocity at the wall, u and fluid kinematic viscosity, ν is as follow:

$$y^+ = \frac{\Delta y.u}{\nu} \qquad \text{Equation 1}$$

According to the SST turbulence model criterion a fine mesh near the wall boundary is needed with a $y^+<1$.

A grid independency study was performed to confirm the independence of the results of the grid size. Six different grids, each with certain number of nodes were generated to find the optimum mesh number. The tabulated result in Table 1 shows a grid independent behavior for a node number of 205400. As it is evident no significant variation are shown in time averaged parameters listed in Table 1. Since Strouhal number converged for the certain value at Model 3 afterwards, this model was chosen in the current work.

Table 1 Grid independency test

Model No	Number of nodes	Cd	Cl	St Number
Model1	153800	1.16	1.574	0.235
Model2	177600	1.14	1.564	0.235
Model3	205400	1.16	1.573	0.240
Model4	237200	1.16	1.566	0.240
Model5	273000	1.15	1.586	0.239
Model6	512000	1.17	1.572	0.239

1.3 Analysis type

Regarding the inherit time dependence of the flow characteristics in present work, e.g., the vortex shedding development, a transient simulation was applied in flow analysis. The time dependent behavior of the flow for the transient simulation is specified through the time step and the time duration. The time step size must be small enough to resolve the time dependent features. The Courant number is often used to estimate a time step and is of fundamental importance for transient flow. This gives the number of mesh elements the fluid passes through in one time step. Typical values are 2...10, but in some cases higher values are allowed, (ANSYS, 2011). For a one dimensional grid a Courant number is defined as follow:

$$Courant = \frac{u.\Delta t}{\Delta x} \qquad \text{Equation 2}$$

where u is the fluid speed, Δt is the time step and Δx is the mesh size. To keep the Courant number within an acceptable range, the desired time step must be small enough. In the present case, the calculated time step for Re=4000 is Δt=0.01 s which keeps the Courant number in the range of 2 to 10. This value is also rechecked to see if it gives enough resolution to study one cycle of vortex shedding. According to the Reynolds and Strouhal number, the frequency of the vortex shedding is the value of 0.65 Hz, so the time needed to pass one frequency is 1.53 s. Now it can be decide how many steps are needed to reach this time. In the case of 10 steps, the calculated time step will be Δt=0.1 s. Comparing this with the former calculated one, Δt=0.01 s was applied in the current transient simulation.

1.4 Boundary conditions

In this work, boundary conditions including inlet, openings, wall and symmetry were applied. Since the flow is predominately directed into the domain an inlet boundary condition was used for inlet surface. A uniform velocity was imposed at domain inlet in such a way that the desired Reynolds number was obtained. The simulation was performed for the both stationary and oscillating cylinder in water channel. In this case in order to compare the results with (Morse & Williamson, 2009), (Williamson & Govardhan, 2004), (Carberry, Sheridan, & Rockwell, 2001) and (Carberry, Sheridan, & Rockwell, 2005)the value of 0.1185 m/s was considered for the water stream velocity corresponding to Re=4000 at which the experimental results were already presented in the literature review.

Since the only known value for domain exit is the pressure, an opening boundary condition was assigned to the domain exit and the value of 1 atm was applied for relative pressure. In a preliminary investigation a no-slip, adiabatic wall boundary conditions had been considered for the upper and lower domain surfaces. In order to minimize the wall effect on the vorticity development, the upper and lower B.C. were changed to an opening boundary condition allowing the fluid flows simultaneously in and out of the domain. The entrainment with the static pressure option was used for determining mass and momentum in the opening boundary tab.

The heat transfer is considered to be isothermal this requires an absolute uniform temperature for the fluid. A value of 25 °c was applied in water stream channel.

Since an accurate simulation of boundary layer, the beginning and the amount of flow separation is highly important in the present study a brief investigation has been performed to choose the most practical turbulence model for present case. The k-epsilon model as one of the most general and prominent turbulence model was the first choice. It is considered the industry standard model and has proven to be stable, numerically robust and of good predictive capability. For general purposes the k-epsilon model shows a good balance between accuracy and robustness, (ANSYS, 2011). Although the k-epsilon model as a standard two equation model, provides good prediction for many engineering flows, there are application for which this model is not suitable. Flows with boundary layer separation and flows over curved surfaces are such examples. In general, turbulence models based on the epsilon model predict the separation point too late.

Currently, the most eminent two-equation model to capture the flow separation phenomenon more accurate is the k-omega based model of (Menter, 1992).The k-omega based shear stress transport (SST) model was designed and developed to accurately predict the beginning and amount of flow separation under adverse pressure gradients. To model transitional effects the Gamma-Re Theta model was applied in this study. To predict transition, ANSYS has designed and developed a locally formulated transport equation for intermittency, which can be used to trigger transition (ANSYS, 2011). Figure 2 shows a comparison between the k-epsilon and the k-omega based shear-stress-transport (SST) model with and without Gamma-Re Theta transitional model. The results are being evaluated based on the experimental work by (Achenbach, 1968).

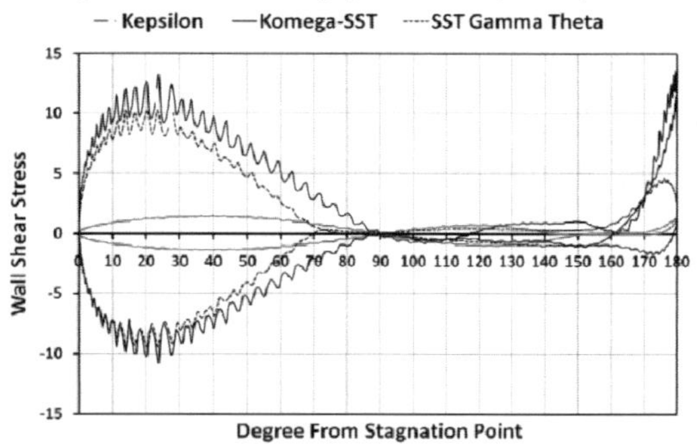

Figure 2 Wall shear stress distribution around cylinder through applying different turbulence model

1.5 Numerical results

Although this work does not mainly focus on flow parameters such as the drag coefficient, the Strouhal number, or the vorticity for the case of stationary cylinder, the results are included here as a basis for comparison with similar results in literature in order to validate the applied CFD model.

Frequency spectra of the fluctuating lift force

(Schewe, 1983) studied the shape of the power spectra at different Reynolds numbers. His wide study shows that at the subcritical Reynolds number, a narrow band of spectrum is taken at Strouhal number of the value of 0.2. The FFT analysis of the fluctuating lift force for the circular cylinder in the current study shows a dominant peak at the frequency of the value of 0.734 Hz. This peak corresponds to the frequency of vortex shedding for the stationary cylinder at St=0.22. Figure 3 shows the spectra analysis for lift and drag force. The frequency of the drag force is twice the vortex shedding frequency while the lift force oscillates at the vortex shedding frequency.

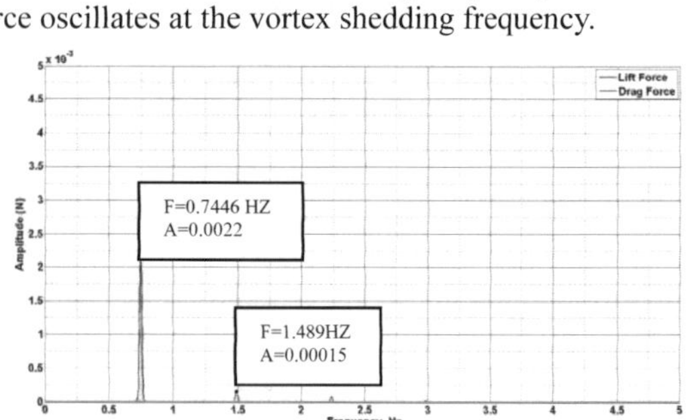

Figure 3 Frequency spectra of the lift and drag fluctuations for stationary circular cylinder at subcritical Reynolds Number of the value of 4000.

Wake mode

Figure 4 and Figure 5 show the alternate shedding of the vortices in the near wake in a classical vortex street configuration. As can be seen in a typical Karman vortex street, the regions of concentrated vorticity are shed into the downstream flow from alternate sides of the body, giving the appearance of an upper row of negative vortices and lower row of positive vortices. As shown in Figure 5, if the vortex becomes strong enough it can draw the vortex of the opposite side across the wake. These two vortices are in contrast with the direction of rotation. The approach of oppositely-signed vorticity cuts off the further supply of the vortices to the vortex of the opposite side. At this stage the vortex is being shed (Gerrard, 1966).

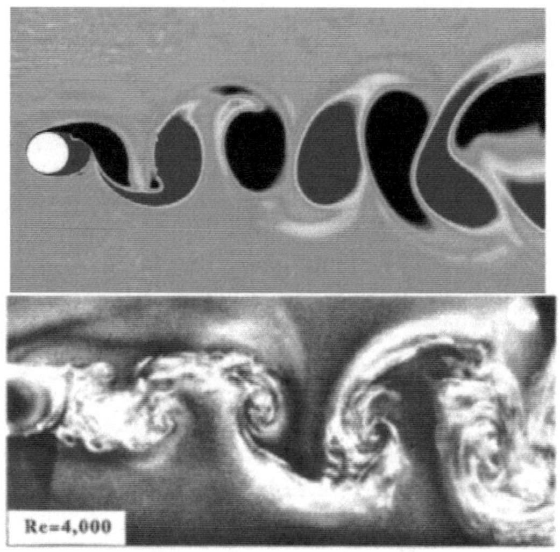

Figure 4 Time development of vortex shedding motion for the stationary cylinder at Re=4000 compared with experimental result. Contour of Vorticiy ,-1<Velocity Curl-z<1 of the current study and experimental result by (Williamson C. H.,1996)

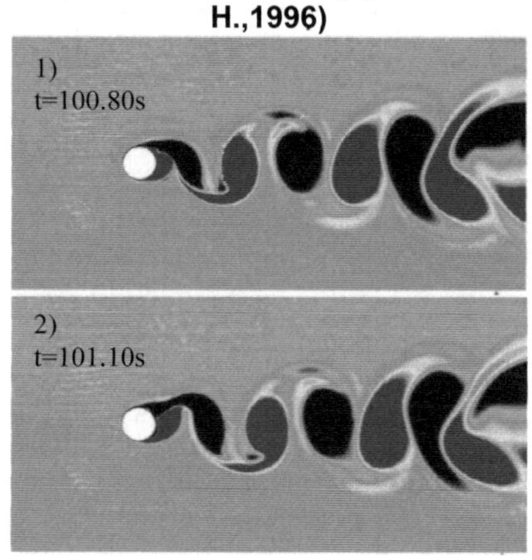

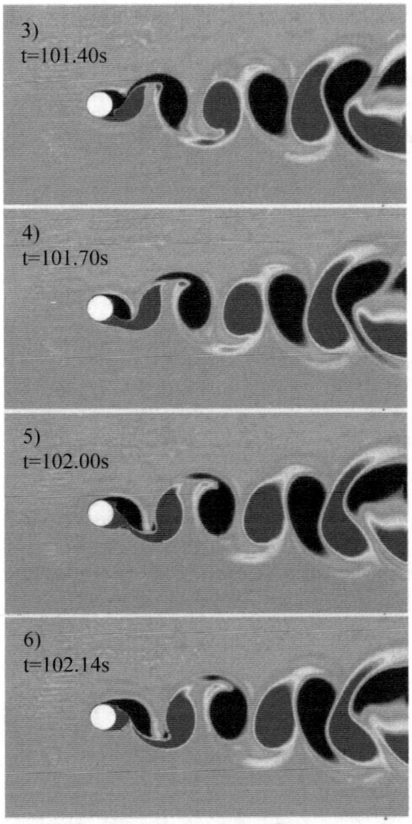

Figure 5 Time development of vortex shedding motion for the stationary cylinder at Re=4000- it takes 1.34 s to shed a vortex to downstream flow.

Considering the vortex shedding timing reveals that 1.34 s is the time of shedding one vortex to the downstream. The calculated Strouhal number is 0.22 and the experimental results show the value of 0.20. So the calculated Strouhal number is in the acceptable range. The wake mode is captured with a desirable resolution compared to the experimental result and the calculated frequency of the fluctuating lift force and Strouhal number in an acceptable range which confirms the accuracy of the applied CFD model. In the next section, the verified CFD model will be implemented to study the response characteristics of the eccentrically rotating cylinder in the flow stream.

2 Eccentricity rotating cylinder

A common approach to further investigate the vortex-induced vibration is to control the motion of the body, allowing the response of the wake to the motion to be studied in isolation. (Morse & Williamson, 2006) concluded that it is very important to carefully match the experimental conditions between the free and the controlled vibration. Regarding the difficulties of the implementation of a vertically moving cylinder in an experimental study, the vortex dynamics and the lock-in behavior of the eccentrically rotating cylinder is studied in this scientific work. This research is set out to answer a number of questions, such as the following: What mode of vortex formation can be observed in this configuration? What is the shape of response diagram? How does the lock-in region looks like in this case? What is the relation between vortex modes and fluid excitation? To what extend can the eccentricity rotating cylinder be used to precisely predict the behavior and response of a freely oscillating cylinder. This work tries to answer these questions by considering the time and hardware limitation. To discuss an aforementioned questions, the lift coefficient and the wake mode are investigated in the following amplitude-frequency ratio plane. The boundary of the synchronization region is being determined through a comprehensive spectral study. To investigate the force characteristics, the wake modes and the transition between the wake modes, it is important to describe the manner in which the wake state was established. In the present study for each simulation the value of fe/fs , ferequency ratio, is constant and the initial value for the transient simulation is the steady solution of stationary cylinder in a free stream. So in fact in each simulation the wake state at t=0 was the Karman wake state and for t>0 the cylinder was oscillated at a constant desired frequency until the end of the simulation to form a new wake state. In this investigation the force characteristics, the wake mode and the range of the synchronization region have been studied through varying the frequency of rotation (forcing frequency) for a range of A/D at Re=4000. Several points were picked for evaluation and some of these points have been identified as a boundary of the lock-in region. Figure 7 presents the space parameter of the current study.

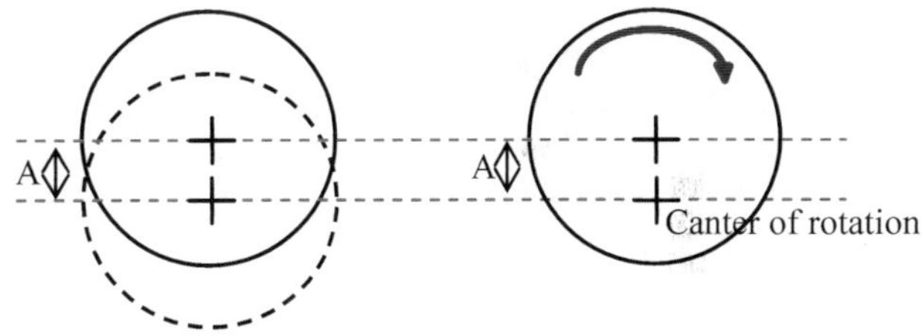

Figure 6 Schematic view of vertically moving cylinder and eccentrically rotating cylinder.

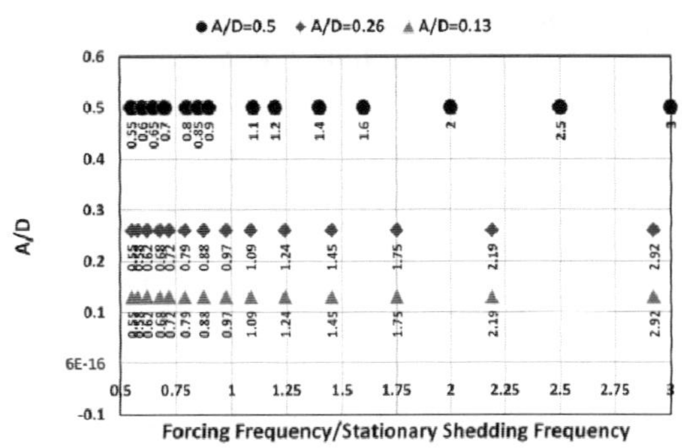

Figure 7 Distribution of simulation points in Amplitude-Frequency plane.

2.1 Lift and drag force characteristics

A vey vast effort has been devoted to characterize the unsteady lift force acting on a cylinder forced to oscillate. (Mercier, 1973), (Staubli, 1983), (Gopalkrishnan, 1993)and (Sarpkaya, 1979) have made a wide study in this area. In some cases, these lift characteristics are represented in terms of amplitude and phase, C_l and φ versus fe/fs. Figure 8 shows results from number of previous studies adopted from (Carberry, Sheridan, & Rockwell, 2005) and Figure 9 shows the results of current investigation. As can be seen at low values of fe/fs the amplitude of the lift force is small and out of phase with the cylinder motion. As oscillating frequency approaches the natural Karman frequency fs, the lift force experiences an abrupt jump in amplitude and phase shift of the order of 180. So at higher value of the frequency ratio fe/fs the lift force has a larger amplitude and the approximately in the phase with the oscillation of the cylinder. (Carberry, Sheridan, & Rockwell, 2005) discussed this change in amplitude and phase as a transition from low-frequency lift force to high-frequency lift force and the frequency experiencing this sudden change is called transition frequency ft. According Figure 9, the CFD model of the eccentrically rotating cylinder shows almost the same general shape in lift coefficient versus frequency ratio compared to the experimental data.

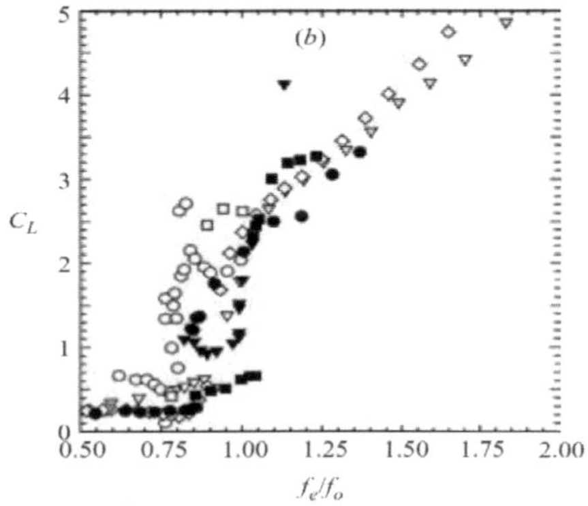

Figure 8 Experimental results of previous studies adopted from (Carberry, Sheridan, & Rockwell, Controlled oscillations of a cylinder: forces and wake modes, 2005). Lift coefficient $C_L$as a function of frequency ratio fe/fs.

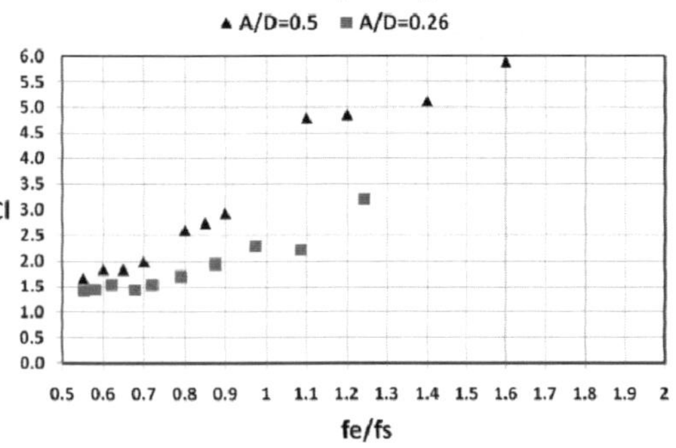

Figure 9 Numerical results of Current study. Lift force coefficient C_L for two different amplitude ratios as a function of frequency ratio fe/fs.

2.2 Vortex shedding frequency spectral analysis

The occurrence of the lock-in is estimated through study the spectral analysis of the fluctuating lift force. Figure 10 and Figure 11 show the fluctuating lift force and their corresponding frequency spectrum.

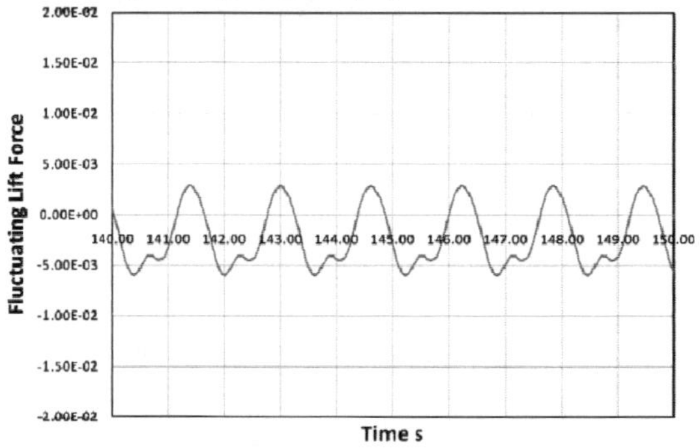

Figure 10 Time tracing of fluctuating lift force - Close view of synchronized lift force for 140<t<150. A/D=0.13, F=0.88 and Re=4000.

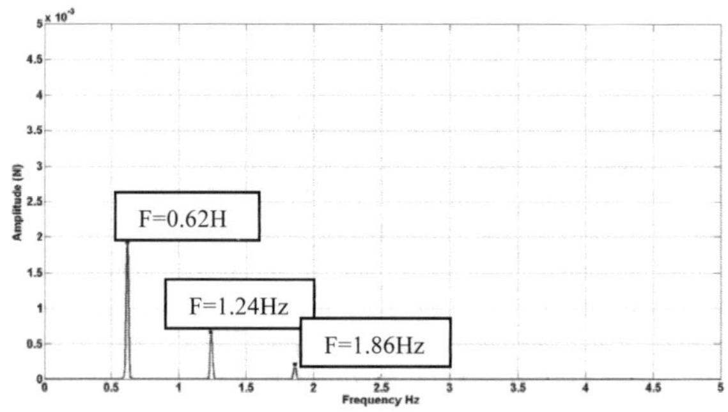

Figure 11 Spectral analysis of fluctuating lift force for 100 s of rotation corresponding to almost 62 cycle of rotations. A/D=0.13, F=0.88 and Re=4000.

When the lock-in does not occur, a peak can be observed at the vortex shedding frequency (Strouhal Frequency). While in the lock-in region the peak at the vortex shedding frequency disappears. Figure 10 and Figure 11 show the time tracing of the fluctuating lift force and the frequency spectra of the lift force, respectively. It becomes clear that for the case at A/D= 0.13 with the frequency ratio value of 0.88 and Re=4000 there is no peak at the Strouhal frequency, 0.73 Hz, and the only visible peaks occurs at the forcing frequency, 0.62Hz, and its multiples. This characteristic along with the synchronized shape of the lift force can be interpreted as an existence of a lock-in or synchronization region in the case under investigation. The same analysis must be done for the other selective points to specify the bounds of the lock-in region.

To develop an understanding about the lock-in region, it is worthy to study the wake mode inside the boundary of the synchronization region. The effect of the lock-in on the flow field and the wake state is investigated in more detail in the following sections.

2.3 Study of wake mode and vortex timing

As discussed in the previous section, according to (Williamson, & Roshko, 1988) and (Carberry, Sheridan, & Rockwell, 2001) the jump in the phase and the amplitude of the lift coefficient corresponds to a change in both phase and mode of the vortex shedding. These changes show a transition from one near wake state to another.

Figure 12 through Figure 14 show the instantaneous vorticity fields. One of the purposes of this section is to study the type of wake synchronization that may occur throughout the lock-in region. In the case of the vorticity field study, (Bearman, 1984), (Williamson, & Roshko, 1988) and (Carberry, Sheridan, & Rockwell, 2005) reported repeatable vortex shedding patterns, each one occurring within a certain range of amplitudes. (Williamson, & Roshko, 1988) found that the process of pairing of vortices is a fundamental one, not only for repeatability of the patterns, but also for enabling vortices to convect away under their own induced vortices. Figure 12 shows the time development of the vortex shedding for the case of A/D=0.13 corresponding to 5 mm of eccentricity at F=0.88 in one cycle of rotation. The low frequency wake, shown in Figure 12 produces a long attached shear layer at the bottom. Due to higher relative velocity at the upper side, the timing of the vortex formation at this side is less than in the stationary case. As can be seen in picture #1 of Figure 12 at t=151.612 s a negative vortex is about to shed and a shear layer is going to form at the lower side. As moving through the time steps, the lower shear layer is growing and shed in picture #9 while a positive vortex has been shed already. It is clear that the pattern of the vorticity field in picture #9 resembles the pattern of picture #1. So it can be concluded that the vortex shedding pattern repeats after each cycle and becomes periodic and synchronized in the specific mode. Since two single vortices have been shed in one cycle the wake mode it is named 2S which means that two single vortices are fed into the downstream wake in each cycle of rotation.

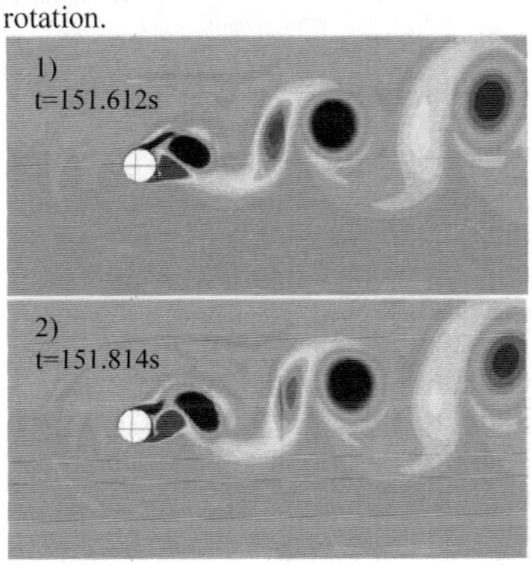

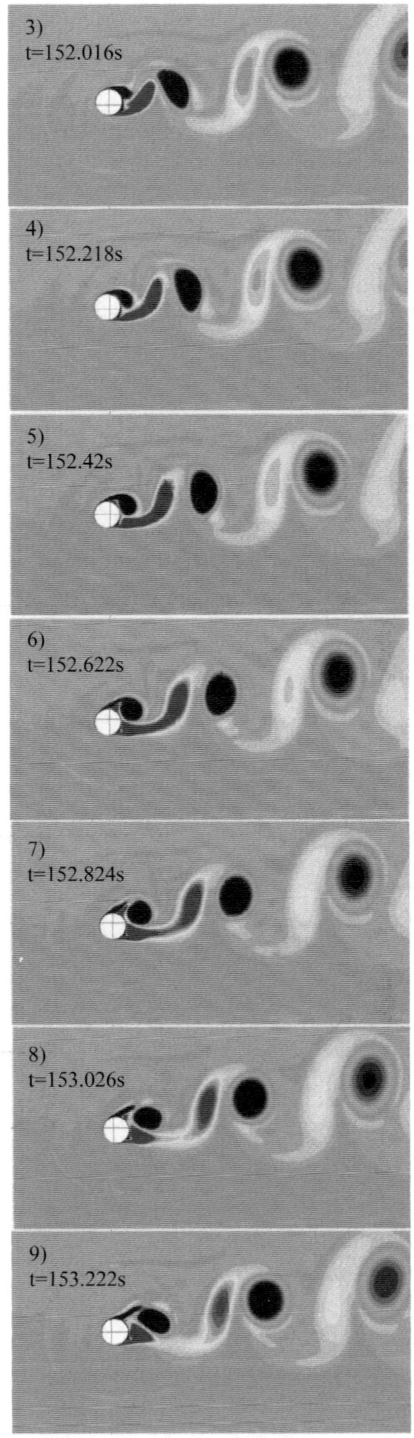

Figure 12Time development of vortex shedding motion, for A/D=0.13, F=0.88 and Re=4000. Vortex street type of wake 2Smode.

The study of the flow at the slightly larger amplitude at A/D=0.26, corresponding to 10 mm eccentricity, shows although the appearance and the general configuration of the near-wake vortices remains very similar to Figure 12, the timing of the vortex formation is smaller in the case with larger an amplitude of oscillation.

Figure 13 and Figure 14 show two consecutive cycles of rotation in the case of A/D=0.5 and F=0.6. The formation of more than two vortices per cycle is apparent in these figures. In each cycle of rotation a vortex pair is fed into the downstream wake with the single vortex. This vortex pattern defines the P+S mode (in accordance with the terminology introduced in (Williamson, & Roshko, 1988)).

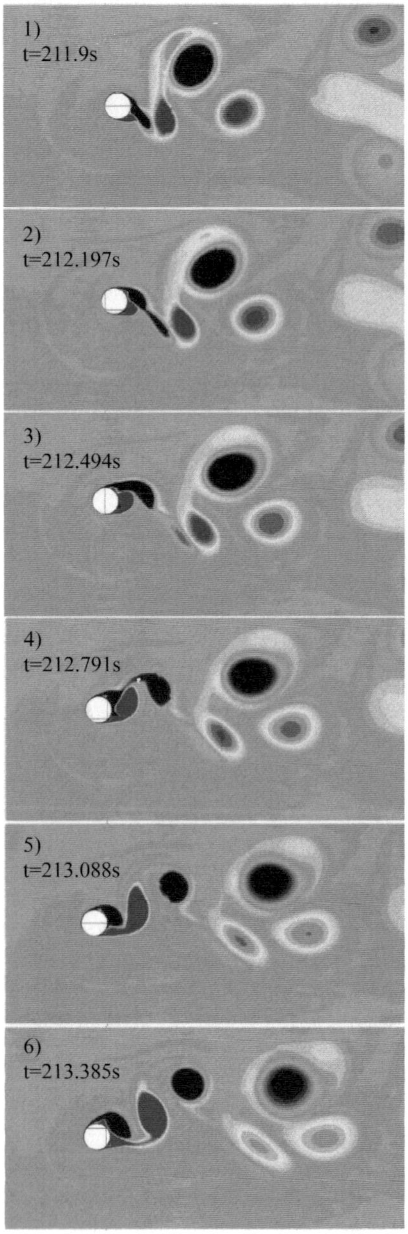

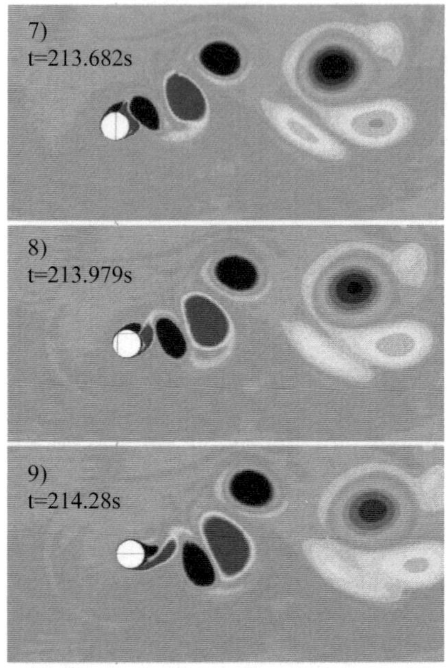

Figure 13 Cycle 1- Time development of vortex shedding motion for A/D=0.5, F=0.6 and Re=4000. Vortex street type of wake P+S mode.

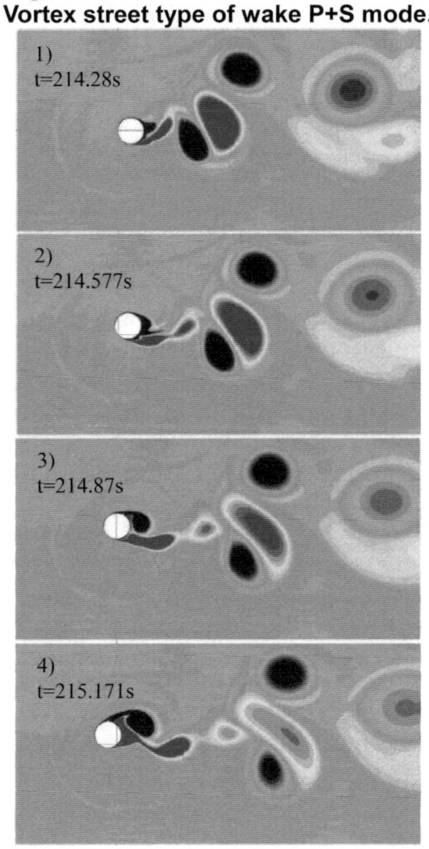

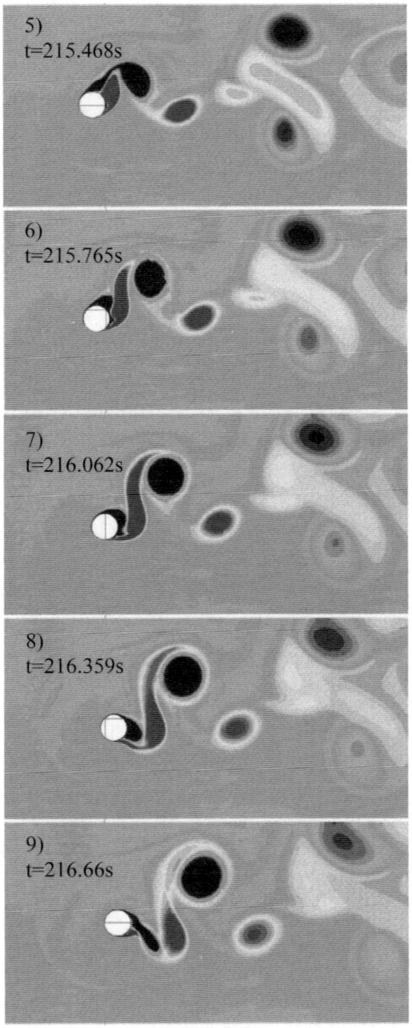

Figure 14 Cycle 2- Time development of vortex shedding motion for A/D=0.5, F=0.6 and Re=4000. Vortex street type of wake P+S mode.

There is a change from one mode to another mode as the frequency ratio increases. Figure 15 was generated by imposing a new larger forcing frequency to the case of**Error! Reference source not found.**and F=0.79. The frequency ratio of the new case is 1.24. Following Figure 15 and Figure 16 becomes apparent that this wake mode repeats itself after two consecutive cycles. This new wake mode is quite distinct from the known modes in vortex induced vibration. This new mode of vortex formation comprises a negative vortex in one cycle followed by two single positive vortices in next cycle.

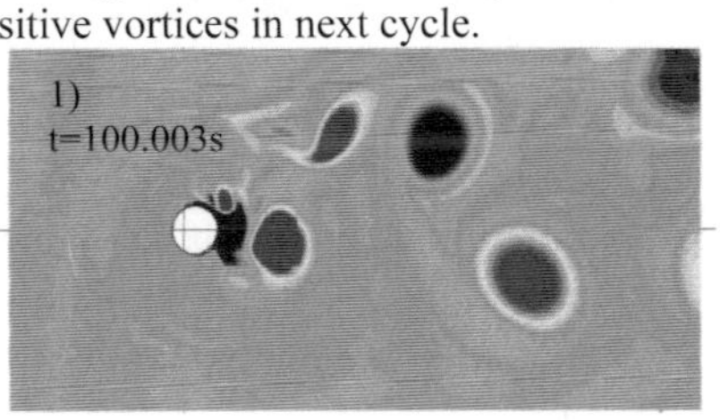

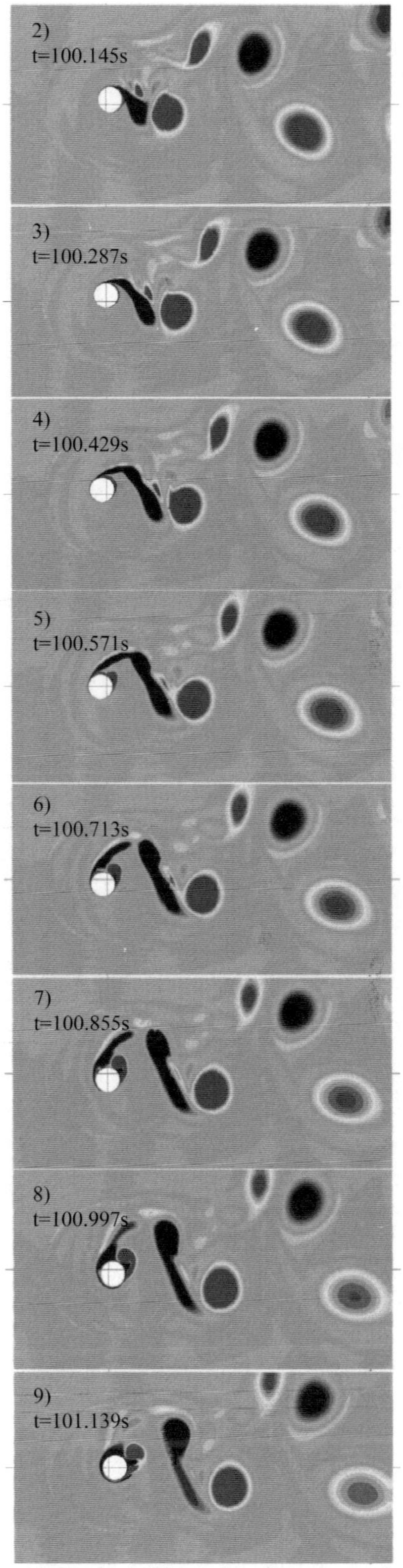

Figure 15 Cycle 1- Time development of vortex shedding motion, for A/D=0.26, F=1.24 and Re=4000.

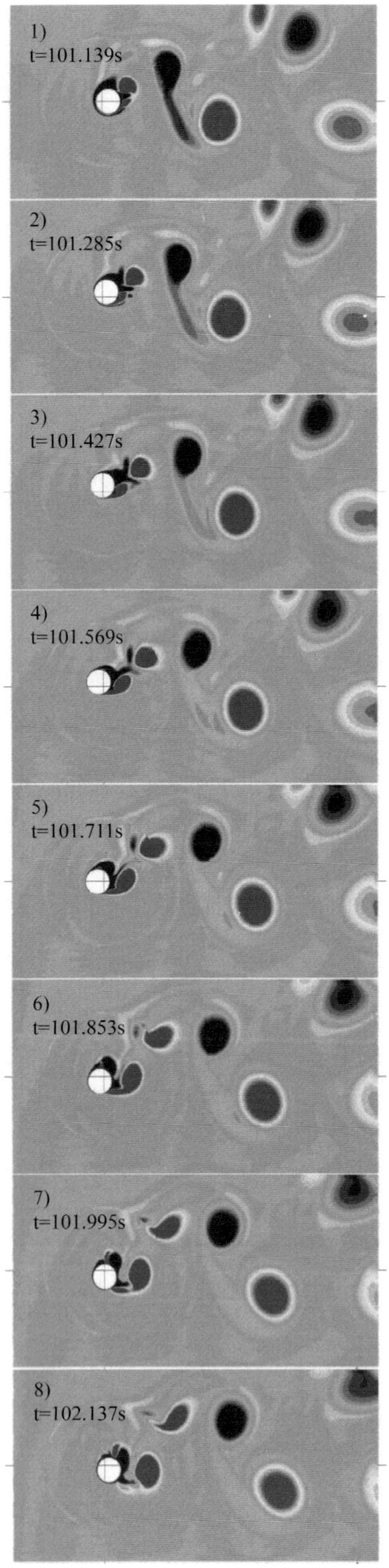

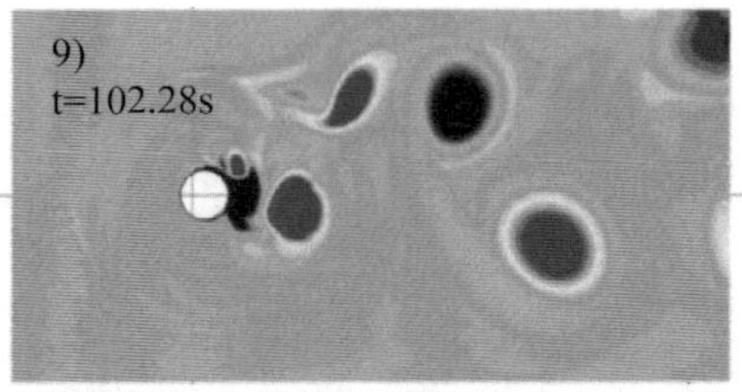

Figure 16 Cycle 2- Time development of vortex shedding motion, for A/D=0.26, F=1.24 and Re=4000.

As discussed in the preceding paragraphs, exploring the existence and shape of the synchronization region has been the main objective of this work. The following figure shows the bounds of the lock-in region in the case of the eccentrically rotating cylinder. It is clear that the range of the synchronization region in the current case is larger than the transversely oscillating cylinder. Simultaneous occurrence of cross-flow and in-line oscillation can justify the large range of synchronization region, since the lock-in happens at the frequency of twice of the Strouhal frequency in the case of in-line oscillation. Figure 18 shows the synchronized wake mode at the bounds of the lock-in region.

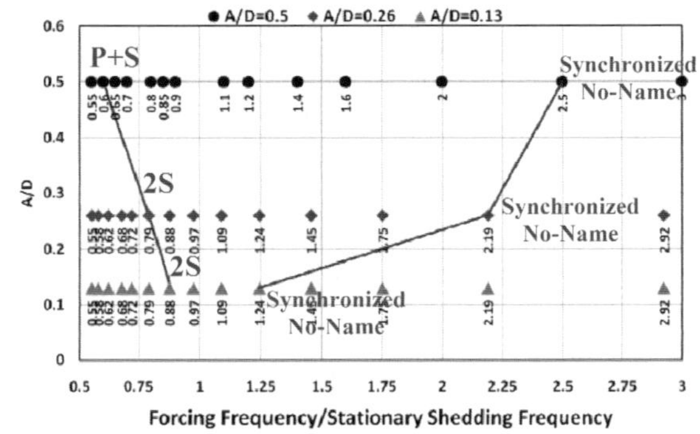

Figure 17 Bounds of synchronization region in eccentrically rotating cylinder

CONCLUSION

In this work the existence of the lock-in or synchronization region for the case of an eccentrically rotating cylinder has been investigated. The lock-in bounds were found through a study of the drag and lift force characteristics, spectral analyses of the fluctuating lift force, and through exploring the vortex synchronization in the frequency-amplitude plane. The manner in which the wake mode varies with the frequency of rotation (forcing frequency) was studied for a range of A/D at Re=4000. This value for Re has been chosen due to available experimental results for transversely oscillating cylinders in water channels. Besides that, over a very wide range of Re, namely $300<Re<3x10^5$, the wake structure is almost similar and the frequency of the vortex shedding and of the fluctuating force coefficient remains constant. For this subcritical range, the boundary layer over the cylinder surface remains laminar and the wake is completely turbulent. This fact suggests an extension of the current work to higher and more practical Re values.

Several points in the amplitude-frequency plane were picked for further study and some points have been identified as boundaries of the lock-in region. In the neighborhood of lock-in bounds, different vortex patterns are synchronized, each within the certain range of the amplitude and frequency ratio and some of them are similar to those patterns found in the literature for a transversely oscillating body. Despite some symmetry wake modes, a couple of very asymmetric wake modes were observed. A typical synchronized vortex wake which has been observed here involves the generation of two single counter-rotating vortices in each cycle of rotation. This so-called '2S' wake mode was observed in two different amplitude ratio, A/D=0.13 and A/D=0.26 at the vicinity of the left boundary of synchronization region. In contrast with the typical '2S' mode in a transversely moving cylinder, feeding a single vortex in each half cycle and in total two counter-rotating vortex in each cycle, the aforementioned observed '2S' mode in the case of an eccentrically rotating cylinder sheds two counter-rotating vortices in each cycle, but not necessarily in each half cycle. At the higher A/D at the left bound of the synchronization region, the formation of more than two vortices per cycle was observed. This vortex pattern defines the P+S mode, in accordance with the terminology introduced in (Williamson, & Roshko, 1988). At constant A/D, increasing forcing frequency through the lock-in region leads to a transition from one wake mode to another wake mode. By increasing the rotating frequency from 0.79 to 2.19 at constant A/D wake mode, this transition is observed. The interesting phenomenon about this transition in the wake mode is that it coincides with an abrupt change in the fluid force that was observed by (Bishop & Hassan, 1964).

According to Figure 9, a sharp jump in the lift force can be interpreted as a resonant synchronization within the lock-in region. It has been concluded that the sudden jump and the large force are induced by the shedding of a more concentrated region of vorticity. To examine the exact position of the resonance between the cylinder motion and wake formation, a finer resolution in the amplitude-frequency plane is needed. It is worthy to study such a wide range in the future, since the highest vibration amplitude occurs at the resonant point for a freely oscillating cylinder.

It is common to study the forced oscillating cylinder at the constant amplitude ratio A/D while the frequency of the motion is varied about the point where fe/fs=1. Although in this work, the bounds of the synchronization region were specified around fe/fs=1, it is possible that further resonance between the motion of the cylinder and the vorticity field and thereby further resonance in the cylinder forces exist in other parts of the frequency-amplitude plane.

One of the major questions for research into this class of fluid-structure interaction is whether the results of the controlled oscillating experiment can be applied to predict the vortex-induced vibration of a body that is free to oscillate. Whilst the current study did not seek to answer this question, a number of studies have been published in this area (Morse & Williamson, 2009). The relationship between an eccentrically rotating cylinder and an elastically mounted cylinder is very complicated. However there are striking similarities in the wake mode and the lift force characteristics for the eccentrically and freely moving cylinder. In both cases, the wake transition is observed around fe/fs=1. Therefore, despite some very good similarities in the wake states of the eccentrically rotating and the freely moving cylinder, further work is required in order to determine how the case of rotation with eccentricity can be used to predict the vortex-induced motion.

REFERENCES

Achenbach, E. (1968). Distribution of local pressure and skin fricsion around a circular cylinder in cross-flow up to Re=5x106. *Journal of fluid Mechanics , 34* (04), 625-639.

ANSYS. (2011). *ANSYS CFX-Solver Modeling Guid;* (14.0 ed.). ANSYS, Inc.

Bearman, P. (1984). Vortex shedding from oscillation bluff bodies. *Annual review of Fluid Mechanics , 16* (1), 195-222.

Belivins, R. D. (1990). *Flow-induced vibration* (2 ed.). New York: Van Nostrand Reinhold Co.

Bishop, R. E., & Hassan, A. Y. (1964). The lift and drag forces on a circular cylinder oscillating in a flowing fluid. *proceedings of the royal society of london. series A. mathematical and physical sciences , 277* (1368), 51-75.

Carberry, J., Sheridan, J., & Rockwell, D. (2005). Controlled oscillations of a cylinder: forces and wake modes. *Journal of fluid Mechanics , 538*, 31-69.

Carberry, J., Sheridan, J., & Rockwell, D. (2001). Forces and wake modes of an oscillating cylinder. *Journal of fluids and structures , 15* (3), 523-532.

Fransson, T. H., n.d. Dynamic Aeroelasticity. In: *Aeroelasticity in Axial-Flow Turbomachines.* s.l.:s.n.

Gerrard, J. H. (1966). The mechanics of the formation region of vortices behind bluff bodies. *Journal of fluid mechanics , 25* (2), 401-413.

Gopalkrishnan, R. (1993). Vortex-induced forces on oscillating bluff cylinders. *(No. WHOI-92-38). WOODS HOLE OSCEANOGRAPHIC INSTITUTION MA .*

Kielb, R. E., Barter, J. W., Jeffrey, P. T., & Kenneth, C. H. (2003). Blade Excitation by aeromechanic instabilities- A compressor Blade Study. *ASME Turbo Expo 2003, collected with the 2003 International Joint Power Generation Conference. American Society of Mechanicsl Engineering*, 399-406.

Kielb, R. E., Hall, K. C., Spiker, M., & Thomas, J. P. (2006). Non-Synchronous Vibration of Turbomachinery Airfoils. *Duke University Durham NC Dept of Mechanical Engineering and Materials Science .*

Menter, F. R. (1992). Improved two-equation k-omega turbulence models for aerodynamic flows. *NASA STI/Recon Technical Report N , 93.*

Mercier, J. A. (1973). Large amplitude oscillations of a circular cylinder in low speed stream. *Diss. Stevens Institude of Technology* .
Morse, T. L., & Williamson, C. H. (2006). Employing controlled vibration to predict fluid forces on a circular undergoing vortex-induced vibration. *journal of fluids and structures* , *6* (22), 877-884.
Morse, T. L., & Williamson, C. K. (2006). Employing controlled vibration to predict fluid forces on a circular undergoing vortex-induced vibration. *journal of fluids and structures* , *6* (22), 877-884.
Morse, T. L., & Williamson, C. K. (2009). Fluid forcing, Wake modes, and transition for a cylinder undergoing controlled oscillations. *Journal of Fluids and Structure* , *25* (4), 697-712.
Sarpkaya, T. (1979). Vortex-induced oscillation. *Journal of Applied Mechanics* , *46* (2), 241-258.
Schewe, G. (1983). On the force fluctuations acting on a circular cylinder in crossflow from subcritical up to transcritical Reynolds bnumbers. *Journal of Fluid Mechanics* , *133*, 265-285.
Staubli, T. (1983). Calcualtion of the vibration of an elastically mounted cylinder using experimental data from forced oscillation. *Journal of Fluids Engineering* , *105* (2), 225-229.
Summer, B. M., & Fredsoe, J. (1997). *Hydrodynamics around cylindrical structure* . Singapore: World Scientifc Publishing Co Pte Ltd.
Williamson, C. H. (1996). Vortex dynamics in the cylinder wake. *Annual review of fluid mechanics* , *28* (1), 477-539.
Williamson, C. H., & Govardhan, R. (2004). Vortex-induced vibration. *Annu. Rev. Fluid Mech.* (36), 413-455.
Williamson, C. H., & Roshko, A. (1988). Vortex formation in the wake of an oscillating cylinder. *Journal of fluid and structures* , *2* (4), 355-381.
Williamson, C. K., & Govardhan, R. (2004). Vortex-induced vibrations. *Annu. Rev. Fluid mech* , *36*, 413-455.
Williamson,, C. K., & Roshko, A. (1988). Vortex formation in the wake of an oscillating cylinder. *Journal of fluid and structures* , *2* (4), 355-381.

Proceedings of the 14th International Symposium on Unsteady Aerodynamics, Aeroacoustics & Aeroelasticity of Turbomachines
ISUAAAT14
8-11 September 2015, Stockholm, Sweden

I14-S12-1

ON LIP STALL SUPPRESSION IN POWERED INTAKE: HIGH AND LOW FIDELITY APPROACH

Mauro Carnevale
Department of Mechanical Engineering
Imperial college of London, SW7 2BX, UK
m.carnevale@imperial.ac.uk

Jeff S. Green
Rolls-Royce plc
Derby, DE24 8BJ,UK

Luca di Mare
Department of Mechanical Engineering
Imperial college of London, SW7 2BX, UK

ABSTRACT
This work describes a computational and analytical study of the flow in a turbofan intake at high incidence. It is well known that lip separation can occur above a certain angle of flight incidence, depending on the flight Mach number, Reynolds number and engine mass flow rate. There has been a steady rise in the lower lip loading in turbofan installations in recent years. This trend is set to continue because of the necessity to reduce the installation weight and drag in view of the introduction of ultra-high bypass ratio engines. This makes designs with shorter and thinner lips more attractive, but at the same time riskier because of the occurrence of lip separation.
Once separation occurs, a distortion with pressure, entropy and vortical components is passed into the fan. While these distortions are potentially detrimental to the life and stability of the fan, it is also known that the presence of the fan itself changes the conditions at which separation occurs. It is important therefore, to investigate and understand the mechanism by which the fan modifies the flow in the intake.
In this study, detailed simulations have been performed of powered and aspirated intakes at the same reduced mass flow rate and spanning a wide range of incidence angles. Geometries representative of real engine configuration have been used. The computational results are shown to be in agreement with wind tunnel results and demonstrate the beneficial effect of the fan on the intake flow.
Furthermore, the distortions at fan face are analysed using wave decomposition.
It is shown that the effect of the fan is carried by the rejected waves generated at fan face - consistently with actuator disc theory - but it is amplified by the nature of the separated flow in the intake.

INTRODUCTION
It is a commonly accepted design practice to provide the Low Pressure Compression system (LPC) with uniform flow under all operating conditions. Under this point of view the intake represents the most critical component. High distortion levels to fan stage and therefore to the OGV and ESS blades, can cause blades instability and have detrimental effects on whole compression system.
At the same time, the presence of the LPC modifies the flow in the intake. A beneficial effect of the presence of the fan has been demonstrated in term of improving the stall margin of fan stage.

In recent years Computational Fluid Dynamic CFD provides a robust and high fidelity approach to investigate the interaction mechanisms in intake-fan coupled systems. The weak point of these kind of approaches lies on the high computational costs and make them not suitable for preliminary design process. Fast tools are therefore necessary to investigate the distortion propagation for early stage of the design process or whenever a quick solution is desirable, when minimal assumptions are made about the studied geometry and nature of the gas.

A large literature review based on low fidelity methods and their application to problem (typically acoustic) involving distortions in boundary conditions is available.

Satyanaryan et al [1] compared theoretical results and experimental data highlight the necessity to overcome the single blade analysis models. In 1969, Kaji & Okazaki [2] have described a model, providing the magnitudes of the reflected and transmitted waves as they approach the blade cascade. This approach represents the widest validation reference for numerical approaches and will exploit in most of the work will be discussed herein. The model has been update in 1970 [3], where Kaji & Okazaki by considering also the blade thickness and camber angle effects. These models are built using the semi-actuator-disk theory and it is very efficient for cases where the wavelength of the incident wave is much longer than the length of the blade chord. Amiet [4] proposed a simplified theory for evaluation of the reflection and transmission coefficients. This approach shows a good comparison with the semi-actuator disk model by Kaji & Okazaki for very long waves (wavelength-to-chord ratios 32 and above). Amiet extended his approach from single blade row to two-blade rows in [5].

Cumpsty and Marble [6] related entropy fluctuations with vorticity and acoustic waves in bladed passages. The model accounts for the small perturbations in comparison with the steady pressure difference across the blade. The approach shows a very good agreement with Kaji & Okazaki for the reflection and transmission coefficients of acoustic waves as functions of the wave incidence for longer wavelengths.

In 1987 Whitehead [7] published an analytical 2D solution for the flat blade cascade forced response and flutter simulations. Semi-actuator disk methods are used for evaluation of reflection and transmission coefficients as well as blade forcing for forced response and first-order vibration modes cases.

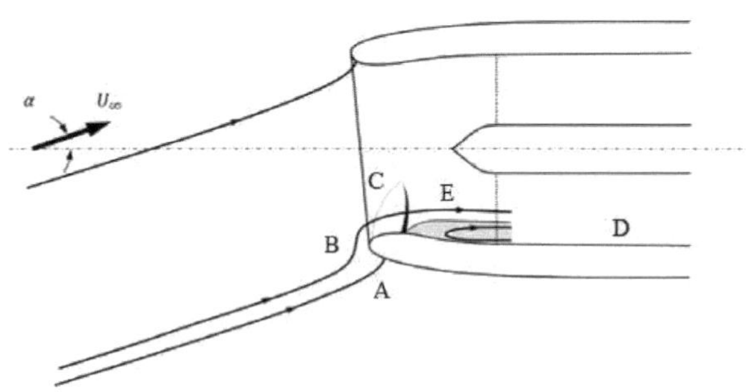

Figure 1 – Intakes flow features

This paper is mainly focused on the problem connected to high incidence flight conditions. Typical scenarios take place during take-off and at beginning of descent phases. When the ingested flow is characterized by high incidence on the intake is exacerbated by the unfavourable combination of relatively high Mach numbers and fan speeds and by the action of high-lift devices which modify the direction of the flow in front of the wing. The flow pattern is determined by the Reynolds number, the Mach number (or equivalently the mass flow rate through the intake) and the incidence.

Fig. 1 shows the main features of the stall formation inside a real civil turbofan intake. Installations where only the intake is considered (i.e. without the presence of the fan) are defined in this paper aspirated intake. The lip stall is caused by the adverse pressure gradient encountered by the flow after turning into the intake around the lower lip. During the design process, the limit conditions for incidence are typically set for the low speed flight and high velocity ratio $V_{intake}/V_{freestream}$ (stream tube capture ratio). Real conditions are characterized by high cruise velocity ($V_{freestream}$) and intake's velocity (V_{intake}) and consequently elevate mass flow rate combined with the incidence result in high flow turning and strong shocks. Typically take off, climb out is in the in the M=0.25 to 0.45 range, where the limit angles of attack can occur. At high speed cruise angle tolerance is much more limited.

The effect of the inlet distortion effect in transonic fan has been numerically investigated by Gunn et al [8]. The rotor inlet non-uniformity creates circumferential and radial variations. Shock position, size and strength vary markedly around the annulus. More recently Perovic et al [10] investigated experimentally the stall effect and its propagation through the whole annulus.

Felderman et al [9] investigated the lip separation at low speed due to the incidence, Wakemann et al. [11] [12] studied intake lip separation at low and high speed and explored strategies for lip stall control based on trips and vortex generator jets. Chen [13] investigated the aspirated intake with a 3D Euler code and compared data with actuator disk codes.

The interaction between the intake flow and the fan is a two-way interaction. Installations which include intake and fan stage are named in this paper powered intake. The importance of coupling in distorted flows has been identified since the 1970s. Callahan and Stenning [14] studied flow redistribution upstream of a low hub-tip ratio fan with combined distortion using a very simple model of compressor behaviour. Henderson and Horlock [15] developed a simplified approach for unsteady flow predictions on compressor blades in distorted flow. Greitzer [16] proved that a compressor would induce an irrotational velocity field ahead of itself in response to incoming distortions. Longley et al [17] studied the interaction between intake and fan flow analytically and show that this interaction takes place at wave lengths of comparable size to the fan diameter. Hawthorne et al [18] investigated the effect of inlet distortions on compressor blades performance and elucidated the mechanisms by which distortions affect fan performance. Hsiao et al [19] performed an actuator disk analyses on powered intakes. Yokum et al [20] performed a study of the effects of rotor-blade geometry, steady loading, and distortion wavelength on the distortion flow field. Peters et al [21] demonstrated the use of body forces as surrogate models of the fan set to design ultra-short nacelles. Fidalgo et al [22] studied the interaction between the flow and a transonic rotor. Carnevale et al [23] investigated how fan and distortion occurring in the intake influence each other when separation occurs at incidence.

This work deals with a study into the flow intake of a civil turbofan at high incidence. The two-way interaction has been assessed by means a CFD calculation referring to a previous work in Carnevale et al [23]. This analysis quantifies the effect of the fan on the intake flow in terms of harmonic content. The response of fan blade has been investigated with the work approach detailed and validated in Romanov et al [24]. Authors propose a generic approach for investigation of arbitrarily-patterned travelling distortion propagation along fan stage domains. Non-reflecting boundary conditions have been implemented using the hyperbolic characteristic theory. The model also includes a set of interchangeable loss- and deviation correlations that account for the total pressure loss across the blade and the corrected value of the Kutta-condition at the trailing edge

Imposed inlet boundary conditions have extracted by CFD aspirated intake. The overlapping of reflected waves by fan blade with the input distortions from CFD for the lower frequencies is comparable with the distortion signal extracted by 3D fully coupled CFD.

NOMENCLATURE

Re	=	Reynolds number [-]
$MRTP$	=	Corrected mass flow-rate $[kg/s \times \sqrt{K}/Pa]$
M_∞	=	Free stream Mach number [-]
P_∞	=	Free stream pressure (total) [Pa]
v_∞	=	Free stream velocity [m/s]
ρ_∞	=	Free stream density [Kg/m^3]
α	=	Angle of wind attack [degree]
C_f	=	Skin friction coefficient [-]
P_{stat}	=	Static pressure [Pa]
P	=	Total pressure [Pa]
T	=	Total temperature [K]
W	=	Real mass flow-rate [kg/s]
u_x	=	Axial velocity [m/s]
u_∞	=	Free stream axial velocity [m/s]
y^+	=	y-plus [-]
u^+	=	u_x-plus [-]
Γ	=	Circulation [m^2/s]
DC_{60}	=	Fan face distortion coefficient [-]
P_{60}	=	Minimum average 60° sector tot pressure [Pa]
P_{ave}	=	Area weighted fan face total pressure [Pa]
q_i	=	Dynamic pressure [Pa]
D_l	=	Highlight diameter [m]
h	=	Cruise altitude [ft]
dt	=	time step [s]
ϕ	=	Mass flow rate function [-]
π	=	Pressureratio[s]
η	=	Efficiency[s]

CFD METHODOLOGY

All computations are carried out using the CFD solver reported in Di Mare et al [25]. The solver uses a cell-centred, unstructured, finite volume scheme. The inviscid fluxes are based on Total Variation Diminishing (TVD) variable extrapolation. The gradients are evaluated using with weighted least square and the numerical fluxes are evaluated using Roe's [26] flux vector difference splitting. Van Leer's flux vector splitting [27], and several implementations of Liou-Steffen (AUSM) [28] flux vector splitting are also available. For low-Mach number flows, a preconditioning procedure based on the AUSM-UP(+) [29] with Choi-Merkle [30] time derivative preconditioning is employed. Convergence is reached by means of Jacobi iteration or, alternatively with GMRES iterations. The solver is fully implicit, and it is second order accurate in space and in time. Different turbulence closures are available in the code in order to face both high and low Reynolds flow configurations: mixing length, Spalart Allmaras (SA) and k-ε, k-ω, k-ω SST. Low Reynolds correction is applied according the approach proposed in Wilcox [31].

The CFD solver has been assessed and validated by means of academic test cases in order to take into account the different flow condition occurring in intake by Carnevale et al [35]. The flow conditions analysed in the present study corresponds to steady flight at 19700 *ft* altitude. The flight Mach number is $M_\infty=0.25$.

The LPC operating condition is representative of a modern fan in civil aero-engine operating at the beginning of landing, or in take-off and climbing manoeuvres. High angle of incidence and high capture stream tube ratios represent a typical condition for landing and take-off manoeuvres. These conditions are next to the limit of angle of attack capabilities of the aircraft.

A number of incidence values have been studied. For the purpose of the present paper, incidence is referred to the highest incidence value for which un-stalled operation can be achieved in a ground test facility. This incidence value will be referred to as α_{ref} in the rest of the paper. The simulations are carried out from $\alpha - \alpha_{ref} = -1^o$, up to $\alpha - \alpha_{ref} = +6^o$.

The computations on domains related to aspirated intake are based on Reynolds Average Navier-Stokes (RANS), whereas the powered intake study uses Unsteady Reynolds Average Navier-Stokes (URANS). Due to the symmetry of the flow in the absence of the fan, aspirated domain only represents one half of the intake.

Unsteady simulations are performed on powered intake, which features a full three-dimensional intake domain and a whole annulus fan model. The exit conditions imposed as Riemann invariant for the fan model are extracted in order to set the same mass flow rate ingested both in aspirated and powered configuration. The time step has been set in order to describe the blade passage with 64 times steps. The distortion signal wavelength and frequency in the rotating frame are sufficiently large to make the chosen time step adequate for the purpose of the current computations. This is consistent with previous observations by Bernardini et al [32], [33] and Carnevale et al [34] that URANS model can be remarkably accurate in predicting low frequency unsteady flow features.

A grid independence study is performed on the intake and on the LPC components. The purpose of this study is to identify the grid sizes required to resolve the flow features of interest in this paper whilst keeping computational expense acceptable. All the grids have been created with an in-house mesh generator.

Table 1 – Intake's meshes features –Fan's meshes features

INTAKE	LEVEL 1	LEVEL 2	LEVEL 3
Typical cell volume near wall	1×10^{-5}	6.5×10^{-6}	1.45×10^{-6}
Typical cell volume inside intake	3.5×10^{-5}	1×10^{-6}	3×10^{-6}
1st layer thickness	2×10^{-4}	5×10^{-5}	1×10^{-5}
B.L. growing ratio	1.2	1.2	1.3
No. layer	20	20	25
Number of cell	12 M	17 M	18 M
Blade			
FAN (single passage)	300 k	400 k	600 k

Intake

The intake domain has been discretized using a hybrid grid, made of prisms in the boundary layer near solid surfaces and tetrahedral in the rest of the volume. The far field boundary is placed 20 intake diameters D_l away from the intake inlet plane. The choice of unstructured mesh is related to the necessity of discretizing the domain with a large range of volume of cells. Very detailed discretization has been performed near the wall and large volumes characterize discretization in the far filed. Three different levels of grid refinement have been considered for the intake domain. Grid's features are provided in Table 1. Details of the grids near the intake lip and the wall are shown in Fig. 2. Mesh levels 1 and 2 differ in terms in the thickness of the first layer as well as number of layers of the boundary grid. Grid refinement is performed at constant aspect ratio and even a small change in the boundary layer grid can cause a considerable variation in the grid cell count. Grid level 2 and the finest level grid level 3 have the same width in term of prisms discretization (in the direction normal to the wall), and a larger number of prism layers allow to impose a lower value of the first layer thickness to grid level 3.

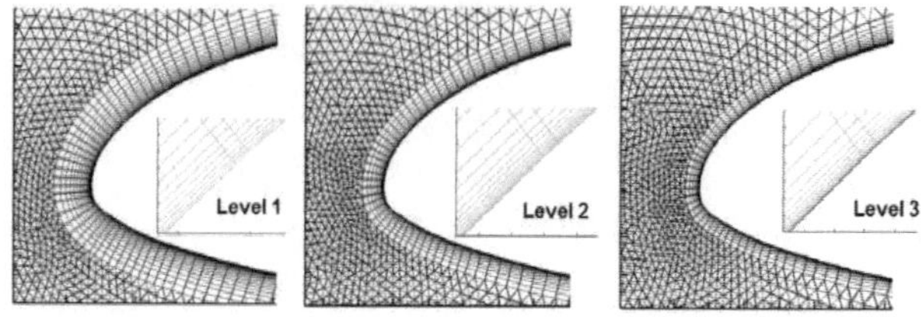

Log-law profiles are plotted in Fig. 3 in two different axial positions in not separated flow.

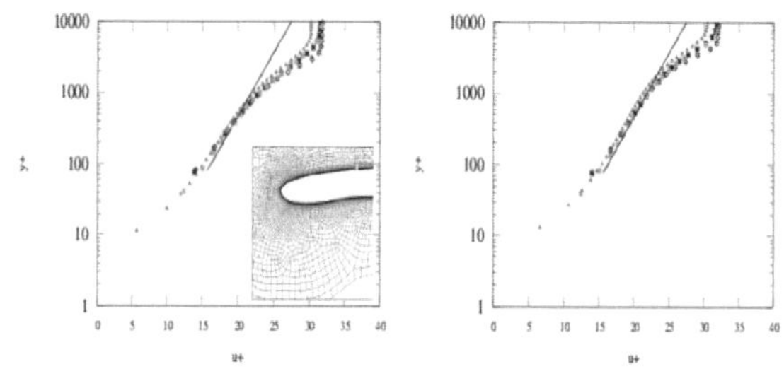

Figure 3 - Log-law profiles near intakes walls

Fan blade

Grid independence has been performed in order to properly discretize blade's geometry. Computations have been carried out with three different grids levels for each blade. The grid parameters are summarized in Table 1. The sensitivity of the mid-span pressure coefficient distribution to the grids is shown in Fig. 4 and details of the fan grid near the trailing edge at the mid-span section are shown in Fig. 5. Based on the results of the grid sensitivity study, the intermediate grid is chosen for the intake and for the turbomachinery domains as the best compromise between accuracy and computational effort.

Meshes have been generated by mirroring the semi-domain of intake with respect to the symmetry plane, and by replicating the single passage fan mesh to the whole annulus. Consequently grid independence has been automatically assessed.

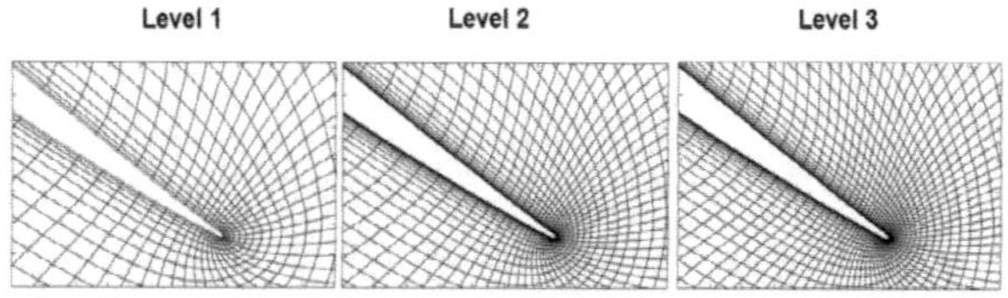

Figure 4: details of fan grid in the blade-to-blade direction

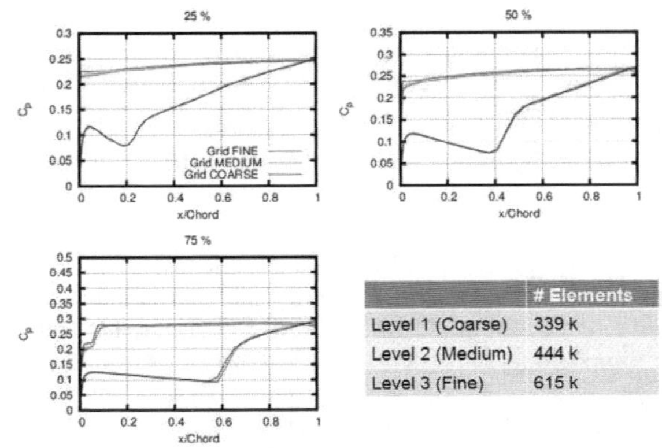

	# Elements
Level 1 (Coarse)	339 k
Level 2 (Medium)	444 k
Level 3 (Fine)	615 k

Figure 5 – Cp distribution on blades – single passages

The selected turbulence model is k-ω model. It has been selected due to the previous experience gained in intake flow field prediction, detailed in Carnevale et al [35].

ACTUATOR DISK APPROACH

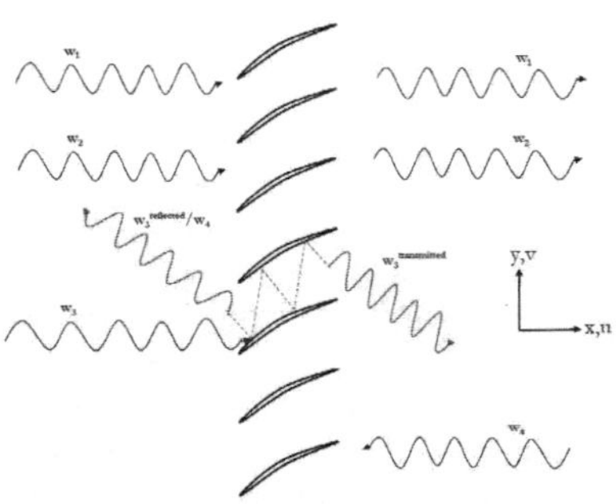

Figure 6 –Actuator disk approach

The modelling of linearized distortions using non-primitive variables finds its roots in the analysis of sound propagation through blade rows.

A schematic representation of actuator disk strategy is shown in Fig. 6. The assumption of a constant mean flow solution and small perturbations, allows for the Euler equations to be greatly simplified. Through a decomposition of the perturbations using Fourier series, four waves appear in the 2-D solution an entropy wave, a vorticity wave, and two acoustic (pressure) waves, one propagating upstream and one downstream.

In general, these waves propagate with spiral wave fronts and as they make their way down or up the blade, they are either rejected or transmitted by the blade rows. Suitable matching conditions must be applied across the cascade in order to link the flow quantities upstream and downstream. This can generally be achieved through a mass continuity relation, an inlet/deviation angle relation and a total pressure loss relation. It is also possible to work in terms velocity potential and shed vorticity. Using these relations, one determines the eigenvalues of the system and thus the stability of the flow field.

Let's define a travelling wave pattern:

$$w_i = \widehat{w}_i\, e^{i[k_i x_i + \omega t]}, \qquad i = 1,2,3,4 \tag{1}$$

As these waves are incident on the blade row at a certain angle, they enter the blade passage and then are diverted due to the fact that there is an enforced outlet flow angle.
The pressure waves are either reflected or transmitted, whilst the entropy and vorticity waves only transmit across and can cause additional pressure waves to occur. The definitions of a reflection and transmission coefficients for each of the waves are:

$$R_j = \frac{\widehat{w}_j^{refl}}{\widehat{w}_j^{inlet}} \qquad , Tr_j = \frac{\widehat{w}_j^{exit}}{\widehat{w}_j^{inlet}} \tag{2}$$

The governing equations for the flow field, assuming an inviscid flow, are the Euler equations.
It can be shown that for the frequencies of interest, the viscous terms are indeed small and can be neglected. The general Euler equations, written in index notation, are:

$$\frac{\partial \rho}{\partial t} + \frac{\partial \rho u_j}{\partial x_j} = 0 \tag{3a}$$

$$\frac{\partial(\rho u_i)}{\partial t} + \frac{\partial(\rho u_i u_j)}{\partial x_j} = -\frac{\partial P}{\partial x_i} \tag{3b}$$

$$\frac{\partial(\rho e)}{\partial t} + \frac{\partial \rho e u_j}{\partial x_j} + \frac{\partial P u_j}{\partial x_i} = 0 \tag{3c}$$

The first equation represents mass continuity, the second is the momentum in the *i-th* direction and the third equation describes the energy. Decomposing each of the flow quantities into their mean and fluctuating components, and neglecting small terms yields:

$$\frac{D\rho'}{Dt} + \frac{1}{\bar{\rho}}\frac{\partial u'_j}{\partial x_j} = 0 \tag{6a}$$

$$\frac{Du'_i}{Dt} + \frac{1}{\bar{\rho}}\frac{\partial P'}{\partial x_i} = 0 \tag{6b}$$

$$\frac{De'}{Dt} + \frac{\bar{P}}{\bar{\rho}}\frac{\partial u'_j}{\partial x_i} = 0 \tag{6c}$$

Where the substantial derivative is defined as:

$$\frac{D'}{Dt} = \frac{\partial}{\partial t} + \overline{u_j}\frac{\partial}{\partial x_j} \tag{7}$$

Equation 6c can be manipulated, assuming an ideal gas to give:

$$\frac{DP'}{Dt} + \bar{\rho}\bar{a}^2\frac{\partial(u'_j)}{\partial x_j} = 0 \tag{8}$$

where $\bar{a}$ is the speed of sound of the mean flow. The above equations can be re-cast in matrix form. For a two-dimensional flow field, where u and v are the velocities in the axial and tangential direction respectively this results in:

$$\frac{\partial}{\partial t}\begin{pmatrix}\rho' \\ u' \\ v' \\ P'\end{pmatrix} = -\begin{pmatrix}\bar{u} & \bar{\rho} & 0 & 0 \\ 0 & \bar{u} & 0 & 1/\bar{\rho} \\ 0 & 0 & \bar{u} & 0 \\ 0 & \bar{\rho}\bar{a}^2 & 0 & \bar{u}\end{pmatrix}\frac{\partial}{\partial x}\begin{pmatrix}\rho' \\ u' \\ v' \\ P'\end{pmatrix} - \begin{pmatrix}\bar{v} & \bar{\rho} & 0 & 0 \\ 0 & \bar{v} & 0 & 0 \\ 0 & 0 & \bar{v} & 1/\bar{\rho} \\ 0 & 0 & \bar{\rho}\bar{a}^2 & \bar{v}\end{pmatrix}\frac{\partial}{\partial y}\begin{pmatrix}\rho' \\ u' \\ v' \\ P'\end{pmatrix} \qquad (9)$$

In more compact form equation 9 can be written as:

$$\frac{\partial q}{\partial t} = -\bar{\bar{A}}\frac{\partial q}{\partial x} - \bar{\bar{B}}\frac{\partial q}{\partial y} \qquad (10)$$

The perturbations in the primitive variables are included inside the vector $\boldsymbol{q}$. It can be written as:

$$\boldsymbol{q} = \hat{\boldsymbol{q}} exp[k\boldsymbol{x} + l\boldsymbol{y} - \omega t] \qquad (11)$$

In equation 11the quantities k, l and ω refer to the axial wavenumber, tangential wavenumber and perturbation frequency respectively. The model attempts to study the growth of perturbations in space.

Losses are assessed exploiting the following correlations:
Tip clearance effects are taken in account following correlation in Storer and Cumpsty [36]; Secondary flows are predicted by means of approach suggested by Wright and Miller [37]; Blade deviation losses are assessed according the correlation of Koening et al [38].

RESULTS AND DISCUSSION

ASPIRATED INTAKE

The sensitivity of the intake flow to incidence has been investigated. Qualitative trend of separation occurrence has been shown in Fig. 7. The flow in the intake is modified by changes in incidence primarily in the bottom part of the intake. When incidence is increased, the stagnation line at the bottom of the intake moves further outside the intake highlight. The flow turning around the lip to enter the intake can accelerate to sonic speed even at relatively low flight Mach numbers. This gives rise to an expansion fan followed by a normal shock. The presence of the shock generates a thickened boundary layer at the bottom of the intake duct. Even in the absence of a shock, the boundary layer is affected by the change in curvature downstream of the lip, which thickens the boundary layer.

This behavior is visible in the axial velocity $\frac{U_x}{U_\infty}$ contours on the mid-plane shown in Fig. 7. When the incidence α_{ref} is exceeded, the nature of the flow changes abruptly, and the flow pattern with a thickened boundary layer at the bottom of the duct is replaced by a sizable separation. It should be noted that the angle α_{ref} is in itself a function of Mach number, Reynolds number and fan face mass flow rate.

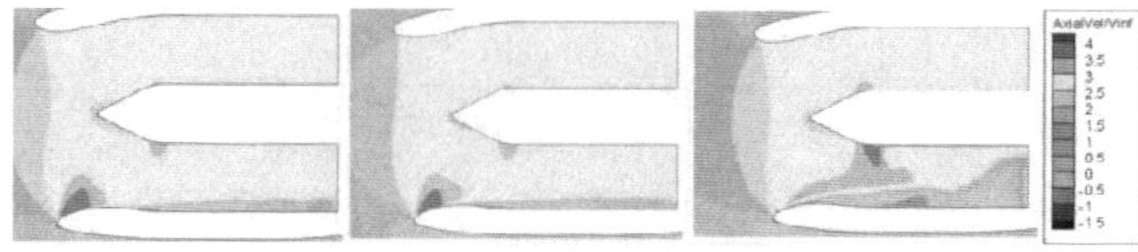

Figure 7: Separation at different angles: (left) $\alpha = \alpha_{ref} - 1^o$; (center) $\alpha = \alpha_{ref}$; (right) $\alpha = \alpha_{ref} + 1^o$

The appearance of a sizeable separation in the lower part of the intake corresponds to a change in the total pressure distribution at fan face. This is shown in Fig. 8 in the form non-dimensional pressure P_{tot}/P_∞. Whereas a total pressure deficit exists in all the conditions examined, the appearance of the separation considerably increases the height and the angular extent of the area of low total pressure.

The distortion at the fan face can be characterized by calculating the DC_{60}, according the UK definition as suggested by Seddon et al [39]:

$$DC_{60} = \frac{P_f - P_{60}}{q} \qquad (10)$$

P_f is the area weighted average total pressure, P_{60} is the area-weighted average of the 60-degree sector with the lowest average total pressure; q is the dynamic pressure. In this work, distortion coefficient has been non-dimensionalized with the DC_{60} corresponding to the highest distortion flow condition.

DC_{60} parameters extracted at fan-face are evaluated for each incidence in Fig. 9. The same graph also shows experimental data obtained in an aspirated intake in conditions matching the present study. The agreement between the two sets is satisfactory and gives confidence in the numerical results. DC_{60} trend show the beneficial effect of the presence of fan stage, which can increase the incidence margin, and consequently installation operability.

Aspirated related data shown in Fig. 9 are related both to Spalart Allmaras and to k-ω turbulence models and are compared to experimental data. Spalart Almaras and K-ω results provide similar responses to not separated flows but a strong under-estimation of losses are provided by Spalart Allmaras models. The main difference in prediction is related to the post separation point. The effect of the turbulence models is small in terms of separation point prediction. Both models evaluate the limit angle of incidence ($\alpha_{ref} + 1$).

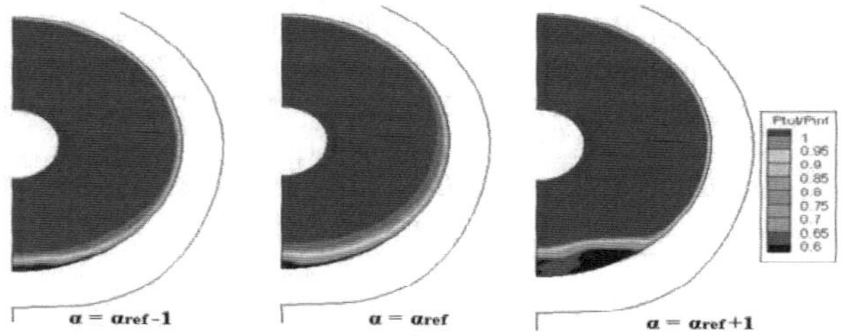

Figure 8: Total pressure contour plots on fan face

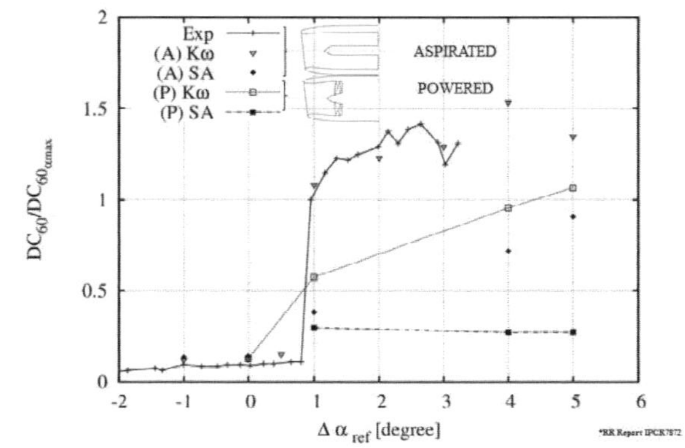

Figure 9: DC_{60} prediction at different axial position

POWERED INTAKE

Unsteady calculations have been performed to assess the effect of the fan on the intake flow. The calculation uses the fan exit condition from the steady state computations as exit boundary condition. The flow is started from steady state solutions and allowed to develop for ten revolutions, when the flow becomes periodic in time. The fan operating condition corresponds to a choked flow. Results shown are related to incidence $\alpha_{ref} + 1^o$.

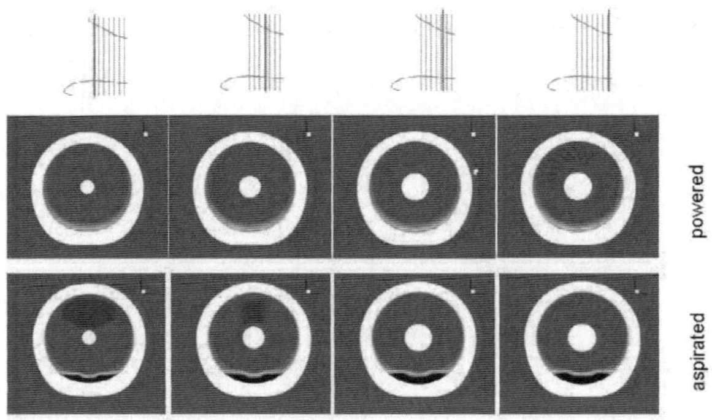

Figure 10 Total pressure distribution on plane parallel to fan face at different axial positions

The effect of the fan on the intake flow and comparison with the aspirated solution is shown in Fig. 10, Fig. 11 and Fig. 12. In Fig. 10 the total pressure contour plot is shown for 4 different section normal to engine axis. The large blue regions characterize the losses due to the evolution of separation traveling inside the intake duct. Fig. 11 and Fig. 12 provide the evolution of axial velocity and static pressure at constant radial section at 92% of the lip. The angle θ provide the angular position. The *θ=0* position corresponds to the *y* positive side axis of Fig.10. Comparison between the aspirated and powered intake highlights the beneficial effect of the presence of the fan: The presence of the fan suppresses the separation at the bottom of the intake. The separation zone (on lower lip) is located around $\theta = -\frac{\pi}{2}$. The extent to which the flow in the intake is cleaned up by a choked fan can be appreciated by the value of the DC_{60} coefficient (powered intake point in Fig. 9). The value of DC_{60} for the powered intake at high incidence is comparable with the DC_{60} evaluated for aspirated configuration at lower incidence. The distortion level is caused by the higher turning at the lip, but it is considerably lower than the value for the stalled aspirated intake. In powered intake distortions on fan face cause a local effect. When the passage is operating in clean flow, the fan stage is working at the nominal point indicated in Fig. 13 by the lower dashed line. In passage is affected by a nominal distortion in the lower lip region the passage is affected by a higher pressure ratio and consequently by lower efficiency. In Fig. 13 the upper dashed line represent the stall limits in terms of pressure ration when the fan is subjected to uniform flow. As result of the increasing of distortions the pressure ratio increase corresponds to the higher flow incidences. Consequently the presence of distortions reduces the limit of fan stage operability.

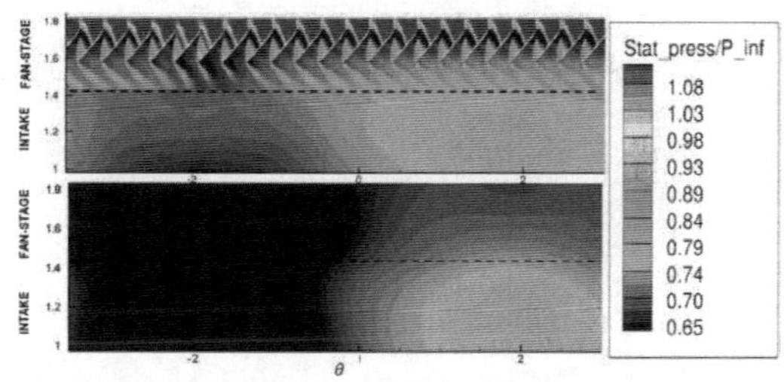

Figure 11 - Static pressure travelling from intake to fan stage: (top) powered configuration, (bottom) aspirated configuration

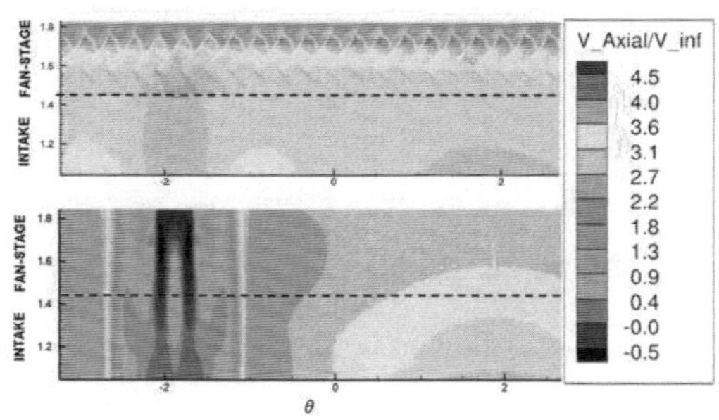

Figure 12 - Axial velocity travelling from intake to fan stage: (top) powered configuration, (bottom) aspirated configuration

ACTUATOR DISK INVESTIGATION

Blade response to incoming distortion has been predicted with actuator disk code. Two radial sections define two different 2D configurations for blade cascade. Radial sections have been extracted at 60% and 90% span-blade. Exit boundary conditions have been determined in order to reproduce the same working point of the whole three dimensional calculation. Fig. 13 shows the performance maps drawn for the relative sections. Inlet boundary conditions correspond to the radial static pressure extracted from the aspirated intake (see Fig. 14).

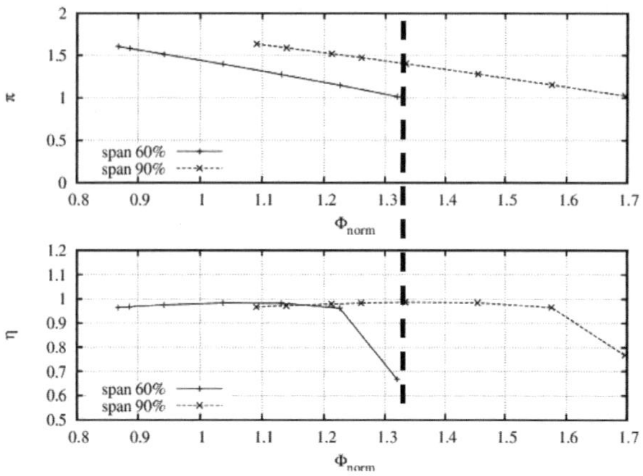

Figure 13 – Steady state performance related to 60% and 90% span fan's section.

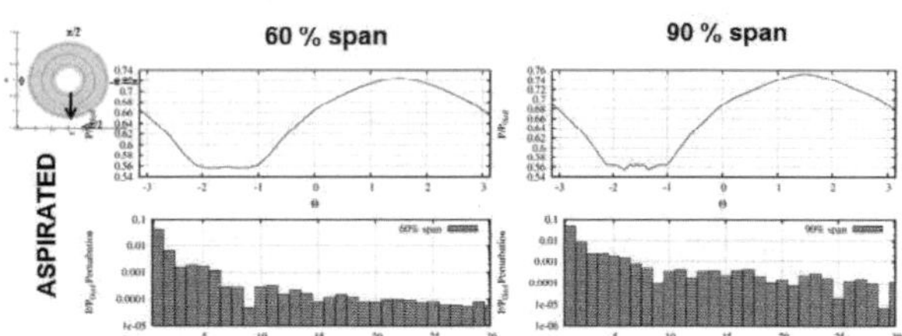

Figure 14 – Distortion pressure signal and relative dft for aspirated test-case.

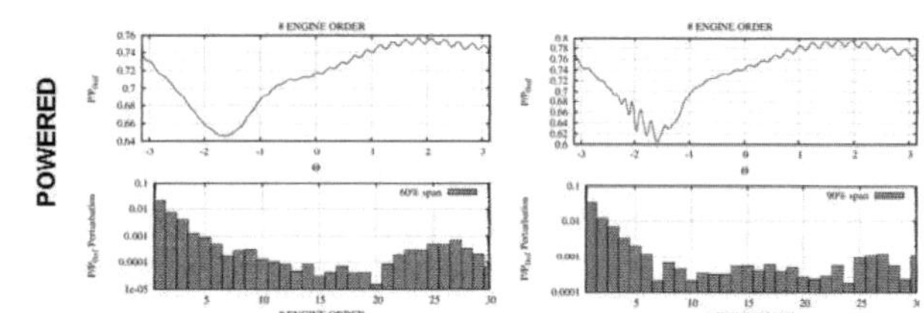

Figure 15 – Distortion pressure signal and relative dft for powered test-case.

Fig. 16 and Fig. 17 provide the comparison of amplitudes related to the each frequency component from order 1 to order 4 among the aspirated, powered intake and blade response for actuator disk. For each frequency a set of 4 histograms are shown. The pure distortion signal (1st histogram) is compared with the signal provided by CFD of powered intake (4th histogram). The 2nd histogram is related to the reflected wave predicted by means of the actuator disk (see Fig.15). The 3rd histogram is relative to the difference of the incoming distortion wave and the reflected wave and it is comparable to the signal provided by powered intake. This demonstrated that the lower frequency distortions are combined linearly with the reflected waves from the blade. This kind of relation has value both for low section perturbed incoming waves (~60 % R) and for high perturbed incoming waves (~90 % R).

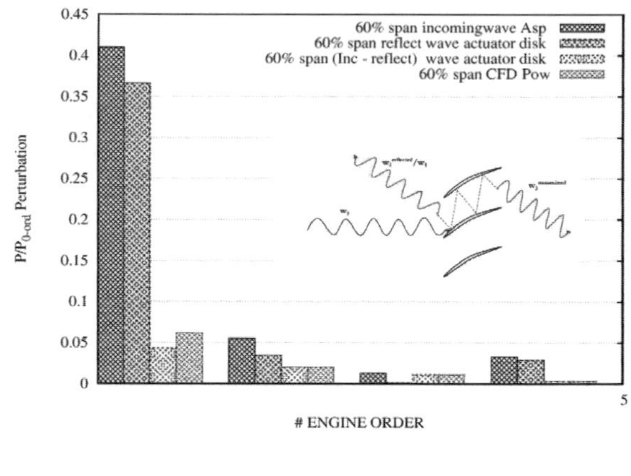

Figure 16 – 60% Radius: Four engine order prediction from CFD and actuator disk: I) Aspirated form CFD – II) reflected wave from Actuator Disk – III) difference I – II – IV) Powered form CFD

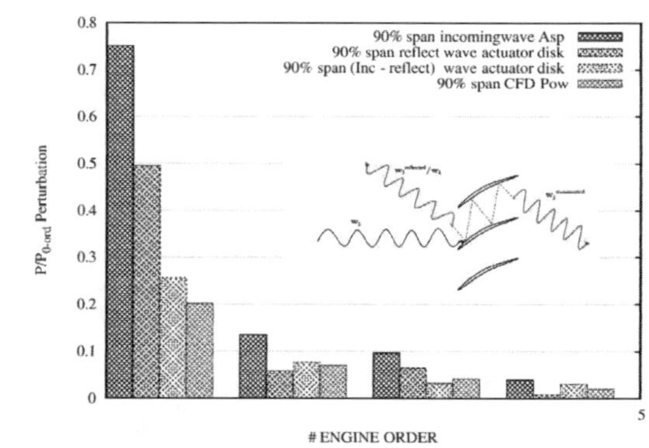

Figure 17 – 90% Radius: Four engine order prediction from CFD and actuator disk: I) Aspirated form CFD – II) reflected wave from Actuator Disk – III) difference I – II – IV) Powered form CFD

CONCLUSION

A computational study into the effect of a choked fan on the flow in an intake at incidence has been presented. The study is based on intake and fan geometry of an in–service configuration. Two studies have been carried out: for the aspirated intake and one for the powered intake. The powered intake computations are based on a choked fan flow.

The data for the aspirated intake are compared to the distortion coefficients measured in an experimental facility. It has been shown that the numerical predictions match the measured data with good accuracy.

The powered intake CFD study has shown the extent to which the fan flow can alter the intake flow, even for incidence value for which the aspirated intake is stalled.

The powered intake study has been demonstrated to increase the margin of separation occurrence and installation operability. When a large separation occurs, the presence of the fan stage has drastically drops distortion. Distortion increase gradually. Powered intake's distortion reaches the same value of the aspirated distorted conditions (corresponding to $\Delta\alpha_{ref} = 1°$) at $\Delta\alpha_{ref} = 5°$. The effect of the presence of the fan affected by distortion incoming waves has been investigated by means of actuator disk approach. Two radial constant sections have been taken in account, and two cascades have been defined. Geometrical features have been extracted for both the blade and cascade configuration. Inlet boundary conditions in terms of perturbations wake have been extracted at the same sections form the aspirated intake.

The effect of the presence of the fan in the powered intake has an effect of reducing the distortion components related to lower engine order component, which corresponds to the higher amplitudes. Higher components which are characterized by weaker distortions are slightly amplified or remain unchanged.

The comparison between the reflected waves from fan stage and the signal extracted just upstream the fan's leading edge, had highlight the linear behavior of the lower disturbance frequency.

ACKNOWLEDGMENTS

The authors gratefully acknowledge Rolls-Royce plc for funding this work and granting permission for its publication. The computations in this paper have been carried out on the facilities of the Imperial College HPC Service managed by Mr Simon Burbidge.

REFERENCES

1. Satyanarayana, B., J. P. Gostelow, and R. E. Henderson. "A Comparison Between Experimental and Theoretical Fluctuating Lift on Cascades at Low Frequency Parameters." ASME 1974 International Gas Turbine Conference and Products Show. American Society of Mechanical Engineers, 1974.
2. Kaji, S., Okazaki, T., "Propagation of Sound Waves Through a Blade Row", Journal of Sound and Vibration, Vol. 11, 1970.
3. Kaji, S., Okazaki, T., "Generation of Sound by Rotor-Stator Interaction", Journal of Sound and Vibration, Vol. 13, 1970.
4. Amiet, R. K., "Transmission and Reflection of Sound by a Blade Row", AIAA Journal, Vol. 9, 1971.
5. Amiet, R. K., "Transmission and Reflection of Sound by Two Blade Rows", Journal of Sound and Vibration, Vol. 34(3), 1974, 399-412.
6. Cumpsty, N. A., Marble, F. E., "The Interaction of Entropy Fluctuations with Turbine Blade Rows; A Mechanism of Turbojet Engine Noise", Proc R Soc Lond A.357, 323-344, 1977.
7. Whitehead, D. S., "Classical Two-Dimensional Methods", AGARDograph No.298, 1987
8. Gunn, E.J., Tooze, S.E., Hall, C.A., Colin, Y., "An experimental study of loss sources in a fan operating with continuous inlet stagnation pressure distortion" Proceedings of the ASME Turbo Expo-2012, Copenhagen, Denmark, Paper No-GT2012-68888, 2012.
9. Felderman, E.J and Albers, J.A. "Comparison of experimental and theoretical boundary layer separation of inlet at incidence angle at low speed conditions" NASA-TM-X-3194, NASA Lewis Research Centre 1975.
10. Perovic, D., Hall C.A., Gunn E. J., "Stall inception a boundary layer ingesting fan" *Proceedings of ASME Turbo Expo 2015: Turbine Technical Conference and Exposition,* Montreal , Canada june 15-19, 2015.
11. Wakelam, C.T., Hynes, T.P., Hodson, H.P., Evans, S.W. and Chanez, P. "Separation control for aeroengine intakes, Part 1: Low speed investigation of control strategies" *Journal of Propulsion and Power*, Vol. 28, No 4 July-August 2012.
12. Wakelam, C.T., Hynes, T.P., Hodson, H.P., Evans, S.W. and Chanez, P. "Separation Control for aeroengine intakes, Part 2: High speed investigation". *Journal of Propulsion and Power*, Vol. 28, No 4 July-August 2012.
13. Chen, H.C., Yu, N.J., Rubbert, P.E. and Jameson, A. "Flow simulations for general nacelle configurations using Euler equations" *AIAA Paper* AIAA-83-0539, 1983.
14. Callahan, G. M. and Stenning, A. H., "Attenuation of inlet flow distortion upstream of axial flow compressors", *Journal of Aircraft,* 8-4 pp 227-233, 1971.
15. Henderson, R.E., and Horlock, J.H., "An approximate analyses of the unsteady lift of airfoils in cascade" *ASME Journal of Engineering for Power*, 94-4 pp 233-240, 1972.
16. Greitzer, E.M. "Upstream attenuation and quasi-steady rotor-lift fluctuations in asymmetric flows in axial compressors". *ASME Paper* No. 73-GT-30, pp. V001T01A030, 1973.
17. Longley, J.P. and Greitzer, E.M. "Inlet distortion effects in aircraft propulsion system integration" In AGARD Steady and Transient Performance Prediction of Gas Turbine 1992.
18. Hawthorne, W.R., Mitchell, N. A. and McCune, J.E. and Tan, C.S."Non-axisymmetric flow through annular actuator disk: inlet distortion problem" *ASME Journal of Engineering for Power*, 100 pp 604-617, 1978.

19. Hsiao, E., Naimi, M., Lewis, J.P., Dalbey, K., Gong, Y. and Tan, B., "Actuator-duct model of turbo-machinery for powered-nacelle Navier-Stokes Calculations" *ASME Journal of Propulsion and Power*, 17-4, pp.919-927, 2001.
20. Yocum, A. M., and R. E. Henderson. "The Effects of Some Design Parameters of an Isolated Rotor on Inlet Flow Distortions." *Journal of Engineering for Gas Turbines and Power* 102.1 (1980): 178-186.
21. Peters, Andreas, et al. "Ultrashort Nacelles for Low Fan Pressure Ratio Propulsors." *Journal of Turbomachinery* 137.2 (2015): 021001.
22. Fidalgo V. Hall C.A., Colin Y. " A study of fan-distotion interaction withinf NASA rotr 67 transonic stage *Proceedings of ASME Turbo Expo 2010.*
23. Carnevale M., Wang F., Green, J., Di Mare , L, " On lip stall suppression in powered intake" Journal of Propulsion and Power, 2015.
24. Romanov, Artyom, and Luca Di Mare. "Extended Turbomachinery Aeromechanical Model." *ASME Turbo Expo 2013: Turbine Technical Conference and Exposition*. American Society of Mechanical Engineers, 2013
25. Di Mare, L., Davendu, Y.K., Wang, F., Romanov, A., Ramar, P.R. and Zachariadis, Z.I. "Virtual gas turbines: Geometry and conceptual description" *Proceeding of ASME Turbo Expo 2011*, June 6-10, Vancouver, Canada, 2011.
26. Roe, P. L., "Approximate Riemann solvers, parameter vectors and difference schemes", *Journal of Computational Physics*, 43, 357-372, (1981).
27. Van Leer, B., "Flux-vector splitting for the Euler equations", *8th International Conference on Numerical Methods in Fluid Dynamics*, Volume 170, 1982, pp 507-512.
28. Liou, M.S., Steffen C.J. Jr. "A new flux splitting scheme" *J. Comput. Phys.*, 107 (1993), pp. 23–39.
29. Liou, M.S. "A Sequel to AUSM, Part II: AUSM+-up for all speeds" *J. Comput. Phys.*, Vol. 214, Issue 1, pp 137- 170, 2006.
30. Choi, Y.H., Merkle, C.L. "The application of preconditioning in viscous flows", *J. Comput. Phys.*, Volume 105, Issue 2, April 1993, Pages 207–223.
31. Wilcox, D.C. *Turbulence modeling for CFD.* 2nd Ed. La Canada, CA: DCW industries, Chap 4, 1998.
32. Bernardini, C., Carnevale, M., Salvadori, S. and Martelli, F. "On the Assessment of an Unstructured Finite-Volume DES/LES Solver for Turbomachinery Application." *WSEAS Transactions On Fluid Mechanics*. 2011
33. Bernardini, C., Carnevale, M., Manna, M., Martelli, F., Simoni, D. and Zunino, P. "Turbine blade boundary layer separation suppression via Synthetic jet: an experimental and numerical study". *Journal of Thermal Science* Vol 21. No (2012) 404-412.
34. Carnevale, M., Montomoli, F., D'Ammaro, A., Salvadori, S., and Martelli, F. "Uncertainty quantification: A stochastic method for heat transfer prediction using LES". *Journal of Turbomachinery* 06/2013; 135(5). DOI:10.1115/1.4007836.
35. Carnevale, M., Green, J.S. and Di Mare, L. "Numerical studies into intake flow for fan forcing assessment". *Proceedings of ASME Turbo Expo 2014: Turbine Technical Conference and Exposition,* Düsseldorf, Germany June 16 – 20, 2014.
36. Cumpsty, N. A., Storer J. A., "An Approximate Analysis and Prediction Method for Tip Clearance Loss in Axial Compressors", ASME Journal of Turbomachinery, Vol. 116, 1994
37. Wright, P. I., Miller, D. C., "An Improved Compressor Performance Prediction Model", IMechE, C423/028, 1991, pp.69-82.
38. König, W. M., Hennecke D. K., Fottner, L., "Improved Blade Profile Loss and Deviation Angle Models for Advanced Transonic Compressor Bladings: Part I – A Model for Subsonic Flow", ASME Journal of Turbomachinery, Vol. 118, 1996
39. Seddon, J. and Goldsmith, E.L. *Intake aerodynamics* Blackwell Science Oxford 1999.

Proceedings of the 14th International Symposium on Unsteady Aerodynamics, Aeroacoustics & Aeroelasticity of Turbomachines
ISUAAAT14
8-11 September 2015, Stockholm, Sweden

I14-S12-3

EFFECTS OF ROTATING INLET DISTORTION ON COMPRESSOR STABILITY WITH A NOVEL CASING TREATMENT

Dong Xu
BeiHang University
Beijing, China

Li Fanyu
BeiHang University
Beijing, China

Sun Dakun
BeiHang University
Beijing, China

Sun Xiaofeng
BeiHang University
Beijing, China

ABSTRACT

This paper conducts an experimental research of rotating inlet distortion on a low-speed large size test compressor with emphasis on the stability problem of axial fan/compressors, and the stall margin enhancement with a kind of stall precursor-suppressed (SPS) casing treatment. Some results on compressor stall margin and pre-stall behavior under the restriction of rotating inlet distortion are presented. The experimental results show that whether the inlet distortion is co-rotating or counter-rotating, the SPS casing treatment can still improve the stall margin without leading to additional efficiency loss caused by such configuration. The experiment results also show that the mechanism of the stall margin improvement with such casing treatment is associated with delaying the nonlinear development of the stall precursor waves and weakening the unsteady flow disturbances in a compression system.

INTRODUCTION

Advanced aircraft are often subjected to high intensity flight maneuvers, which place great operational demands on the compression system (fan/compressor) of air-breathing gas turbine engines (e.g., [1]). It is desired that the compressor should operate in a stable manner during all aspects of flight. However, rapid flight transients cause the inlet to produce a highly distorted flow field to the compressor inlet, whereas high distortion levels may generally cause a decrease in performance and, more importantly, a lessening of the stall margin of the compressor. Meanwhile, certain inflow distortions may even lead to catastrophic damage to the entire engine. Therefore, how to enhance compressor stall margin under inlet distortion condition has long been a problem of concern (e.g., [2-9]). Inlet distortion may be categorized as two kinds of different inflow conditions, the first is relatively static inlet non-uniform flow like total temperature distortion and total pressure distortion, the second deals with time-correlated non-uniform flow and rotating inlet distortion. Especially, the latter often occurs in the high pressure compressor of multispool engines due to the rotating stall happening in upstream stage (low pressure compressor). Compared to the investigation on the static inlet distortion (e.g., [2-6]), there are relatively small amount of works and open published literature about the effect of rotating inlet distortion on compressor's performance and stability (e.g., [7-9]). From a practical point of view, further work in this regard is required to know how to extend compressor stall margin under the condition of rotating inlet distortion.

Aiming at the rotating inlet distortion, Longley et al. (e.g., [7]) first conducted such investigations on a 4-stage low-speed research compressor with 4 different kinds of rotors in 1996. Their experiments clearly show the influence caused by rotating inlet distortion on stall margin decrease, compression ability decrease and efficiency loss. Especially, they have shown very meaningful results about the different effect of co-rotating and counter-rotating distortion. In addition, Peters et al. (e.g., [8-9]) did some further investigations about the effect of co-and counter-rotating inlet distortion on a 5-stage high pressure compressor, and some similar conclusions were drawn about the different effect of co-rotating and counter-rotating distortion. Furthermore, they gave experimental results of the compressor's response to the rotating inlet distortion frequency, which is helpful to understand the possible resonant peak encountered in some specific rotating inflow conditions. Recently, Salunkhe et al. (e.g.,

[10]) carried out an experimental investigation on a single stage axial flow fan about the stall inception mechanism under clean and distorted inlet conditions, which can help to design robust actuators and control schemes to delay/suppress the stall onset. And then Hah et al. (e.g., [11]) presented a study about the pre-stall behavior in a modern transonic fan by using both unsteady flow calculation and experiment methods, with the same purpose as Salunkhe's. It is known that the real purpose of studying the rotating inlet distortion is to make the compressor more robust by any means. Therefore, an exploration is justified to extend compressor stall margin under the restriction of rotating inlet distortion.

It is noted that among all stability control methods, casing treatments are well known as an effective approach to extend stall margin (e.g., [12-16]). In the wake of developments in experimental technology, there are more and more elaborate works to study the effect of casing treatment and the mechanism of stall margin enhancement caused by various casing treatments (e.g., [17-18]). The existing research shows that casing treatment can enhance the stall margin of both subsonic and transonic compressors in varying degrees along with some efficiency loss as cost. There are also some investigations on the performance and effectiveness of various casing treatments with inlet distortion flow. Madden et al. (e.g., [19]) once carried out both experimental and numerical studies about the effects of inlet distortion on the stability of a fan stage with casing treatment, and they thought that the casing treatment could restrain the influence on surge trigger caused by inlet distortion and recognized the stability enhancement of casing treatment under the inlet distortion. Later on Huang and Wu (e.g., [20]) did a numerical investigation of inlet distortion on an axial flow compressor rotor with circumferential groove casing treatment, and the relevant comparison was made on the stability enhancement effects of casing treatment under both clean and distorted inlet conditions. However, up to now there has been no work concerning the combined effect of the casing treatment under the restriction of rotating inlet distortion on compressor performance and stability.

Recently, a kind of stall precursor-suppressed (SPS) casing treatment was proposed (e.g., [21-22]). Compared with the above conventional casing treatments aimed at the improvement of the blade tip flow structure, the objective of the SPS casing treatment is to affect the inception or evolution of the stall precursors to realize stall margin enhancement. More importantly, the geometrical parameters of the SPS casing treatment can be determined based on the application of flow stability model (e.g., [21-25]) instead of trial-and-error method. In fact, although there are many theoretical models available for predicting the flow stability related to isolated airfoil (e.g., [26-27]) and compressor cascade (e.g., [28-29]), it is found that a three-dimensional flow stability model (e.g., [21-25]) is more suitable to study on such casing treatment. In addition, the casing treatment configuration has an annular air chamber and a low perforated ratio about 6%, which is far smaller than that in traditional casing treatments with over 50% open area ratio. The results show that the SPS casing treatment can obviously improve the stall margin, accompanied by less efficiency loss (e.g., [22]). Because of this, it is reasonable to judge that the SPS casing treatment may also influence the change of compressor pre-stall behaviour caused by rotating inlet distortion, and realize the purpose of stall margin enhancement. Therefore, there is a strong motivation to extend such casing treatment to the case of rotating inlet distortion.

The present work is devoted to the experimental study of the mechanism of stall margin variation with rotating inlet distortion and the relevant control strategy with SPS casing treatment. All experimental work is finished in a low speed compressor with controllable rotating distortion device. The approach adopted is to rotate a distortion screen upstream of a low-speed compressor. The experimental facility used enables the distortion screen to be rotated in either direction (co- or counter rotational) at various fractions of rotor speed. Both time-mean and time-resolved measurements are recorded for this compressor with the results analyzed in terms of the given data processing approach. Comparisons are also made with the performance and stability limit corresponding to both solid casing and SPS casing treatment. The experimental results show that the SPS casing treatment can not only improve the stall margin by 8–12% for this subsonic compressor under the clean inlet condition, but also make up the stall margin loss caused by rotating inlet distortion. Especially, in some situation the stall margin benefit can be even obtained under certain distorted inlet condition with lower rotating degree. In addition, the results also show that the distorted inlet conditions will accelerate the nonlinear development of the stall process by stimulating the propagation of the stall precursor waves, as compared with the evolution of the precursors without inlet distortion. Comments on the distinguishing features of the SPS casing treatment under the restriction of rotating inlet distortion will then be given, along with discussion and analysis of the broader implications of the present work.

EXPERIMENTAL FACILITY, INSTRUMENTATION AND DATA ACQUISITION

The configuration of the SPS casing treatment is shown in Fig. 1, which mainly contains a back-chamber and a perforated plate (5, 1 in Fig. 1). The perforated plate can be closed, the SPS casing treatment will thus be an ordinary solid wall casing. In this paper, the perforated ratio is about 6%.

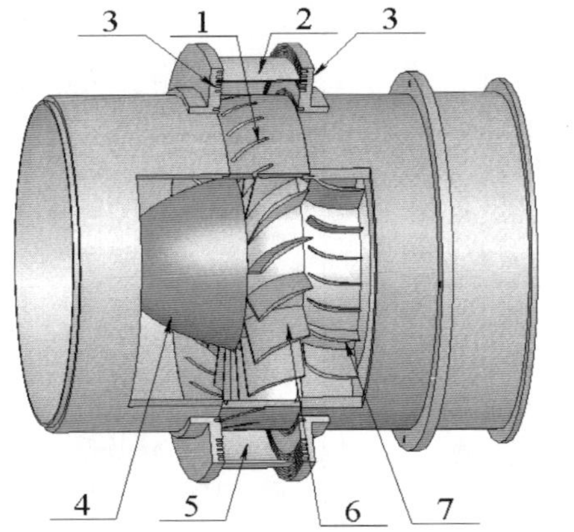

Fig. 1 Configuration of the SPS casing treatment

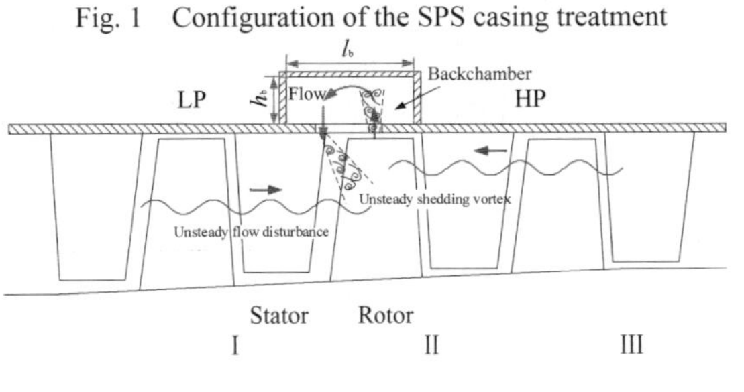

Fig. 2 The vortex-pressure wave interaction mechanism of the SPS casing treatment

Figure 2 shows the mechanism of the vortex-wave interaction with SPS casing treatment, the unsteady shedding vortex from the edge of orifice or slot will interact with the perturbation wave like stall precursor wave, which will delay its nonlinear development. Now, based on this mechanism, it has been found that the SPS casing treatment can suppress the unsteady flow disturbance wave in uniform inflow condition (e.g., [20]). Therefore, it is reasonable to assume that such configuration may also be used to suppress the stall precursor wave with higher energy triggered by inlet distortion. As shown in Fig. 2, no matter where the stall precursor wave originates, the unsteady flow disturbance wave coming from upstream or downstream will physically interact with the casing treatment, the additional system damping will be naturally formed and the nonlinear evolution of the wave will be then delayed. Under the condition of rotating inlet distortion, it is worth noting that whether such SPS casing treatment can still take effect on the stall margin improvement. The following experiment will be designed to study these problems and concerns.

The tests are carried out with a low-speed large size test compressor TA36 (Fig. 3). It contains a one-stage fan (4 in Fig. 3) with 20 blades, 300mm radium and 1.2067 pressure ratio; a fixed throttle cone (10 in Fig. 3); an asynchronous ac-motor (7 in Fig. 3) as the power of this rig; a distortion generator (Fig. 4) will be placed on position 11 in Fig. 3 and some pressure sensors (2, 3, 6 in Fig. 3).

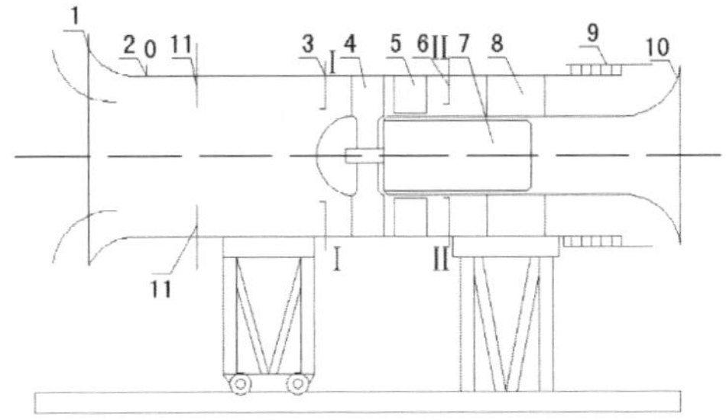

Fig. 3 Configuration of TA36

The distortion generator is shown in Fig. 4, it contains 8 screens circumferentially placed on the shaft of a servo motor. The screens can rotate in the speed from 0 to 3000r/min, two directions. So this distortion generator can simulate rotating inlet distortion in different intensities. Besides, only two screens placed in 180 degrees are used in this present work.

Fig. 4 Rotating Inlet Distortion Generator

The instrumentation of the TA36 rig mainly contains high-frequency response pressure sensors. The total pressure sensors are placed at the position of 3,6 in Fig. 3 to measure the inlet/outlet total pressure; and the static pressure sensors are placed on the casing wall at the position 2,6 in Fig. 3 to measure the inlet/outlet static pressure. Eight high-frequency response pressure sensors are mounted at the plane about half a chord length upstream of the rotor inlet to measure the static pressure perturbations, which are equally spaced around the circumference of the machine.

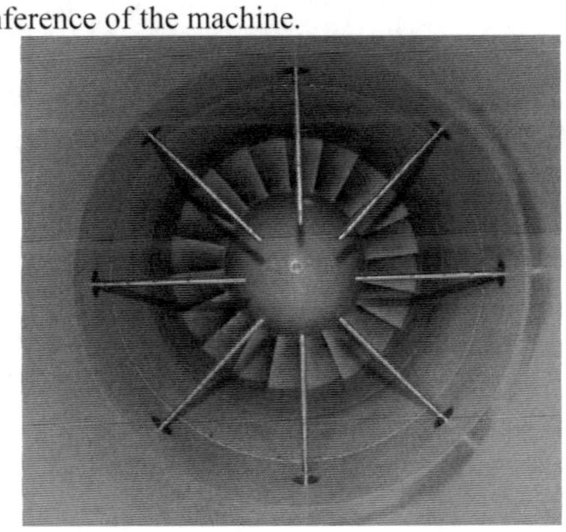

Fig. 5 8 total pressure combs

There are also 8 total pressure combs with 40 Kulite high-frequency response pressure sensors shown in the Fig. 5, those pressure sensors are used to measure the inlet pressure field structure and analyze the characteristic of inlet distortion.

EXPERIMENTAL RESULTS

Degree of Inlet Distortion

In this paper, PXXX/NXXX means the direction and speed of rotating inlet distortion generator. For example, P200 means the rotating direction of screens is the same as the rotor's, and rotating speed is 200r/min; N500 means that the rotating direction of screens is opposite, and the rotating speed is 500r/min.

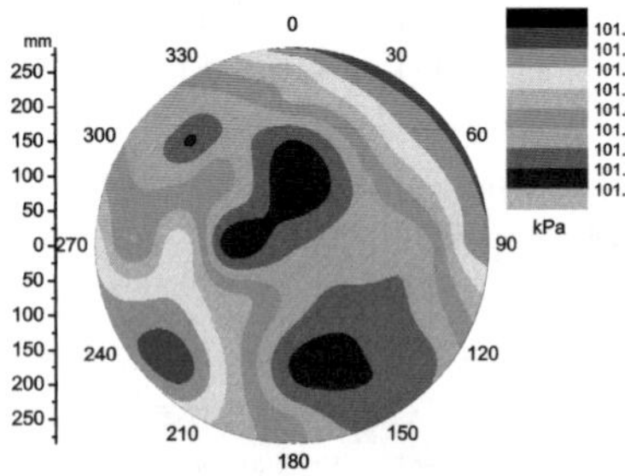

Fig. 6 Inlet total pressure field structure on 70% working speed under the P200 rotating inlet distortion

As shown in the previous section, there are 40 high-frequency response pressure sensors used to measure the inlet total pressure field, so the rotating inlet distortion can be described by plotting a pressure contour in some moments, and the relevant comparison with each other can be also made in a constantly time domain. Figure 6 shows the inlet total pressure field structure on 70% working speed under the P200 rotating inlet distortion in one moment, the two screens of distortion generator make the pressure field in three districts, one high-total-pressure districts and two low-total-pressure districts. And if the pressure field structure information is picked up in the time step length of 0.05s, the movement of inlet total pressure field structure can be captured in two cycle periods, as shown in Fig. 7. From this figure, there is an obvious rotating movement, the total pressure loss and rotating movement are the main characteristics of rotating inlet distortion.

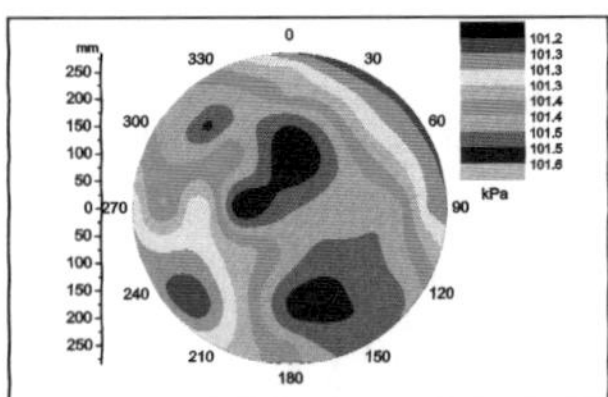	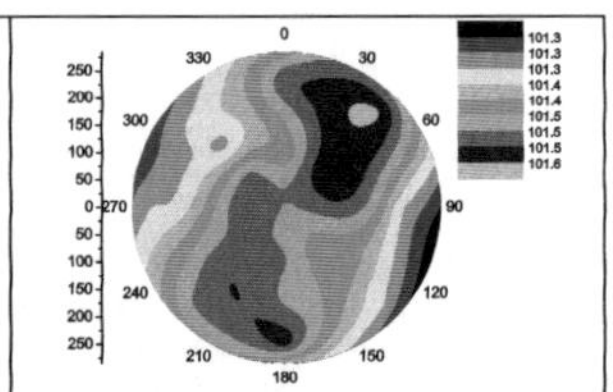
0.00s	0.20s
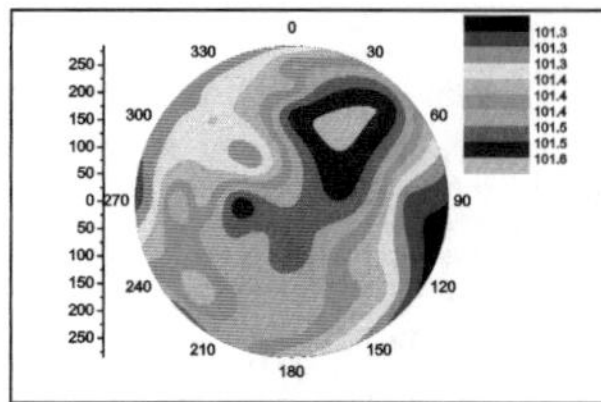	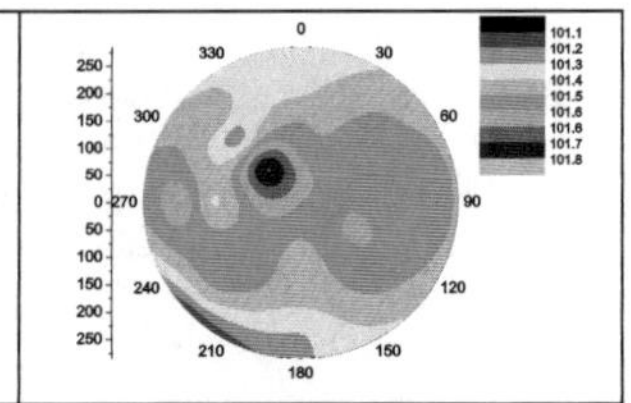
0.05s	0.25s
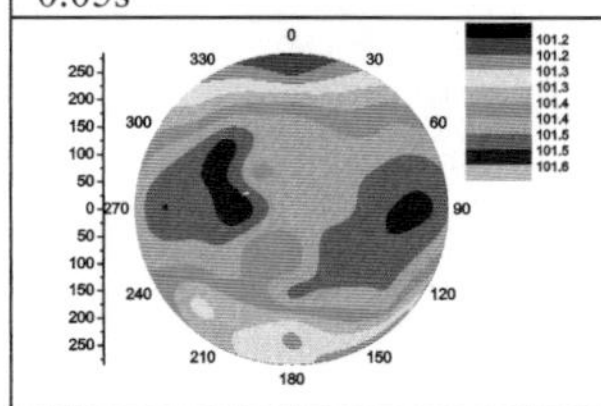	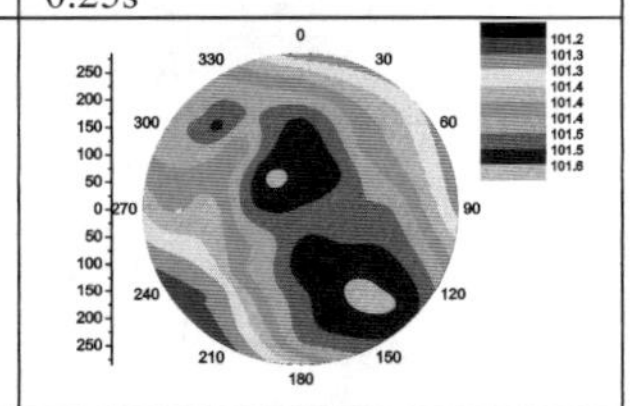
0.10s	0.30s
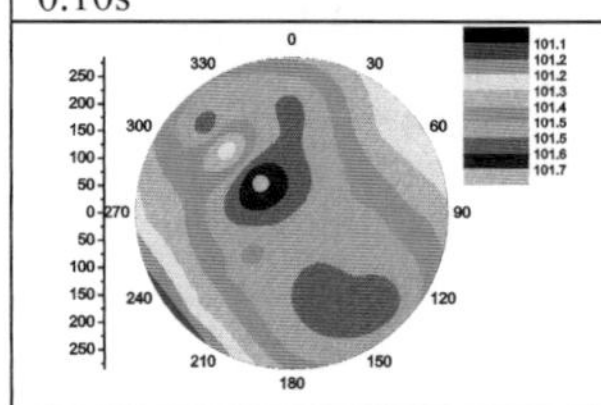	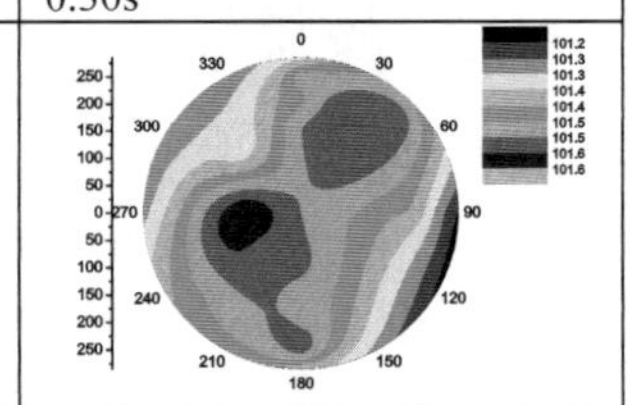
0.15s	0.35s

Fig. 7 Inlet total pressure field structure on 70% working speed under the P200 rotating inlet distortion

The present work also contains the experiments under the inlet distortion generator rotating in two directions, but the inlet total pressure field structure are same, the only difference is that the rotating speed of inlet flow, so the figure of the total pressure field structure in that condition is not displayed. Figure 8 also shows the frequency characteristic of distorted inlet flow, the rotating inlet distortion can input an obvious disturbance with a main characteristic frequency which is the same as the rotation frequency of distortion generator's screens.

Total pressure field structure	Normalized frequency
P200	
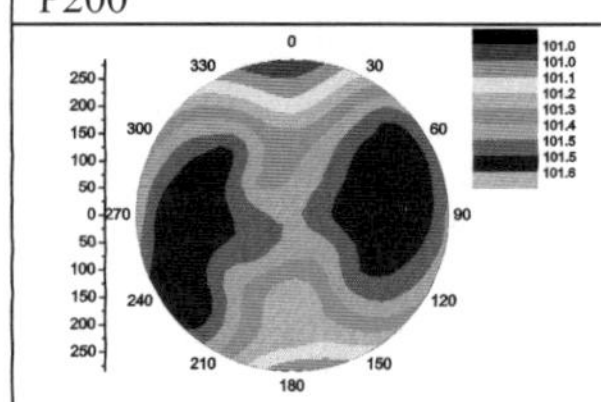	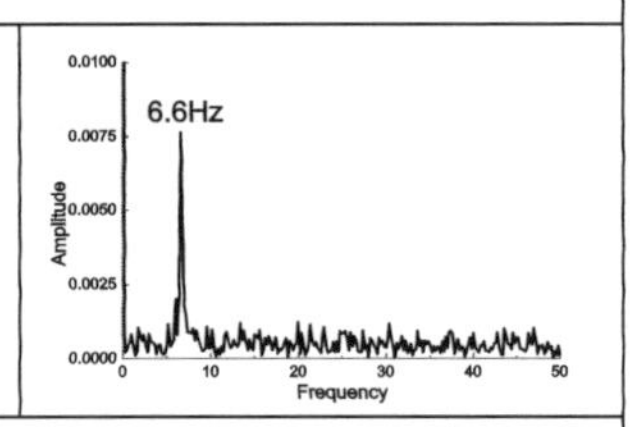
P500	
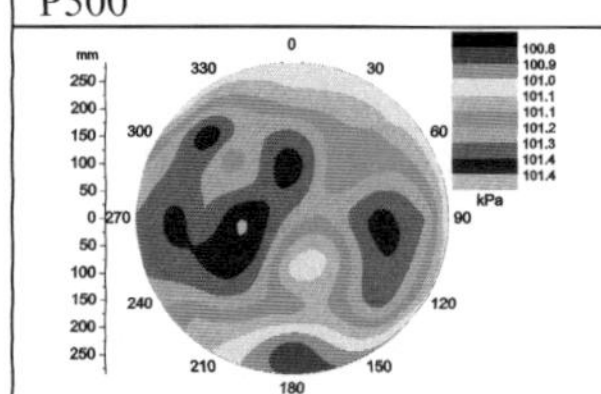	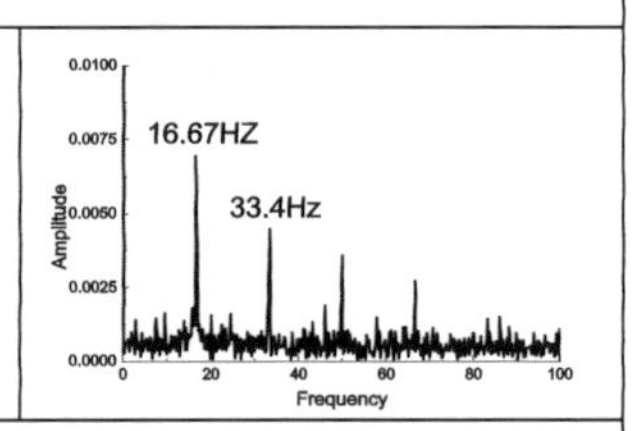
P800	

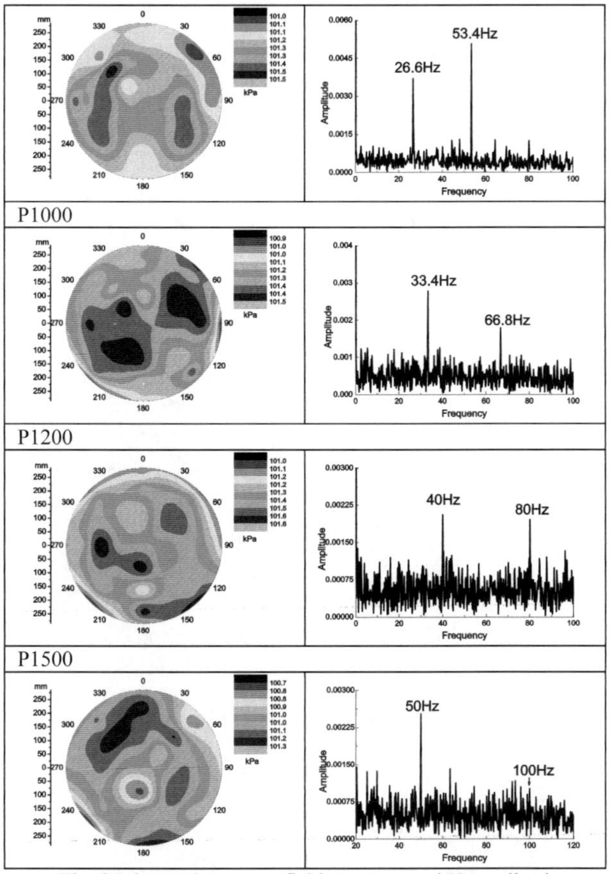

Fig. 8 Inlet total pressure field structure and Normalized frequency on100% working speed in different degrees inlet distortion

Enhancement of SM by SPS Casing Treatment

Performance of test compressor under the rotating inlet distortion with SPS casing treatment.

There are a series of experiments designed in different inlet conditions and different working speeds to prove the SPS casing treatment's ability of stall margin enhancement. In this paper, CT means using the SPS casing treatment with about 6% perforated ratio, SC means that the perforated plate is closed, which just corresponds to an ordinary solid wall casing.

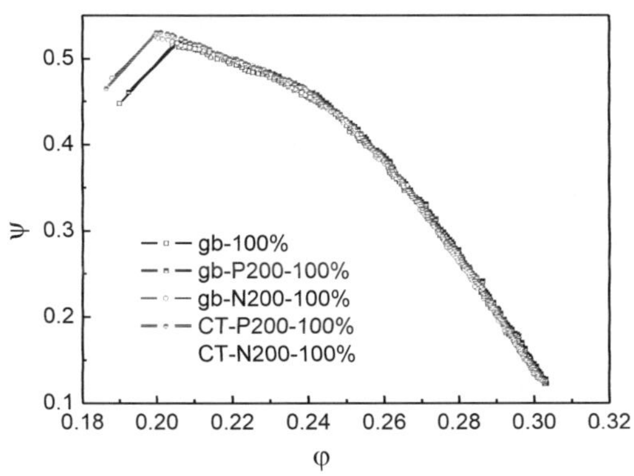

Fig. 9 The pressure rise curve under the 200r/min distortion speed

Figure 9 shows the pressure rise curve under the 200r/min distortion speed and 100% design working speed. In this figure, the SPS casing treatment can move the stall point to the left under the inlet distortion of this degree, and not only make up the stability decrease which is caused by inlet distortion but also enhance the stall margin compared to the solid wall casing with clean inlet condition. At the same time, the rotating inlet distortion with counter-rotating speed can also enhance the stall margin by 1%~5% compared to the distortion with co-rotating speed.

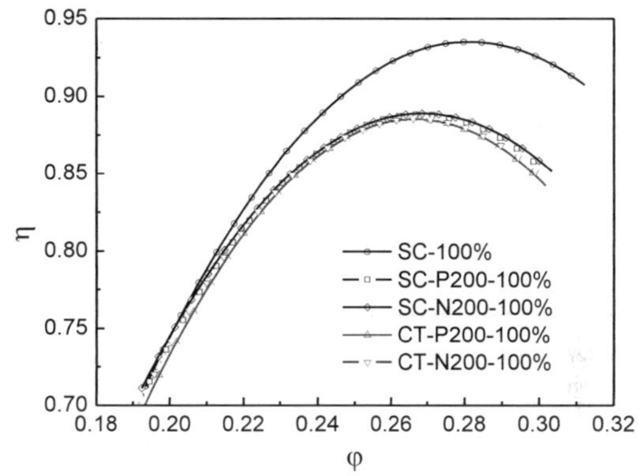

Fig. 10 Efficiency curves with 200r/min distortion speed under rated speed

The efficiency curves with 200r/min distortion speed under rated speed is shown in Fig. 10. The inlet distortion causes a big loss on the efficiency, and using the SPS casing treatment also causes a little additional loss on efficiency. But from this figure, the efficiency loss caused by this SPS casing treatment is much smaller than that caused by the inlet distortion. In fact, using the static data, the additional efficiency loss caused by SPS casing treatment is less than 1%, which is almost within the error range.

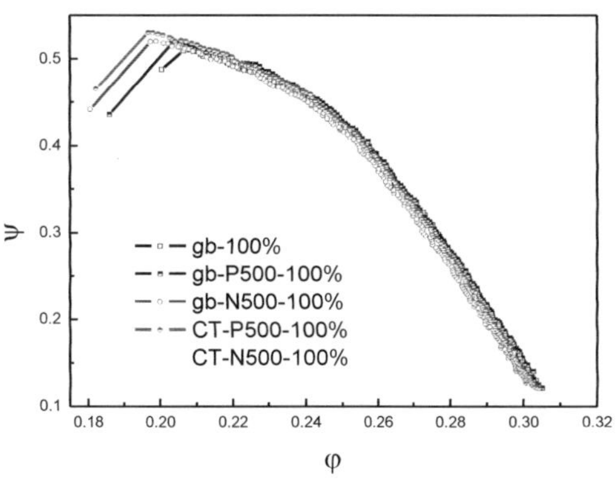

Fig. 11 The pressure rise curve under the 500r/min distortion speed

Figure 11 shows the pressure rise curve under the 500r/min distortion speed and 100% design working speed. It is easily seen that with the increase of distortion speed, the difference between the influences caused by counter- and co-rotating speed become more obvious. And with this kind of inlet distortion, the SPS casing treatment still has a good ability in enhancing the stall margin. And in this figure, the stall margin benefit obtained from counter-rotating inlet distortion becomes more obvious. Perhaps, this is why some modern aircraft engines make use of multi-shaft strategy with counter-rotating system.

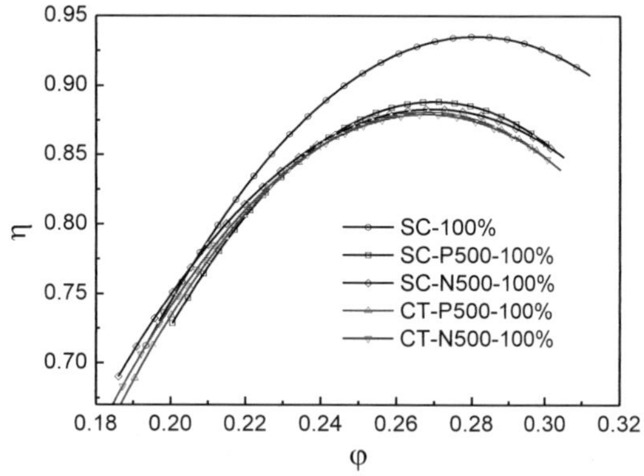

Fig. 12 Efficiency curves with 500r/min distortion speed under rated speed

Figure 12 shows the efficiency curve with 500r/min distortion speed under rated speed. The implications understood from this figure are the same as Fig. 10, the efficiency loss caused by this SPS casing treatment is less than 1%, which is still much smaller than that caused by inlet distortion.

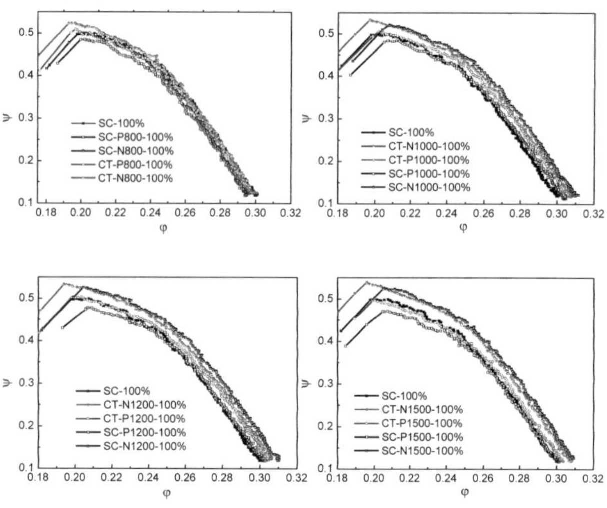

Fig. 13 The pressure rise curve under the 800r/min, 1000r/min, 1200r/min, 1500r/min distortion speed

If further raising the rotating speed of distortion generator, the compressor's pressure rise characteristic is shown in Fig. 13. This figure shows four different distortion speeds; in those four cases, the inlet condition becomes worse and worse with the rising of distortion speed, and the stall margin also becomes lower and lower. While the screens rotate in high speed and even bring certain influence on the measurement of the flow rate of this compressor, the differences between the curves of counter-rotation speed and co-rotation speed become even bigger. In spite of this condition, the SPS casing treatment still has the ability to enhance the stall margin.

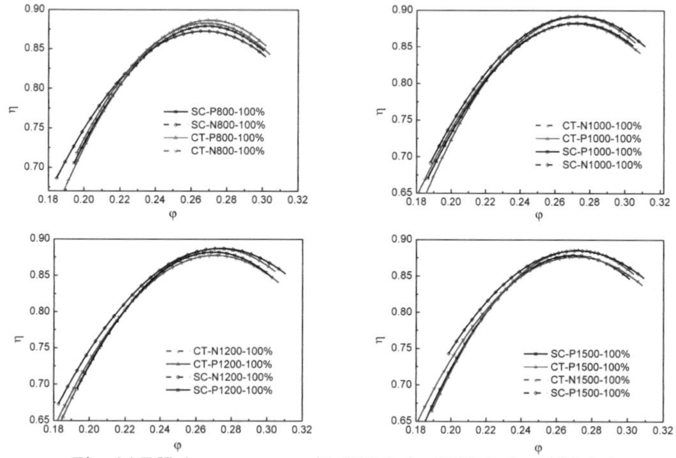

Fig. 14 Efficiency curves with 800r/min, 1000r/min, 1200r/min, 1500r/min distortion speed under different working speed

Figure 14 shows the efficiency curves with four different distortion speeds under the rated speed. Compared with Fig. 10 and Fig. 12, it is found that no matter what kind of inlet distortion speed is, the additional efficiency change caused by SPS casing treatment is always so small as to be neglected.

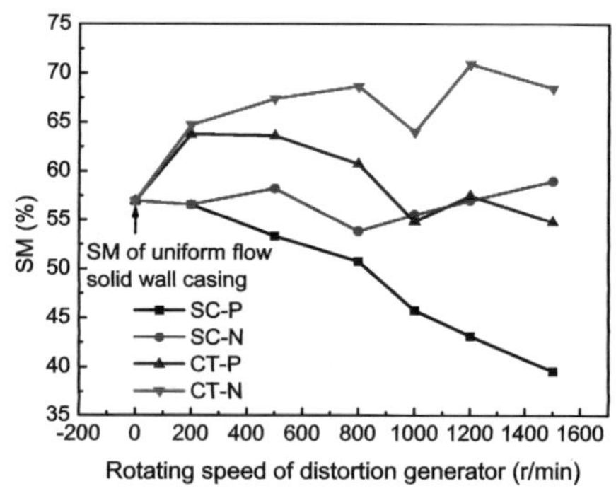

Fig. 15 The comparison of SM in different inlet and casing conditions

These compressor characteristic curves in the previous section obviously show the stall margin enhancement caused by SPS casing treatment, and if comparing the stall margin in different conditions, as shown in Fig. 15, the rotating inlet distortion with co-rotating speed may make the stall margin lower and lower with the increase of the rotating speed of distortion generator. In fact, the black line in Fig. 15 just represents monotonically decreasing curve. In contrast, the blue line in Fig. 15 shows a relatively smooth SM variation along with the increase of the rotating speed. This means that the SPS casing treatment can effectively delay the deterioration of stall margin even in the condition of the co-rotating inlet distortion. And for the situation of the rotating inlet distortion with counter-rotating speed, the rotating inflow may not result in the deterioration of flow stability in solid casing condition. Especially, in some specific situation, such non-uniform inflow even realize stability benefit in some degrees, as shown in the red line of Fig. 15. Furthermore, it is also noted that whether the rotating inlet distortion is co-rotational or counter-rotational, as shown in the blue and green lines in Fig. 15, the ability of stall margin enhancement with SPS casing treatment is quite obvious in comparison with the solid casing.

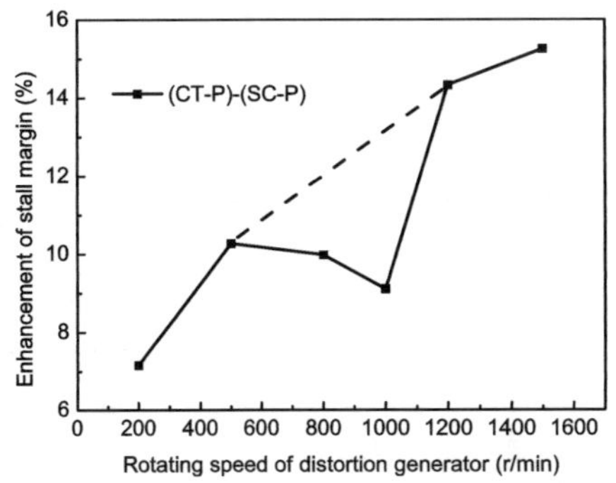

Fig. 16 The comparison of SM enhancement within using SPS casing treatment under the different degrees rotating inlet distortion

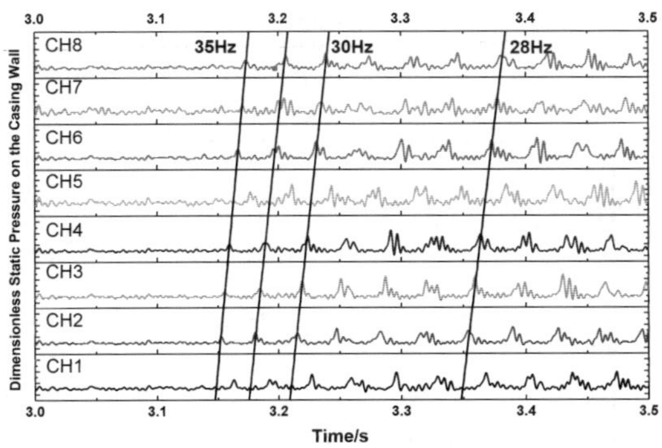

Fig. 17 Wall static pressure on the case 0.5 chord length upstream of the rotor measured by 8 high-frequency response pressure sensors under the 100% working speed with P1000 distortion

Furthermore, the comparison of the stall margin enhancement caused by SPS casing treatment in different inlet conditions with co-rotational speed is presented in Fig. 16 It is noted that the SPS casing treatment can improve the stall margin by 7% -15%. However, the stall margin enhancement does not monotonically increase with the increase of the rotating speed of distortion generator. In Fig. 17, the frequency of stall precursor waves in 100% working speed is about 35Hz, and the propagating frequency of fully developed stall cell is about 28Hz, these two characteristic frequency can be exactly stimulated by the inlet distortion generator rotating in the speed of 1050r/min and 840r/min, respectively, as shown in Fig. 8. It means that the distortion generator rotating in the speed about 1000r/min can continuously generate a disturbance in the frequency near to the stall precursor wave frequency, and make the compressor always work with the trigger factor which cannot be restrained by SPS casing treatment in time. This is the reason why the curve of the stall margin enhancement in Fig. 16 has a large decrease during the 800~1000 r/min. From some degree, this result may prove that the mechanism of the stall margin improvement with such casing treatment is associated with delaying the nonlinear development of the stall precursor waves.

Analysis and discussion on the effect of SPS casting treatment.

In order to simulate the most rotating inlet distortion situations in a multistage compressor, the distortion generator used can be set to rotate in co-and-counter rotational directions, and in different rotating speeds. While from the results shown above, the counter-rotating inlet distortion almost never deteriorates the stall margin of this compressor, so this paper has paid main attention to the co-rotating inlet distortion, which can lead to obvious stall margin and compression ability decrease.

Aimed at the situations under the co-rotating inlet distortion, using SPS casing treatment can make up the stall margin loss caused by distorted inlet conditions (Fig. 15) and can almost draw the stall point and other operation points to be back to the positions under clean inlet condition, as shown in Fig. 13. Meanwhile, the efficiency curve shows that using SPS casing treatment nearly doesn't cause additional efficiency loss, and its variation is about 1%, which is equal to the experimental error range.

Next, the dynamic time-resolved experimental results will be carried out for the purpose to uncover the mechanism of stall margin enhancement ability with the SPS casing treatment.

Experimental Results of Dynamic Pre-stall Signals

The experimental results shown in section 3 and section 4 indicate that the mechanism of the stall margin improvement with such casing treatment is associated with delaying the nonlinear development of the stall precursor waves. And based on the vortex-pressure wave interaction theory, it is theoretically shown that the SPS casing treatment can not only delay the nonlinear development of the stall precursor waves but also can weaken the unsteady flow disturbance all the time (e.g., [22]). In order to check its effectiveness under rotating inlet distortion, a technical way to measure and analyze the pre-stall behavior in the compressor system is carried out. As shown in Fig. 19, wall static pressure is measured by using eight high-frequency response pressure sensors (Kulite XT-190M) mounted at the plane about half a chord length upstream of the rotor inlet.

The filtered time histories of the individual sensors can be used to calculate the modal information by taking a discrete Fourier transform (DFT) in space about the circumference of the compressor at each point in time. All the information on the wave position and amplitude are contained in the Fourier coefficients as a function of time. According to the method proposed by Garnier (e.g., [30]), given N measurements about the machine, the complex Fourier coefficients for each mode k are expressed by

$$C_k(t)=\frac{1}{N}\sum_{n=0}^{N-1}p_n(t)\exp\left[-\frac{2ikn\pi}{N}\right] \tag{8}$$

In order to observe the pre-stall behavior, the compressor operating point is first stabilized close to stall point and then the throttle is closed very slowly so that the machine will stall within 10 to 20 seconds. Data are taken during this entire period from the eight static pressure sensors about the compressor annulus. The period of a rotor revolution is used as the unit of time, since this is a characteristic time scale for the phenomena. The analysis is performed using signal processing methods from Refs. [30]. And in order to observe the pre-stall behavior as much as possible, all the showing PSD information is ended at 0.25 second before the happening of stall. Figure 18 shows the PSD of the normalized frequency at 100% design working speed during a period of pre-stall, within P500 inlet distortion, the normalized frequency is the ratio over shaft frequency, in this figure there are three obvious characteristic frequency, i.e., distortion frequency in 0.34, second order distortion frequency 0.68 and the shaft frequency 1.

Normalized Frequency from 0.2 to 1.2:

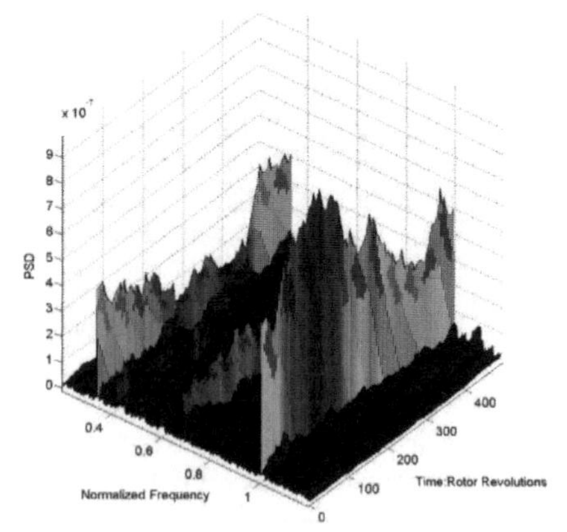

Normalized Frequency from 0.2 to 0.8:

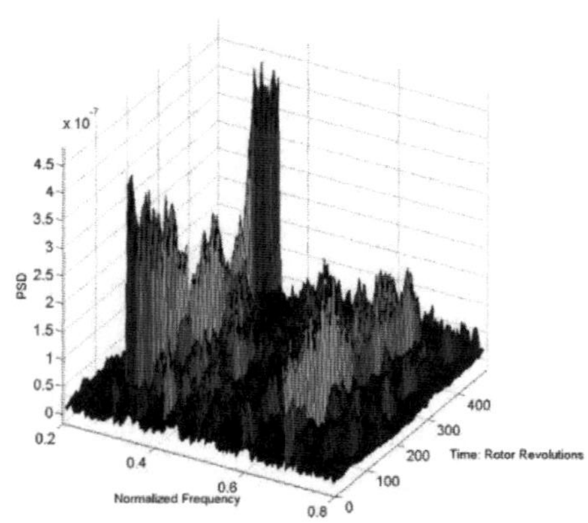

Fig. 18 PSD of the Normalized Frequency at 100% design working speed during a period of pre-stall, within P500 inlet distortion

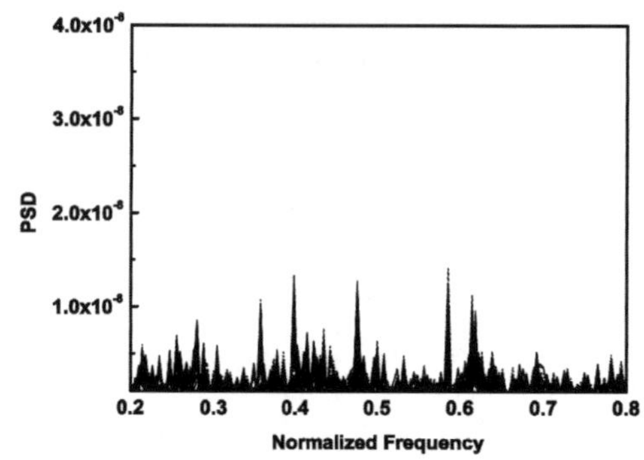

Fig. 19 PSD of the Normalized Frequency at 100% design working speed during a period of pre-stall, without inlet distortion

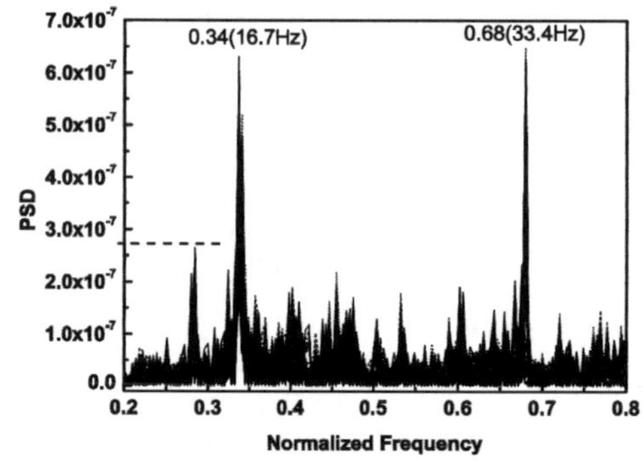

(SC)

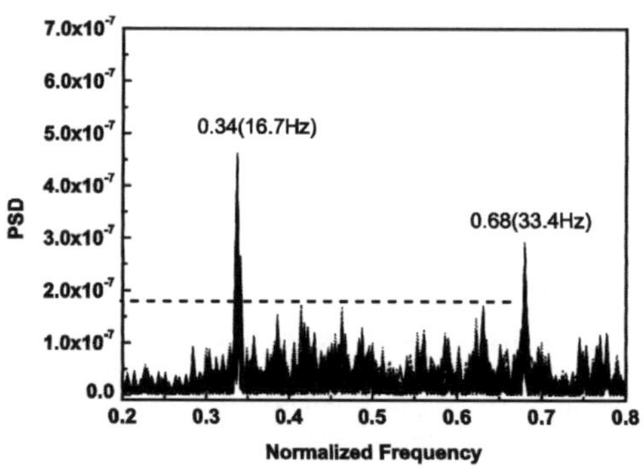

(CT)

Fig. 20 Comparison of PSD with and without casing treatment at 100% design working speed during a period of pre-stall, within P500 inlet distortion

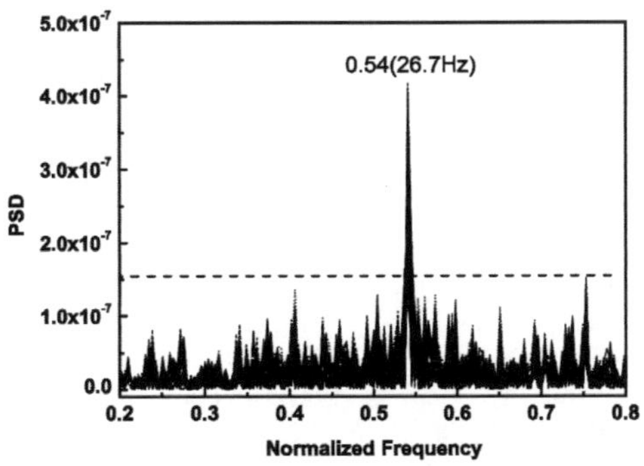

(SC)

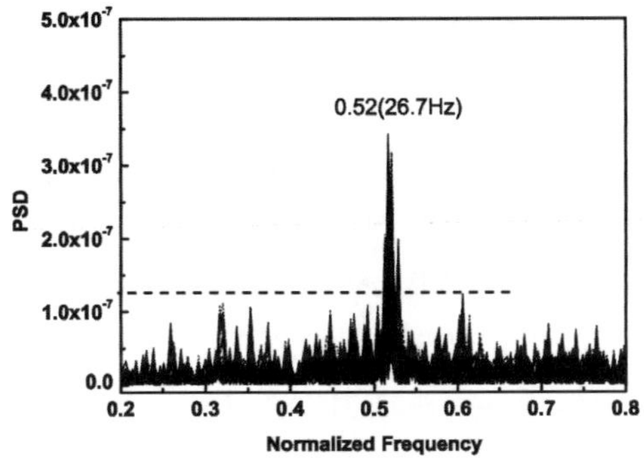

(CT)

Fig. 21 Comparison of PSD with and without casing treatment at 100% design working speed during a period of pre-stall, within P800 inlet distortion

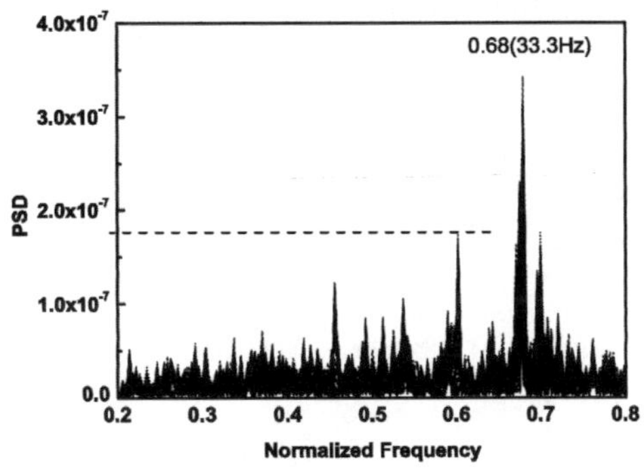

(SC)

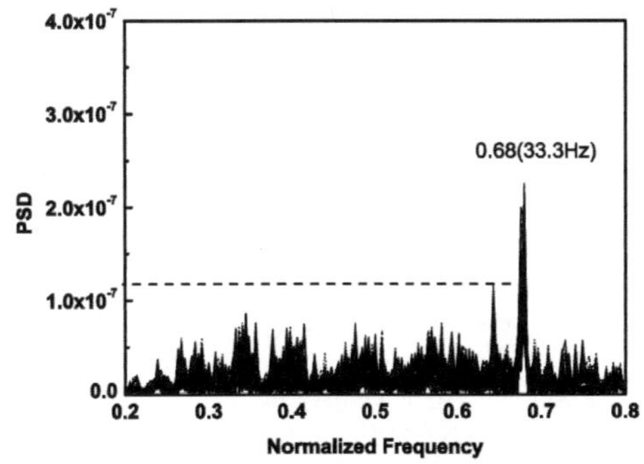

(CT)

Fig. 22 Comparison of PSD with and without casing treatment at 100% design working speed during a period of pre-stall, within P1000 inlet distortion

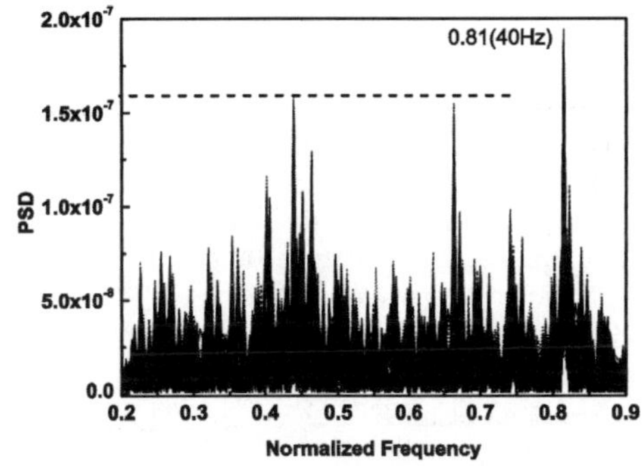

(SC)

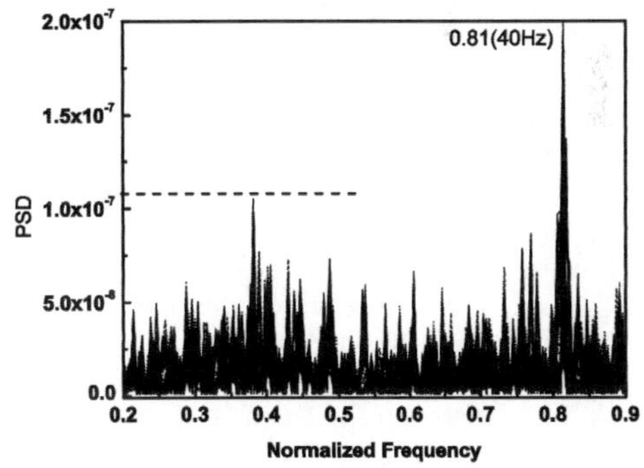

(CT)

Fig. 23 Comparison of PSD with and without casing treatment at 100% design working speed during a period of pre-stall, within P1200 inlet distortion

Because of the 3-D PSD figure can't show out the differences with or without SPS casing treatment obviously, so the time axis is cut down and the curves of different moments are plotted in one single moment, as shown in Figs. 19-23. In these figures, the working speed is 100% design working speed, which is about 2941r/min, so the shaft frequency is about 49 Hz. Figure 28 shows the PSD of the normalized frequency at 100% design working speed during a period of pre-stall under the uniform flow inlet condition, compared with the cases with inlet distortion shown in Figs. 20-23, the PSD amplitudes of disturbances in the range of 0.2 to 0.8 normalized frequency are much smaller, and this is why the distorted inlet conditions can amplify the flow disturbances. More importantly, it is such high amplitude disturbances that more rapidly lead to the happening of compressor stall inception.

Figures 20-23 show the comparison of PSD with or without casing treatment at 100% design working speed within four kinds of different inlet distortion. It is obvious that the SPS casing treatment can weaken the disturbances in broadband, and can also decrease the probability of the amplification of disturbance in flow.

CONCLUSIONS

This work presents various experimental data concerning the combined effect of the casing treatment under the restriction of rotating inlet distortion. The main results of rotating inlet distortion on compressor stability with SPS casing treatment can be summarized as follows.

In order to simulate the rotating distorted inlet conditions, a distortion generator has been designed, and from the experimental results, the distortion generator used in this paper can produce non-uniform inlet condition with an obvious total pressure decrease region and specific characteristic frequency, including both counter-rotating and co-rotating inlet distortions. It is found that the rotating inlet distortion with counter-rotating speed can enhance the stall margin by 1%~5% compared to the distortion with co-rotating speed, which is consistent with the results observed by Longley et al. (e.g., [7]) and Peters et al. (e.g., [8-9]). More importantly, the present experiment shows that counter-rotating inlet distortion has a completely different mechanism from co-rotating inlet distortion, the latter often leads to obvious stall margin loss.

Since the rotating inlet distortion with counter-rotating speed can enhance the stall margin, this investigation has paid more attention on the performance of test compressor and the effect of SPS casing treatment. Under the co-rotating inlet distortion, the SPS casing treatment can move the stall point to the left, and not only make up the stall margin loss caused by inlet distortion but also enhance the stall margin compared to the solid wall casing under the clean inlet condition. Meanwhile, the efficiency variation caused by the SPS casing treatment is less than 1%, which is within the range of the experimental error, therefore, the SPS casing treatment almost does not lead to additional compressor efficiency loss with various non-uniform inflow conditions.

The mechanism of the stall margin improvement with SPS casing treatment is to weaken the unsteady flow disturbances in broadband and delay the nonlinear development of the stall precursor waves.

ACKNOWLEDGMENTS

This work has been supported by National Natural Science Foundation of China (Nos. 51236001, 51106154 and 51006005), and National Basic Research Program of China (No. 2012CB720201) and Aeronautical Science Foundation of China (No. 2014ZB51018) and Fundamental Research Funds for the Central Universities.

REFERENCES

[1] Bowditch, D. N., and Coltrin, R. E., 1983, "A Survey of Engine Inlet Distortion Capability," NASA TM-83421.

[2] Hynes, T. P., and Greitzer, E. M., 1987, "A Method for Assessing Effects of Circumferential Flow Distortion on Compressor Stability," ASME Journal of Turbomachinery, 109(3), pp. 371-379.

[3] Spakovszky, Z. S., Weigl, H. J., Paduano, J. D., Van Schalkwyk, C. M., Suder, K. L., and Bright, M. M., 1999, "Rotating Stall Control in a High-Speed Stage With Inlet Distortion, Parts I-II," ASME Journal of Turbomachinery, 121(3), pp. 508-522.

[4] Hah, C., Rabe, D. C., Sullivan, T. J., Wadia, A.R., 1998, "Effects of Inlet Distortion on the Flow Field in a Transonic Compressor Rotor," ASME Journal of Turbomachinery, 120(2), pp. 233-246.

[5] Lesser, A., and Niehuis, R., 2014, "Transonic Axial Compressors with Total Pressure Inlet Flow Field Distortions," ASME Paper No. GT2014-26627.

[6] Lucas, J. R., O'Brien, W.F., Ferrar, A. M., 2014, "Effect of BLI–Type Inlet Distortion on Turbofan Engine Performance," ASME Paper No. GT2014-26666.

[7] Longley, J. P., Shin, H. W., Plumley, R. E., Silkowski, P. D., Day, I. J., Greitzer, E. M., Tan, C. S., Wisler, D. C., 1996, "Effects of Rotating Inlet Distortion on Multistage Compressor Stability," ASME Journal of Turbomachinery, 118(2), pp. 181-188.

[8] Peters, T., Fottner, L., 2002, "Effects of Co-and Counter-Rotating Inlet Distortions on a 5-Stage HP-Compressor," ASME Paper No. GT-2002-30395.

[9] Peters, T., Bürgener, T., Fottner, L., 2001, "Effects of Rotating Inlet Distortion on a 5-Stage HP-Compressor," ASME Paper No. 2001-GT-0300.

[10] Salunkhe, P. B., Pradeep, A. M., 2010, "Stall Inception Mechanism in an Axial Flow Fan under Clean and Distorted Inflows," ASME J. Fluids Eng., 132(12), p. 121102-1-14.

[11] Hah, C., Shin, H. W., 2012, "Study of Near-Stall Flow Behavior in a Modern Transonic Fan with Compound Sweep," ASME J. Fluids Eng., 134(7), p. 071101-1-7.

[12] Koch, C. C., 1970, "Experimental Evaluation of Outer Case Blowing or Bleeding of a Single-Stage Axial Flow Compressor," No. NASA CR-54592.

[13] Takata, H., and Tsukuda, Y., 1977, "Stall Margin Improvement by Casing Treatment—Its Mechanism and Effectiveness," ASME Journal of Engineering for Gas Turbines and Power, 99(1), pp. 121-133.

[14] Muller, M.W., Schier, H. P., Voges, M., and Hah, C., 2011, "Investigation of Passage Flow Features in a Transonic Compressor Rotor with Casing Treatments," ASME Paper No. GT2011-45364.

[15] Kang, C. S., McKenzie, A. B. and Elder, R. L., 1995, "Recessed Casing Treatment Effects on Fan Performance and Flow Field," ASME Paper No. 95-GT-197.

[16] Kroeckel, T., Hiller, S. J. and Jeschke, P., 2011, "Application of a Multistage Casing Treatment in a High Speed Axial Compressor Test Rig," ASME Paper No. GT2011-46315.

[17] Voges, M., Willert, C., Mönig, R., Schiffer, H. P., 2013, "The Effect of a Bend-Slot Casing Treatment on the Blade Tip Flow Field of a Transonic Compressor Rotor," ASME Paper No. GT2013-94006.

[18] Pixberg, C. T., Schiffer, H. P., Ross, M. H., Cameron, J. D., Morris, C. S., 2013, "Stall Margin Improvement by Use of Casing Treatments," ASME Paper No. GT2013-95842.

[19] Madden, D. S., West, M. A., 2005, "Effects of Inlet Distortion on the Stability of an Advanced Military Swept Fan Stage with Casing Treatment," ASME Paper No. GT-2005-68693.

[20] Hung, J., Wu, H., 2008, "Numerical Investigation of Inlet Distortion on a Compressor Rotor with Circumferential Groove Casing Treatment," Chinese Journal of Aeronautics 21(6), pp. 496-505.

[21] Sun, X., Sun, D., Liu, X., Yu, W. and Wang, X., 2014, "Theory of Compressor Stability Enhancement Using Novel Casing Treatment, Part I: Methodology," AIAA Journal of Propulsion and Power, 30(5), pp. 1224-1235.

[22] Sun, D., Liu, X., Jin, D., Gui, X. and Sun, X., 2014, "Theory of Compressor Stability Enhancement Using Novel Casing Treatment, Part II: Experiment," AIAA Journal of Propulsion and Power, 30(5), pp.1236-1247.

[23]Dakun Sun, Xiaohua Liu, and Xiaofeng Sun*, "An Evaluation Approach for the Stall Margin Enhancement with SPS Casing Treatment," ASME Journal of Fluids Engineering, Vol. 137, 2015.

[24] Sun, X., 1996, "On the Relation between the Inception of Rotating Stall and Casing Treatment," 32th AIAA/ASME/SAE/ASEE Joint Propulsion Conference, No. AIAA-96-2579.

[25] Sun, X., Sun, D., and Yu, W., 2011, "Model to Predict Stall Inception of Transonic Axial Flow Fan/Compressors," Chinese Journal of Aeronautics, 24(6), pp. 687-700.

[26] Rusak, Z., and Morris, W. J., 2011, "Stall Onset on Airfoils at Moderately High Reynolds Number Flows," ASME J. Fluids Eng., 133(11), p. 111104.

[27] Morris, W. J., and Rusak, Z., 2013, "Stall Onset on Aerofoils at Low to Moderately High Reynolds Number Flows," Journal of Fluid Mechanics, 733, pp. 439–472.

[28] Sun, X., Liu, X., Hou, R. and Sun, D., 2013, "A General Theory of Flow Instability Inception in Turbomachinery," AIAA Journal, 51(7), pp. 1675-1687.

[29] Liu, X., Sun, D., Sun, X., 2014, "Basic Studies of Flow-Instability Inception in Axial Compressors Using Eigenvalue Method," ASME J. Fluids Eng., 136(3), p. 031102-1-9.

[30] Garnier, V. H., Epstein, A. H., and Greitzer, E. M., 1991, "Rotating Waves as a Stall Inception Indication in Axial Compressors," ASME Journal of Turbomachinery, 113(2), pp. 290-302.

Proceedings of the 14th International Symposium on Unsteady Aerodynamics, Aeroacoustics & Aeroelasticity of Turbomachines
ISUAAAT14
8-11 September 2015, Stockholm, Sweden

I14-S12-4

INFLUENCE OF UNSTEADY EFFECTS ON INTEGRAL AND LOCAL PERFORMANCES OF COUNTER ROTATING FANS

Victor Mileshin
CIAM
Moscow, Russia

Yaroslav Druzhinin
CIAM
Moscow, Russia

Alexander Stepanov
CIAM
Moscow, Russia

ABSTRACT

3D viscous flow in blade-to-blade channels of CRTF2A counter-rotating fan developed under the European VITAL program [1] is simulated by NUMECA FINE TURBO in two problem statements: steady state (Mixing Plane approach) and Non-Linear Harmonic Method. NLH gives an opportunity to investigate an influence of rotor-rotor interaction on integral and local characteristics of CRTF2A.

Analysis of the results shows that computed and experimental data are in good agreement. Maximum difference is observed between values of maximum flow rate and maximum efficiency. Computed curves are shifted to the right for massflow by ≈2%; the difference between max. efficiencies is about the same.

Comparison of values computed by Mixing Plane and NLH procedures reveals a slight difference in integral performances. The solution is more sensitive to the choice of a turbulence model.

After finding the solutions in the NLH problem statement, a time-reconstruction is made to get unsteady flow fields in blade-to-blade channels.

Static pressure fluctuations at tip section of R1 and R2 were assessed.

INTRODUCTION

One of the perspective approaches in efficient turbomachine design is taking into account unsteady blade row interaction. Conventional methods of optimization are generally based on steady approach implemented in calculations with Mixing Plane technology. The core of this method is pitchwise averaging of the flow solution at the rotor-stator interface, and the decision gives reliable results for great variety of applications. But modern trends in compressor and fan design are directed to reducing the number of stages, number of blades per row, etc. This leads to unavoidable increasing of aerodynamic load. At the same time, new phenomena, which could not be described by classical computational methods occur in turbomachines.

Full unsteady calculations became available and more widely spread nowadays due to development of high performance computer systems. It gives an opportunity to carry out more complicated and resource consuming simulations, investigate some kind of new effects, for example, clocking effect originated from wake-blade interaction between stages in multistage compressors and turbines.

Despite of all the advantages of full unsteady statement with sliding mesh row-row interfaces, it is still too much time consuming and can be applied mostly at final stage of design procedure to check and prove the quality of previous statements.

Non-Linear Harmonic Method turns to be a kind of ways, which make possible to achieve unsteady solution many times faster than in case of full unsteady simulation. Particularly it is known to be applicable for the flows without strong non-periodical effects like surge and stall.

CRTF2A counter rotating ducted fan depicted on fig.1 was studied experimentally and numerically using many different approaches, and some features caused by problem statement were revealed. It gives background to conduct a series of calculations using steady Mixing Plane and NLH methods implemented in NUMECA FINE software package.

Fig.1. General view of CRTF2A fan model

NOMENCLATURE

NLH – Non-Linear Harmonic Method;
N_h – number of frequencies;
R1 – first rotor;
R2 – second rotor;
π^* – total pressure ratio;
η – isentropic efficiency;
G, kg/s – massflow rate;
N_{cor} – corrected rotational speed;
LE – leading edge;
TE – trailing edge

NUMERICAL METOD

Numerical simulation of the flow in ducted counter rotating fan CRTF2A was carried out using NUMECA FINE TURBO software [2] for solving Reynolds averaged 3D Navier-Stokes equations (3D RANS, URANS)

Two differential turbulence models: SST and Spalart-Allmaras with extended wall function were used for simulation. Navier-Stokes equations solutions were obtained by Jameson scheme [3].

A block structured O-H mesh was generated by automatic grid generator Numeca Autogrid 5. The computational domain is divided to 22 grid blocks. The grid contains totally 2 989 128 cells (fig.2).

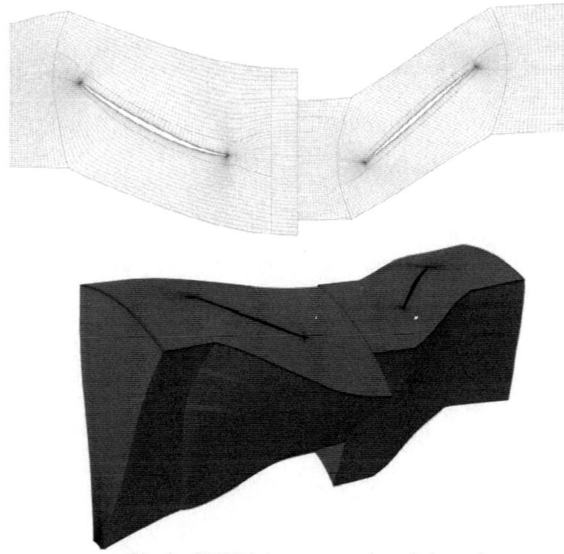

Fig.2. CRTF2A computational domain

Computations of viscous flow in CRTF2A have been performed with corrected inlet conditions T*=288.15K, P*=101325Pa at N_{cor}=90% and N_{cor}=100%. Radial flow equilibrium is applied at the outlet section imposing static pressure at tip area. Non-slip adiabatic conditions are specified on the solid walls.

Steady flow computations in CRTF2A have been carried out using Mixing Plane approach. Standard algorithm takes into account mutual influence of adjacent rows correctly defining radial distributions of pitchwise averaged parameters of flow at upstream and downstream side of row-row interface.

Unsteady simulation is carried out using Non-Linear Harmonic Method. This method proposed by He and Ning (1998) [4] provides unsteady solutions for the rotor-rotor system, but it requires much less computational resources and time as compared with full-unsteady problem statement (3D URANS) [5], because the NLH approach needs no simulation of several blade channels to reach common periodicity at the rotor-rotor interface that is necessary for the unsteady interface of sliding grids. Furthermore, convergence time in NLH statement is much less than it is in case of full unsteady simulation.

The main idea of NLH is that the flow perturbations that make the flow unsteady are written about a time-averaged value of the flow and are Fourier decomposed in time. Converting the unsteady Navier-Stokes system into the frequency domain, transport equations are obtained for each time frequency.

The accuracy of the unsteady solution is defined by the order of the Fourier series.

Computations of the flow in NLH statement were conducted with number of frequencies N_h=6 per one perturbation. Results of steady simulations with standard Mixing Plane were used as an inital solution for NLH.

Time-reconstruction of NLH solutions has been performed for the analysis of unsteady flow fields and investigation of rotor-rotor interaction in CRTF2A. Time-reconstruction takes into account numbers of blades in rows (10 and 14 for 1st and 2nd rotor correspondingly) and operates with minimum common period consisting of 5 blade passages of 1st rotor and 7 blade passages of 2nd rotor. Time discretization is limited by 10 instantaneous solutions.

NUMERICAL RESULTS

Fig. 3 and Fig. 4 show the computed data for CRTF2A integral performances obtained by Mixing Plane and NLH methods in comparison with experimental data and calculations completed earlier by CIAM using in-house 3D RANS "3D_IMP MULTI" software.

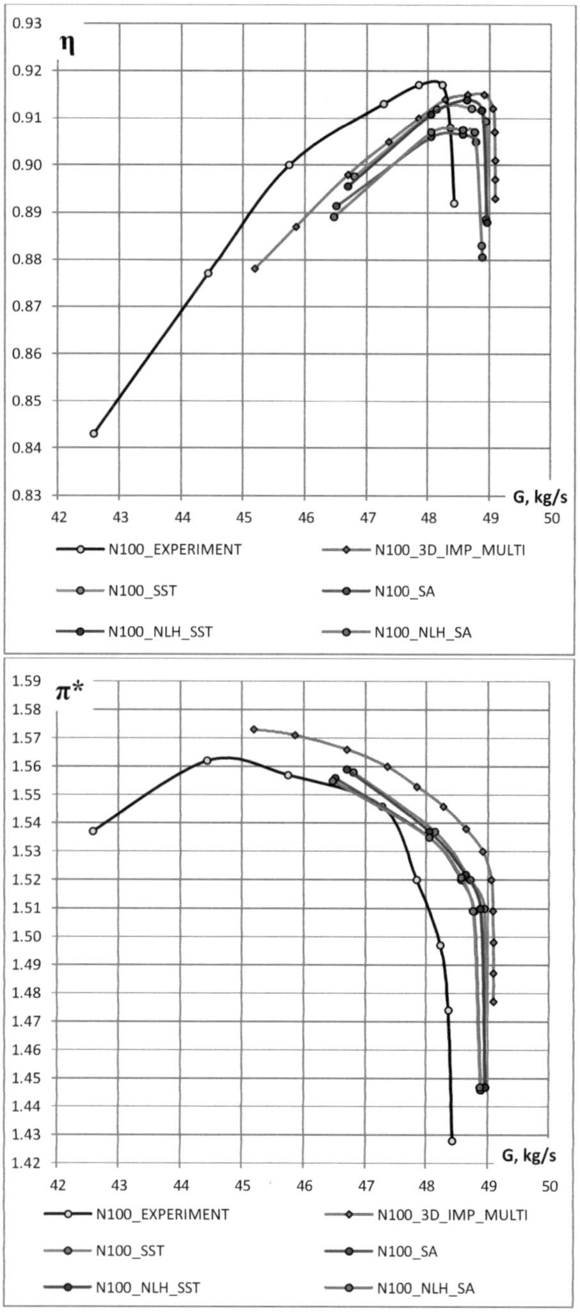

Fig.3. Integral performances at N_{cor}=100%

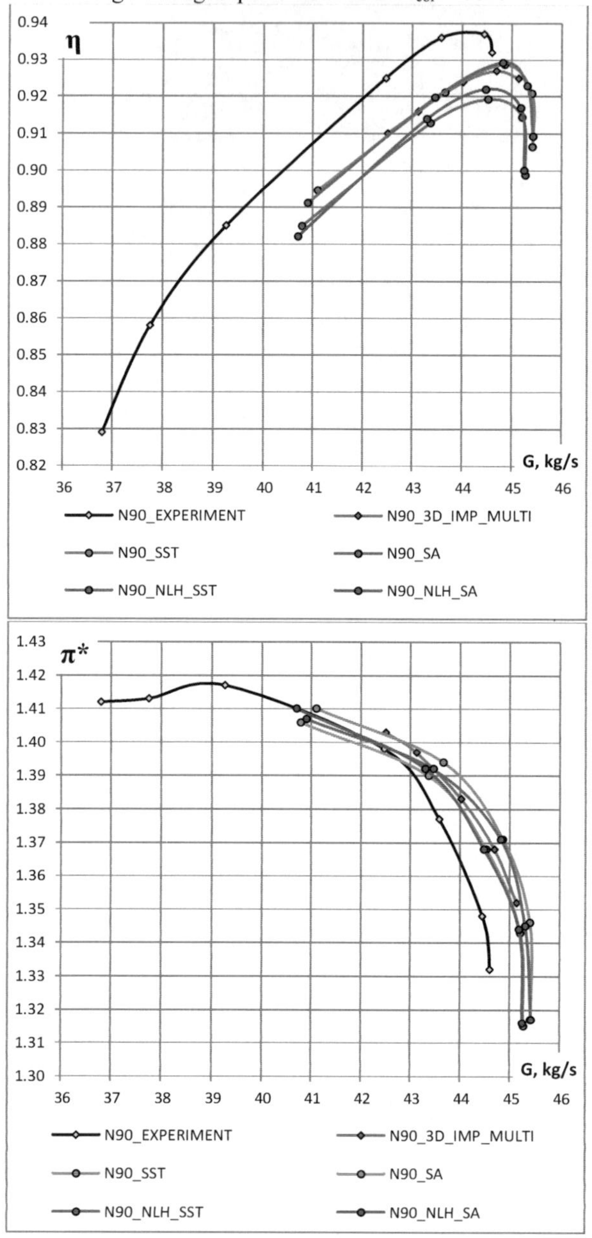

Fig.4. Integral performances at N_{cor}=90%

Analysis of the results shows that computed and experimental data are in good agreement. Maximum difference is observed between values of maximum flow rate and maximum efficiency. Computed curves are shifted to the right for massflow by ≈2%. The difference between maximum efficiencies varies from 0.5% to 1.0% depending on turbulence model at 100% rpm and from 1.0% to 2.0% at 90% rpm. Results obtained by SST model demonstrate higher values of isentropic efficiency in comparison with Spalart-Allmaras model.

Moreover, it is worth noting that computed values of surge margin are noticeably lower than experimental value for both rpm.

Comparison of values computed by Mixing Plane and NLH procedures reveals a slight difference in integral performances. NLH approach shows higher by ≈0.3% value of efficiency for 100% rpm and by ≈0.2% for 90% rpm keeping almost the same values of massflow.
The solution is more sensitive to the choice of a turbulence model. The reason of such effect might be explained by a difference in modelling local separation zone near R1 hub.

ANALISYS OF THE FLOW

Fig.5-6 show the most evident distinction in flow fields obtained in steady statement depending on turbulence model. Local separation occurs in hub area (5% blade span) of 1st rotor at 100% rpm as well as at 90%. The choice of turbulence model has an influence on boundary layer behaviour. SST model demonstrates at fig.5 more intensive separation.

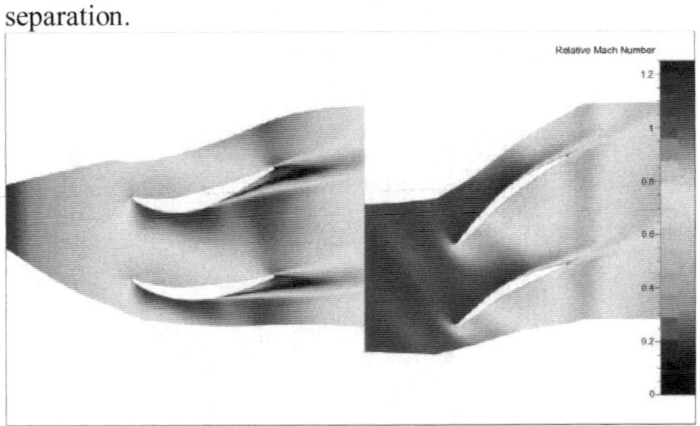

Fig.5. Relative Mach number distribution at 5% span. N_{cor}=100%. Mixing Plane approach. SST turbulence.

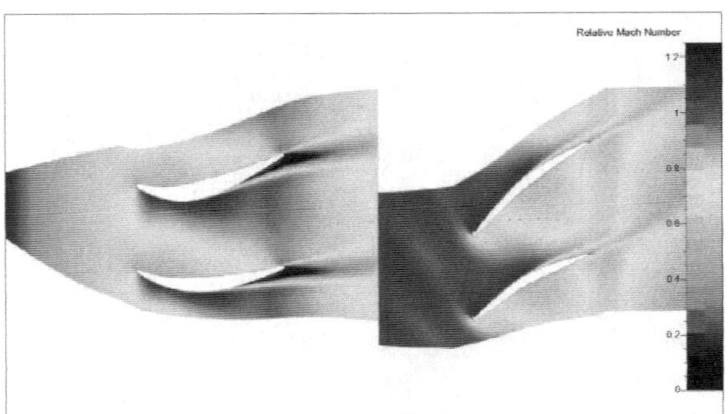

Fig.6. Relative Mach number distribution at 5% span. N_{cor}=100%. Mixing Plane approach. SA turbulence.

Fig.7-12 demonstrate the results of time-reconstruction of harmonic solution at 100% rpm. The most important phenomena to be analyzed are viscous wakes propagation through the axial gap between rotors and unsteady interaction of wakes with shock waves and boundary layer at 2nd rotor airfoils.

Static pressure distributions show sufficient quality of shock waves resolution. Distributions in near hub sections show presence of thick wakes generated by 1st rotor blades caused by separation of boundary layer. The wakes propagates through the axial gap almost without dissipation and influence the flow field in 2nd rotor blade passages.

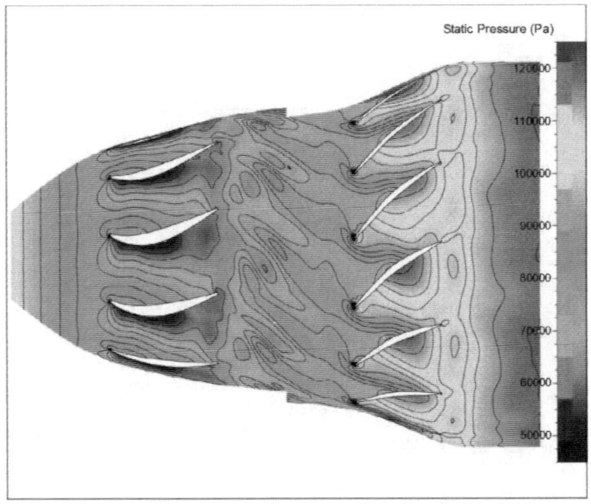

Fig.7. Instantaneous Static Pressure distribution at 5% span. N_{cor}=100%.

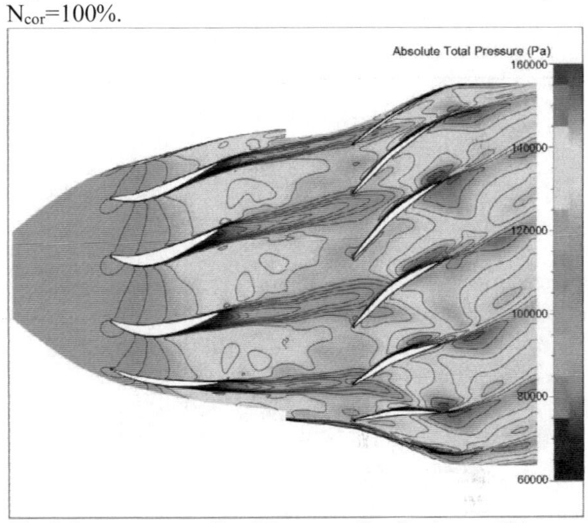

Fig.8. Instantaneous Total Pressure distribution at 5% span. N_{cor}=100%.

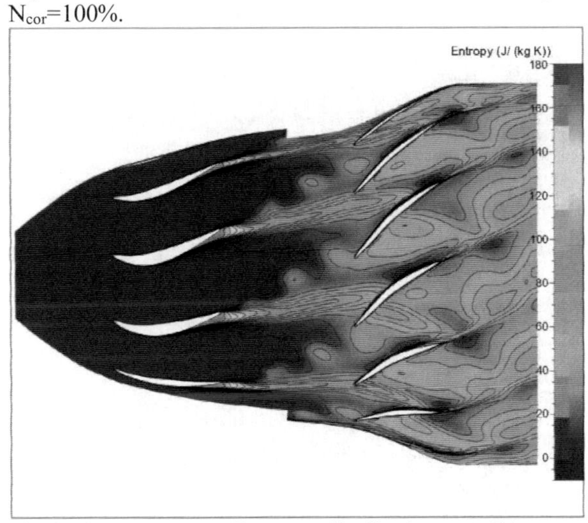

Fig.9. Instantaneous Entropy distribution at 5% span. N_{cor}=100%.

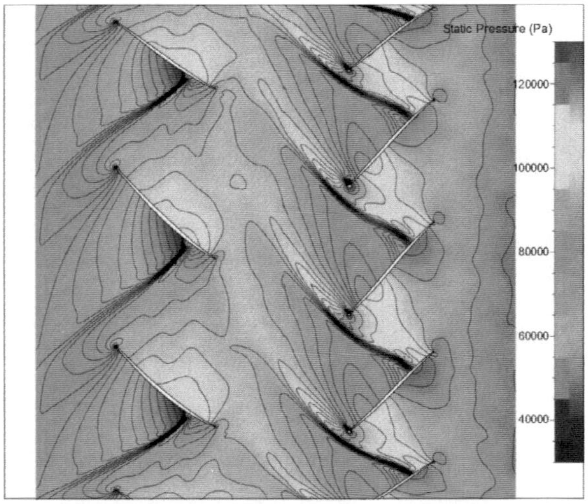

Fig.10. Instantaneous Static Pressure distribution at 95% span. N_{cor}=100%.

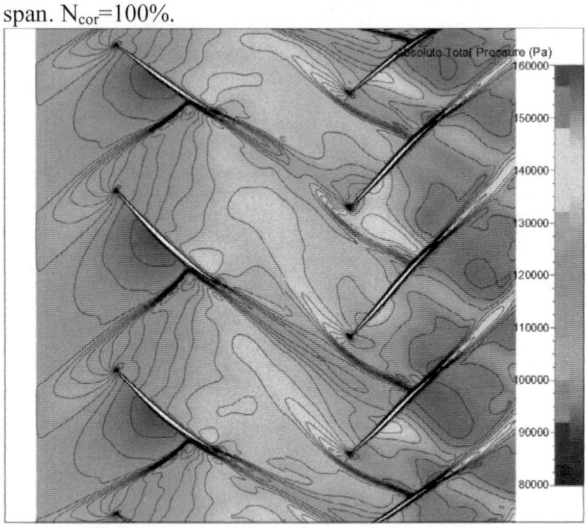

Fig.11. Instantaneous Total Pressure distribution at 95% span. N_{cor}=100%.

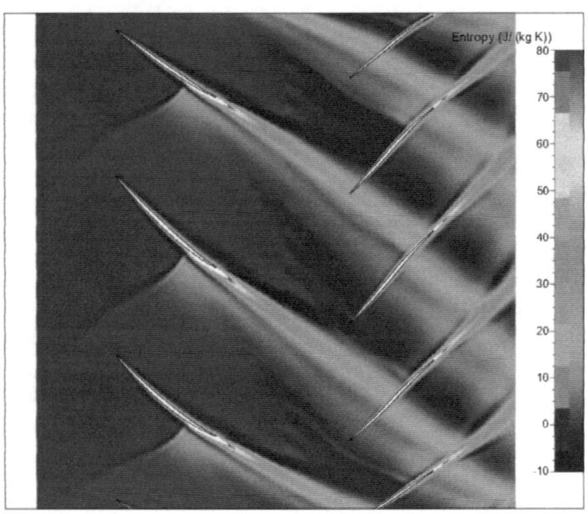

Fig.12. Instantaneous Entropy distribution at 95% span. N_{cor}=100%.

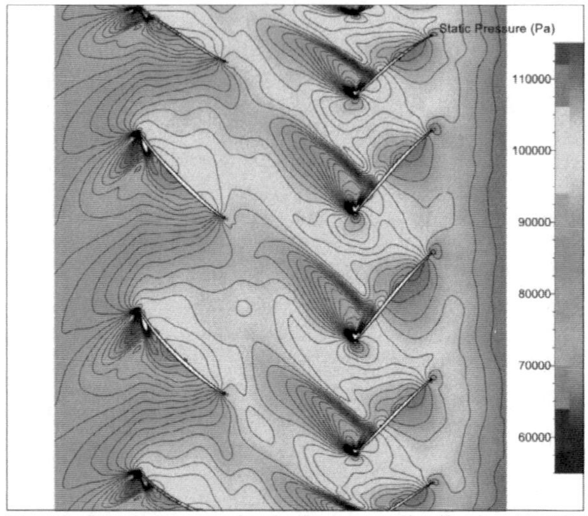

Fig.13. Instantaneous Static Pressure distribution at 95% span. N_{cor}=90%.

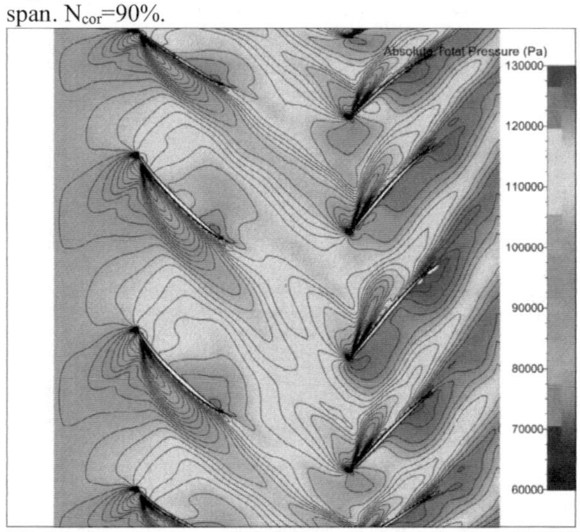

Fig.14. Instantaneous Total Pressure distribution at 95% span. N_{cor}=90%.

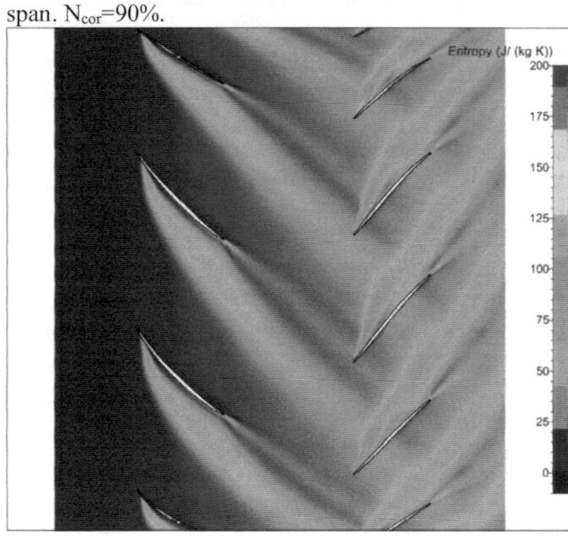

Fig.15. Instantaneous Entropy distribution at 95% span. N_{cor}=90%.

Fig.10, 11 demonstrates presence of supersonic zones ended by strong shock waves near trailing edges in tip sections. Nevertheless, the value of axial distance between blade rows is big enough to reduce the intensity of interaction between shock waves and trailing edges of 1st rotor. Shock disturbances become week while reaching 1st rotor blades.

Fig. 13-15 shows Pressure and Entropy distributions at 95% blade span section at 90% rpm. This fields reveal a significant difference of flow in comparison with 100% rotational speed. Supersonic zones in both rotors are located near leading edges and shocks take form of detached waves in this case. Total Pressure and Entropy distributions indicate vortexes produced by tip leakage interacting with shock waves. The vortexes originate strong disturbances which propagate downstream.

Fig. 16, 17 show instantaneous Entropy distribution in outlet section. Wakes generated by the 1st rotor and interfering with 2nd rotor can be distinguished. It should be noted that R1 wakes have skewed shape and cover more than two R2 blade passages.

Wake disturbances look more distinctive at N_{cor}=100% (fig.16) in comparison with N_{cor}=90%. It may cause higher level of fluctuations for the design point.

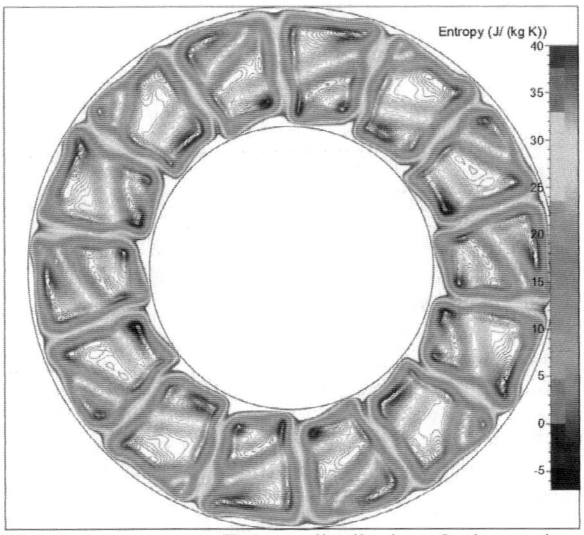

Fig.16. Instantaneous Entropy distribution. Outlet section. N_{cor}=100%.

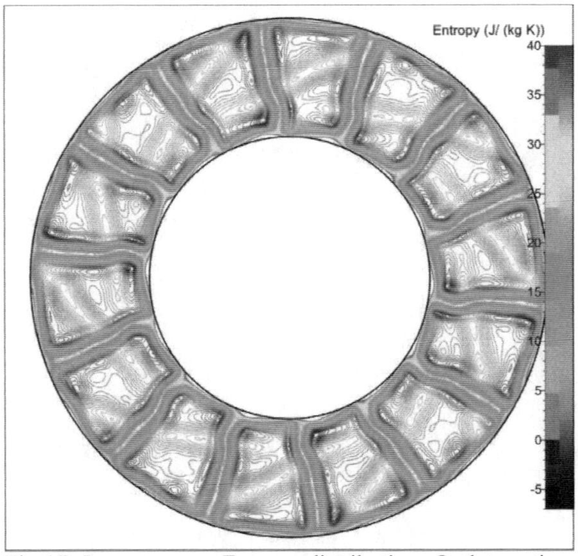

Fig.17. Instantaneous Entropy distribution. Outlet section. N_{cor}=90%.

ANALYSIS OF PRESSURE FLUCTUATIONS

Instantaneous pressure distributions at blade airfoils were extracted from time-reconstructed NLH solutions. Fig.18-21 show the distributions at tip sections (95% span) of both rotors. The graphs demonstrate the difference between unsteady load distributions along the airfoil of 1st and 2nd rotors. Rotational speed has an influence on location of maximum pulsation intensity at 1st rotor. Comparison of fig.18 and fig.20 displays that maximum value of pulsation is about 5% and it is shifted to the left part (closer to LE) of the profile at N_{cor}=100% but at N_{cor}=90% to the right (closer to TE). It is worth to note also, that the distribution at suction side of 1st rotor is almost steady for both rpm. It means that the position and intensity of shock wave in R1 is not sensitive to unsteady rotor-rotor interaction.

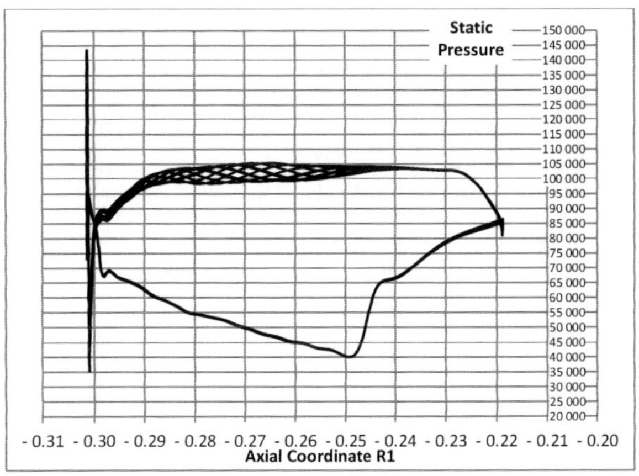

Fig.18. Instantaneous Pressure distribution at pressure and suction side of R1 blade at 95% span. N_{cor}=100%.

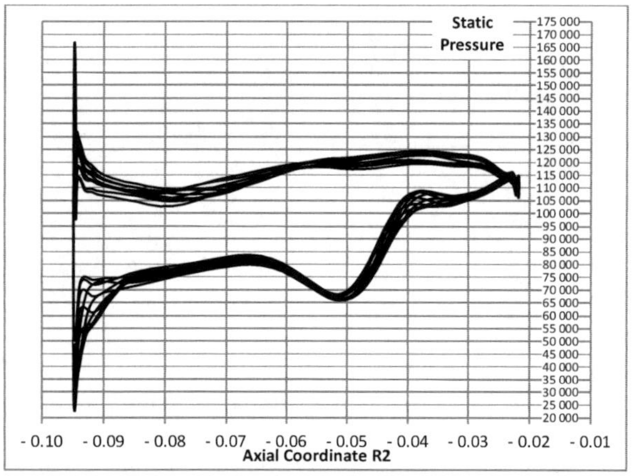

Fig.19. Instantaneous Pressure distribution at pressure and suction side of R2 blade at 95% span. N_{cor}=100%.

On the contrary, analysis of R2 pressure fluctuations shows presence of unsteadiness at both side of the airfoil. Absolute value of fluctuation is higher at N_{cor}=100% and it is distributed uniformly along the profile excepting the increase of ≈20% deviation near LE of R2. Pressure drop at shock wave is also oscillating. These results prove that wake-blade and wake-shock-wave interaction has strong influence on the flow in R2. In spite of big value of axial gap between 1st and 2nd rotor (≈150% of CRTF2A blade tip chord axial projection), wake disturbance has not enough free path for dissipation.

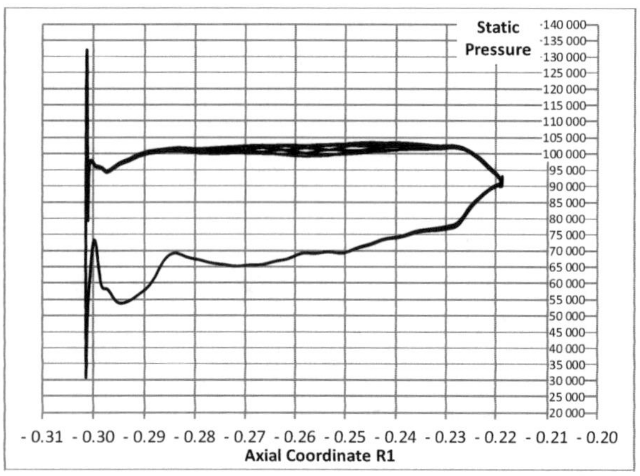

Fig.20. Instantaneous Pressure distribution at pressure and suction side of R1 blade at 95% span. N_{cor}=90%.

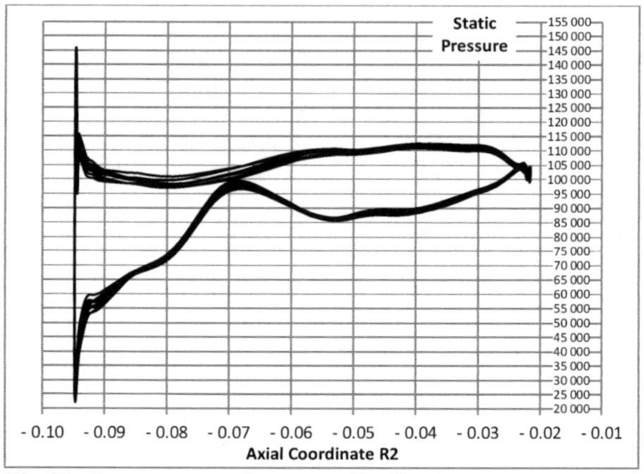

Fig.21. Instantaneous Pressure distribution at pressure and suction side of R2 blade at 95% span. N_{cor}=90%.

CONCLUSION

1. A numerical study of flow peculiarities in ducted counter rotating fan CRTF2A was carried out in two different problem statements: steady Mixing Plane approach and Non-Linear Harmonic method.
2. Results of calculations were compared with experimental data for two rpm.
3. Slight difference between steady and NLH results was found for integral performances.
4. An influence of turbulence modelling on integral and local performances was investigated. Calculations with SST model found to be closer to the experimental results in values of isentropic efficiency.
5. Time-reconstruction was carried out for analysis of instantaneous flow fields in CRTF2A.
6. Rotor-rotor interaction was studied. Viscous wakes propagation and interaction of them with blades and shock waves was analysed.
7. Unsteady pressure distributions at R1 and R2 airfoils near tip were studied. The intensity of pressure fluctuations was assessed.

REFERENCES

1. V.I. Mileshin, S.V. Pankov, V.A. Fateev, S.K. Shchipin. "Ducted Counter–Rotating Fan Blades Optimization Based on 3D Inverse Problem Solution aiming at fan gasdynamics improvement". Proceedings of ISABE International Conference, Montreal, Canada, ISABE-2009-1334, 2009.
2. Hirsch Ch., "Numerical Computation of Internal and External Flows". Volume 2. John Wiley & Sons, 1990
3. Jameson, A., "Time dependent calculations using multigrid, with applications to unsteady flows past airfoils and wings". AIAA-91-1596, 1991
4. He, L. and Ning, W., 1998, "Efficient Approach for Analysis of Unsteady Viscous Flows in Turbomachines", AIAA Journal, Vol. 36, No. 11
5. I.A. Brailko, V.I. Mileshin, A.M. Volkov, V.N. Korznev. "Numerical and experimental investigations of CRF with simulation of flow non-uniformity in the basic flight conditions". 27th International Congress of Aeronautical Sciences, Nice, France, 2010

Proceedings of the 14th International Symposium on Unsteady Aerodynamics, Aeroacoustics & Aeroelasticity of Turbomachines
ISUAAAT14
8-11 September 2015, Stockholm, Sweden

I14-S13-1

DEPENDENCE OF FAN TONE NOISE ON AXIAL GAP BETWEEN FAN ROTOR AND STATOR IN ROW-TO-ROW ACOUSTIC INTERACTION

Hidekazu Kodama
IHI Corporation
Nishitama-gun, Tokyo, Japan

Masanobu Namba
Professor Emeritus, Kyusyu University
Yokosuka-shi, Kanagawa, Japan

ABSTRACT

This paper describes the investigation of dependence of fan tone noise on axial gap between fan rotor and stator in row-to-row acoustic interaction. In this work, a three-dimensional lifting surface theory for a fan stage model composed of a rotor and a stator is used to calculate the sound field under acoustic interaction between rotor and stator. In the parametric study with the fan rotor-stator axial gap varied as an independent variable, the predicted acoustic power level of rotor wakes-stator interaction noise shows no monotonous variation with the axial gap, but sharp and round peaks at certain axial gaps. The investigation of the relevance of axial wave lengths of acoustic duct modes to the gaps between the rotor and the stator indicates that the peaks might be caused by a resonance mechanism similar to the acoustic response in a resonator.

INTRODUCTION

The primary source of the fan tone noise is the unsteady loadings on the stator vanes generated by interaction with oncoming fan rotor wakes. Therefore, it has been believed that an increase of the axial gap between the fan rotor and the stator giving interaction with highly decayed wakes is a useful strategy of reducing the fan tone noise. However, increase in the axial gap has a penalty of increase in a fan axial length. The selection of the axial gap between the fan rotor and the stator is an important key point for an engine design, because it determines not only the level of the engine noise but also the engine weight. So far, a number of experimental and analytical studies have been made of the dependence of fan tone noise on axial gap between fan rotor and stator.

In the earlier works, some experimental efforts to study the trends with axial gap between fan rotor and stator on fan tone noise level appeared. Woodward and Glaser (1981) conducted fan noise tests using a single stage test fan with a transonic fan rotor. Two stator configurations were used. One was cut-on and another was cut-off design with respect to propagation of the fundamental blade passing tone due to interaction between fan rotor wakes and stator. For both stator configurations, the fan noise was measured at three fan rotor-stator axial gaps. The results showed that there was no clear consistency in the trends with the axial gap between the two stator configurations. It is interesting that the fundamental tone level of the cut-on stator configuration for the largest axial gap showed a higher value than that for the next to largest axial gap. Kantola and Gliebe (1981) conducted a similar fan noise test also using two cut-on and cut-off stator configurations. The trends of the fan tone noise level with the fan rotor-stator axial gap were compared between two fan pressure rise levels. It can be seen that widening the axial gap from the next to largest value to the largest value resulted in an increase of the third blade passing tone level at the higher pressure rise condition for each of the two stator configurations. These increases in the tone noise levels with the axial gap cannot be explained by the ordinary mechanism of noise generation by interaction between rotor wakes and stator.

Most of analytical efforts appeared so far were based on analytical models composed of only stator cascade with oncoming rotor wakes. There has been no work that predicted the above unexpected trend in the tone noise level. Namba (2006) developed a three-dimensional lifting surface theory for prediction of the acoustic interaction of two annular cascades in counter rotation. This theory was applied to investigate the effect of axial gap between rotor and stator on fan tone noise level and it was suggested that the variation of the tone noise level with axial gap is not monotonous but highly complicated (Kodama et al. (2009)).

This paper investigates the dependence of the fan tone noise on the axial gap between fan rotor and stator. In this study, the prediction method developed by Namba (2006) is used to evaluate the contribution of acoustic interaction between fan rotor and stator to the dependence on the axial gap. The parametric studies of axial gap are made on two sets of the two fan noise test conditions. The previous study (Kodama et al. (2013)) clearly indicated a significant influence of rotor-stator acoustic interaction on the sound power in these two sets of the test conditions. The computation results with rotor-stator acoustic interaction showed sharp and round peaks in the variation in the acoustic power level of rotor wakes-stator interaction noise with the axial gap.

NOMENCLATURE

C_{a1}	axial chord length of rotor
C_{a2}	axial chord length of stator
G	axial gap between rotor and stator
G_r	resonant axial gap between rotor and stator
h	hub to tip ratio
$k_\ell^{(n)}$	radial eigenvalue
ℓ	radial mode order
M_a	axial Mach number
N_1	number of rotor blades
N_2	number of stator vanes
n	circumferential wave number
p	disturbance pressure
r_H^*	inner radius of the annular duct (dimensional value)
r_T^*	outer radius of the annular duct (dimensional value)
W_a^*	axial flow velocity of undisturbed fluid (dimensional value)
ρ_0^*	undisturbed fluid density (dimensional value)
Ω_1^*	angular velocity of rotor rotation (dimensional value)
ω	reduced frequency of wake interaction

ROTOR-STATOR ACOUSTIC INTERACTION

Experiment

The cross-section sketch of the model fan is shown in Fig. 1. The fan is composed of 20 rotor blades and 30 stator vanes. In this case, even when the relative inlet flow at rotor tip is subsonic, the blade number/vane number ratio gives fundamental tone cut-on condition. The fan design parameters are shown in Table 1.

As seen in Fig. 1, the flow path downstream of the stator leading edge is composed of a cylindrical duct. At the stator trailing edge, the flow is still swirled with a tangential flow angle of about 15 degree to prevent high diffusion. Such flow path is not usual in a real machine, but it satisfies the assumption about the duct geometry used in the unsteady lifting surface theory.

The fan noise tests were conducted in the IHI anechoic test facility. Inflow Control Devices (ICD) was mounted on the fan duct to prevent additional noise sources caused by inflow turbulences. The test facility is not capable of measuring aft far field noise. Therefore aft propagating sound pressure was measured on the casing wall at the axial position downstream of the stator as shown in Fig. 1. The pressure transducers were mounted at 10 circumferentially spaced locations.

Figure 2 shows the variation in the measured SPL for the fundamental blade passing frequency (1BPF), the second harmonic blade passing frequency (2BPF) and the third harmonic blade passing frequency (3BPF) with the mechanical fan speed. The measured pressures at the aft outer casing wall were decomposed into the BPF components and transformed into a SPL for each BPF. The SPLs acquired at circumferentially 10 locations were arithmetically averaged.

The SPL of each BPF shows the same overall trend that the SPL increases with the fan speed, but the slope of the increment is different. The higher harmonic order BPF has the steeper slope. It can be observed that, on the underlying variation, there is an up-and-down behaviour in the variation, which is larger for a higher BPF. This behaviour cannot be explained by the ordinary mechanism of noise generation by interaction between rotor wakes and stator. The mechanism assumes that the amplitude of upwash velocity of the rotor wakes at the stator vane surface determines the unsteady vane loading, and as a consequence, the magnitude of the acoustic power levels of the propagating sound waves. However it cannot account for the up-and-down behaviours of the noise level.

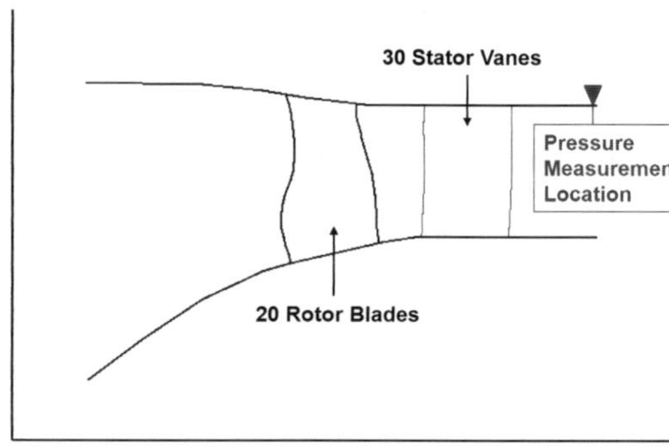

Figure 1. Cross section sketch of fan

Number of rotor blades	20
Number of stator vanes	30
Rotor tip speed (m/sec)	423
Stage pressure ratio	1.61
Tip radius at rotor inlet (m)	0.161
Hub to tip ratio at rotor inlet	0.43
Hub to tip ratio of stator inlet	0.535

Table 1. Design parameters of fan

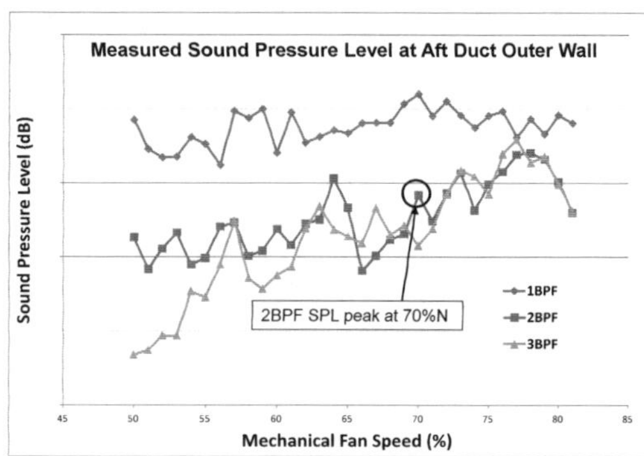

Figure 2. Measured SPL variation with the fan rotation speed

Analytical Investigation

The cause of up-and-down behaviour in the SPL variation was investigated by using the analytical method in which prediction can be conducted with and without acoustic interaction between fan rotor and stator. Figure 3 shows the fan stage model for the fan rotor-stator acoustic interaction analysis. The analytical model is a stage of blade rows composed of a rotor and a stator in an annular rigid walled duct of infinite axial extent. The fluid is assumed to be an inviscid and non-heat-conducting perfect gas. The oncoming undisturbed flow is of a uniform axial velocity at a Mach number $\boldsymbol{M_a}$ and a uniform density $\boldsymbol{\rho_0}^*$. It is further assumed that the rotor blades and stator vanes are of no lean, no sweep and no thickness, and therefore they have no steady loading. The unsteady disturbances are assumed to be small. In the figure, axial chord length of rotor and stator, $\boldsymbol{C_{a1}}$ and $\boldsymbol{C_{a2}}$ respectively, are normalized by the duct outer radius $\boldsymbol{r_T}^*$ and hub to tip ratio $\boldsymbol{h}$ is defined by the ratio of the duct inner radius $\boldsymbol{r_H}^*$ to the duct outer radius. The dimensionless angular velocity of rotor rotation corresponds to the rotor tip speed normalized by the axial velocity $\boldsymbol{W_a}^*$ ($\boldsymbol{\Omega_1} = \boldsymbol{\Omega_1}^* \boldsymbol{r_T}^* / \boldsymbol{W_a}^*$). Reduced frequency of wake interaction $\boldsymbol{\omega}$ is expressed as $\boldsymbol{\nu N_1 \Omega_1}$ (the fundamental blade passing frequency ($\nu = 1$) and higher harmonics ($\nu = 2,3,..$)). The upwash velocity of oncoming wakes is derived from the wake model of Schlichting (1955) in which the profile of the velocity defect is represented by cosine function.

In Figs. 4(a), 4(b) and 4(c), the predicted SPL variations with the fan speed are compared between conditions with (red line) and without (green line) rotor-stator acoustic interaction for 1BPF, 2BPF and 3BPF, respectively. The noise sources of the prediction without rotor-stator acoustic interaction are restricted to the unsteady vane loadings caused by hydrodynamic interaction with rotor wakes. Therefore the difference between the predictions with and without rotor-stator acoustic interaction is a measure to represent the influence of the acoustic interaction.

From the comparisons, it can be said that a chief cause of up-and-down variations observed in the measurements is the influence of the fan rotor-stator acoustic interaction. In the figures, the measured variations (blue line) are also plotted for comparison. There are observed definite differences in the noise level between the experiment and the prediction. However, it should be noticed that the predictions with the acoustic interaction reasonably capture the overall and local variations obtained in the measured SPL for the three BPF's. This suggests that the present analysis reasonably predicts the acoustic behaviours occurred in a real machine.

DEPENDENCE ON AXIAL GAP

In Figure 4, the prediction with rotor-stator acoustic interaction shows a steeper drop in the SPL from 50 to 51 percent rotation speed than the prediction without rotor-stator acoustic interaction for all the three BPF's. The measured SPL also shows a relatively steep drop from 50 to 51 percent rotation speed except for the 3BPF. From 71 to 72 percent speed, the predicted SPL with rotor-stator acoustic interaction shows an increase for all the three BPF's, while the predicted SPL without rotor-stator acoustic interaction shows rather a decrease. The trends with the fan rotation speed in the measurements agree with those of the predictions with rotor-stator acoustic interaction for all the three BPF's.

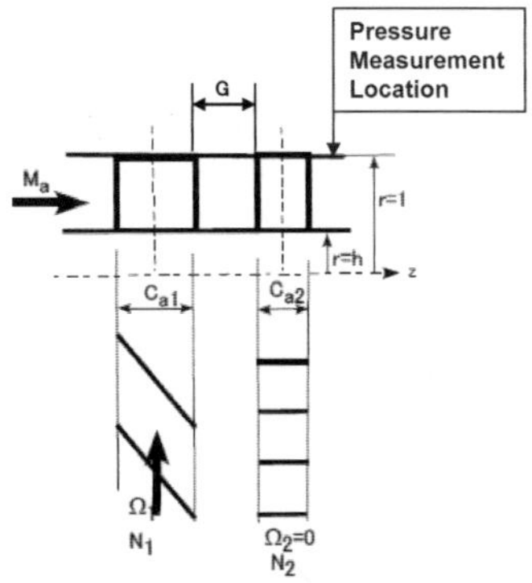

Figure 3. Fan stage analytical model

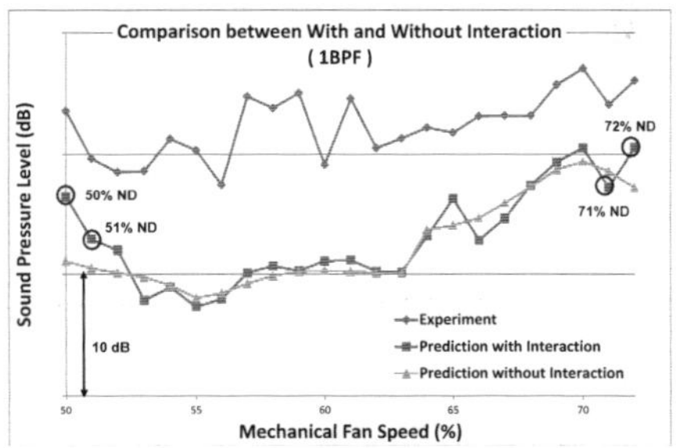

(a) 1BPF

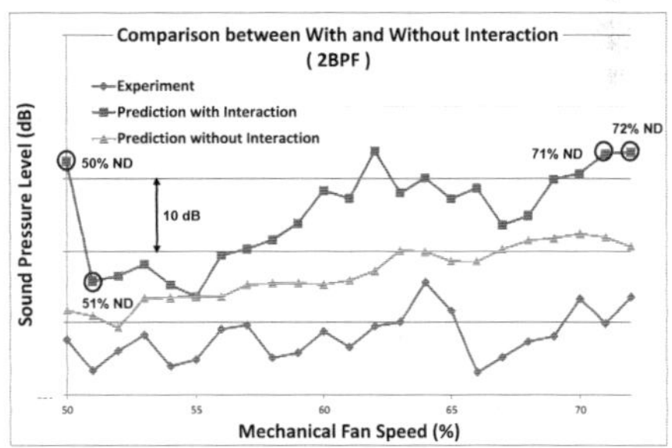

(b) 2BPF

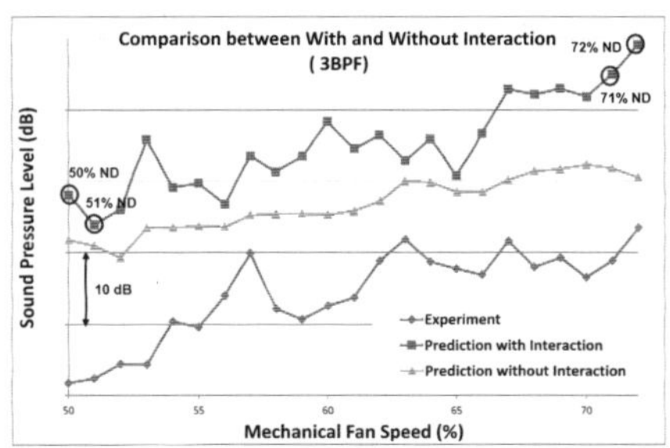

(c) 3BPF

Figure 4. Comparison between with and without interaction

We focus our attention to these two sets of the two adjacent fan rotation speeds in the following study of the contribution of the acoustic interaction to the dependence on the axial gap.

In the present study, the axial gap is varied parametrically while the other parameters are kept the same as those used in the previous study (Kodama et al. (2013)) for each fan rotation speed. We also use the same wake model as used in the previous study in which a wake decays with axial gap in accordance with the diffusion formula by Schlichting (1955). In a real machine, radial distribution of the circumferential phase of the wake varies with the axial gap. However, it is assumed to be constant in this study.

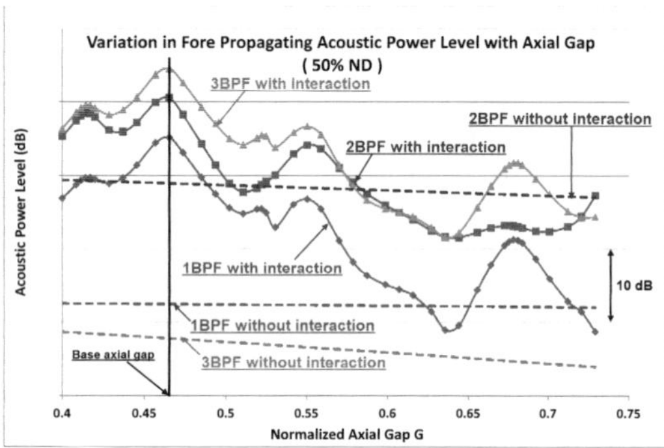

(a) Fore propagation

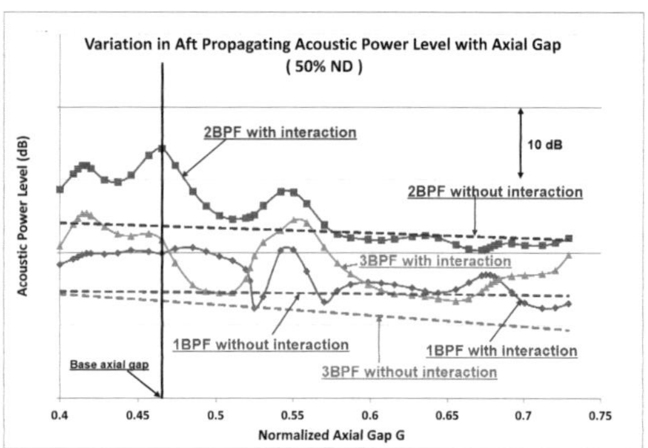

(b) Aft propagation

Figure 5. Variation in APL with axial gap at 50% rotation speed

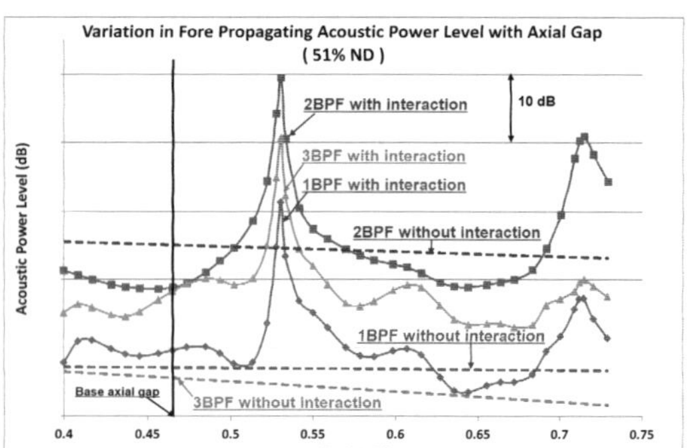

(a) Fore propagation

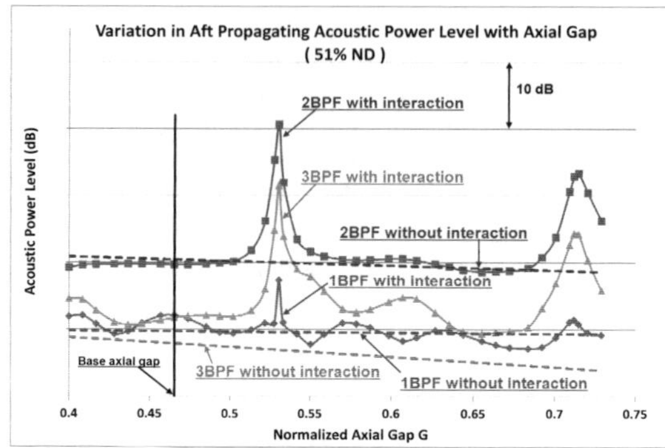

(b) Aft propagation

Figure 6. Variation in APL with axial gap at 51% rotation speed

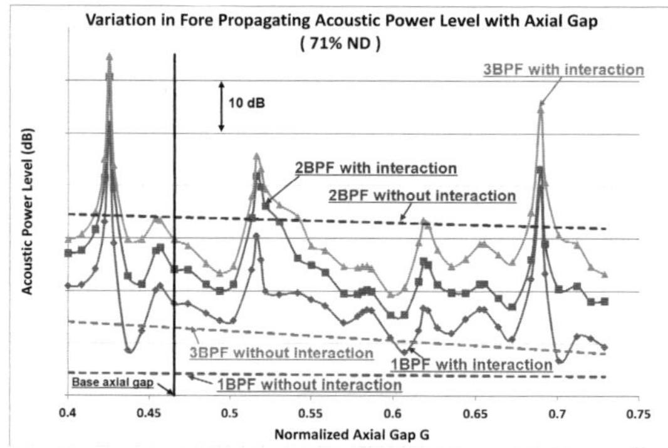

(a) Fore propagation

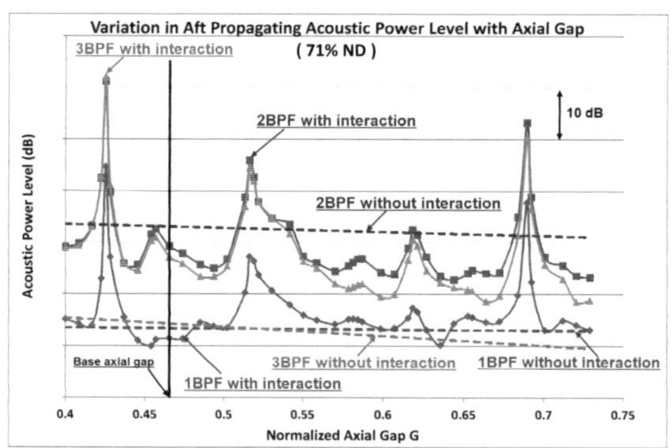

(b) Aft propagation

Figure 7. Variation in APL with axial gap at 71% rotation speed

Figures 5, 6, 7 and 8 show the predicted variations in acoustic power level (APL) with the axial gap for 1BPF, 2BPF and 3BPF at 50, 51, 71 and 72 percent fan rotation speeds, respectively. In each Figure, the APL's for fore propagation (a) and aft propagation (b) are shown. Here the axial gap G (shown in Fig. 3) is normalized by the duct outer radius, and the base axial gaps shown in the Figures are the values used in the predictions (Kodama et al. (2013)) for the comparisons with the experiments (see Fig. 4). There are two predictions for each BPF, one is the prediction with rotor-stator acoustic interaction (solid line) and another is that without interaction (dotted line). Each prediction without interaction indicates a monotonous variation in accordance with decay in the amplitude of upwash velocity of the wake at stator vane location. On the other hand, the predictions with the rotor-stator acoustic interaction show a large variation independent of the wake amplitude. Also the significant difference in the overall levels of the APL can be seen between the predictions with and without interaction for each BPF. This is attributed to a scattering of one BPF component into other BPF components as a result of coupling among the BPF's in the rotor-stator acoustic interaction.

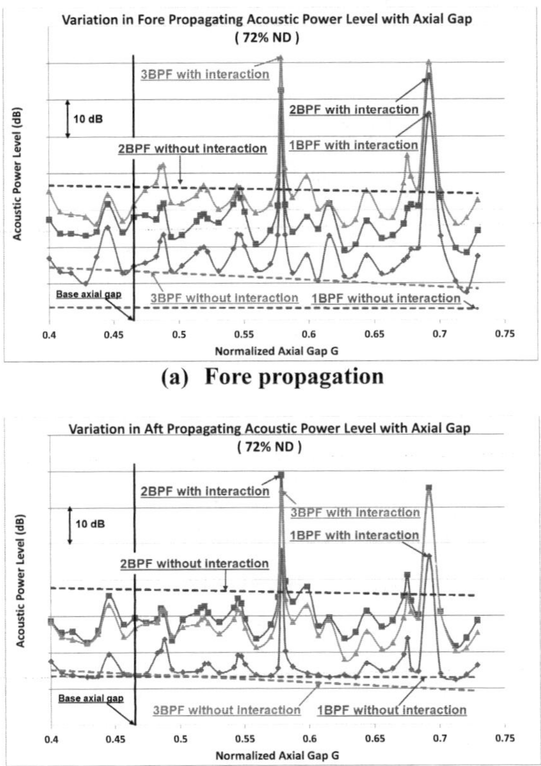

(a) Fore propagation

(b) Aft propagation

Figure 8. Variation in APL with axial gap at 72% rotation speed

As seen in Fig. 5(a), in the fore propagating APL at 50 percent rotation speed, a similar variation can be seen for the three BPF's, in which round peaks appear at certain axial gaps. The base axial gap coincides with one of the round peaks. A cycle of increase and decrease in APL is repeated with the maxima declining with increasing axial gap. In the aft propagating APL at 50 percent rotation speed (Fig. 5(b)), some round peaks can be also seen at the same axial gaps, but the variations differ among the BPF's.

The predicted APL with the interaction at 51 percent rotation speed (Fig. 6) shows the same features as seen in Fig. 5, in which the peaks appear at certain axial gaps for the three BPF's. However, the variations themselves are quite different from those in Fig. 5 despite the slight change in the rotation speed.

The axial gaps at which the peaks exist are quite different from those at 50 percent rotation speed. The base axial gap is far away from the gaps giving the peaks. This explains the reason why a big drop in the APL variation appear from 50 to 51 percent rotation speed in Fig. 4, especially for 2BPF. In the case of 51 percent fan rotation speed, the APL variations with axial gap are similar among the BPF's and also between the fore and the aft propagation in contrast to those at 50 percent rotation speed.

At 71 percent rotation speed, significantly high peaks appear in the APL variations with the axial gap as seen in Fig. 7(a) and 7(b), and some low peaks exist between the high peaks. The variations themselves are similar among the BPF's and also between the fore and the aft propagation.

The same features as seen in Fig. 7 are observed in the APL variation with axial gap at 72 percent rotation speed (Fig. 8(a) and (b)). However, again the axial gaps at which the peaks exist are quite different from those at 71 percent rotation speed despite the slight change in the rotation speed. In the case at 72 percent rotation speed, there appear a larger number of low peaks between the high peaks than at 71 percent rotation speed. These differences result in the different trends in the APL variation from 71 to 72 percent rotation speed between the predictions with and without rotor-stator acoustic interaction.

SOUND AMPLIFICATION IN A DUCT BETWEEN ROTOR AND STATOR

As seen in Figs. 5 through 8, some of the peaks at certain discrete axial gaps are significantly high. The levels are towering by 20 to 30 dB. This reminds us of a resonator that shows large amplitude response to excitation at the resonance frequencies. It might be possible that the duct between the fan rotor and the stator plays a role like a resonator when the axial gap corresponds to a multiple of an axial wave length of an acoustic mode.

Axial Wave Length of Duct Modes at BPFs

Axial wave length normalized by the duct radius of a duct mode of circumferential wave number $\boldsymbol{n}$ and radial mode order $\boldsymbol{l}$ at a frequency $\omega = \nu N_1 \Omega$ is given by

$$\lambda_{\pm}(\nu, n, l) = \frac{2\pi}{\omega \dfrac{M_a^2}{\beta_a^2} \mp \sqrt{\omega^2 \dfrac{M_a^2}{\beta_a^4} - \dfrac{k_l^{(n)2}}{\beta_a^2}}} \tag{1}$$

Here

$$\beta_a^2 = 1 - M_a^2 \tag{2}$$

and $\boldsymbol{k_l^{(n)}}$ is the radial eigenvalue.

A resonator would maximize the amplitude of acoustic wave when an axial wave length of the duct mode ($\boldsymbol{n}$, $\boldsymbol{l}$) satisfies the following equation.

$$G_r = m\left(\frac{\lambda_{\pm}(\nu,n,l)}{2}\right) \quad (3)$$

$m = 0, \pm 1, \pm 2, \pm 3, \cdots$

Subscript

+ : propagating to downstream

− : propagating to upstream

Here G_r is resonant axial gap normalized by the duct outer radius.

Calculation of Resonant Axial Gap

For all of the cut-on acoustic modes of 1BPF, 2BPF and 3BPF in the duct between fan rotor and stator, the resonant axial gaps that satisfy Equation (3) are calculated and compared with the axial gaps at which sharp and round peaks appear in the APL variation.

In Figs. 9, 10, 11 and 12, the resonant axial gaps which are close to the sharp and round peaks are shown with vertical lines in the variation in fore propagating APL at 50, 51, 71 and 72 percent rotation speeds, respectively. In the Figures, 'Up' and 'Down' denote upstream propagating and downstream propagating sound waves, respectively. It can be seen that at least one resonant axial gap always exists near the peaks. This indicates that the peaks might be caused by a similar mechanism to the acoustic response in a resonator.

There are many resonant axial gaps for both upstream propagating and downstream propagating acoustic waves in the range of G from 0.4 to 0.73. However, the amplification of the sound wave does not always occur at each resonant axial gap. The level of the amplification or attenuation will essentially be dependent on the interaction behaviours of the sound waves with the rotor blades and the stator vanes. In particular interaction with the moving rotor blades will be highly influential from the standpoint of sound energy modification.

As seen in Figs. 9 through 12, the resonant axial gaps which are close to the peaks are those of the duct modes for 3BPF $(\nu = 3)$. However we should note that the sound waves of the other BPF's are also amplified at the same axial gaps. It will be because, due to coupling among different frequency components in the rotor-stator acoustic interaction, the sound waves of the other BPF's are also influenced.

In Fig. 10 for the fore propagating APL at 51 percent rotation speed, two of the three resonant axial gaps are those of the duct modes of circumferential wave number 0 at 3BPF $(\nu = 3)$, but the remaining one resonant axial gap is that of the duct mode of circumferential wave number 20 at 2BPF. In spite of the resonant gap of 2BPF, however, there is almost no amplification for the 2BPF at this axial gap. It will mean that the interaction of the 2BPF sound waves with the rotor blades results in a scattering of the 2BPF component into other frequency components. The other reason will be that the sound level of 2BPF is reduced during the transmission through the fan rotor.

In general, the number of resonant axial gaps increases as the number of the cut-on modes increases. In the range of the present computation, we have no change of the number of cut-on modes between 71 and 72 percent rotation speeds. Nevertheless, however, the number of high and low peaks increases from 71 to 72 percent rotation speed as seen in Figs. 11 and 12. This will be because the interaction behaviours for each cut-on mode are substantially changed by the change from 71 to 72 percent rotation speed.

It seems that the appearance of the peaks is greatly influenced by the interaction behaviours of the sound waves with the blade rows, and the behaviours, e.g., the phase condition of the sound waves at the interfaces of the blade rows, seem to be very sensitive to the flow conditions. Therefore the APL variation with axial gap is significantly changed by a slight change in the rotation speed.

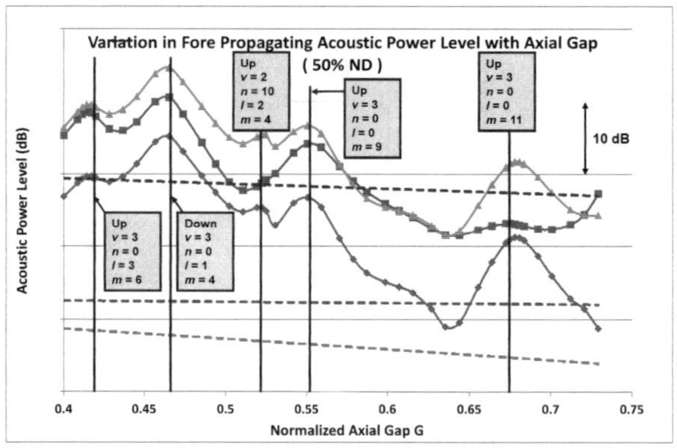

Figure 9. Relation between resonant axial gaps and the peaks in the variation in fore propagating APL (50% rotation speed)

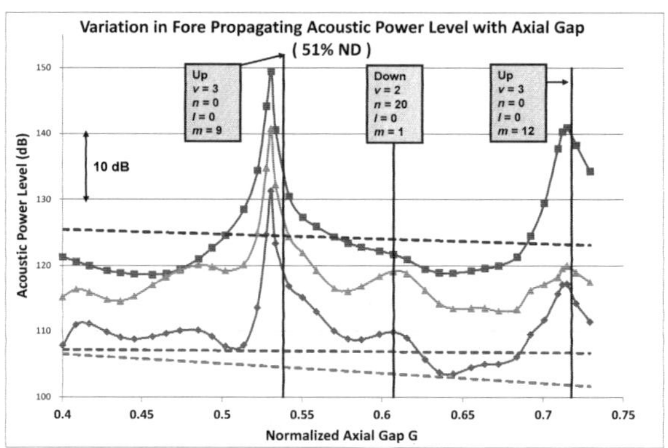

Figure 10. Relation between resonant axial gaps and the peaks in the variation in fore propagating APL (51% rotation speed)

Row-to-Row Acoustic Interaction Analysis with Restricted Frequency and Mode Coupling

So far in the present study, we have dealt with the three BPF's, 1BPF, 2BPF, and 3BPF under calculation with frequency and mode coupling up to third harmonics in the row-to-row acoustic interaction. As a result of the coupling, the APL variation with axial gap becomes highly complex. In order to examine the evidence of the complexity arising from the coupling, we conduct a row-to-row acoustic interaction analysis by computation with restricting the highest harmonic order in coupling.

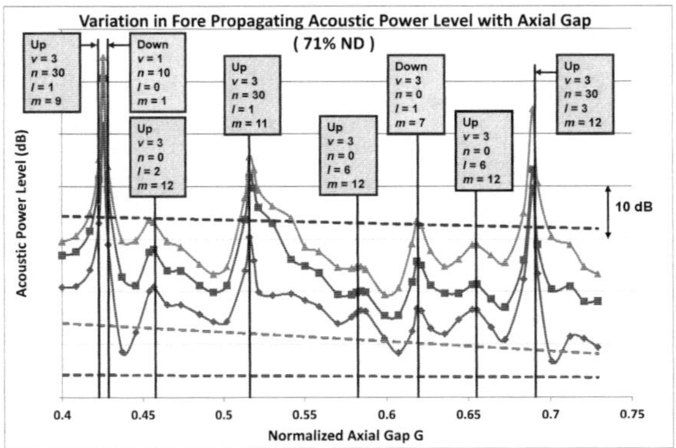

Figure 11. Relation between resonant axial gaps and the peaks in the variation in fore propagating APL (71% rotation speed)

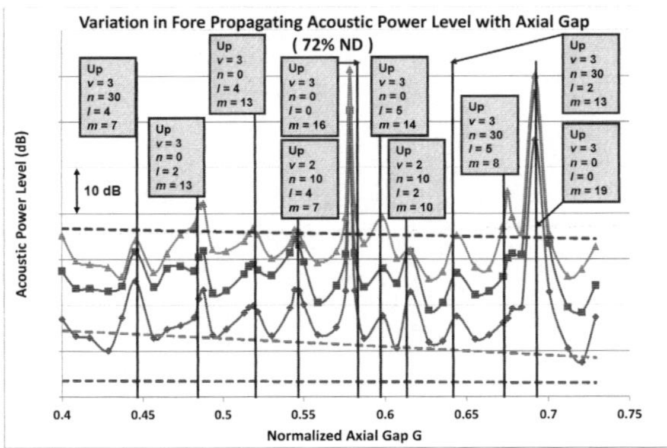

Figure 12. Relation between resonant axial gaps and the peaks in the variation in fore propagating APL (72% rotation speed)

Figure 13 compares the fore propagating APL variations with axial gap for the 1BPF at 70 percent rotation speed between the interaction analysis with coupling of the three BPF's and that with restricted coupling of only 1BPF (no scattering from higher harmonics). The low and high peaks seen in the analysis with the three BPF's disappear in the analysis with only 1BPF, because there is no influence of the resonant duct modes of 3BPF $(\nu = 3)$ shown in Fig. 11. The resonant axial gaps of both 1BPF and 3BPF exist near the high peak at the axial gap of 0.426 as seen in Fig. 11. This peak is considered to be attributed to the resonant duct mode of 3BPF $(\nu = 3)$, because there is no peak in the analysis with only 1BPF in Fig. 13.

In Fig. 14, the predicted APL variations with axial gap by row-to-row acoustic interaction analysis with restricted frequencies to 1BPF and 2BPF (coupling up to the second harmonics) are shown for the fore propagation. The analysis is conducted for the fan test condition at 72 percent rotation speed. In the variations, the peaks which correspond to the resonant axial gaps of the duct modes for 3BPF disappear and those which correspond to the duct modes for 2BPF remain. This suggests that the root cause of the peaks appeared in the APL variation with axial gap is sound amplification of a resonant mode of one of the BPF's and the other BPF sound waves are substantially influenced by the resonant mode through the row-to-row acoustic interaction.

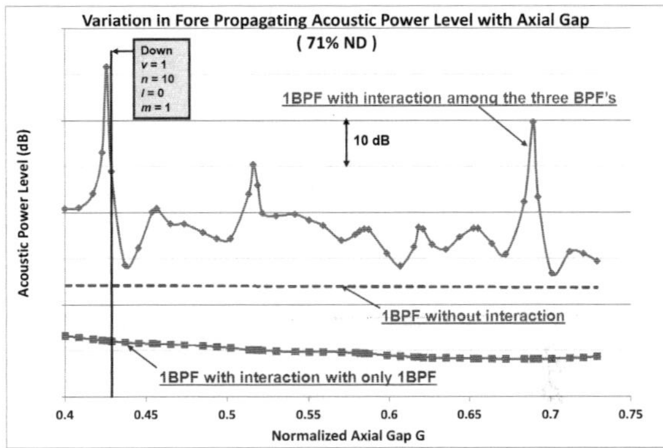

Figure 13. Row-to-row acoustic interaction analysis with only 1BPF

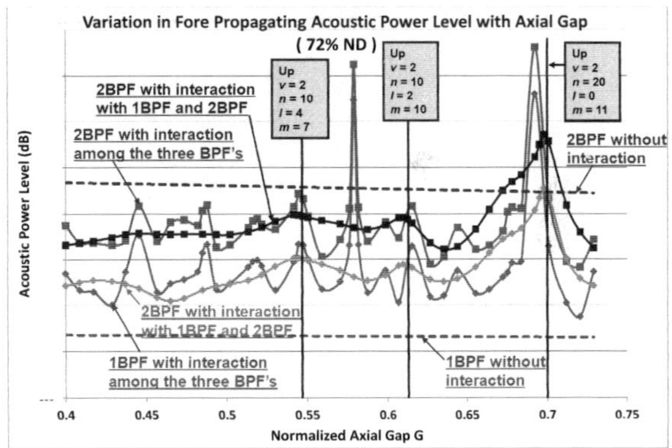

Figure 14. Row-to-row acoustic interaction analysis with coupling of 1BPF and 2BPF

CONCLUSION

The dependence of the fan tone noise on the axial gap between fan rotor and stator is investigated by using a three-dimensional lifting surface theory for a prediction of rotor-stator acoustic interaction.

It is found that, when the rotor-stator axial gap is varied parametrically, round and sharp peaks appear in the variation in both fore and aft propagating acoustic power levels at certain axial gaps. With a slight change in a flow condition, the peaks appear in quite different ways.

The relation between the axial gaps of the sound level peaks and the axial wave lengths of cut-on duct modes suggests that the cause of the peaks will be amplification of the sound waves due to a similar mechanism to the acoustic response in a resonator.

REFERENCES

Kantola, R.A. and Gliebe, P.R., "Effects of Vane/Blade Ratio and Spacing on Fan Noise," AIAA-81-2033, 1981

Kodama, H., Kusuda, S., Ooishi, T. and Namba, M., "Application of Unsteady Three-dimensional Lifting Surface Theory for Coupled Rotor/Stator Cascades to Fan Noise Prediction System," International Symposium on Air Breathing Engines, ISABE-2009-1144, 2009

Kodama, H., Kusuda, S., Ooishi, T., and Namba, M., "Experimental and Analytical Investigation of Acoustic Reflection and Scattering for Fan Tone Noise," *International Symposium on Air Breathing Engines*, ISABE-2013-1313, 2013.

Namba, M. and Nishino, R., "Unsteady Aerodynamic Response of Oscillating Contra-Rotating Annular Cascades, Part 1: Description of Model and Mathematical Formulations," *Transaction of Japan Society for Aeronautical and Space Sciences of Sound and Vibration*, Vol. 49, No. 165, pp. 181-186, 2006.

Schlichting, H., "Boundary Layer Theory," Chapter 24, McGraw-Hill, 1955.

Woodward, R.P. and Glaser, F.W., "Effects of Blade-Vane Ratio and Rotor-Stator Spacing on Fan Noise with Forward Velocity," AIAA-81-2032, 1981

Proceedings of the 14th International Symposium on Unsteady Aerodynamics, Aeroacoustics & Aeroelasticity of Turbomachines
ISUAAAT14
8-11 September 2015, Stockholm, Sweden

I14-S13-2

COMPUTATION OF THE RADIATED NOISE OF A CONTROLLED-DIFFUSION AIRFOIL VIA THE HYBRID METHODS OF AEROACOUSTICS

RABEA MATOUK
UNIVERSITE LIBRE DE BRUXELLES
rabea.matouk@ulb.ac.be
Brussels, Belgium

GERARD DEGREZ
UNIVERSITE LIBRE DE BRUXELLES
gdegrez@ulb.ac.be
Brussels, Belgium

ABSTRACT

The accurate prediction of the broadband or trailing-edge noise and the determination of its sources are essential to reduce fans and propellers noise. This noise component is due to the scattering of the turbulent boundary layer into acoustic waves by the trailing edge. In this paper, the noise emanating from a CD (Controlled-Diffusion) airfoil, belonging to an automotive fan, is simulated and computed via the hybrid methods of aeroacoustics. In these methods, the aerodynamic and acoustic fields are computed separately. The flow data are obtained using the in-house LES solver SFELES [1]. ACTRAN acoustic solver [2] has been used to solve the acoustics and to provide the near and far fields propagation via Lighthill [3] and Mohring [4] analogies. Curle's analogy [5] is applied as well in its integral compact formulation which takes the presence of walls into account. Curle's formulation is applied proposing two approaches; the first approach is to implement the volume and surface integrals in SFELES and to calculate them simultaneously with the flow in order to avoid the storage of noise sources which requires a huge space. In the second approach, the fluctuating aerodynamic forces, already obtained during the aerodynamics simulation, are used to compute the noise considering just the surface sources. As results, tonal noise that arises from vortex shedding generated by laminar boundary layer is found the main acoustic source for the laminar flow where a dominant peak is found at the frequency of the vortex shedding (369 Hz). Turbulent boundary layer and wake are the more efficient sources for the turbulent flow around the airfoil. Volume sources contribution (quadrupole) to the overall radiated noise is found not to be negligible even if for low Mach number whereas surface sources (dipole) are dominant for the studied airfoil. The convectional effects studied via Mohring's analogy are important for mid and high frequencies, the effect up to 5dB at the peak corresponding to the frequency 1360 Hz for the laminar flow case. Curle's analogy is found a simple and reliable formulation and can give a fast idea of the airfoil radiated noise without any additional sources cost.

1. INTRODUCTION

Hybrid methods of aeroacoustics consist in two-steps simulations, where the determination of the aerodynamic field is decoupled from the computation of the acoustic waves' propagation as shown in Fig. 1.

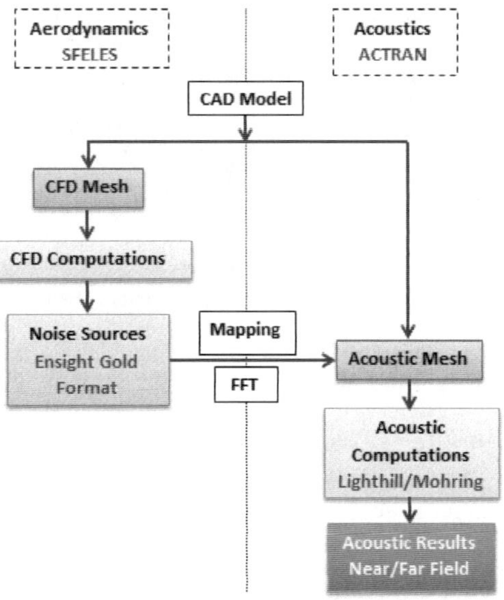

Figure 1: Hybrid approach of aeroacoustics

This decoupling gives the possibility to adapt the numerical techniques to the constraints of each computational step, such as the mesh refinement and boundary conditions. So the first step in this methodology is to determine a space-time evolution of a turbulent aerodynamic field from a solution of Navier-Stokes equations, either by direct calculation (DNS), Large Eddy Simulation (LES), used in this study, or unsteady Reynolds Averaged Navier-Stokes (URANS). Pressure and/or velocity fluctuations are stored and then used to build the acoustic source terms, according to the used analogy, which are integrated into a propagation model, leading to the radiated noise in the far field. These hybrid approaches still lack information on the interactions between the aerodynamic and acoustic fields. The effect of sound waves on the fluid flow and the potential couplings between the aerodynamic and the acoustic fields are neglected due to the large difference in energy content where the feedback from acoustics to the flow is very small for subsonic flows. In this study the refraction effects of the mean flow on the acoustic waves propagation is taken into account through Mohring's analogy.

2. AEROACOUSTICS ANALOGIES APPLIED IN THIS STUDY:

2.1 LIGHTHILL'S ANALOGY:

Lighthill's analogy is derived from the Navier-Stokes equations without any approximations. Morover, it is an exactly valid equation, which admits an acoustic interpretation.

The acoustic domain is decomposed into a flow region containing the source region in which the turbulent fluctuations generate the sound and an acoustic quiescent and uniform region where the acoustic waves propagate to the far field listener's position as shown in Fig. 2.

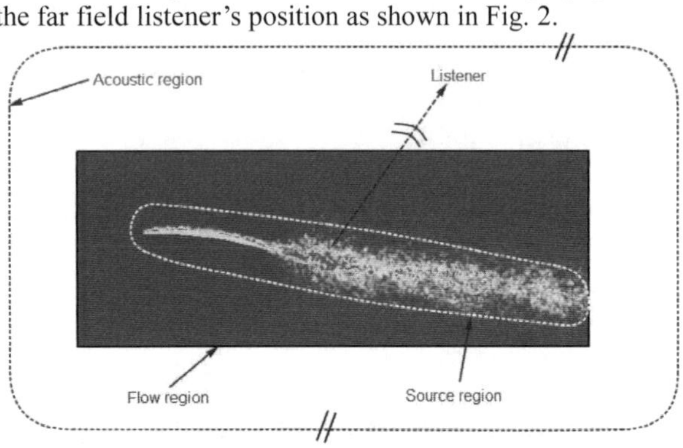

Figure 2: Regions in an aeroacoustic problem

The final formulation of Lighthill's analogy is:

$$\left\{\begin{array}{c} \frac{\partial^2 \rho_a}{\partial t^2} - c_0^2 \nabla^2 \rho_a = \frac{\partial^2 T_{ij}}{\partial x_i \partial x_j} \\ T_{ij} = \rho u_i u_j + [(p - p_0) - c_0^2(\rho - \rho_0)]\, \delta_{ij} - \tau_{ij} \end{array}\right\} \quad (1)$$

Where $\rho_a = \rho - \rho_0$, the acoustic density and T_{ij} is known as Lighthill's tensor. In this formulation it is clear that outside the flow region T_{ij} is null. In numerous cases, Lighthill's tensor's expression can be simplified as in this study: for high Reynolds number, the viscous stress tensor τ_{ij} is negligible in comparison with the Reynolds stresses $\rho u_i u_j$ because viscous effects are much smaller than inertial effects. The term $[(p - p_0) - c_0^2(\rho - \rho_0)]\, \delta_{ij}$ is related to entropy variations inside the source region which is zero for isentropic flows. Lighthill's tensor is eventually reduced for low Mach number flow to: $T_{ij} = \rho u_i u_j$. It corresponds to quadrupole sources; this term has just been considered in this study. For the complete derivation, the reader is referred to [3].

2.2 MOHRING'S ANALOGY:

Mohring's analogy is based on the stagnation enthalpy B, which obeys a linear convected wave equation, as an acoustic variable. Starting from Navier-Stokes equations (Crocco's form) and using the energy conservation equation and other thermodynamic relations, it is arrived to the following scalar formulation:

$$\frac{\partial}{\partial t}[\frac{\rho}{\rho_T^2 c^2}\frac{\partial b}{\partial t} + \frac{\rho v_i}{\rho_T^2 c^2}\frac{\partial b}{\partial x_i}] + \frac{\partial}{\partial x_i}[\frac{\rho v_i}{\rho_T^2 c^2}(\frac{\partial b}{\partial t} + v_i \frac{\partial b}{\partial x_i}) - \frac{\rho}{\rho_T^2}\frac{\partial b}{\partial x_i}] = R$$

$$R = -\frac{\partial}{\partial x_i}\left[\frac{\rho}{\rho_T}(\vec{v} \times \vec{\omega})_i - \frac{\partial \tau_{ij}}{\partial x_j}\right] + \frac{\partial}{\partial x_i}\left[\frac{v_i}{\rho_T}\left(\frac{\partial \rho}{\partial s}\right)_p \frac{\partial s}{\partial t} - \frac{\rho T}{\rho_T}\frac{\partial s}{\partial x_i}\right]$$
$$+\frac{\partial}{\partial t}\left[\frac{1}{\rho_T}\left(\frac{\partial \rho}{\partial s}\right)_p \frac{\partial s}{\partial t} - \frac{\rho v_i}{\rho_T^2}\frac{\partial \rho_T}{\partial x_i}\right] \quad (2)$$

With v_i the mean velocity, ρ the mean density, c the sound speed, $\vec{\omega} = \nabla \times \vec{v}$ the vorticity, ρ_T the total density and b the scaled total enthalpy fluctuations defined by: $\frac{Db}{Dt} = \rho T \frac{DB}{Dt}$, where $B = h + \frac{v_i v_i}{2}, \frac{D}{Dt} = \frac{\partial}{\partial t} + v_i.\nabla$

In the equation (2) the left-hand side corresponds to an acoustic wave operator in the presence of a heterogeneous flow. The right-hand side is considered as acoustic sources where it contains the flow fluctuations represented by three terms. The first term represents the turbulent noise and the other two terms represent the combustion noise. Neglecting the viscous effects and the total density fluctuations, for an isentropic flow the sources related to turbulent noise are eventually reduced to: $R = -\frac{\partial}{\partial x_i}\left[\frac{\rho}{\rho_T}(\vec{v} \times \vec{\omega})_i\right]$. As mentioned, the acoustic variable in this analogy is the fluid stagnation enthalpy B which is related to the acoustic pressure via the energy equation: $\frac{DB}{Dt} = \frac{1}{\rho}\frac{Dp_a}{Dt}$. The application field of this analogy is more general than Lighthill's analogy especially in the case of convectional effects cannot be neglected; they may cause refraction of the acoustic waves.

2.3 CURLE'S INTEGRAL FORMULATION:

Curle extended the Lighthill's analogy in order to incorporate the influence of the presence of solid boundaries. Therefore, the overall radiated sound field is here a contribution of three origins:

1. Quadrupole sources (Lighthill), which are present in the vicinity of the solid boundaries and due to the turbulence such as the dispersion of turbulent boundary layer in the wake of a body, an airfoil or a cylinder etc.
2. Dipole sources, generated by fluctuating aerodynamic forces on the solid boundaries, acting on the fluid.
3. Monopole sources, due to the mass flux through the surface S.

The acoustic field is often calculated by the contribution of the dipole sources only, and neglecting the quadrupole sources due to its high costs and small contribution. However, in this study the contribution of the quadrupole sources is taken into account. Curle's formulation is given by the following relation:

$$\rho(X,t) = \frac{1}{4\pi c_0^2}\frac{\partial^2}{\partial x_i \partial x_j}\int_V \frac{T_{ij}}{r} dV(V) - \frac{1}{4\pi c_0^2}\frac{\partial}{\partial x_i}\int_S \frac{n_j}{r}(p\delta_{ij} - \tau_{ij}) dS(y) \quad (3)$$

Changing the spatial derivatives into temporal and applying $p' = \rho * c_0^2$ yield in a far-field approximation:

$$p'(X,t) = \frac{x_i x_j}{4\pi \|X\|^2 c_0^2}\frac{\partial^2}{\partial t^2}\int_V [\frac{T_{ij}}{r}]_{t_e} dV(V) -$$
$$\frac{1}{4\pi}\frac{\partial}{\partial t}\int_S [\frac{\rho v_i}{r}]_{t_e} n_i dS(y) -$$
$$-\frac{x_j}{4\pi \|X\| c_0}\frac{\partial}{\partial t}\int_S [\frac{P_{ij+\rho v_i v_j}}{r}]_{t_e} n_i dS(V) \quad (4)$$

With $P_{ij} = p\delta_{ij} - \sigma_{ij}$ is the stress tensor. Assuming that the source region is compact, the retarded time t_e can be neglected and $r = |x|$ can be written if the origin is $y = 0$ inside the body. This assumption leads for our study to:

$$L < \lambda \rightarrow chord < \frac{c_0}{f_{max}} \rightarrow f_{max} < 2507 \text{ [Hz]}$$

Where L is a typical dimension of the solid boundaries (chord of the airfoil $C = 0.1356\ m$ in our case), λ is the sound wave length. This assumption is quite suitable for frequencies of interest (less than 2500 [Hz]). Furthermore, the second integral refers to a monopole-like sound field, can be omitted because the studied airfoil is not permeable. The viscosity term σ_{ij} will not be considered further. It is also assumed that the position of the receiver is large compared to L (chord) which means far field sound. After all these simplifications, the relation is eventually written:

$$p'(X,t) \cong \frac{x_i x_j}{4\pi\|X\|^3 c_0^2}\frac{\partial^2}{\partial t^2}\int_V [T_{ij}]dV(Y) - \frac{x_j}{4\pi\|X\|c_0}\frac{\partial}{\partial t}\int_S pn_i dS \quad (5)$$

3. METHODOLOGY:

3.1 THE APPLICATION OF LIGHTHILL AND MOHRING ANALOGIES BY SFELES/ACTRAN:

In order to extract the primary variables (velocity and/or pressure) during the aerodynamic simulation to use them for building the Lighthill and Mohring sources via ACTRAN, the transient format Ensight Gold [7], supported by ACTRAN, is implemented in SFELES. This format builds 3-D fields and geometry from the 2-D modes solutions and the original CFD mesh. Figure 1 shows the processes applied for the complete hybrid approach.

First, the aerodynamic simulations have been carried out, on the well refined mesh ($y^+ = 1$) shown if Fig. 3, using the SFELES solver. When the simulation arrived to the statistically converged solution, the velocity components are extracted in the Ensight Gold format on every point of the reconstructed 3-D CFD mesh for more than 5 airfoil flow-through times (almost 0.1 sec) to cover frequencies until 2500 [Hz]. Second, the sources are projected from the CFD mesh on the acoustic mesh shown in Fig. 4 using the integration method which conserves the sources by integrating them on the acoustic mesh elements [8]. This is performed by a tool combined with Actran called iCFD. Then the Lighthill and Mohring volume source terms are computed in the time domain and a Fourier transform (DFT) is applied to change them to frequency domain.

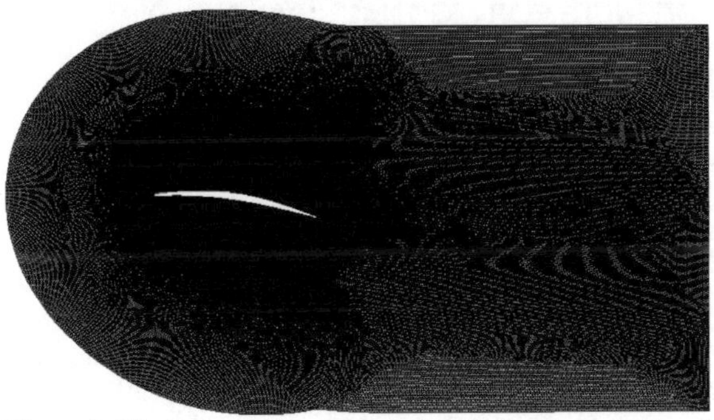

Figure 3: The 2-D aerodynamic mesh with the computational domain

Figure 4: The 3-D acoustic mesh

Third, the acoustic simulation is performed via ACTRAN to compute the radiated noise, represented by the sound pressure level (SPL) in the near and the far fields. The SPL in far field is computed using the Infinite Elements [9, 10] which provide a direct numerical estimate of the acoustic pressure at all points in the outer domain. ACTRAN uses them to model the unbounded acoustic domain and to prohibit the reflection of the acoustic waves into the finite element domain; they act as a non-reflective boundary condition.

The acoustic mesh is smaller in dimensions and coarser than the CFD mesh, the criterion 6 quadratic elements by an acoustic wave is taken in consideration basing on the maximum required frequency to be solved (2500 Hz).

3.2 THE APPLICATION OF CURLE'S ANALOGY:

Two approaches have been proposed to apply Curle's integral formulation.

The first approach is the implementation of volume and surface source terms of the relation (5) in SFELES. They are calculated simultaneously with the flow in order to avoid the storage of noise sources. The second using the fluctuating aerodynamic forces directly to compute the noise considering just the surface sources of the equation (5) as the following:

$$p'(X,t) \cong -\frac{x_j}{4\pi\|X\|c_0}\frac{\partial}{\partial t}\int_S pn_i dS$$

$$\frac{\partial}{\partial t}\int_S pn_i dS = \frac{\partial F_i}{\partial t}(Y,t) \Rightarrow p'(X,t) \cong -\frac{x_j}{4\pi c_0 r^2}\frac{\partial F_i}{\partial t}(Y,t)$$

$$p'(X,t) \cong -\frac{1}{4\pi c_0 r^2}[x.\frac{\partial D(t)}{\partial t} + y.\frac{\partial L(t)}{\partial t}]$$

$x = r.\cos(\theta), y = r.\sin(\theta)$. r and θ are the receiver coordinates. In this study the receiver is placed in the mid-span plane above the trailing edge ($r = 2\ m$ and $\theta = 90^\circ$).

The aerodynamic forces are written:

$D(t) = 0.5 * \rho\ * U_\infty^2 * S * C_d(t)$: The drag force.

$L(t) = 0.5 * \rho\ * U_\infty^2 * S * C_L(t)$: The lift force.

The temporal history of the aerodynamics coefficients C_d and C_L is already stored during the aerodynamic simulation so this relation is applied by a simple post-processing procedure by Matlab. The aerodynamic coefficients consist of two parts, one related to the pressure and the other to the viscosity which is neglected here.

$C_d(t) = C_{dp}(t) + C_{dv}(t)$ and $C_L(t) = C_{Lp}(t) + C_{Lv}(t)$

4. SPATIAL FILTER:

Truncation of the Lighthill and/or Mohring source terms at the domain boundaries can create spurious sound especially at low frequencies. This is called truncation phenomenon which pollutes the acoustic field with artificial sources and may causes an overprediction of the acoustic pressure field.

Figure 5 (above) presents this phenomenon obtained in our study for the laminar case where an artificial dipole has appeared near the domain outlet for the frequency 369 Hz.

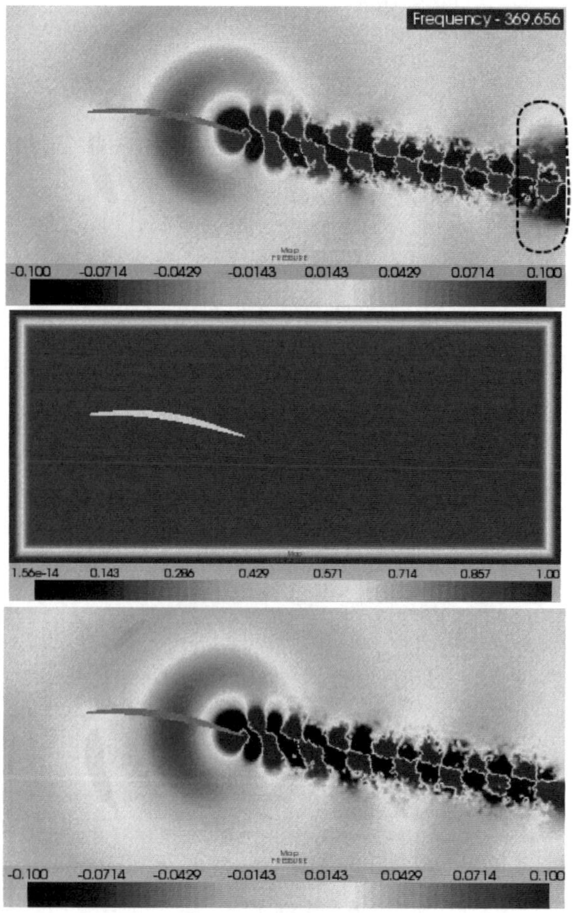

Figure 5: (Above) Truncation phenomenon at the outlet for the frequency 369 Hz, (middle) the applied Cosine Filter, (bottom) the same acoustic field after the application of the filter

In order to prevent this phenomenon, a spatial filter of type cosine is applied near the borders. The source terms are damped out slowly toward the sources boundaries by the applied filter as shown in Fig. 5 (middle). In this Figure an amplitude value of 1 (red color) means that the sources in that area are 100% taken into account whereas 0 value (blue color) means that the area is not taken into account at all. The result of the application of is filter is shown in Fig 5 (bottom), it is clear that the artificial dipole has been removed from the acoustic field. The spatial filtering is applied in all acoustic simulations performed via ACTRAN in this paper.

5. THE PERFORMED SIMULATIONS:

Two cases have been addressed in this study; the first one is a laminar flow at a Reynolds number of 12000 and the second is a turbulent flow at a Reynolds number of 160000 as shown is Table 1.

Case	Reynolds number	Fourier modes	Flow regime	Time step (sec)	Acoustics results
1	12000	1 (2d)	Laminar	10^{-5}	**converged**
2	160000	64 (3d)	Turbulent	10^{-5}	**under development**

Table 1: Performed simulations

As mentioned before, SFELES solver is used. SFELES is a hybrid spectral/finite elements code capable of simulating 3-D unsteady incompressible viscous flows over axisymmetric and planar geometries. It is a large eddy simulation solver uses the stabilized Galerkin finite element discretization in the plane and a spectral discretization in the periodic transverse direction. The three analogies have been applied for the two cases.

However, detailed acoustics results will be presented for the laminar flow case whereas for the turbulent flow case, preliminary results will just be presented in this paper considering Lighthill's analogy for the near and far field to show the noise generation mechanism. That is because the sources need to be improved to obtain converged results for the far field acoustic pressure spectra for high frequencies.

6. RESULTS FOR LAMINAR CASE STUDY AT REYNOLDS 12000:

The aerodynamic simulation has been carried out at a Reynolds number of 12000 and at a Mach number of 0.047 in order to compute the tonal noise corresponding to the Von Karman vortex shedding shown in Fig. 6.

Figure 6: Vorticity contours (15-50) colored by longitudinal velocity

6.1 RESULTS FOR LIGHTHILL'S ANALOGY:

Lighthill's analogy has been applied according to the methodology shown in Fig. 1 and explained in section 3.1. Lighthill's sources are computed on the acoustic mesh in the time domain and shown in Fig 7. Figures 8, 9 and 10 show the Lighthill's sources for the frequencies 369 Hz, 628 Hz and 731 Hz respectively. These frequencies have been chosen to show that the sources corresponding to the dominant frequency 369 Hz, which is the Von Karman vortex shedding frequency and its harmonics, are strong and important whereas for other frequencies the sources are weak.

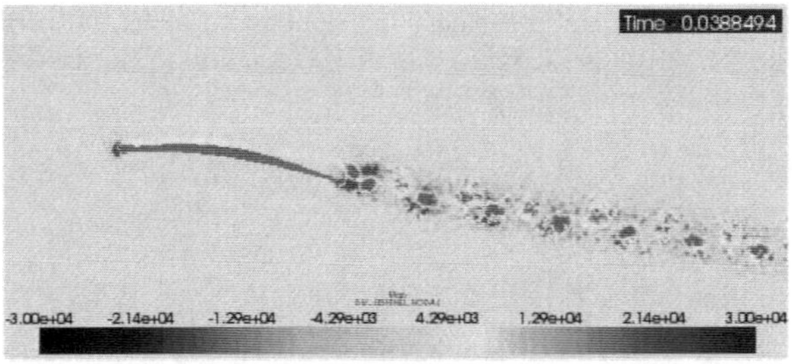

Figure 7: Lighthill sources on the acoustic mesh in the time domain

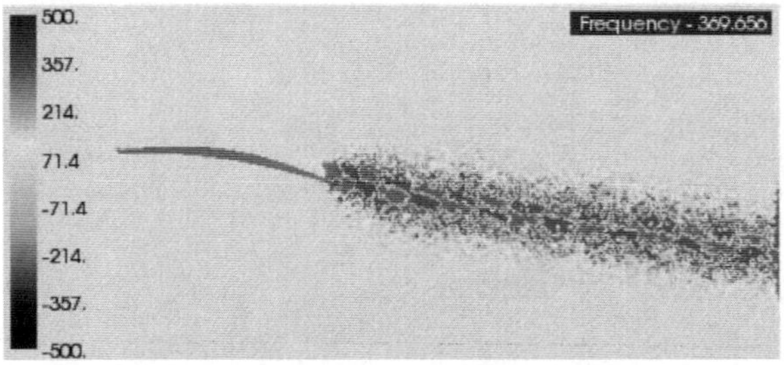

Figure 8: Lighthill sources on the acoustic mesh for the frequency 369.656 Hz

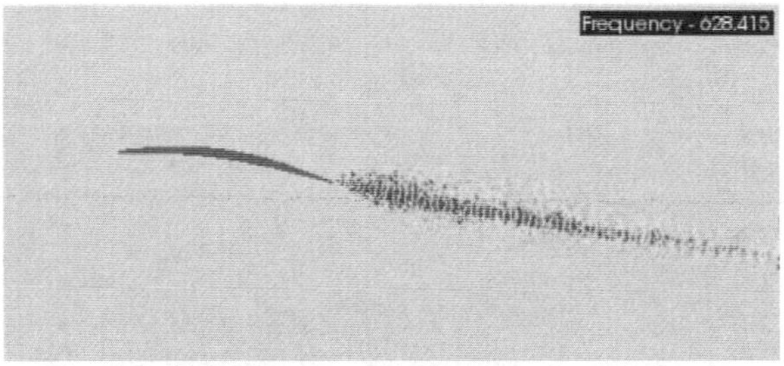

Figure 9: Lighthill sources on the acoustic mesh for the frequency 628.415 Hz

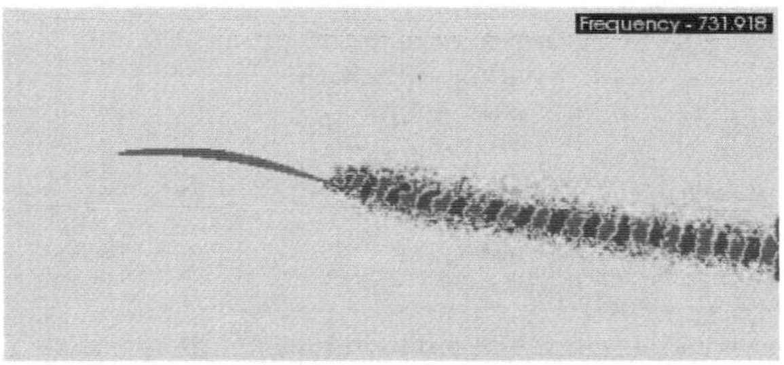

Figure 10: Lighthill sources on the acoustic mesh for the frequency 731.918 Hz

Figures 11, 12 and 13 present the Lighthill's acoustic pressure maps for the near field for frequencies 369 Hz, 628 Hz and 731 Hz respectively. These results show that the acoustic pressure is important for the frequency 369 Hz which corresponds to the Von Karman vortex shedding and this shedding is the main source of this tonal noise. For the frequency 628 Hz, the acoustic pressure is very small whereas for the frequency 731 Hz, the acoustic pressure is important because this frequency corresponds to the first harmonic of the dominant frequency.

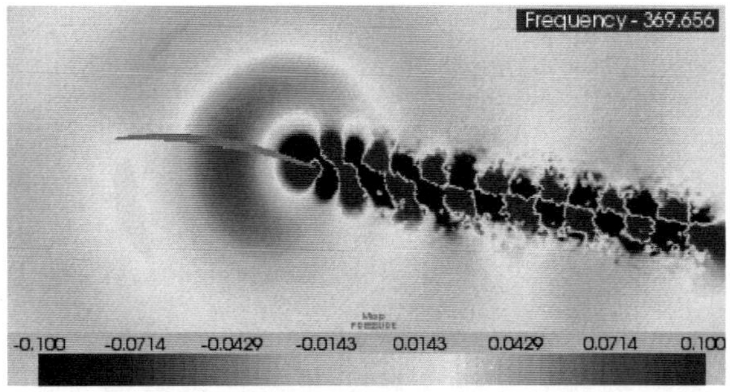

Figure 11: Lighthill near field acoustic pressure for the frequency 369.656 Hz

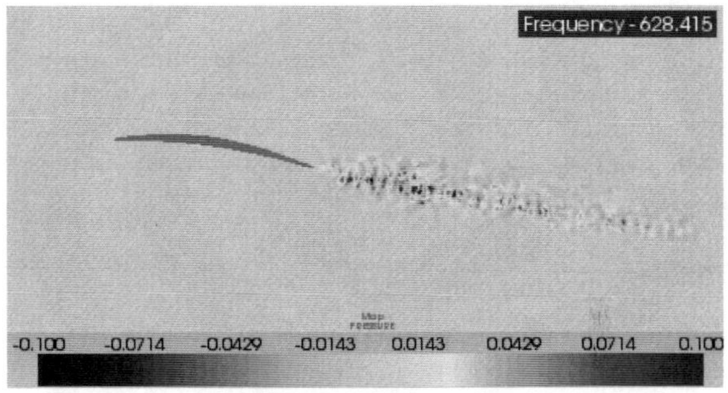

Figure 12: Lighthill near field acoustic pressure for the frequency 628.415 Hz

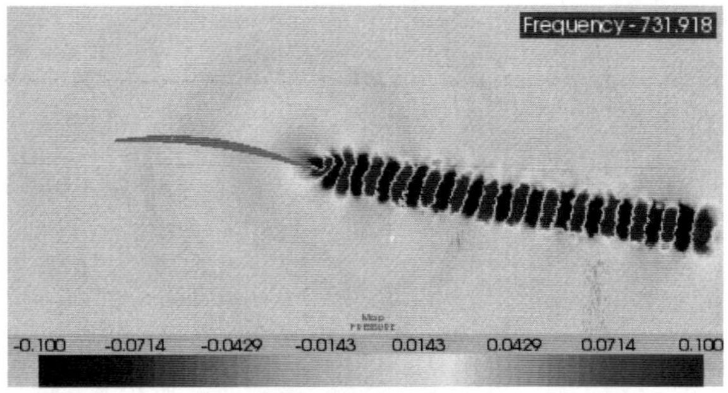

Figure 13: Lighthill near field acoustic pressure for the frequency 731.918 Hz

In order to visualize the acoustic pressure outside the finite elements domain where the infinite elements are used to model the unbounded domain, ACTRAN gives the possibility to create field maps as a post-processing. So a plane of dimensions (2.7 m*2.7 m) is added in the mid-span to show the acoustic waves' propagation from the sources to the far field. Results are shown in Fig. 14 for the frequency 369 Hz. It is clear that the acoustic waves leave the computational domain boundaries smoothly without any reflections; this is thanks to the use of the Infinite elements.

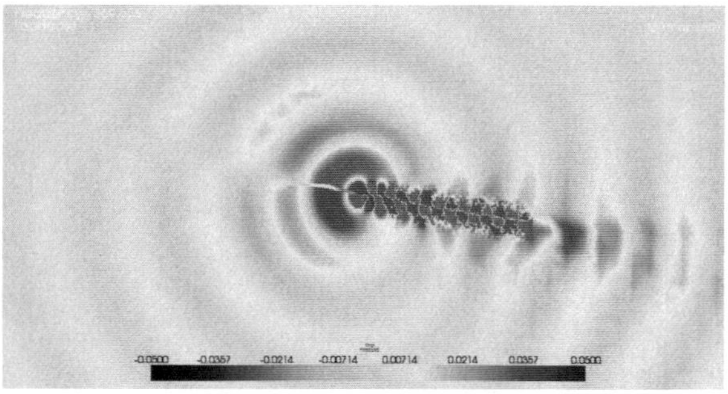

Figure 14: Lighthill, acoustic waves' propagation to far field for the frequency 369.656 Hz

Far field acoustic pressure spectra for the considered receiver are depicted on Fig. 15. The result presents a main peak corresponding to the Von Karman vortex shedding frequency which is 369 Hz. Other tones exist and they correspond to harmonics of the dominant frequency. Apart from these frequencies, the acoustic pressure is really negligible.

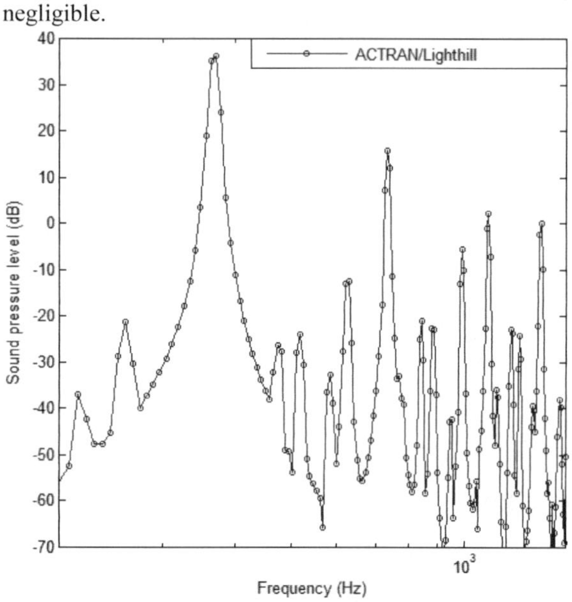

Figure 15: Lighthill far field acoustic pressure spectra for the receiver located at mid-plane, 2 m above the trailing edge

6.2 RESULTS FOR MOHRING'S ANALOGY:

Mohring's analogy has been applied using the same steps as Lighthill's analogy according to the methodology shown in Fig. 1. The importance of this analogy that it allows to study the convectional effects of the mean field, shown in Fig. 16, on the acoustic waves' propagation. Figures 17 and 18 show Mohring sources mapped on the acoustic mesh in the time domain and the frequency domain (at the frequency 369 Hz). The vorticity appears clearly in the source in the time domain because as shown in equation (2) the source term contains it. Mohring acoustic pressure maps for the near field for frequencies 369 Hz and 731 Hz are shown in Figures 19 and 20.

They look more complex than that obtained by Lighthill. Figure 21 show the acoustic waves propagation to the far field for the frequency 369.656 Hz.

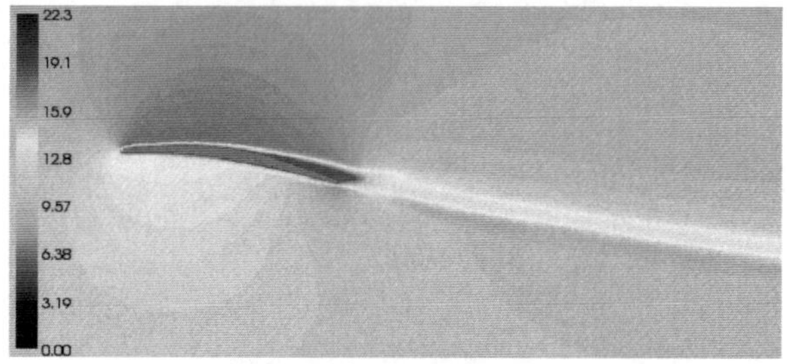

Figure 16: The mean flow

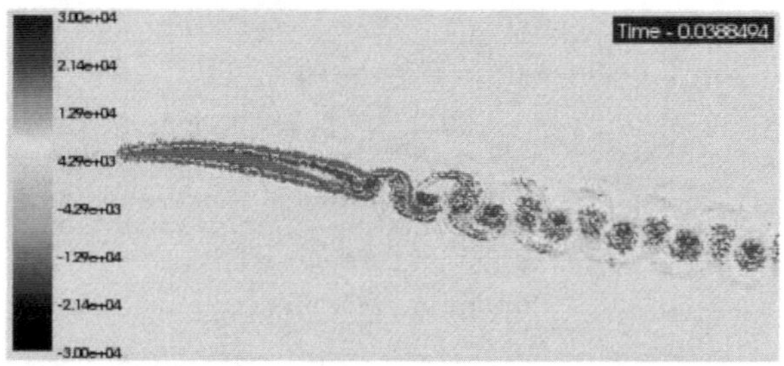

Figure 17: Mohring sources on the acoustic mesh in the time domain

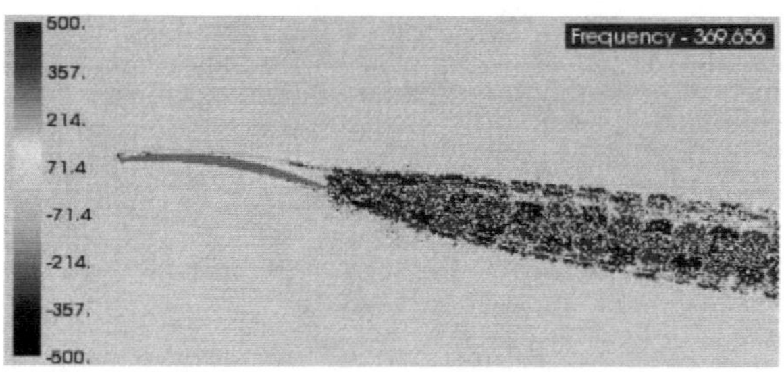

Figure 18: Mohring sources on the acoustic mesh for the frequency 369.656 Hz

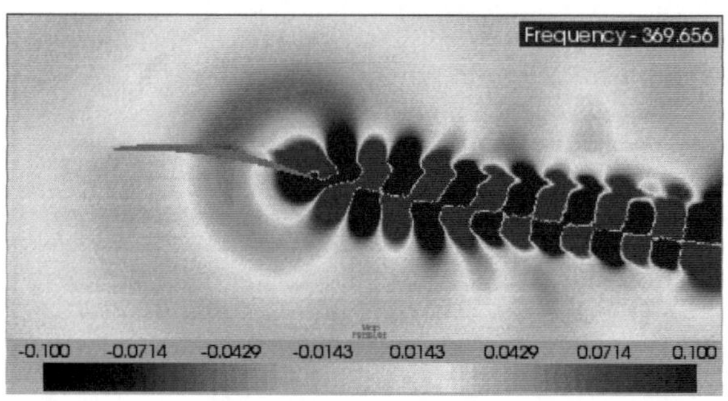

Figure 19: Mohring near field acoustic pressure for the frequency 369 Hz

Figure 20: Mohring near field acoustic pressure for the frequency: 731 Hz

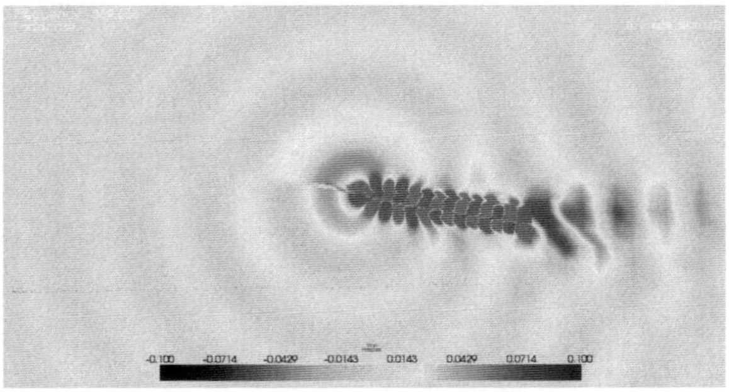

Figure 21: Mohring, acoustic waves' propagation to far field for the frequency 369.656 Hz

The far field pressure is predicted and compared with that obtained using Lighthill'a analogy in Fig. 22. The tonal noise is very similar; the acoustic pressure corresponding to the main frequency (369 Hz) is almost the same. However for mid and high frequencies the acoustic pressure is more important for Mohring's analogy than for Lighthill. This is because of the diffraction of the acoustic waves caused the mean flow. To demonstrate that more clearly, a zoom on the important peaks in this domain is made and depicted in Fig. 23.

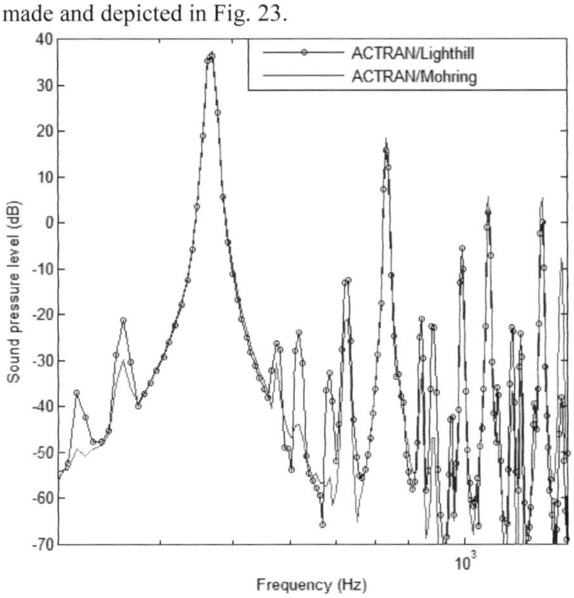

Figure 22: Comparison between Lighthill and Mohring far field volume contribution

As we can see, the effect at the frequency 731 Hz is almost 2.5 dB and 3.7 dB at the frequency 1102 Hz whereas it is up to 5.5 dB for the frequency 1360 Hz.

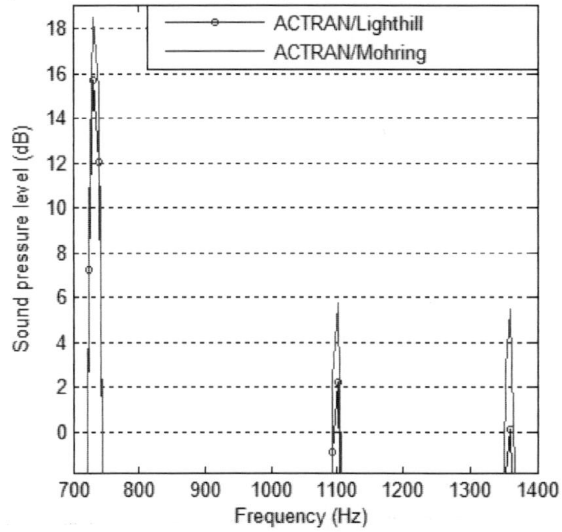

Figure 23: Comparison between Lighthill and Mohring, Zoom for the peaks corresponding to frequencies 731 Hz, 1101 Hz and 1360 Hz

6.3 RESULTS FOR CURLE'S ANALOGY:

Curle's analogy is applied using the two approaches proposed in this paper and explained above. The effect of solid boundary (airfoil) is taken into account as well as the volume sources in the first approach. Figure 24 presents the far field acoustic pressure considering the surface and volume sources contributions where these terms are implemented in SFELES. It is clear that the surface source contribution is dominant but the volume contribution is important as well and not negligible.

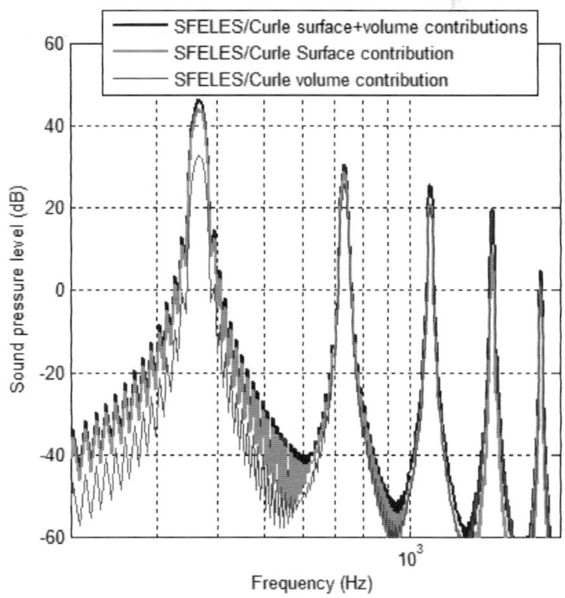

Figure 24: Far field acoustic pressure at the considered receiver via Curle's analogy

A comparison between volume source contributions of the far field acoustic pressure obtained by the three analogies: Lighthill, Mohring and Curle is made and shown in Fig. 25. This Figure shows that the dominant frequency, its harmonics and the corresponding acoustic pressure are well captured by the proposed approach for Curle's analogy. For mid and high frequencies peaks, Curle's acoustic pressure has higher values.

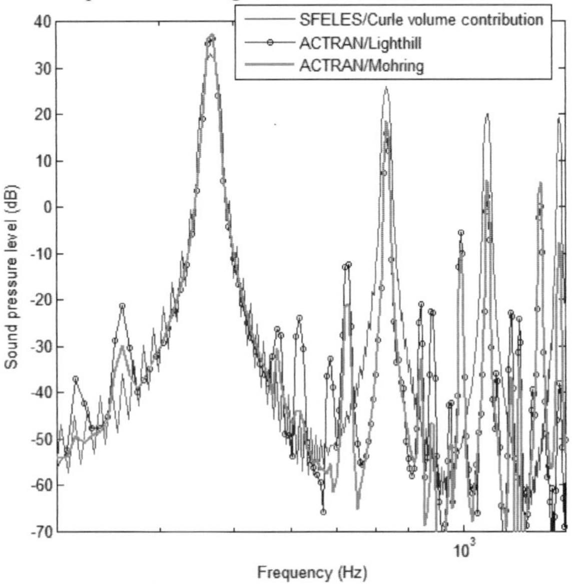

Figure 25: Comparison between volume source contributions obtained by the three analogies: Lighthill, Mohring and Curle

The second approach of Curle's analogy, which is based on the aerodynamics forces and takes into account just the surface source contribution, is then applied. The acoustic pressure spectra obtained from this approach are compared with that obtained by the first approach. The result is shown in Fig. 26. The acoustic pressure is very similar which means that it is a valid and fast approach.

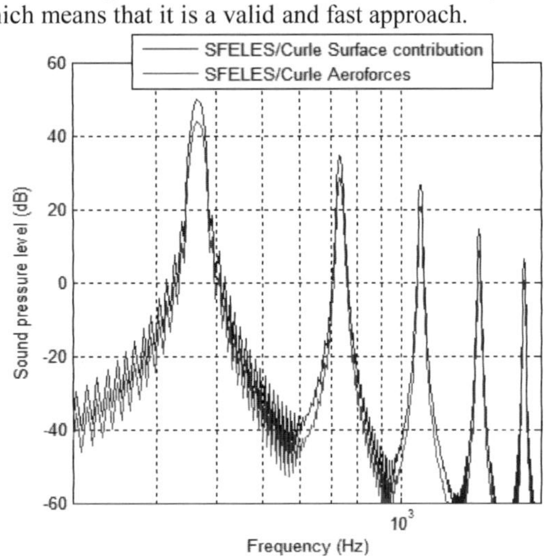

Figure 26: Comparison between the surface source contributions obtained by the two approaches proposed for Curle's analogy

These results show that Curle's analogy is a simple and reliable formulation and can give a fast idea about the far field radiated noise from airfoils.

7. RESULTS FOR THE TURBULENT CASE STUDY AT REYNOLDS 160000:

7.1 AERODYNAMICS SIMULATION AT A GLANCE:

The turbulent simulation has been carried out at a Reynolds number of 160000 and at a Mach number of 0.047. In this simulation, the static Smagorinsky SGS is used with the constant $C_s = 0.17$. The imposed time step is $\Delta t = 10^{-5}$ sec. In the spanwise direction, 64 Fourier modes have been imposed. Figure 27 shows the flow topology described by the contours of the Q criterion where $Q = \frac{1}{4}(\omega^2 - 2S_{ij}S_{ij})$ with $S_{ij} = 0.5(\frac{\partial u_i}{\partial x_j} + \frac{\partial u_j}{\partial x_i})$. This Figure depicts the level of vorticity and the size of turbulent structures in the flow at a given instant.

Figure 27: Q criterion contours (75000) colored by longitudinal velocity

The velocity components have been extracted on every point of the reconstructed 3-D CFD mesh for a period 0.1 sec after that the simulation arrived to the statistically converged solution. For more details about the aerodynamic results, the reader is referred to [6].

7.2 PRELIMINARY ACOUSTICS RESULTS FOR LIGHTHILL'S ANALOGY:

Lighthill analogy has been applied according to the methodology shown in Fig. 1. Figures 28, 29, 30 and 31 show Lighthill sources computed on the acoustic mesh for the frequencies 345 Hz, 494 Hz, 815 Hz and 1012 Hz respectively. It is clear that main noise sources come from the turbulent boundary layer and wake. For high frequencies, the sources are less efficient.

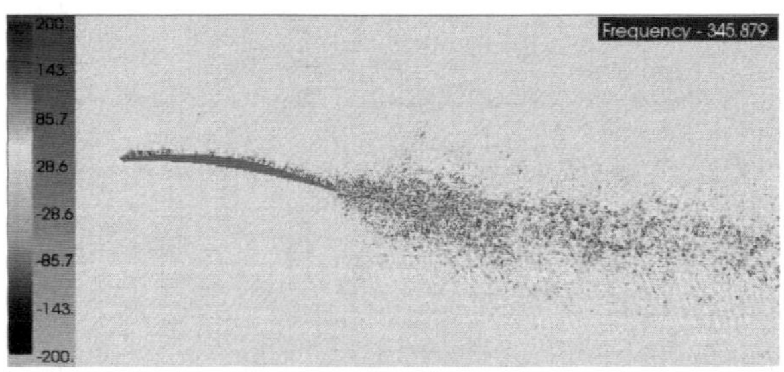

Figure 28: Lighthill sources on the acoustic mesh for the frequency 345 Hz

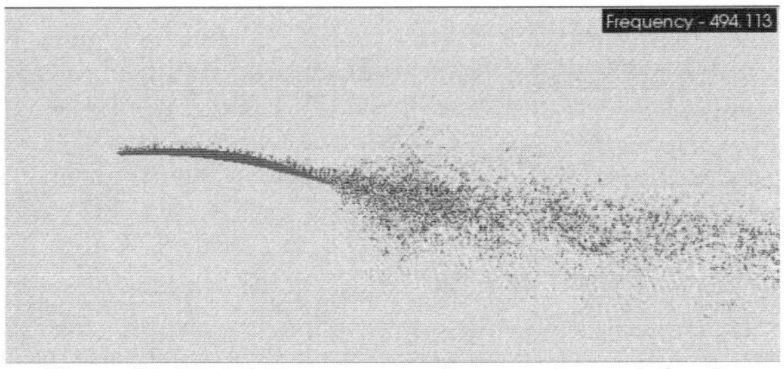

Figure 29: Lighthill sources on the acoustic mesh for the frequency 494 Hz

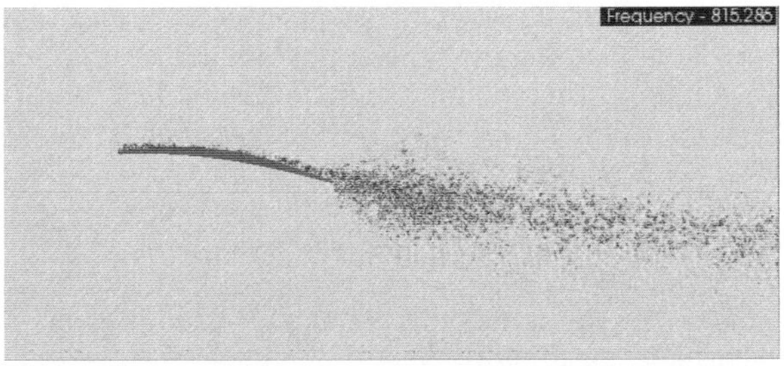

Figure 30: Lighthill sources on the acoustic mesh for the frequency 815 Hz

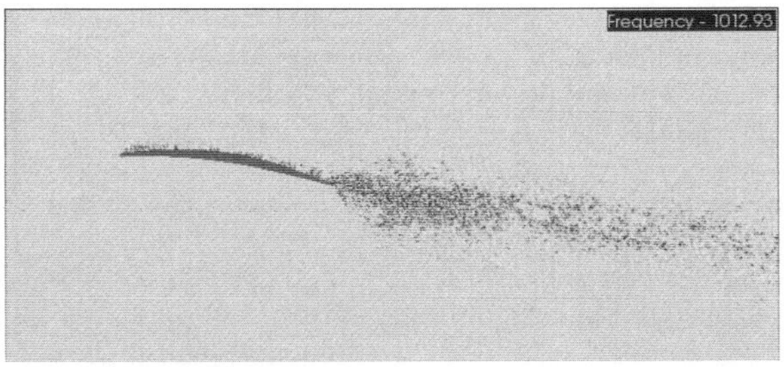

Figure 31: Lighthill sources on the acoustic mesh for the frequency 1012 Hz

The corresponding near field acoustic pressure maps are shown in Figures 32, 33, 34, and 35.

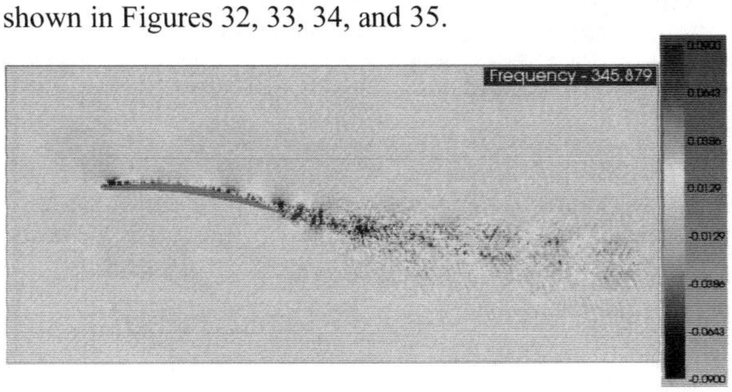

Figure 32: Lighthill near field acoustic pressure for the frequency 345 Hz

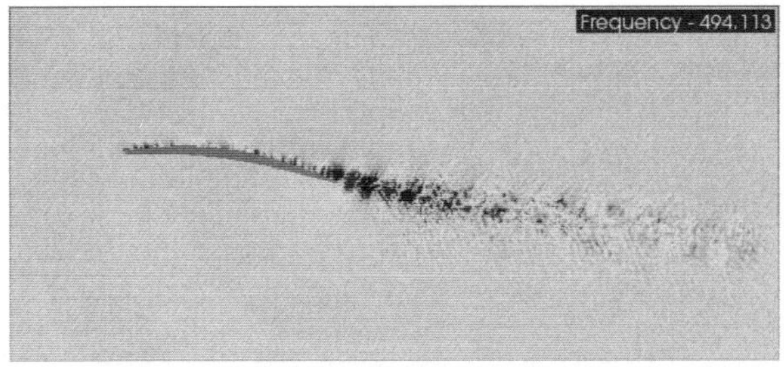

Figure 33: Lighthill near field acoustic pressure for the frequency 494 Hz

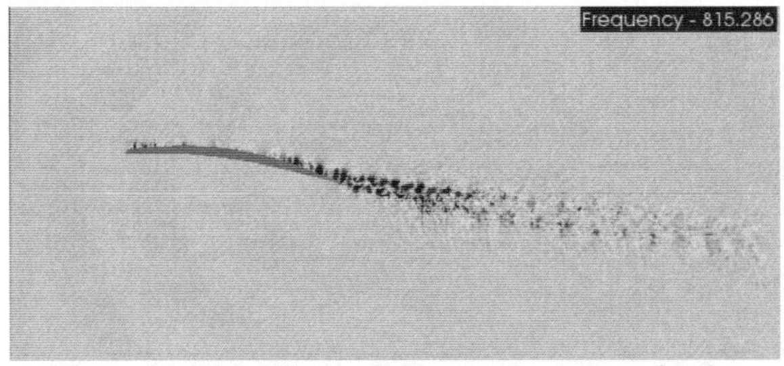

Figure 34: Lighthill near field acoustic pressure for the frequency 815 Hz

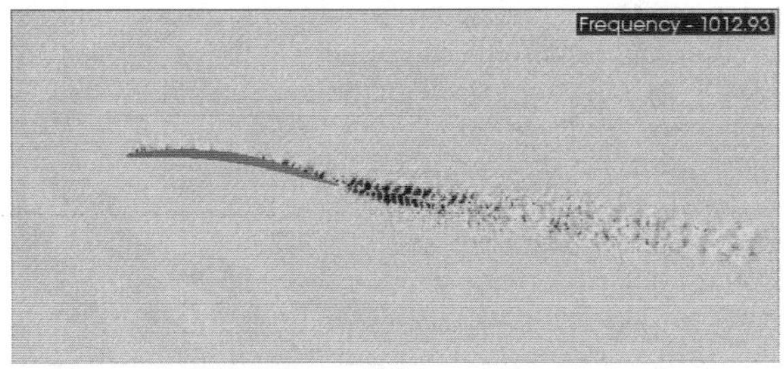

Figure 35: Lighthill near field acoustic pressure for the frequency 1012 Hz

The acoustic waves' propagation from the airfoil to the far field is shown in Figures 36 and 37 for frequencies 345 Hz and 494 Hz respectively. The acoustic waves leave the computational domain boundaries smoothly and without any reflections. A higher Infinite Elements order is needed to assure that in comparison with the laminar case.

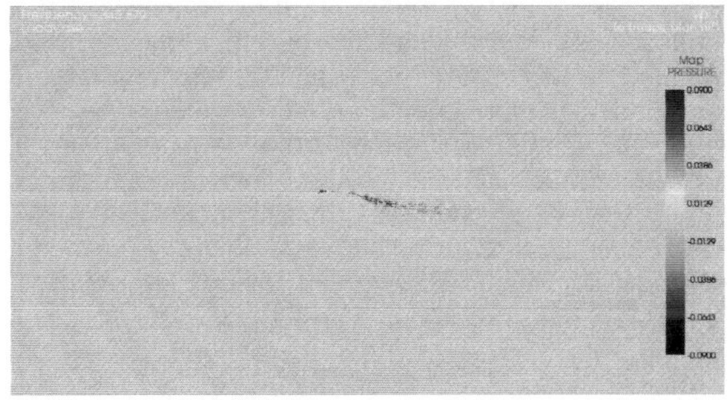

Figure 36: Lighthill field map acoustic pressure for the frequency 345 Hz

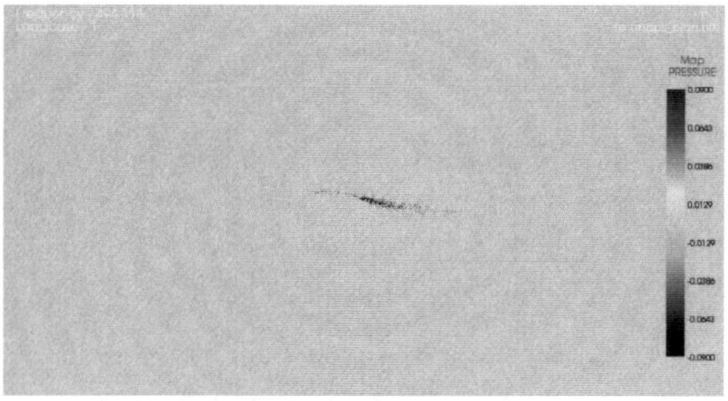

Figure 37: Lighthill field map acoustic pressure for the frequency 494 Hz

These results are preliminary because the time resolution of the snapshots is not sufficient to resolve high frequencies but can give an idea about the acoustic emission from the turbulent flow. The acoustic pressure spectra in the far field as well as Mohring and Curle analogies results will be published later.

8. CONCLUSIONS

In this paper, the radiated noise from a CD airfoil is simulated and computed via the hybrid methods of aeroacoustics. Three analogies, namely those of Lighthill, Mohring and Curle, have been applied and their results are compared. In Lighthill's analogy, the quadrupole sources are computed. Maps of the corresponding radiated noise are demonstrated. The convectional effects of the mean flow on the acoustic waves propagation has been accounted for via Mohring analogy whereas via Curle integral formulation, the dipole sources caused by the presence of the airfoil are computed. Broadband noise is studied at Reynolds number of 160000 whereas a tonal noise case is studied for a laminar flow at Reynolds number of 12000. Tonal noise that arises from vortex shedding generated by laminar boundary layer is the main acoustic source for the laminar flow where a dominant peak is found at the frequency of the vortex shedding (369 Hz). Turbulent boundary layer and wake are the more efficient sources for the turbulent flow around the airfoil. The dipole sources are dominant whereas the quadrupole contribution is found important despite of the low Mach number considered in this study. The convectional effects are important as well for mid and high frequencies, the effect up to 5dB at the peak corresponding to the frequency 1360 Hz for the laminar flow case. Finally, the results of the proposed approaches for Curle's analogy match well the results obtained by Lighthill and Mohring analogies. The vortex shedding frequency, its harmonics and the corresponding acoustic pressure are well captured for the laminar case. So we can say that Curle' analogy is a simple and reliable formulation and can give a quick idea of the radiated noise of the studied airfoil with the limitations and restrictions mentioned in this paper such as compactness source region, utilizable only for exterior noise problem in free field, etc… .

ACKNOWLEDGMENTS

Authors gratefully acknowledge the FFT Company in Belgium (Prof. Jean Louis Migeot, Dr. Yves Detandt and all staff) for providing ACTRAN, for their permanent support and for giving the opportunity for an internship in the company.

REFERENCES

[1] Snyder, D., A parallel finite-element/spectral LES algorithm for complex two dimensional geometries, 2002, PhD Thesis, von Karman Institute for Fluid Dynamics, Belgium, Utah State University, USA.

[2] Actran Aeroacoustic Solutions: Free-Field-Technologies, Belgium, User's Manual, 2014.

[3] Lighthill, M. J., On sound generated aerodynamically: I. General theory, Proceedings of the Royal Society of London. Series A, 1952, 211,564-587.

[4] Mohring, W., A well posed acoustic analogy based on a moving acoustic medium, Aeroacoustic workshop, SWING. Dresden 1999.

[5] Curle, N., The Influence of Solid Boundaries upon Aerodynamic Sound, Proceedings of the Royal Society of London. Series A, Mathematical and Physical Sciences, 1955, 231(1187):505-514.

[6] Matouk, R., Degrez, G. and Christophe, J., Aerodynamics and aeroacoustics study of the flow around an automotive fan airfoil, Proceedings of WCCM 2014, ECCM 2014 and ECFD 2014. July 2014, pages 5677-5688.

[7] EnSight User Manual for Version 10.0, Computational Engineering International, Inc.

[8] Caro, S., Detandt, Y., Manera, J., Toppinga, R., and Mendonca, F., Validation of a New Hybrid CAA strategy and Application to the Noise Generated by a Flap in a Simplied HVAC Duct, 15th AIAA/CEAS Aeroacoustics Conference and Exhibit, 11 - 13 May 2009, Miami, FL.

[9] Jeremy Astley R., Infinite Elements, Institute of Sound and Vibration Research, University of Southampton, Highfield, Southampton, SO17 1BJ, United Kingdom

[10] Caro, S., Ploumhans, P., and Gallez, X., Implementation of Lighthill's Acoustic Analogy in a Finite/Infinite Elements Framework, AIAA paper, 2004-2891.

[11] Moreau, S. and Roger, M., Effect of Airfoil Aerodynamic Loading on Trailing-Edge Noise Sources, AIAA Journal, January 2005, 43(1):41-52.

[12] Moreau, S., Neal, D. and Foss., J., Hot-Wire Measurements Around a Controlled Diffusion Airfoil in an Open-Jet Anechoic Wind Tunnel, J. Fluids Eng., 2006, 128:699-706.

[13] Palumbo, D. Determining correlation and coherence lengths in turbulent boundary layer fight data. Journal of Sound and Vibration, (2012), 331(16):3721{3737.

[14] Wang, M. and Moin, P., Computation of Trailing-Edge Flow and Noise Using Large-Eddy Simulation. AIAA Journal, 38(12):2201, 2209, December 2000.

[15] Corcos, G. M., The Structure of Turbulent Pressure Field in Boundary-Layer Flows, J. Fluid Mech., 1964, 18(3):353-378.

Proceedings of the 14th International Symposium on Unsteady Aerodynamics, Aeroacoustics&Aeroelasticity of Turbomachines
ISUAAAT14
8-11 September 2015, Stockholm, Sweden

I14-S13-3

ASSESSMENT OF ACOUSTIC CHARACTERISTICS OF COUNTER-ROTATING OPEN ROTORS

Anton Rossikhin
Central Institute of Aviation Motors
Moscow, Russia

Igor Brailko
Central Institute of Aviation Motors
Moscow, Russia

Victor Mileshin
Central Institute of Aviation Motors
Moscow, Russia

ABSTRACT

The results of computational analysis of tonal noise of three counter-rotating open rotors are presented. The analysis was performed using 3DAS (3 Dimensional Acoustics Solver) CIAM in-house solver, designed for three dimensional calculation of turbomachinery tone noise generation, propagation and radiation in the near and the far acoustic fields. The results of the computations were compared with the results of experiments performed in the wind tunnel of Central Aerohydrodynamic Institute (TsAGI), Russia, within the project DREAM (Vali**d**ation of **r**adical **e**ngine **a**rchitecture syste**ms**) of European Seventh Framework Programme (FP7).

INTRODUCTION

One of the perspective schemes of next generation aviation engines is the scheme of gas-turbine engine with counter-rotating open rotors [1]. This type of engines is expected to have high fuel efficiency. On the other side they have some disadvantages. One of the most important - is the relatively higher noise radiation in comparison with conventional high and ultrahigh bypass ratio turbofans with similar thrust. In order to solve this problem it is important to have efficient numerical methods with high fidelity and robustness for calculation of aerodynamic and acoustic characteristics of open rotors. That's why significant efforts were applied for the development of such methods. Examples can be found in [2,3,4,5,6,7]. The 3DAS solver was also used for the first time for calculation of open rotor noise [8].

The present paper is devoted to the comparison between the results of aeroacoustic calculations done using the 3DAS solver, and measurements from low speed wind tunnel tests, performed in TsAGI test facility for open rotors in low Mach number working conditions. These assessments were conducted within the European FP7 project DREAM [9] Sub-project 3 (SP3) - Direct-Drive Pusher Open Rotor led by Snecma and more specifically in the concept studies carried out along the Work Package (WP) 3.2. Partners include Airbus, CIAM, Cenaero, DLR, Dowty and ONERA.

In the framework of the SP3 several open rotors were tested. In this study we will describe our results for three model open rotors named V2.0 [3,10], V2.1a [10] and V2.3. They were designed on the second stage of WP3.2 using different design strategies.

All open rotors under consideration contain twelve blades in the first blade row and ten blades in the second. Rotation speeds of the front and the rear blade rows are equal in the absolute value. The radius of the second blade row for all three open rotors is noticeably smaller than the radius of the first one (clipping). This design feature can lead to the reduction of interaction noise in some working conditions, because in this case tip vortexes from the blades in the first blade row can at least partly pass above the tips of the blades in the second row.

The paper is the direct continuation of the work [11], devoted to the calculation of tone noise of two model open rotors V0 and V1.1 also tested in the framework of SP3.

NOMENCLATURE

B_f - the blade passage frequency of the first rotor
B_a - the blade passage frequency of the second rotor

3DAS SOLVER

3DAS [8,12] solver is designed for the calculations of tone noise, generated by inter-row interaction in a turbomachine, in the near and far acoustic fields.

The method of inter-row interaction calculation, implemented in 3DAS solver, is based on the solution of three-dimensional linear or nonlinear Euler equations for disturbances over viscous steady mean flow in the reference frames of blade rows. The system of equation is derived from the Navier–Stokes equations by division of unsteady flow on predefined steady mean flow and unsteady disturbances under the assumption that the effects of viscosity and heat conduction normally control mean flow distribution rather than acoustic fluctuations.

Interaction between blade rows is provided by special “sliding grid” interfaces [8]. The motion of one blade row relative to the other is accomplished by "sliding" the grid system of the blade row past the grid system of the other blade row. Interfaces transfer unsteady acoustic (and also entropy and vorticity) disturbances and non-uniformities of mean steady flow field (for example rotor wakes) from one blade row to another.

The main features of the 3DAS solver are the following.

The input data for the solver is a steady mean flow field in the computational domain and a body-fitted computational grid. For turbomachinery tone noise applications the steady mean flow field in the blade rows must be calculated using a Reynolds-Averaged Navier–Stokes (RANS) finite volume steady solver of the second or higher orders. The steady coupling between rows must be provided by "mixing-plane" type interface.

Calculation can be performed either using nonlinear Euler equations for disturbances or with linearized Euler equations dependent on the problem under consideration. For open rotor noise calculation nonlinear approach is preferable due to strong pressure pulsations observed on the blades.

3DAS solver admits calculations either in time domain or in frequency domain for specified set of harmonics [8,13]. In the current work we use the time domain method of calculation.

The fourth order DRP (Dispersion Relation Preserving) scheme proposed by K. Tam [14] was taken as a base of a spatial difference scheme used in our code. The scheme was generalized to the finite volume method (see [8] for details) for convenience of the work with mean flow fields, provided by finite volume solvers.

Second-order (fourth order in the linear case), four stages Runge-Kutta LDDRK (Low Dissipation and Dispersion Runge-Kutta Scheme) scheme (F. Hu et al. [15]) was used for time discretization.

To suppress high frequency spurious waves, artificial selective damping was used, based on the fourth-order central filtering coefficients of O. Vasilyev [16].

With the scheme parameters, used in this work, 3DAS program package needed resolution of 10 - 12 points per wave length to give appropriate results. This concerns not only linear perturbations, but also highest resolved frequency components of nonlinear perturbations. The interfaces, implemented in the program package, accurately describe propagation of waves having at least 8 grid points per spatial period in circumferential direction.

The method based on the Ffowcs-Williams and Hawkings equation with a penetrable data (integration) surface [17] is used to predict tone noise in the far field, using results of the calculations in the near field.

The 3DAS solver was applied for the assessment of tonal noise of SV-27 counter-rotating open rotor [8]. Also it was used for the calculations of tone noise of ducted counter-rotating fans [18] in the framework of European Project VITAL (EnVIronmenTALly friendly aeroengine). In both cases the results were quite successful. Other applications of the solver were the calculations of acoustic characteristics of high bypass ratio fans of turbofan engines and last stages of low pressure turbines [12].

Detailed description of 3DAS solver can be found in [12], descriptions of time domain method implementation - in [8] and [18].

DESCRIPTION OF THE TEST RIG

The tests, described in this work were performed in TsAGI low speed wind tunnel T-104. This wind tunnel has an open test section of 7 m diameter and 13 m length [19]. The test rig, which was used in the experiments, is shown in the fig. 1.

Fig. 1. TSAGI wind tunnel T-104

The T104 facility is a conventional wind tunnel without anechoic open chamber. In such facility acoustic measurements will be contaminated with reverberations and refractions effects from surrounding reflective surfaces. To obtain satisfactory results acoustic improvement [19] of the facility was performed - reflecting surfaces were covered with sound absorbing material and sound absorbing coulisses were mounted. The reverberation measurements in the facility showed that such actions, in conjunction with the use of directional microphones, allowed obtaining in the test sections conditions approaching those of the free field.

Other results, concerning the comparison between calculations and experiment within SP3, can be found in [3,4].

a

b

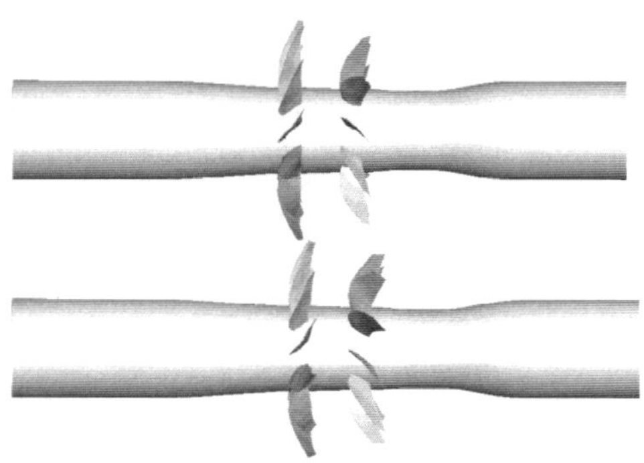

c

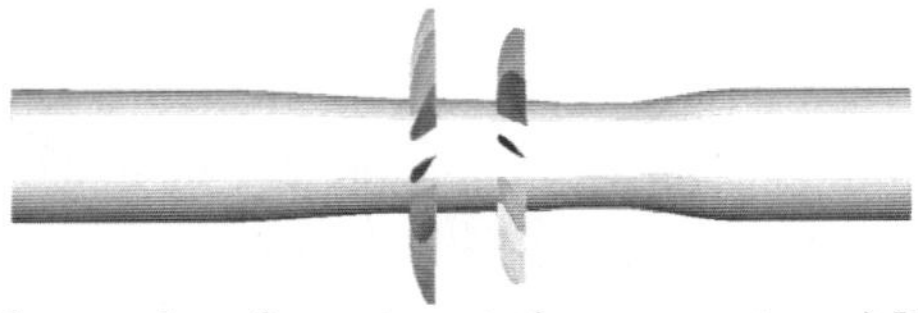

Fig. 2. Geometries of counter-rotating open rotors a) V2.0, b) V2.1a, c) V2.3

SIMULATION OF TONE NOISE OF COUNTER-ROTATING OPEN ROTORS

The counter rotating open rotors considered in this work are presented in the fig. 2a – V2.0, in the fig. 2b – V2.1 and in the fig. 2c – V2.3. In our calculations we didn't took into account the interaction of rotors with wakes produced by the rig supporting construction. The inclusion of this construction into the computational domain greatly complicates the problem. The geometry of the rig in the calculation was taken axisymmetric. It was evident from the beginning that such initial statement would prevent us from obtaining good correspondence between calculation and experiment for noise on the harmonics of blade passage frequencies (BPFs) of rotors. Nevertheless we hoped to obtain consistent data for interaction tones which were known to be less sensitive [20] for incoming distortions of flow parameters.

For all open rotors calculations were performed for working conditions correspondent to the take-off working conditions of full-sized open rotors (free stream Mach number – 0.2). It is not a certification point, but just one of the design points.

First of all we performed aerodynamic calculations using CIAM's 3DFS (3 Dimensional Flow Solver) RANS solver [8]. Comparison between the results of computations and experimental data for steady flow field parameters showed good correspondence, especially taken into account that we neglected the influence of supporting structures. In the fig. 3 are shown distributions of the static pressure on the hub. Also in the fig. 3 are shown positions of pressure sensors on the hub and positions of upstream and downstream rakes relative to the open rotor. In the fig. 4 are shown distributions of the Mach number downstream the open rotors.

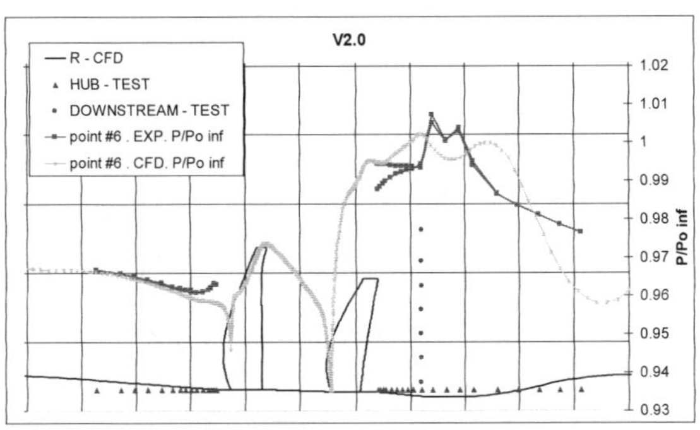

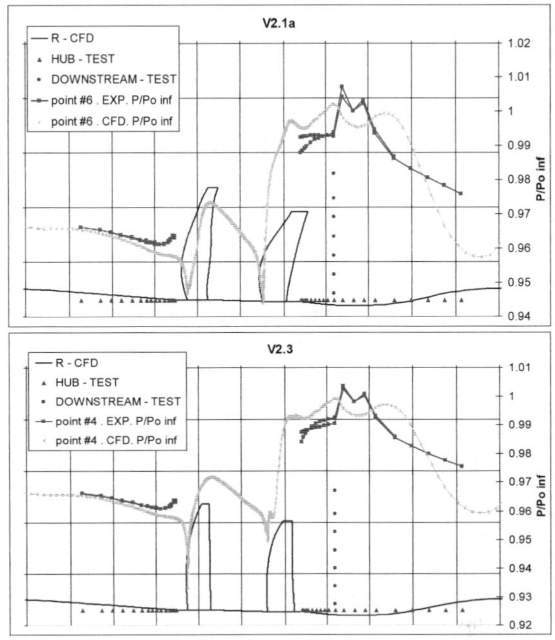

Fig. 3.Distributions of the static pressure on the hub for V2.0, V2.1a and V2.3 (Po inf – static pressure in infinity).

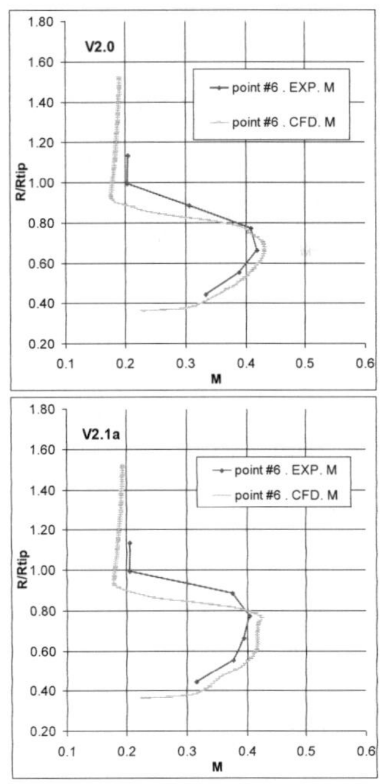

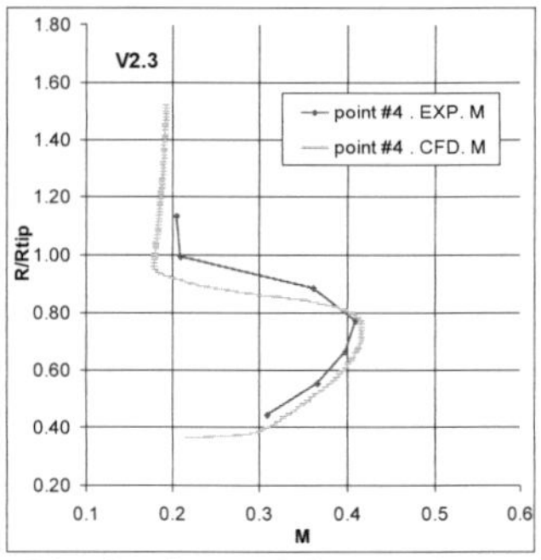

Fig. 4. Distributions of the Mach number downstream the open rotors for V2.0, V2.1a and V2.3 (R_{tip}-radius of the first rotor).

Calculations of steady flow field for aeroacoustic calculations were performed for simplified geometries with zero thicknesses of blade tips. This simplification was made for acceleration of calculations. Computational grids in all cases contained approximately 19 million cells.

For the unsteady calculation the grids were additionally changed. They were refined in the regions between the rotors and the positions of the integration surfaces for the method of Ffowcs-Williams and Hawkings. The grids were made dense enough to compute propagation of waves with frequencies up to $3(B_f+B_a)$ for correspondent open rotor. In addition, the grids were stretched near the outer boundaries of the computational domain in order to prevent unphysical reflections. Also blocks of the first rotor were periodically duplicated 6 times and of the second - 5 times to cover total spatial period. New computational grids contained approximately 19.8 million cells for all configurations.

Unsteady calculations were performed advancing solution by time with the constant time step, using nonlinear Euler equations for disturbances. Period of rotation contained 25400 time steps for V2.0, 22000 time steps for V2.1a and 32000 time steps for V2.3.

In the fig. 5 unsteady fields of static pressure on the curvilinear cylindrical surface surrounding the rotors are presented for the geometries under consideration (normalized at the static pressure in infinity). The most intensive pulsations are observed for V2.1a.

V2.0

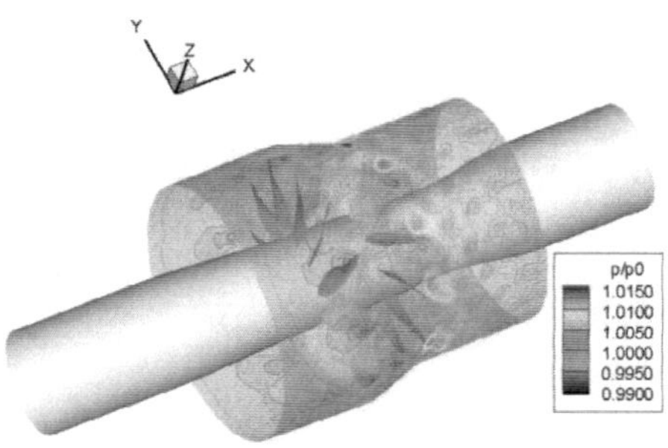

V2.1a

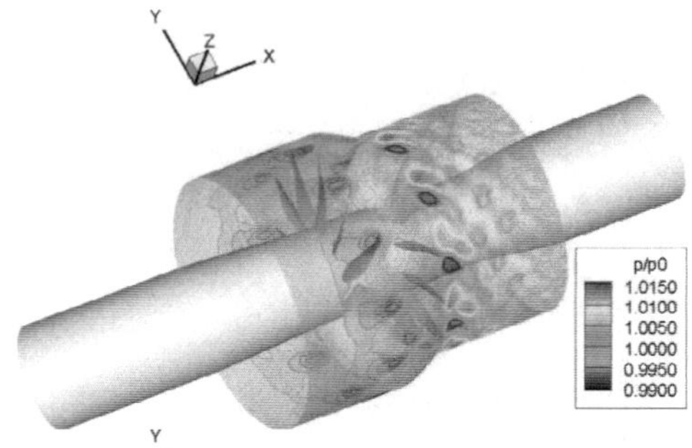

V2.3

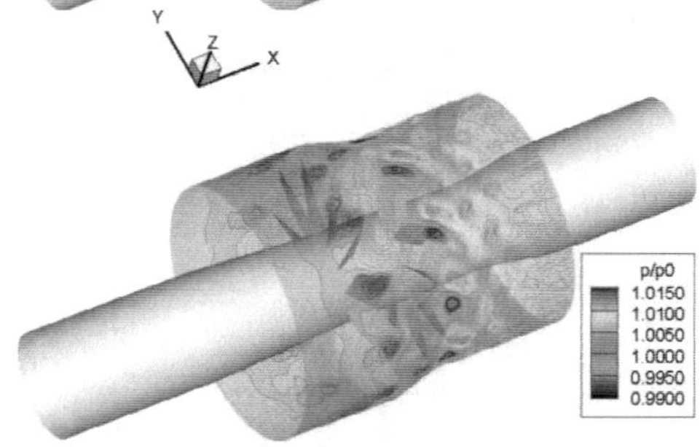

Fig. 5. Unsteady fields of static pressure (normalized) on the surfaces surrounding the rotors

V2.0

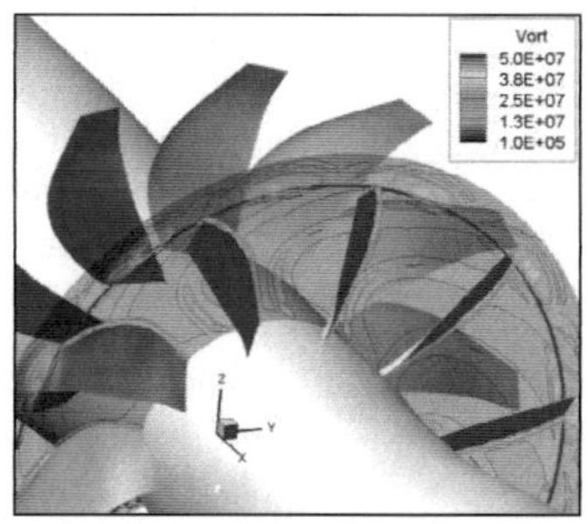

V2.1a

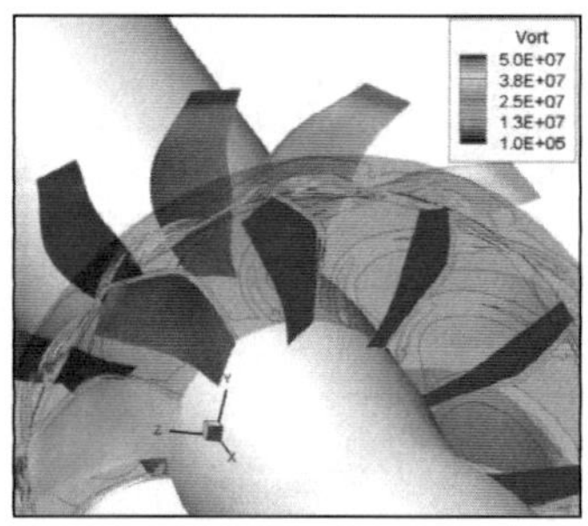

V2.3

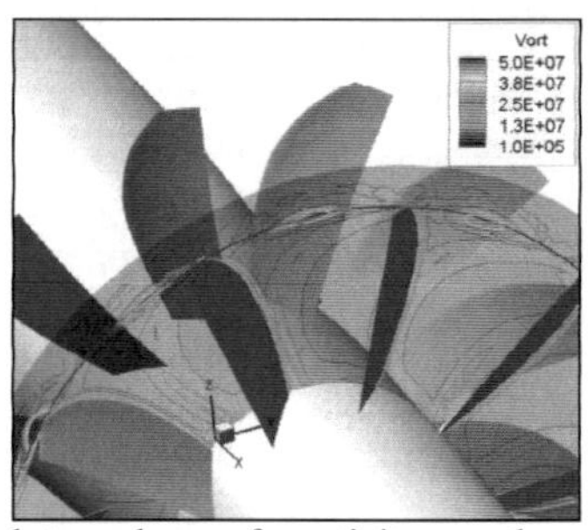

Fig. 6. Absolute values of vorticity on the surfaces, placed in front of the second rotors.

In order to make clear how clipping of the second rotor influences on the rotor-rotor interaction we presented in the fig. 6 the distributions of the absolute values of vorticity - |rotV| (where V is the velocity of flow) on the surfaces placed in front of the second rotors. Red circle shows the height of the second rotor blades. We see that some interaction between tip vortexes and second rotor is unavoidable in all three cases. Also, it is possible to conclude, that the most intensive interaction between rotors should be observed for V2.3.

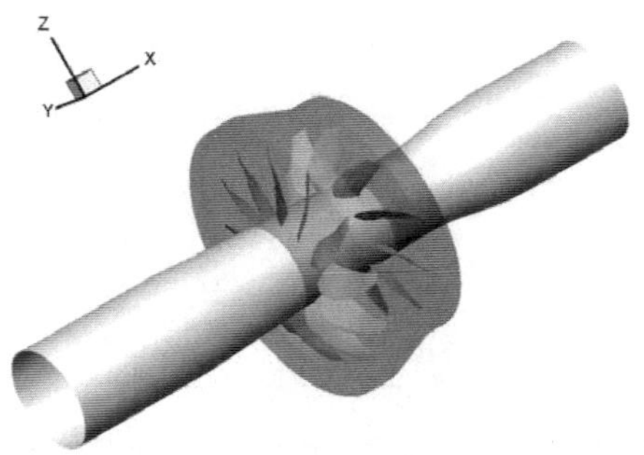

Fig. 7. Integration surface – V2.0

On the last stage of computations we performed the calculations of sound pressure pulsations in the far field using method based on the Ffowcs Williams – Hawking equation. In the fig. 7 is shown the position of the integration surface in the computational domain for V2.0, the pictures for other configurations are analogous. The positions of microphones are placed on the line parallel to the engine axis. Microphones are distributed within the angle range from 20° to 60° with step 2°. The angle 0° corresponds to the upstream direction.

COMPARISON BETWEEN EXPERIMENTAL AND COMPUTATIONAL DATA

Experimental data were presented as a set of narrow band spectra for 12 microphones (fig. 8). The frequency bands of the spectra were equal to 12.5 Hz. As in the calculation the positions of microphones were distributed on the line parallel to the engine axis and located on the same distance from it. In the test rig microphones were placed outside the jet of the wind tunnel, while in the computation they were placed in the incident flow. Before the comparison, the experimental data obtained were preliminary processed to subtract effects caused by refraction on the boundary of the jet using Amiet method [19].

In the fig. 9 integral spectra (sum of the spectra for all 12 microphones) for experimental and computational data are presented. Here for experimental data we showed narrowband spectra. For computational data we showed the main BPFs harmonics and the combinational harmonics, for which in the expression, determining the frequency of the combinational harmonic – $F_{nm} = | n B_f + m B_a |$, integer multipliers satisfy condition n,m>=1. Other combinational harmonics are omitted, because, as it is known from the theory of open rotor noise [20], the levels of noise for the harmonics with n,m>=1 must be much higher than for the combinational harmonics with m or n <0. In the fig. 9 we see, that the peaks, which are correspondent to the harmonics of tone noise in the experimental data have significant width. This widening of peaks can be caused by variability in the speeds shafts rotation and haystacking (reduction of tone peaks in favour of a more distributed spectral hump around each tone frequency arising due to the interaction of tones with the unsteady turbulent jet shear layer). That's why we used in our comparison the intensities of peaks obtained by integration of spectrum intensity over a width of the experimental peaks (after the extraction of the broadband noise contribution).

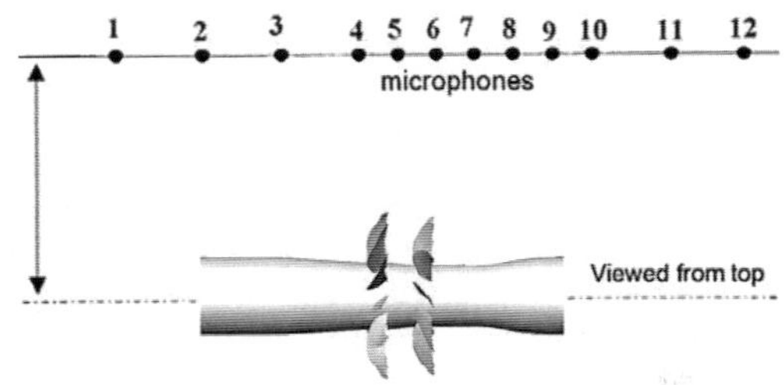

Fig. 8. Positions of microphones in the experiment (the fan is not in proper scale)

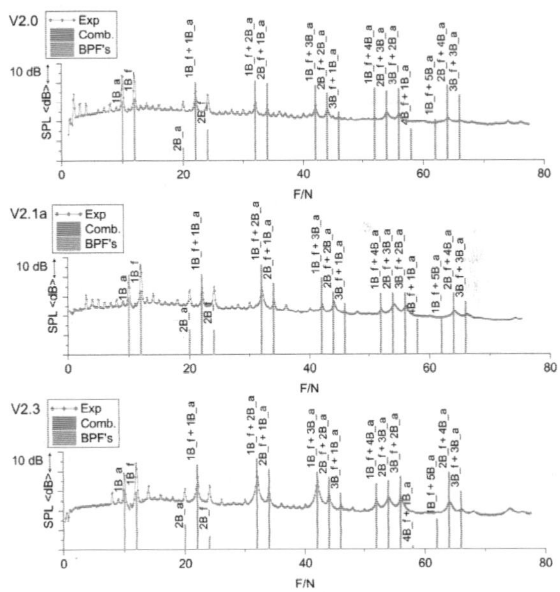

Fig. 9. Comparison of integral spectra for experimental and computational data (N – frequency of shafts rotation)

In the fig. 10 we presented comparison between experimental and computational directivity diagrams for the first two harmonics (B_a and B_f) of BPFs. The shapes and the levels of computational and experimental directivity diagrams are quite different. Computational directivity diagrams for frequencies B_a and B_f have the shapes, typical for the noise of rotation, while correspondent experimental diagrams – the shapes typical for the interaction noise [20]. We can conclude that in the experiment the main contributions to the noise on these harmonics were made by the interaction of rotors with nonuniformities of the incident flow (caused by installation effects and turbulence). Results of the acoustic evaluation, performed in [4], showed that inclusion of the test rig design into the computation can significantly improve the results for BPFs harmonics.

In the fig. 11 we presented comparison between experimental and computational directivity diagrams for the first five combinational harmonics, with n,m>=1 (for all three open rotors). For higher harmonics it becomes too hard to unambiguously extract tonal noise from the experimental data – the peaks correspondent to harmonics of tonal noise are too flat (see fig. 9).

We see satisfactory correspondents between computation and experiment for the first combinational harmonic B_f+B_a for all three open rotors. Computational and experimental curves are close to each other, and we see that the difference between them in some positions of microphones is less than 3 dB. Also satisfactory correspondence between the curves is observed for the fifth combinational harmonic $2B_f+2B_a$.

The situation is quite different for the other three combinational harmonics. We again see appropriate correspondence between the computational and experimental curves for the V2.3 open rotor. However the results of computation and experiment for the V2.0 and V2.1a open rotors are quite distant (except for some positions of microphones). Computational curves lays higher than experimental.

The differences between computation and experiment in these cases are caused from our point of view by the underestimation of the influence of clipping in our calculations. We used simplified geometry of second rotor blades tips. As we can see from fig. 6 in all three cases we cannot avoid interaction between tip vortexes of the first rotors blades and second rotors blades. The interaction should depend on the geometry of rotor tips. This dependence is less significant for the V2.3 geometry, for which the regions on the second rotor blades, where pressure pulsation are induced by the interaction with the tip vortexes, occupy larger regions of blades.

In general the discrepancy between computation and experiment for the noise of the V2.3 open rotor on combinational harmonics is comparable with the results presented in the works [3,4] and work [11]. The results of the work [4] show that the inclusion of the rig construction into the computational domain makes discrepancy smaller.

In the future we are planning to perform calculations for the open rotors under consideration using exact geometries of blades tips. The following step of our work will be the calculation of tone noise of counter-rotating open rotor taking into account installation effects.

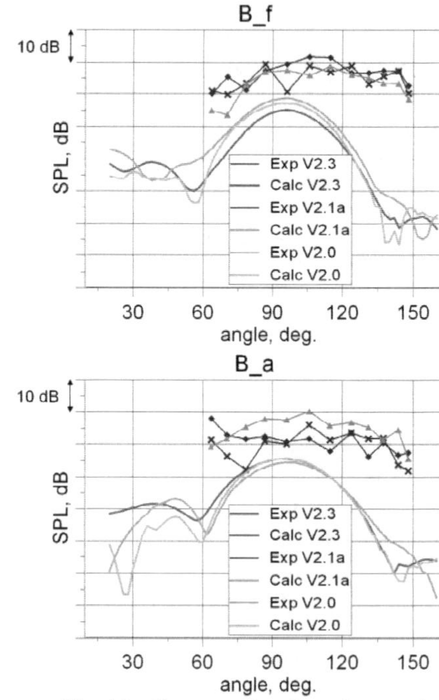

Fig. 10. Computational and experimental directivity diagrams for the first two BPFs harmonics

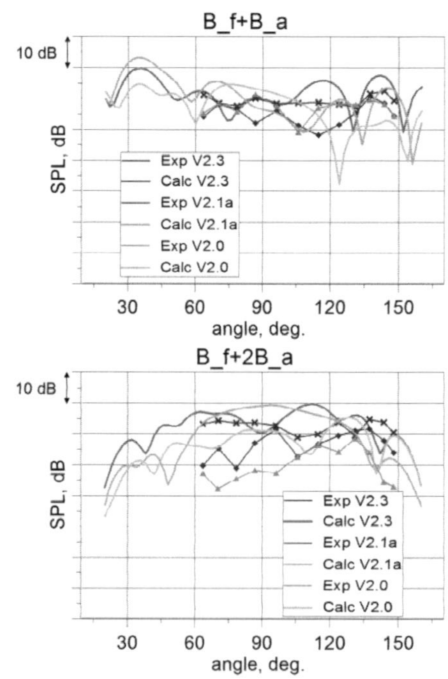

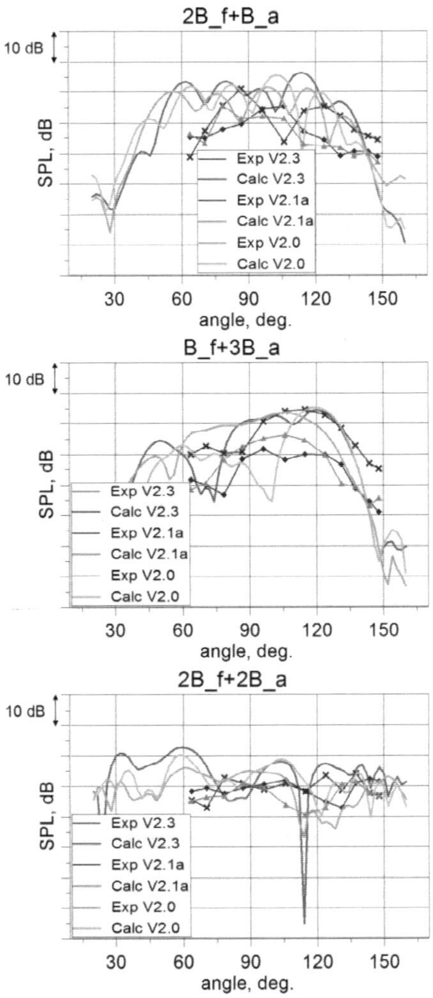

Fig. 11. Comparison between computational and experimental directivity diagrams for the combinational harmonics.

CONCLUSION

The comparison between the results of aeroacoustic calculations and the results of measurements obtained in the low speed wind tunnel tests for three open rotors in take-off working conditions is provided. The calculations were made using 3DAS aeroacoustic solver.

The comparisons between the results of computations and experiments for the tone noise on the first five combinational harmonics (main source - inter-row interaction) showed satisfactory qualitative and quantitative correspondence of directivity diagrams for one of the open rotors – V2.3. For the other two open rotors (V2.0 and V2.1a) correspondence is observed only for the first and the fifth harmonics. The authors expect, that the differences for the other harmonics are caused by the simplifications of the geometry they made in the calculations.

For the noise on the harmonics of blade passage frequencies of blade rows the results of computations and experiments were qualitatively different. The differences can be explained taking into account design features of the test rig.

ACKNOWLEDGMENTS

The work described in the paper concludes a part of the CIAM effort in the European funded project DREAM (FP7/2007-2013).

REFERENCES

[1] Dron, S. – Toward Acare 2020: Innovative Engine Rchitectures to Achieve the Environmental Goals?, 26-th International Congress of the Aeronautical Sciences, ICAS 2008, Anchorage, USA, 2008

[2] Hoffer, P.A., Deconinck, T., Hirsch, Ch., Ortun, B., Canard-Caruana, S., Rahier, G., Pascal, S., Caruelle, B. – Aeroacoustic Computations Of Contra-Rotating Open Rotors Using The Nonlinear Harmonic Method And A Chorochronic Approach, GT2012-68982, ASME Turbo Expo 2012, Copenhagen, Denmark, June 11-15, 2012

[3] Schnell, R., Yin, J., Funke, S., Siller, H. – Aerodynamic and Basic Acoustic Optimization of a Counter Rotating Open Rotor with Experimental Verification, AIAA 2012-2127, 18th AIAA/CEAS Aeroacoustics Conference, Colorado Springs, CO, June 04 - 06, 2012

[4] Boisard, R., Falissard, F., Delattre, G. – HPC capabilities of the elsA CFD software applied to a counter rotating open rotor test rig, FP08-2012-boisard, 47th Symposium of Applied Aerodynamics, Paris, 26-28 March 2012

[5] Busch, E.R., Keßler, M, Krämer, E. – Numerical Investigation Of Counter-Rotating Open Rotor Noise, GT2012-68625, ASME Turbo Expo 2012, Copenhagen, Denmark, June 11-15, 2012

[6] Envia. E. – Open Rotor Aeroacoustic Modelling, (NASA/TM—2012-217740) 15th International Conference on Fluid Flow Technologies (CMFF'12), Budapest, Hungary, September 4–7, 2012

[7] Sharma, A. and Chen, H., "Prediction of Tonal Aerodynamic Noise from Open Rotors," AIAA-2012-2265, 18th AIAA/CEAS Aeroacoustics Conference, Colorado Springs, CO, CO, June 04 - 06, 2012

[8] Nyukhtikiv, M., Brailko, I., Mileshin, V., Pankov, S., – Computational and Experimental Investigation of Unsteady and Acoustic Characteristics of Counter – Rotating Fans, HT-FED2004-56435, ASME Heat Transfer / Fluids engineering Summer Conference 2004, Charlotte, North Carolina USA, July 11-15, 2004

[9] Detailed description of the program DREAM – "Validation of radical engine architecture systems" can be found in: http://www.dream-project.eu

[10] Lepot, I., Leborgne, M., Schnell, R., Yin, J., Delattre, G., Falissard, F., Talbotec, J. Aero-mechanical optimization of a contra-rotating open rotor and assessment of its aerodynamic and acoustic characteristics European Turbomachinery Conference March 21-25, Istanbul, Turkey, 2011

[11] Rossikhin, A.A., Brailko, I.A., Mileshin, V.I., "Numerical investigations of tonal noise of counter-rotating open rotors," International Journal of Aeroacoustics, vol. 13, num. 3 & 4, pp. 303 – 320, 2013
[12] Nyukhtikov, M.A., Rossikhin, A.A., Sgadlev, V.V., Brailko, I.A. – Numerical Method for Turbo-Machinery Tonal Noise Generation and Radiation Simulation Using CAA Approach, GT2008-51182, ASME Turbo Expo 2008: Power for Land, Sea and Air, GT2008, Berlin, Germany June 9-13, 2008
[13] Nyukhtikov M.A., Rossikhin A.A. – Numerical method for calculating three-dimensional fan tonal noise due to rotor-stator interaction, P. 268-280, International Congress «Turbomachines: Aeroelasticity, Aeroacoustics, and Unsteady Aerodynamics», TORUS PRESS Ltd. Moscow, 2006.
[14] Tam, C.K.W., Webb, J.C. – Dispersion-Relation-Preserving Schemes for Computational Acoustics, Journal of Computational Physics, Vol. 107, pp. 262-281, 1993
[15] Hu, F.Q., Hussaini, M.Y., Manthey, J. – Low-Dissipation and Low-Dispersion Runge-Kutta Schemes for Computational Acoustics, Journal of Computational Physics, Vol. 124, pp. 177-191, 1996
[16] Vasilyev, O.V. – A General Class of Commutative Filters for LES in Complex Geometries, Journal of Computational Physics, Vol. 146, pp. 82–104, 1998
[17] Brentner, K.S., Farassat, F. – Analytical Comparison of the Acoustic Analogy and Kirchhoff Formulation for Moving Surfaces, AIAA Journal, Vol. 36, No. 8, pp. 1379–1386, 1998
[18] Rossikhin, A., Pankov, S., Brailko, I., Mileshin, V. – Numerical Method For 3D Computation Of Turbomachinery Tone Noise, Paper 035, International Conference Fan 2012, Senlis, France, April 18-20, 2012
[19] Pankratov, I.V., Samokhin, V.F., Vlasov, E.V., Chevagin, A.F. – Acoustic tests of contra rotating propellers in the Dream project, AIAA 2011-2761, 17th AIAA/CEAS Aeroacoustics Conference, Portland, Oregon, 05 - 08 June 2011
[20] Lewy, S. – Semi-empirical prediction of tone noise due to counter-rotating open rotors, 20th International Congress on Acoustics, ICA 2010, Sydney, Australia, 23-27 August 2010

Proceedings of the 14th International Symposium on Unsteady Aerodynamics, Aeroacoustics & Aeroelasticity of Turbomachines
ISUAAAT14
8-11 September 2015, Stockholm, Sweden

I14-S14-1

TURBOCHARGER TURBINE PERFORMANCE: A SENSITIVITY TO THE BOUNDARY CONDITIONS

Shyang Maw Lim
KTH Royal Institute of Technology
Competence Center for Gas Exchange (CCGEx)
Stockholm, Stockholm, Sweden

Mihai Mihaescu
KTH Royal Institute of Technology
Competence Center for Gas Exchange (CCGEx)
Stockholm, Stockholm, Sweden

ABSTRACT

Deviation of turbocharger's performance under engine-like pulsating flow conditions from that estimated on a test rig under steady-flow conditions motivated this study. This numerical study aimed to look into the sensitivity of turbocharger turbine performance to different pulse profiles for the exhaust gases which are feeding the turbine through the exhaust manifold. Exhaust pulse profiles which are corresponding to an automotive engine working with wide open throttle and with divided exhaust period respectively, as well as corresponding to a test rig pulse generator were investigated by using Detached Eddy Simulations (DES) approach under adiabatic condition. The study found that the level of phase shift in mass from upstream to downstream in the turbine is different. Furthermore, rotor torque was found to be strongly affected by the pulse profile, considering the same injected mass flow per engine cycle. No direct correlation between rotor's upstream total pressure loss and the rotor torque was observed. The instantaneous relative flow field under pulsating flow condition was found to be dissimilar qualitatively when compared with the corresponding non-pulsating flow condition. When compared with literatures, this study observed difference rotor behaviour in handling period with no flow. Also, different direction of phase shift in mass at rotor downstream was also observed, as compared with past study. These differences might suggest that the effects of pulsating flow to turbine performance is geometry dependence and further study is necessary to find the optimum matching of turbine and inflow pulse.

1. INTRODUCTION

Stricter legislations to improve fuel economy and to reduce pollutants of automotive vehicles have been imposed in the last decade. This has made the automotive industry to constantly seek for innovative technologies to meet the legislations requirements. Downsizing of the Internal Combustion Engine (ICE) and the use of turbocharging is considered to be a mature technology, which will be in used for the incoming years to reduce fuel consumption and maintain or improve engine efficiency.

The flow in the exhaust manifold, downstream of which the turbocharger's turbine is mounted, is highly unsteady and pulsating. The performance of the turbocharger is closely related to the efficiency with which the kinetic energy in the exhaust flow can be extracted by the turbine (see e.g. (1)). It is known that the performance of the turbocharger's turbine under pulsating conditions deviates from the steady state conditions (2,3) and that the turbine operates more efficiently under continuous flow conditions (1).

Much effort has been made to understand the underlying physics of turbine operating under pulsating conditions. Experimentally, flow field measurement with Laser Doppler Velocimetry (LDV) under steady and pulsating conditions showed that under pulsating conditions, the velocity triangles at the inlet and outlet of rotor of a mixed flow turbine change during the pulse period, and it deviates from an optimum (4). A detailed flow field assessment within an experimental set-up might be difficult and costly. Computational Fluid Dynamics (CFD) can be an alternative or a complementary approach to expensive measurements. It has been applied to investigate the unsteady flow features in three-dimensional (3D) space to help understanding turbine performance under pulsating conditions, see e.g. (5).

Many studies have been devoted to understand the relationship between the flow parameters and turbine performance. Pulse frequency is the most studied variable among flow parameters and it has a strong influence on the turbine efficiency since it affects the incident angle at the rotor inlet (4,6,7). A recent study suggested that pulse frequency might also affect the friction loss in the rotor, which then influences the turbine efficiency (7). Moreover, the turbine speed, the profile of the pulsating flow and the developed secondary flow within manifold also have great influence on turbine performance. Turbine speed affects the matching time of optimum incidence angle with high energy instances at the rotor inlet, thus influencing the average power output (7). Also, the torque generated by the rotor is affected by the pulse profile (8). The developed secondary flow upstream of the turbine caused by geometrical complexity of the manifold (resulting in e.g. vortical flow structures, possible flow separation and swirling flow motion), influences the flow distribution at the rotor. This has also a direct impact on turbine performance (9,10).

The sensitivity of turbocharger's turbine performance to different pulse profiles imposed for the exhaust gases at the corresponding cylinders exit ports was investigated in this study. Different pulse profiles were investigated, which are corresponding to an automotive engine working with wide open throttle and with divided exhaust period, respectively. Furthermore, a typical pulse profile generated by a pulse generator in a test rig was also investigated. Injected mass per engine cycle was kept constant for all cases.

2. METHODOLOGY

The target of this study was a turbocharger radial turbine used in a passenger car with a four cylinders, two litres spark-ignition (SI) engine. The turbine is nozleless and waste-gated, with a rotor of twelve blades. The rotor is characterized by 55mm inducer diameter and a trim of 70.

The study was divided into two parts. The first part of the study focused on computational mesh sensitivity study and validation of computed performance parameters with steady gas stand experimental data. The second part of the study investigated the effects of pulse profiles to the performance of turbocharger turbine by incorporating exhaust manifolds in the computational domain. The commercial finite-volume based CFD solver StarCCM+ by CD-Adapco was used to generate the computational mesh and to perform the numerical simulations. It has been previously validated and used in the assessment of flow behaviour within a centrifugal compressor under design and off-design operating conditions (11,12).

2.1 Mesh sensitivity and validation study

2.1.1 Continuous flow set-up

Figure 1(a) shows the computational domain used for the validation study. The domain consists of an inlet pipe, the scroll, the rotor and an outlet pipe, as according to the steady gas stand setup. On the other hand, Figure 1(b) corresponds to the integrated exhaust manifold-turbine geometry, the computational domain considered under pulsatile flow conditions. The waste gate and its by-pass channel were not modelled because steady gas stand experiment was performed with the waste gate closed. The gap between the rotor and backplate, which was neglected in some literatures (9,10,13), was included in this study to account for windage loss. Tip clearance between the rotor blade and the shroud wall was modelled as tip leakage flow is well known to have great impact on the aerodynamic performance of turbomachinery.

The fluid domain was discretized using a polyhedral computational mesh suitable for such a complex geometry. Polyhedral mesh has more neighbouring faces, and it has been shown that it can predicts flow gradient with less numerical diffusion at fewer cells (14). At regions far upstream from the scroll and downstream from the rotor such as the inlet pipe and lower part of outlet pipe, a hexahedral mesh was extruded from the polyhedral cells up to the inlet and outlet boundaries. The computational grid near-by the rotor was refined locally to capture the blade curvature and to resolve the complex flow. Thus, prism layers were used to generate orthogonal prismatic cells to capture flow and temperature field near to the wall. Five prism layers were employed in all near-wall regions. In order to ensure smooth transition of mesh size from near-wall region to the core-flow region, stretching factor of 1.5 was used in prism layers.

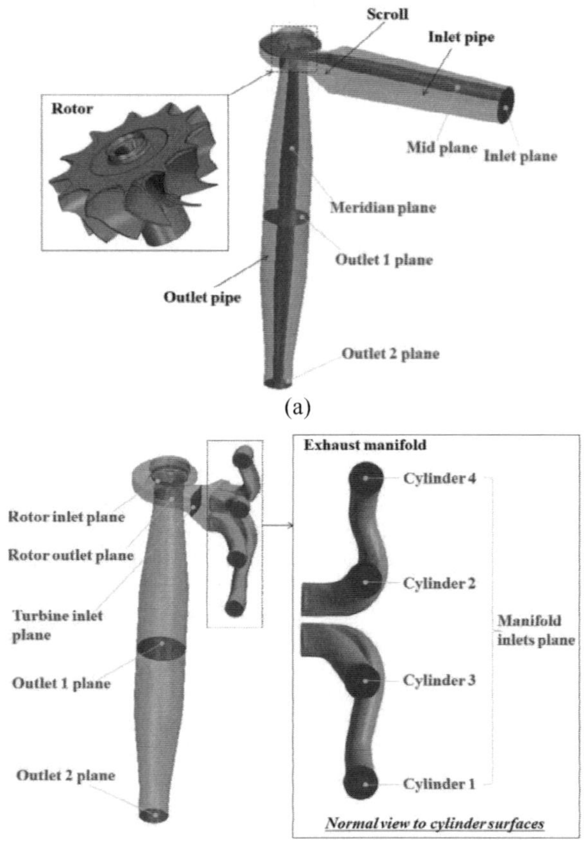

Figure 1. Computational domains: (a) Simplified geometry corresponding to the gas-stand experiments under continuous flow; (b) Integrated exhaust manifold-turbine under pulsating flow conditions.

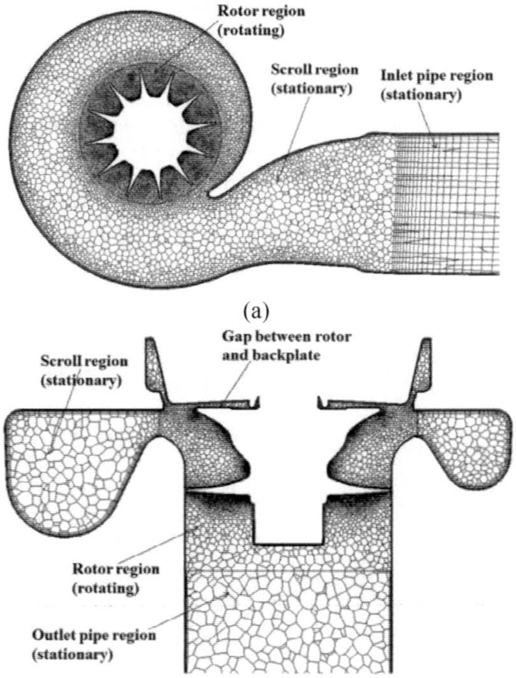

Figure 2. Computational mesh (medium resolution) used, part of the sensitivity and validation studies: (a) Meridian plane; (b) Mid-longitudinal cut through the turbine's rotor.

The mesh sensitivity study was performed according to (15). The sensitivity of global performance parameters such as pressure ratio and rotor torque to the mesh size was evaluated on three successively refined meshes of approximately 1.7 million (coarse), 2.4 million (medium) and 3.8 million (fine), as shown in Table 1. The refinement factors, as calculated by using Eq. 1 and Eq. 2, from coarser mesh to finer mesh are 1.12 (r_{fm}) and 1.18 (r_{mc}). Note that the mesh refinement factors used in this study are lower than the recommended value of greater than 1.3 (15). This is because with a refinement factor greater than 1.3, the computational cost for such a complex geometry will become too high for application purpose. However, the refinement factor employed in this study falls within 1.1~1.2, i.e. typical range found when treating turbomachinery corresponding flows (see e.g. (13,16)).

$$h = \left[\frac{1}{N}\sum_{i=1}^{N}(\Delta V_i)\right]^{1/3} \quad (1)$$

Here h is the representative mesh size, ΔV_i is the volume of ith cell, and N is the total number of cells used in the computational domain. The refinement factor r is:

$$r = \frac{h_{coarser}}{h_{finer}} \quad (2)$$

The boundary conditions imposed correspond to the maximum efficiency operating point on a low speed line, according to the turbine map obtained from the steady gas stand experiment. Mass flow rate of 0.0587 kg/s and total temperature of 1173.15 K were imposed at the inlet of the computational domain. The fluid was assumed to enter normally into the inlet boundary, and thus no secondary flow was imposed at the inlet. 10% turbulence intensity and turbulent length scale corresponds to the radius of inlet pipe was imposed at the inlet, as according to literature (8). At the outlet of the computational domain, a constant atmospheric pressure and static temperature of 972K was imposed. The rotor's rotation (constant rotational speed of 84533 rpm) was handled by using the sliding mesh technique, previously tested and found superior to the. Multi Reference Frame (MRF) approach (9,13). The time step used was about 2 x 10^{-6} seconds. It corresponds to one degree rotation of the rotor. All the walls in the computational domain were treated as smooth, adiabatic and no-slip walls. Standard wall functions were used to approximate the near-wall velocity profiles in the present study. Further studies will assess the solution's sensitivity to this aspect.

The working medium (air) in the turbine was assumed to be a thermally perfect gas. Thus, the behaviour of the air was modelled by using the ideal gas law, with Sutherland's Law approximation for its dynamic viscosity and thermal conductivity. The specific heat of air was modelled by using polynomial functions to reflect its dependency on temperature.

Initially, a steady-state Reynolds Averaged Navier-Stokes (RANS) simulation based on k-ω SST turbulence model was carried out, using the mixing plane interface method. This implies that the transfer of conserved quantities across the interface (between stationary and rotating domains) is done by using circumferential averaging technique. The RANS data was then used to initialise the flow field for the unsteady flow calculation, namely the DES with Smagorinsky Subgrid Scale (SGS). The DES calculations were run until statistically convergence of global performance parameters was reached.

A density-based couple solver was used to solve the mass, momentum and energy conservation equations. Implicit hybrid bounded-central differential scheme was used to discretize the convective and diffusion terms in the governing equations, as recommended in Star-CCM+ documentation (17). The temporal discretization used with DES was second order.

2.1.2 Continuous flow results

Table 1. Mesh sensitivity analysis

	Number of cells, N (-)	Pressure ratio, π (-)	Rotor torque, τ (N-m)
Experiment	-	1.30	-
Coarse, c	1697398	1.30	0.4087
Medium, m	2371663	1.30	0.4101
Fine, f	3848317	1.30	0.4109

Table 1 shows the results of mesh sensitivity and validation study. Pressure ratio is calculated as the ratio of mass flow averaged total pressure at the inlet plane to the mass flow averaged static pressure at outlet 1 plane; see Figure 1(a) for locations of planes. Rotor torque is obtained by integrating the pressure and shear forces on the rotor surface. The computational data reported in Table 1 are the mean values computed from statistically converged DES simulation. Pressure ratio and rotor torque were found to be statistically converged after about 120 revolutions, or 0.08 second. From Table 1, it can be observed that the sensitivity of pressure ratio prediction to mesh size is small. Since the predicted pressure ratio value is 1.30 and it agrees well with the experimental value, it is thought that exact solution has been attained for pressure ratio. Although no experimental data is available for rotor torque validation, discretization errors for the fine mesh in term of Grid Convergence Index (GCI) for rotor torque is estimated to be small, i.e. 0.12%.

2.2 Pulsating flow set-up

The time scale of exhaust pulse is larger than blade passing event of a rotor at an order of three (18). In order to solve frequencies as lowest as 10 Hz (assuming 1250 rpm engine speed as in this study) with a time-step equivalent to one degree of rotor rotation, a physical simulation time of about 0.1 second is required (to solve for one engine cycle in pulsating flow computations). Furthermore, to get a statistically converged cycle to cycle average data (e.g. rotor torque estimations) with accuracy better than 1%, the minimum physical time, Γ_{min} (see Eq. 3 (19)) is estimated to be about 0.6 second, or equivalent to six engine cycles. This implies that the computational time of pulsating flow is about eight times longer than the continuous flow computation replicating the gas stand conditions.

$$\Gamma > 20000\Lambda\left(\frac{\sigma_x}{\mu_x}\right)^2 \qquad (3)$$

Here Λ is the integral time scale, σ_x and μ_x are the standard deviation and the mean of evaluation parameter (e.g. rotor torque, mass flow rate, total pressure, etc.), respectively. In the estimation discussed above, the integral time scale, Λ was taken as ~0.1 second, the physical time of an engine cycle. The ratio of $\left(\frac{\sigma_x}{\mu_x}\right)$ was taken as ~0.017 second, a value estimated from rotor torque data over four engine cycles.

As shown in Figure 1(b), for the pulsating flow computations, the inlet pipe was replaced by four exhaust ports merging into a manifold that feeds the turbine. Since the predicted pressure ratio is unaffected by the mesh size, and the discretization error from medium to fine mesh is small (see Section 2.1.2), the mesh characteristics for the scroll, rotor and outlet pipe were chosen to be the same as the medium mesh described in Section 2.1. As for the exhaust pipes, two prism layers were used. This was an effort to reduce the computational time. The selection of medium mesh integrated with exhaust manifold resulted in ~3.5 million polyhedral cells.

Figure 3(a) shows the pulse profiles investigated in this study. The profile named "Normal" and "Short" correspond to an automotive engine working with wide open throttle and with divided exhaust period, respectively. These profiles were adopted from literature (8). Furthermore, typical pulse profile generated by a pulse generator in a test rig (20) was also investigated. Pulse frequency was the same for all profiles, i.e. corresponds to 1250 rpm engine speed. Injected mass per engine cycle was kept constant for all cases, focusing the study on the influence of pulse profiles to the turbine performance. Note that this resulted in different pulse amplitudes, as can be seen in Figure 3(a).

As for the boundary conditions, time dependent mass flow and total temperature were imposed according to the engine firing order (i.e. 1-3-4-2) at the four inlets to the computational domain. The total temperature profile was adopted from literature (8) and it was the same for all cases. The fluid was assumed to enter the domain normally with 5% turbulence intensity and turbulent length scale equivalent to the radius of inlet, as according to literature (8). Constant atmospheric pressure and static temperature of 1090 K were assigned to the outlet. Otherwise, the computational setup was the same as described in Section 2.1.1 The simulations were run for several engine cycles and it was found that maximum cycle to cycle variations of peak rotor torque is less than 2% after four engine cycles (2880 CAD). Due to this and due to the fact that obtaining statistically converged mean value in pulsating flow simulations will require a much longer computational time, the data presented hereinafter is the cycle to cycle averaged data using four engine cycles to give relative comparison of the effects of different pulse profiles.

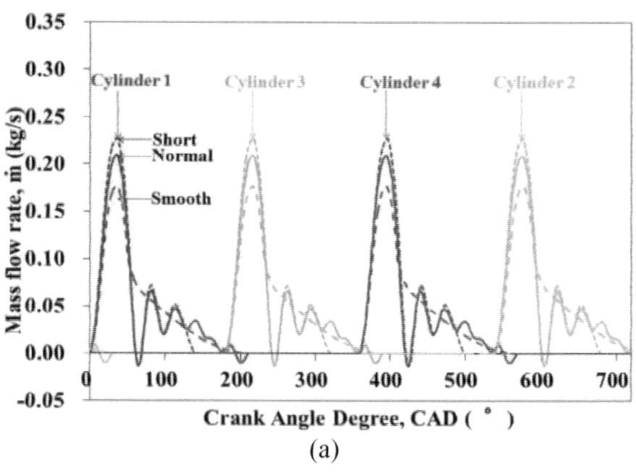

(a)

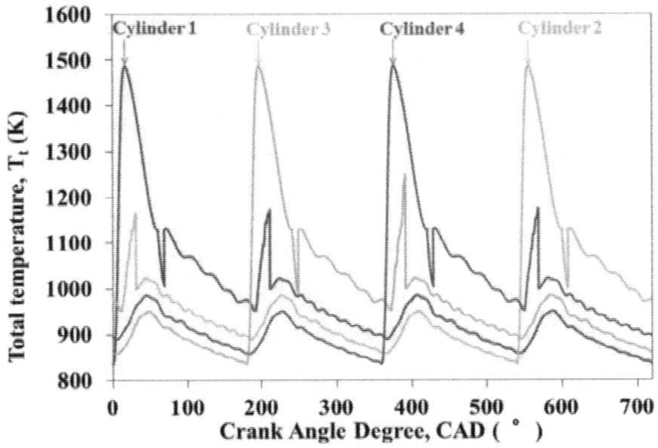

(b)

Figure 3. Inlet boundary conditions at exhaust manifold inlets for pulsating flow study: (a) Mass flow rate; (b) Total temperature.

3. PULSATING FLOW DATA

3.1 Mass flow characteristics

Figure 4 shows the change of mass flor rate at different locations (see Figure 1(b) for plane locations) over an engine cycle for different pulse profiles. For Normal profile, it can be observed from Figure 4(a) that the peak mass flow at turbine inlet is shifted with approximately 10 CAD compared with the peak mass flow at the manifold inlets. Mass phase shift is also observed from turbine inlet to rotor inlet, but it is not as much when compared to manifold inlets to turbine inlet. This is because the mean path travelled by a gas particle is longer in the exhaust manifold due to bended pipe geometry, as compared with the scroll. Since the mean path between rotor inlet and rotor outlet is short, no significant mass phase shift is observed in the rotor. Furthermore, an increase in mass flow between rotor outlet and outlet 1 plane is observed. This phenomenon is also reported in another study and it was attributed to the effect of rotor's downstream plenum acting as resonator (13). However, in this study, the mass flow at outlet 1 plane is observed to be in-phase again with the mass flow at manifold inlets. This observation is different from the literature, which reporting a continuous phase shift from upstream to downstream (13). The difference is due to the expansion in the outlet pipe geometry downstream of the rotor, which is used here to resemble the gas-stand experimental set-up. This turned out to affect the phase-shifting phenomenon. Similar observations apply to the Short and Smooth profiles, as depicted in Figure 4 (b) and (c).

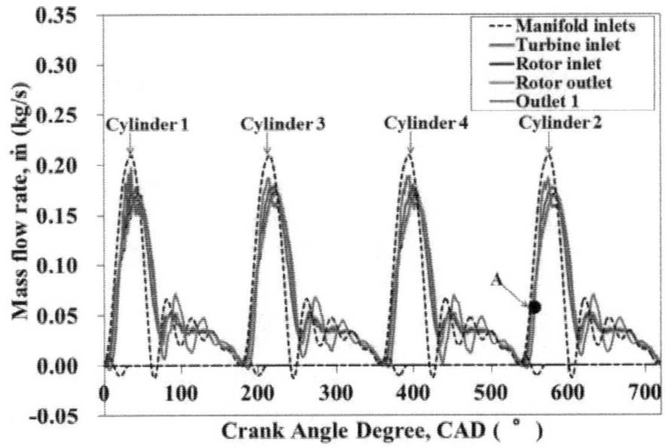

(a)

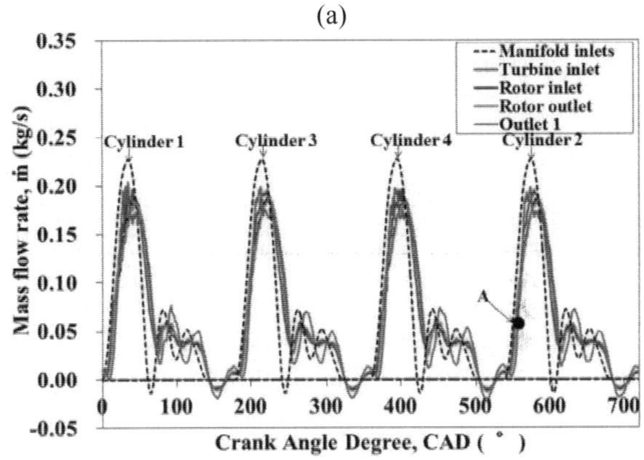

(b)

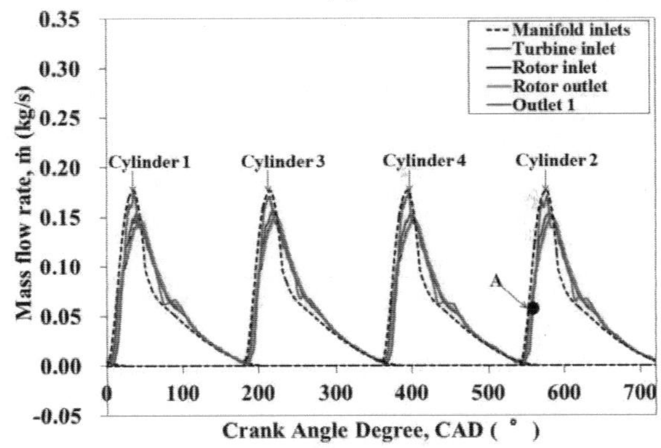

(c)

Figure 4. Comparisons of cycle to cycle averaged mass flow rate at different locations in turbine. (a) Normal; (b) Short; (c) Smooth. Point A correspond to $\dot{m}$=0.0587 kg/s.

3.2 Rotor torque characteristics

Figure 5 shows the variations of rotor torque over an engine cycle. From Figure 5, it can be seen that Short profile has highest peak in rotor torque, followed by Normal and respectively Smooth profile. The rotor torque goes to zero or even negative when there is very low mass flow at the manifold inlets. This means that the rotor cannot handle the long period of time with no flow entering the geometry. This trend is different from the result published in other studies, which observed non-zero rotor torque for a long period of no flow entering the geometry (8,10). Also, from Figure 5, it can be seen that peak rotor torque is the highest when the flow comes from cylinder 1, and almost equal peak rotor torque is obtained when the flow comes from cylinder 2, 3 and 4. The observed trend is independent of pulse profiles. Again, this result is different from past literature, which observed higher peak rotor torque when the flow comes from outer cylinders 1 and 4, compared with the instance when the flow comes from inner cylinders 2 and 3 (8). As can be seen from Figure 1(b), in this study, the exhaust pipes coming from the outer cylinders 1 and 4 have different length and bend, and these geometrical differences might affect the behaviour of rotor torque. Differences with the literature observed in this study suggest that there exist a best match between pulse profile and turbine geometry for optimum turbine performance in pulsating flow conditions.

The variation of absolute total pressure over an engine cycle at different locations (See Figure 1(b) for location of planes) is shown in Figure 6. It can be seen that total pressure loss in the exhaust manifold is almost unaffected by the pulse profile. Furthermore, in the scroll, the total pressure loss is also unaffected by the pulse profile and it is small, as compared with total pressure loss in the exhaust manifold. From the observations above, the correlation between the rotor torque and upstream total pressure loss is almost insignificant in this study. The variation in rotor torque seems to be dictated by the applied mass flow at the manifold inlets, i.e. the higher the mass flow peak, the higher the rotor torque peak (See Figure 3(a) also).

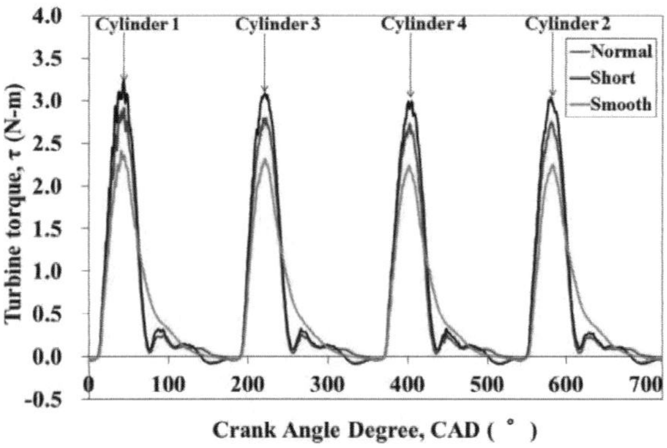

Figure 5. Comparisons of cycle to cycle averaged rotor torque for different pulse profiles.

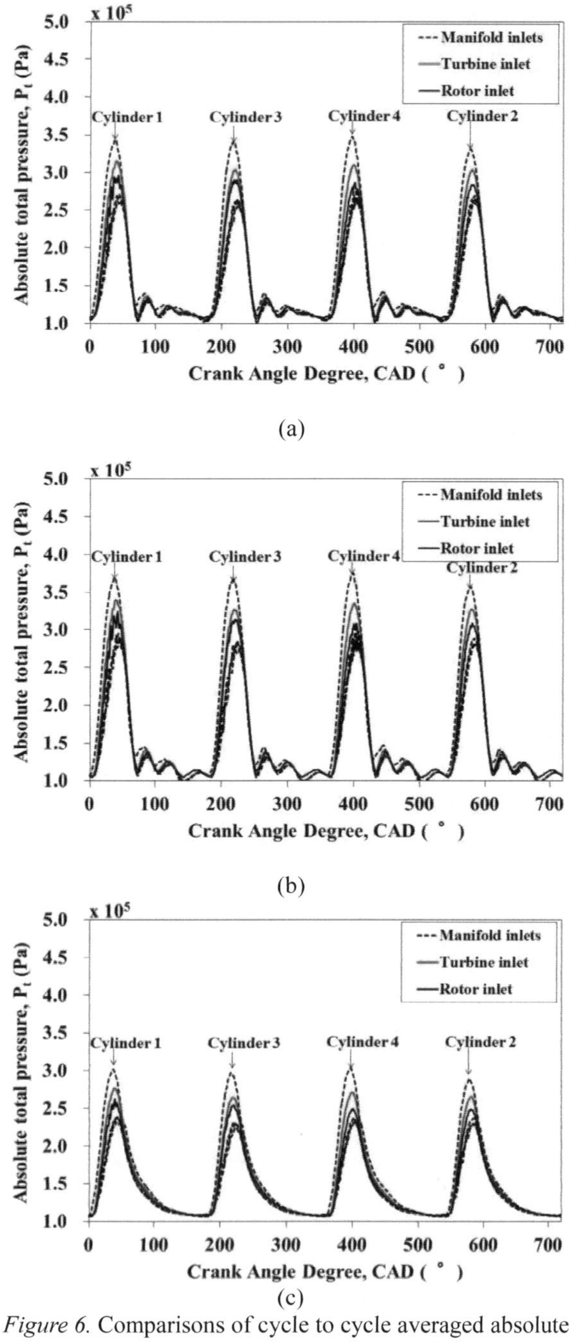

Figure 6. Comparisons of cycle to cycle averaged absolute total pressure at different locations in turbine. (a) Normal; (b) Short; (c) Smooth.

3.2 Flow field evaluation

Figure 7 compares the instantaneous relative flow field at the mid span of rotor between steady flow condition and pulsatile flow conditions under the same mass flow rate of $\dot{m}$=0.0587 kg/s at the rotor inlet (corresponding to point A in Figure 4). Figure 7(b), (c) and (d) are the instantaneous flow field at the fourth engine cycle. From Figure 7(a), it can be seen that the flow is attached to the rotor blades under steady flow condition. The operating point studied corresponds to maximum efficiency operating point in the steady turbine performance map and hence secondary flow structures are not expected to present under steady flow conditions. In contrast, as for pulsating flow, flow separation is observed at the pressure surface of rotor blade near to the rotor inlet, as depicted in Figure 7(b), (c) and (d). The flow enters the rotating domain at a smaller incidence angle instantaneously for pulsating flow condition, which cause mismatch of relative incoming flow angle with the blade angle. This phenomenon suggests that there is no similarity in the flow field between the steady flow and pulsating flow conditions, although the mass flow rate into the rotor is the same.

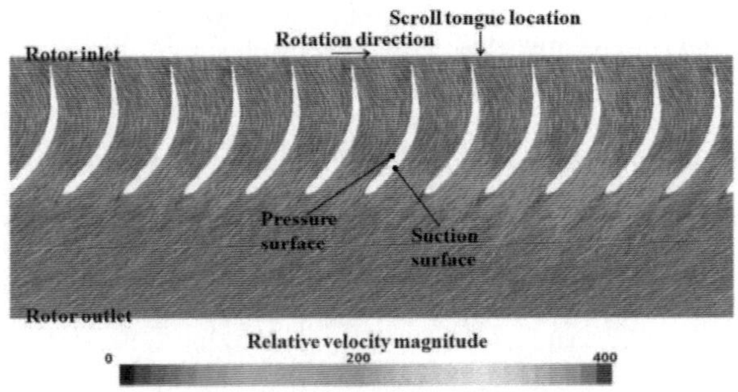

(a)

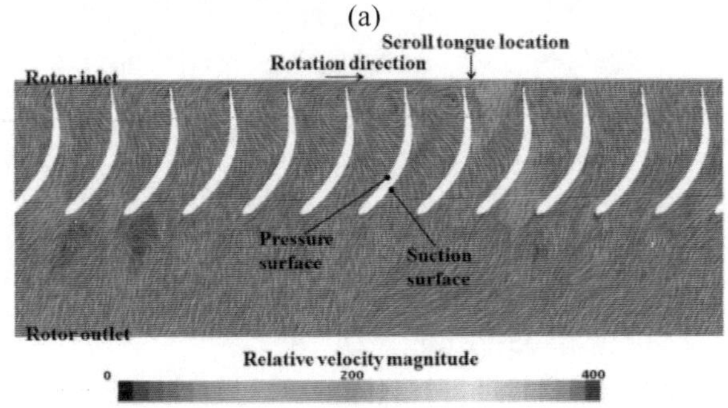

(b)

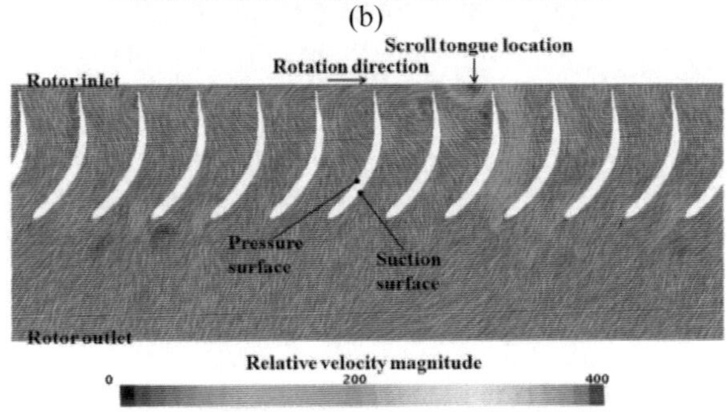

(c)

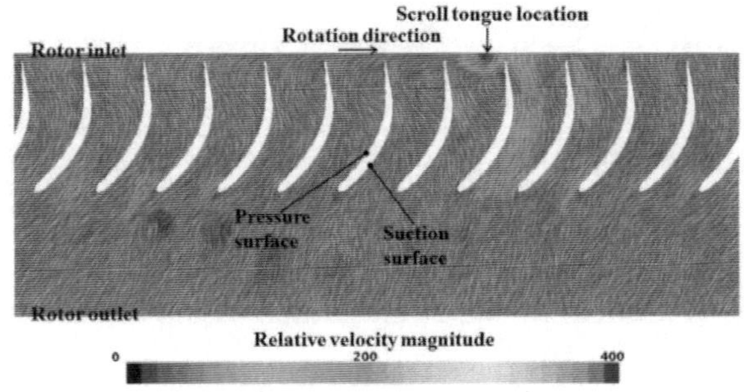

(d)

Figure 7. Comparison of instantaneous relative velocity flow field at midspan of rotor when $\dot{m}$=0.0587 kg/s at rotor inlet. (a) Steady flow; (b) Normal; (c) Short; (d) Smooth. Flow field for pulsating cases correspond to point A in Fig. 5.

4. CONCLUSIONS

This study represents an effort in understanding the effects of exhaust pulse profile to the performance of turbocharger turbine under pulsating flow conditions. Phase shift in mass with different levels was observed at different locations in the turbine. Rotor torque was found to be strongly affected by the pulse profile, under the condition of same injected mass per engine cycle. No direct correlation between rotor's upstream total pressure loss and the rotor torque was observed. Furthermore, instantaneous relative flow field under pulsating flow condition was found to be dissimilar with the corresponding steady operating point.

Several differences with regards to the effects of pulsating flow to the performance of turbine were observed between this study and published literatures. For example, phase recovery in mass was obtained at the downstream of rotor in this study, whereas other literature observed continuous phase lagging from upstream to downstream of turbine (13). Also, the behaviour of rotor in handling period with no flow was also observed to be different from literatures (8,10). All these differences might be due to different turbine geometry and future study is necessary to investigate the optimum matching between pulse profile and turbine performance over a wide range of engine speed. Since cycle to cycle averaged data published in this study is not fully statistically converged, there is a need to run the pulsating computations for a longer time to get better quantitative representation of the turbine performance data in the future.

NOMENCLATURE

Notations

h	characteristic mesh size	(-)
$\dot{m}$	mass flow rate	(kg/s)
N	total number of cells	(-)
H	representative mesh size	(-)
P	absolute pressure	(Pa)
r	mesh refinement factor	(-)
T	temperature	(K)
ΔV_i	Volume of ith cell	(m^3)

Greek

τ	rotor torque	(N-m)
π	total to static pressure ratio	(-)
Γ	Physical time	(s)
Λ	integral time scale	(s)
σ_x	standard deviation	
μ_x	mean	

Subscripts

mc	coarse to medium
fm	medium to fine
t	total
min	minimum

ACKNOWLEDGMENTS

The work was carried out within the framework of the Competence Center for Gas Exchange (CCGEx) at KTH. The financial support from the Swedish Energy Agency and the contributions from Scania, Volvo Trucks and Borg Warner are greatly acknowledged. The Swedish National Infrastructure for Computing via Parallel Computing Center at KTH is acknowledged for providing the necessary computational resources to perform this study.

REFERENCES

1. Heywood J. Internal Combustion Engine Fundamentals. McGraw-Hill Education; 1988.

2. Wallace FJ, Blair GP, Engineers. AS of M. The pulsating-flow performance of inward radial-flow turbines. New York, N.Y.: ASME; 1965.

3. Benson RS, Scrimshaw KH. An experimental investigation of non-steady flow in a radial gas turbine. Proceedings of the Institution of Mechanical Engineers, Conference Proceedings. 74–85, 1965. http://pcp.sagepub.com/content/180/10/74.abstract

4. Karamanis N, Martinez-Botas RF, Su CC. Mixed Flow Turbines: Inlet and Exit Flow under Steady and Pulsating Conditions. ASME-AMER SOC MECHANICAL ENG; 359–371, 2001.

5. Palfreyman D, Martinez-Botas RF. The pulsating flow field in a mixed flow turbocharger turbine: An experimental and computational study. ASME-AMER SOC MECHANICAL ENG, 144–155, 2005. http://dx.doi.org/10.1115/1.1812322

6. Hellström F, Fuchs L. Numerical computations of pulsatile flow in a turbo-charger. AIAA paper, AIAA 2008-735, 2008.

7. Padzillah MH, Rajoo S, Martinez-Botas RF. Influence of speed and frequency towards the automotive turbocharger turbine performance under pulsating flow conditions. Energy Convers Manag. 2014 Apr;80:416–28.

8. Fjällman J. Large Eddy Simulations of Complex Flows in IC-Engine's Exhaust Manifold and Turbine. PhD Thesis, KTH Royal Institute of Technology, ISBN 978-91-7595-270-3, 2014. http://www.diva-portal.org/smash/get/diva2:748463/FULLTEXT01.pdf

9. Hellström F, Fuchs L. Effects of inlet conditions on the turbine performance of a radial turbine. ASME paper, GT2008-51008, 1985-2001, 2008.

10. Hellström F, Fuchs L. Numerical computation of the pulsatile flow in a turbocharger with realistic inflow conditions from an exhaust manifold. ASME paper, GT2009-59619, 1317-1329, 2009.

11. Semlitsch B, Jyothishkumar V, Mihaescu M, Fuchs L, Gutmark E, Gancedo M. Numerical Flow Analysis of a Centrifugal Compressor with Ported and without Ported Shroud. SAE technical paper, 2014-01-1655, 2014.

12. Sundström E, Mihaescu M. Centrifugal Compressor: The Sound of Surge. AIAA paper, AIAA 2015-2674, 20154.

13. Galindo J, Fajardo P, Navarro R, García-Cuevas LM. Characterization of a radial turbocharger turbine in pulsating flow by means of CFD and its application to engine modeling. Appl Energy. 2013 Mar;103:116–27. http://www.sciencedirect.com/science/article/pii/S0306261912006538

14. Peric M. Flow simulation using control volumes of arbitrary polyhedral shape. ERCOFTAC Bulletin, no 62, September 2004.

15. Procedure for Estimation and Reporting of Uncertainty Due to Discretization in CFD Applications. J Fluids Eng. ASME; 2008 Jul 22;130(7):78001. http://dx.doi.org/10.1115/1.2960953

16. Sundström E, Semlitsch B, Mihaescu M. Assessment of the 3D Flow in a Centrifugal compressor using Steady-State and Unsteady Flow Solvers. SAE Technical paper, 2014-01-2856, 2014.

17. CD-adapco. STAR-CCM+ User Guide 10.02.012. 2014.

18. Copeland C, Newton P, Martinez-Botas RF, Seiler M. A comparison of timescales within a pulsed flow turbocharger turbine. Proc IMechE, Pap. 2012;(C1340/086):389–404.

19. Segalini A, Alfredsson PH. Uncertainty Analysis. KTH Mechanics; 2014.

20. Laurantzon F, Örlü R, Segalini A, Tillmark N, Alfredsson PH. Experimental analysis of turbocharger interaction with a pulsatile flow through time-resolved flow measurements upstream and downstream the turbine. 10th Int Conf on Turbochargers and Turbocharging (iMechE). 2012. p. 405–15.

Proceedings of the 14th International Symposium on Unsteady Aerodynamics,
Aeroacoustics & Aeroelasticity of Turbomachines
ISUAAAT14
8-11 September 2015, Stockholm, Sweden

I14-S14-2

UNSTEADY NUMERICAL INVESTIGATION OF TIP CLEARANCE FLOW AND ROTOR-STATOR INTERACTION IN THE CENTRIFUGAL COMPRESSOR

Alireza Ameli
Laboratory of Fluid Dynamics
School of Energy Systems
Lappeenranta University of Technology
Lappeenranta, Finland
Email: Alireza.Ameli@lut.fi

Teemu Turunen-Saaresti
Laboratory of Fluid Dynamics
School of Energy Systems
Lappeenranta University of Technology
Lappeenranta, Finland
Email: Teemu.Turunen-Saaresti@lut.fi

Ahti Jaatinen-Värri
Laboratory of Fluid Dynamics
School of Energy Systems
Lappeenranta University of Technology
Lappeenranta, Finland
Email: Ahti.Jaatinen-Värri@lut.fi

Jonna Tiainen
Laboratory of Fluid Dynamics
School of Energy Systems
Lappeenranta University of Technology
Lappeenranta, Finland
Email: Jonna.Tiainen@lut.fi

ABSTRACT

Tip clearance is one of the most important factors which causes losses in centrifugal compressors. The flow inside centrifugal compressors is three-dimensional and unsteady and the tip clearance is one of the major flow characteristics affecting the flow. In this study, the effect of the tip clearance and the width of the diffuser on the diffuser and impeller unsteady performance is studied. Moreover, the mechanisms of the rotor-stator interaction on the flow field in the impeller are discussed. Special attention is paid on the effect of rotor-stator interaction on the tip clearance flow. Six different axial tip clearances with two pinched and one unpinched vaneless diffusers are modeled in 3D and numerically solved in steady and some of them in transient states. The pinched diffusers has 15% and 10% reduction in width on the shroud. One blade passage and corresponding sections of the vaneless diffuser and inlet cone were modelled.

The flow field over the tip clearance is studied in order to find out different ways about how the tip clearance flow increases the losses in the impeller passages and vaneless diffuser. Unsteady results reported in this study also show that the unsteady flow fields at the diffuser inlet and in the impeller are different with different types of diffusers. Diffuser type affects the tip clearance flow and losses. Beside, by comparing the mass averaged pressure along the tip clearance, leading edge to trailing edge, different influences of pinched and unpinched diffusers on the tip flow are demonstrated. Different time steps are compared in order to illustrate that tip clearance flow is unsteady. Numerical results were compared to experimental measurements which were conducted using the test compressor in the Laboratory of Fluid Dynamics in Lappeenranta University of Technology.

INTRODUCTION

Centrifugal compressors are widely used in many industrial applications such as air conditioning, refrigeration, small gas turbines, turbochargers, and in gas and oil industries. In small applications, such as small gas turbines, centrifugal compressors are more commonly used because of the lower manufacturing cost, in comparison with multistage axial compressors.

The interaction between rotor and stator has two major aspects. The most pronounce aspect is the effect of impeller on diffuser flow field. The effects of the vaneless diffuser on the impeller flow was studied in the literature [1]. Complex flow field develops in the impeller and strong fluctuations in the velocity caused by the vaneless diffuser could be seen in the circumferential and axial directions [2].

Moreover, the influence of the tip clearance on the centrifugal compressors performance has been studied [3] .However, there is no single correlation taking account the tip clearance loss in different compressors and phenomena related to tip clearance [4].

In the studies of Pampreen [5] and Mashino et al. [6], tip clearance was sized by changing the location of the shroud axially and it was concluded that by decreasing the tip clearance, its effect on the main flow can be decreased. Also, it was found out by moving the shroud of the impeller axially, better performance can be achieved in comparison with movement in both axial and radial directions at the same time.

The interaction of rotor-stator has been studied by many researchers. Ziegler et al. [7] performed studies on the effect of the rotor-stator interaction on the unsteady and transient flow field inside the impeller, vaned and vaneless diffusers. It was concluded that smaller radial gap between the impeller and diffuser is leading to more uniform and steady flow field inside the vaned diffuser and also it leads to a higher pressure recovery in the diffuser. Also it was seen that, a huge reduction of the wake region at the impeller outlet is shaped with vaned diffuser compared to the vaneless.

The effect of the rotor-stator interaction on the performance of a centrifugal compressor by using the unsteady calculation was evaluated by Shum et al. [8].In the study it was observed that, blockage and slip associated with the unsteady tip flow inside the impeller were reduced by effects of the rotor-stator interaction. Also it was concluded that the impeller exit unsteadiness has some effects on the pressure recovery in the diffuser and in magnitude point of view is similar to the axial distortion of velocity from diffuser inlet.

Steady state and unsteady simulations by Dawes [9] showed that the most of the pressure losses occur inside the diffuser and in the total of diffuser losses, only 5% were related to unsteady losses. Numerical investigation by Feng et al. [10] showed that the change of radial gap in a vaned diffuser does not have significant effect on the time averaged impeller performance and pressure field.

Sato and He [11] used a three dimensional unsteady code to study the flow inside a centrifugal compressor with vaned diffuser. Their result indicated the significant effect of the downstream flow near to the trailing edge of impeller on the centrifugal compressor performance. More stable flow field was generated by the effects of the rotor-stator interaction in comparison with the steady single row simulations.

In investigations by Krain [12], unsteady and time averaged results for impeller with vaned and vaneless diffusers were compared. It was seen that flow at the impeller exit is highly distorted for both diffusers. An unsteady numerical investigation by Leboeuf et al. [13] shows that by using transient RANS simulations, prediction of compressor performance with higher accuracy can be achieved compared to the steady simulation. It was concluded that the shock deformation and the backward and forward waves are the main responsible on the local losses at impeller outlet in transient simulations. Also it was seen that the effect of the unsteadiness on the total performance in the centrifugal compressor is much higher than axial compressor. An intensive research about unsteady phenomena in vaneless diffuser of a centrifugal compressor was done by Turunen-Saaresti [14]. It was seen that pressure fluctuations are less in the impeller compared to the vaneless diffuser in unsteady simulations.

In this study, effect of tip clearance flow and diffuser width on the unsteadiness in the diffuser and impeller is studied in more detail.

STUDIED CASES

The modelled compressor is an industrial high-speed, variable speed driven compressor, equipped with magnetic bearing and design pressure ratio of 1.78. The unshrouded impeller has seven full and seven splitter blades with 40° backsweep. Three different vaneless diffusers, varying in diffuser width were studied. The pinched diffusers has 15% and 10% reduction in widths $b/(b_2 + t_2)$. In the previous study by authors it was concluded that further reduction in the width of the vaneless diffuser is ended up in lower performance [15].

The pinch was implemented at diffuser inlet at a radius of $r_{2'}/r_2 = 1.01$. Diffuser outlet ratio is $r_3/r_2 = 1.68$. Figure 1 illustrates a schematic of the centrifugal compressor. The modeled compressor has previously been studied numerically and experimentally [16] [17] [18].

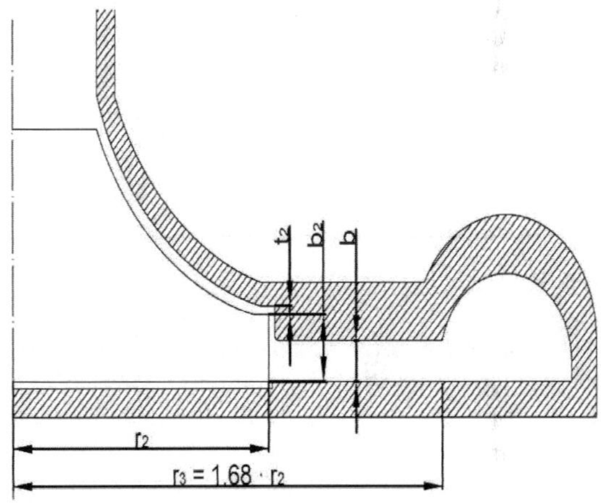

Figure 1. A schematic view of the compressor

A total of six different tip clearance were investigated in steady state and due to time consuming and difficulty of convergence in transient mode, only three different tip clearances with all types of vaneless diffusers were studied in unsteady mode. In all cases, pinch was implemented on the shroud of the diffuser because this it has been found to be more beneficial according to previous study by authors [16]. The studied cases are summarized in Table 1.

Table 1. Studied cases.

Case	t_2/b_2	$b/(b_2+t_2)$	*Steady*	*Unsteady*
Unpinched	0.026	1.00	Done	------
	0.052	1.00	Done	Done
	0.077	1.00	Done	Done
	0.103	1.00	Done	Done
	0.129	1.00	Done	------
	0.155	1.00	Done	------
Pinched 1	0.026	0.899	Done	------
10% pinch	0.052	0.902	Done	Done
	0.077	0.904	Done	Done
	0.103	0.907	Done	Done
	0.129	0.909	Done	------
	0.155	0.911	Done	------
Pinched 2	0.026	0.849	Done	------
15% pinch	0.052	0.853	Done	Done
	0.077	0.856	Done	Done
	0.103	0.860	Done	Done
	0.129	0.863	Done	------
	0.155	0.866	Done	------

NUMERICAL METHODS

The commercial solver ANSYS CFX 15.0 [19] was used for all computations in this study. The computational approach is based on finite volume method with second order spatial discretization. RANS equations were closed through the two equation $k-\omega$ SST turbulence model of Menter [20].

The convergence criteria of all calculations in this study were based on the reduction of RMS momentum, mass and energy residual below 10^{-5}, imbalance percentage below 5×10^{-3} percent and stability in the stage polytropic and isentropic efficiencies.

According to the pitch angle (51.4 degree), rotational speed (18000 rpm) and number of blades (14 in total), time step was calculated (10^{-5}second), but in some cases because of the difficulty of the convergence and some instabilities, smaller time steps were selected (5×10^{-6} second).

As it was mentioned previously, in order to reduce simulation costs, one passage of each compressor was modelled and simulated. Three zones were modelled, guide pipe which guides the air into the rotational part, impeller and diffuser. As the volute is not modelled, there is a short exit section after the diffuser. This approach was used to ensure that the outlet boundary would not affect the result. Similar approach was used by Oh et al. [21].

Figure 2 demonstrates computational domain of the centrifugal compressor. Total pressure and temperature were the initial condition of inlet boundary. Reference pressure was set at zero and mass flow rate is assumed to be outlet initial condition. Frozen Rotor was selected for interface between stationary and rotational zones in steady simulation, and Transient Rotor-Stator (sliding mesh) was selected for unsteady calculations.

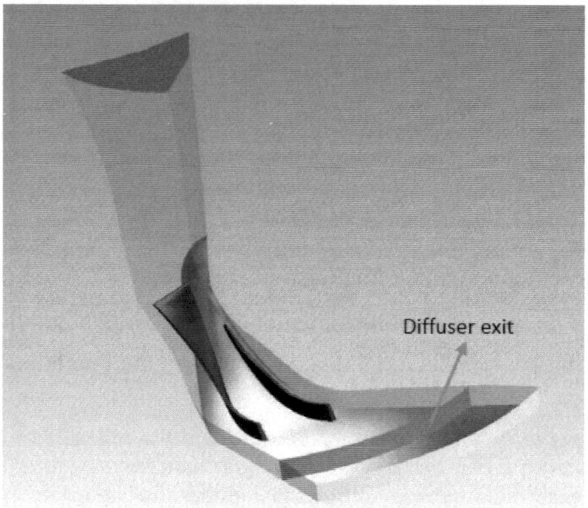

Figure 2. Boundary conditions

GRID DEPENDENCY

An accurate grid dependency check was done. All diffusers were modelled with the second smallest tip clearance ($t_2/b_2=0.052$) with three different structured grids. The grids had approximately 0.81, 1.9 and 4.28 million cells. Non-dimensional total efficiency by reference efficiency with respect to the number of cells can be seen from Fig. 3. It can be seen that when the cell number is increasing from 2 million to 4 million cells, total efficiency is hardly changed. It can be concluded that roughly 2 million cells is sufficient. Total efficiency was calculated with mass flow averaged values as follow:

$$\eta_t=\frac{T_{t1}(\pi_t^{R/C_p}-1)}{T_{t3}-T_{t1}}$$

Moreover, quality of structured mesh was tested according to Skewness Centroid and Jacobian. Although a few cells near the sharp and complex locations had Skewness Centroid near 0.87, but all other cells had Skewness Centroid below 0.8, which shows that the accuracy is acceptable according to the complexity of the geometry.

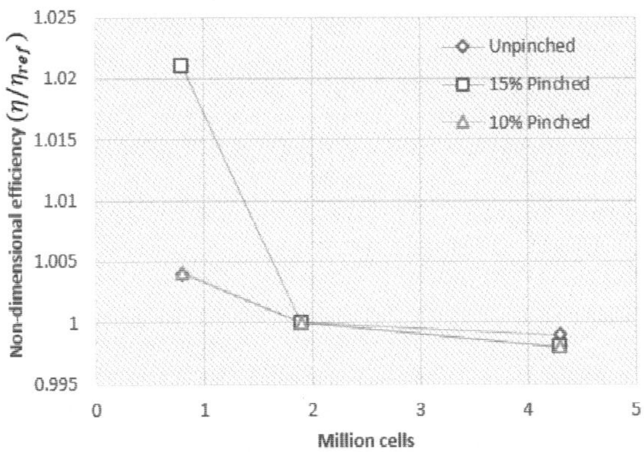

Figure 3. Grid dependency

INTERACTION EFFECT ON THE DIFFUSER

Total pressure loss through the diffuser should be concerned by designers. For calculating the total pressure loss coefficient, integrated mass averaged pressure was used across the diffuser inlet and outlet. In Fig. 4, results of unsteady simulations for unpinched and pinched diffusers for three different tip clearances are compared. Trends of all tip clearances are mostly the same although the amounts of each point has a bit difference.

Mass flow averaged values were used for calculating the performance parameter.

The diffuser total pressure loss coefficient can be calculated as follow:

$$K_{pr} = \frac{p_{s3} - p_{s2}}{p_{t2} - p_{s2}}$$

It should be mentioned that, each pitch size is 51.4 degree and in all cases, results are reported in each 10 degree. Also, each pitch takes approximately 4.76×10^{-4} second.

From Fig. 4 it can be derived that by increasing the tip clearance, fluctuation and unsteadiness of non-dimensional pressure loss coefficient is increased specially in unpinched case. But in the pinched diffusers pressure loss coefficient does not change much by increasing the tip clearance.

In other words, in vaneless unpinched diffuser the effect of the tip clearance on the unsteady behaviour of the impeller and unsteadiness of the performance is more significant compared to the pinched diffusers.

In the pinched diffusers, non-dimensional pressure loss coefficient is more unstable according to the time and degree of rotation. It can be achieved for the slope of each trend which in the unpinched is less than others. But averaged pressure loss in pinched diffusers is smaller than in unpinched one. In the unpinched diffuser, there is less difference and fluctuations between each time step, and results seem to be steadier.

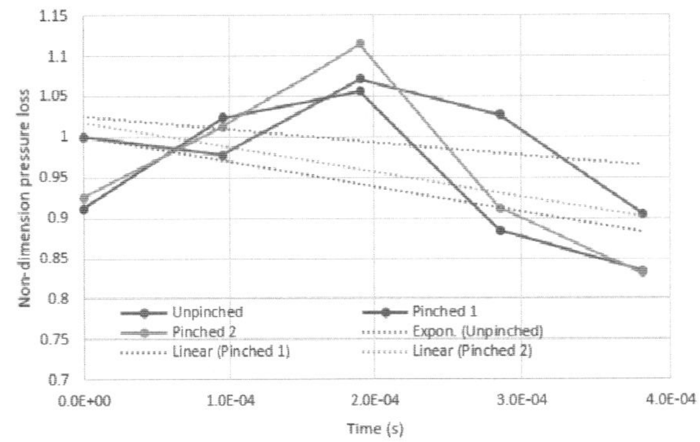

(a) $t_2/b_2 = 0.052$

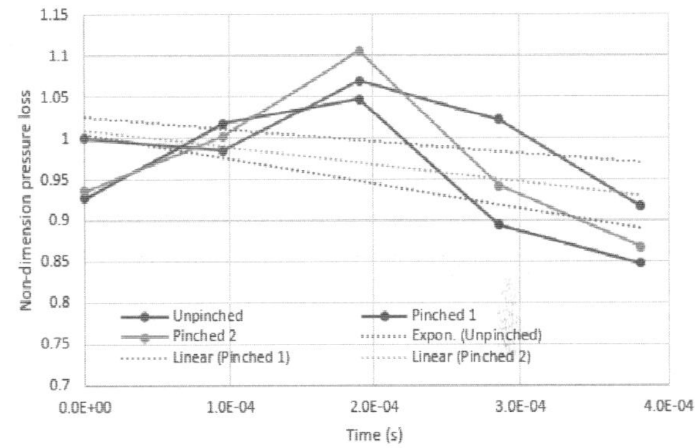

(b) $t_2/b_2 = 0.077$

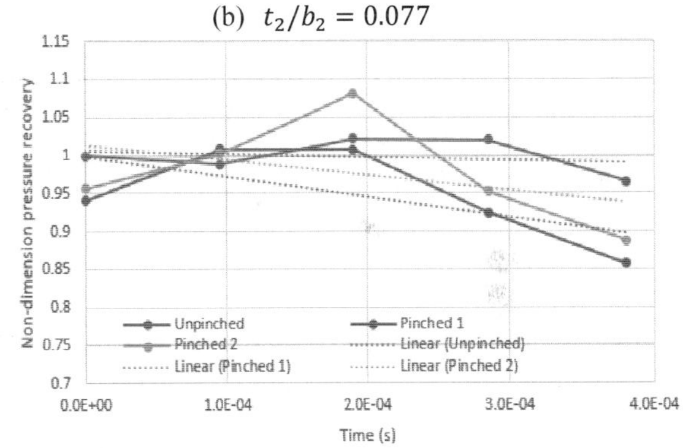

(c) $t_2/b_2 = 0.132$

Figure 4. Non-dimensional pressure loss coefficient

Furthermore, mass averaged pressure distribution from hub to the shroud inside the diffuser at the radius of($r/r_2 = 1.1$) is used to compare and investigate the unsteady behavior and effect of the pinch inside the diffuser. Figure 5 demonstrates pressure distribution from hub to the shroud of all diffusers at the same tip clearance ($t_2/b_2 = 0.077$). In Fig. 5, each color shows a specific time of rotation.

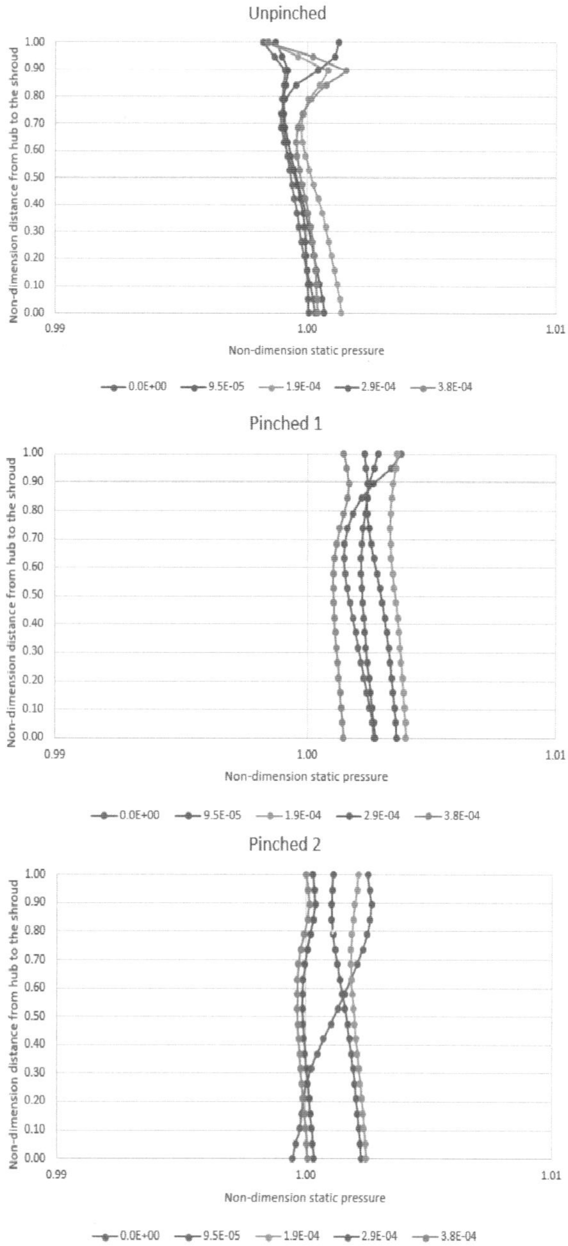

Figure 5. Non-dimensional pressure distribution from hub to the shroud. From top: Unpinched, pinched 1 and pinched 2

The axial pressure distribution from hub to shroud in the diffusers were calculated. In unpinched diffuser, there is a slight increase in the pressure near the shroud, in front of the tip clearance caused by transient tip flow. With the pinched diffusers, there is no such sharp increase in the pressure but periodic flow unsteadiness can be seen clearly. Periodic tip flow unsteadiness was more clearly observed in the pinched diffusers. Difference between trends of pressure distribution from hub to the shroud between pinched and unpinched diffusers implies the effects of the tip clearance flow and pinch on the unsteadiness of pressure distribution.

Also unsteadiness of the pressure is more significant in the axial direction from hub to the shroud in comparison with span wise direction in the diffuser.

It should be mentioned that y axis in above plot is non-dimensional distance from hub (0.0) to the shroud (1.0).

INTERACTION EFFECT ON THE IMPELLER AND UPSTREAM FLOW

According to the Fig. 6, influence of rotor-stator interaction of pinched diffuser is more sensible compared to the unpinched. In the pinched diffuser, this interaction also affects the pressure stability at the impeller inlet. In unpinched diffuser, more stable and steady pressure distribution can be achieved respect to time. Impeller-diffuser interaction affects the stability of the flow field inside the impeller as well as the diffuser.

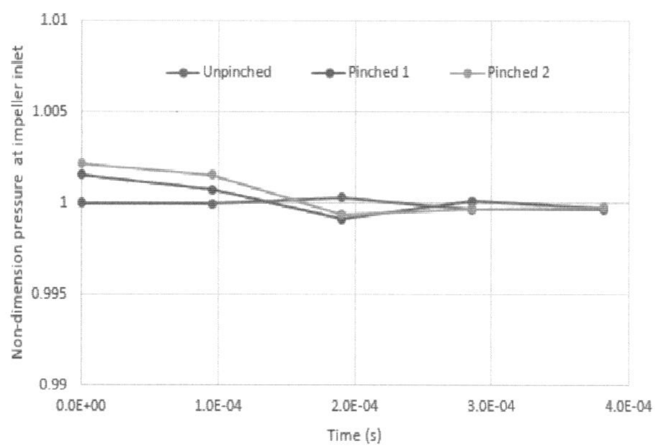

Figure 6. Non-dimensional pressure at impeller inlet

According to Fig. 7, performance parameters like pressure ratio inside the impeller in both pinched and unpinched diffusers are unsteady, however in pinched diffuser performance parameters are more significant compared to the unpinched due to non-uniform flow and pressure distribution. Figure 7 shows the total to total pressure ratio inside the impeller. Mass averaged total pressure at inlet and outlet of impeller was calculated in different time intervals. Rotor-stator interaction also has an influence on the impeller pressure ratio with pinched diffuser. The pressure ratio with unpinched diffuser is hardly affected by the impeller, although for pinched diffusers, this change is weak.

Total loss of the unpinched compressor is higher than pinched one. Inside the pinched diffusers, flow wake goes faster to the outlet and losses are smaller. Inside the impeller, in unsteady simulations, due to the periodic flow at the inlet and tip clearance which comes from the effect of the pinch on the upstream stability and uniformity, unsteady losses are more significant.

Periodic flows can generate more losses due to the shear and mixing of the wake with the flow. Unsteady losses especially at the trailing edge of the blades and periodic flow unsteadiness are more significant in the pinched diffuser compressor which come from rotor-stator interaction compare to the unpinched diffuser, but amount of unsteady losses is low compared to the total loss in the impeller and diffuser.

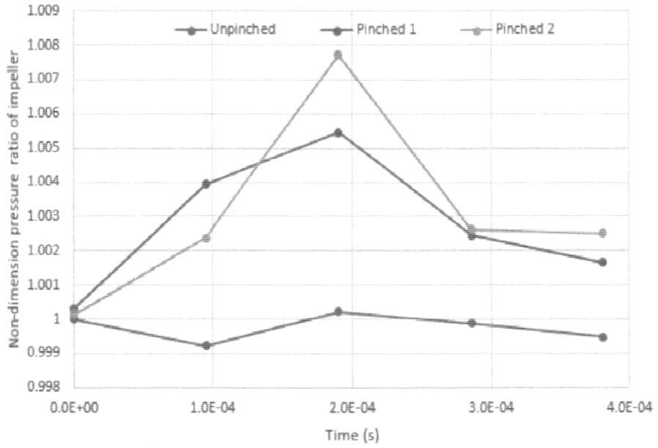

Figure 7. Non-dimensional pressure ratio of impeller

Furthermore, effect of the pinch and rotor-stator interaction on the tip clearance flow is studied. Figures 8 demonstrates the mass averaged entropy distribution along the tip clearance of the main blade for pinched and unpinched diffusers of the same tip clearance ($t_2/b_2 = 0.052$).

As it can be derived from Fig. 8, there is not any significant difference of entropy field between two types of pinched diffusers which used in this study. However, by comparing pinched diffusers with the unpinched one it can be noted that the pinch affects the transient upstream and tip clearance flows. By using pinch at the inlet of the diffuser, tip clearance loss can be reduced especially close to the trailing edge of the blades. Close to the trailing edge, static entropy in unpinched case is higher than pinched diffusers.

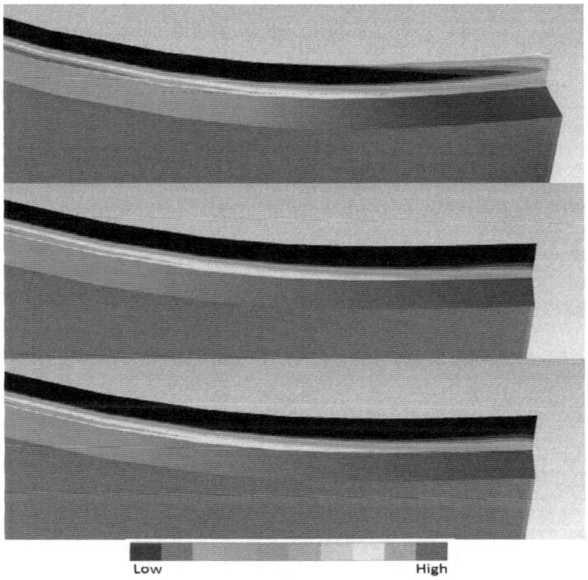

Figure 8. Entropy distribution along tip clearance. From top, unpinched, pinched 1 and pinched 2

By studying the back flow from diffuser into the impeller, it was concluded that moderate pinch can reduce back flow from diffuser into the impeller and due to this reason, entropy/loss close to the trailing edge can be reduced. Effect of the pinch on the upstream flow was studied in previous chapter, this effect can be investigated in detail by studying the mass averaged pressure on the tip gap in all types of diffusers.

Figure 9 illustrates the mass averaged static pressure along the tip clearance of the main blades for all diffusers. Y axis is non-dimensional static pressure and x axis is time of the rotation. Difference between unpinched and pinched diffusers is noticeable. Rotor-stator interaction especially at the pinched diffusers can affect the upstream and tip clearance flows in this centrifugal compressor. Mass averaged pressure at the tip clearance in pinched diffusers is higher than in unpinched.

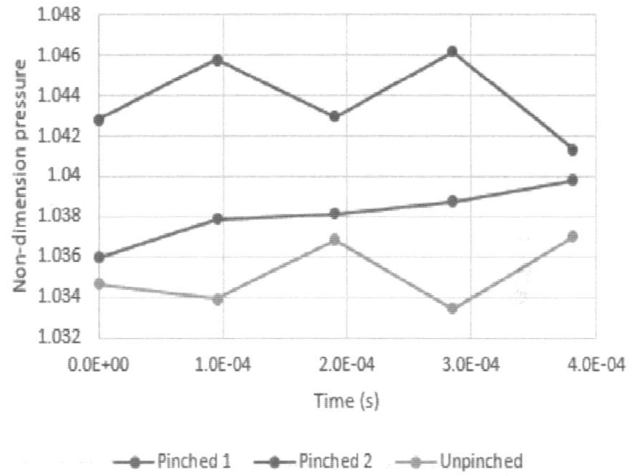

Figure 9. Non-dimensional mass averaged pressure of tip clearance

CONCLUSION

In this study, the influence of width of the diffuser and tip clearance on the performance and flow inside the diffuser and impeller has been discussed. First, the grid independence check was done and the results showed that the number of grids was sufficient. Unsteady simulations for three different tip clearance and a vaneless diffuser with three different reduction in width were conducted.

By analyzing the effects of the rotor-stator interaction on the performance of the impeller and diffuser, it can be concluded that pinch has a significant effect on unsteadiness inside the diffuser and the impeller. In unpinched diffuser, total pressure loss coefficient did not change significantly with respect to time. Mass averaged total and static pressures at the impeller inlet and in the diffuser did not change significantly with the unpinched diffuser versus time. It can be concluded that in the unpinched diffusers, unsteady behaviors are less important compared to the pinched diffusers.

Unsteady effect on the compressor performance can be divided in two parts, impeller and diffuser performances. Unsteady behavior has less effect on the impeller performance according to the results which are presented in last section. The unsteady effects are more sensible inside the vaneless diffuser especially in the pinched diffusers, also by increasing the tip clearance, unsteady behavior is more announced in unpinched diffuser. But in total, by comparing the total performance of the compressor, it can be conducted that unsteady effects are rather weak.

The unsteady simulations also showed that the rotor-stator interaction has some significant effects on the flow field at the impeller inlet. In the pinched diffuser, interaction affects the pressure stability and uniformity at the impeller inlet, while in the unpinched one, more stable pressure distribution was achieved.

By comparing the pressure distribution inside the diffusers, from hub to the shroud after the pinch radius, it can be concluded that in the pinched diffuser, unsteadiness and significant differences according to the time can be seen while in the unpinched case, unsteady behavior is less significant. Therefore, it could be suggested that, in unpinched diffusers it would be enough to only calculate the case in steady mode, because there is not any significant difference between different time intervals, and it helps to save time and cost. On the other hand, with pinched diffusers, transient simulation is suggested. More complex and unsteady flow before and after the diffuser is observed. Also by transient simulation, effects of the rotor-stator interaction and losses inside the impeller and diffuser can be analyzed in more details.

By studying the entropy distribution and mass averaged pressure along the tip clearance, it can be concluded that pinch can reduce loss close to the trailing edge of the impeller by decreasing the back flow from diffuser to the impeller, and can increase the pressure along the tip clearance simultaneously. This phenomena has been seen in both steady and unsteady simulations.

Further studies are conducting to study and compare the vaned diffusers and see the effects of the vane on the upstream flows in unsteady simulations.

NOMENCLATURE

LATIN ALPHABET

b	Blade height/channel width
C_p	Specific heat capacity
K_{pr}	Diffuser total pressure loss coefficient
p	Pressure
r	Radius
R	Specific gas constant
t	Tip clearance
T	Temperature

GREEK ALPHABET

η	Efficiency
π	Pressure ratio

SUBSCRIPTS

1	Compressor inlet
2	Impeller outlet
$2'$	Start of pinch
3	Diffuser outlet
in	Inlet
ref	Reference
t	Total
s	Static

ACKNOWLEDGMENTS

The authors would like to acknowledge the financial contribution of Academy of Finland and Graduate School of Lappeenranta University of Technology.

REFERENCES

[1] R. C. Dean, "On the Unresolved Fluid Dynamics of the Centrifugal Compressor," *Advanced Centrifugal Compressors, ASME,* pp. 1-55, 1971.

[2] R. Dean and Y. Senoo, "Rotating Wakes in Vaneless Diffuser," *Basic Engineering,* vol. 82, no. 3, pp. 563-570, 1960.

[3] S.-S. Hong and a. R. S. Abhari, "Effect of Tip Clearance on Impeller Discharge Flow and Vaneless Diffuser Performance of a Centrifugal Compressor," *Journal of Power and Energy,* vol. 226, pp. 963-974, 2012.

[4] T. Turunen-Saaresti and A. Jaatinen, "Influence of the Different Design Parameters to the Centrifugal Compressor Tip Clearance Loss," *Journal of Turbomachinery,* vol. 135, no. 1, 2013.

[5] R. Pampreen, "Small Turbomachinery Compressor and Fan Aerodynamics," *ASME Journal of Engineering for Power,* vol. 95, pp. 251-256, 1973.

[6] T. Mashino, I. Watanbe and I. Ariga, "Effect of Fluid Leakage on Performance of a Centrifugal Compressor.," *ASME J. Eng. Power,* vol. 101, pp. 337-342, 1979.

[7] K. U. Ziegler, H. E. Gallus and R. Niehuis, "A Study on Impeller-Diffuser Interaction-Part I: Influence on the Performance," *Journal of Turbomachinery,* vol. 125, pp. 173-182, 2003.

[8] Y. K. P. Shum, C. S. Tan and N. A. Cumpsty, "Impeller-Diffuser Interaction in a Centrifugal Compressor," *ASME, Journal of Turbomachinery,* vol. 122, pp. 777-786, 2000.

[9] W. Dawes, "A Simulation of the Unsteady Interaction of a Centrifugal Impeller With Its Vaned Diffuser: Flow Analysis," *Journal of Turbomachinery,* vol. 117, pp. 213-221, 1995.

[10] J. Feng, F.-K. Benra and H. J. Dohmen, "Numerical Study on Impeller-Diffuser Interactions with Radial Gap Variation," in *Proceedings of the 4th WSEAS International Conference on Fluid Mechanics and Aerodynamics*, Elounda, Greece, August 21-23, 2006.

[11] K. Sato and a. L. He, "Effect of Rotor-Stator Interaction on Impeller Performance in Centrifugal Compressors.," *Journal of Rotating Machinery,,* vol. 5, pp. 135-146, 1999.

[12] H. Krain, "A Study on Centrifugal Impeller and Diffuser Flow," *ASME J. Eng. Power,* vol. 103, pp. 688-697, 1981.

[13] F.Leboeuf, I.Trébinjac, X.Ottavy and N.Gourdain, "Aerodynamic Studies in High-Speed Compressors Dedicated to Aeronautical Applications," in *27th International Congress Of The Aeronautical Sciences*, Nice, France, September 19-24, 2010.

[14] T. Turunen-Saaresti, Computational and Experimental Analysis of Flow Field in the Diffusers of Centrifugal Compressors, Doctoral Thesis, Lappeenranta: Lappeenranta University of Technology, 2004.

[15] A. Jaatinen-Värri, A. Grönman, T. Turunen-Saaresti and J. Backman, "Investigation of the Stage Performance and Flow Fields in Centrifugal Compressor with a Vaneless Diffuser," *International Journal of Rotating Machinery,* no. ID 139153, p. 10, 2014.

[16] A. Jaatinen, A. Grönman, T. Turunen-Saaresti and P. Röyttä, "Effect of Vaneless Diffuser Width on the Overall Performance of a Centrifugal Compressor," *Journal of Power and Energy,* vol. 225, pp. 665-673, 2011.

[17] A. Jaatinen, T. Turunen-Saaresti, A. Grönman, P. Röyttä and J. Backman, "Experimental Study of the Effect of the Tip Clearance to The Diffuser Flow Field and Stage Performance of a Centrifugal Compressor," in *ASME Turbo Expo 2012*, Copenhagen, Denmark, June 11-15, 2012.

[18] A. Jaatinen-Värri, T. Turunen-Saaresti, P. Röyttä, A. Grönman and J. Backman, "Experimental Study of Centrifugal Compressor Tip Clearance and Vaneless Diffuser Flow Fields," *Journal of Power and Energy,* no. 227, pp. 885-895, 2013.

[19] ANSYS, Theory Manual, 15.0 , Canonsburg: ANSYS.

[20] F. R. Menter, "Two-Equation Eddy-Viscosity Turbulence Model for Engineering Applications," AIAA, 1994, pp. 1598-1605.

[21] J. S. Oh, C. W. Buckley and A. G. Agrawal, "Numerical Investigation of Low Solidity Vaned Diffuser Performance in a High-Pressure Centrifugal Compressor Part II: Influence of Vane Stagger," in *In Proceedings of ASME Turbo Expo*, Berlin, Germany, June 9-13, 2008.